Maleev and Hartman's
Machine Design
in SI Units
SIXTH EDITION

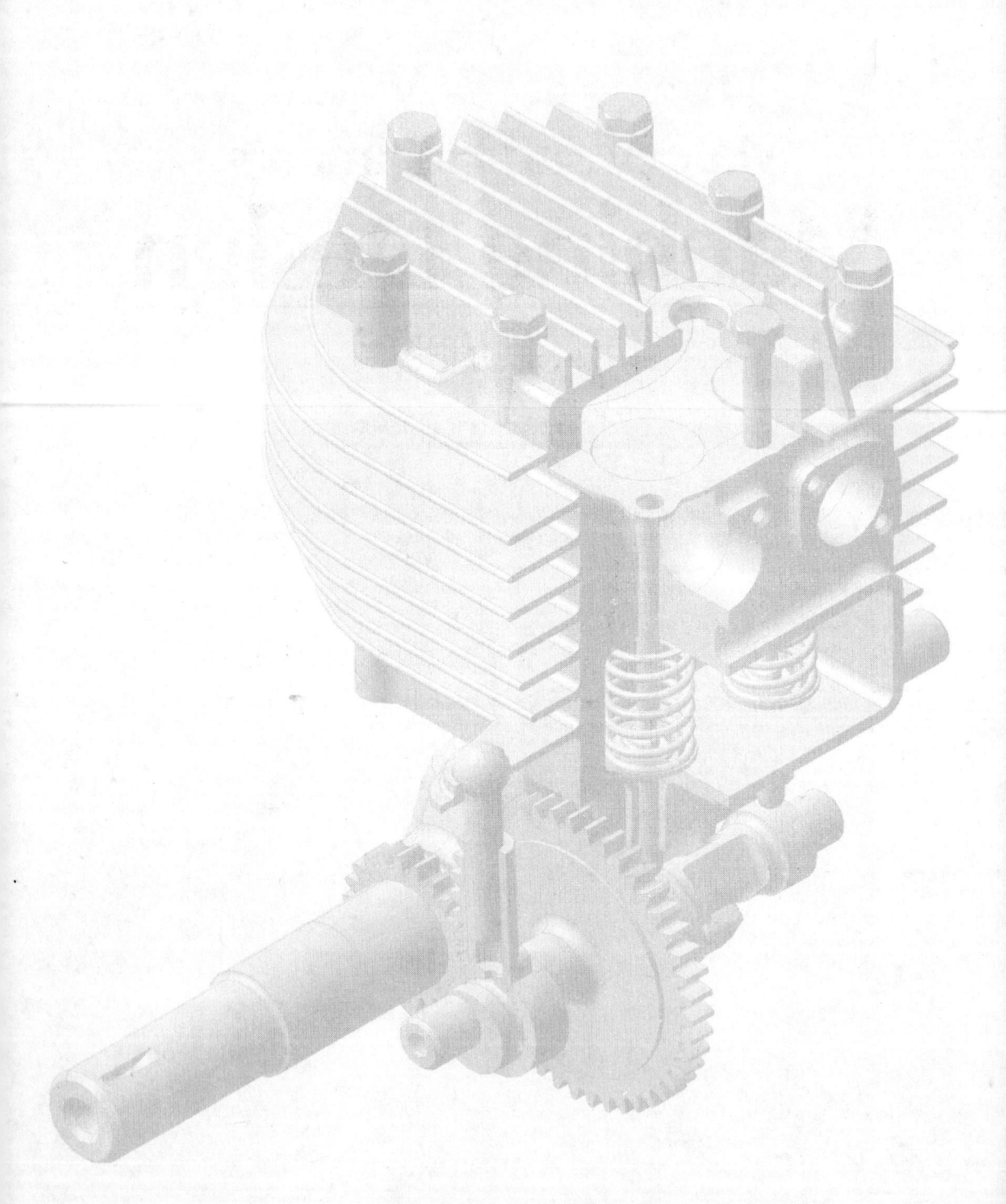

Maleev and Hartman's
Machine Design
in SI Units

SIXTH EDITION

Revised and Edited by

OP Grover
BE Mech, MTech, PhD, FIE, MISME, FIIET

Professor and Dean (Academic)
Maharaja Agrasen Institute of Technology, Delhi

Ex-Professor, Department of Mechanical Engineering
Delhi College of Engineering, Delhi

Ex-Professor and Head, Department of Mechanical Engineering
and Dean, Faculty of Technology
University of Delhi, Delhi

Past Hon. Secretary, Institution of Engineers (India)
Delhi State Centre

CBS

CBS Publishers & Distributors Pvt Ltd

New Delhi • Bengaluru • Chennai • Kochi • Mumbai • Pune
Hyderabad • Kolkata • Nagpur • Patna • Vijayawada

Maleev and Hartman's

Machine Design

in SI Units

SIXTH EDITION

ISBN: 978-81-239-2632-2

Copyright © Publisher

Sixth edition: 2015 , Reprint : 2018

Fifth edition (revised and enlarged): 1999

Published by Satish Kumar Jain and produced by Varun Jain for

CBS Publishers & Distributors Pvt Ltd

4819/XI Prahlad Street, 24 Ansari Road, Daryaganj, New Delhi 110 002, India.
Ph: 23289259, 23266861, 23266867   Website: www.cbspd.com
Fax: 011-23243014   e-mail: delhi@cbspd.com; cbspubs@airtelmail.in.
Corporate Office: 204 FIE, Industrial Area, Patparganj, Delhi 110 092
Ph: 4934 4934   Fax: 4934 4935   e-mail: publishing@cbspd.com; publicity@cbspd.com

Branches

- **Bengaluru:** Seema House 2975, 17th Cross, K.R. Road,
  Banasankari 2nd Stage, Bengaluru 560 070, Karnataka
  Ph: +91-80-26771678/79   Fax: +91-80-26771680   e-mail: bangalore@cbspd.com
- **Chennai:** 7, Subbaraya Street, Shenoy Nagar, Chennai 600 030, Tamil Nadu
  Ph: +91-44-26260666, 26208620   Fax: +91-44-42032115   e-mail: chennai@cbspd.com
- **Kochi:** Ashana House, No. 39/1904, A.M. Thomas Road, Valanjambalam, Ernakulam 682 016, Kochi, Kerala
  Ph: +91-484-4059061, 65, 67   Fax: +91-484-4059065   e-mail: kochi@cbspd.com
- **Mumbai:** 83-C, Dr E Moses Road, Worli, Mumbai 400 018, Maharashtra
  Ph: +91-22-24902340/41   Fax: +91-22-24902342   e-mail: mumbai@cbspd.com
- **Pune:** Bhuruk Prestige, Sr. No. 52/12/2+1+3/2 Narhe, Haveli
  (Near Katraj-Dehu Road Bypass), Pune 411 041, Maharashtra
  Ph: +91-20-64704058, 64704059, 32392277   Fax: +91-20-24300160 e-mail: pune@cbspd.com

Representatives

- **Hyderabad** 0-9885175004
- **Nagpur** 0-9021734563
- **Vijayawada** 0-9000660880
- **Kolkata** 0-9831437309, 0-9051152362
- **Patna** 0-9334159340

Printed at Swastik Packaging, Patpharganj, Delhi-92

to
all engineers

Preface to the Sixth Edition

Without engineering, the world would not have been as it exists today. The role of engineer is visible from a needle to an aeroplane. Every organization indicates the contribution of engineering skills. From the first discovery of a bar, i.e. a lever used as a machine for shifting stones, to the most modern machines in all fields including medicine, agriculture, chemicals, textiles, etc. engineers are the backbone of development. Quoting Theodore Van Karman "scientists discover the world that exists; engineers create the world that never was". For the engineer, the design is the ultimate goal. To simplify the design, steps have been provided not only for the elements but also for the systems so that uncertainties are eliminated.

Design steps of the elements with iterative procedure are easily adopted for making computer simulation. Design of crane hook block, ICE piston, gear box, external and internal shoe brake, and the choice of transmission are added. Information about latest development of ball and roller bearings has been given.

The book now covers the entire course of machine design for all institutions and also public examinations. **A special feature of the book is practical short answer questions. The answers are provided in the text itself in italics with question numbers given in the bracket at the end of the answer.** These will be helpful in understanding the fundamentals of the subject and to answer any type of questions in competitive examinations. These questions have been highly appreciated by experts including Prof BC Nakra, former Director, IIT Delhi.

I am thankful to my colleagues for giving suggestions for improvement.

I am highly thankful to Mr YN Arjuna and his team at CBSPD for bringing out the book in the present form.

Any suggestions for improvement will be welcome and acknowledged.

O P Grover

Preface to the Fifth Edition

All that exists in the present world in various fields is due to engineering and all engineering activities are a design activity. *Maleev and Hartman's* **Machine Design** has been a classic book and a very popular one all over the world and had been used as a reference book by practising engineers. The revision has been done to the SI system (ISO 1000–1973) and many additional topics, problems and objective type questions have been added keeping the same classical style and enhancement of the aim of the original authors, retaining also the general overall acceptability.

Competency in design depends on many factors such as imagination, curiosity, creativity, judgment, assumptions, empirical data, codes, standards and many others. Illustrations have been given to make use of such factors. First chapter has been revised to give a proper definition of system design concept including the iterative design process. The iterative concept has been used in many solved problems. The present trend of computer aided design is also presented in the design procedures of various elements leading to making of computer programmes. The objective type of problems given at the end of each chapter will be highly useful in understanding the design concepts and also for various competitive examinations and interviews, the likes of which has been missing in the textbooks. This book is useful for a two-semester course of machine design of engineering undergraduate courses or equivalent examinations. I hope the present edition will be appreciated by all the users.

OP Grover

The Editor

Dr **OP Grover** graduated in 1957 in mechanical engineering from Delhi Polytechnic, now Delhi College of Engineering. After one year at Bhakra Dam Project, he joined the College and rose to the level of Professor in 1977. He obtained MTech from IIT Kanpur and PhD from IIT Delhi. He served as Head, Mechanical Engineering and Dean of Faculty of Technology, University of Delhi, Delhi. He was awarded KF Antia Memorial Award by the Institution of Engineers (India) in 1773. He was Hon. Secretary, Institution of Engineers, for two terms and President, Indian Society of Mechanical Engineers, for two terms. He has published a large number of papers, and attended and organized national and international conferences. He has been an expert member of DRDO, NTPC, UPSC and many universities. He was also a member of the Board of Technical Education, Delhi. He has been responsible for establishing two private engineering colleges — Institute of Technology and Management, Gurgaon, and Lingaya's Institute of Management and Technology, Faridabad (as its Director), affiliated to MD University, Rohtak, and presently its Chairman, Advisory Board. He has been a Professor and Dean (Academic) at Maharaja Agrasen Institute of Technology, Delhi.

Contents

PART V: HOISTING MACHINERY

PART VI: PARTS TRANSMITTING ROTARY MOTION

Abbreviations and Symbols

ABBREVIATIONS

Abbreviation	Meaning
abs	absolute
AGMA	American Gear Manufacturers Association
AISI	American Iron and Steel Institute
ASA	American Standards Association
ASME	American Society of Mechanical Engineers
Bhn	Brinell hardness number
BIS	Bureau of Indian Standards
cu	cubic (metre or mm)
DFA	Drop Forging Association
ISO	International Standards Organisation
kN	kilonewton
kW	kilowatt
m	metre
min	minute(s)
MN	meganewton
mpm	metre per minute
mps	metre per second
N	newton
Nm	newtonmetre
N/m^2	newton per square metre
rms	root mean square value
rpm	revolutions per minute
s	second(s)
SAE	Society of Automotive Engineers
sq	square (metre or mm)
vibr	vibrations

SYMBOLS
[With dimensions in the International System (F, L, T)]

Symbol	Units	Dimensions	Name of Quantity
a	m/s^2	LT^{-2}	acceleration, linear
A	sq (mm or cm or m)	L^2	area, total or of a cross section
b	mm	L	breadth, width
c	mm	L	distance from neutral axis to extreme fibre
C	–	–	constant (may have various subscripts)
C	N	F	centrifugal force
d, D	m or mm	L	diameter
e	m or mm	L	deformation, total
e	m or mm	L	eccentricity, as of force application
e	1	0	efficiency (mostly with a subscript)
E	N/m^2 or N/mm^2	FiL^{-2}*	modulus of elasticity, direct (tension or compression)
f	1	0	coefficient of sliding friction
f	vibr per sec	T^{-1}	frequency of vibration
F	N or kN or MN	F	force, concentrated load
g	m/s^2	LT^{-2}	acceleration due to gravity, 9.81 m/s
G	N/m^2	FL$^{-3}(\theta)^{-1}$	modulus of elasticity, transverse (shear or torsion)
h	m or mm	L	height, depth, or thickness
i	1	0	number of elements if n is used for rpm
I	m^4 or mm^4	L^4	moment of inertia, rectangular (for areas)
I	Nms^2	FLT^2	moment of inertia, rotating mass
J	m^4 or mm^4	L^4	moment of inertia, polar (for areas)
k, K	–	–	coefficients in empirical formulas, mostly with subscripts
K	1	0	stress-concentration factor due to discontinuities (also with subscripts)
k	m	L	radius of gyration, rectangular or ratio
k_o	m	L	radius of gyration, polar
l, L	mm or m	L	length, distance
m	kg	FL^{-1}T^2	mass
M	Nm	FiL	moment of a force couple
n	1	0	factor of safety, design
n'	1	0	factor of safety, actual
n	1	0	number of elements or parts
n	rpm	T^{-1}	number of revolutions per minute
N	N	F	force, normal
p	mm or m	L	pitch (also a ratio without denomination)
p	N/m^2 or N/mm^2	FiL^{-2}*	pressure
P	kW	FLT^{-1}	power, 1000 watts/s

* The notation i indicates that the two terms are at right angles.

(Contd...)

(Contd...)

Symbol	Units	Dimensions	Name of Quantity
Q	N	F	load, total
r, R	mm or m	L	radius, radius of curvature
R	N	F	force of reaction
s	N/m² or N/mm²	$FiL^{-2}*$	stress, direct or normal, tensile or compressive
s_c	N/m² or N/mm²	$FiL^{-2}*$	stress, compressive only when necessary to differentiate from tensile
s_n, s_o	N/m² or N/mm²	$FiL^{-2}*$	stress, nominal
s_s	N/m² or N/mm²	FL^{-2}	stress, shear or tangential
S_d	N/m² or N/mm²	$FiL^{-2}*$ ⎤	design stress
S_e	N/m² or N/mm²	or	elastic limit
S_{en}	N/m² or N/mm²	FL^{-2} ⎦	endurance limit
t	deg C	0	temperature
t	sec, min, hr	T	time
T	Nm	$FiL*$	torque, torsional moment
T	sec	T	period (harmonic motion)
u	Nm per m³	$FiL^{-2}*$	modulus of resilience
U	Nm	FL	resilience
v	m/s	LT^{-1}	velocity, linear
V	cu (m or mm)	L^3	volume
w	N/m³	FL^{-3}	weight, specific
W	N or kN or MN	F	weight, total, load
W	Nm	FL	work
y	mm or m	L	deflection
y	1	0	Lewis factor in gear computations
z	1	0	Number of teeth
Z	m³	L^3	section modulus, rectangular (for areas)
Z_o	m³	L^3	section modulus, polar (for areas)
Z	Ns/m²	$FL^{-2}T$	viscosity, absolute
Z_k	m²/s	L^2T^{-1}	viscosity, kinematic
α	1	0	coefficient of thermal expansion, linear
α, β, γ	deg	0	angle between two lines
γ	1	0	specific gravity as compared with water
ε	1	0	unit deformation
θ	radian	$LiL^{-1}*$	angular distortion
θ	radian	$LiL^{-1}*$	angle
λ	deg	0	lead angle of worm or screw threads
μ	1	0	Poisson's ratio
μ_o	Ns/m²	$FL^{-2}T$	viscosity, absolute, Reynolds
π	1	0	ratio of circumference to diameter, 3.1416
ρ	kg/m³	$FL^{-4}T^2$	density, mass per unit of volume
ϕ	deg	0	angle of friction
ω	radian per sec	$(\theta)T^{-1}$	angular velocity

* The notation *i* indicates that the two terms are at right angles.

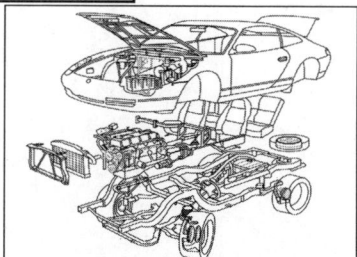

Introduction

1

1.1 MACHINE DESIGN

Machine design is the art of developing new ideas for the construction of machines and expressing those ideas in the form of plans and drawings. The idea may be almost entirely new, as in the case of an invention or an improvement upon existing machinery; or it may be only partially new, as when a machine or a machine part is to differ in size, load, or materials from those already existing.

For a machine to be well-designed the parts must be strong enough for the duty required of them and must be adequate for the functions they must perform, but they must not involve unnecessary expenditure of material or prohibitive cost of construction.

To design well any machine or part, the designer must have a working knowledge of the elements of machine construction; must know how to analyze the applied forces and their reactions and how to determine the resulting stresses; must possess sufficient information about materials; and must understand the influence of shape, method of assembling, and working conditions of parts upon the operation and maintenance of the machine. Thus modern machine design involves the application of the principles of three fundamental engineering subject: mechanisms, mechanics, and strength of materials, including elements of the theory of elasticity. In addition, possession of or access to experimental data on the performance of similar machines already existing is of great value.

1.2 DESIGN PROCEDURE

The procedure for designing a machine involves various steps. Figure 1.1 gives a systematic design procedure. The order in which these steps are taken in a certain sense are not too important since the engineering design is always an iterative decision making process. One has to traverse these principles several times before the final design emerges.

1.3 NEED

Depending upon the need of a particular goal, e.g. there may be a basic need like the need of a fastener or a cheaper mode of transport or to protect a spacecraft from destruction due to heat of friction or to reduce pollution from automobiles, etc. one can think of possible solution to such problems. The beginning of any design process is the recognition of need. The designer may recognize a need on his own or it may be communicated to him by his employer or customer or society. The process of design for finding solutions starts if and only if need is required to be fulfilled or if the need is not necessarily to be satisfied, there is no development. The recognition of need is a creative process through imagination or problems being faced.

The identification of need is a crucial step in the design process. Once there is an agreement on the preliminary need statement, this statement is to be analysed for getting a complete definition.

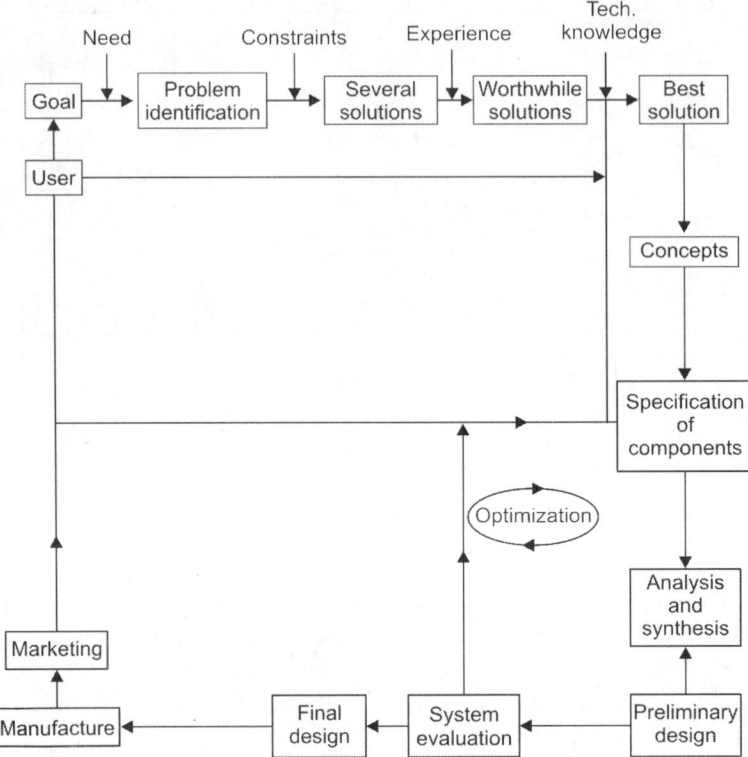

Fig. 1.1: Systematic design procedure

1.4 PROBLEM IDENTIFICATION

After a predicted or imagined goal, the first step is the problem identification or stating or defining the problem. To state a problem means to write down all data as to what is required. A clear definition and statement of the problem is absolutely necessary before the design process can begin. The problem of course will be identified as a result of market information which will specify the expected application of the product, working of the product in the particular environment, its purpose, frequency of use, etc. Identification of the problem as a result of market information is one aspect which is easily compared with the task of identifying the problem when the user himself is not aware of the need for the product. Depending upon the statement of the problem, the solution varies. Transport or a ground-effect machine, etc. may not be readily apparent without a detailed analysis.

In a certain sense, the market information will indicate the necessity for a solution but in several cases, the user himself will not be aware of the need for a particular item. Once the item is put in the market, the need for a better and new design might become imperative. Taking the example of a hoola-hoop, the public was not aware of its need when it was put in the market. Once it was marketed, industries multiplied overnight to meet the demand.

1.5 SOLUTIONS

The statement of the problem is followed by solutions. Solutions are obtained by various ideas, while several solutions may be plausible, all of them may not be possible or worthwhile. Only those solutions that are realizable or feasible have to be selected for which constraints and restrictions play a major role. The constraints could be of various types like cost, size, time to develop, means of production, materials, weight, etc.

The same problem will have different solution depending upon constraints, shortage or non-availability of materials, may call for a unique solution for the same common problem. While electric motor may be a solution at one place, diesel engine may be a solution at another place where no electricity is available. The environment like weather and temperature may demand the use of different materials. The design of a space vehicle has to be different from that of a terrestrial vehicle. Lack of proper appreciation to the constraints in all its various aspects—material, human, economic, social, political—lead in many cases to faulty designs. From a set of realizable solutions, depending on the experience of the designer, technical available know-how, etc. the best solution is selected.

1.6 CONCEPTS

The selected solution is then given a form. This stage is called the conceptual stage. This form exists in most of the cases in abstract. The designer will have an approximate idea as to the shape of the system, the components and subsystems performing respective operations resulting in the desired function of the system. Some drawings will depict these concepts.

1.7 SPECIFICATIONS

The functional requirements of these components and subsystems are then stated, i.e. specifications are laid down regarding the scope and operation of the components. These specifications have to be compatible with one another. The compatibility of the specifications may depend on various parameters, like strength, weight, wear, sensitivity, range, etc. Achieving compatibility may not always be very easy due to certain restrictions. Certain sacrifices may have to be made because after laying down the specifications for the various components, one has to go through a decision making called the buy or make decisions.

1.8 SYNTHESIS AND ANALYSIS

Then comes the process of synthesizing as a result of which the preliminary design emerges. This process involves putting together the various components and subsystems with their assigned functional requirements so that a system emerges.

1.9 ANALYZING THE PROBLEM

A thorough analysis of all external and internal forces acting upon a machine part is essential. These forces may be classified as follows:

(a) Useful loads due to the energy to be transmitted by the part
(b) Deadweight forces
(c) Forces due to frictional resistance
(d) Inertia forces due to changes in velocity
(e) Centrifugal forces due to changes in the direction of motion
(f) Forces due to changes of temperature
(g) Forces due to procedures in manufacturing
(h) Forces due to the shape of the part

Determination of Forces: Forces must be determined with an accuracy consistent with the importance of the part. Complex loads should be resolved into combinations of simple forces or forces and couples. When forces and their relations change, as during the motion of the piston of an engine, the procedure must be repeated for two or more phases of the change in order to find the maximum stresses in the dangerous sections of the parts to be designed.

Assumptions: If the designer does not have sufficient information to evaluate accurately all the factors in an analysis, he should make certain assumptions based on similar, simpler, or more fully investigated conditions. No definite rules can be set for making such assumptions. However, the fewer the assumptions, the quicker and better will be the results. An assumption should be made only when there is no means of acquiring the necessary data. When an assumption is made it should always be on the safe side. If possible, it should be checked in the course of preparing the design.

Experience is of great help in making necessary assumptions. In the beginning designer should not become discouraged when he sees that he does not have the desired experience. He will acquire it by working systematically and by keeping a record of good and poor results.

Sketches: The result of an analysis of loads should be presented in the form of clear sketches showing all forces, with the direction, magnitude, and point of application of each, and the amount and position of each couple.

Selecting the mechanism, materials and stresses. The three most important selections to be made in starting to design a machine are the mechanism, the materials to be used, and the allowable stresses.

Mechanism: The selection of the proper mechanism, or kinematic arrangement, for a machine part is usually indicated by the purpose for which the part is to be designed. Consideration is given to strength, wear, accuracy of motion, efficiency, and cost.

Materials: In the design of a certain machine part the selection of the most suitable materials, from the large variety of modern materials available, is determined first by the shape of the part and secondly by the condition of loading to which the part is subjected. Resistance to corrosion and wear is often an important consideration also, since both, greatly affect the life of various parts, particularly those subjected to repeated loads.

A designer must also keep costs in mind. If two materials are equally good for a specific application, the one that involves the least cost for the part should be selected. It must also be remembered that the cost of a manufactured object is the sum of the cost of the material and the cost of the labor necessary to make the object. Sometimes it is so much easier to machine a more expensive material that the saving in machining more than offsets the extra cost of the material. Since ordinarily a more expensive material is also a better one, the part so produced may be improved without an increase in cost.

Naturally, experience helps greatly in selecting the best materials. In discussing various machine details throughout this book, the proper material to be used will be mentioned.

Stresses: The selection of proper allowable stresses to be used in designing a part requires a working knowledge of materials. At the same time this selection must be based on an analysis of load variations, the stress distribution due to abrupt changes of sections, and other working conditions. Thus, in order to obtain uniform allowable stresses, it may be necessary in the calculations to use different allowable stresses for the same materials in different parts of a machine or even in different sections of the same part.

1.10 PRELIMINARY DESIGN

The purpose of the preliminary design is to determine the probable dimensions of all the parts that will form the machine. The chief consideration in establishing these dimensions is to provide sufficient strength, but rigidity and resistance to wear may also be important for certain parts.

Frequently the shape of a machine part depends on the operating conditions and the shape of adjoining parts to such an extent that the useful design procedure should be reversed. First the shape of the part is determined in accordance with service requirements; then the stresses in various sections are checked and necessary alterations are made in the dimensions.

If strength alone is the aim, the proper procedure is to design all sections of a part, and all parts of a machine, to be equally strong. An increase in a dimension of a section beyond the necessary size may mean not only a useless expenditure of material but also the weakening of an adjoining section and of the whole piece.

The designer should remember that the actual stresses in a part cannot always be computed by merely inserting numerical values for factors in stress formulas. The proper selection of values for these factors and coefficients is possible only if the load application and the working conditions in general are clearly understood.

Parts that may develop resonance, whether through lateral or torsional vibration, must be checked for critical speeds.

Excessive wear must be avoided by using special materials with sufficiently large and rigid bearing areas; by making proper provision for lubrication; and by eliminating eccentric application of loads, which always tends to cause uneven wear.

If rigidity (that is, absence of excessive distortion) is essential for the operation of a machine or a part, the deflection of each part under the applied forces must be determined, and materials that have a high modulus or rigidity must be selected. Thus a construction of welded steel will often be preferable to one of cast iron. However, with our present knowledge of the strength and elasticity of materials and our knowledge or stress distribution, the determination of deflections by analysis can be carried out with sufficient accuracy only for simple sections. In determining deflections in parts with more involved sections, either former experience with similar cases must be relied upon or special studies and experiments must be carried out to obtain relations between acting forces, dimensions, and deflections.

Sketches: The shapes decided upon in the preliminary design must be put on paper in the form of freehand sketches. A sufficient number of views and sections should be shown, and all the dimensions that can be computed should be included. Occasionally, oblique projections may help to present the design more clearly.

Calculations: All calculations must be as complete as possible and be recorded in a neat and legible form. This is just as essential in a classroom as in an engineering office. Legibility facilitates the checking of results, whether it is done by the designer himself or by another person. Neatness in writing calculations should become a habit of the designer.

All assumptions and the reasoning behind them must be stated clearly so that person who will check the calculations will know on what basis they were made.

1.11 SYSTEM EVALUATION

When the product of this preliminary design is put into evaluation—analytical or experimental it will satisfy all the specified constraints and restrictions and is a solution to the problem. But it will not have gone through one of the most important steps in engineering design that is the process of optimization. In this processes the utility function and the pay off functions are maximised. This is an iterative process where the specifications are reviewed and functions of the components are analysed in view of the desired requirements. Because of this process the engineering design has been defined as an iterative decision making process.

The utility function varies from problem to problem. In some cases it can be expressed mathematically.

During the evaluation stage, there will be individual evaluation and compound evaluation of the subsystems and the system. The interaction effect between the components are considered carefully.

The cost economics are evaluated for the optimum design. This gives a feedback to the best solution or to the revision of specifications.

1.12 REVISING THE DESIGN

Before working drawings are made from the sketches of the preliminary design, these sketches must be revised to take into consideration even practical requirement and contingency.

Manufacturing requirements: The first step in revision is to consider the problems of manufacturing, such as those involved in the making of patterns, in methods of forging, and in machining. The way a pattern can be simplified, or can be made with less defects or made more cheaply, for instance, may suggest a change in the shape of a cast part. Sometimes the division of a complicated casting into two or more simpler ones may be advisable in order to decrease the danger of having to scrap an entire defective piece. Due consideration might be given to the question of whether an improvement can be attained by changing a part ordinarily made of cast iron to a welded steel construction. Forged pieces might be made less expensively by some changes in forging methods which would alter the shape. Sometimes it may be desirable to change the shape of a part so that the machining operation may be less expensive. The availability of existing patterns, fixtures, or tools may suggest other changes. In a manufacturing plant an experienced designer often consults the various foremen of the pattern shop, foundry, forge shop, and machine shop.

The desirability of using standard commercial parts or of complying with the recommendations of the various committees on standardization may require changes of still another character.

Operation requirement: The second step in revision is to consider the requirements of the operation of the machine, such as provisions for lubrication and adjustment for wear. It may be found advisable to divide a part that is subject to severe wear into two pieces, one of which provides the strength while the other takes all the wear. The part which takes the wear should be made inexpensive and easily replaceable. The designer must also consider the safety problem, as by avoiding protruding parts in rotating members and by providing guards for gears and reciprocating parts.

Assembly requirements: The third and final step in revision begins when the designer starts to make assembly drawings. It may be found necessary to modify the shape of certain parts in order to avoid interference with adjoining parts or to connect them properly. Ease of assembly, maintenance, and dismantling should be considered carefully.

It may be advisable to change the shape of a part to give it a more pleasing appearance. Fanciful curves used in machinery fifty years ago are no longer in vogue. Simple lines that are appealing to the eye, and a general impression of well-balanced proportions, represent good design and help sell the finished machinery.

The assembly drawing should give all the dimensions and information needed for assembling and installing the machine, but it should show no unessential detail dimensions.

1.13 FINAL DESIGN AND DRAWINGS

After the assembly drawing has been made and all possible phases of revision have been taken into consideration, working drawings, or shop drawings, are made. The modern practice is to prepare them directly on vellum paper ready for photocopying or directly on computer and to show each part of the machine on a separate sheet to facilitate eventual alterations. Despite the careful revision of the preliminary design, as just outlined, more changes may be found desirable after a piece has been made and used.

A working drawing must be clear, concise, and complete. It must have enough views and cross sections to show all details. The main view of a part should show it in the position it is designed to occupy. Every dimension must be given, so that there will be no occasion for guesswork and no necessity for scaling the drawing.

Tolerances: All dimensions that are important for correct assembly and interchangeability must be given in micrometre; and the tolerance, or limit of permissible deviation, should also be indicated. The use and magnitude of tolerances is discussed in Chapter 6.

Information and notes: A working drawing for a piece should contain, in the form of sections, dimensions, and notes, all the information that may assist the men in the shops in making the pattern for molding and casting the piece, in welding parts together, in machining it, or in heat-treating it. The sequence of machining operations may be given if it affects the accuracy of the final dimension. The kind of finish should be indicated in order to obtain a piece with a finish required by its duty. This is explained in Chapter 6.

Bill of material: When the design of a mechanism or a subassembly of a machine has been completed, a bill of material, or list of all parts, is made. This list must show the number of every drawing, the name of the part on the drawing, the material of which the part is made, and the number of parts in the assembly. The bill of material must also show a list of all standard parts such as bolts, nuts, washers, and cotter pins, their sizes, and the quantity of each needed.

1.14 MANUFACTURE

From the drawings made, the product can be manufactured. After testing the prototype, full production is possible and goes to the user. Even at that final state, a come back may have to be made depending upon the reaction of the user. Also depending on the research and development carried out, further revision may have to be done. Finally, there is a very important point which is generally neglected in engineering design—that is, the roll of aesthetics in design. This is not a simple problem, because personal reactions and emotions are involved. The engineering design should not merely be a utilitarian design. It should satisfy the whole man. To quote Dr. Mackinnon "For the truly creative person it is not sufficient that a problem be solved, there is the further need that the solution be elegant."

1.15 TESTING AND MARKETING

After the manufacture of the system, the trial runs and final testing is carried out before marketing. The user purchases the system from the market. The user's and markets' reactions are again responsible for changes to be incorporated in the design. It is the user who justifies whether the goal is achieved or not.

1.16 BASIC PRINCIPLE OF DESIGN

The following considerations should be made in designing an element or a system.

1. Functional requirement is the first and foremost. The function that the element or the system has to perform is the most important one and correspondingly, links, gears, belt drive, etc. can be chosen to perform a particular function in combination with the prime-movers.

2. The strength requirements should never be sacrificed. Depending upon the forces, the dimension of the parts should be so chosen that the strength of the materials of parts is taken care of. The strength also includes the static stiffness and dynamic rigidity of the parts.

3. Effect of temperature: Temperature affects all parts. It has to be taken into consideration during manufacture and also the effect of temperature during actual working, e.g. parts of any internal combustion engine.

4. Alignment of parts: Depending upon the mating parts or the rotating parts, the dimension of parts or the distances between parts should aim at the proper alignment of bearings or sliding parts, etc.

5. Use of indigenous materials and their proper choice. As far as possible, indigenous materials should be selected to avoid undue delay in procurement and in production and in maintenance.

6. Use of standard sizes of components, standard speeds and preferred numbers. The use of standard size of components, e.g. bolts, nuts, standard speeds of prime movers and preferred numbers of various sizes, speeds, etc. result in reduced cost.

7. Existing facilities for manufacture. In the already existing works, the existing facilities for manufacture should be utilized for a new product design to start with.

8. Facilities for wear adjustment and protection from environment conditions. In the case of mating parts with relative motion, the wear adjustment facility results in longer life. Wherever necessary, environmental conditions (e.g. in mines, chemical plants) should be taken care of in design.

9. Ease of assembly, inspection and maintenance. Assembly requirements should be considered in the design to reduce wastage. The inspection of various parts subjected to wear, fracture or failure should be convenient and so should be the deassembly.

10. Lubrication and cooling: Wherever necessary the lubrication and cooling of parts with relative motion be provided for in the design.

11. Location of controls should be conforming to standards and should be within the reach of operator. This requires the proper selection of direction, force and distance.

12. Built-in-safety devices: The design must include built-in safety devices for overload protection and for the safety of the operator. "Safety should not be by accident".

13. Overall weight and available space. As far as possible, efforts to be made to reduce the weight of the element or the system be made to the size for a given available space.

14. Manufacturing tolerance is a necessary evil. For all parts, depending upon the specific use, proper tolerances need to be decided.

15. Aesthetic appeal: An effort should be made to give the product a modern appearance which is best described by words such as 'smooth', 'rounded', 'uncluttered', or 'streamlined', etc. Any design which looks bad to the eye has something wrong with it. It may be due to wrong calculation or wrong decisions, etc. A design which is good to look at, may not have anything wrong with it.

16. Human aspects of design: To consider the physiological, psychological, anatomy, anthropological factors in design, as every system is made by man, operated by man and used by man.

17. Reliability: The reliability of the performance of the machine depends on a number of factors and should be optimized.

18. Cost: The optimized design is obtained by compromising with conflicting requirements, combined with reasonable profits. It results in the least cost.

19. Transportation and erection facilities: For proper safe transportation and erection of the machine or subassemblies provision needs to be made in the design for handling purpose like eye bolts, etc.

20. Aim at basic principle of prosperity of the organisation, of the community and of the country.

1.17 ERGONOMICS

Ergonomics is the sequence of human characteristics relevant to industrial performance. It deals with various aspects such as:

1. Anatomical factors in workplace layout, i.e. placement of machines/equipment and components to suit human body measurements and design of seats, etc.

2. For accurate perception of various display panels and presentation of all types of instruments and control dials.

3. Design of control wheels and control levers to suit human mental and physical characteristics, viz force, direction and distance.

This science has to correlate the activities of man and machine as man–machine system. The performance of any system does not only depend on the system but also on human being who is to operate even when the system may be fully automatic. The human being has to control and operate. A car driver will not be able to operate the car if his seat is not only comfortable but also be able to see the motion of the car and also see everything—other vehicles, animals, obstructions, etc. on the road along with conveniently placed hand wheel, gear shift, brake pedal, acceleration pedal and many other controls. Here also a compromise is made between the human needs, comforts of man, cost and performance. All controls and displays are dependent on:

(a) Structural body dimensions, i.e. dimensions recorded with body in fixed position: Standing and/or sitting. This is useful for designing car seats, aeroplane seats, pilot seats, earphones, glasses, wrist watches, chairs, desks, tables, etc.

(b) Functional body dimension, i.e. those dimensions which result from the movement or motions of the body and limbs: Legs, feet, arms, hand and fingers. Many activities need to be performed by movement of individual limbs, e.g. considering arm does not only involve just its length but also it is dependent on the trunk, the shoulder and the hand. All the controls of machines are dependent on the forces which can be applied and also their dimensions.

The size of controls and dials on the capability of a normal eye to see the pointer and read the scale.

The consistency and compatibility, the usual preset ideas in the mind of man need to be considered. Turning a lever or hand wheel clockwise results in increase in temperature, speed, quantity and vehicle turning right side. Similarly turning anticlockwise should result in decrease of speed, temperature and vehicle taking left turn. Imagine if turning the hand wheel anticlockwise were to turn the vehicle to the right side, there would have been difficulties to control the vehicles. The force that a finger can apply to operate a switch, the force the hand can apply to operate a hand wheel/lever and the force a foot can apply to control the brake and acceleration, need to be known for the proper operation and also how much energy is available to run a system for a definite length of time, i.e. a man can dissipate or has 4500 k.cal which is equivalent to 1200–6000 Nm to accomplish a task or run a system for a certain work to be performed and for how long so that he does not feel tired. To take care of all such activities, the following principles need to be considered for design.

Basic principles of human aspects:

1. For proper operation of a system the sitting position should be preferred. The standing position may be preferred only when it is not possible to work in the sitting position.

2. Normal or natural body position is the best. Unnatural body positions such as cramped body position, stoopping positions should be avoided.

3. In the static work position, rests may be provided for elbows, arms, hands and feet. Holding a small weight with stretched arms for a few minutes even, is not easy.

4. Various activities for controlling or working should be divided between different limbs of the body, i.e. arms, hands, legs and feet.

5. The working tables or platform should be of proper height for sitting and standing positions.

   For standing posture, the work table be 50 mm below elbow height for accurate work and 100 mm for light work and for heavy work 150 mm.

Elbow height from ground	1040 mm for man
	980 mm for woman

6. Torsional movement of body, arms and legs should be avoided.

7. For performing an operation, the number of groups of muscles required should be minimized.

8. For various operations to be performed, implements, levers, hand wheels, foot pedals, etc. should be as close to the body as possible for frequent operations.

9. All the limbs of body have different load carrying capacity or the loads which

can be applied by the limbs. Maximum load capacity may be used only for short duration. To operate a switch, a finger is to be used, the force which can be applied by a finger needs to be known. The following table serves some guide:

Limbs	Capacity (N)	Size (mm)
Push button	5–15	12–30
Toggle switch: One finger	2.5–15	5
Knobs: One hand	5–10	25–100
Levers	150–300	25–75
Hand wheels	10–150	100–150
Pedals	50–600	100–150*

* Movement up to 125 mm

10. Working area should be provided with adequate light. Reflecting surfaces should be avoided. Glare and naked light is also not desirable

11. For reducing eye strain, headache, etc. tasks requiring constant visual observation control should be so located as to allow a comfortable head position. For most comfortable head position, the line of sight is 38° ± 2° below horizontal for sitting position; 30° ± 2° for standing position.

12. Operation of controls: The clockwise rotation should cause increase in speed, increase in intensity of light, increase in temperature and turning right side. The reverse of these require movements in anticlockwise rotation.

13. The size of dials and the least count should be accurately readable so the pointer has to be near the scale. The size and number of graduations has to be matched with the distance from the operator. Minimum mental effort may be required to read the scales. The subdivisions should be 1, 2, 5 or 10. The size of dials and scales should be readable at convenient distance from the operator. Some guidelines are given below:

Distance from operator	Diameter of dials	No. of gradients
700	25	50
1000	40	35
2000	70	15
3000	100–110	10

The shape of scales or dials preferably should be either circular or straight.

14. Visual aspects of design: The appearance of the design should be such that it looks good to the eye. A guiding principle is "what is not looking alright to the eye— is not a good design or has something wrong with it" the reverse may or may not be true. The appearance is based on the following factors:

 (a) Shape which should be smooth, streamlined and should have good proportions.

 (b) Special features which account for appearance are: Overall shape, controls, displays, handles, tone and colour.

PROBLEMS

1.1 Define engineering design.

1.2 Compare talent and creativity. How does talent enter in design process?

1.3 List the tools used by (a) physicians, (b) lathe operators, (c) dentists, (d) butchers, (e) architects and (f) engineers.

1.4 Trace the evolution of (a) automobile, (b) radios, (c) airplanes, (d) shirts, (e) washing machines and (f) bicycle.

1.5 Describe a systematic design process by making a block diagram.

1.6 List the basic principles of a good design.

1.7 Safety should not be by accident. Discuss.

1.8 Consider the design process as it applies to the design of a machine for shelling peanuts. List the consideration, criteria and alternatives to be considered. Set up your solution in block diagram.

1.9 Explain optimization in design.

1.10 Design is an iterative process. Explain.

1.11 Discuss human aspects in design.

2 Failure Criteria and Static Stresses in Machine Parts

2.1 INTRODUCTION

This book is primarily concerned with the mechanical design of the elements which, functioning together, comprise a machine. In order to compare design alternatives, discussed in Chapter 1, the mechanical elements must be given form. They are analyzed for strength and rigidity, the materials are selected, and they are properly shaped and distributed. It is only at this stage that fabrication methods, cost, weight, and other pertinent details can be compared and a decision reached.

The most commonly used machine elements are:

Screws and bolts	Levers
Keys	Gears
Cylinders	Bearings and lubricants
Springs	Flywheel
Brakes	Chains and wire rope
Clutches	Belts
Shafts	Motors

Almost every type of machine consists of a combination of these elements.

2.2 BASIS OF GOOD DESIGN

One of the necessary conditions for good design requires that an element functions without failure for its required life. It is not always necessary, economically feasible, or possible to design for infinite life.

Under certain conditions it is economically possible to design a part to have infinite life, whereas some parts are designed with a finite life. A shaft can be designed to operate without breaking. Gears and bearings, however, can only be designed with a limited working life, since there is no way of avoiding wear.

Aircraft transmissions are designed to meet economic and weight restrictions, whereas maintenance is critical for buses. Aircraft parts are designed with a much shorter useful life. It is more economical to pay increased maintenance costs and carry larger pay loads.

2.3 FAILURE OF MACHINE PARTS

Failure of machine parts are of two types: Functional failure and fracture.

Machine parts fail in their function in many ways. Failure is not always overt or permanent in nature; it may be reversible or irreversible. A shaft which fails to function owing to excessive deflection may perform its function at reduced loads. A spring which deforms permanently under overload has failed permanently if, as a result, it fails to meet the preload of free-length requirements.

Failure may appear as change of the kinematic relationships, noise above a tolerable limit, decrease in mechanical efficiency, increase in heat generation, fire, explosion, or inability of the machine to respond to its controls by either stopping or starting.

Fracture of a machine part is due to one or more primary causes: *Deformation, wear, corrosion* or it may break in one or more pieces. It is possible, by proper design, to prevent excessive deformation at normal temperatures and, in many cases, to eliminate corrosion; the effect of the wear and deformation that are due to high-temperature creep, however, can only be mitigated and taken into account in design, but never avoided. Since a machine always includes moving parts, wear always exists as the life-limiting factor.

2.4 DEFORMATION

All loads are associated with deformation, and all stresses are associated with strain. Deformation is the sum of strains in a structure. Deformations are also caused by temperature change.

Deformation as a cause of failure of machine part is related to the type of element and the machine in which it is used. Excessive deformation may interfere with function by using up clearances, as in the case of a turbine blade; changing operation characteristics, as in the case of a cam; increasing the noise level, as in the case of a gear; or catastrophic failure, as in the case of fracture of connecting rod or pressure vessel. Table 2.1 lists the types of deformation at normal operating temperatures (below one half the absolute melting point of most materials).

Table 2.1: Types of deformation

Type	Cause
Elastic	a. Stress below the yield point
	b. Temperature change
	c. Instability (buckling of columns or plates)
Plastic	a. General stress above the yield point
	b. Local stress above the yield (stress raisers)
	c. Creep
Fracture	a. Stress above the ultimate strength
	b. Cyclic stress above the endurance limit

At elevated temperature, deformation is primarily plastic and is due to time related creep which terminates in rupture. This type is discussed under high temperature creep and rupture.

Elastic Deformation

The simplest types of elastic deformation is associated with stresses below the yield point of the material. Most rigid members are designed in this stress range. Figure 2.1 shows the load deflection relationship for a material obeying Hooke's law. The limiting condition is the yield point. The yield stress as a criterion may not be sufficient to avoid failure in flexible members or where clearances are very small. An element should always be checked to be sure that, in service, it is compatible with the adjacent elements in the machine.

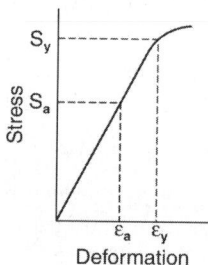

Fig. 2.1: $S-\varepsilon$ curve for a Hooke's law material

Example 2.1: Find the required area of the aluminium arm a of the centrifuge shown in Fig. 2.2. The carriage weight is 2 kN, and its center of gravity is 5 m from the center of rotation. The clearance between the end of the carriage and the retaining wall is 12 mm, and angular velocity is 50 rad/sec. The allowable stress of the material is 275 MN/m² and $E = 69000$ MN/m². Neglect the weight of the arm.

(a) Find the centrifugal force. (b) Find the area of the arm required to support the load. (c) Find the area of the arm required to give a deflection of less than 12 mm.

(a) $C = \dfrac{w}{g}rw^2 = \dfrac{2000}{9.81} \times 5 \times \dfrac{(50)^2}{10^6} = 2.548$ MN

(b) Based on the allowable stress,

$$A = \frac{C}{S_d} = \frac{2.548 \times 10^6}{275 \times 10^6} = 0.00926 \text{ m}^2$$

(c) Based on the 12 mm clearance,

$$A = \frac{Cr}{eE} = \frac{2.548 \times 10^6 \times 5}{12 \times 10^{-3} \times 69000 \times 10^6} = 0.0154 \text{ m}^2$$

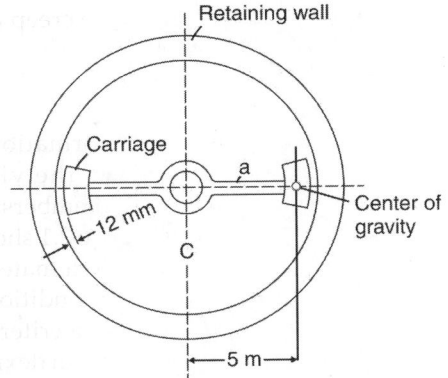

Fig. 2.2: Centrifuge

The arm area must be at least 0.0154 cm² to avoid contact between the carriage and retaining wall, due to elastic deflection, although a 0.00926 m² area would satisfy the strength requirement.

Temperature strains: Change in length or warpage (such as in a bimetallic element) results from temperature changes. Deformation due to temperature change may be added to or subtracted from strains resulting from applied loads.

The change of length

$$e_t = \alpha \, l \, (t_2 - t_1) \qquad \ldots (2.1)$$

where, α is the linear coefficient of thermal expansion.

Deformation due to temperature may result in failure in the same manner as deformation due to load, or they may both contribute to failure.

Example 2.2: Find the minimum area of the arm a in Example 2.1 if, owing to air friction during operation, the temperature increase is 10°C. Assume the arm length to be 5 m.

(a) Calculate the elongation e_t due to the temperature rise. (b) Calculate the net elongation. (c) Calculate the required area.

α for aluminium = 2.36×10^{-6} (from Table 3.2)

$$e_t = 23.6 \times 10^{-6} \times 5 \times 10$$

$$= 1180 \, \mu m \qquad \ldots (a)$$

$$e = 12 - 1.18 \qquad \ldots (b)$$

$$= 10.82 \, mm$$

$$A = \frac{Cr}{eE}$$

$$= \frac{2.548 \times 10^6 \times 5}{10^{-3} \times 10.82 \times 69000 \times 10^6}$$

$$= 0.017 \, m^2$$

Owing to the effect of temperature, the arm area must be larger than indicated by considering only the deformation caused by the centrifugal load.

Instability: Slender compression members and thin plates and shells are subject to sudden large deflections and changes in shapes at stresses above specific critical values of load or stress. These critical stresses are below the yield point of the material. At stresses above critical, the structure is said to become unstable. Compression members subject to instability are called columns. Column failure is discussed in Section 2.15. Plates and shells buckle above critical values of stress. Evaluations of members subject to instability are available in numerous texts dealing with the specific problem.

The stress–strain diagram of a member subject to instability is shown in Fig. 2.3. Large strains occur at stresses above S_a.

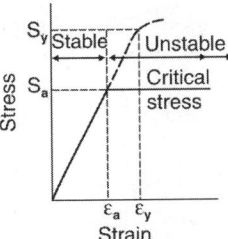

Fig. 2.3: Member subject to instability failure

Slender compression members, fins in bending, thin shear members, and thin flexural members must always be analyzed to determine the critical load and stresses. These may be much lower than limitations set by other criteria.

Plastic Deformation

Plastic strains and deformation are irreversible. When the load is removed, only the elastic portion of the deformation is recoverable. Plastic strains are very large compared to elastic strains for the same stress increment. In many types of structures, large deformation are associated with stresses in the plastic range.

Figure 2.4 shows a ductile material loaded in the plastic range. The slope of the elastic portion is large compared to the plastic portion of the curve.

$$\frac{S_y}{\epsilon_y} >> \frac{S_a - S_y}{\epsilon_a - \epsilon_y} \qquad \ldots (2.2)$$

or $\qquad E >> E_p \qquad \ldots (2.3)$

where, E_p is the modulus of plasticity. Very little advantage is gained in load carrying capacity for the large increase in strain.

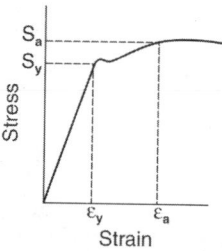

Fig. 2.4: Loading in the plastic range

It is primarily because of the large deformations that rigid machine elements are designed with stress levels below the yield stress. The permanent nature of plastic deformation is also a factor in the design of springs or other members which must retain their original shape or size.

Local plastic strains: Elements made of ductile material where there are large stress gradients, such as beams, thick walled pressure vessels, and torsion members, are often permitted to develop plastic strains. In effect, the plastic strain is localized at a surface, and the resultant total deformation is not increased abruptly, even when the local strains are plastic.

Figure 2.5 shows the deflection y of the cantilever beam with an end load Q. The maximum stress due to bending is

$$S_b = Ql \times \frac{c}{I} \qquad \ldots (2.4)$$

and occurs only at the wall at point a where yielding starts. As the load is increased, the volume of material subjected to plastic strain

increases gradually, as does the total deformation. The fact that an element a on the beam strains plastically is obviously not particularly a cause of failure if the material is ductile.

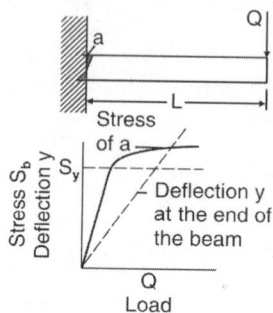

Fig. 2.5: Local yielding

A similar situation occurs in the area of a stress raiser in a ductile material. Plastic flow generally occurs near a notch or a hole or under a rivet. Rather than being a cause of failure, this plastic flow prevents failure by distributing the load more evenly and preventing fracture.

Elements such as hooks, chains, and pressure vessels are often overloaded by 50 to 100 per cent of their working load. Yielding may take place. The resultant work hardening and residual stresses are protection against overload in service. This is called proofing.

Fracture

Fracture of a machine part is the most severe type of failure. The fracture can take place due to tensile or shear stresses. Under certain conditions fracture may be observed as a crack which has not completely traversed its section to cause complete separation. Usually, fracture is represented by a complete separation of the structure into two or more parts. Fracture may be considered as an infinite strain. It is an irreversible failure and very often is catastrophic, causing failure of other elements or of the total mechanism. Fracture of an automobile engine connecting rod usually damages the engine block. The explosion associated with the fracture of a pressure vessel is well known. Fracture of a single gear tooth will play havoc with an entire transmission.

Fracture due to static stress: Figure 2.6 illustrates S–ε diagrams for typical ductile and brittle materials. The basic difference lies in the relative strains associated with ultimate strength. The more brittle the behaviour of the material, the closer the ultimate strain is to the yield strain. The ideal brittle material does not have a yield point; either it is elastic or it fractures. Glass behaves like this.

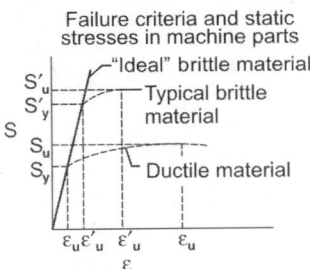

Fig. 2.6: Ductile and brittle materials

It is important, when using brittle materials, to anticipate possible overload conditions and maintain stress levels below the ultimate strength of the material.

Owing to the proximity of the ultimate strength and yield point in brittle materials, fracture may be caused by the presence of a stress raiser such as a hole, notch, or keyway. Since the stress in the area of the stress raiser cannot be distributed by yielding locally, fracture may start at the point of concentration and proceed through the entire section. Stress raiser in ductile materials subjected to static loads have no deleterious effect and can be ignored.

Fracture due to cyclic stress: Fatigue is the name commonly used to describe progressive fracture due to cyclic loading. The effect of a fatigue fracture is the same as failure due to fracture under static load.

Fracture under cyclic loading is much more common than static fracture. It is progressive in nature and requires time to nucleate and grow. It cannot be prevented by a simple overload test prior to use. The endurance limit, which is the lowest alternating stress causing fatigue failure, is lower than the yield point of the material and much less quantitatively

predictable than the yield stress or ultimate stress. Fatigue failure affects ductile and brittle materials alike, and it is very sensitive to the condition of the surface, the presence of stress raisers, and residual stresses.

This type of fracture invariably leads to a service failure during the life of the part. It is therefore sensitive to environmental factors such as corrosion, temperature, wear and handling. Very often, the cyclic nature of the load is not apparent, as when it is due to vibrations.

The endurance stress, or the quantitative effect of a particular stress raiser or design situation, is not always available. More than for any other type of loading, the prevention of fatigue failure depends on qualitative as well as quantitative judgments.

Example 2.3: Find the required clearance and area if the arm *a* of Example 2.2 is to be designed to have a minimum cross sectional area.

$$e = \frac{sl}{E} = \frac{275 \times 5}{69000} = 0.0199 \text{ m} \qquad \text{... (a)}$$

$$e_{total} = e + e_t = 0.0199 + 0.0118 = 0.021 \text{ m} \qquad \text{... (b)}$$

This is the minimum clearance required

$$A = \frac{2.548 \times 10^6}{275 \times 10^6} = 0.00926 \text{ m}^2 \qquad \text{... (c)}$$

This area is based on yielding as a failure criteria.

High temperature, creep and rupture: At temperatures above one-half the absolute melting point of a material, permanent deformation under load starts to occur with time as a factor. Strain increases with no increase in stress. This phenomenon is called *creep*. The rate of deformation, known as the creep rate, is primarily a function of the stress level and temperature. Figure 2.7a illustrates deformation as a function of time and temperature for given stress level. Figure 2.7b illustrates deformation as a function of time and stress level for a given temperature.

Creep continues at a relatively constant rate, then starts to increase rapidly, and culminates in rupture. There is no way of preventing either creep or rupture. Design for elevated temperature is based on two criteria. Deformation

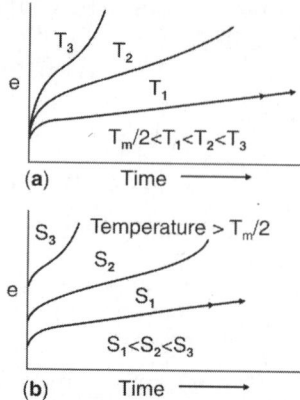

Fig. 2.7: Deformation versus time at high temperature, (a) Constant stress, (b) Constant temperature

due to creep must remain within permissible limits, and rupture must not occur during the design service life. The lower the stress level and/or temperature, the longer the life of the part. Creep of bolts and piping has been a problem in high-temperature power-generating systems for a long time. In recent years, knowledge of this characteristic of materials has increased in importance with the development of gas turbines, jet and rocket engines, and nuclear power systems.

The amount of creep and rupture data available has increased since World War II. In many areas, particularly on long time creep and rupture, several methods are used to extend existing data to cover longer time periods. Analysis in this area is very complex, and, in critical applications, tests for verification are required.

2.5 WEAR

Wear is the unintentional change of shape of the surface of a solid due to the motion of another material relative to it by removal of material. The impinging material may be solid, liquid or gas, or a combination of all (2.1). The word unintentional is stressed. The intentional change of shape defines a machining process. When a gear is formed, both the gear blank and the gear cutter change shape; the gear blank is being machined, but the gear cutter is wearing.

The wear of safety-valve seats is due to the flow of gas or steam. One of the most severe wear situations results from the flow of liquid, with hard particle (such as sand) in suspension, impinging on metal or rock.

Wear and wear resistance are not material properties but are related to the service. Wear can result from either distortion of or removal of material from the surface. Distortion, or "Brinelling," may occur on a surface subjected to high impact loads or high static loads. Proper selection of the surface area and material can reduce this form of wear.

Wear due to material removal is much more complex. It is a function of many parameters in varying degrees. As a result, wear cannot be measured in absolute units. Even under laboratory conditions it is impossible quantitatively to repeat results in wear experiments.

Some of the parameters are listed below:

Surface finish	Hardness
Load	Relative velocity
Environmental conditions	Corrosion resistance
	Notch sensitivity
Temperature	Yield strength
Rate of heat conduction	Ultimate strength
Melting point	Lubricant
Specific heat	System of lubrication
Fatigue strength	Shock

Wear is not static. It varies with time, and thus factors which were not effective initially may have a prominent effect as wear progresses. For example, a coarse cast-iron surface may, in time, become glazed, resulting in reduced friction and heat generation.

All the factors operate cooperatively to determine the rate at which wear occurs. A change in any one will affect all the others.

Under certain conditions or for particular materials, various combinations of factors play more prominent roles than others. Wear resulting from particular combinations of factors is recognized by the appearance of the wear surface at failure.

Wear of machine parts is known as scoring, galling, scuffing, or pitting, depending on the appearance and the type of part on which it occurs. Wear occurs in a number of ways, but falls primarily into one or a combination of four mechanisms, namely adhesion, abrasion, surface fatigue, or fretting.

Adhesion

No matter how well a surface is finished, it consists of microscopic irregularities in the form of hills and valleys. When a pair of surfaces are loaded, the hills in contact are very highly stressed. As shown in Fig. 2.8a, these stresses are many times higher than the average stress on the surfaces. The movement of one surface relative to the other generates heat and increases the temperature at the point of contact.

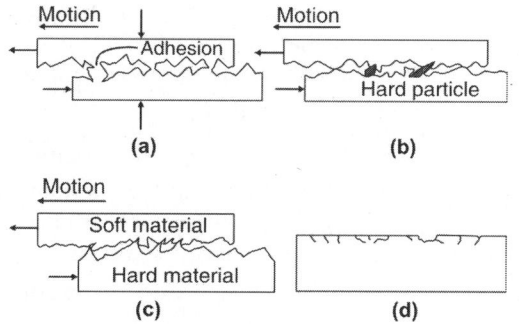

Fig. 2.8: Various wear mechanisms, (a) Adhesive wear, (b) Abrasive wear, (c) Abrasive wear, (d) Surface fatigue

Under certain conditions the hills in contact will fuse. As movement is continued, they will be pulled apart, and particles of material will be removed from the surfaces. When a particle is removed, a new hill and valley are formed. Depending on the materials involved, the resultant valleys may be deeper or shallower than the original surface. The former case will lead to an increased wear rate, as with ductile and tenacious materials; the latter case will result in a decreasing wear rate, with the surface taking on a more polished appearance. This occurs in "breaking-in" operations, particularly in the case of cast iron and other hard and brittle materials.

This type of wear can be mitigated by using materials of different melting points, providing for good heat conduction, providing a large lubricant supply and using hard materials.

Galling (Scoring) is a problem with stainless steels (particularly the 300 series–austenitic), titanium, and the precious metals. When using soft ductile materials in a wear situation, surface should have a good finish and be well lubricated. Under high load conditions adhesion may terminate in seizing, or freezing, of the two surfaces, which would result in the machine becoming inoperative.

Abrasion

Wear due to abrasion is caused by the presence of hard particles between the rubbing surfaces or by the action of a hard surface on a soft surface, resulting in a grinding or machining of material from the surfaces, as shown in Figs 2.8b and 2.8c.

Hard particles are present in the air, on surfaces prior to assembly, and in the lubricating system. This is particularly true in an industrial environment. Hard particles are also generated as a result of adhesive wear, so that abrasive wear may result from adhesive wear. In an extreme instance, large particles may increase the pressure sufficiently to cause galling or seizing.

In general, abrasive wear is continuous and results in a constant wear rate. Failure occurs when the surface thickness or the shape has been changed sufficiently to warrant replacement or repair.

The rate of wear can be reduced by using one soft surface in which hard particles can become embedded, making the hard surface extremely smooth, and preventing contamination of the lubrication system by the use of filters (2.3).

Surface Fatigue (Pitting)

Surfaces of machine elements subjected to high cyclic contact stresses (such as ball and roller bearings or gear teeth) wear because of surface fatigue.

Contact stresses reach high intensities close to the surface and dissipate quickly. Their proximity to the surface causes cracks to nucleate at surface stress raisers. The cracks grow with time, but growth away from the surface cannot be sustained, because of the low stress level. The crack either stops growing or continues to grow beneath and parallel to the surface. Finally, it either turns back toward the surface or consolidates with another crack. In either case, a particle of the surface is released. As this process continues, the surface grows rough in the area where the material has left the surface, since the space becomes a stress raiser which accelerates crack nucleation and growth. The wear resulting from this mechanism is called *pitting*. Figure 2.8d illustrates the growth of fatigue cracks under repeated loads.

This type of wear can be reduced by using a highly polished surface, free of stress raisers, and by developing a high endurance limit at the surface. A high endurance limit may be attained by using hardened alloy material; by surface treatment such as carburizing, nitriding, flame hardening, or chrome plating; and by work-hardening the surface, as with shot peening or rolling.

Fretting

Fretting is sometimes called fretting corrosion because the wear products, when observed, are oxidized. Fretting is a type of wear which occurs at surfaces of tight-fitting parts that are subjected to small-amplitude oscillations. A gear or pulley pressed onto a shaft, which is rotating under load, may, in time, lose its press fit. When the parts are disassembled, red rust is observed. The rust is composed of oxided particles which have been worn off the surface, with a resultant loss of fit, the wear is due to the relative motion between the hub and the shaft, which is flexing as it rotates under load. This relative motion, with pressure between the surfaces, causes a combined abrasive and surface fatigue type of wear, followed by heat generation and the resultant oxidized particles. The situation can be corrected by plating one of the surfaces or by introducing a thin, soft insert (such as paper) between the surfaces.

2.6 CORROSION

Corrosion is a chemical or electrochemical action on a material, causing an undesirable change in shape, colour, or texture. Generally, we think of corrosion in relation to metal parts, although nonmetallics, such as plastics, rubber and neoprene, are also subject to chemical attack.

Chemical Corrosion

General chemical attack, in which a machine part will react with a particular chemical, can easily be anticipated. The reactions of various metals and nonmetals with acids, hydroxides, fluorides, chlorides, and other active chemicals are well established. These reactions increase with temperature. Good design requires a proper choice of materials to attain the desired low corrosion rate. Where data are not available, tests must be run to establish corrosion rates.

In situations where a nonreactive material is not available, or its use is not feasible economically, coatings are used. Chrome plating, cadmium plating, gold plating, galvanizing, glass lining, teflon coating, lead oxide, and other surface finishes permit the use of less expensive and more technically desirable materials in chemical environments.

Electrochemical Corrosion

Electrochemical corrosion is a suitable type of reaction which causes many types of damage, depending on atmospheric conditions, temperature, and state of stress. It may result in general corrosion or pitting of a surface, or damage between two surfaces which are in contact. The latter is the more dangerous, since it occurs unobserved and may terminate in fracture and failure, without warning. This is the case, for example, when corrosion takes place on the shank of a bolt under the head or nut. It is usually the bolt failure which discloses the existence of corrosion.

Electrochemical corrosion is also known as galvanic, or wet, corrosion. It requires an electrically conductive material to be in contact with an electrolyte in the form of water with some salt, acid, or alkali in solution.

This type of corrosion occurs because of a potential difference which develops between any two points electrically in contact in the presence of the electrolyte. One point becomes anodic and the other cathodic. Metal leaves at the anode and goes into solution in the electrolyte, and a current flows between the anode and the cathode. A potential difference is caused by any difference existing between two points or areas. The difference may be due to variations in the composition of the electrolyte or to variations in the metal.

Table 2.2 lists some of the variables which affect electrochemical corrosion. Proper design and maintenance can prevent this kind of corrosion. Most often, these variable can be controlled, if they are anticipated.

Table 2.2: Factors influencing electrolytic corrosion

Variation in the metal

Different metals or alloys in contact

Chemical variations in single alloy

Presence of inclusions

Variations in stress, or presence of residual stresses

Variation in the electrolyte

Temperature variation

Oxygen-ion concentration variation

Metal-ion-concentration variation

Chemical variations in the electrolyte

Different velocities of the electrolyte relative to the metal

---

Wear failure can be reduced by using hard materials of different melting points, providing for good heat conduction by supplying abundant lubrication. Surfaces can be hardened and highly polished free of stress raisers and having high surface endurance limit. To reduce corrosion fatigue, surfaces can be chromium or cadmium or gold plated or coated with teflan, lead oxide, etc. In certain situations extreme pressure (EP) additives, viz. graphite or molybdenum disulphide are recommended for temperatures below 100°C as in rolling bearings.

Antiwear (AW) additives are having similar effect by providing a protective layer on the asperities so that surfaces slide over each other. Calcium sulphonate complex provides EP/AW effect (2.2).

2.7 DEFORMATIONS AND STRESSES

When an external force, or load, is applied to a body, the body will move in the direction of the applied force unless that force is balanced by an opposite force called a *reaction*. Such forces always act in pairs (although mention of the reaction is often omitted). The action of this pair of forces produces a *deformation* of the body, also called a *strain*. The deformation may be only longitudinal, symbolized by *e* and by a plus sign (+) to indicate lengthening of the body or a minus sign (−) to indicate shortening. Sometimes the deformation is also angular, symbolized by α, denoting a change in the angle between adjacent elements of a body.

Unit deformation: The deformation per unit length in the direction of the load is called *unit deformation*, or *strain*. It is symbolized by $\in$, and the units are mm per mm or m/m. Thus

$$\in = \frac{e}{l} \qquad \ldots (2.5)$$

Unit deformation α produced by angular distortion is called *shear strain* and is represented in Fig. 2.9a. Because the angle is very small,

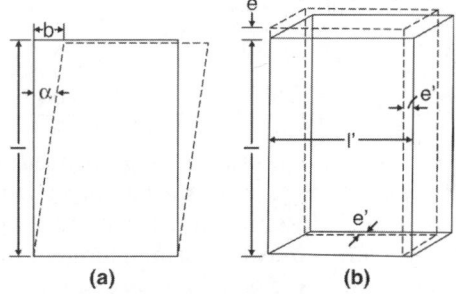

Fig. 2.9: Deformations

$\alpha = \sin \alpha = \tan \alpha$ and the shear strain may be expressed by the relation

$$\alpha = \frac{b}{l} \qquad \ldots (2.6)$$

Poisson's ratio: When a bar is subjected to tension or compression and its length changes from l to $(l \pm e)$, as indicated in Fig. 2.9b, its transverse dimension l' changes to $(l' \mp e')$, the deformations e and e' having opposite signs. The ratio of the transverse strain $\in' = e'/l'$ to the corresponding longitudinal strain $\in = e/l$ is called *Poisson's ratio* and is designated μ. Thus

$$\mu = \frac{\in'}{\in} \qquad \ldots(2.7)$$

Values of μ for various materials, found from tests, are given in Table 2.3.

Table 2.3: Poisson's ratio (μ)

Material	μ
Malleable cast iron	0.230
Nickel	0.239
Gray cast iron	0.210–0.27
Cast steel	0.265
Wrought iron	0.278
Steel, high-carbon	0.295
Steel, mild	0.303
Aluminium, cast	0.330
Monel metal	0.32–0.37
Zinc	0.331
Copper	0.340
Brass	0.340
Ductile iron	0.34–0.37
Aluminium, drawn	0.348
Inconel x	0.41
Lead	0.431
Rubber	0.45–0.49

Stresses: Any deformation of body produced by external forces causes internal forces within its material. These internal forces are called *stresses*. A stress may be normal, tangential, or oblique to the plane on which it acts. A normal stress or the normal component of an oblique stress is referred to as a direct stress. Such a stress may be a tensile stress (+) or a compressive stress (–). A tangential stress is called *shear*. It may be produced either by a transverse load or by torsion.

Normal and tangential stresses seldom exist alone; any load usually produces a combination of both. The relative magnitudes of the stresses are determined by the type of load and by the location of the reference plane with respect to the direction of the load.

Unit stress, used in computations and usually called simply stress, is the amount of the internal force per unit area of the section. In the SI system of units a stress is expressed in Newton per square mm. Direct unit stresses are designated by S, although S_c is sometimes used for compression. Shear is conveniently designated by S_s.

When several stresses act at a point, normal stresses that act on a plane through the point along which there is no shear are called *principle stresses*.

To bring out more clearly the meaning of stresses in various formulas, stresses *created* in a member will be designated by small letters, such as s, s_c, and s_s, while stresses that *characterize* a certain material will be designated by capital letters, such as S_e, S_u, and S_{en}.

Stress–strain diagram: The relation between loads on a specimen of the material and its deformations is determined by tests, and the results may be presented in a diagram like that in Fig. 2.10, where stresses are ordinates and strain as abscissas. The general form of the diagram is the same of either tensile or compressive stresses. Up to point e the strain is

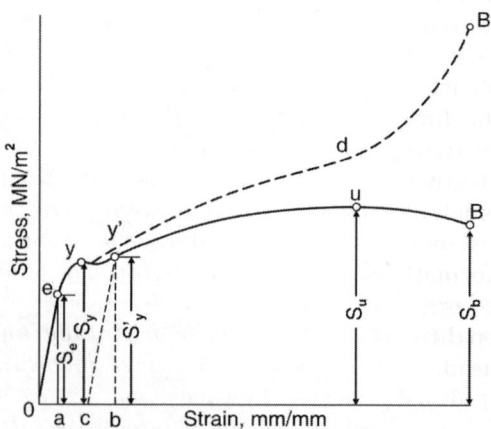

Fig. 2.10: Stress–strain diagram

directly proportional to the stress (Hooke's law), and when the load is removed the body assumes its original shape and size. *The tendency of the material to resume its original shape and size is called elasticity, and the stress S_e is called the elastic limit. The maximum stress up to which the strain of a material remains proportional to the stress is called the proportional limit S_p (not shown in Fig. 2.10).* For the majority of metals S_p is very close to S_e. Therefore they are often considered equal and termed the *proportional elastic limit. An increase of the stress above S_e is accompanied by a faster increase of deformation until point y is reached. At this point the resistance of the molecules begins to break down rapidly, and a sudden and large increase of deformation occurs without an increase in the load.* The corresponding stress S_y is called the *yield point.* After the removal of a load causing a stress greater than S_e the body does not return to its original shape and a permanent set Oc remains. *The maximum stress S_u, is called the ultimate strength. The stress at the breaking point is called breaking stress S_b.*

For the purposes of drawing a stress–strain diagram the stresses are referred to the original area of the test piece. If the actual areas, which gradually become smaller as the piece stretches, are used, the corrected stress curve would rise more rapidly, as shown by curve d in Fig. 2.10. The actual breaking stress at B', S_b' is highest.

In order that a piece may retain its shape after the load has been removed, its stress must not exceed the elastic limit, a knowledge of which is therefore very important. In many instances it is easier to determine the proportional limit S_p. But some materials, such as cast iron or copper, do not have a proportional limit and do not follow Hooke's law. The deformation of such a material increases faster than the stress increases, even within the limits of elastic deformations. Accurate determination of the proportional elastic limit of the material is then difficult, and the yield point is often used instead.

The yield point is therefore also known as the commercial elastic limit. However, the difference between the proportional elastic limit and the yield point may be considerable– up to 20 to 30 per cent; and so whenever possible the design should be based on the elastic limit and not on the yield point. In machine design the elastic limit is one of the most important properties of a material. *For a material that does not have a proportional elastic limit or for one for which a figure has not yet been established, the apparent elastic limit may be used. This means the limit stress which, when the load is removed, leaves such a small permanent deformation (0.02 per cent or less) that for all practical purposes it may be neglected.* It is also called *proof stress.*

2.8 THE MODULI OF ELASTICITY

A modulus of elasticity is defined as the ratio of the stress to the strain below the elastic limit.

The modulus of elasticity in tension, which is symbolized by E, is also called the *direct modulus,* or *Young's modulus.* It may be expressed by the relation

$$E = \frac{s}{\epsilon} \qquad \qquad \ldots (2.8)$$

The value of E, which is expressed in N/m^2 or N/mm^2 or MN/m^2 or GN/m^2, is represented in Fig. 2.10 by the tangent of the angle eOa.

The modulus of elasticity in compression is symbolized by E_c. It is determined in the same way as E. For most ductile materials the values of E_c and E are practically equal.

The modulus of elasticity does not depend upon the ultimate strength or elastic limit of material, but it is a function of the stiffness of the material. Therefore, it is often called the *modulus of rigidity* or the *coefficient of rigidity.* In Fig. 2.11 stress–strain diagrams are given for several grades of steel. These diagrams show that, regardless of the greatly varying values of ultimate strengths and elastic limits, the values of moduli of elasticity represented by the slopes of the lines of elastic deformation are almost the same for all these steels. The modulus decreases only slightly with an increase of the carbon content. A machine part made of soft steel therefore is practically as rigid as one

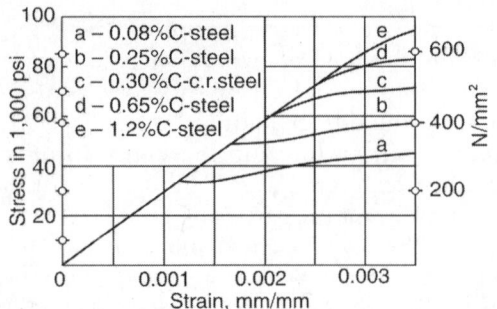

Fig. 2.11: Stress–strain diagram for several steels

made of high-carbon steel, although the one made of high-carbon steel may be two or more times as strong.

The modulus of elasticity in shear or torsion, which is symbolized by G, is also called the *transverse modulus*. It is equal to the ratio of the shear stress S_s and the deformation per unit length $c\theta/l$. Thus (as modulus of rigidity)

$$G = \frac{S_s l}{c\theta} \qquad \ldots (2.9)$$

where, l is the length of the twisted bar, in meters, c is the distance from the surface to the center line, in m; and θ is the angle of distortion, in radians. In accordance with equation 2.6, $c\theta/l = \alpha$ and if $c = 1$, then

$$G = \frac{S_e}{\alpha} \qquad \ldots (2.10)$$

Theoretically, G and E are related as follows:

$$G = \frac{E}{2(1+\mu)} \qquad \ldots (2.11)$$

where, μ is Poisson's ratio. Values of G computed by equation 2.11 are slightly smaller than those found from actual tests, but they may be used in the absence of information about G. Conversely, if E and G are known μ may be computed with sufficient accuracy from the relation

$$\mu = \frac{E}{2G} - 1 \qquad \ldots (2.12)$$

2.9 SIMPLE STRESSES

Tension, compression, and shear are called *simple stresses* when they can be considered

singly. A simple stress is considered to be distributed uniformly over the cross section of the part to which the force is applied.

Tensile stress: The stress s in simple tension in a machine member is equal to the external force F, in Newton, divided by the cross-sectional area A of the member, in square meters, Thus

$$s = \frac{F}{A} \qquad \ldots (2.13)$$

From equation 2.5, the total elongation of a member l meters long is $e = \varepsilon l$. Using for ε its values from equation 2.8 gives

$$e = \frac{sl}{E} \qquad \ldots (2.14)$$

Then, using for s the value given by equation 2.13 results in

$$e = \frac{Fl}{AE} \qquad \ldots (2.15)$$

Compressive stress: For a straight short compression members in which the resultant load F acts through the centre of gravity of every cross section, the stress s is also found from equation 2.13.

If the length of a member is more than twenty times that least radius of gyration of its normal cross section, the member must be treated as a column, and the stresses are determined by one of the column formulas.

Shear stress: A shear stress is induced in a plane between two adjacent parts of a body by two equal external forces acting on opposite sides of the plane and in opposite directions along lines of action parallel to the plane considered. In Fig. 2.12 the load F acting downward upon plate a is transmitted to plate b through the dowel pin c and creates a reaction $R = F$ in plate b. The simultaneous action of the forces F and R induces shear stress in the dowel pin c. Ordinarily it is assumed that the shear stress is distributed uniformly over the cross section of the pin normal to its axis, in which case

$$S_s = \frac{F}{A} \qquad \ldots (2.16)$$

The actual conditions are probably different. The stresses are greatest near the points of action of forces F and R and gradually decrease toward the center of the pin c. The assumption of uniform distribution of the shear stress over the whole area of the cross section is used only to simplify the computations. Such an assumption should never be made where shear is due to bending.

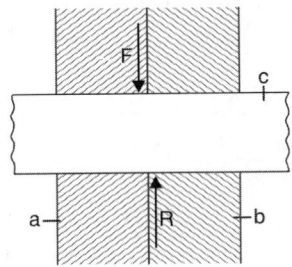

Fig. 2.12: Part in shear

2.10 TORSION

Stresses produced by torsion and bending are termed *compound stresses*, because torsion and bending produce normal and tangential stresses in the same plane simultaneously.

Torsional stress: The stress produced in a member twisted by a couple is pure shear, but its intensity on any fiber depends on the distance to the fiber from the center line of the twisted member. For a circular section, the shear stress is proportional to the distance from this center line.

Equating the external torsional moment T to the internal resisting moment gives

$$T = s_s Z_o \qquad \dots (2.17)$$

where, Z_o is the polar section modulus and s_s is the maximum shear stress. For a solid or hollow circular section, $Z_o = J/c$, in which J is the polar moment of inertia and c is the distance from the axis to the most remote fiber where the stress is s_s.

For a solid round bar, $Z_o = \dfrac{1}{16}\pi D^3$ (*see* Table 2.4) and the maximum shear stress is expressed as follows:

$$s_s = \frac{16T}{\pi D^3} \qquad \dots (2.18)$$

Equation 2.17 does not apply for a non-circular section. Because of the uneven distortion

of such a section, the greatest intensity of shear stress occurs in the outer fibers *closest* to the center line. The value of the maximum shear stress in torsion of one of the sections shown in Table 2.4 can be computed from the data in the table.

The relation between the angular deflection in radians and the torsional moment T for a section can be obtained by substituting for s_s in the equation for torque for the section (such as equation 2.18) the value of s_s from equation 2.9, and solving the resulting equation for θ. Thus, for a round shaft,

$$\theta = \frac{32lT}{\pi D^4 G} \qquad \dots (2.19)$$

Values of θ for other sections are given in the next-to-last column of Table 2.4.

The angular deflection θ of a bar may be expressed also as a function of the stress by applying equation 2.19. Thus for a round bar for which $c = D/2$,

$$\theta = \frac{2ls_s}{DG} \qquad \dots (2.20)$$

Values of θ for other sections are given in the last column of Table 2.4.

Torque: It may be well to recall here the relation between the torque T, the power P in kW, and the speed n in revolutions per minute. If we designate by D, the diameter at the end of which the tangential force F is applied and we express values in meters m and Newton N, the definition of power is

$$P = \frac{\pi D n F}{60 \times 1000} = \frac{2\pi n T}{60 \times 1000} \qquad \dots (2.20a)$$

or if n is rps, $P = \dfrac{2\pi n T}{1000}$

Since $\dfrac{FD}{2} = T$, solving for T gives

$$T = \frac{60 \times 1000 P}{2\pi n} = \frac{30000 P}{\pi n} \qquad \dots (2.21)$$

where, T is expressed in Nm.

Example 2.4: A shaft made of a seamless steel pipe 3 m long, with an outside diameter of 100 mm and wall thickness of 5 mm transmits 190 kW at 370 rpm.

Table 2.4: Torsion of shafts of various cross sections

Type	Cross section	Polar section modulus $Z_o = \dfrac{I}{c}$	Polar radius of gyration k_o	Angular deflection θ	
				In terms of torsional moment T	In terms of maximum Stress s_o
a		$\dfrac{\pi D^3}{16}$	$\dfrac{D}{\sqrt{8}} = 0.354\,D$	$\dfrac{32l}{\pi D^4}\dfrac{T}{G}$	$\dfrac{2l}{D}\dfrac{s_s}{G}$ s_s at circumference
b		$\dfrac{\pi(D_1^4 - D_2^4)}{16 D_1}$	$\sqrt{\dfrac{(D_1^2 + D_2^2)}{8}}$ $= 0.354\sqrt{D_1^2 + D_2^2}$	$\dfrac{32l}{\pi(D_1^4 - D_2^4)}\dfrac{T}{G}$	$\dfrac{2l}{D_1}\dfrac{s_s}{G}$ s_s at outer circumference
c		$\dfrac{\pi b^2 h}{16}$ * $h > b$	$\dfrac{1}{4}\sqrt{b^2 + h^2}$	$\dfrac{16(b^2 + h^2)l}{\pi b^3 h^3}\dfrac{T}{G}$	$\dfrac{(b^2 + h^2)l}{bh^2}\dfrac{s_s}{G}$ s_s at A †
d		$\dfrac{2b^2 h}{9}$ * $h > b$	$\sqrt{\dfrac{(b^2 + h^2)}{12}}$ $= 0.289\sqrt{b^2 + h^2}$	$\dfrac{m(b^2 + h^2)l}{b^3 h^3}\dfrac{T}{G}$ $\dfrac{h}{b} =$ 1 $\quad$ 2 $\quad$ 4 $\quad$ 8 $m =$ 3.56 $\;$ 3.50 $\;$ 3.35 $\;$ 3.21 $n =$ 0.79 $\;$ 0.78 $\;$ 0.74 $\;$ 0.71	$\dfrac{n(b^2 + h^2)l}{bh^3}\dfrac{s_s}{G}$ s_s at A‡
e		$\dfrac{b^2}{20}$ *	$0.289b$	$\dfrac{46.2l}{b^4}\dfrac{T}{G}$	$\dfrac{2.31l}{b}\dfrac{s_s}{G}$ s_s at center of side
f		$0.92b^3$ *	$0.645b$	$\dfrac{0.967l}{b^4}\dfrac{T}{G}$	$\dfrac{0.9l}{b}\dfrac{s_s}{G}$ s_s at center of side

* This value is not a ture value of Z_o but is a value of Z_o for a circular section of equal strength and may be used for determining the maximum stress by the formula $s_s = T/Z_o$.

† At B, shear stress $= 16T/\pi bh^2$.

‡ At B, shear stress $= 9T/2bh^2$.

Assume that the material is 10 C8 (SAE 1010 steel). Determine (a) the maximum shear stress, (b) the shear stress at the inner surface of shaft and (c) the angular deflection of the shaft in a length of 3 m.

The torque is, by equation 2.21

$$T = \frac{30000P}{\pi n} = \frac{30000 \times 190}{\pi \times 370} = 4903.7 \text{ Nm}$$

(a) The polar section modulus of shaft is determined from Table 2.4, for case b, when $D_1 = 100$ mm, $D_2 = 100 - 2 \times 5 = 90$ mm.

$$Z_0 = \frac{\pi(100^4 - 90^4)}{16 \times 100} = 67525 \text{ mm}^3$$

Then by equation 2.17

$$S_S = \frac{4903.7 \times 1000}{67525} = 72.6 \text{ N/mm}^2$$

(b) In round shaft the stress at any point is proportional to the distance from the centre line.

Since $\dfrac{D_2}{2} = 45$, the shear stress at the inner surface is

$$\frac{72.6}{50} \times 45 = 65.34 \text{ N/mm}^2$$

(c) From Table 4.2 the transverse modulus of elasticity of 10 C8 (SAE 1010 steel) is $G = 80000 \text{ MN/m}^2$. By substituting known values in the expression in the fifth column of Table 2.4 for case b, using $l = 3$ m, we get,

$$\theta = \frac{32 \times 3 \times 10^3 \times 4903.7 \times 10^3}{\pi(100^4 - 90^4) \times 80000} = 0.054 \text{ radian}$$

This is equivalent to $\alpha = 0.054 \times \dfrac{180}{\pi} = 3.09°$

2.11 BENDING

When a load is applied to a machine member transversely, it produces stresses of several kinds.

Tensile and compressive stresses: A cantilever beam carrying a concentrated load F at the end, Fig. 2.13a, may serve as an example of tensile and compressive stresses due to bending. The nature of the stresses at a section located at any distance x for the load F may be made evident by imagining that two opposite forces F_1 and F_2, each equal to F, are applied at this section. The forces F_1 and F form a couple whose moment is $M = Fx$. This couple bends the beam, producing a tensile stress on its upper, or convex, side and a compressive stress on the lower, or concave, side, as shown to a larger scale in Fig. 2.13b.

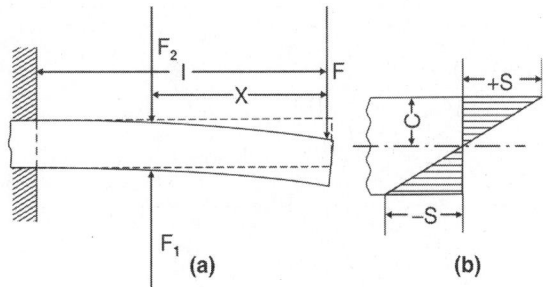

Fig. 2.13: Bending

Shear stress: The remaining force F_2, Fig. 2.13a, produces a shear stress at right angles to the tensile and compressive stresses. This shear stress due to a transverse force, unlike a simple shear stress, is not distributed uniformly over the cross section of the beam.

Its intensity is zero at the outer fibers and increases toward the neutral plane, where it is a maximum.

The distribution of shear stress in a beam can be best explained as follows:

1. If a shear stress exists on a plane at a certain point in a beam, there always exists a shear stress of equal intensity at that point on a plane at right angles to the first plane.
2. When a beam in bending assumes a curved shape, the fibers in any two adjoining planes have a tendency to slide over one another, and shear stresses are created in planes parallel to the neutral plane. This is clearly shown in a simple beam composed of several layers, as indicated in Fig. 2.14. In this case the axial shear stress is zero at the outer fibers and its value gradually increases until it becomes maximum in the neutral plane.

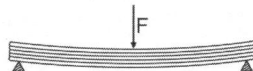

Fig. 2.14: Beam made up of several layers

In accordance with statement 1 in the preceding paragraph, the transverse shear stress at any section follows the same pattern, and has the same value at each point, as the axial shear stress. A general expression for the shear stress at any point i in a beam, both in the axial plane and at right angles to it, may be written by using the designations of Fig. 2.13. This expression is

$$S_s = \frac{F_s}{Iz_1} \int_y^e y \, dA \qquad \dots (2.22)$$

where, F_s is the shear force, in Newtons I is the moment of inertia of the full cross section with respect to the neutral axis x-x, in m^4.

Table 2.5 gives the values of the shear stress found in equation 2.22 for several of the more important beam section.

For an irregular cross section it may be convenient to use a modified equation. With the designation of Fig. 2.15,

$$S_s = \frac{F_s A'c'}{Iz_1} \qquad \text{...(2.23)}$$

where, A' is the cross-sectional area between the plane of the computed stress and the outer fiber of the beam, as the area shown cross-hatched in Fig. 2.15.

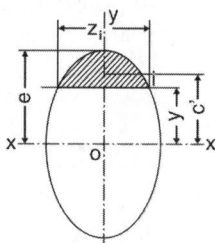

Fig. 2.15: Notation for shear caused by bending

c' is the distance from the center of gravity, or *centroid*, of the cross-hatched area to the neutral axis or neutral plane.

In relatively long metal beams, the actual shear stress is always small compared with the allowable value when the beam can safely withstand the tensile and compressive stresses, and shear is often neglected. In short beams, however, the shear stress may be large enough to be taken into account. Values of the maximum

shear stress for various types of beam and kinds of loads can be computed by using data from Table 2.5.

A valuable feature of shear diagrams, such as those shown in Table 2.4, in Fig. 2.23c, and elsewhere in the text, is that the bending moments have their highest values where the shear changes its sign. The change may be abrupt, as in cases c and d in Table 2.6, or it may be gradual, as in cases f and j.

The relation between the bending moment M, the normal tensile or compressive stress s, and the dimensions of the cross section of a beam may be obtained by equating the external moment to the internal-stress moment. Thus,

$$M = \frac{sI}{c} \qquad \text{...(2.24)}$$

where, I is the moment of inertia of the cross section with respect to the neutral axis normal to the direction of the load F, and where c is the distance from its centre of gravity to the outermost fiber. The quantity I/c is called the *section modulus*. This definition is expressed by the equation

$$\frac{I}{c} = Z \qquad \text{...(2.25)}$$

Table 2.5: Shear stress in beam, caused by bending

Type	Section	Shear stress at a distance y from neutral axis N/mm^2	Maximum shear stress N/mm^2
a		$\dfrac{3F}{2bh}\left[1-\left(\dfrac{2y}{h}\right)^2\right]$	$\dfrac{3F}{2bh}=1.5\dfrac{F}{A}(\text{for } y=0)$
b		$\dfrac{4F}{3\pi r^2}\left[1-\left(\dfrac{y}{r}\right)^2\right]$	$\dfrac{4F}{3\pi r^2}=1.33\dfrac{F}{A}(\text{for } y=0)$
c		$\dfrac{F\sqrt{2}}{b^2}\left[1+\dfrac{y\sqrt{2}}{b}-4\left(\dfrac{y}{b}\right)^2\right]$	$1.591\dfrac{F}{A}\left(\text{for } y=\dfrac{c}{4}\right)$
d		–	$\dfrac{3F}{4a}\left[\dfrac{bc^2-(b-a)d^2}{bc^3-(b-a)d^3}\right](\text{for } y=0)$

Table 2.6: Beams of uniform cross section, loaded transversely

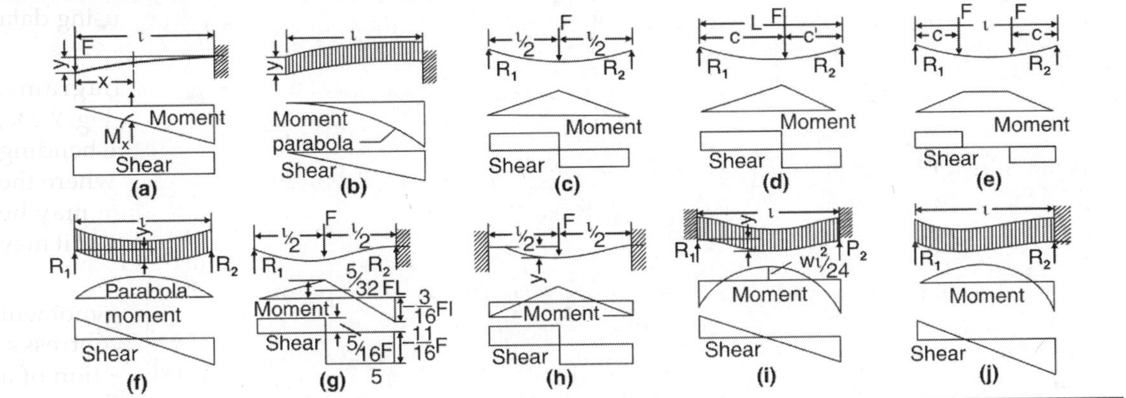

Beam case	Maximum bending moment M	Maximum shear force F_s	Dangerous section	Maximum deflection y	Reactions R_1	Reactions R_2
a	$-Fl$	$-F$	At support	$\dfrac{Fl^3}{3EI}$	F	–
b	$\dfrac{-wl^2}{2}$	$-wl = -F$	At support	$\dfrac{Fl^3}{8EI}$	F	–
c	$\dfrac{Fl}{4}$	$\pm\dfrac{F}{2}$	At center	$\dfrac{Fl^3}{48EI}$	$\dfrac{F}{2}$	$\dfrac{F}{2}$
d	$\dfrac{Fcc'}{l}$	$-\dfrac{Fc}{l}$	At load	$\dfrac{Fc'}{3EIl}\left[\dfrac{c(l+c')}{2}\right]^{\frac{3}{2}}$	$\dfrac{Fc'}{l}$	$\dfrac{Fc}{l}$
e	F_c	$\pm F$	Between the loads	$\dfrac{Fc(3l^2-4c^2)}{24EI}$	F	F
f	$\dfrac{wl^2}{8}$	$\pm\dfrac{wl}{2}=\pm\dfrac{F}{2}$	At center	$\dfrac{5Fl^3}{384EI}$	$\dfrac{F}{2}$	$\dfrac{F}{2}$
g	$-\tfrac{3}{16}Fl$	$-\tfrac{11}{16}F$	At fixed support	$\dfrac{7Fl^3}{768EI}$	$\tfrac{5}{16}F$	$\tfrac{11}{16}F$
h	$\pm\dfrac{Fl}{8}$	$\dfrac{F}{2}$	At support and at center	$\dfrac{Fl^3}{192EI}$	$\dfrac{F}{2}$	$\dfrac{F}{2}$
i	$-\dfrac{wl^2}{12}$	$\pm\dfrac{wl}{12}=\pm\dfrac{F}{2}$	At support	$\dfrac{Fl^3}{384EI}$	$\dfrac{F}{2}$	$\dfrac{F}{2}$
j	$\dfrac{wl^2}{8}$	$\tfrac{5}{8}wl=\tfrac{5}{8}F$	At fixed support	$\dfrac{Fl^3}{185EI}$	$\tfrac{3}{8}F$	$\tfrac{5}{8}F$

By substituting this value in equation 2.24 and solving for the stress, we get

$$s = \frac{M}{Z} \qquad \text{... (2.26)}$$

Rigidity: When a load is placed on a beam, the beam is bent and every portion of the beam is moved in a direction parallel to the direction of the load. The distance that a point on the beam moves is called the *deflection* of the beam at that point. The deflection increases as the distance to the point from the support becomes greater, until a maximum value is reached.

The rigidity of a beam is measured by its deflection, which can be determined by means of the equation

$$\frac{d^2y}{dx^2} = \frac{M}{EI} \qquad \ldots (2.27)$$

where, y is the deflection of the beam and x is the distance from the support. In order to use equation 2.27, an expression for the bending moment M in terms of x for the case considered must first be established and substituted in that equation; the resulting equation is then integrated twice to find y. The maximum bending moments and deflections for the more frequently encountered types of bending loads are given in Table 2.6.

Table 2.7 gives properties of cross sections commonly adopted for beams.

Example 2.5: Using the pipe of example 2.1 as a simple beam with a span of 3 m between the supports, determine (a) the load which applied at the middle of span, will produce a maximum bending stress of 100 N/mm², (b) the corresponding maximum deflection.

(a) According to Table 2.7, the section modulus in bending is,

$$Z = \frac{\pi(100^4 - 90^4)}{32 \times 100} = 33762 \text{ mm}^3$$

Also from Table 2.6 case c, the maximum bending moment is $M = F \times \dfrac{3000}{4} = 750\,F$, substituting the value of s, Z and M in equation 2.26 and solving the resulting equation for F, we get

$$F = 100 \times \frac{33762}{750} = 4502 \text{ N}$$

(b) The maximum deflection can be computed by substituting proper value in the expression in column 5 of Table 2.6 for case c, namely $F = 4502$ N, $l = 3000$ mm, $E = 207$ GN/m², $I = Z.c = 337620 \times 50$

$$y = \frac{4502 \times (3000)^3}{48 \times 207000 \times 33762 \times 50} = 7.25 \text{ mm}$$

2.12 DEFLECTION OF BEAMS

When a beam is loaded, the neutral plane becomes a curved surface. Its intersection with a vertical plane drawn lengthwise through the center of the beam is called the *elastic curve*. The distance to any point on the elastic curve from the initial straight neutral axis of the unloaded beam is a measure of the deflection of the beam at that point.

Radius of Curvature of Elastic Curve

The radius of curvature at any point on a curve is the radius of the circle that is tangent to the curve at the point. This radius is normal to the straight line that is tangent to the curve at the point. In general the elastic curve of a beam is not a circle. However, a very short length of the curve, such as $CD = \Delta l$ in Fig. 2.16 may be considered to be the arc of a circle. Before the load was applied to the beam, the section AE was in the position GH and was parallel to the section BF. After the load has been applied, the sections AE and BF make an angle θ with each other but remain perpendicular to the elastic curve CD. They are parts of radii of the same circle, and the intersection of these radii at O is the *center of curvature*. Because of the bending of the beam, the fiber HF lengthens to EF. The increase in length is $EH = \lambda$, and the unit elongation, or strain, is

$$\epsilon = \frac{\lambda}{\Delta l} \qquad \ldots (2.28)$$

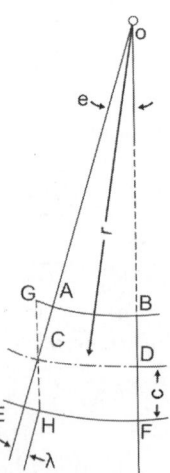

Fig. 2.16: Element of a bent beam

Table 2.7: Properties of various cross sections

Type	Section	Moment of inertia I	Distance to farthest point c	Section modulus $Z = \dfrac{I}{c}$	Radius of modulus gyration $k = \sqrt{\dfrac{I}{A}}$
a		$\dfrac{bh^3}{12}$	$\dfrac{h}{2}$	$\dfrac{bh^2}{6}$	$0.289h$
b		$\dfrac{b}{12}(H^3 - h^3)$	$\dfrac{H}{2}$	$\dfrac{b(H^3 - h^3)}{6H}$	$\sqrt{\dfrac{H^3 - h^3}{12(H - h)}}$
c		$\dfrac{BH^3 - bh^3}{12}$	$\dfrac{H}{2}$	$\dfrac{BH^3 - bh^3}{6h}$	$\sqrt{\dfrac{BH^3 - bh^3}{12(BH - bh)}}$
d		$\dfrac{Bc_1^3 - bh^3 + ac_2^3}{3}$ $h = c_1 - d$	$c_1 = \dfrac{aH^2 + bd^2}{2(aH + bd)}$ $c_2 = H - c_1$	$\dfrac{I}{c_1}$ and $\dfrac{I}{c_2}$	$\sqrt{\dfrac{I}{Bd + a(H - d)}}$
e		$\dfrac{BH^3 + bh^3}{12}$	$\dfrac{H}{2}$	$\dfrac{BH^3 + bh^3}{6H}$	$\sqrt{\dfrac{BH^3 - bh^3}{12(BH + bh)}}$
f		$\dfrac{(6b^2 + 6bb_o + b_o^2)h^3}{36(2b + b_o)}$	$\dfrac{(3b + 2b_o)h}{3(2b + b_o)}$	$\dfrac{(6b^2 + 6bb_o + b_o^2)h^2}{12(3b + b_o)}$	$\sqrt{\dfrac{I}{A}}$
g		$\dfrac{\pi D^4}{64} = \dfrac{\pi R^4}{4}$	$\dfrac{D}{2} = R$	$\dfrac{\pi D^3}{32} = 0.098D^3$	$\dfrac{D}{4} = \dfrac{R}{2}$
h		$\dfrac{\pi}{64}(D_1^4 - D_2^4)$ $= \dfrac{\pi}{4}(R_1^4 - R_2^4)$	$\dfrac{D_1}{2} = R$	$\dfrac{\pi(D_1^4 - D_2^4)}{32D_1}$	$\dfrac{\sqrt{D_1^2 + D_2^2}}{4}$ $\dfrac{\sqrt{R_1^2 + R_2^2}}{2}$
i		$\dfrac{\pi bh^3}{64}$	$\dfrac{h}{2}$	$\dfrac{\pi bh^2}{32}$	$\dfrac{h}{4}$

Since the triangles COD and ECH are similar,

$$\frac{OD}{CD} = \frac{CH}{EH}$$

or

$$\frac{r}{\Delta l} = \frac{c}{\lambda} \qquad \ldots (2.29)$$

Equating the values of e from 2.8 and 2.28, and replacing s by the expression obtained by solving equation 2.24, gives.

$$\frac{\lambda}{\Delta l} = \frac{s}{E} = \frac{Mc}{EI}$$

Also, from equation 2.29,

$$\frac{\lambda}{\Delta l} = \frac{c}{r}$$

Finally, by equating the two values of $\lambda / \Delta l$ and solving for r, these result

$$r = \frac{EI}{M} \qquad \ldots (2.30)$$

From Fig. 2.16 it is evident that an increase in the deflection of a beam makes the radius of curvature smaller. Thus, the deflection is inversely proportional to the radius of curvature r. Therefore the deflection is directly proportional to M and inversely proportional to E and I.

This analysis also permits the establishment of a rule for determining the sign of the bending moment by noting the location of the center of curvature of the bent beam: If the center of curvature is below the beam, as in the case of a cantilever beam, the bending moment is considered *negative*; if the center of curvature is above the beam, as in the case of a simple beam, the bending moment is considered *positive*.

Deflection formulae: In Fig. 2.17 let OU be the neutral axis of an unloaded cantilever beam supported at O and having its end U free; and let $OKRS$ represent to an exaggerated scale the elastic curve of the beam carrying a uniform load from O to P and having no load from P to U. For practical purposes it may be assumed that the point U moves downward vertically and its deflection is the distance US. The problem is to find the deflection QR of any point Q.

Suppose that $BCDB$ is the moment diagram. Divide the part OK of the elastic curve into a

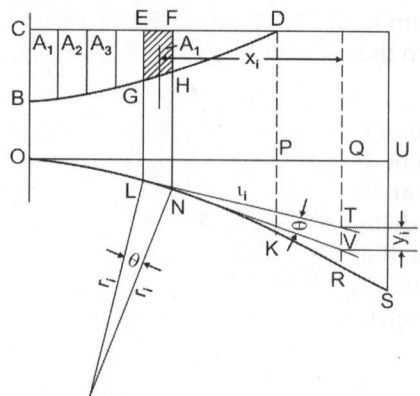

Fig. 2.17: Notation for deflection of a beam

number of small segments, one of which is LN. Since the part PU of the beam does not carry any load, the part KS of the elastic curve is a straight line. At L and N are drawn radii of curvature having a length r_i and forming an angle θ; the tangents LT and NV also form the angle θ with each other. If the length a_i of the arc LN is very small, it may be assumed that $LT = NV = l_i = x_i$. The bending of the portion of the beam from L to N causes point Q to deflect through the distance $TV = y_i$. But $y_i = l_i \theta$; $\theta = a_i / r_i$; and $r_i = EI / M_i$. Therefore,

$$y_i = \frac{l_i a_i}{r_i} = \frac{l_i a_i M_i}{EI} \qquad \ldots (2.31)$$

Now the bending moment M_i at L is equal to the ordinate EG of the moment diagram. Also a_i is practically equal to the distance GH. Therefore $a_i M_i$ is equal to the area A_i of the portion $EFHG$ of the moment diagram, and

$$y_i = \frac{x_i A_i}{EI} \qquad \ldots (2.32)$$

The product $x_i A_i$ is the moment of the moment area A_i with respect to point Q of the beam.

The total deflection QR is the sum of the deflections y_i caused by the bending of the small parts into which the beam was divided. Thus the deflection of point Q is

$$y = y_1 + y_2 + \ldots + y_n = \frac{x_1 A_1 + x_2 A_2 + \cdots + x_n A_n}{EI}$$

$$= \frac{\sum x_i A_i}{EI} \qquad \ldots (2.33)$$

From a study of theoretical mechanics it is known that

$$\Sigma x_i A_i = \bar{x} A \qquad \ldots (2.34)$$

where, A is the entire area of the portion of the moment diagram between the support and the point at which the deflection is to be found, $\bar{x}$ is the distance from the center of gravity of that area to that point for a cantilever beam, and to the nearest support for a simple beam with symmetrical loading. Thus

$$y = \frac{\bar{x} A}{EI} \qquad \ldots (2.35)$$

Example 2.6: Determine the expression for the deflection of a cantilever beam with a concentrated load on its free end.

The moment diagram for this beam is shown in Fig. 2.18, where the free end is at A and the support is at B. To find the deflection of a point C at a distance x from the free end, the moment area $CBDE$ between the point C and the support is divided into the rectangle $CBGE$ and the triangle EGD. Since the ordinate CE is equal to $-Fx$, the area of the rectangle and its moment arm are, respectively,

$$A = -Fx(l-x) \qquad \bar{x} = \frac{l-x}{2}$$

For the triangle EGD, the height $GD = BD - BG = -Fl - (-Fx) = -F(l-x)$. Hence,

$$A = \frac{-F(l-x)(l-x)}{2} = \frac{-F(l-x)^2}{2} \text{ and } \bar{x} = \frac{2}{3}(l-x)$$

The sum of the moments of the rectangle and the triangle is

$$\Sigma x_i A_i = \frac{Fx(l-x)^2}{2} - \frac{F(l-x)^3}{3} = -\frac{F(2l^3 - 3l^2 x + x^3)}{6}$$

Hence, the deflection at C, by equation 2.31, is

$$y = -\frac{F(2l^3 - 3l^2 x + x^3)}{6EI}$$

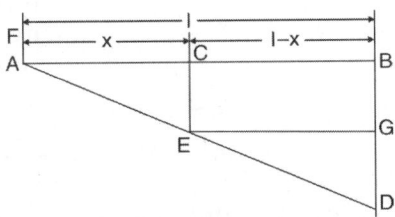

Fig. 2.18: Moment diagram

The maximum deflection is at the free end and is found by making x equal to zero. Thus

$$y_{max} = -\frac{F(2l^3)}{6EI} = -\frac{Fl^3}{3EI}$$

2.13 STATICALLY INDETERMINATE STRUCTURES

Stress in a machine part can be determinate only if all external forces, both loads and reactions, acting upon the part are known. The external loads must be given or found from data that are furnished. The reactions are determined, either completely or partially, from the following three conditions of equilibrium of a free body:

1. The summation of all horizontal forces must be zero.
2. The summation of all vertical forces must be zero.
3. The summation of the moments of all forces with respect to any point must be zero.

If the laws of equilibrium are sufficient for the determination of the reactions, the structure is said to be *statically determinate*.

Redundant elements: If a structure has more supports or members than are necessary for stability, the three equilibrium conditions are not sufficient for determining the reactions. Additional equations must then be set up by taking deformations of the structure into account. Such a structure is said to be *statically indeterminate*; and the supports or members that can be removed without destroying the stability of the structure are called *redundant elements*.

Figure 2.19a shows a beam that is statically indeterminate externally because if the support at A were removed, the beam would remain a stable cantilever beam. Therefore the support is redundant. The structures shown in Fig. 2.19b and c are statically indeterminate internally

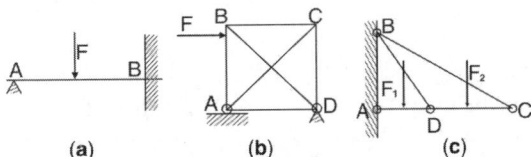

Fig. 2.19: Structures with redundant elements

because if one member, as *BD*, were removed from either structure, the remaining members could carry the loads. Each structure therefore has one redundant member.

A general rule that helps in the design of a statically indeterminate structure may be stated as follows: In transmitting a load to its final support the stresses in the members take the *most rigid path*.

There are many ways of analyzing statically indeterminate structures. In the simplest method, deformations are used explicitly in setting up equations that give relations between the forces, moments, and stresses.

A typical method of analysis will be illustrated by the following examples.

Example 2.7: A machine weighing 100 MN is supported on three very short columns. The columns have equal lengths and cross-sectional areas and are spaced as shown in Fig. 2.20a. Assuming that the machine frame and the supporting foundations are so rigid that their deformations can be neglected, find the load on each column.

If the loads on the columns are designated as F_1, F_2, and F_3, the conditions for equilibrium of vertical forces and equilibrium of moments about C.G. may be written as follows:

$$F_1 + F_2 + F_3 = 100 \qquad \ldots \text{(a)}$$

and $\quad 1.8F_1 + 0.6F_2 = 1.8F_3$ or $3F_1 + F_2 = 3F_3 \quad \ldots$ (b)

Because of the rigidity of the foundation and of the supported body, the tops of the columns, after deformations have taken place, will be on a straight line, as shown in Fig. 2.20b. From the similar triangles, it follows that

$$\frac{\Delta l_2 - \Delta l_1}{1.2} = \frac{\Delta l_3 - \Delta l_2}{2.4} \qquad \ldots \text{(c)}$$

The general expression for the total vertical deformation of any column is

$$\Delta l = \frac{Fl}{AE} \qquad \ldots \text{(d)}$$

where, *A* is the cross-sectional area of the column in square mm.

Since all three columns have the same length *l*, the same cross-sectional area *A*, and the same modulus of elasticity *E*, we will introduce the designation $l/AE = C$. Then

$$\Delta l_1 = F_1 C, \quad \Delta l_2 = F_2 C, \quad \Delta l_3 = F_3 C \qquad \ldots \text{(e)}$$

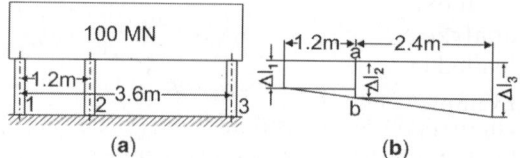

Fig. 2.20: Conditions for example 2.4

Substituting values of Δl from equation *e* in equation *c* gives the equation

$$2F_1 - 3F_2 + F_3 = 0 \qquad \ldots \text{(f)}$$

Solving equations a, b and f simultaneously, these result

$$F_1 = 28.6 \,\text{MN} \quad F_2 = 32.1 \,\text{MN} \qquad F_3 = 39.3 \,\text{MN}$$

Example 2.8: Determine the loads on the columns described in example 2.7 if column 2 has been machined 0.05 mm shorter than the others, the normal length of all three columns is 300 mm. The area A of each cross-section is 30 sq.cm., and the steel has a modulus of elasticity of $207 \times 10^3 \,\text{N/mm}^2$.

As in example 2.7

$$F_1 + F_2 + F_3 = 100 \qquad \ldots \text{(g)}$$

and from b above $3F_1 + F_2 = 3F_3 \qquad \ldots$ (h)

Again the tops of the columns after the deformations have taken place, will be on a straight line, as in Fig. 2.20b since columns 2 was originally 0.05 mm short, however, the distance corresponding the ab in Fig. 2.20b will be $\Delta l_2 + 0.05$, and the equation based on the deformation is

$$\frac{\Delta l_2 + 0.05 - \Delta l_1}{1200} = \frac{\Delta l_3 - \Delta l_2 - 0.05}{2400} \qquad \ldots \text{(i)}$$

In this case, the vertical deformation of each column may be taken as

$$\Delta l = \frac{1000\,F \times 300}{3000 \times 207 \times 10^3} = 0.483 . \times 10^{-2}\,F \qquad \ldots \text{(j)}$$

or $\Delta l_1 = 0.00483\,F_1$, $\Delta l_2 = 0.00483\,F_2$, and $\Delta l_3 = 0.00483\,F_3$ substituting these values in equation g gives

$$2\,F_1 - 3\,F_2 + F_3 = 41.4 \qquad \ldots \text{(k)}$$

Solving equations g, h, and k simultaneously, these result

$$F_1 = 34.48 \,\text{MN}, F_2 = 23.27 \,\text{MN}, F_3 = 42.25 \,\text{MN},$$

Comparison of the loads in example 2.8 with the corresponding loads of example 2.7 shows what a great difference a small change of length can make.

In example 2.9, which follows, we shall analyze a statically indeterminate structure loaded in tension.

Example 2.9: Determine the loads on the rods in Fig. 2.21a. The rods are hinged at both ends and are loaded by a suspended weight of 12000 N. The cross-sectional area of the vertical rod is $A_2 = 6$ sq.cm and that of the rods 1 and 3 is $A_1 = A_3 = 3$ sq.cm.

The forces acting on the rods will be designated by F_1, F_2, and F_3, Since cos 30° = 0.866 and cos 45° = 0.707, summation of the vertical forces is

$$0.866\,F_1 + F_2 + 0.707\,F_3 = 12000 \qquad \ldots (1)$$

The relation involving the summation of horizontal forces should not be used before checking whether rod 2 remains absolutely vertical. However, the necessary second and third equation may be obtained by considering the elongations of the rods indicated in Fig. 2.21b. If it is assumed that the end of rod 2 moves vertically downwards from point a to b, the end of rod 1 must swing from c to b and the end of rod 3 must swing down from d to b. Then

$$\frac{\Delta l_1}{\cos 30°} = \Delta l_2 = \frac{\Delta l_3}{\cos 45°} \qquad \ldots (m)$$

The elongations must be expressed as follows:

$$\Delta l_1 = \frac{l\,F_1}{A_1 E\,\cos 30°}$$

$$\Delta l_2 = \frac{l\,F_2}{A_2 E}$$

$$\Delta l_3 = \frac{l\,F_3}{A_2 E\,\cos 45°} \qquad \ldots (n)$$

We can now substitute these expressions in equation m. Replacing cos 30° by 0.866, cos 45° by 0.707, A_1 and A_3 by 3 and A_2 = 6 and simplifying the results, we obtain

$$F_1 = 0.375\,F_2 \qquad \ldots (o)$$

$$F_3 = 0.25\,F_2 \qquad \ldots (p)$$

Solving equations l, o, and p simultaneously, we get the required answers;

$$F_1 = 3000\ \text{N}, F_2 = 8000\ \text{N}, F_3 = 2000\ \text{N}$$

It is interesting to check the possible angle β formed by rod 2 with the original vertical position. A summation of horizontal components gives

$$3000 \sin 30° - 8000 \sin \beta - 2000 \sin 45° = 0$$

From this

$$\sin \beta = \frac{3000 \times 0.5 - 2000 \times 0.707}{8000} = 0.01075$$

Hence β = 0.61°, a value so small that it did not effect the accuracy of equation j.

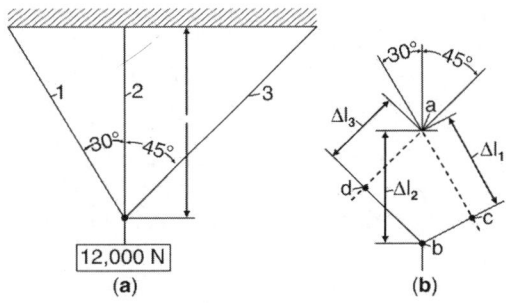

Fig. 2.21: Indeterminate structure loaded in tension

2.14 INDETERMINATE BEAMS

Basically, the procedure for analyzing a statically indeterminate beam is the same as that used for a structure consisting of parts loaded in tension and compression. The main difference is in the greater use of moment equations, of the relations between the reactions, and of shear diagrams. The redundant elements are supports that are not necessary for stability of the beam. Again the procedure can be best illustrated by analyzing some typical cases.

Beams fixed at one end and supported at the other. In Fig. 2.22a is shown a beam fixed at one end and supported at the other, and carrying a uniformly distributed load. The full line in Fig. 2.22b represents the elastic curve. If the left-hand support were removed, the beam would become a cantilever and the elastic curve would be represented by the broken line. In this event, the loading would correspond to that for case b in Table 2.6 and the maximum deflection would be

$$y_{\max} = \frac{Fl^3}{8EI}$$

The reaction of the left-hand support may be considered a concentrated load R_1 on the free end of the assumed cantilever. This reaction must deflect the beam upward a distance equal to y_{max}. According to case a in Table 2.6.

$$y_{max} = \frac{R_1 l^3}{3EI}$$

Evidently

$$\frac{R_1 l^3}{3EI} = \frac{Fl^3}{8EI}$$

and

$$R_1 = \frac{3}{8}F \qquad \qquad \text{... (2.36)}$$

With R_1 known, the shear and moment diagrams are constructed as shown in Fig. 2.22c and d. At point A in Fig. 2.22d the bending moment is zero. This point is called a *point of inflection*. The only stress here is shear.

The following example will illustrate how to take into account the conditions that the supports are not on the same level.

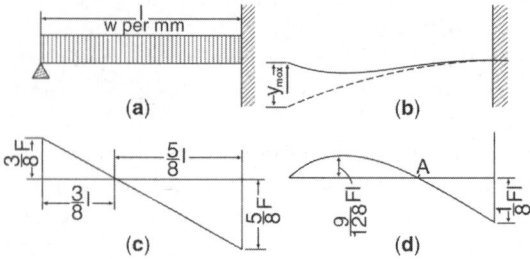

(a)   **(b)**   **(c)**   **(d)**

Fig. 2.22: Beam with one end fixed and the other simply supported

Example 2.10: A 150 mm by 170 N I beam with a 2 m span are fixed at one end, are simply supported at the other, and carry a uniformly distributed load of 32 kN. The supports settle 3 mm under the action of load. Find the reactions on supports and the maximum fibre stress in the beam I = 839.1/cm⁴.

If the supports are removed, the deflection at the end would be

$$y_1 = \frac{R\,l^3}{8EI} = \frac{32000 \times (2000)^3}{8 \times 20700 \times 839.1 \times 10^4} = 18.42 \text{ mm}$$

But the reaction of the support brings the end up a distance equal to

$$y_2 = \frac{R'\,l^3}{3EI} = \frac{R' \times (2000)^3}{3 \times 207000 \times 839.1 \times 10^4} = 1.53 \times 10^{-3}\,R'$$

Equating the deflection gives

$$18.42 = 1.53 \times 10^{-3}\,R' + 3$$

$$R' = 10078 \text{ N}$$

If the supports did not settle, the reaction R_1 of one beam would be

$$R_1 = \frac{3}{8}R = \frac{3}{8} \times \frac{64000}{2} = 12000 \text{ N}$$

Because settling of supports reduced their reactions, the reaction on the fixed ends are increased from

$$R_2 = 32000 - 12000 = 20000 \text{ N}$$

$$R'_2 = 32000 - 10078 = 21922 \text{ N}$$

The dangerous bending moment is at the fixed end and is

$$M = R_1\,l = \frac{Fl}{2}$$

Without settling, this moment for one beam would be

$$M = 12000 \times 2000 - 32000 \times \frac{2000}{2} = -8 \times 10^6 \text{ N mm}$$

Because of settling, it is

$$M' = 10078 \times 2000 - 32000 \times \frac{2000}{2} = -9.844 \times 10^6 \text{ N mm}$$

The maximum stress, with

$$Z = \frac{I}{c} = \frac{839.1}{7.5} = 111.88 \text{ cm}^3 \,,$$

is

$$S = \frac{9.844 \times 10^6}{111.88 \times 10^3} = 88 \text{ N/mm}^2$$

Beam on three supports: In Fig. 2.23a a beam is shown with two equal spans each l m, long and uniformly loaded with w N/m. The supports are on the same level. The curved full line in Fig. 2.23b represents the elastic curve. If the center support is removed, the beam becomes a simple beam with a span $2l$, as shown by the broken line. Its deflection in the middle would be, by case f in Table 2.6

$$y = \frac{5}{384}\frac{F(2l)^3}{EI}$$

where, $F = 2wl$. Now, to bring the beam back to its original shape, a single force R_2 must be applied at the center. The upward deflection, by case c in Table 2.6, is

$$y = \frac{1}{48}R_2\frac{(2l)^3}{EI}$$

Equating the expressions for the deflections and solving for R_2 gives

$$R_2 = \tfrac{5}{8}F = \tfrac{5}{4}wl \qquad \text{... (2.37)}$$

Since $R_2 + 2R_1 = F$,

$$R_1 = \frac{F - R_2}{2} = \frac{2wl - \frac{5}{4}wl}{2} = \frac{3}{8}wl \quad \ldots (2.38)$$

With the reactions determined, the shear force diagram can be drawn as in Fig. 2.23c, and the points A and B at which the shear is zero can be located. The moment diagram is shown in Fig. 2.23d in which at points A and B moment is zero. They are points of contraflexure.

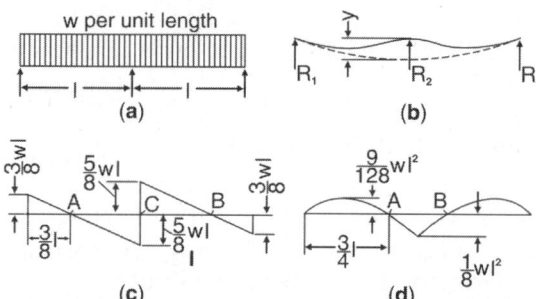

Fig. 2.23: Beam on three supports equally spaced

Continuous beams with more than three supports are seldom used in machine design and therefore will not be discussed here.

General moment equation: The beam shown in Fig. 2.24 rests on two supports and carries several types of loads. There is uniformly distributed load of w/m length and there are also two concentrated loads, F_1 to the left of left-hand support and F_2 between the supports. The reactions on the supports from all loads are R_1 and R_2.

Taking moments about point A of all the forces to the left of that point, we have

$$M_A = -F_1(a + x) + R_1 x - \tfrac{1}{2}wx^2 - F_2(x - b)$$
$$= -F_1 a + (R_1 - F_1)x - \tfrac{1}{2}wx^2 - F_2(x - b)$$

But $-F_1 a$ is the moment of all loads to the left of support 1 about point 1 and may be designated M_o. Furthermore, $R_1 - F_1$ is the shear force at the right-hand edge of the support. If this shear is designated V_1,

$$M_A = M_o + V_1 x - \tfrac{1}{2}wx^2 - F_2(x - b) \quad \ldots (2.39)$$

This is called the *general moment equation*. By means of it, when the moment M_o at a support

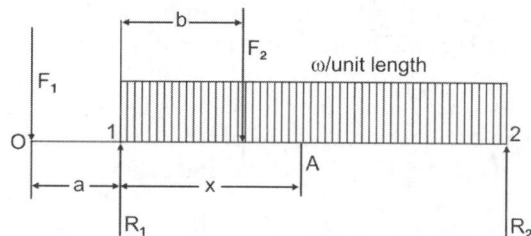

Fig. 2.24: General case of beam loading

is known, the shear and bending moment at any point to the right of that support due to any given loading may be determined.

Example 2.11: Using the general moment equation, find the bending moment at the points A and C of the beam in Fig. 2.23, also draw the moment diagram.

Applying equation 2.39, in which $M_o = 0$, $V_1 = R_1 = \frac{3}{8}wl$, and $x = \frac{3}{8}l$, we get for the bending moment at A

$$M_A = \tfrac{3}{8}wl \times \tfrac{3}{8}l - w\frac{\left(\tfrac{3}{8}l\right)^2}{2} = \frac{9}{128}wl^2$$

For point C, where $x = l$,

$$M_C = \tfrac{3}{8}wl \times l - \tfrac{1}{2}wl^2 = -\tfrac{1}{8}wl^2$$

By calculating the moments at several points of the beam in this manner, the moment diagram in Fig. 2.23d can be drawn.

If the beam in Fig. 2.23 were cut in two at the center support so as to form two simple beams, the maximum bending moment in each simple beam would be $M = -\tfrac{1}{8}wl^2$. Thus, the maximum bending moment for a continuous beam on three supports equally spaced is the same as that for two simple beams on the same supports.

2.15 STRESSES IN COLUMNS

A cast-iron compression member whose length is more than 6 times as great as its least lateral dimension, or a compression members of a ductile material whose length is 8 times as great, is likely to buckle and should be treated as a column.

End conditions: The resistance of a column to failure depends upon the condition of its ends. The least favorable case is when one end of the column is fixed and the other is free, as in Fig. 2.25a. Smaller stresses result when both ends are round-ended and guided or hinged, as in Fig. 2.25b. Still more favorable is the case when one end is fixed and the other is round-ended and guided or hinged, as in Fig. 2.25c. The smallest stresses are produced when both ends are fixed rigidly, as in Fig. 2.25d.

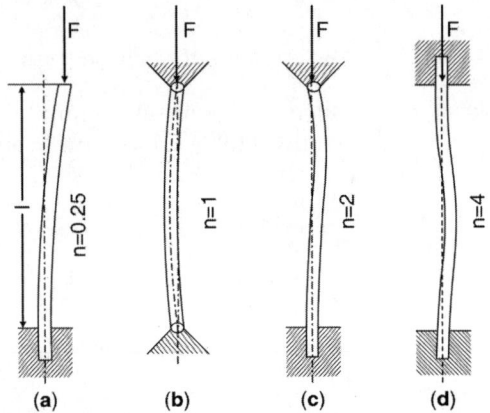

(a) (b) (c) (d)

Fig. 2.25: End conditions of columns

Depending upon the *slenderness ratio*, which is the ratio of the length l of the column to the least radius of gyration k of the cross section, columns are divided into two groups: *Short columns* and *long columns*.

The radius of gyration of any section with regard to any axis through the center of gravity of the section can be found by the relation

$$k = \sqrt{\frac{I}{A}} \qquad \ldots (2.40)$$

where, I is the moment of inertia of the section with regard to the same axis, and A is the area of the section. Expressions for the radius of gyration for various sections commonly used as columns are given in the last column of Table 2.7. In determining the slenderness ratio of a column the least radius of gyration of the section must be used.

Short columns: Cast-iron columns with a slenderness ratio l/k not greater than 80, and columns of steel and other ductile materials for which l/k is not greater than 100, are considered short columns. In a short column the intensity of the simple compressive stress on its concave side is augmented by the stress

that arises from bending under the axial load. In order to prevent failure, the combined stress should always be below the elastic limit.

Several formulas have been proposed for determining the relation between the external load F and the induced stress s_c in short column.

Gordon-Rankine formula gives the crippling load:

$$F = \frac{S_e A}{1 + a\left(\frac{l}{k}\right)} \qquad \ldots (2.41)$$

The value of a for hinged ends for steel is $\dfrac{1}{6250}$, $CI = \dfrac{1}{1250}$ and its value is $\dfrac{1}{4}$ of these values for both ends fixed.

The *Ritter formula*, which is a modification of the Gordon-Rankine formula, has the advantage that it can be applied for any material for which the elastic limit S_e and the modulus of elasticity E are known. It is

$$S_c = \frac{F}{A}\left[1 + \left(\frac{l}{k}\right)^2 \frac{S_e}{\pi^2 n E}\right] \qquad \ldots (2.41a)$$

Where, in addition to the designations already mentioned, A is the cross sectional area of the member, in m^2 and n is the coefficient of end conditions. Values of n for various conditions are given in Fig. 2.25.

Example 2.12: A steel bar 500 mm long, acting as a column with both ends hinged, supports a weight of 55 kN. The cross-section of the bar is a rectangle 45 mm × 25 mm. The elastic limit of steel is 300 N/mm² and its modulus of elasticity is 207 × 10³ N/mm², determine the stress in the outer fibres.

From case a, in Table 2.7, the least radius of gyration is

$$k = 0.289 \times 25 = 7.225 \text{ mm}$$

Hence, $\dfrac{l}{k} = \dfrac{500}{7.225} = 69.2$ which is less than 100.

From Ritter's formula (eq. 2.41a) and with end condition coefficient n as 1

$$S_e = \frac{55000}{45 \times 25}\left[1 + \left(\frac{500}{7.225}\right)^2 \times \frac{300}{\pi^2 \times 207000}\right]$$

$$= 83.27 \text{ N/mm}^2$$

Long columns: Long columns are those with a slenderness ratio l/k greater than 100 for ductile materials and greater than 80 for cast iron. Long columns fail by buckling due to instability. The ultimate or breaking load for a long column may be found from Euler's rational formula, which is

$$F_u = \frac{n\pi^2 E}{l^2} = \frac{n\pi^2 AE}{\left(\dfrac{l}{k}\right)^2} \qquad \text{...(2.42)}$$

in which the coefficient n theoretically has the same values as in Ritter's formula. However, test data indicate that in long columns the end conditions do not exert so great an influence.[1]

Because of the instability of long columns when subjected to maximum loads, they are avoided in machines unless the maximum load is known very accurately and the design is based on that load.

2.16 INDUCED STRESSES

Because of the deformations produced in a machine part by the applied loads, other stresses known as *induced stresses* or *secondary stresses* are created.

Shear caused by tension or compression: When a force F produces a normal tensile or compressive stress $s = F/A$ in a body, it also produces on every oblique plane of the body a shear stress s_s', as indicated in Fig. 2.26. The magnitude of this shear stress is

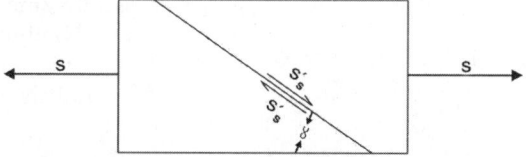

Fig. 2.26: Shear produced by a direct stress

$$s_s' = 0.5\, s \sin 2\alpha \qquad \text{...(2.43)}$$

This stress reaches a maximum in a plane forming an angle $\alpha = 45°$ with the main plane. In this case

$$s_s' = 0.5\, s \qquad \text{...(2.44)}$$

Normal stress caused by shear: Similarly, as indicated in Fig. 2.27, a force producing a shear stress s_s in a body produces also a normal stress s' on every oblique plane of the body. The magnitude of this stress is

$$s' = s \sin 2\alpha \qquad \text{...(2.45)}$$

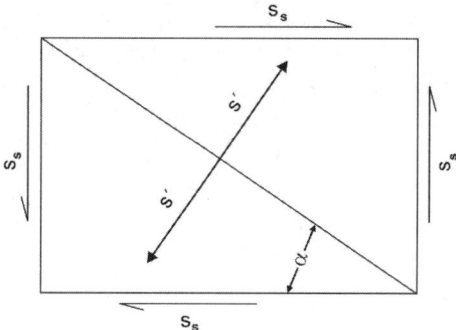

Fig. 2.27: Normal stress caused by shear

This stress reaches a maximum when $\alpha = 45°$. Then

$$s' = s_s \qquad \text{...(2.46)}$$

Lateral stress: When a force produces a normal stress s, it also causes a lateral strain ϵ', which may be considered as caused by a stress s' normal to s. The magnitude of this fictitious stress s' is

$$s' = -\mu s \qquad \text{...(2.47)}$$

where, μ is Poisson's ratio and the minus sign shows that the nature of the stress s' is opposite to that of s. If s is tensile stress, s' is a compressive one; and vice versa.

2.17 COMBINATION OF STRESSES

When the loads applied to a member induce stresses of several kinds, the material is affected by the combination of all the stresses. The stress that results from the combined simultaneous action of several stresses is called the *resultant stress*.

The resultant of several stresses acting on the same plane at a point of a body is the geometric sum of their individual actions. But stresses acting at a point on different planes cannot be combined or resolved like forces.

[1] B. Kirsh, "Knickfestigkeit langer Stable," *Zeitschrift Verein Deutscher Ingenieure*, Vol. 49 (1905), p. 907

Superposition: Where several loads acts in the same plane, their combined effect may be visualized as the separate effects of all the individual loads superimposed one on top of the other. This method, called *superposition,* can be applied to a variety of effects, such as stresses, deformations, and reactions.

It should be noted that superposition gives correct results only under two conditions:

1. The stresses and deformations must be directly proportional to the loads.

2. The deformations must not be so great as to change appreciably the configuration of the system or the point of application of the load.

Cases in which superposition cannot be used are shown in Fig. 2.28. A large increase of the load on the beam in Fig. 2.28a changes the horizontal distance from the support to the point of its application and hence changes the moment arm from l_1 to l_2. In Fig. 2.28b the spring grid of a Falk flexible coupling is shown (the coupling itself is shown in Fig. 21.16). The spring d is laid in grooves g cut in the disks h. The grooves have curved sides and widen toward each other. The free beam length of the spring elements is l_1. When a torque is applied the springs are bent along the arcs of the grooves, and the span is reduced to some smaller values, such as l_2.

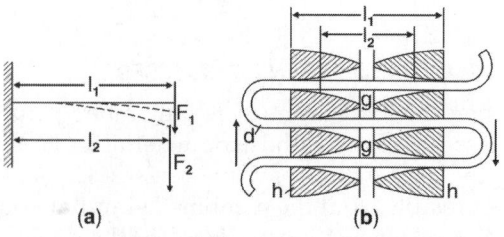

(a) **(b)**

Fig. 2.28: Variable load conditions

Normal stresses at right angles: If at a certain point there exist simultaneously a normal stress s_1 and another normal stress s_2 whose line of action is at right angles to the line of action of s_1, then there will exist an additional stress in the direction of s_1. According to equation 2.47

the magnitude of the additional fictitious stress will be $-\mu s_2$. The resultant stress, found by superposition, is

$$s' = s_1 - \mu s_2 \qquad \ldots (2.48)$$

The maximum shear stress at this point is in a plane making an angle of 45° with the main plane. Its magnitude, found from equation 2.44 by superposition, is

$$s' = 0.5\,(s_1 - s_2) \qquad \ldots (2.49)$$

Normal and shear stresses: The stress analysis of an element subjected to the simultaneous action of a normal stress s and a shear stress s_s shows that the combined action of these stresses produces an internal normal stress s' and a tangential stress s_s' on an interior plane, as indicated in Fig. 2.29. The ratio s/s_s determines the value of the angle β. The stress s' is maximum when $\tan 2\beta = -2s_s'/s$, and s_s' is maximum when $\tan 2\beta = +s/2s_s$. The principal direct stresses are

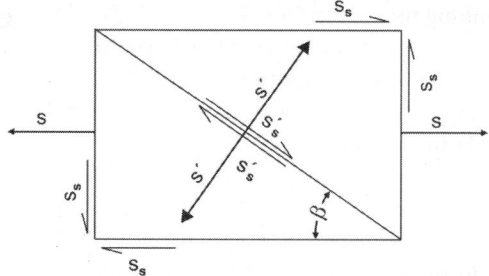

Fig. 2.29: Combined tensile and shear stresses

$$s' = \tfrac{1}{2}s \pm \sqrt{\left(\tfrac{1}{2}s\right)^2 + s_s^2} \; = \tfrac{1}{2}\left[s \pm \sqrt{s^2 + 4s_s^2}\right] \ldots (2.50)$$

The maximum shear stresses are

$$s_s' = \sqrt{\left(\tfrac{1}{2}s\right)^2 + s_s^2} \; = \pm\tfrac{1}{2}\sqrt{s^2 + 4s_s^2} \qquad \ldots (2.51)$$

The two stresses s' from equation 2.50 are at right angles to each other. The two shear stresses s_s' from equation 2.51 are also at right angles to each other, but they act in planes that make an angle of 45° with the planes of the principle stress s'.

The greater principal stress determined by equation 2.50 is usually considered to be the maximum stress within the material. However, according to the maximum-strain theory of failure the design should be governed by the stress that correspond to the maximum strain, which takes into account the lateral deformation. The normal stress s'' which will produce the maximum strain is

$$s'' = (1-\mu)\tfrac{1}{2}s + (1+\mu)\sqrt{\left(\tfrac{1}{2}s\right)^2 + s_s^2}$$

$$= \frac{1}{2}\left[(1-\mu)s + (1+\mu)\sqrt{s^2 + 4s_s^2}\right] \quad \text{...(2.52)}$$

In practice, equation 2.50 is usually preferred to equation 2.52, although equation 2.52 gives a slightly higher value for the normal stress, especially when the shear stress s_s is relatively high.

Example 2.13: Determine the resultant stresses created in the journal of the crankshaft shown in Fig. 2.30. The diameter D = 75 mm, the crank radius is 125 mm, the distance l = 95 mm the tangential force F = 24.5 kN.

Then stress in bending in fibres 1 and 2 due to bending moment M = Fl, by eq. 2.26 is

$$s = \frac{24500 \times 95 \times 32}{\pi \times 75^3} = 56.19 \text{ N/mm}^2$$

The average shear stress due to bending is, according to Table 2.6

$$s_s = \frac{24500}{\dfrac{\pi}{4} \times 75^2} = 5.54 \text{ N/mm}^2$$

In fibres 1 and 2, the shear stress $s_s = 0$, in the middle section 3.4, it reaches maximum value, which according to Table 2.5 is

$$1.33\, S_s = 1.33 \times 5.54 = 7.368 \text{ N/mm}^2$$

the shear stress due to torsion produced by the torque Fr is

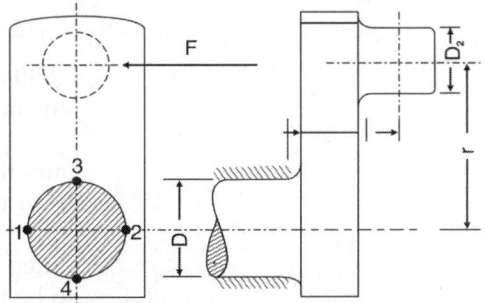

Fig. 2.30: Crankshaft with side crank

$$s_s = \frac{24500 \times 125 \times 16}{\pi \times 75^3} = 36.97 \text{ N/mm}^2$$

Thus in fibres 1 and 2, the principal stresses are, by eq. 2.50

$$s'' = 0.5 \times 56.19 \pm \sqrt{(0.5 \times 56.19)^2 + (36.97)^2}$$

$$= 82.367 \text{ N/mm}^2$$

and –26.176 N/mm²

The maximum shear stress in fibres 1 and 2 is by eq. 2.51

$$s_s = \sqrt{(0.5 \times 56.19)^2 + (36.97)^2} = 54.271 \text{ N/mm}^2$$

by equation 2.48 in which m = 0.303, the normal stress is

$$s'' = (1 - 0.303) \times \frac{56.19}{2} + (1 + 0.303) \times 54.271$$

$$= 90.295 \text{ N/mm}^2$$

The resultant shear stress in fibre 3 due to bending and torsion is

$$s_s'' = 7.368 + 36.97 = 44.338 \text{ N/mm}^2$$

The resultant shear stress in fibre 4, because of the opposite directions of shear stresses, is

$$s_s'' = 36.97 - 7.368 = 29.6 \text{ N/mm}^2.$$

2.18 BIAXIAL STRESS CONDITION

If in a system referred to three coordinate axes the components of all stresses parallel to one of the axes are zero, the stress condition is called *biaxial*. This condition is particularly important in machine design, because it is the most common condition on the surface of a part where the stresses are greatest.

The relationship between the stresses on various planes passing through a point may be determined by applying the equations of equilibrium to an element of the material at this point taken as a free body. For analyzing a biaxial-stress condition the most convenient free body is a triangular wedge such as that in Fig. 2.31, in which the angle α may have any value assigned to it. Since the shear stresses at right angles are equal, the resultant normal stress is found to be

$$s_n = s_x \sin^2 \alpha + s_y \cos^2 \alpha + s_s \sin 2\alpha \quad \text{...(2.53)}$$

The shear stress s_{sn} on the inclined plane is

$$s_{sn} = \tfrac{1}{2}(s_x - s_y)\sin 2\alpha + s_s \cos 2\alpha \quad \text{...(2.54)}$$

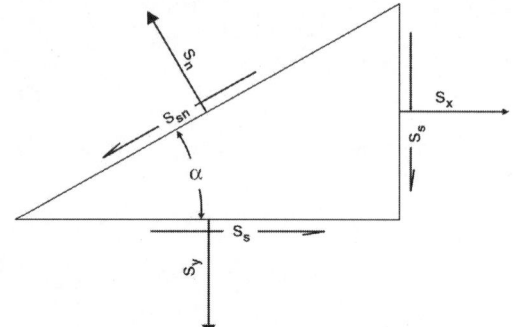

Fig. 2.31: Biaxial-stress condition

Principal stresses: Differentiating equation 2.53 with respect to α and equating the derivative to zero gives the direction of the plane on which the resultant normal stress becomes a maximum or minimum. The result is

$$\tan 2\alpha_m = -\frac{2s_s}{s_x - s_y} \qquad \ldots (2.55)$$

Equation 2.55 defines two planes which are at right angles to each other. The normal stress is a maximum on one of the planes, and it is a minimum on the other. The angle α_m is the angle between either of the axes x and y; to which the original stresses s_x and s_y are parallel, and one of the axes u and v, to which the two principal stresses are parallel.

For the value of α determined by equation 2.55, the expression for the principal stresses given in equation 2.53 becomes

$$s_{u,v} = \tfrac{1}{2}(s_x + s_y) \pm \tfrac{1}{2}\sqrt{(s_x - s_y)^2 + 4s_s^2}$$

$$= \tfrac{1}{2}\left[(s_x + s_y) \pm \sqrt{(s_x - s_y)^2 + 4s_s^2}\right] \ldots (2.56)$$

where, s_u is the maximum principal stress, which is parallel to the axis u, and s_v is the minimum principal stress, which is parallel to the axis v.

From equations 2.54 and 2.55 it can be shown that the shear stresses on the planes of principal stresses are equal to zero.

Maximum shear stresses: By a similar procedure the angles of the planes in which the shear stresses are greatest are determined by the relation

$$\tan 2\alpha_s = \frac{s_x - s_y}{2s_s} \qquad \ldots (2.57)$$

Equation 2.57 defines a pair of planes that make angles of 45° with those determined by equation 2.55. For the angle determined by equation 2.57, the maximum shear stresses found by equation 2.54 are

$$s_s' = \pm \tfrac{1}{2}\sqrt{(s_x - s_y)^2 + 4s_s^2} \qquad \ldots (2.58)$$

Equation 2.58 shows that the two maximum shear stresses are equal in magnitude and opposite in direction.

Example 2.14: At a point on the vertical side of a beam the horizontal stress s_x is 24 N/mm², tension, the vertical stress s_y, is 9 N/mm², compression, and the shear stress s_s due to a positive vertical shear is 12 N/mm². Determine the maximum shear stress and the principal stresses.

From eq. 2.58, the maximum shear stress is

$$s_s' = \pm \tfrac{1}{2}\sqrt{(24 + 9)^2 + 4 \times 12^2} = 20.4 \text{ N/mm}^2$$

Also from eq. 2.56, the two principal stresses are,

$$s_u = \frac{24 - 9}{2} + 20.4 = +27.9 \text{ m}^2 \text{ N/mm}^2$$

$$s_v = \frac{24 - 9}{2} - 20.4 = -12.9 \text{ N/mm}^2$$

Direction of stresses: A convenient way to show the relative directions of the stresses whose magnitudes are determined by equations 2.56 and 2.58 is to compute the values of the pairs of angles defined by equations 2.55 and 2.57, and to draw the corresponding free-body diagram. The procedure is best explained by an example.

Example 2.15: Determine the directions of the four stresses found in example 2.14.

The angles that the principal stresses make with the x axis are found from equation 2.55. For the given stresses,

$$\tan 2\alpha_m = -\frac{2 \times 12}{24 + 9} = -0.727$$

$$\alpha_{m1} = -\frac{36° \ 2'}{2} = -18° \ 1'; \ \alpha_{m2} = 71° \ 59'$$

In Fig. 2.32a the original block with the acting stresses is shown; and in Fig. 2.32b the block is shown cut along the line O_a making an angle $\alpha_{m2} = 71° \ 59'$ with the x axis. Of the fives forces, N_x, N_y, T_{xy}, T_{yz}, and N_u, acting on the block, the directions of the first four are known. From the diagram it is evident that the resultant of N_x and T_{xy} is a force that acts to the right and must be balanced by the horizontal component of N_u. Thus the force N_u must represent a tensile stress, and the stress with this direction is s_u. The directions of the axes are shown in Fig. 2.32c. The u axis is perpendicular to the line O_a, the directions of the principal stresses are shown in Fig. 2.32d.

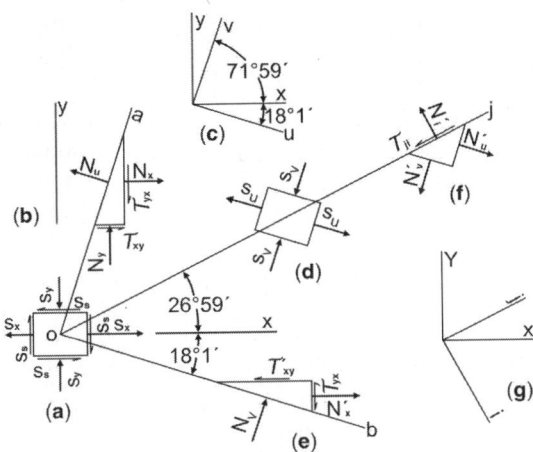

Fig. 2.32. Determination of direction of stresses

The same result is obtained if the free body is assumed to be cut along the line O_b, Fig. 2.32e, making an angle $\alpha_{m1} = 18° 1'$ with the x axis.

The maximum shear stresses make an angle of 45° with, the principal stresses and are found in the manner just described from a diagram of the block cut along the line O_j, as in Fig. 2.32f. The components of N_u', N_v' parallel to the line O_j act in the same direction; therefore the component of T_{ji} must be to the left to keep the free body in equilibrium, and the directions of the maximum shear stresses will be as shown in Fig. 2.32g.

Other stresses: Let s_u and s_v denote the principal stresses at a point. Since the shear stress in each of these planes is equal to zero, the normal stress s_n on any plane that is perpendicular to the u–v plane and that intersects the $u-v$ plane in a line which makes an angle α with the u axis is, as found from equation 2.53,

$$s_n = \tfrac{1}{2}(s_u + s_v) - \tfrac{1}{2}(s_u - s_v)\cos 2\alpha \qquad \text{... (2.59)}$$

The normal stress s_t acting on a plane at right angles to s_n is

$$s_t = \tfrac{1}{2}(s_u + s_v) + \tfrac{1}{2}(s_u - s_v)\cos 2\alpha \qquad \text{... (2.60)}$$

The magnitude of the shear stress s_{sn} acting on either of two inclined planes is, from equation 2.54,

$$s_{sn} = \tfrac{1}{2}(s_u - s_v)\sin 2\alpha \qquad \text{... (2.61)}$$

Also, from equations 2.58 and 2.56, the maximum shear stress is

$$s_s' = \pm \tfrac{1}{2}(s_u - s_v) \qquad \text{... (2.62)}$$

When $\alpha = \alpha_m$, it is evident that $s_n = s_y$ and $s_t = s_x$.

Use of the Mohr circle: The Mohr circle provides a graphical relation between the biaxial stresses s_x and s_y, the shear stress s_s, the principal stresses s_u and s_v, and the maximum shear stress s_s', all acting at a certain point in a body. To explain the procedure, data from example 2.14 will be used. First the normal stresses are laid off to scale as horizontal vectors from the origin o, Fig. 2.33a. Tensile stresses, since they are considered positive, are laid off to the right from o; thus, $s_x = oa$. Compressive stresses, considered negative, are laid off to the left; thus, $s_y = ob$. Shear stresses are laid off from the points a and b as vertical vectors. Those which tend to rotate a free-body element clockwise are considered positive and are laid off upward; thus, $s_s = ad$. Shear stresses tending to rotate the element counter-clockwise are considered negative and are laid off downward; therefore, $-s_s = be$. With the line ed as a diameter and its middle point c as the center, a circle $dfeg$ is described. From Fig. 2.33a, keeping in mind the signs of s_x and

$$oc = \tfrac{1}{2}(s_x + s_y), \quad ca = \tfrac{1}{2}(s_x - s_y),$$

and $cd = \tfrac{1}{2}\sqrt{(s_x - s_y)^2 + 4s_s^2}$. According to equation 2.56, $of = oc + cf = oc + cd = s_u$ and $og = cg - oc = cd - oc = s_v$. By equation 2.55, $\tan\varphi = ad/ca = s_s / \tfrac{1}{2}(s_x - s_y) = -\tan 2\alpha_m$.

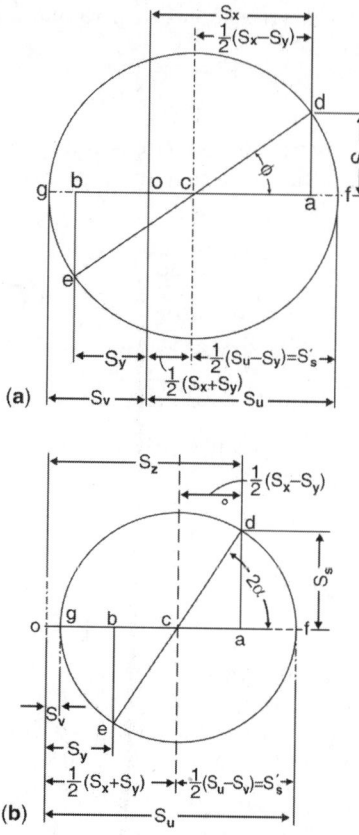

(a)

(b)

Fig. 2.33: Mohr's circle for stresses

The foregoing explanation shows how Mohr's circle provides a simple way of finding, for a biaxial stress condition, the principal stress, the maximum shear stresses, and the angle α between the principal stresses and the biaxial stresses. On the other hand, if the principal stresses are known, Mohr's circle permits one to find very easily the maximum shear stresses and also, for any desired angle, the corresponding stresses s_x, s_y, and s_s. In general this circle gives a clear picture of the relationship of stresses. If at any point there are biaxial stresses s_x and s_y, and the shears stress s_s is zero, these biaxial stresses are the principal stresses.

Mohr's circle also shows the influence of any change in the biaxial stresses. As an illustration let it be assumed that the sign of the vertical stress s_y in example 2.14 and Fig. 2.32a is

changed. Mohr's circle for this condition is shown in Fig. 2.33b. Angles of the Mohr's circle are double those on the actual body. The values of α_m show the angles between the biaxial stresses and the principal stresses. Thus the planes on which the maximum principal stress, s_u acts make a clockwise angle α_m with the planes on which s_x acts.

2.19 COMBINED LOADS

When a machine part is subjected to several loads simultaneously, the governing stresses are best found by first determining the stresses from each load and then combining them by one of the following methods: Superposition; application of equations 2.50, 2.51 or 2.52; application of equations 2.56 and 2.58; or construction of a Mohr circle. To illustrate the procedure several typical cases will be discussed.

Bending and Axial Load

The main stresses in bending are direct tensile and compressive stresses, each of which is s_1; and they are simply added algebraically to the direct stress s_2 from the axial load. By superposition the resultant stress is where a tensile stress is given a plus sign and a compressive stress is given a minus sign.

$$s = \pm s_1 + s_2 \qquad \ldots (2.63)$$

In Fig. 2.34 is represented a cantilever beam that is bent and stressed in tension. Shear from bending does not affect the stresses that must be taken into account. In the outside fibers, s_s is zero; and in the center fibers, where s_s is greatest, the bending stress s_1 is zero. If the axial stress s_2 is large, the combination of s_s and s_2 may be critical.

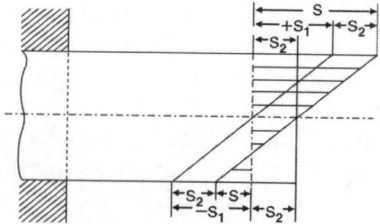

Fig. 2.34: Bending and axial load

Bending by several loads: If several loads are imposed on a beam, their actions are superimposed in every respect—to find bending moments, reactions, shears, and deflections.

Example 2.16: Work example 2.4 by using the principle of superposition and taking into account the influence of the weight of the pipe on both the stresses and the deflections.

Since the specific weight of steel is 77100 N/m^3, the weight of the pipe is

$$\frac{\pi}{4}\frac{(100^2 - 90^2)}{1000^2} \times 3 \times 77100 = 345.16 \text{ N}$$

By case f in Table 2.4, the bending moment from this weight is,

$$M = \frac{Wl}{8} = \frac{345.16 \times 3}{8} = 129.435 \text{ Nm}$$

The stresses produced by this moment with $Z = 33762$ mm^3, from example 2.5 is

$$s = \frac{129.435 \times 1000}{33762} = 3.83 \text{ N/mm}^2$$

(a) The stress due to the weight of the pipe must be deducted from the allowable stress of 100 N/mm^2. With this correction, the expression for F in example 2.4 becomes

$$F = (100 - 3.83) \times \frac{33762}{750} = 4329 \text{ N}$$

(b) By case f in Table 2.4, the deflection from uniformly distributed load is

$$y = \frac{5Fl^3}{384EI} = \frac{5 \times 345.16 \times (3000)^3}{384 \times 207000 \times 33762 \times 50} = 0.347 \text{ mm}$$

The deflection due to force $F = 4329$ N is

$$y = \frac{4329 \times (3000)^3}{48 \times 207000 \times 33762 \times 50} = 6.97 \text{ mm}$$

The combined deflection is, $y = 0.347 + 6.97 = 7.317$ mm. Thus while the outside load is decreased by $\frac{100 \times (4502 - 4329)}{4502} = 3.8\%$, the total deflection is increased by $\frac{100 \times (7.317 - 7.25)}{7.25} = 0.92\%$. However, the combined load is $4329 + 345 = 4674$ N which is greater than 4502 N.

Torsion and axial load: A combination of torsion and axial load may be encountered in a propeller shaft of a ship, in a shaft used for driving a heavy worm gear, or in a bolt when its nut is tightened. The primary shear stress s_s is found from equation 2.17 for the torsional moment involved. If the axial load is tension, the direct stress s is found by equation 2.13 if the axial load is compression and the piece is short, the stress is found by equation 2.13.

The combined normal stress is found by equation 2.50, and the combined shear stress by equation 2.51.

If the piece must be treated as a column, equations 2.41 and 2.42 may be used.

Torsion and bending: Torsion and bending are combined in all shafts used for transmission of power by means of pulleys, sprockets, or gears. The primary shear stress is found from equation 2.17; the bending moment causes a direct stress (tension or compression), which is found by equation 2.26. The compound stresses are found in the same way as in the case of a normal stress and a shear stress by using equations 2.50 and 2.51.

Eccentric load application: If, as in Fig. 2.35, the line of action of a force F does not go through the center of gravity of the cross section of a body but is parallel to the axis of the body, the body will be subjected simultaneously to direct tension and bending. If we apply at the center of gravity of any cross section two imaginary opposite forces F_1 and F_2, each equal to the force F and parallel to it, the system of forces acting on the part of the body to the left of the action may be considered as the combination of a couple formed by the forces F and F_1 with an arm e and a residual axial force F_2. The stresses produced in the body by the couple with a moment Fe are

$$s_1 = \pm\frac{Fe}{Z}$$

Also, the stress produced by the force $F_2 = F$ is pure tension, and its magnitude is

$$s_2 = \frac{F}{A}$$

The combined stress, as found by superposition, is

$$s = \frac{F}{A} \pm \frac{Fe}{Z} \qquad \dots (2.64)$$

When the section is not symmetrical, the proper value of Z must be used with the corresponding sign (+ or –).

In Fig. 2.35 the force F produces axial tension; but equation 2.64 applies also if the force produces compression, provided the member is short. In general an eccentric load is equivalent to simultaneous action of bending and an axial load.

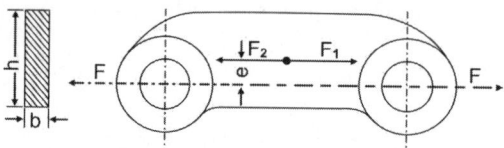

Fig. 2.35: Eccentric loading

Example 2.17: Determine the maximum stress created in the offset link in Fig. 2.35 by a force F = 4 kN if the height h of its cross section is 50 mm, its width b = 10 mm and the arm e of the force is 10 mm.

The section modulus as shown in Table 2.5 is

$$Z = \frac{bh^2}{6} = \frac{10 \times 50^2}{6} = 4166 \text{ mm}^3$$

and the bending stresses are

$$s_1 = \pm \frac{4000 \times 10}{4166} = \pm 9.6 \text{ N/mm}^2$$

The stress in tension is

$$s_2 = \frac{4000}{50 \times 10} = 8 \text{ N/mm}^2$$

The tensile stress in the outer fibres nearest to the line of action of the force F is

$$s = s_1 + s_2 = 9.6 + 8 = 17.6 \text{ N/mm}^2$$

the stress in the outer fibres at the other edge is

$$s = s_1 + s_2 = -9.6 + 8 = -1.6 \text{ N/mm}^2 \text{ compression}$$

Eccentrically loaded columns: If the member in compression is of such length that it can be classed as a short column, the maximum primary stress s_2 is determined by equation 2.41 and the bending stress s_1 is added to it. To simplify the result, substitute in equation 2.64 the expression for Z in equation 2.25 and replace the moment of inertia I by its equivalent value k^2A. If Ritter's formula (equation 2.41a) is used, the maximum combined stress becomes

$$s_c = \frac{F}{A}\left[1 + \left(\frac{l}{k}\right)^2 \frac{S_e}{\pi^2 nE} + \frac{ce}{k^2}\right] \quad \dots (2.65)$$

The influence of eccentric loading on the maximum stress in a long column is small and can usually be disregarded.

2.20 THICK-WALLED CYLINDER

In Fig. 2.36a is shown a cross section of a thick-walled cylinder subjected to uniform normal pressure p_i and p_o on the inside and outside surfaces, respectively, but not subjected to any external force parallel to its axis. Since the body and the loading are symmetrical with respect to the cylinder axis, there will not be any shear stresses in the tangential or radial directions; and any element in a thin cylindrical section will be subjected only to normal stresses s_t and s_r as shown. In order to find the relation between these stresses and the pressures p_i and p_o, consider the equilibrium condition of a semicircular element like that in Fig. 2.36b. The axial thickness of the element is taken as unity. The vertical component of the resultant of all the inward radial stresses across the diameter of the element is equal to $2s_r r$, and the vertical component of the resultant of the outward stresses is $2(s_r + ds_r)(r + d_r)$. Since the sum of downward stresses on the ends of the element is $2s_t dr$, the equilibrium equation for the element becomes

$$2s_r r + 2s_t dr = 2(s_r + ds_r)(r + dr)$$

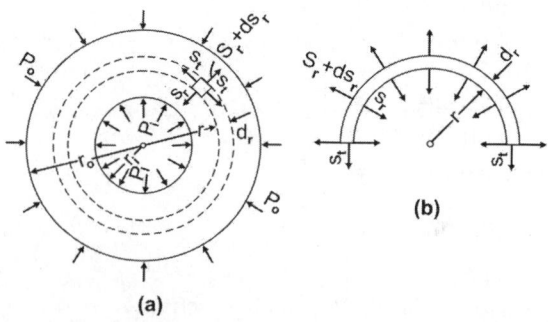

(a)

(b)

Figs 2.36a and b: Thick-walled cylinder subjected to inside and outside pressures

After expanding the right side of this equation and neglecting the infinitesimals of higher order than the first, we obtain

$$s_t - s_r = r\frac{ds_r}{dr} \qquad \ldots (2.66)$$

If Poisson's ratio is denoted by μ and the modulus of elasticity is denoted by E, the axial deformation ϵ caused by the stresses s_t and s_r can be written as

$$\epsilon = -\frac{\mu s_t}{E} - \frac{\mu s_r}{E}$$

Hence,

$$s_t + s_r = -\frac{\epsilon E}{\mu} \qquad \ldots (2.67)$$

The right-hand side of this equation is a constant, which can be conveniently designated as $2C_1$. Subtracting equation 2.67 from equation 2.66 gives

$$r\frac{ds_r}{dr} + 2s_r = 2C_1$$

Multiplying each term of the last equation by r and rearranging gives

$$\frac{d(r^2 s_r)}{dr} = 2rC_1$$

After integration we get

$$r^2 s_r = C_1 r^2 + C_2$$

where, C_2 is the constant of integration. From this equation,

$$s_r = C_1 + \frac{C_2}{r^2} \qquad \ldots (2.68)$$

Substituting this value of s_r in equation 2.67, we find that

$$s_t = C_1 - \frac{C_2}{r^2} \qquad \ldots (2.69)$$

At the inner boundary, $r = r_i$ and the radial stress $s_r = -p_i$. Thus equation 2.68 becomes

$$-p_i = C_1 + \frac{C_2}{r_i^2} \qquad \ldots (2.70)$$

At the outer boundary $r = r_o$ and the radial stress $s_r = -p_o$. So equation 2.68 now becomes

$$-p_o = C_1 + \frac{C_2}{r_o^2} \qquad \ldots (2.71)$$

Simultaneously solving equations 2.70 and 2.71 results in

$$C_1 = \frac{r_i^2 p_i - r_o^2 p_o}{r_o^2 - r_i^2} \qquad \ldots (2.72)$$

and

$$C_2 = \frac{r_i^2 r_o^2 (p_o - p_i)}{r_o^2 - r_i^2} \qquad \ldots (2.73)$$

Substitution of these values in equations 2.68 and 2.69 gives

$$s_r = \frac{r_i^2 p_i - r_o^2 p_o + \dfrac{r_i^2 r_o^2 (p_o - p_i)}{r^2}}{r_o^2 - r_i^2} \qquad \ldots (2.74)$$

and

$$s_t = \frac{r_i^2 p_i - r_o^2 p_o - \dfrac{r_i^2 r_o^2 (p_o - p_i)}{r^2}}{r_o^2 - r_i^2} \qquad \ldots (2.75)$$

These equations were first developed by Lamè.

PROBLEMS

2.1 How does wear cause failure in a machine? Give examples.

2.2 Is it possible to prevent wear? Explain.

2.3 Explain the role of lubricants in reducing wear.

2.4 Can corrosion be prevented? Explain.

2.5 Calculate how much a standard 25 mm. Steel pipe (33.40 mm OD, 26.66 mm ID) 3 m long will stretch when carrying a load in tension of 40 kN. Assume $E = 2 \times 10^5 \text{ N/mm}^2$.

2.6 After the nut on the M27 steel bolt with SI thread in Fig. P2.1 is drawn snug, it is given one-quarter of a turn. The diameters of the round spacers on the bolt are $d_o = 50$ mm and $d_i = 32$ mm. The shorter one has a length l_1 to 100 mm and is of steel for which $E = 2 \times 10^5 \text{ N/mm}^2$.

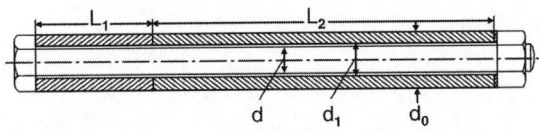

Fig. P2.1

The longer one has a length l_2 of 300 mm and is of cast iron for which $E = 0.8 \times 10^5$. Data for the screw thread are given in Table 11.2. Find the force set up in the bolt.

2.7 A M22 steel bolt a (Fig. P2.2) with SI threads holds two plates b in place, a 25 mm standard wrought-iron pipe being used as a spacer. After the nut is tightened just enough to take up the slack, it is given an additional one-fifth of a turn. Determine the stresses set up in the shank of the bolt and in the spacer c, assuming that $E = 2 \times 10^5$ N/mm^2 for the steel and that $E = 1.8 \times 10^5$ N/mm^2 for the wrought iron. Also, $l = 30$ mm and $h = 16$ mm. Neglect the deformation of the plates b.

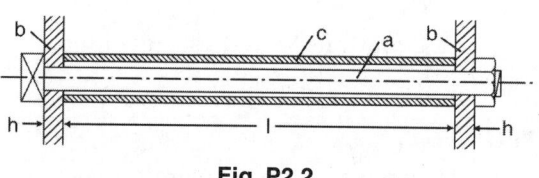

Fig. P2.2

2.8 Determine the theoretical value of the transverse modulus of elasticity of a high carbon steel bar having a diameter of 22 mm which, when subjected to a tensile load of 50 kN, stretched from 200 mm and 200.02 mm between the marks.

2.9 A tension member in a steel frame (Fig. P2.3) made of flat steel having the dimensions $b = 100$ mm and $h = 12$ mm is fastened to the main plate by four 16 mm rivets. Assuming that the load F of 5 kN is distributed equally between the rivets, determine the tensile stresses in the sections 1–1, 2–2, and 3–3 and also the shear and bearing stresses in the rivets.

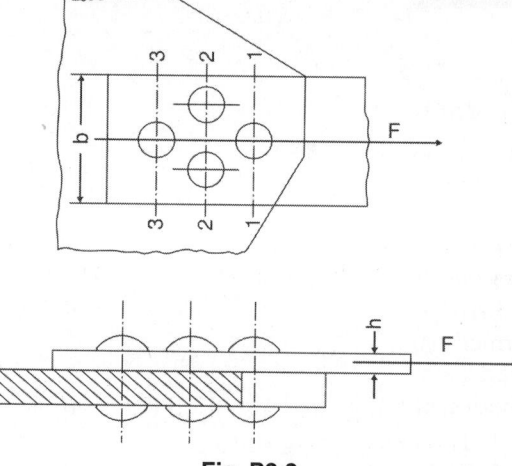

Fig. P2.3

2.10 Find the maximum stress and the degrees of torsion of a hollow steel shaft, 125 mm outside diameter, 50 mm inside diameter, and 7 m long, transmitting 440 kW at 240 rpm. The modulus of elasticity in shear of the material is 0.8×10^5 N/mm^2.

3 Dynamic Stresses and Stress Concentration

3.1 VARIABLE LOADS

Only a few machine parts are subjected to constant, or static, loads. *The loads on most parts vary, sometimes in magnitude, sometimes in direction, and frequently both in magnitude and direction. Such loads are called dynamic loads (3.1).*

Stresses produced by variable loads are termed *dynamic stresses*. The variable, or dynamic, loads which produce such stresses may be divided into two groups:

1. Those produced by outside forces.
2. Those resulting from the inertia of the mass of the member itself when it moves with a variable speed.

   Loads produced by variable outside forces are also called live loads, as contrasted with *static loads*, or *dead loads*.

A live load may be produced either by the gradual change of a force already applied or by the sudden application of an outside force. When it is produced by a sudden outside force, it is called a *shock load*, or *impact load*.

The influences of these various types of dynamic loads upon the internal stresses of a member differ considerably and must be investigated separately.

3.2 INERTIA STRESSES

Inertia loads are caused by acceleration that acts when a change of velocity, either linear or angular, takes place. The acceleration may be due to a change in the direction of the motion (as of a rotating crank), a change in the magnitude of the velocity (as of a connecting road, one end of which rotates while the other moves back and forth).

Centrifugal loads: The general expression for the centrifugal force created in a mass *m* is

$$C = \frac{mv^2}{R} \qquad \dots (3.1)$$

in which v is the circumferential velocity, in metres per second, and R is the radius of curvature of the path of the motion of the mass, in metres. Considering a mass of 1 m^3 and expressing it as ω/g where ω is the specific weight of the material, in N per cubic metre and $g = 9.81$ m per sec^2. Also, the radius of the curvature r in metres, equation 3.1 then becomes

$$C = \frac{\omega v^2}{gr} \qquad \dots (3.2)$$

Centrifugal force can create tensile stresses, as in the case of the rim of a rotating pulley; or it can create bending stresses, as in the case of a locomotive coupling rod or a connecting rod.

A change of rotary speed, as in starting or stopping an engine, creates additional stresses, such as bending in the arms of a flywheel or torsion in the shaft which carries it.

Pulley or flywheel: In the case of a pulley or a flywheel, the velocity can be expressed by the relation

$$v = \frac{2\pi rn}{60} = \frac{\pi rn}{30} \qquad \dots (3.3)$$

where, n is the number of revolutions per minute.

Consider an element A of a pulley rim Fig. 3.1 included by $d\theta$, the effect of arms is negligible.

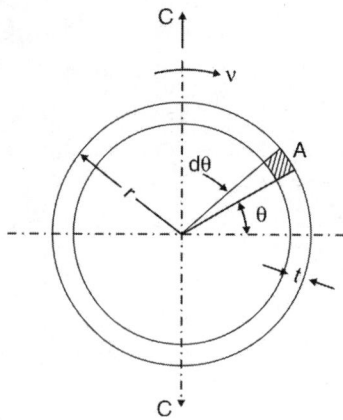

Fig. 3.1: Rotating pulley

Area of the element $= rd\theta.t$

If b is the width of the pulley, volume of the element $= tbrd\theta$

The centrifugal force acting on this element is

$$C = \frac{\omega v^2}{gr} tbrd\theta$$

$$= \frac{\omega v^2 tb}{g} d\theta$$

It has a horizontal component $= \dfrac{\omega v^2 tb\, d\theta}{g}\cos\theta$

and a vertical component $= \dfrac{\omega v^2 tb\, d\theta \sin\theta}{g}$

Total horizontal component of force in top half

$$= \int_0^\pi \frac{\omega v^2 tb \cos\theta d\theta}{g} = 0$$

Total vertical component of force in top half

$$F_v = \int_0^\pi \frac{\omega v^2 tb \sin\theta d\theta}{g}$$

$$F_v = \frac{\omega v^2 tb \times 2}{g} \qquad \ldots (3.4)$$

An equal vertical component will act in the lower half of pulley. The stress across horizontal diameter or any diameter is tensile stress

$$= \frac{F_v}{Area} = \frac{\omega v^2 tb \times 2}{g \times 2 tb} = \frac{\omega v^2}{g} \qquad \ldots (3.5)$$

Coupling rod: In the case of a coupling rod like that shown in Fig. 3.2 the centrifugal force acts as a load uniformly distributed on a simple beam of length l. Its bending component has a maximum value expressed by equation 3.6 when the rod ends pass through the points b or through the point b'. It becomes zero when the rod ends are at the points a or a'.

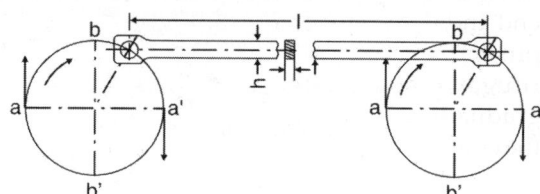

Fig. 3.2: Locomotive coupling rod

The maximum uniformly distributed bending load is

$$C = \frac{\omega htlv^2}{gr} \qquad \ldots (3.6)$$

where, t is the thickness of the rod, in metres.

The axial component of the centrifugal force is zero when the rod ends are at points b or at points b'. It has its maximum value given by equation 3.6 when the rod ends are at point a or a'. This force creates compressive and tensile stresses on the cross section which vary from zero at one end of the rod to a maximum value at the other end.

Example 3.1: Find the stress due to centrifugal force in the locomotive coupling rod shown in Fig. 3.2. The wheels turn at 200 rpm, $r = 230$ mm, $l = 1.83$ m, $h = 64$ mm, and $t = 40$ mm. Use $w = 74660$ N/m^3.

$$v = \frac{2\pi \times 230 \times 200}{60 \times 1000} = 4.82 \text{ m/s}$$

The bending load is found by equation 3.6. It is

$$C = \frac{74660 \times 64 \times 40 \times 1.83 \times 4.82^2}{9.806 \times 230 \times 1000} = 3603 \text{ N}$$

If the rod is treated as a simple beam uniformly loaded, the bending moment is, according to Table 2.4,

$$M = \frac{3603 \times 1.830}{8} = 824 \text{ Nm}$$

The section modulus, according to Table 2.5, is

$$Z = \frac{40 \times 64^2}{6} = 27307 \text{ mm}^2$$

Hence, the bending stress is

$$s = \frac{824 \times 1000}{27307} = 30 \text{ N/mm}^2$$

The direct stress is very small. It is

$$s_2 = \frac{3603}{40 \times 64} = 1.4 \text{ N/mm}^2$$

Connecting rod: The motion of a connecting rod is a combination of the rotation of the crank end and the reciprocating motion of the wrist-pin end, or small end. Accordingly, the centrifugal force acting on an element of the rod gradually changes from a maximum value at the crank end to zero at the small end, as shown by the shaded triangle in Fig. 3.3. The force of inertia that produces bending acts normally to the rod axis. Without any appreciable error so far as the stress is concerned, it may be assumed that the rod has a uniform cross section and that the inertia force is represented by the shaded triangle. The magnitude of the inertia force perpendicular to the rod will be one-half that given by the basic equation 3.2, or

$$F_i = \frac{Wv^2 \sin \alpha}{2gr} \qquad \ldots (3.7)$$

where, W is the weight of the rod itself, not including the ends.

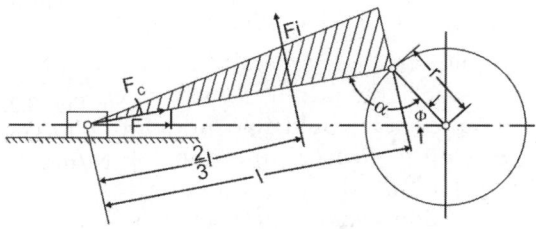

Fig. 3.3: Forces acting on a connecting rod

This force acts at a distance of $\frac{2}{3}l$ from the wrist pin. The maximum bending moment produced by this force is at a distance of $\frac{1}{\sqrt{3}}l$ from the wrist pin, and its magnitude is[1]

$$M = \frac{2F_i l}{9\sqrt{3}} \qquad \ldots (3.8)$$

Substituting for F_i its value from equation 3.7 gives

$$M = \frac{Wv^2 l \sin \alpha}{9\sqrt{3}\, gr} \qquad \ldots (3.9)$$

The bending stress can be found as usual by equation as

$$s = \frac{M}{z}$$

Naturally, in a connecting rod the main stress is caused by the force F acting on the piston (see Fig. 3.3), or more specifically by its component F_c acting along the axis of the rod. This force is also variable and depends upon the position of the piston, or angle α. The connecting rod is a short column subjected to buckling effect. The compressive stresses must be found for several values of the angle α, and these stresses must be combined with stresses caused by bending due to the inertia force for the same angles.

Reciprocating motion: The usual case of reciprocating motion is that of a piston and a part of the connecting rod due to gas force. The force of inertia F in this case is expressed approximately by the formula

$$F = \frac{Wv^2}{gr}\left(\cos\phi + \frac{r}{l}\cos 2\phi\right) \quad \ldots (3.10)$$

where, W is the weight of the reciprocating masses, in N
v is the crank velocity, in metre per second
g is the acceleration of gravity $= 9.81$ metre per sec^2.
ϕ is the angle between the crank and the center line of the piston
l is the length of connecting rod in metres

The angle ϕ is measured from the head-end dead center. Substituting the value of v from

---

[1] Expressions for triangle load distribution are given in Lionel S. Marks, ed., *Mechanical Engineers' Handbook*, 5th ed. (New York: McGraw-Hill Book Company, Inc., 1951), p. 427, and in R. T. Kent, Mechanical Engineers' Handbook, 12th ed., Vol. II, Design and Production, ed. by Colin Carmichael (New York: John Wiley & Sons, Inc., 1950), p. 8–13.

equation 3.3 and carrying through the numerical calculations, we obtain

$$F = \frac{\pi^2 Wrn^2}{900g}\left(\cos\phi + \frac{r}{l}\cos 2\phi\right) \quad \dots (3.11)$$

The maximum value of F occurs when $\phi = 0$, or when the crank is at the head-end dead center, and is

$$F_{max} = \frac{\pi^2 Wrn^2}{900g}\left(1 + \frac{r}{l}\right) \quad \dots (3.12)$$

At the crank-end dead center, where $\phi = 180°$, F attains the maximum negative value, acting in the opposite direction. Then

$$F_2 = \frac{\pi^2 Wrn^2}{900g}\left(1 - \frac{r}{l}\right) \quad \dots (3.13)$$

The inertia forces of reciprocating masses produce in machine parts stresses which must be added to those produced by outside forces. The inertia stresses may exceed the stresses due to the outside forces.

3.3 IMPACT STRESSES

If a moving body strikes another body, the second body is subjected to an impact which is equal to the kinetic energy of the moving body.

Impact energy: If the weight of a moving body is W, and its velocity is v m/s its kinetic energy in Nm is

$$K_i = \frac{Wv^2}{2g} \quad \dots (3.14)$$

For stress calculations it is convenient to express the impact energy as if it were produced by a falling body. The height of fall, in m, that would develop the velocity v is

$$h = \frac{v^2}{2g} \quad \dots (3.15)$$

Hence, the impact energy of body falling from a height h is

$$K_i = Wh \quad \dots (3.16)$$

where, K_i refers to the instant when the falling body strikes the other body.

The total work done by a falling body can also be computed by equation 3.16 if h is understood to include both the height of the fall before contact and the deformation of the body that is struck.

Impact stress: A general expression for impact stress may be derived by considering the normal stress created by a falling weight W, which is allowed to drop through a distance h before it strikes axially a steel rod whose cross-sectional area is A and whose length is l. When the weight W strikes the rod, the rod will be compressed a distance e', and the external work performed by the weight will be $W(h + e')$.

The stress in the rod was zero before the weight struck it; and after it has been compressed a distance e', the stress become s'. If s' is within the elastic limit, the internal work of the variable compressive force is $\frac{1}{2}As'e'$. If the internal work is equated to the external work and the equation is solved for s', the result is

$$s' = 2W\frac{(h+e')}{Ae'} \quad \dots (3.17)$$

But $W/A = s$, where s is the static stress; and for elastic deformations

$$e' = \frac{es'}{s}$$

where, e is the static deformation under the action of the weight W. Substituting these values in equation 3.17 and solving it for s', we get

$$s' = s\left(1 + \sqrt{1 + \frac{2h}{e}}\right) \quad \dots (3.18)$$

Equation 3.18 is general, and s' is the impact stress due to the force W which, when applied as a static load, would give the deformation e.

Since the stresses and deformations are proportional, the general equation for the deformation under impact action is

$$e' = e\left(1 + \sqrt{1 + \frac{2h}{e}}\right) \quad \dots (3.19)$$

Impact from a direct load: For tension or compression, $e = sl/E$ and the formula for the impact stress is

$$s' = s\left(1 + \sqrt{1 + \frac{2hE}{sl}}\right)$$

$$= \frac{W}{A}\left(1+\sqrt{1+\frac{2hEA}{Wl}}\right) \qquad \ldots (3.20)$$

Impact and bending: When a beam is subjected to impact, the deformation e means the deflection y, which is computed for the type of beam and the loading in the particular case. Equations 3.18 and 3.19 apply without any other changes.

Impact and torsion: Equations 3.18 and 3.19 also be used, when a force causing torsion is applied with impact. In this case, however, the deformation e means the angular deflection θ, in radians; and the travel h of the force must also be expressed in radians. If the impact is given in energy units, the stress s' is found more easily from the following equations

Shear stress $s'_s = s_s\left[1+\sqrt{1+\frac{2h}{r\theta}}\right] \qquad \ldots (3.21)$

Angular deflection

$$\theta' = \theta\left[1+\sqrt{1+\frac{2h}{r\theta}}\right] \qquad \ldots (3.22)$$

θ' and θ are angular deflection in radians r is the moment arm.

Sudden load: If the weight W is applied suddenly but it does not have an appreciable velocity before it strikes the body, the value to be substituted of h in equation 3.18 is zero and

$$s' = 2s \qquad \ldots (3.22a)$$

The stress produced by a suddenly applied load is therefore double that produced by the same load if it were applied statically or gradually. The deformation e' is $2e$.

Inertia effect: When a body having a weight W strikes another body that has a weight W', some of the impact energy is used to overcome the inertia of the weight W' and to accelerate the weight. Therefore, because of the inertia of the body that is struck, that body is subjected to less impact and the resulting stresses and deformations in it are reduced. According to the laws of collision of two perfectly inelastic bodies, it may be assumed that the impact

energy Wh is reduced to nWh. The value of n is less than 1 and may be found by the formula

$$n = \frac{1+am}{(1+bm)^2} \qquad \ldots (3.22b)$$

where, $m = W'/W$, and a and b are the coefficient whose values are given in Table 3.1 for the main cases of impact action.

Table 3.1: Coefficients in the inertia equation 3.22b

Type of impact	a	b
Longitudinal impact on bar	$\frac{1}{3}$	$\frac{1}{2}$
Center impact on simple beam	$\frac{17}{35}$	$\frac{5}{8}$
Center impact on beam with fixed ends	$\frac{13}{35}$	$\frac{1}{2}$
End impact on cantilever beam	$\frac{4}{17}$	$\frac{3}{8}$

When the inertia effect is considered, stresses and deformations may be computed by substituting nh for h in equations 3.18, 3.19 and 3.20. However, when one body comes in contact with another, the local stress near the area of contact is increased by the inertia of the struck body; and the impact may produce a local stress that exceeds the elastic limit and thus may cause a permanent deformation.

3.4 RESILIENCE

When external forces deform a body and create internal stresses in it, the body can absorb some of the energy. Also, the body can give up energy when the forces are removed. This property of a material is generally called *resilience*. In other words, the resilience of a body is the potential energy stored up in it when it is deformed. Since resilience is measured by the amount of deformation, or strain, it is also known as *strain energy*. Resilience is a function of internal stresses, regardless of whether they are created by static loads or dynamic loads. However, since resilience is important chiefly in cases of dynamic loads, it is discussed here rather than in Chapter 2.

Numerically, the resilience U of body is equal to the work required to produce a deformation which sets up a stress s, when s does not exceed the elastic limit. This work is equal to the product of the average applied force and the final deformation. Thus,

$$U = \frac{1}{2}Fe \qquad \ldots (3.23)$$

If a load W is applied with impact, the force F may be taken as

$$F = \frac{We'}{e} = \frac{Ws'}{s} \qquad \ldots (3.24)$$

where, e' and s' have the same meaning as in the preceding section.

The resilience U indicates the amount of impact energy that a body can withstand without permanent deformation when acting as a rigid spring. Thus the resilience of a body is a measure of its ability to resist impact. Equation 3.23 is general and applies to all types of impact.

Modulus of resilience: The ability of a material to withstand impact without permanent deformation is indicated by its modulus of resilience. This modulus, which is designated by u, is a measure of the elastic energy stored up in a unit volume of the material at the elastic limit. It is usually expressed in Nm/m^3.

Resilience in tension or compression: By substituting for F and e in equation 3.23 the values given by equations 2.13 and 2.14, we can obtain the following general formula:

$$U = \frac{1}{2}\frac{As \times sl}{E} = \frac{s^2 V}{2E} \qquad \ldots (3.25)$$

where, U is the resilience, in Nm
s is the direct stress invoked by the load, in N/m^2
A is the constant cross-sectional area of the body, in m^2
$V = Al$ is the volume of the body, in m^3

The modulus of resilience may be found from equation 3.25 by substituting S_e for s and 1 for V. Thus,

$$u = \frac{S_e^2}{2E} \qquad \ldots (3.26)$$

For a direct stress, u may be found graphically by determining the area Oea in Fig. 2.10. For a deformation resulting in a stress s_y', the unit resilience is represented by the area $cy'b$.

It should be noted that the modulus of elasticity of a given metal is practically a constant characteristic. The elastic limit, on the other hand, depends to a great extent upon the method of shaping the piece, which may be casting, forging, or rolling; upon the addition of certain chemical elements, such as carbon, chromium, nickel, and vanadium; and finally upon the subsequent heat treatment. By proper heat treatment the elastic limit of certain alloy steels can be made 2.4 times as great as that of untreated metal, and the resilience can be made 5.7 times as great.

Table 3.2 gives moduli of resilience in tension for typical materials used in machinery. These values may serve as guides for selecting the proper material when shock loads must be dealt with.

Example 3.2: A 16 mm 6 × 19 construction steel rope runs at a speed of 5.6 km/h between the rails of a narrow-gage road. The weight of a loaded car which must be connected to and pulled by this rope is 8 kN; the weight of the rope is 9.316 N/m, the length of the rope between the driving pulley and the point where the car is hooked in, is 30 m. Determine the stress created in the rope by the impact of hooking-in the car, assuming that the modulus of elasticity of the rope is $207 \times 10^9 \, N/m^2$.

$$v = \frac{5.6 \times 1000}{60 \times 60} = 1.56 \text{ m/s}$$

The impact energy, by equation 3.14, is

$$K_i = \frac{8 \times 1000 \times 1.56^2}{2 \times 9.806} = 992.7 \text{ Nm}$$

If it is assumed that the flexible core weighs 5 per cent, the net weight of the steel wires is $9.316 \times 0.95 = 8.85 \, N/m$.

With a specific weight of $74660 \, N/m^3$ the cross-sectional area is

$$A = \frac{8.85}{74660} = 0.00011854 \text{ m}^2$$

By equation 3.25 the resilience of the cable is

$$U = \frac{s^2 \times 0.00011854 \times 30}{2 \times 207 \times 10^9} = 8.589 \times 10^{-15} s^2 \text{ Nm}$$

Equating this resilience to the impact energy and solving for s gives

$$s = \sqrt{\frac{992.7}{8.589 \times 10^{-15}}} = 3.4 \times 10^8 \text{ N/m}^2$$

$$= 340 \text{ N/mm}^2$$

Table 3.2: Resilience in tension

Material	Elastic limit S_e N/mm²	Modulus of elasticity E N/mm² × 10³	Modulus of resilience u Nm/m³ × 10⁻³	Impact strength (Izod number)
Cast iron				
Class 20 (ordinary) FG 150	45*	69	15	–
Class 25 FG 200	70*	90	27	–
Nickel, Grade II	120*	125	58	–
Malleable	140	175	56	7.9
Aluminum alloy, SAE 33	48	67	17	–
Brass, SAE 40 or SAE 41	70	83	29	2.7
Bronze, SAE 43	193	110	169	–
Monel metal				
Hot-rolled	210	175	126	120
Cold-rolled, normalized	480	175	658	110
Steel				
7C4 (SAE 1010)	210	210	105	–
30C8 (SAE 1030)	250	210	149	20
50C4 (SAE 1050), annealed	335	206	272	–
98C6 (SAE 1095), annealed	415	206	418	–
98C6 (SAE 1095), tempered	520	206	656	–
20Ni3 (SAE 2320), annealed	310	206	233	52
20Ni3 (SAE 2320), tempered	690	206	1156	40
50Ni2 Cr80 (SAE 3250) annealed	550	214	707	–
50Ni2 Cr80 (SAE 3250) tempered	1400	214	4579	30
50Cr6 V23 (SAE 6150), annealed	430	214	432	–
50Cr6 V23 (SAE 6150), tempered	1100	214	2827	–
Rubber	2.06	1.035	2050	–

* Cast iron has no well-defined elastic limit, but the values may be safely used for all practical purposes

Resilience in bending: A cantilever beam with a concentrated load at its free end will be used as a typical case to illustrate the procedure in computing the resilience of a beam. When a load F is applied at the end of a cantilever beam having a uniform cross section and a length l, the deflection of the free end is

$$y = \frac{Fl^3}{3EI}$$

By the general equation 3.23 the resilience is

$$U = \frac{Fy}{2} = \frac{F^2 l^3}{6EI}$$

If s denotes the maximum stress in the extreme fiber at the support,

$$Fl = \frac{sI}{c}$$

Substituting Ak^2 for I gives

$$U = \frac{s^2 k^2 Al}{6Ec^2} \qquad \ldots (3.27)$$

The maximum resilience per unit volume of the beam is

$$u = \left(\frac{k}{c}\right)^2 \frac{S_e^2}{6E} \qquad \ldots (3.28)$$

It is easy to prove that equations 3.27 and 3.28 hold also for a simple beam loaded at the middle or for a beam clamped at the ends and also loaded at the middle. Equations for other beams have the same general form but the numerical coefficients are different.

It should be remembered that the stress in bending in a beam varies from end to end and in each transverse section from the outer fiber to the neutral plane and further on to the other outer fibre. In equation 3.27, s is the maximum stress produced in the outer fibres of the dangerous section.

Resilience in shear: By starting with the general equation 3.23 and proceeding in the manner described for deriving equation 3.25, the resilience in transverse shear of a member with the volume V can be determined. The result is

$$U_s = \frac{S_s^2 V}{2G} \qquad \ldots (3.29)$$

The modulus of resilience is

$$u_s = \frac{S_s^2}{2G} \qquad \ldots (3.30)$$

Resilience in torsion: Assume that a round bar is held at one end and twisted at the other by a couple whose moment is Fa; and that the angular deflection of the free end is θ. The work of the couple is then $\frac{1}{2} Fa\theta$, and the resilience of the bar is

$$U_s = \tfrac{1}{2} Fa\theta$$

If s_s denotes the maximum shear stress at the fixed end, equation 2.17 gives

$$Fa = \frac{s_s J}{c}$$

Replacing J by Ak_o^2, where k_o is the polar radius of gyration, and eliminating θ by use of equation 2.5, we can obtain the equation

$$U_s = \frac{s_s^2 k_o^2 Al}{2Gc^2} \qquad \ldots (3.31)$$

The resilience per unit volume can be found by taking Al as 1. Thus,

$$u_s = \left(\frac{k_o}{c}\right)^2 \frac{S_s^2}{2G} \qquad \ldots (3.32)$$

For a round hollow shaft with an outside diameter D_1 and an inside diameter D_2, the polar radius of gyration is $k_o = \sqrt{(D_1^2 + D_2^2)/8}$ and $c = \frac{1}{2} D_1$. Therefore the unit resilience is

$$u_s = \left[1 + \left(\frac{D_2}{D_1}\right)^2\right] \frac{S_s^2}{4G} \qquad \ldots (3.33)$$

For a round solid shaft, $D_2 = 0$ and the unit resilience is

$$u_s = \frac{S_s^2}{4G} \qquad \ldots (3.34)$$

Example 3.3: Determine the maximum torsional impact that can be withstood without permanent deformation by a 100 mm cylindrical shaft 5 m long and made of 30C8 [SAE 1030] steel. For this steel, $S_e = 180$ MN/m^2 and $G = 82000$ MN/m^2. By equation 3.34, the unit resilience is

$$u_e = \frac{(180 \times 10^6)^2}{4 \times 82000 \times 10^6} = 98.78 \times 10^3 \text{ Nm/m}^3$$

$$V = \frac{\pi}{4} \times (0.1)^2 \times 5 = 0.03927 \text{ m}^3$$

Hence, the maximum torsional impact, which is equal to the resilience, is

$$k_{it} = U_s = 98.78 \times 10^3 \times 0.03927 = 3879 \text{ Nm}$$

Values of the maximum resilience per unit volume for various types of loading and the more commonly used cross sections are given in Table 3.3.

3.5 STRESS CONCENTRATION

When the cross section of a part changes abruptly at a certain place, and when the part has a hole, a notch, or a groove, or where a small section joins a larger one, then the stress in the part ceases to follow the elementary equations established in sections 2.9 to 2.11. In such a part the fibers closest to the abrupt change in the cross section are affected more than those farther from the place. *This phenomenon of increased stress is called stress concentration, or localized stress, and change of section is called a discontinuity or a stress raiser.* As shown in Fig. 3.5, the stress in the fibers nearest to the discontinuity is increased most. The increase tapers off as the distance to the fibers from the discontinuity becomes greater until at the most remote part of the section that stress decreases even below the nominal, or average, value. This must be true because the total internal resistance of a section with a stress raiser is the same as that of an equal section without a stress raiser.

Table 3.3: Maximum resilience per unit volume

Type of loading	Modulus of resilience (N/m^3)
Tension or compression	$\dfrac{S_e^2}{2E}$
Shear, simple transverse	$\dfrac{S_s^2}{2G}$
Bending in beams with simply supported ends:	
Concentrated central load and rectangular cross section	$\dfrac{S_e^2}{18E}$
Concentrated central load and circular cross section	$\dfrac{S_e^2}{24E}$
Concentrated central load and I beam section	$\dfrac{3S_e^2}{32E}$
Uniform load and rectangular section	$\dfrac{3S_e^2}{56E}$
Torsion:	
Solid round bar	$\dfrac{S_s^2}{4G}$
Hollow round bar with D_1 greater than D_2	$\left[1+\left(\dfrac{D_2}{D_1}\right)^2\right]\dfrac{S_s^2}{4G}$
Springs:	
Laminated with flat leaves of uniform strength	$\dfrac{S_e^2}{6E}$
Flat spiral with rectangular section	$\dfrac{S_e^2}{24E}$
Helical with round section and axial load	$\dfrac{S_s^2}{4G}$
Helical with round section and axial twist	$\dfrac{S_e^2}{8E}$
Helical with rectangular section and axial twist	$\dfrac{S_e^2}{6E}$

Stress concentration due to a discontinuity occurs whether the stress comes from a direct load, from bending, or from torsion.

If s_o is the nominal stress in a section with a discontinuity, as determined by an elementary formula, and if s_1 is the maximum or significant stress at the discontinuity, then

$$s_1 = s_o K' \qquad \ldots (3.35)$$

In Fig. 3.4a, bar is having a V notch and effective area is A. The stress at the section of notch is

$$S_1 = \frac{F}{A}.K'$$

In Fig. 3.4b, the bar is subjected to bending moment M and effective moment of inertia is I, the maximum bending stress is

$$S_1 = \frac{My}{I} \, K'$$

In Fig. 3.4c, a twisting moment T is applied. The effective polar moment of area is J and r is the effective radius, the maximum stress is

$$S_1 = \frac{Tr}{J} \, K'$$

K' is called the *stress-concentration factor*. Thus the stress-concentration factor indicates the maximum increase of the stress over the nominal stress computed without the influence of the discontinuity, i.e.

$$K' = \frac{\text{Maximum stress at discontinuity}}{\text{Average stress}}$$

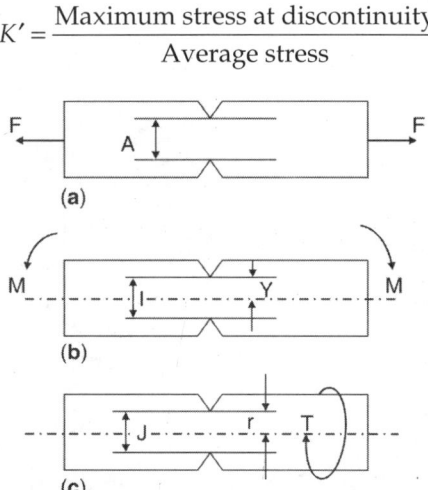

(a)

(b)

(c)

Fig. 3.4: Stress concentration in direct load, bending and torsion

The increase of the stress as expressed by the stress-concentration factor depends on three things: the type and size of the discontinuity, the material of the part, and the character of the load. Only the influence of the discontinuities as such will be discussed here. The other two factors will be discussed in Chapter 5.

Form stress factor: The theoretical magnitudes of stress-concentration factors are determined either by mathematical analysis or by special experimental methods, the more important of which are the photoelastic,[2] the plaster-model,[3] and the soap-film[4] methods.

By the nature of its determination the theoretical stress factor depends only on the type and relative size of the discontinuity or on its geometrical form. Therefore it is commonly called the *form stress factor* and is designated by K in order to distinguish it from the actual stress-concentration factor, which depends also on the material and type of loading.

Most engineering materials are elastic and have the ability to yield at the point of excessive stress. These characteristics spread the localized high stress and lower its intensity. As a result the actual stress-concentration factor K' is always lower than the corresponding form stress factor K. *However, in a brittle material like glass a line scribed with diamond point is sufficient to separate the two parts by giving a small blow or impact load [3–5].* The form stress factor K may be considered as the high limit of stress concentration under the most adverse conditions.

The values of the form stress factors for various types of discontinuities given in this book, mostly in the form of curves, are based on the most nearly reliable data available at present, but should not be considered absolutely correct. New investigations are being conducted and published continuously, and as a result of these the data which follow may require revision.

Causes of Stress Concentration

1. Variation in the properties of material from point to point, e.g. internal cracks and flaws, cavities in welds, air holes in steel and concrete, nonmetallic inclusions, etc.

2. Pressure at points and areas at which loads are applied, e.g. contact between wheel

[2] M. M. Frocht, "Recent Advances in Photoelasticity," *Transactions of the American Society of Mechanical Engineers*, Vol. 53, APM-53-11 (1931), p. 135.
[3] F. B. Seely and T. J. Dolan, "*Stress Concentration at Fillets, Holes, and Keyways as Found by the Plaster Model Method*," Bulletin No. 276, University of Illinois Engineering Experiment Station (June, 1935)
[4] F. B. Seely, Advanced Mechanics of Materials (New York: John Wiley & Sons, Inc., 1932), pp. 189 ff.

and rail, balls and races of bearings, contact between beam and its supports, contact between gear teeth, contact between threads of nut and screw, etc.

3. Abrupt changes in section. These changes occur in two classes.

(a) Surface conditions such as scratches due to machining and handling, indent marks, tool marks and grinding marks, etc.

(b) Form of member, i.e. when a member has abrupt changes in section as in crankshaft, machine shaft or spigot or socket of a colter joint.

The effect of discontinuities is explained below:

A small hole means usually a hole less than 3 mm diameter.

3.6 STRESS CONCENTRATION FROM DIRECT LOADS

In the case of a round hole the stress concentration depends on the condition of the hole, as whether it is free, empty, or filled with a bolt or pin.

Free round holes: The stress distribution in a plate of infinite width containing a round hole, as found by a theoretical analysis, is illustrated in Fig. 3.5. At a point 1 at one edge of the hole, $K = 3$. From there the stress decreases at first very rapidly, and then gradually, asymptotically approaching s_o.

If the width of the plate is b and the diameter of the hole is d, the value of K decreases with the increase of the ratio d/b, as shown in Fig. 3.6. Theoretically, a small hole has a greater effect than a large one. Actually, the value of stress concentration factor is given below:

Mathematically expressed as $\dfrac{r}{b} \to 0, K \to 3$

At a point 2, Fig. 3.5, a transverse stress is induced. Its intensity is

$$s_2 = N s_o \qquad \ldots (3.36)$$

Values of the coefficient N are also given in Fig. 3.6, which shows that the transverse stress has a negative sign. If the direct stress is tension, the transverse stress is compression, and vice versa.

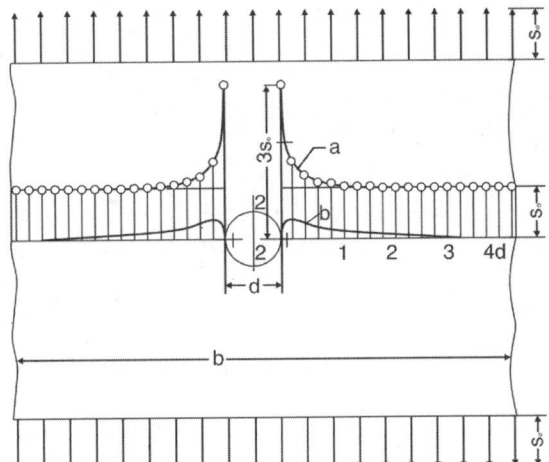

Fig. 3.5: Stress concentration due to hole in wide plate.

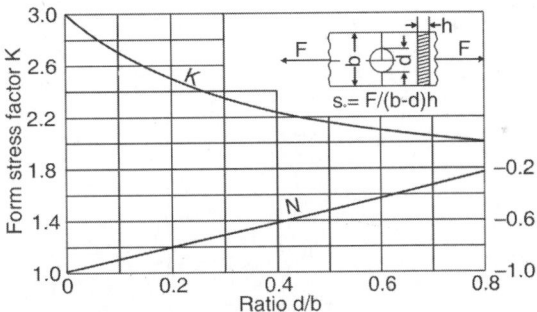

Fig. 3.6: Form stress factor due to hole in narrow plate

When $s_1 = 3\, s_o$ at 1
$$s_2 = -s_o \text{ at } 2$$

Shaft with hole: The form stress factor for a shaft containing a hole can be found from Fig. 3.7.[5] As may be seen by comparing this curve with Fig. 3.6, the value of K for a shaft are about 10 per cent lower.

---

[5] C. Lipson, G. C. Noll, and L. S. Clock, Stress and Strength of Manufactured Parts (New York: McGraw-Hill Book Company, Inc., 1950), p. 68, and M. M. Frocht, "Factors of "Stress Concentration Photoelastically Determined," Trans. ASME, Vol. 57, APM (1935), p. A-67.

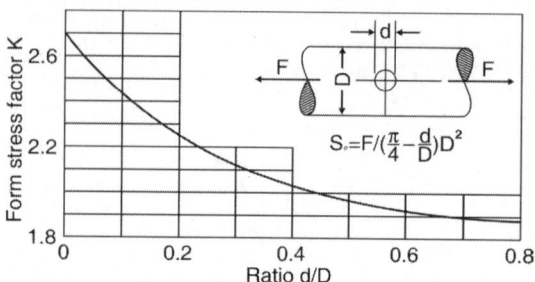

Fig. 3.7: Form stress factor due to through hole in shaft loaded in tension

Filled holes: The relative values of the axial stresses found by the photoelastic method in a plate with a hole filled by a pin, where the load is applied at the ends of the plate, are given in Fig. 3.8. It shows that the value of K is 2.5, which is practically the same as that for a free hole when d/b is 0.2. The lower curve gives the transverse stresses s_r in the main section. At first there is compression amounting to $-0.75 s_0$, where the plate touches the pin. The stress then decrease rapidly to zero; and after that it reaches a maximum of about 0.25 s_0 in tension.

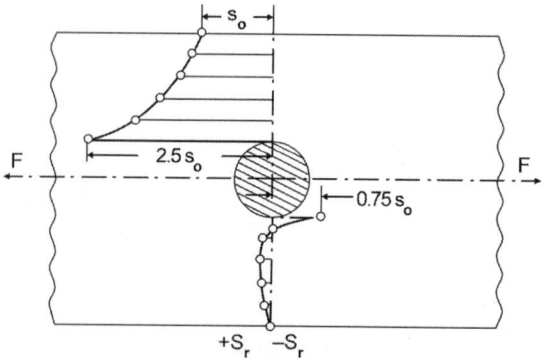

Fig. 3.8: Hole filled with pin

The stress distribution in a plate when the hole is filled by a pin and the load is applied through the pin is shown in Fig. 3.9. The maximum tensile stress in the main section parallel to the line of action of the load occurs at point c and is about the same as before; thus, $s = 2.5 s_0$. All the radial, or transverse, stresses s_r in this section are tension and the maximum is about 0.5 s_0. The curve for s_c shows the radial compressive stresses in the section through the center line of the plate, under the pin. The stress where the pin presses against the hole edge is very great, probably of the order of 6.5 s_0. At a distance 1.5 mm lower, it is about 3 s_0, and it becomes zero about 10 mm from the hole. All the transverse stresses in the section b–b are compressive, the maximum[6] being at point b and equal to 6.5 s_0. This gives a value of $N = -6.5$ in equation 3.36. This stress is very high; but being a compressive one, it produces a plastic deformation in actual machine parts and does not cause failure.

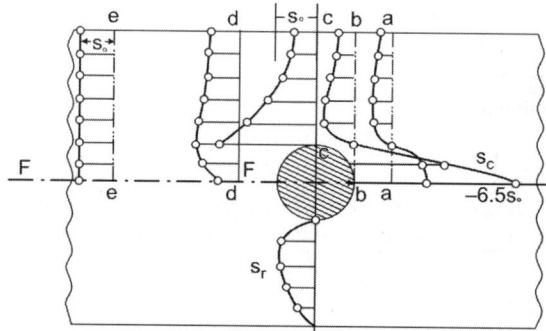

Fig. 3.9: Hole filled with pin which transmits load

The values of K for such a plate are shown in Fig. 3.10. The curves[7] show that K is affected both by the relative size of the hole, or the ratio d/b, and by the distance from the free end of the plate, or the ratio d/b.

Stresses between two holes: Stress concentrations from two adjacent holes are superimposed, as shown in Fig. 3.11 for two free holes.

Eye bar: In Fig. 3.12 are shown stresses created in the tangential direction around the edge of the hole in an eye bar due to a load applied by means of a pin which fills the opening. The stresses are those found by photoelastic measurements and are represented by the polar diagram *gbiej* with the hole edge

[6] E. Lehr, Spannungsverteilung in Konstructionselementen (Berlin: Vercin Deutscher ingenieure, 1934), p. 43.
[7] Lipson, Noll, and Clock, op, cit., p. 93.

as a base line.[8] The greatest stress is not in the section ck where it is expected, but in the section dl at point d. It is about 4.2 times as great as the nominal stress s_0 in the eye section ck.

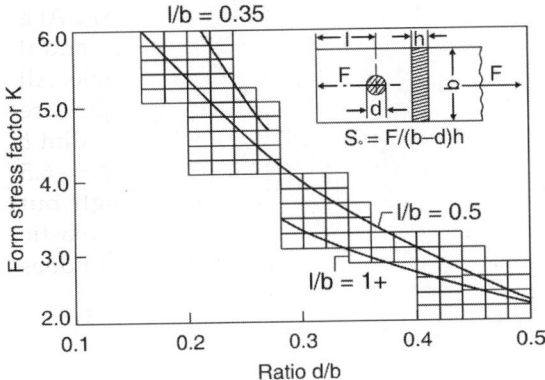

Fig. 3.10: Form stress factor for hole filled with pin which transmits load

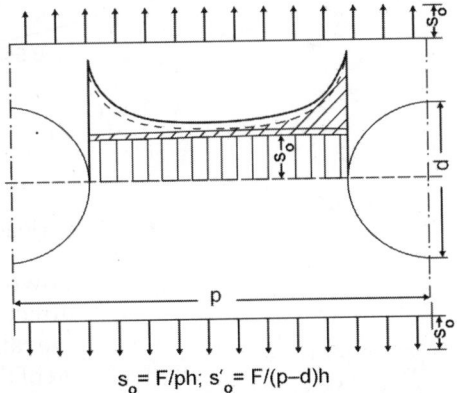

$$s_0 = F/ph; \quad s'_0 = F/(p-d)h$$

Fig. 3.11: Stresses between two holes

Curve mno gives the stress variation in the section dl from the inner edge to the outer edge. It shows that the outer third of the eye section is under compression instead of tension.

Curve pqr gives the compressive stresses normal to the inner surface of the hole. This curve shows that the maximum bearing pressure is not at the center line. Instead, because of the deformation of the eye bar, the greatest pressure occurs about 45° below the center line.

[8] Seely, op. cit., p. 220.

It should be noted that the stress curves and their scales apply only to a bar having the relative dimensions shown in Fig. 3.12. With other dimensions the numerical values for the stresses may change somewhat. However, dimensions used in Fig. 3.12 are typical for various machine part of similar shape.

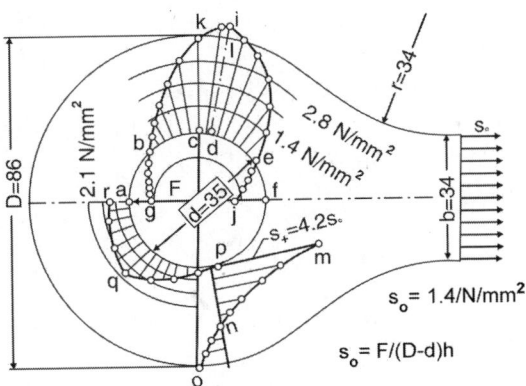

Fig. 3.12: Stresses in eye bar

Biaxial loading: If a plate with a hole is subjected to stresses s_0' and s_0'' acting at right angles, the combined maximum stress is found by superposition and is

$$s'' = Ks_0' + Ns_0'' \qquad \dots (3.37)$$

Example 3.4: Determine the stresses at the edges of a cylindrical vessel subjected to an internal pressure. The conditions are represented in Fig. 3.13. The hoop stress in an unweakened section is s_0.

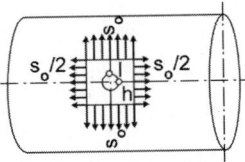

Fig. 3.13: Hole in cylindrical shell

The stresses at points h and l can be found from equation 3.37 by using the corresponding values for s_0' and s_0''. Another method is to proceed step by step. In a closed vessel the longitudinal stress is $0.5\,s_0$. It invokes at point l a concentrated stress $0.5\,Ks_0$ and at point h a transverse stress $0.5\,Ns_0$. The hoop stress

s_o invokes at point h a concentrated stress Ks_o and at point l a transverse stress Ns_o. From Fig. 3.4, for a infinitely wide plate, $K = 3.0$ and $N = -1$. Combining the stresses acting at the same points, we obtain for the longitudinal stress at l,

$$s_1 = 1.5\,s_o - s_o = 0.5\,s_o$$

The hoop stress at h is

$$s_1 = 3\,s_o - 0.5\,s_o = 2.5\,s_o$$

Inner holes: The effect of an inner blowhole in a casting or of a flaw in a forging is similar to that of a hole in a plate. Theoretically the maximum stress produced at the edge of a spherical hole is

$$s = 2s_o \qquad \dots (3.38)$$

Elliptical holes: The effect of an elliptical hole depends on the relative values of the plate width B, the major half-axis a, and the minor half-axis b, and also on the direction of the axis of the ellipse with respect to the direction of the applied force.

For a plate having a width B greater than $12a$ and containing an elliptical hole with its major axis normal to the direction of the stress, the form stress factor may be computed by the expression[9]

$$K = 1 + 2\frac{a}{b} \qquad \dots (3.39)$$

If the minor axis is normal to the direction of the stress,

$$K = 1 + 2\frac{b}{a} \qquad \dots (3.40)$$

In Fig. 3.14 is indicated the influence of the orientation of an ellipse for which $a = 3b$.[10] The general effect is similar to that of notches of different sharpness, as may be seen by comparing Fig. 3.14 with Fig. 3.15. A microcrack is a special case of an elliptical hole with a large major axis and length of minor axis approaching

zero. So the value of $K \to \infty$. *The only way to stop the propagation of a crack is to drill a small hole at each end of the crack. This will reduce the large stress concentration factor to 3 (Fig. 3.3).*

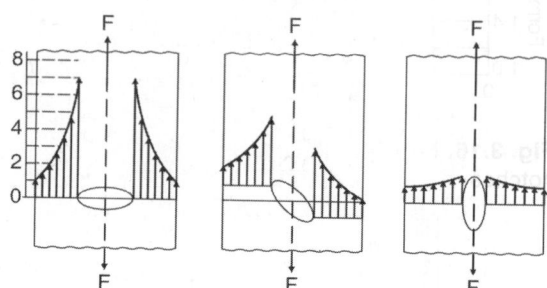

Fig. 3.14: Stress concentration caused by elliptical holes

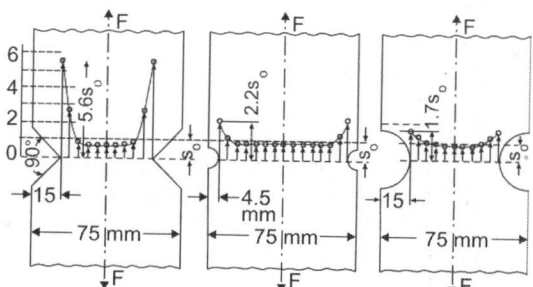

Fig. 3.15: Stress concentration caused by notches

Notches and grooves: Notches (re-entrant) in flat pieces and cylindrical grooves in round bars have an effect similar to that of holes. While protruding notches do not produce stress concentration. The form stress factor depends on the depth h_1 of the notch or groove as well as on its shape. The curves in Fig. 3.15 show typical stress concentration due to tension. The chart in Fig. 3.16 gives numerical data for various relative sizes of semicircular notches,[11] and the curves in Fig. 3.17 are for notches of various relative depths h_1 and radii r at the bottom.[12]

[9] S. Timoshenko and J. N. Goodier, *Theory of Elasticity*, 2nd ed. (New York: McGraw-Hill Book Company, Inc. 1951), p. 86.

[10] Battelle Memorial Institute, *Prevention of the Failure of Metals Under Repeated Stress* (New York: John Wiley & Sons, Inc., 1941), pp. 53, 55.

[11] A. M. Wahl and R. Beeuwkes, Jr., "Stress Concentration Produced by Holes and Notches," *Trans.* ASME, Vol. 56, APM–56–11 (1934), p. 621.

[12] Lehr, op. cit., Table 4, Fig. 18.

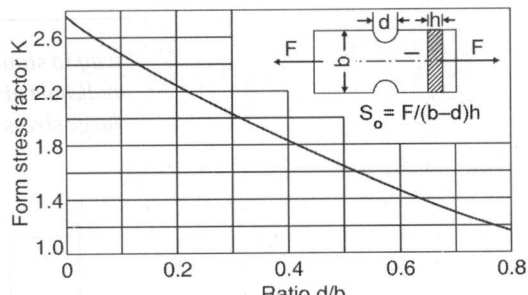

Fig. 3.16: Form stress factor due to semicircular notches

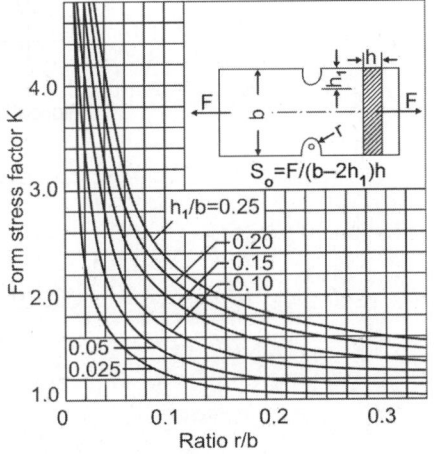

Fig. 3.17: Form stress factor due to various notches

For small notches the form stress factor can be computed from the equation[13]

$$K = 1 + 2\sqrt{\frac{h_1}{r}} \qquad \ldots (3.41)$$

Values of K for notches having various shapes, particularly different angles between the side, are given in Fig. 3.18. These curves are based on an infinite width b but may help to estimate the correction of the value of K found from the curves in Figs 3.16 and 3.17 when the sides of the notch are not parallel.

Fillets: The influence of a fillet depends on the relative size of the fillet, as expressed by the

ratio r/b, as shown in Fig. 3.19; but it does not depend on the relative size of the rib, as expressed by the ratio B/b, if r is less than $\frac{1}{2}(B-b)$.[14] The form stress factor may be determined by the curve a. While Fig. 3.19 shows a plate, the same values of K may also be used with sufficient accuracy for a round bar, shaft, bolt, or similar part.

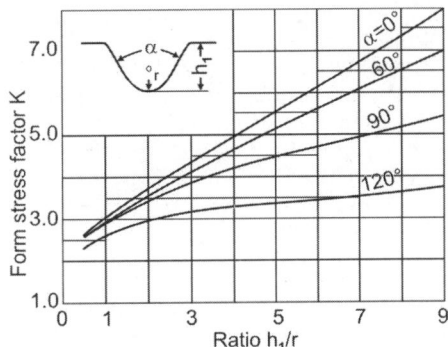

Fig. 3.18: Form stress factor due to notches of various shapes

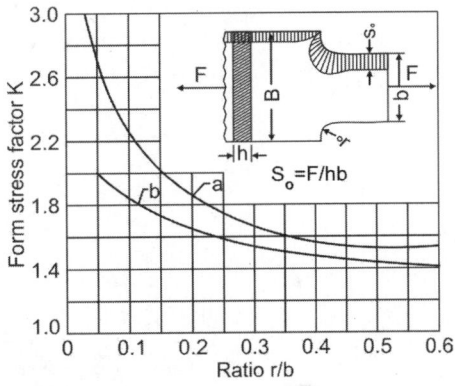

Fig. 3.19: Form stress factor due to fillets, for tension

For a full fillet for which r is not less than $\frac{1}{2}(B-b)$, such is encountered in a shaft or a hub of a disk, the form stress factor is lower and may be presented by the curve b.[15]

[13] R. E. Peterson, "Stress Concentration Phenomena in Fatigue of Metals," *Trans.* ASME, Vol. 55, APM–55–19 (1933), p. 161.

[14] E. E. Weibel, "Studies in Photoelastic Stress Determination," Trans. ASME, Vol. 56, APM–56–13 (1934), p. 641.

[15] S. Timoshenko and W. Dietz, "Stress Concentration Produced by Holes and Fillets," Trans. ASME, Vol. 47 (1925), p. 210. See also Lipson, Noll, and Clock, op. cit., p. 59.

Screw thread: Screw threads are similar in their effect to circular grooves, but the helical shape reduces the form stress factor somewhat. The shapes of various threads were established before the effect of stress concentration was recognized. Value of the factor K are therefore rather high. Thus, for the American National thread, $K = 5.62$; and for the British Whitworth thread, $K = 3.86$.[16]

Example 3.5: Determine the form stress factor for a plate 96 mm wide that is loaded in tension and has two symmetrical notches in the main cross section. Each notch is 12 mm deep and has a radius of 3 mm at the bottom, and its sides are at right angles.

From Fig. 3.17, for $h_1/b = \dfrac{12}{96} = 0.125$ and $r/b = \dfrac{3}{96} = 0.031$, the factor K is 3.6.

The influence of the 90° flareout can be found from Fig. 3.18. For $h_1/r = 12 \div 3 = 4$, the ratio of the values for 90° and 0° is $4.3 \div 5.0 = 0.86$. Therefore the corrected value of the form stress factor is

$$K = 3.6 \times 0.86 = 3.10$$

Protrusions: Even such a discontinuity as a protrusion on the edge of a plate causes a certain stress concentration, as shown in Fig. 3.20.[17]

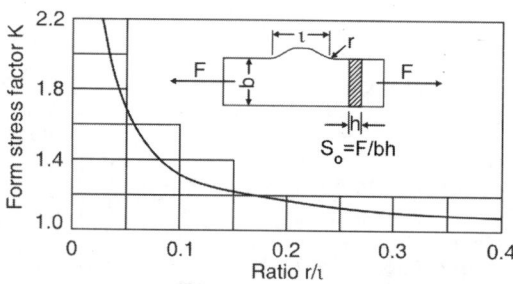

Fig. 3.20: Form stress factor due to protrusion on plate

3.7 STRESS CONCENTRATION IN BENDING

Abrupt changes in cross section may occur in various parts of beams for a number of reasons.

Holes in beams: The influence of holes on the stresses in a bar subjected to bending may be found by superimposing the hole concentration stresses, determined by use of Figs 3.5 or 3.8, on the bending stresses, represented in Fig. 2.13b. A hole with its axis located in the neutral plane of a beam, as in Fig. 3.21, has practically no effect on its strength. A hole located nearer the beam edge may result in high stresses at the hole edge, these stresses exceeding those in the outer fibers of the beam. Thus, holes located halfway between the neutral plane and the edge of a beam, as in Fig. 3.22, show maximum stresses about 1.5 times as great as those at the outer fibers.

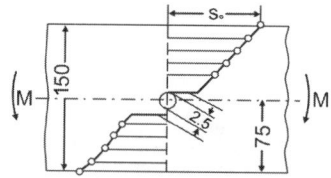

Fig. 3.21: Stress in beam with one central hole

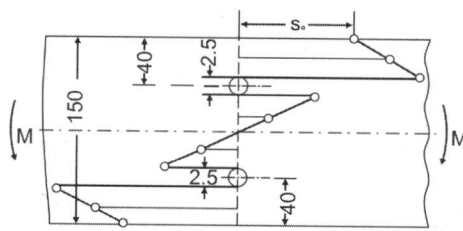

Fig. 3.22: Stresses in beam with two holes

Notches: Values of the form stress factor, as found experimentally for rapid bending of beams with notches,[18] are given in Fig. 3.23. The curves for constant r have a parabolic shape, and those for constant h have a hyperbolic shape. With the decrease of r in sharper notches, the values of K increase very fast as they approach the limit values. The values

[16] H. F. Moore and P. K. Henwood, *The Strength of Screw Threads Under Repeated Tension*, Bulletin No. 264, University of Illinois Engineering Experiment Station (1934), p. 15.

[17] G. H. Neugebauer, "Stress Concentration Factors and Their Effect on Design," Product Engineering, Vol. 14 (1943), p. 82.

[18] F. Rötscher, "Die Ermittlung der Spannungsverteilung in Konstructionsteilen durch Dehnungsmessungen," *Z. VDI*, Vol. 77 (1933) p. 375. The article does not give the depth b of the beam; h/b seemingly was very small.

from Fig. 3.23 can also be used with sufficient accuracy for tension and compression.

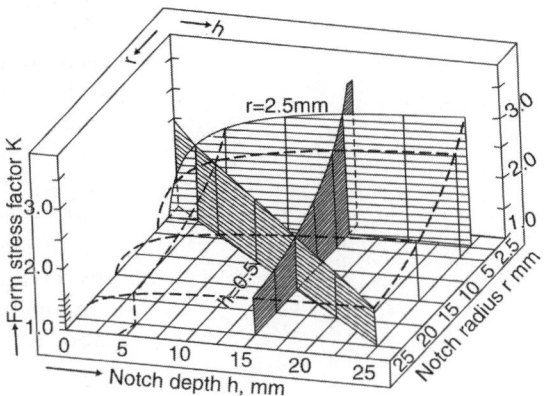

Fig. 3.23: Form stress factor due to notches, for rapid bending

The curves in Fig. 3.24 give values of the form stress factors for beams with notches of the shape indicated. These curves take into account the dimension b of the bar, the depth h of the notch, and the radius r of its bottom. In Fig. 3.25 is pictured the effect of a sharp notch in a beam. This diagram shows both the stress distribution and the distance along the axis through which the notch exerts its influence.

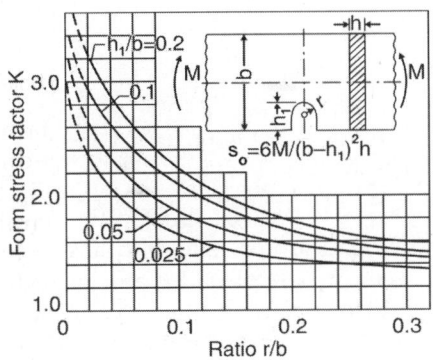

Fig. 3.24: Form stress factor due to notches in beam

Fillets: For a fillet in symmetrical beam sections the influence of the ratio of the fillet

radius r to the plate of width b is represented by curves similar to curves a and b in Fig. 3.19 for tension. For pur bending, however, the numerical values of K are about 20 per cent lower than those given in Fig. 3.19 for tension. The form stress factor K shown in Fig. 3.19 applies both for flat sections, such as ribs, and for junctures of shafts of different diameters. The influence of a one-sided fillet, as in an angle-shaped section, is shown in Fig. 3.26, which gives values of K found by photoelastic measurements.[19]

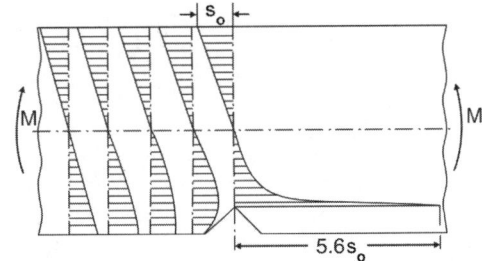

Fig. 3.25: Stress distribution in bending by sharp notch

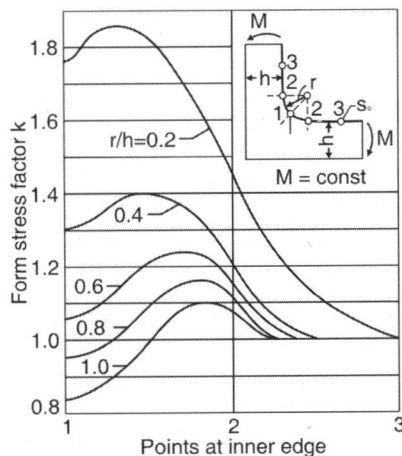

Fig. 3.26: Increase in bending stress caused by fillet

The form stress factor for an angle shaped section for which the thickness ratio H/h is 4 is given in Fig. 3.27 as a function of the ratio r/h and the ratio l/h.

[19] L. Föppl, "Fortschritte auf dem Gebiet des Spannungsoptischen Untersuchung von Konstructionen," *Z. VDI*, Vol. 76 (1932), p. 507.

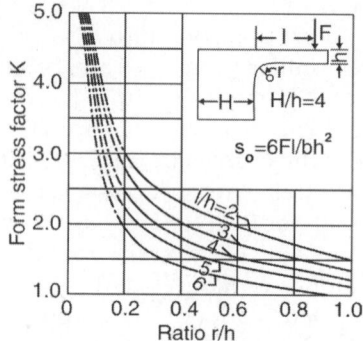

Fig. 3.27: Form stress factor for angle shaped section bending

Enlarged central section: The curves in Fig. 3.28[20] give values of the form stress factor for pure bending of a flat bar with an enlarged central section. The same curves may be used with sufficient accuracy for bending of a shaft with an enlarged section. It may also be added that an increase of the width to $D = 3d$ raises the

values of K very little; a decrease of the length to $L = D$ reduces K very little; and a decrease in length to $L = 0.25 D$ reduces K to the values given by the curves c and d for $D/d = 2.0$ and $D/d = 1.25$, respectively.

Gear teeth: The stresses at the roots of gear teeth are greatly influenced by the fillet.[21] In Fig. 3.29 the stress distribution around the fillet is shown; and the curve in Fig. 3.30 gives the factor K as a function of the fillet radius r. With a small fillet the stress concentration is considerable, as K ranges from 2 to 2.5.

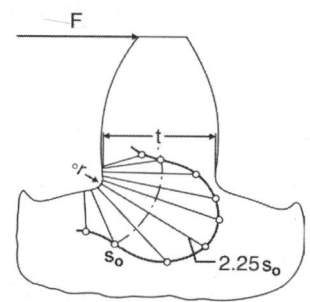

Fig. 3.29: Stress concentration at root of gear tooth

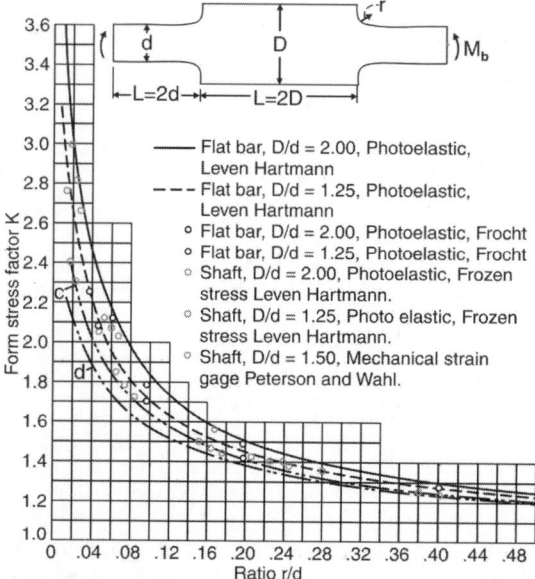

Fig. 3.28: Form stress factor for bar with enlarged section in bending

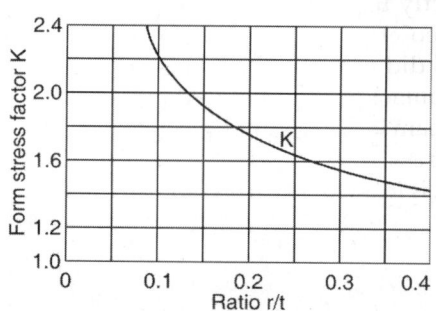

Fig. 3.30: Influence of fillet on stress concentration at root of gear tooth

Concentrated load: The application of a concentrated load produces high local stresses, as may be seen in the explanation for Fig. 3.9, where the form stress factor reached a value of 6.5.

It can be shown mathematically[22] that a concentrated load F, Fig. 3.31, applied to a

---

[20] J. B. Hartman and M. M. Leven, "Factors of Stress Concentration for the Bending Case of Fillets in Flat Bars and Shafts with Central Enlarged Section," Proceedings of the Society of Experimental Stress Analysis, Vol. IX, No. 1 (1951), p. 61.

[21] S. Timoshenko and R. U. Baud, "The Strength of Gear Teeth," Mechanical Engineering, Vol. 48 (1926), p. 1105.

[22] Seely, op. cit., p. 233.

beam with a narrow rectangular cross section creates radial compressive stresses s_r. The magnitude of the stress at any point a is

$$s_r = \frac{2F\cos\theta}{\pi r} \qquad \dots (3.42)$$

where, F is the load per unit of beam width. This stress combined with the flexure stress $s = M/Z$ will give the resultant stress in the proximity of the load. The maximum value of s_r will be for the angle $\theta = 0$, as at point b, where

$$s_r = \frac{2F}{\pi r} = \frac{0.637F}{r} \qquad \dots (3.43)$$

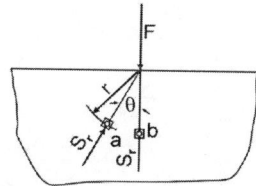

Fig. 3.31: Action of a concentrated load

Directly under the load, where $r = 0$, the stress theoretically would become infinite if the material did not yield locally and change the line of contact to a small area of contact.

In a beam loaded at the center, the maximum shear stress s_s' near the top, as at point c in Fig. 3.32, is in accordance with equation 2.49,

$$s_s' = 0.5\,(s_r - s) \qquad \dots (3.44)$$

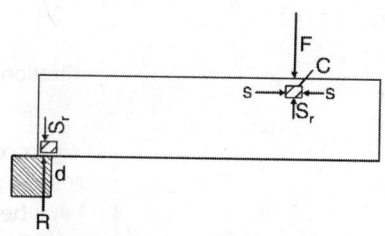

Fig. 3.32: Stresses in beam caused by concentrated load

At a point d near the support the bending stress s is negligible, but a proper bearing area must be provided to avoid a high stress concentration. The accuracy of this analysis has been confirmed experimentally.

---

[3] *Ibid.*, p. 228.

3.8 STRESS CONCENTRATION IN TORSION

When a shear stress is caused by torsion, the form stress factor generally is smaller than that for direct stress or flexure. If the cross sections of a straight shaft are circular but vary in diameter, a plane section remains plane after twisting, but its intersection with an axial plane does not remain a straight line after twisting. The change in the diameter results in a concentrated shear stress, as illustrated in the upper corner of Fig. 3.33. The form stress factor is a function both of the ratio of the two adjacent diameters and of the relative size of the fillet as shown by the curves in the same figure.

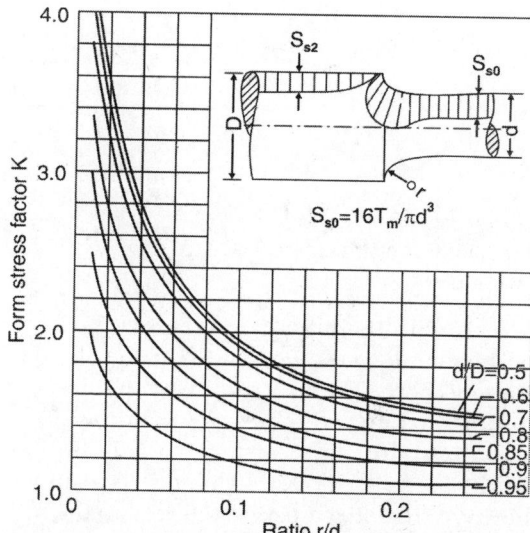

Fig. 3.33: Form stress factor due to enlargement of shaft in torsion

Keyways: A keyway in a shaft has a stress-concentration effect which is particularly noticeable at its end. A few numerical values are available. The relation between the form stress factor K in torsion and the ratio of the size of the fillet r in the keyway corner c to the width b, determined by the soap-film method, is shown in Fig. 3.34.[23] Numerically K is very high, and rectangular keyways should be avoided in highly stressed shafts. The influence of the keyway extends to the center e of the bottom of the keyway.

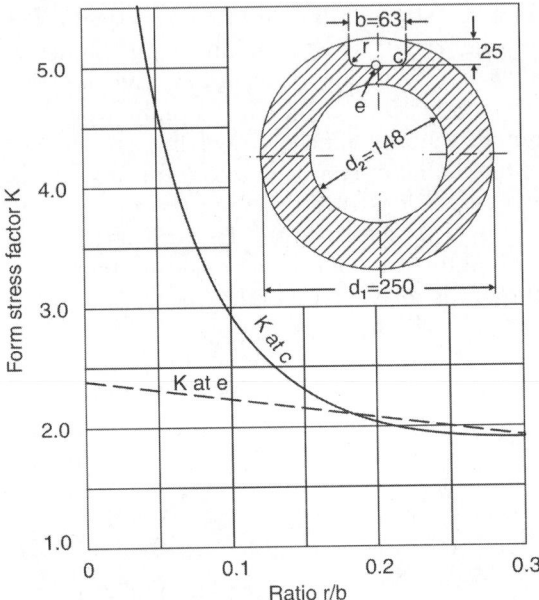

Fig. 3.34: Form stress factor due to keyway in shaft in torsion

Tests with the plaster-model method give considerably lower values for K than do tests with other methods. Thus, for a keyway of standard proportions, for which $b = d/4$ and $h = b/2$, carefully conducted tests gave $K = 1.68$ for a profiled keyway and $K = 1.44$ for a sled-runner keyway.[24] In both cases K is referred to the full shaft section. In bending the same tests gave values for K that were lower by about 15 per cent.

Radial holes in shafts: In the case of a shaft subjected to torsion the influence of a hole passed radially through the shaft is given in Fig. 3.35. The data found by the plaster-model method are referred to the unweakened shaft.[25] Therefore, the values of K cannot be compared with values given by other curves, such as those shown in Fig. 3.7. However, data furnished by Fig. 3.35 are really what a designer needs. They show directly how much a certain hole will weaken the shaft.

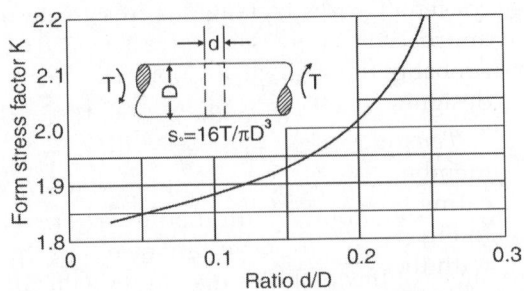

Fig. 3.35: Form stress factor due to hole in shaft in torsion

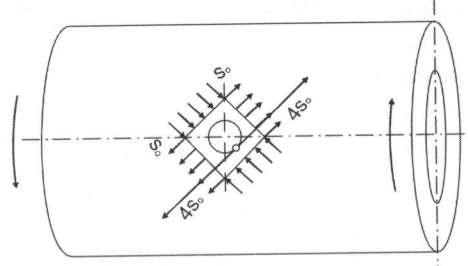

Fig. 3.36: Stress concentration due to torsion

The influence of a radial hole on a hollow shaft subjected to torsion can be found by applying the equation of stresses for biaxial loading to a hollow shaft, Fig. 3.36, in which two main stresses have different signs.

To determine the stress distribution shown in Fig. 3.36, we assume that $d/b \to 0$ and we use equation 3.37, in which $K = 3$ and $N = -1$. The result is

$$s'' = 3s_o + (-1) \times (-s_o) = 4s_o \qquad \dots (3.45)$$

Thus, a small hole, such as an oil hole, theoretically causes a great concentration of stress, i.e. $K = 4$.

A similar effect occurs in the wing of an aeroplane caused to flutter which has the same effect as that of torsion (3.10).

3.9 QUALITATIVE EVALUATION OF STRESS CONCENTRATION

The discontinuities previously discussed and the resulting stress-concentration effects cover only the most typical cases. For other cases and for a clear understanding of stress

---

[24] Seely and Dolan, loc, cit., p. 24.

[25] *Ibid,* p. 21.

concentration in general, there exist qualitative approaches that help the designer to visualize probably stress concentration in the part he is designing.

Tension: Stress concentration in tension members may be represented by lines indicating the direction of the principal stresses. In Fig. 3.37 is shown an axially loaded plate with these direction lines. At both ends of the plate the lines are parallel, indicating uniform stresses; but at the right end they are closer together, indicating a higher stress. Where the lines join near the discontinuity, they are more crowded. A local stress increase is thus indicated.[26] In practice these lines are more descriptively called *force-flow lines.*

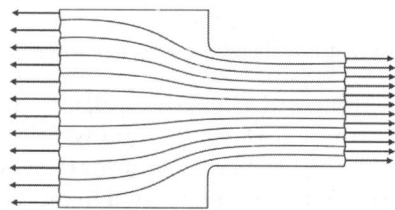

Fig. 3.37: Force flow in tension member

The usefulness of the method of drawing force-flow lines can be illustrated by the example of stress concentration created by an elliptic hole, as shown in Fig. 3.38. The force-flow lines indicate that there is a lower stress concentration in Fig. 3.38b than in *a*, although $2b = 2a$ and the cross-sectional area of the plate is reduced by the same amount. That the stress in *b* is lower can be confirmed by applying equations 3.39 and 3.40.

If in both cases $a/b = 2$, the form stress factor in Fig. 3.38a is $K_a = 1 + 2 \times 2 = 5$ and that in Fig. 3.38b is $K_b = 1 + 2 \times \frac{1}{2} = 2$.

Torsion: A circular shaft with two different diameters can be considered to be divided into a number of concentric hollow shafts and a solid central shaft of such sizes that each of these elementary shafts carries an equal share

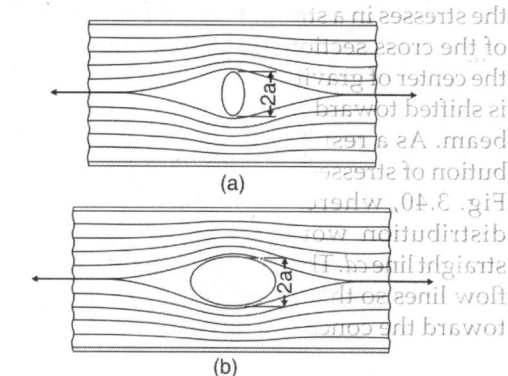

Fig. 3.38: Force flow in tension member with elliptic hole

of the torque transmitted by the original solid shaft.[27] In Fig. 3.39 is shown a typical stepped shaft having both, the part with the large diameter and the part with the small diameter divided into four concentric hollow shafts and a solid central shaft. The two sets of shafts of constant diameter are joined by shafts of variable diameter along smooth curves, as indicated in the drawing. Since the thickness of the outer shaft just at the discontinuity is very small, high stresses are indicated at this place. The thicknesses of the other shafts are also reduced at the discontinuity, but to a lesser degree. Using a larger fillet will give smoother junctures and will lower the localized stresses.

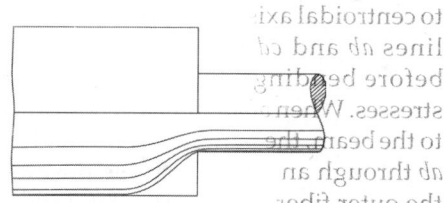

Fig. 3.39: Division of shaft into equitorque hollow shafts

3.10 CURVED BEAMS

Many machine parts, such as frames of shears, punch presses, crane hooks, and various levers loaded in bending, have the basic shape of a curved beam.

The normal stresses created in a curved beam when it is subjected to bending differ from

[26] R. V. Baud, "Avoiding Stress Concentration by Using Less Material," *Product Engineering,* Vol. 5 (1934), p. 170.
[27] L. S. Jacobsen, "Torsional-Stress Concentration in Shafts of Circular Cross Section and Variable Diameter," *Trans.* ASME, Vol. 47 (1925), p. 619.

the stresses in a straight beam. The neutral axis of the cross section, instead of going through the center of gravity, or centroid, of the section, is shifted toward the center of curvature of the beam. As a result there is a nonlinear distribution of stresses, as shown by the curve ab in Fig. 3.40, whereas for a straight beam the distribution would be represented by the straight line cd. The curvature changes the force-flow lines so that there is stress concentration toward the concave side of the beam.

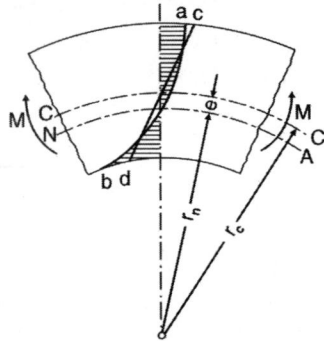

Fig. 3.40: Bending stresses in curved beam

The equation for a curved beam can be derived by assuming, as for a straight beam, that planes normal to the centroidal axis before bending remain plane and also remain normal to centroidal axis after bending. In Fig. 3.41 the lines ab and cd represent two such planes before bending, that is, when there are no stresses. When a bending moment M is applied to the beam, the plane cd rotates with respect to ab through an angle $d\theta$ to the position fg, and the outer fibers are shortened while the inner fibers are elongated. The original length of a strip at a distance y from the neutral line is $(r_n + y)\theta$. It is shortened by the amount $y\,d\theta$, and the stress in this fiber is, by equation 2.4,

$$s = E\epsilon = -\frac{Ey\,d\theta}{(r_n + y)\theta} \qquad \ldots (3.46)$$

Then the load on a strip having a thickness dy and a cross-sectional area dA is

$$dF = s\,dA = -\frac{Ey\,d\theta\,dA}{(r_n + y)\theta}$$

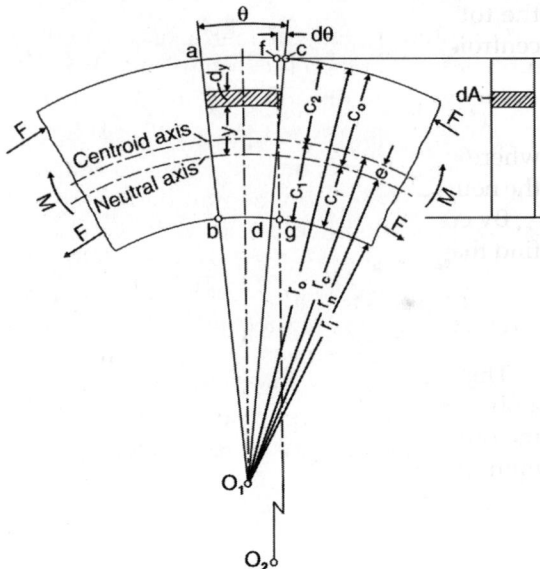

Fig. 3.41: Analysis of stresses in curved beam

From the conditions of equilibrium, the summation of forces over the whole cross section is zero, and the summation of the moments of these forces is equal to the applied bending moment. Therefore

$$\int dF = -\frac{E\,d\theta}{\theta} \int \frac{y\,dA}{r_n + y} = 0$$

Since $E\,d\theta/\theta$ is not equal to zero, it follows that

$$\int \frac{y\,dA}{r_n + y} = 0 \qquad \ldots (3.47)$$

For a given cross section, dA can be expressed in terms of y and dy, and the radius of curvature r_n can be determined from this equation.

If moments are taken about the neutral axis,

$$M = -\int y\,dF$$

By substituting for dF, we can obtain

$$M = \frac{E\,d\theta}{\theta}\int \frac{y^2\,dA}{r_n + y} = \frac{E\,d\theta}{\theta}\int\left(y - \frac{yr_n}{r_n + y}\right)dA$$

$$= \frac{E\,d\theta}{\theta}\int y\,dA$$

Since $\int y\,dA$ represent the statical moment of area, it may be replaced by Ae, the product of the total area A and the distance e from the centroid axis to the neutral axis. Then

$$M = \frac{AeE \, d\theta}{\theta} \qquad \ldots (3.48)$$

where e is the distance from the centroid axis to the neutral axis, or $e = r_c - r_n$, as in Fig. 3.41.

By combining equations 3.46 and 3.48, we find that

$$s = -\frac{M}{Ae}\left(\frac{y}{r_n + y}\right) \qquad \ldots (3.49)$$

This is the general equation for the stress in a fiber at a distance y from the neutral axis. At the outer fiber, y is equal to c_o and the maximum compressive stress due to bending is

$$s_o = -\frac{Mc_o}{Aer_o} \qquad \ldots (3.50)$$

At the inner fiber, y is $-c_i$ and the maximum tensile stress due to bending is

$$s_i = \frac{Mc_i}{Aer_i} \qquad \ldots (3.51)$$

The value of r_n depends on the shape of the beam. For the more commonly used cross sections the value of r_n may be computed from data given in Table 3.4. The values of c_i and c_o in Fig. 3.41 may be determined from data given in Table 2.5, since

$$c_i = r_c - r_i - e = c_1 - e, \quad c_o = r_o - r_c + e = c_2 + e \ldots (3.52)$$

Final remarks: The stresses determined by equations 3.49, 3.50 and 3.51 represent pure bending stresses produced by a couple. Curved machine members usually are bent by an eccentrically applied force. In such a case the normal stress from the force applied at the centroid must be added to the bending stresses.

If holes must be put in a curved beam they should be located on the neutral axis rather

Table 3.4: Values of radius to neutral axis for curved beams

Type	Section	Radius of Neutral surface r_n
a		$r_n = \dfrac{(\sqrt{r_o} + \sqrt{r_i})^2}{4}$
b		$r_n = \dfrac{h}{\ln(r_o/r_i)}$
c		$r_n = \dfrac{\frac{1}{2}h(b_i + b_o)}{\frac{b_i r_o - b_o r_i}{h}\ln(r_o/r_i) - (b_i - b_o)}$ If $b_o = 0$, this section reduces to a triangle.
d		$r_n = \dfrac{A}{b_i \ln \frac{r_i + a_i}{r_i} + b_2 \ln \frac{r_o - a_o}{r_i + a_i} + b_o \ln \frac{r_o}{r_o - a_o}}$ If $a_o = 0$, the section reduces to a $\perp$ section; r_n is the same for a box section shown in dotted lines with each side panel $\frac{1}{2}b_2$ thick.

than on the centroid axis, to decrease the effect of stress concentration as explained in section 3.7 and shown in Figs 3.21 and 3.22.

Example 3.6: Determine the maximum stress in the frame of the 100 kN punch press shown in Fig. 3.42.

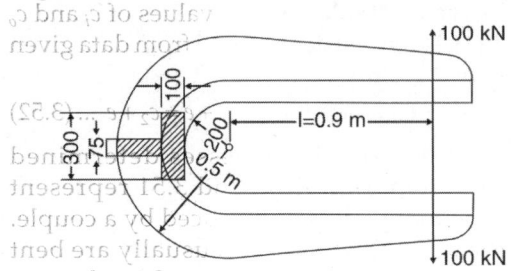

Fig. 3.42: Frame of punch press

The radius of curvature of the neutral axis may be determined by using the expression given for case d in Table 3.4 and letting $a_o = 0$. Thus

$$r_n = \frac{75 \times 200 + 100 \times 300}{300 \ln\left(\frac{200+100}{200}\right) + 75 \ln\left(\frac{500}{200+100}\right)}$$

$$= 281.4 \text{ mm}$$

The distance from the centroidal axis to the inner fibre may be computed by applying the formula given for case d in Table 2.7. The result is

$$c_1 = \frac{75 \times 300^2 + 225 \times 100^2}{2[75 \times 300 + 225 \times 100]} = 100$$

$$c_2 = 300 - 100 = 200$$

The distance from the centroid axis to the neutral axis is

$$e = (200 + 100) - 281.4 = 18.6 \text{ mm}$$

From equation 3.52, $c_i = 100 - 18.6 = 81.4$ mm and $c_o = 200 + 18.6 = 218.6$ mm.

The bending moment referred to the neutral axis is

$$M = F(r_n + l) = 100000\,(281.4 + 900)$$

$$= 118 \times 10^6 \text{ Nmm}$$

The compressive stress in the outer fibres due to bending is, by equation 3.50

$$s_o = \frac{118 \times 10^6 \times 218.6}{45000 \times 18.6 \times 500} = -61.6 \text{ N/mm}^2$$

The direct stress from force $F = 100,000$ N is tension. Its magnitude is

$$s_2 = \frac{100,000}{45,000} = 2.2 \text{ N/m}^2$$

The tensile stress in the inner fibres due to bending is by eq 3.51

$$s = \frac{118 \times 10^6 \times 81.4}{45000 \times 18.6 \times 200} = 57.3 \text{ N/mm}^2$$

The combined stress in the outer fibre is

$$s = -61.6 + 2.2 = -57.3 \text{ N/mm}^2$$

and that in the inner fibre is

$$s = 57.3 + 2.2 = 59.5 \text{ N/mm}^2$$

3.11 REPEATED STRESSES

Experiments show that test specimens of metal fail when loads are repeated or reversed several million times, even though the unit stresses do not reach the elastic limit. Until recently this phenomenon was called *fatigue of material*.[28] It has now been established that the failure is due to a crack, which occurs at a point of the surface of the part where the highest tensile stress exists.[29] The fissure gradually spreads until failure takes place. Therefore the proper name is failure by progressive fracture.[30] This phenomenon is illustrated in Fig. 3.43 by means of lines of force flow. The dotted line b in Fig. 3.43a connects the points of maximum stresses. The beginning of a crack is shown in Fig. 3.43b; and the gradual increase of the crack depth is indicated in Figs 3.43c and 3.43d. The fracture follows the surface of maximum stresses and is always normal to the nearest line of force flow. Absence of elongation and of reduction of the area at the break is characteristic of a failure by progressive fracture. In general, ductile metals do not resist repeated stresses better than brittle metals.[31]

[28] J. B. Johnson, M. O. Withey, and James Aston, Materials of Construction, 8th ed. (New York; John Wiley & Sons, Inc., 1939), p. 771, and R. E. Peterson, op. cit., p. 157.

[29] J. O. Almen, "Shot Peening to Increase Fatigue Resistance," SAE Journal, Vol. 51, No. 7 (July, 1943), p. 248.

[30] Seely, op. cit., p. 56.

[31] H.W. Gillett, "Effect of Alloying and Heat Treatment on Endurance Limit of Steel," Proceedings of the American Society for Testing Materials, Vol. 30, Part I (1930), p. 291.

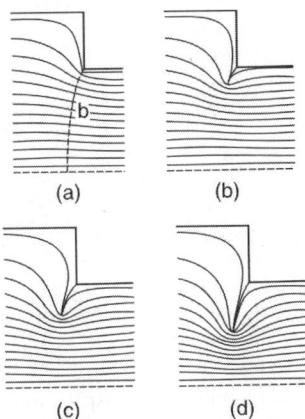

Fig. 3.43: Force flow and failure due to progressive fracture

If a thin ductile steel wire is bent by 180°, then straightened and bent by 180° in the reverse direction and process repeated a few times, results in the breaking of wire due to repeated stresses (3.6).

Repeated loading is especially dangerous for parts with discontinuities, which always cause stress concentrations. The maximum local tensile stress may produce a minute crack which will cause a failure of the part after a sufficiently large number of changes in the load.

The crack propagation depends on the fracture toughness which is the property of the material that defines its ability to resist stress at the tip of a crack. It also depends on the stress intensity factor which is a measure of stress at the section of the crack. The stress intensity factor is proportional to the applied nominal stress and also square root of the width of crack. If the stress intensity factor is less than the fracture toughness, crack is not affected. When the stress intensity factor reaches fracture toughness, the crack propagates with velocities reaching higher than 1.5 km/s.

Endurance limit: The maximum stress which, with a complete reversal, can be repeated many millions of times (infinite times) without causing failure by progressive fracture is called the endurance limit. There is no numerical relation between the elastic limit and the endurance limit. However, there seems to exist

[32] Supplement to Z. *VDI*, Vol. 77, No. 42 (1933), p. 1

a certain relation between the endurance limit S_{en} and the ultimate tensile strength S_u. For ferrous materials S_{en} is about $0.5\,S_u$, and for nonferrous materials S_{en} is between $0.25\,S_u$ and $0.35\,S_u$.[32]

S-N curves: The endurance limit of a material is determined by applying to several specimens repeated loads which cause completely reversed stresses of known values, and recording the number of stress reversals each specimen did endure before it failed. Each specimen is subjected to a lower stress than was the preceding one, and it breaks after a greater number of cycles. Finally, a stress is reached that does not cause failure, regardless of how many reversed cycles are applied. The results are plotted, as shown in Fig. 3.44, with the stresses S as ordinates and the numbers of cycles N as abscissas. Values of S_{en} and N are laid off on logarithmic scales in order to shorten the diagram and increase its accuracy. Such plots are called S-N curves.

Fracture starts by initiating a small crack which forms the nucleus. When the stress at the crack is tensile, it opens up and crack has a tendency to increase as the stress at the end of crack is very high. When there is compressive stress, the two sides of the crack try to close there by smoothening the surfaces at every compressive stress. Finally as the crack propagates, the resultant resisting area becomes so small that it fractures. *The fractured parts will be characterised by a smooth area or*

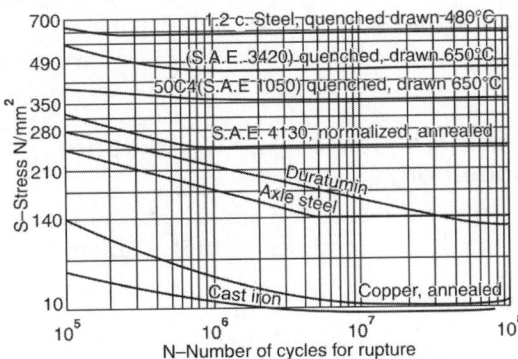

Fig. 3.44: S-N curves

bright area indicating progressive compressive stresses and a dull area (rough area) indicating a sudden fracture (3.9).

For steels, the S-N curves flatten out abruptly and show a knee when the stress is approaching the endurance limit. With a nonferrous metal, the S-N curve does not show such a pronounced knee; and for some materials, such as some aluminum alloys, the curve does not level off even after 10^9 cycles. For such a material no endurance limit exists, and it is possible to give only an endurance strength indicating a stress and the number of cycles withstood at that stress.

Endurance strength or endurance limit is the maximum stress that can be subjected to a test specimen or a machine part by repeated loading or reversed loading infinite times without causing failure by progressive fracture. For most materials 10^7 is considered a sufficient number of cycles to define the endurance limit. An alternating or completely reversed stress is one which reverses completely, from tension to an equal value in compression, or from clockwise shear to an equal value of shear in the opposite direction.

In testing, the method of applying an alternating load is by using a rotating bending member. The maximum fiber stress alternates between equal values of tension and compression as the test member rotates.

The term fatigue strength is commonly applied to the highest repeated stress to which a material can be subjected without failure due to a infinite number of cycles. The repeated stress need not be alternating but may cycle between any upper and lower limits.

Designing for Fatigue Conditions— Soderberg's Diagram

To determine the limiting stress conditions when variable and steady stresses are present, Soderberg suggested a method for defining the limiting stress condition with the known values of endurance limit and the yield strength of a member.

The ordinate of Fig. 3.45 represents the variable stress S_r. The point A represents the

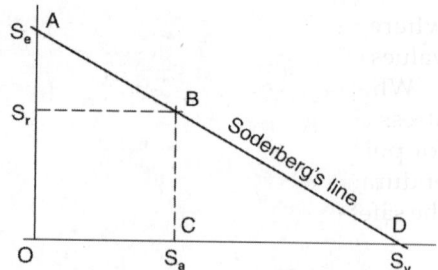

Fig. 3.45 Soderberg's diagram

magnitude of endurance limit s_e, the steady or the average stress s_a is represented on the abscissa. The point D represents the yield strength s_y.

From similar triangles BCD and AOD,

$$\frac{S_r}{S_y - S_a} = \frac{S_e}{S_y}$$

From which

$$S_r = S_e \left[1 - \frac{S_a}{S_y} \right] \qquad \dots (3.53)$$

This equation is called the Soderberg's law and it gives the limiting values of the variable stress S_r for the given value of the endurance limit, and the yield strength and for various values of the average stress S_a.

Figure 3.46 shows the use of factor of safety in the case of fatigue loads.

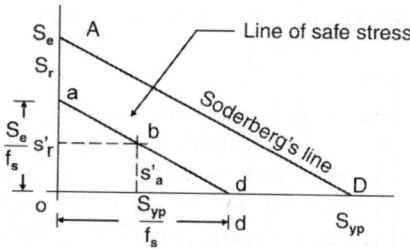

Fig. 3.46: Soderberg's diagram with factor of safety

The factor of safety is usually the same for the endurance limit and the yield strength and the safe line *ad* is obtained instead of AD for any point *b*, with the factor of safety f_s, the variable stress

$$S_r = \frac{S_e}{f_s} \left[1 - \frac{S_a'}{S_y / f_s} \right] \qquad \dots (3.54)$$

where, S_r' and S_a' are respectively the safe values of the varying stress and average stress.

When the load is fluctuating, a suitable stress concentration factor k must be applied for points of stress concentration. If S_e is the endurance stress and range of stress S_r, then the safety factor n is

$$n = \frac{S_e}{KS_r} \qquad \ldots (3.55)$$

Similarly n can be written for completely reversed stress.

In many actual situations, a part of the stress is static and with less accurately known alternating stresses superimposed. This can be considered as being made up of two parts: (i) the steady or the average load F_{av} and (ii) the variable or range load F_r. The maximum load F_{max} is

$$F_{max} = F_{ar} + F_r$$

and minimum load $F_{min} = F_{ar} - F_r$ correspondingly the stress S_{av} and S_r can be found.

To take care of various types of combinations, line of failure is used for various combination of loads. If results are plotted in Fig. 3.47 with S_u on the x-axis and S_e on the y axis, the line AB indicates the Failure points as Goodman's line where point A indicates the failure for a completely reversed stress and point B is the failure corresponding to ultimate stress.

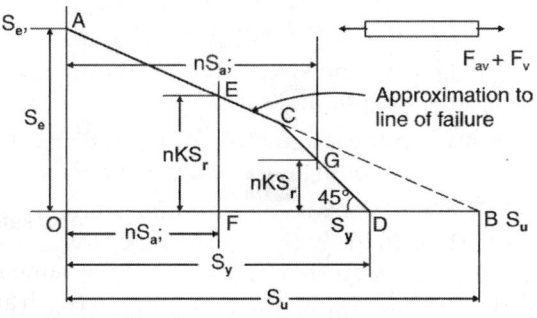

Fig. 3.47: Modified Goodman's diagram

For any combination, failure points lie between A and B. A modified Goodman's diagram is drawn as the curve ACD. If stress concentration is present, it is applied to the varying component.

A safety factor n is used for both components S_a and KS_r to avoid failures.

A point on the curve such as E to be safe consists of static stress $OF = n\,Sa$ and varying component $FE = nKS_r$.

By similar triangles

$$\frac{EF}{FB} = \frac{OA}{OB}$$

Equivalent static stress OB, $S_u = OF + FB$

$$OB = OF + \frac{OB}{OA} \times EF$$

$$= nS_a + \frac{S_u}{S_e} \times nKS_r$$

$$S = \frac{S_u}{n} = S_a + K\frac{S_u}{S_e}S_r \qquad \ldots (3.56)$$

$$\frac{1}{n} = \frac{S_a}{S_u} + \frac{KS_r}{S_e} \qquad \ldots (3.57)$$

Maximum shear theory of failure (Guest) can be applied when normal and shear stresses in a shaft are fluctuating. It is assumed that normal and shear stresses reach their maximum values simultaneously. Similar to equation 3.56, the equivalent shear stress S_s

$$S_s = S_{sa} + \frac{K.S_u}{S_e}.S_{sr} \qquad \ldots (3.58)$$

Substituting equations 3.56 and 3.58 into equation 2.51

$$S' = \sqrt{\frac{1}{4}\left(S_a + \frac{KS_u}{S_e}S_r\right)^2 + \left(S_{sa} + \frac{KS_u}{S_e}S_{sr}\right)^2} \qquad \ldots (3.59)$$

$$= \frac{16}{\pi d^3}\sqrt{\left(M_a + \frac{KS_a}{S_e}M_r\right)^2 + \left(T_a + \frac{KS_u}{S_e}T_r\right)^2} \qquad \ldots (3.60)$$

where, M_a is the average bending moment
M_r is the varying bending moment
T_a is the average torque
T_r is the varying torque

Example 3.7: A shaft is subjected to a bending moment of 2000 Nm and average torque of 4500 Nm. If the torque fluctuates by ± 20% and stress concentration for bending and torsion is 1.5; the ultimate strength of the shaft material is 720 N/mm², yield strength 420 N/mm^2 and endurance strength 276 N/mm^2, find the shaft diameter for a factor of safety of 2.

$$\frac{kS_u}{Se} = \frac{1.5 \times 720}{276} = 3.913 \text{ and 20\% of } 4500 = 900 \text{ Nm}$$

Taking $M_a = 0$

$$d^3 = \frac{16}{\pi \times \frac{420}{2}} \times$$

$$\sqrt{(0 + 3.913 \times 2000 \times 10^3)^2 + (4500 \times 10^3 + 3.913 \times 900 \times 10^3)^2}$$

$$= \frac{16 \times 10^6}{\pi \times \frac{420}{2}} \sqrt{61.246 + 64.347}$$

$d = 64.7$ mm say 65 mm.

Bending: During bending, only the outer fibers on both sides of the neutral plane are subjected to the maximum tensile and compressive stress. The adjoining fibers are stressed less and, in general, fibers located nearer to the neutral plane can take part of the load off the more highly stressed fibers located further from the neutral plane. This fact explains why accurate tests show that the elastic limits in tension and compression in the case of bending are higher than the elastic limit found for direct tension and compression. This difference is about 20 per cent for carbon steels and about 11 per cent for alloy steels. However, in beam calculations these differences are neglected, as this practice is on the side of safety. On the other hand, endurance limits in pure tension and compression are probably lower by a similar amount. However, no reliable data relating to this feature are yet available, the only information being that given in the endurance diagram of Chapter 4.

Summary: Progressive fracture can be the result of the repetition of direct loads, bending moments, or torques. The endurance limits in tension or compression are lower than those in bending, averaging for steels about 65 per cent of the bending limit; and those in torsion are still lower, averaging about 55 per cent of the

bending limit.[33] For cast iron the endurance limit in torsion is about 92 per cent of that in bending.[34] The endurance limits given for various materials are usually found by subjecting the test specimens to repeated bending in opposite directions or reverse stresses. At present most of the important materials have been sufficiently investigated, and diagrams similar to that in Fig. 3.46 can be drawn for them for use in design calculations.

PROBLEMS

3.1 What are dynamic loads?

3.2 What is the effect of a small circular hole in a beam when it is a situation (a) at neutral axis (b) away from neutral axis?

3.3 How can you prevent the propagation of a crack in a part subjected to fatigue load?

3.4 If the dimensions of the parts are same, which is more critical: (a) circular hole (b) semicircular groove (c) fillet?

3.5 A glass sheet is cut by scribing a line by a diamond point and then giving it a blow in bending. Explain why does it split along the scribed line.

3.6 What is the cause of fracture of a wire when it is bent, straightened and bent on the reverse side and reversed and so on?

3.7 If the size of standard specimen in fatigue testing be increased, what is the effect on endurance strength?

3.8 How is the strength of a plate with a rectangular hole be increased?

3.9 Why is a fatigue fracture characterised by a smooth area and a dull area?

3.10 What is stress concentration factor for a hole in the wing of an aeroplane when it flutters due to wind?

3.11 (a) Determine the stress caused by the centrifugal force in the rim of a cast iron belt pulley. The pulley has an outside diameter of 0.6 m, a face width of 200 mm,

[33] F. B. Seely, Resistance of Materials, 3rd ed. (New York: John Wiley & Sons, Inc., 1947), p. 292.
[34] W. Herold Wechselfestigkeit Metallischer Werkstoffe (Vienna: Julius Springer, 1934), p. 68.

and average rim thickness of 8 mm and it turns at 800 rpm. The specific weight of cast iron is 75 kN/m³. (b) Find the stress if the speed is increased to 1000 rpm.

3.12 (a) Determine the stress due to centrifugal action in the rim of a cast-iron belt pulley, the face of the pulley is 250 mm, the thickness of the rim is 22 mm, the pulley has an outside diameter of 1.5 m and it turns at 240 rpm. Assume the weight of cast iron to be 75 kN/m³. (b) Find the stress if the pulley speed is increased to 400 rpm.

3.13 A steel shaft 56 mm diameter is supported on two bearings and has four 900 N counterweights attached as shown in Fig. P3.1. The shaft rotates at 300 rpm. Assuming that the supports are at the centre of bearing, find the bending stress caused by centrifugal forces.

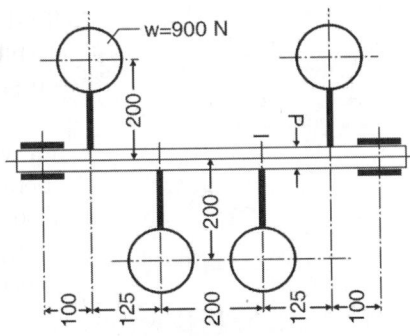

Fig. P3.1

3.14 A locomotive coupling rod connecting three wheels has a rectangular section 40 mm thick and 76 mm deep, and it is 1 m long between the each pair of centers. The cranks connected by the rod have a radius of 265 mm and a maximum speed of 250 rpm. The maximum power developed is 890 kW and is applied to the rear axle. Assume that from there the power is distributed evenly between the six wheels. Determine (a) the tensile and compressive stresses created in the rod by the useful efforts, (b) the stresses

created by the forces of inertia, and (c) the maximum resultant stress.

3.15 A weight W of 1.3 kN falls through a height h of 1.5 mm before the load strikes the square head a of the round steel bar 20 mm in diameter and 450 mm long as shown in Fig. P3.2. (a) Determine the stress induced, assuming a modulus of elasticity of 2.2×10^5 N/mm² and neglecting the interia of the bar. (b) Compare it with the stress from the same load acting statically. (c) Find how much the rod length can be changed before the impact stress exceeds 125 N/mm². (d) Determine how much the clearance can be increased before the impact stress exceeds 125 N/mm². (e) Compare the influence of the inertia of the bar upon the impact stress, using $w = 80$ kN/m³.

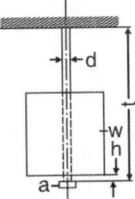

Fig. P3.2

3.16 Determine the maximum impact load which a 200 × 60 and 99 N/m standard beam can take without receiving a permanent deformation. The span is 1.8 m, the ends are clamped and the load acts in the center. Use $S_e = 201$ N/mm² and $E = 2.2 \times 10^5$ N/mm².

3.17 A steel clamp, as shown in Fig. P3.3, has the following dimensions: $a = 65$ mm, $b = 16$ mm, $h = 56$ mm, and $r = 12$ mm. Assuming for both the maximum tensile and compressive stresses, a value of

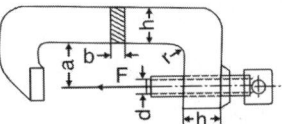

Fig. P3.3

110 N/mm² and taking into account the theortical value of stress concentration in the corners, determine the force F which can be applied by the screw.

3.18 (a) Determine the maximum stress in a cast steel angle bracket, Fig. P3.4, if $F = 5$ kN, $l = 76$ mm, $h = 22$ mm, $H = 76$ mm, $r = 6$ mm, and the width of the bracket is $b = 100$ mm. (b) Find how much this stress will be changed with $r = 10$ mm.

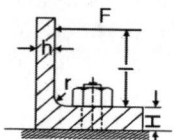

Fig. P3.4

3.19 The main dimensions of a crane hook are designated in Fig. P3.5. Determine the stresses in the inner and outer fibers in the dangerous section 1–1 for a 90 kN hook for which $r = 65$ mm, $h = 110$ mm, $b_i = 85$ mm, and $b_o = 25$ mm.

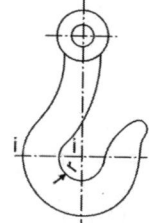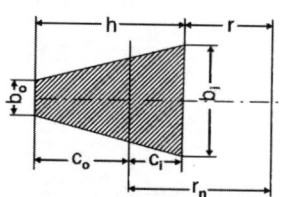

Fig. P3.5

3.20 (a) Determine the force F which will produce a maximum stress of 70 N/mm² in a cast iron punch frame Fig. P3.6 with the following dimensions, $r = 76$ mm, $l = 250$ mm, $b_i = 76$ mm, $b_o = 40$ mm, $h = 110$ mm and $a_i = 40$ mm. (b) Determine the corresponding compressive stress in the outer fibres.

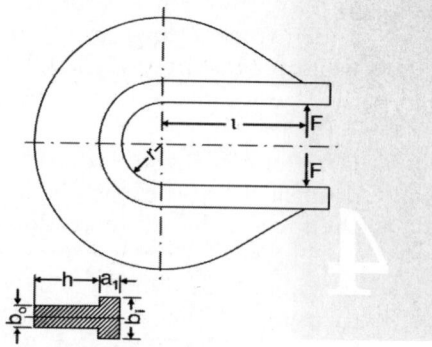

Fig. P3.6

3.21 A frame of a small drill press has the shape shown in Fig. P3.7. The dimensions of the cross section, with the designation of Table 3.4, Type d are $r_i = 50$ mm, $a_i = 25$ mm, $a_2 = 50$ mm, $a_o = 12$ mm, $b_1 = 50$ mm, $b_2 = 12$ mm, $b_o = 40$ mm, if $l = 150$ mm, determine (a) The force F that will produce a maximum tensile stress of 55 N/mm² in section 1–1. (b) The corresponding compressive stress in section 1–1. (c) The stresses in the outer fibres of section 1–1, the stresses in the outer fibres of section 2–2.

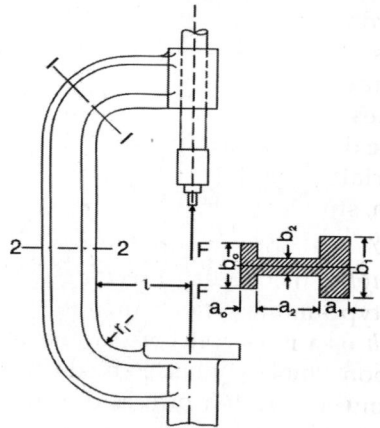

Fig. P3.7

4

Engineering Materials

4.1 TECHNICAL PROPERTIES

The technical properties of materials used in machinery may be classified as: the physical, such as composition, structure, homogeneity, specific weight, thermal conductivity and expansibility, and resistance to corrosion; the technological, or those relating to manufacturing, such as fusibility, forgeability, malleability, bending to shape, machinability; and the mechanical, which are established by tests and used in machine design. The basic mechanical properties are: elastic limits, moduli of elasticity, ultimate strength, endurance limits, and hardness. Secondary mechanical characteristics, determined from the basic ones or simultaneously with them, are: resilience, toughness, ductility, and brittleness. For machine design the mechanical properties of a material that are of prime importance are strength, stiffness, or rigidity, hardness, and ductility.

Strength: The strength of a part depends on the type and nature of loading. The *static strength* of a material is expressed by the corresponding elastic limits stress S_e or the yield point S_y. The static strength of a material is also expressed by ultimate strength S_u and breaking stress strength by S_u or S_b. The *impact strength* is measured by the corresponding modulus of resilience u, i.e. the amount of energy, Nm, required to cause failure. The endurance strength is expressed by the corresponding endurance limit s_{en}.

Stiffness, or *rigidity*, is measured by the modulus of elasticity, E being used for tension or compression and G being used for shear.

Hardness is a relative characteristic. There are several methods of measuring it, all of an arbitrary nature. The *Brinell hardness number* is obtained as follows: A hardened steel ball D mm in diameter is pressed under a load of F kgf into the smooth surface of the material to be tested; the diameter d of the indentation is measured in millimeters; and the hardness number Bhn, in kilogram force per square millimeter, is then computed by the relation

$$\text{Bhn} = \frac{2F}{\pi D(D - \sqrt{D^2 - d^2})} \qquad \ldots (4.1)$$

The standard ball has a diameter D of 10 mm, and the standard load F is 3000 kgf for steels and irons and 500 kgf for the softer, nonferrous metals.

The *Rockwell hardness number* is based upon the additional depth to which a test point is driven by a heavy major load beyond the depth to which the same penetrator has been driven at first by a lighter, minor load. The indentor is either a hardened steel ball or a spherical tipped conical diamond, called a *Brale*. The hardness number is read directly from a scale which is calibrated corresponding to penetration.

Several combinations of indentor and load are possible. Those most commonly used are for softer materials: the R-B scale, which assumes the use of a 3 mm steel ball and a major load of 100 kgf; for harder metals, the R-C scale, which assumes the use of a Brale and a major load of 150 kgf.

The *Vickers method* of determining hardness is similar in principle to the Brinell method. It expresses the pressure under the indentor in kilograms per square millimeter. The indentor is a diamond having the shape of a square pyramid, the loads are much lighter, and the indentation is measured with a medium-power microscope.

When the *Shore Scleroscope* is used, a small cylinder of steel with a hardened point is allowed to fall upon the smooth surface of the materials, and the height of the rebound of the cylinder is taken as the measure of hardness.

Comparison of hardness determinations. The Brinell method of determining hardness is used quite commonly because it gives accurate results at reasonable cost. However, it cannot be applied to brittle materials; neither can it be applied to a finished product without leaving a mark.

The Rockwell test is simple, fast, and accurate. It can be used with either flat or round surfaces, and the indentor leaves a smaller mark than does that of the Brinell tester. Its use is gradually increasing, especially in production work.

The Vickers test is considered to be more accurate than either the Brinell test or the Rockwell test, but the equipment is almost four times as expensive.

The Scleroscope test is simple, rapid, and definite for materials for which it is suited, but the results vary somewhat with the size and thickness of the sample. It is sufficiently accurate as a comparative measure for different specimens of the same material, but it is not reliable for comparing different metals.

There is no exact relation between the scales of different hardness tests, because each measures a somewhat different kind of hardness. The Brinell and Vickers numbers are identical up to 250. Beyond 250 the Brinell numbers begin to be increasingly lower because of the flattening of the steel ball. At

420 Vhn the difference is about 5 per cent; at 540 it is 8.3 per cent; at 675 it is 14.3 per cent; and at 960 it is 26 per cent.

Industry everywhere uses both the Brinell and Rockwell tests. Therefore both figures are given in tables in this book where hardness data are shown.

The following expressions[1] may be useful for changing from a Brinell hardness number B to a Rockwell-C number R_C or to a Rockwell-B number R_B:

$$R_C = 88B^{0.162} - 192 \qquad \dots (4.2)$$

and $$R_B = \frac{(B-47)}{0.0074B + 0.154} \qquad \dots (4.3)$$

Ductility: A materials is ductile if it is capable of undergoing a large permanent deformation without rupturing. There is no absolute measure of ductility. The percentage of *elongation* or the percentage of reduction of area during a tensile test carried to rupture is used as a relative measure. Ductility helps to relieve localized stress concentration through local yielding. It is a necessary characteristic of a material used to sustain live loads, especially where concentrated stresses may occur.

Toughness: It is the ability of the material to absorb strain energy before fracture. The area under load extension curve gives the strain energy absorbed by the specimen up to the breaking point.

Brittleness is characteristic opposite to ductility and toughness. A material may be considered brittle if its elongation at rupture through tension is less than 5 per cent in a specimen 50 mm long. Usually brittleness and hardness are closely associated, and very hard materials are brittle.

Typical properties: Some mechanical properties have already been discussed in the preceding chapter. Tables given in this chapter contain some of the physical properties and most of the basic mechanical properties, as far as they are established, of the materials used in

---

[1] I. H. Cowdrey and R. G. Adams, *Materials Testing,* 2nd ed., rev. ptg. (New York: John Wiley & Sons, Inc., 1944) p. 75.

modern machine construction. The values given in the tables are average values, not minimum ones. Average values are more representative; and with the use of proper safety factors, they give entirely satisfactory results. At the same time, the use of average values is an incentive for manufacturers of materials to improve the quality of their products, while the use of minimum values would be favorable only to manufacturers who do not keep abreast of progress.

4.2 CAST IRON

Ordinary cast iron is the most used of all materials employed in machines. The reasons are its strength, particularly in compression; the ease with which it can be cast into any desired shape; the facility with which its strength and hardness can be varied; its machinability; and its cheapness. Because of its structure it also has the valuable characteristic of *damping out* vibration.

Composition: Cast iron is iron containing so much carbon, from about 2 to 4 per cent that normally it is not malleable at any temperature. In addition, cast iron containing other elements in varying quantities, the more important ones being silicon, manganese, phosphorus, and sulfur.

Classes: Gray cast irons are produced with various strengths to suit different design requirements. According to ASTM specifications, gray-iron castings are listed by classes. The class number gives the minimum tensile strength of the cast test bar: No. 20, FG 140 cast iron has a strength of 140 N/mm². No. 60, FG 420, the cast iron of the highest test specifications, has a strength of 420 N/mm² only a few representative classes are listed in Table 4.1.

Meehanite: Meehanite is the name of several strong, uniform cast irons produced by a special patented process and having different combinations of physical and mechanical properties. These products are of four main types: (a) general engineering, (b) wear-resisting, (c) heat-resisting, and (d) corrosion resisting. Tensile strengths vary from 170 to 380 N/mm²

and with oil-quenching and tempering 530 N/mm² can be obtained.

White cast iron: It contains 1.75–2.30% carbon 1.2% silicon and small quantities of manganese, sulphur and phosphorus. There is no free graphite and has a white appearance. It is wear resistant and tough.

4.3 HEAT-TREATMENT OF CAST IRON

Several types of heat treatment are used to alter or enhance some properties of cast irons and thus to increase their usefulness.

Aging is applied to relieve casting stresses without materially affecting physical properties. It is carried on during 1 to 5 hr. in the temperature range of 425°C to 537°C. *Annealing* is carried out for 1 to 5 hr at 593°C to 870°C, the temperature depending on the size of the part. It is intended to reduce hardness and to facilitate machining. However, annealing is accomplished at the expense of some strength.

Baking is applied to castings that have been pickled in acid to remove sand and scale. Pickling makes castings brittle, but baking for a few hours at 150°C removes this brittleness.

Quenching cast iron in oil or water after it has been heated above the critical range increases its hardness and also its brittleness.

Drawing, or *tempering*, accomplished by reheating the quenched metal to a temperature below the critical temperature, reduces the brittleness but sill leaves an increase of hardness. By such a treatment a Brinell hardness number from 200 to 400 can be attained, the value depending on the quenching and drawing temperatures.

Malleableizing is also heat-treating process, but the final product has properties entirely different from those of the original product.

Malleable-iron parts are made of clean, white-iron castings, preferably having a low sulphur content and a certain manganese content. The castings are heated in an annealing furnace in contact with some iron oxide, which absorbs part of the combined carbon from the cast iron.

Table 4.1: Mechanical properties of ferrous cast alloys

No.	Name	Material Characteristic*	Ultimate tensile strength S_u N/mm²	Elastic limit, average N/mm² — Tension S_e	Elastic limit, average N/mm² — Compression S_e	Elastic limit, average N/mm² — Shear S_{es}	Modulus of elasticity — Direct, or young's $E \times 10^3$	Modulus of elasticity — Transverse, or shear $G \times 10^3$	Endurance limit S_{en} N/mm²	Hardness minimum — Brinell (Bhn)	Hardness minimum — Rockwell B	Hardness minimum — Rockwell C
1	Cast iron, Class No. 20 (ASTM) [FG - 150]**	Medium section, 12 mm	140	45	345	45	69	30	45	163	85	3
		Light section, < 12 mm	165	50	410	50	83	35	52	180	89	8
2	Cast iron, Class No. 25 [FG - 200]	Medium section, 12 mm	175	70	520	70	90	38	66	180	89	8
		Light section, < 12 mm	210	83	620	83	97	42	76	200	94	14
3	Cast iron, Class No. 35 FG220	Semi steel, 12 mm	240	100	700	100	110	45	90	200	94	14
4	Cast iron, Class No. 50 FG350	Electric-furnace, 12 mm	350	210	830	210	140	55	124	230	98	21
5	Malleable iron, ASTM 32310	Standard	350	225	225	160	172	69	165	120	70	–
6	Malleable iron, ASTM 48005	Pearlite	480	330	330	240	186	76	207	190	91	11
7	Iron-cooper alloy	Special heat treatment	400	210	275	165	165	69	207	280	104	29
8	Meehanite, M40, Type B	General use	310	130	470	125	131	52	130	196	93	13
		Heat-treated	410	275	620	–	–	–	–	237	99	22
9	Meehanite, N50, Type A	General use	350	145	520	140	145	55	152	207	95	16
		Heat-treated	520	350	700	–	193	76	193	250	102	24
10	Nickel cast iron, Grade II	Ni, 1.25	240	120	620	105	110	45	90	180	89	8
11	Nickel cast iron, Grade III	Ni, 1.75	300	124	700	110	124	52	110	180	89	8
12	High test cast iron	Ni. 1.8 ± 0.5; Mo, 0.3	310	70	655	175	160	55	138	220	96	19
13	Ni-Tensyliron	Ni, 2.1 ± 0.2; Mo, 0.3	350	97	970	210	160	55	152	320	108	34
14	Ni Resist cast iron	As cast	170	55	620	55	83	35	55	120	70	–
15	Nitralloy cast iron	Nitrided, heat-treated	450	210	1210	240	160	66	207	1050	–	68
16	Ductile iron, 90–65–02***	As cast	620	450	–	400	172	64	276	225	97	20
17	Ductile iron, 80–60–05	Heat-treated	550	410	–	370	172	64	235	195	93	13
18	Ductile iron, 60–45–15	Annealed	410	310	–	275	172	64	165	140	78	–
19	Ductile iron, 80–60–00	As cast	550	410	–	370	172	64	235	230	98	21

*Specific weight of all cast irons is about 70600 N/m³ in.

**Number after FG is the tensile strength in N/mm²

***90–65–02 indicates S_u – 90,000 psi (620), S_y = 65000 psi (450), Elongations is 02%

The annealing process converts the combined carbon of the original hard casting into a free nodular form called *temper carbon*.

4.4 ALLOY CAST IRONS

In order to obtain a material stronger than ordinary gray cast iron but having all its desirable characteristics, certain metals may be added, such as nickel, vanadium, molybdenum, copper, and aluminum. An alloy cast iron can be improved by heat-treating, just as ordinary cast iron. The annealing results in increase of strength to about 350 N/mm^2 and ductility of 18%. It can replace steel forgings and is much cheaper.

The addition of 1 to 2.5 per cent of nickel increases the tensile strength of cast iron to 420 N/mm^2. At the same time, the nickel makes the cast iron easier to machine but tougher and harder, thus increasing its resistance to wear.

The addition of *aluminum* to cast iron allows it to be nitrided like nitralloy steel. This process gives the iron a very hard surface. Adding *molybdenum*, even in small amount (0.2 to 1.5 per cent) raises the strength and the elastic and endurance limits of cast iron, as may be seen from Table 4.1. It also increases creep resistance.

Tests show that copper can be used instead of nickel with practically the same results. Another interesting development is an alloy of cast iron and copper which is used for making automobile-engine crankshafts.[2] Its composition is C, 1.3; Si, 2.0; Mn, 0.5; Cu, 2.6; Cr, 3.5. As cast the metal is hard, white, and brittle. After a special heat treatment, the material obtains the properties of open-hearth steel and has even a higher endurance limit in bending than steel.

Semisteel is a term often applied to a cast iron produced in the cupola by adding to the usual cast iron mixture from 20 to 40 per cent of low-carbon steel scrap. This mixture, when properly handled in the cupola and while being poured into the mold, produces strong, tough, close-grained castings with good machinability. However, the metal has none of the characteristics of steel and is only a high-grade gray cast iron.

Cast iron produced in an *electric furnace*, from mixtures similar to those used in a cupola, has considerably better mechanical properties, as well as greater hardness and good machinability. It is particularly suitable for impact loads.

4.5 DUCTILE IRON

Recently it has been found that small amounts of magnesium added to cast iron convert flake graphite to spherical, or nodular, graphite. This conversion results in a great increase of strength, elastic limit, hardness and elongation. Since the new material is ductile and on the basis of its mechanical properties resembles steel, it has received the name *ductile iron*.[3] Ductile iron is not a single material, but is rather a group of materials. As in steel, the matrix structure can be modified by alloys and heat treatment. From an industrial viewpoint the most interesting grades are the pearlitic and ferritic ductile irons.

The four main types of ductile iron produced at present are shown in lines 16 to 19 of Table 4.1. The 90–65–02 grade has a pearlitic structure, is used as cast, and has good resistance to mechanical wear. The 80–60–05 grade has a pearlitic-ferritic structure and is used to obtain a combination of strength and toughness. The 60–45–15 grade has a fully ferritic structure and is obtained by a short anneal of either of the first two grades. It has exceptional machinability and is very tough. By proper heat treatment its mechanical

---

[2] P. Dwyer, "Ford Foundry Casts V-8 Engine Crankshaft," *Foundry*, Vol. 62, No. 4 (April, 1934), p. 14.

[3] Based on information furnished by the International Nickel Company. See also A. P. Gagnebin, "The Industrial Status of Ductile Iron," *Mechanical Engineering*, Vol. 73 (1951), pp. 101–18, and A. P. Gagnebin, K. D. Millis, and N. B. Pilling, "Engineering Application of Ductile Iron," *Machine Design*, Vol. 22, No. 1 (January, 1950), pp. 108–14.

properties S_u, S_e, and its Brinell hardness number can be more than doubled; however, there is a loss of elongation down to 1 per cent. The 80–60–00 grade contains more manganese and phosphorus than the other grades. It is used for parts that require high strength and toughness but are subject only to moderate shock.

The relationship between tensile strength, elongation, and hardness in cast condition is shown in Fig. 4.1. A very important feature of ductile iron is that it combines the fluidity and castability of gray iron with properties resembling and in some instances exceeding those of steel. It also has an exceptional resistance to wear. Ductile iron can be used for parts which are too intricate to be cast of steel and which cannot be made strong enough when cast in the best grades of gray iron. In addition there are a great number of other parts that so far have been made of other materials but will have better properties or will be lighter if made of ductile iron. Among these are frames, gears, sprockets, diesel-engine pistons, crankshafts, generator shafts, dies for press forming, and various intricate castings. It can be expected that in time this new material will become the third ranking industrial metal on a tonnage basis, exceeding all other of cast steel and malleable iron.

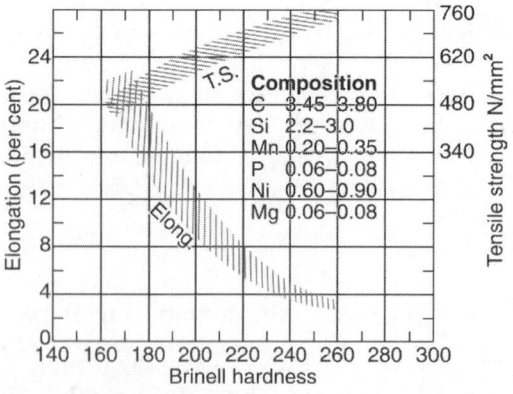

Fig. 4.1: Properties of ductile iron in cast condition

4.6 CARBON STEEL

Simple steel, often called *carbon steel*, consists chiefly of iron, carbon, and manganese. Other elements, such as silicon, phosphorus, and sulphur, are always present but are not essential to the formation of the steel. A steel that contains one or more elements other than carbon and manganese, such as nickel, chromium, vanadium, molybdenum, or tungsten, in sufficient quantity to modify or noticeably improve some of its useful properties, is called an *alloy steel*.

SAE specifications: The Society of Automotive Engineers has adopted specifications for various steels based on their chemical composition.[4] Each specification is given a four-figure or five-figure number. The first two digits indicate the class, while the last two or three indicate the carbon content in hundredths of one per cent. Thus, a carbon steel the carbon content of which is 0.50 per cent has 1050 as its specification number, 10 being the class number of carbon steels. Ordinary steel castings are designated also by carbon content, with 00 in front. An example is SAE 0030 a general purpose steel with a maximum carbon content of 0.30 per cent. High-strength carbon steels and alloy steels are designated by the tensile strength with one zero in front, as in SAE 090 a steel which has S_u = 90 kpsi, i.e. 620 N/mm^2.

AISI specifications: The American Iron and Steel Institute has developed specifications which are intended to limit the number of standard grades. A capital letter prefix indicates the steel making process. The symbol B indicates a Bessemer carbon steel; C, an open-hearth carbon steel; and E, an electric-furnace alloy steel. An index of four numerals, identical with the index of the SAE specifications, indicates the composition. The first two show the type of alloy and the last two give the average carbon content. Thus C1020 is an open-hearth carbon steel with a carbon range from 0.18 to 0.23 per cent; E2515 is a 5 per cent nickel steel having a carbon

[4] *SAE Handbook*, published each year with changes and additions.

content of 0.13 to 0.18 per cent, i.e. average of 0.15% and is made in an electric furnace.[5]

ASTM specifications: Specifications of the American Society for Testing Materials for steel forgings divide all steels into 12 classes numbered from A to M.[6] The ultimate strength is taken as the basis of classification.

BIS Specifications: For the purpose of code designations steels shall be classified as follows:

(a) Steels designated on the basis of mechanical properties.

(b) Steels designated on the basis of chemical composition.

Steels designated on the basis of mechanical properties: Steels in this category are carbon and low carbon steels where the main criterion in the selection and inspection of steel is the tensile strength or yield stress. The code designations consist of the following in the order given:

(a) Symbol 'Fe' or Fe E depending on whether the steel has been specified on the basis of minimum tensile strength or yield stress.

(b) Figure indicating the minimum tensile strength or yield stress in N/mm^2. If no minimum tensile or yield stress is guaranteed, the figure shall be 00.

(c) Chemical symbols for elements, the presence of which characterize the steel.

(d) Symbol indicating special characteristics covering method of deoxidation, (R for rimming steel, K for killed steel), Steel Quality (Q_1 Non-aging, Q_2 freedom from flakes, Q_3 grain size controlled, etc.); degree of purity (P25 phosphorus and sulphur 0.025% P35–0.035 per cent and so on); surface condition, etc. e.g. Fe 410 P35K–killed steel with S & P = 0.035% and minimum tensile strength 410 N/mm^2.

Steel designated on the basis of chemical composition is as follows:

(a) Figure indicating 100 times the average percentage of carbon content.

(b) Letter C.

(c) Figure indicating 10 times the average percentage of manganese content, e.g. 30C8 indicates 0.3% average percentage carbon and 0.8% manganese.

Methods of producing steel: Steel is made from pig iron by burning out the carbon, silicon, manganese, and other impurities, then restoring the carbon content to the desired percentage, and in the case of alloy steels, adding the desired additional elements. The three main processes of producing steel are: (a) the *Bessemer method*; (b) the *open-hearth method*, and (c) the *electric furnace method*.

Bessemer and open-hearth steels do not differ materially in composition and properties. Bessemer steel was formerly used for all rolled products, such as plates, rails, shafting, and structural shapes. Open-hearth steel has largely superseded Bessemer steel for all rolled products and forgings because of its lower production cost.

Electric furnaces are of the induction type and the arc type. The induction furnace produces steel similar in quality to crucible steel obtained from a charge of pig iron and pure scrap steel. The arc furnace produces a high-grade steel from a charge of steel scrap covered with a slag containing a high percentage of lime. The reaction eliminates the impurities, and the proper composition is obtained by introducing the necessary alloys.

The main advantages of electric furnaces are the purity of the product and the ease of controlling its composition. Their disadvantage is a higher production cost.

Methods of manufacturing: Regardless of the process by which steel products are originally obtained, they can be divided into hot-rolled bars, beams, and plates; hot-rolled ingots used for forgings; cold-rolled bars; and castings.

Cold-rolled steel bars and strips are rolled hot to approximately the required size. The surface is then cleaned, usually by chemical means, and the material is rolled cold to very

[5] Data on chemical composition and mechanical properties of AISI steels are given in Lionel S. Marks, ed.,
[6] *Ibid* pp, 582–584

Table 4.2: Mechanical properties of steels and wrought irons

No.	Material Name	Characteristic	Specific weight N/m³	Ultimate tensile strength N/mm²	Elastic limit* N/mm² Tension	Compression	Shear	Modulus of elasticity 10³×N/mm² Direct, or young's	Transverse, or shear	Endurance limit bending N/mm²	Elongation in 50 mm (%)	Brinell (Bhn)	Hardness minimum Rockwell B	C
1	Steel casting, 0.20% C (SAE 0022)	Annealed	76700	410	175	230	115	200	76	169	30	120	70	–
2	Steel casting, 0.30% C (SAE 0030)	Annealed	76700	500	210	270	120	200	76	200	27	140	78	–
3	Steel casting, 0.40% C (SAE 0050)	Annealed	76700	550	220	300	140	200	79	220	22	160	84	2
4	Alloy steel casting (SAE 090, ASTM A-142)	Normalized	76000	620	420	420	250	200	77	276	20	187	–	10
5	Stainless steel: C, 0.10; Cr, 12, Ni, 1	Drawn at 760C	76000	620	380	380	230	200	76	485	–	180	89	8
		Drawn at 430C	76000	1250	900	900	550	200	76	485	–	380	112	40
6	Stainless steel: C, 0.10; Mn, 0.4; Si, 0.35; Cr, 12; Ni, 0.6	Quench 955C, draw 650 F	76700	720	420	420	250	207	83	276	–	200	94	14
7	Stainless steel, SAE 30905	Nonhardenable	76700	660	330	330	210	207	83	276	24	190	92	11
8	Carbon steel, SAE, 1010, 7C4	Hot-rolled	76700	375	215	215	140	228	80	165	36	110	65	–
9	Carbon steel, SAE 1020, 20C8	Hot-rolled	76000	430	240	240	150	220	79	172	30	125	71	–
10	Carbon steel, SAE 1030, 30C8	Soft	76000	520	290	290	180	207	78	220	26	150	81	–
11	Carbon steel SAE 1040, 30C8	Annealed	76000	620	345	345	210	205	77	255	22	180	89	8
12	Carbon steel, SAE 1050, 40C8	Annealed	76000	655	360	360	250	204	77	290	20	190	92	11

(Contd...)

Table 4.2: Mechanical properties of steels and wrought irons (*Contd...*)

No.	Name	Characteristic	Specific weight N/m³	Ultimate tensile strength N/mm²	Elastic limit* N/mm² Tension	Compression	Shear	Modulus of elasticity 10³ × N/mm² Direct or young's	Transverse or shear	Endurance limit bending N/mm²	Elongation in 50 mm (%)	Hardness minimum Brinell (Bhn)	Rockwell B	Rockwell C
13	Carbon steel, SAE 1095 98C6	Annealed	76000	850	425	425	345	204	77	325	20	240	100	23
		Drawn at 480°C	76000	1050	550	550	450	204	77	415	16	300	107	32
14	Carbon steel, SAE 1120, 20C12S14	Screw stock	76000	430	255	255	150	220	86	180	20	125	72	–
15	Nickel steel, SAE 2320 20Ni3	Annealed	76000	480	310	310	190	204	83	276	29	140	78	–
		Drawn at 540°C	76000	830	550	550	345	204	83	415	25	260	103	26
16	Nickel steel, SAE 2340, 40Ni3	Annealed	76000	655	380	380	220	207	83	310	26	190	32	11
		Drawn at 540°C	76000	850	655	700	415	207	83	345	22	220	96	19
17	Cr-Ni steel, SAE 3140, 35Ni1Cr60	Not treated	76000	590	415	415	250	210	86	345	26	170	87	6
		Drawn at 540°C	76000	1100–1520	660	700	395	210	86	550	17	270	104	28
18	Cr-Ni steel, SAE 3240, 40Ni2Cr10	Not treated	76000	760	550	550	335	210	86	345	26	235	99	22
		Drawn at 540°C	76000	1100	830	960	500	210	86	550	19	325	109	35
19	Cr-V steel, SAE 6150, 50Cr6V23	Oil-temper,	76000	1400	1200	1300	700	214	90	550	13	390	112	41
		Drawn at 425°C	76000	1580	1448	1500	792	214	90	620	12	440	120	46
20	Cr-Ni-V steel	Oil-tempered	76000	1100	900	900	550	210	86	485	–	340	109	36
21	Nitralloy steel	Quench, draw hardened	73900	850	620	820	380	200	79	550	13	400	113	42
22	Wrought iron (A41–30)†	Double-refined	73300	320	180	165	110	186	70	140	25	100	60	–
23	Armco ingot iron: Fe, 99.94%	Hot-rolled	75700	300	175	175	105	210	86	165	30	100	60	–
24	Maraging steel	Aged												
	Ni 20%, Ti 1.6%, Al 0.35% Nb 0.5%			1900	1800	–	–	210	84		12			
	Ni 26%, Ti 1.6%, Al 0.35% Nb 0.5%			2130	1900	–	–	210	84		15			
	Ni 19%, Ti 0.25%, Al 0.15% Co 8.9% Mo 3.5%			1600	1400	–	–	210	84		16			
	Ni 19%, Ti 0.75%, Al 0.15% Co 9.5% Mo 5.2%			2170	2100	–	–	210	84		12			

*For sections 12 mm in diameter; gradually decrease with size.
† ASTM.

accurately gaged dimensions. The characteristics of cold-rolled bars are:

(a) The surface is hard, smooth, and bright.

(b) The dimensions are very accurate.

(c) The elastic limit and ultimate strength are increased, but the ductility is decreased.

(d) Inner stresses are set up in the surfaces that cause a twisting or warping of the bar if its skin is removed, especially on one side only, as by cutting a keyway.

Composition: In commercial rolled and forged steels the carbon content varies from 0.05 to 1.50 per cent; the manganese content is from 0.25 to 0.90 per cent; silicon, from 0.08 to 0.16 per cent; phosphorus, from 0.06 to 0.13 per cent; and sulphur, from 0.05 to 0.16 per cent.

Carbon content: As may be seen from Table 4.2[7] an increase of the carbon content of steel raises its ultimate strength, elastic limit, endurance limit, and hardness, both in the annealed state and the heat-treated state. At the same time, an increase of the carbon content decreases the ductility, as is indicated by the lowering of the percentage of elongation.

Carbon steels can be subdivided into three main groups: (a) low-carbon steels, having a carbon content of 0.05 to 0.25 per cent and suitable where only moderate strength but considerable ductility is required; (b) machinery steels, or medium carbon steels which have a carbon content of 0.30 to 0.55 per cent and can be heat-treated to develop high strength; and (c) high-carbon steels, which contain 0.60 to 1.30 per cent and are widely used for tools and springs, mostly with heat treatment.

Steel castings: Steel castings are used for parts that require strength and can be molded to shape so that machining is required only on surfaces where the casting is assembled with other machined parts. There are several classes of commercial steel castings, which differ in carbon content and in alloy content. However, castings of medium-carbon steel, with a carbon content of 0.25 to 0.50 per cent, represent the bulk of the steel-casting output, and this class is considered the regular-grade product.

Steel in the cast form does not differ much in its properties from rolled and forged steel. In comparison with gray cast iron, steel castings have higher strength, much higher ductility, and greater toughness. However, they cannot be used for intricate shapes because of lower fluidity in the molten state. They are also more expensive. They are used practically in every industry, mostly for medium-weight and heavy parts where strength is one of the main requirements.

4.7 HEAT TREATMENT OF CARBON STEEL

The elastic limit, ultimate strength, and hardness of carbon steel can be changed by heating the steel and then allowing it to cool rapidly or slowly.

The various processes that may be used are the following:

Annealing consists in heating steel to a temperature above the critical temperature and cooling it by a relatively slow process as in furnace. Annealing may be used to remove stresses that are produced in forgings and castings, to refine the crystalline structure of steel castings, or to alter ductility or toughness.

Normalizing is a special kind of annealing in which the material is heated to a temperature above the critical range and is subsequently cooled to a temperature below the range, in still air at room temperature.

Hardening consists in heating steel to a temperature above its critical range and cooling it suddenly by quenching it in water, oil, or some other cooling medium that absorbs heat rapidly. Quenching increases the hardness of steel having a carbon content of 0.20 per cent or higher. It also raises the elastic limit and ultimate strength and reduces the ductility. However, it induces internal stresses, and the metal is apt to become brittle.

Drawing, also called *tempering*, consists in reheating hardened or quenched steel to a

---

[7] Supplements to *Zeitschrift Verein Deutscher Ingenieure*, Vol, 77, No. 42 (October 21, 1933), p. 3, and Vol. 77, No. 50 (December 16, 1933), p. 1.

temperature below the critical range in order to remove the internal stresses and restore some of its ductility. The elastic limit and ultimate strength are slightly reduced by drawing, but they are still higher than they were before hardening.

Carburizing is a process for increasing the carbon content of soft steel by heating it below its melting point in contact with carbonaceous material, such as charcoal, leather, or barium carbonate.

Casehardening consists in carburizing a metal product and subsequently heating and quenching it to harden all or part of its surface. The case is that outside portion of a piece in which the carbon content is increased by carburizing; the core is the inner portion of the piece in which the carbon content has not been markedly increased. Casehardening is applied to soft steels with a carbon content of 0.20 per cent or less which cannot be hardened by simple heating and quenching.

Cyaniding is hardening of all or part of the surface of a steel piece by heating it at a suitable temperature in contact with a cyanide salt and then quenching it. Cyaniding gives a thin but very hard case in a very short time.

Nitriding is the introduction of nitrogen into the outer surface of a part made of any one of several special steel alloys called *nitralloys*. The treatment consists in heating the part to a temperature of about 560°C inside a chamber through which a stream of ammonia gas is passed. The part is rough-machined before being nitrided, and is then heated to the nitriding temperature without ammonia, in order to produce whatever distortion may occur. After that the part is finish-machined and nitrided. The nitrided surface has a hardness from 730 to 1100 Bhn and resist corrosion. The hardness is not lost when the nitrided surface is heated up to 540°C. Nitriding also raises the endurance limit of the steel.

Flame hardening is a process in which a steel part is heated locally above the critical temperature and then quenched. The depth of the flame hardened layer can be varied from 1.5 mm to about 3 mm, the exact depth depending on the service requirements. Since flame hardening produces only a very small distortion, this process is useful for hardening surfaces of large steel parts. After being quenched the piece must be stress-relieved by tempering; a temperature of 200°C is usually sufficient. Plain carbon steels with a carbon content of 0.35 to 0.70 per cent are best suited for this method. Flame hardening may be used with castings, forgings, and rolled sections, irrespective of their size. It is often applied to gear teeth and cams.

Local hardening is similar to flame hardening, but the heating is done by an electric current. High-frequency induction heating is well adapted for surface-hardening of cylindrical parts. Crankshaft journals and crankpins are hardened in this manner by the *Tocco process*. The extent of the heated zone can be so closely controlled that the fillets will remain soft while the bearing surface is hardened. Such control reduces the danger of failure by progressive fracture.

4.8 ALLOY STEELS

The elements most often alloyed with steel, singly or two or more together, are—besides carbon—nickel, chromium, silicon, manganese, molybdenum, vanadium, tungsten, and aluminum.

Increase of carbon content increases the hardness and strength of steel but decreases its ductility.

Nickel increases the hardness, toughness, corrosion resistance, and (up to a 12 per cent content) also the elastic limit of steel, but decreases its ductility slightly.[8] The ratio of elastic limit to ultimate strength increases gradually with increasing nickel content up to about 4 per cent, and with a further increase it begins to decrease.

Nickel steels are the most important of the commercial alloy steels. Their nickel content

[8] J. W. Sand, "Nickel in Steel," Metals Handbook (Cleveland: The American Society for Metals, 1948), p. 473.

varies from 0.50 to 5.25 per cent, while the usual carbon content ranges from 0.30 to 0.60 per cent. The chief uses of nickel steels are for structural shapes, rails, steel castings, engine forgings, and automotive parts.

Chromium increases the elastic limit and hardness of steel. It is added either alone or in conjunction with nickel or vanadium. It also increases the resistance to corrosion.

Chrome steels are used for ball and roller bearings, gears, and other machine and automotive parts where hardness is essential. Chrome steels are always heat-treated.

Steels with a chromium content of 11 per cent or more—in some cases over 20 per cent— are known as stainless steels because of their resistance to corrosion. Steel No. 6, Table 4.2, is used for bolts subjected to high temperature and for large studs in steam turbines.[9] At 430°C its elastic limit is still 420 N/mm^2.

Chrome-nickel steels combine high strength with high hardness. They are produced with a chromium content from 0.6 to 1.2 per cent, a nickel content from 1.5 to 3.5 per cent, and various carbon contents. Steels with a carbon content up to 0.2 per cent are used only when case-hardened; those having a content of 0.25 to 0.6 per cent are used for structural parts of automobiles; and those having a content of 0.5 per cent and more are used for gears and automotive parts in place of plain chrome steel.

Silchrome steel contains 0.5 per cent C, 0.30 per cent Mn, 3.50 per cent Si, 8.0 per cent Cr, not more than 0.02 per cent P, and not more than 0.02 per cent S. This steel is used for exhaust valves in internal combustion engines. It is tough, is hard (even at high temperatures), and has a high resistance to scaling and corrosion. Its drawback is the difficulty of machining it.

Silicon-manganese steel contains 0.45 to 0.65 per cent C, 0.60 to 0.90 per cent Mn, and 1.8 to 2.0 per cent Si. It has a high elastic limit and is used extensively for springs and gears. This steel must be given a suitable heat treatment,

the type depending on the kind of service for which the steel is intended.

Vanadium added to carbon steel, nickel steel, or chrome steel, even in such as small amount as 0.15 to 0.25 per cent, increases the elastic limit and resilience. *Chrome-vanadium steel* is used especially for automotive springs. Steel that has an average composition of 0.47 per cent C, 0.84 per cent Mn, 0.032 per cent S, 0.026 per cent P, 0.10 per cent Si, 1.06 per cent Cr, and 0.15 per cent V and has undergone a special heat treatment and drawing has the following properties: If drawn to 260°C, the tensile strength is 2700 N/mm^2 and the elastic limit is 1750 N/mm^2; if drawn to 480°C, the values are 1400 and 1300 N/mm^2, respectively.

Tungsten and *molybdenum* are sometimes added to machine steels for strength and toughness, but their chief use is in the production of high-speed cutting tools. Physical and mechanical properties of a chrome-nickel-tungsten steel are particularly suitable for highly stressed crankshafts.[10]

Aluminum in small amounts 0.10 per cent, increases the fluidity of steel. Aluminum is also added to a ferrous alloy used as stock for nitriding. This alloy, known as nitralloy, has an approximate composition of 0.2 to 0.4 per cent C, 0.5 per cent Mn, 0.2 to 0.5 per cent Si, 0.5 to 0.6 per cent Ni, 1.5 to 1.7 per cent Cr, 0.2 per cent Mo, 0.9 to 1.3 per cent Al, and about 95.5 to 96 per cent Fe.

4.9 MECHANICAL PROPERTIES OF STEEL

The ultimate tensile strength of steel is a very important general characteristic. As shown in Fig. 4.2, it determines within sufficiently close limits the hardness of either carbon steel or alloy steel that has not been heat-treated.

The ratio S_e/S_u of the elastic limit or yield point in tension to the tensile strength also increases. However, it is expressed by a rather wide band, as shown in Fig. 4.2.

The endurance limit in bending for most steels is equal to about 0.45 S_u. However, this

---

[9] W. J. Kerr, et al., Symposium on Effects of Temperature of Metals (New York: McGraw-Hill Book Company, inc. 1932), p. 37

[10] *Ibid.*

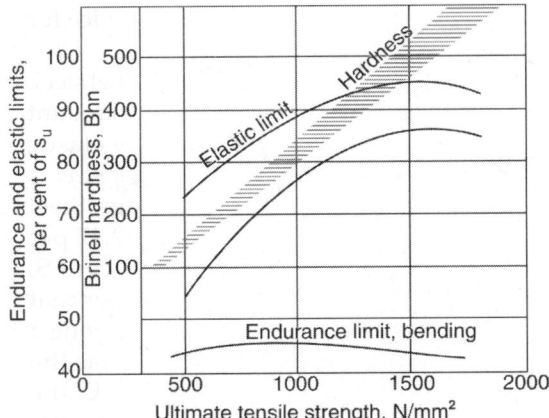

Fig. 4.2: Relation between mechanical properties and tensile strength of steel

factor decreases slightly for high-strength, hard steels. Average mechanical properties of various steels are given in Table 4.2.

4.10 WROUGHT IRON

Wrought iron is a malleable ferrous metal which has been produced from a pasty condition. It contains a small proportion of manganese, 0.07 to 0.30 per cent. When single-refined it has an ultimate tensile strength from 280 to 320 N/mm^2, when double-refined its strength range from 330 to 380 N/mm^2. It is more resistant to corrosion than steel. Being almost pure iron, it cannot be hardened. It is used in the manufacture of pipes, boiler tubes, rivets, staybolts, and crucible steel; for general forging purposes, especially where welding is required; and for parts of electrical machinery. At present wrought iron is not used so extensively, steel having taken its place to a great extent. Table 4.2 contains data pertaining to its mechanical properties.

Armco ingot iron is made in a basic open-hearth furnace similar to that used in producing steel, but it is purified to a much greater degree by the addition of a very pure iron ore. Its composition is especially uniform, and freedom from inclusions, insures mechanical uniformity. Its chief use in machinery is for boilers, tanks, large pipes, tubing, and welding rods.

4.11 COPPER ALLOYS

Copper used in machine parts is alloyed with zinc, tin, lead, aluminum, and nickel. Copper–zinc alloys are as a rule termed *brasses*, although some brasses are called *bronzes*. A true *bronze* is an alloy of copper and tin only, the tin content being 4 to 25 per cent. Copper alloys with tin, lead, or aluminum usually are also termed bronzes.

Industrial brass may be divided into two classes: cast brass and wrought brass.

Cast brass has zinc content from 30 to 40 per cent, the most desirably proportion being 35 per cent. A small percentage of tin is added to increase the hardness, and 1 to 2 per cent of lead gives good machinability. This alloy is commonly called *yellow brass*.

Wrought brass has a zinc content of 15 to 44 per cent. Alloys of copper and zinc, with a zinc content of 38 to 44 per cent, become plastic when heated to redness. The properties of these brasses can be altered by heat treatment. *Muntz metal*, composed of 60 per cent copper and 40 per cent zinc, is rolled at temperature above 590°C and is hardened by quenching.

Modified hot-working brass is brass with small amounts of tin and iron. *Tobin bronze* (SAE 73) is practically the same as *naval brass*. It consists of 59 to 62 per cent Cu, 0.5 to 1.5 per cent Sn, not more than 0.10 per cent Fe, and not more than 0.3 per cent P, the remainder being zinc. It is very resistant to corrosion.

Delta metal is similar in composition and properties to naval brass but it contains from 1 to 3 per cent iron and has a tensile strength of 305 N/mm^2.

Cold-working brass is an alloy of copper and zinc with a zinc content of 28 to 38 per cent. It has considerable strength and great ductility, and it can be worked cold to such forms as rods, tubes, and wires. The addition of a small percentage of tin increases its hardness and resistance to corrosion, and the addition of 1 to 2 per cent of lead increases its machinability.

Cold-rolled admiralty metal, consisting of 71 per cent Cu, 28 per cent Zn, and 1 per cent

Sn, has a tensile strength of 350 to 620 N/mm^2, the actual value depending upon its hardness.

Red brasses, so called because of their colour, are composed of 76 to 88 per cent Cu, 4 to 16 per cent Zn, 2 to 6 per cent Sn, and 1.5 to 6 per cent Pb. They are easily cast and machined.

Aluminum brass consists of 76 per cent Cu, 22 per cent Zn, and 2 per cent Al. Its tensile strength ranges from 280 to 350 N/mm^2, and its elastic limit is from 210 to 270 N/mm^2. Aluminum, when added to brass in the form of aluminized zinc, makes the brass flow freely, so that castings made from it has smooth surfaces and are free from blowholes. Such castings are used in marine service because of their resistance to impingement pitting.

Gun metal is copper combined with 8 to 11 per cent tin and sometimes small percentages of zinc, lead, and iron. It is hard and has a tensile strength of 210 N/mm^2 but its strength decreases very rapidly if the temperature exceeds 260°C. By proper heat treatment, with quenching above 500°C, this alloy, like other true bronzes, can be made stronger and more malleable.

Phosphor bronze in castings contains from 80 to 90 per cent Cu, from 20 to 10 per cent Sn, and up to 1 per cent P. Being hard and tough, it is particularly suitable for gears. In the wrought form it contains only 3 to 9 per cent tin and not over 0.5 per cent of phosphorus. It is hard, tough, and corrosion resistant and is used for springs and wire.

Manganese bronze is in reality a brass, consisting of copper and zinc. Manganese is used only as a deoxidizer, and the finished product often contains only traces of it. The term bronze is applied generally because of the alloy's red colour.

Manganese bronze has high strength, is ductile, and resists corrosion in salt water. Cast manganese bronze is used for such parts as ship propellers, heavy gears, and heavy-duty bearings. Manganese bronze can be wrought like Tobin bronze, and when drawn it has a tensile strength of 520 N/mm^2.

Super strength bronze is also a special brass, similar to manganese bronze, in which iron and aluminum have been substituted in part for zinc. When sand-cast, it has a tensile strength of 750 to 810 N/mm^2 and an elastic limit of 210 to 225 N/mm^2 in tension and about 5 per cent less in compression. Heat treatment and forging or rolling do not change its properties materially.

Aluminum bronze is copper with 2 to 11 per cent aluminum. Its tensile strength increases, as the aluminum content becomes greater, from 210 to 650 N/mm^2. A 10 per cent aluminum bronze is used in cast and hot-worked condition for parts requiring high tensile strength and resistance to corrosion, wear, and alternating stresses. Its mechanical properties are improved by quenching from 900°C and reheating to 590°C. Iron up to 0.5 per cent increases the strength and hardness without reducing the ductility appreciably.

An alloy containing 11 per cent Al, 5 per cent Ni, 5 per cent Fe, and 79 per cent Cu has extreme hardness and wear resistance at high temperatures, a quality essential in aircraft-engine valve seats.

Monel metal is the most used of the copper-nickel alloys, all of which are very strong and highly resistant to corrosion. Monel metal is a natural alloy composed of 68 to 70 per cent Ni, 1.5 per cent Fe, and copper. It retains its strength at temperatures up to 590°C to a greater degree than most brasses and bronzes, and even steels.

Ambrac is another copper-nickel alloy, with properties similar to those of Monel metal. An alloy composed of 65 per cent Cu, 30 per cent Ni, and 5 per cent Zn, and either rolled or drawn, has a tensile strength of 460 N/mm^2 when soft and 560 to 840 N/mm^2 when hard.

Other copper alloys: The number of copper alloys adapted to special purposes is very great. One of the latest developments is *beryllium copper*. Alloys containing from 1.5 to 2.5 per cent beryllium take on exceptionally high characteristics of strength, endurance, and hardness through cold working and heat treatment. Thus the tensile strength of untreated beryllium copper in a soft state is about 420 N/mm² but its strength can be raised to 770 N/mm² by cold working, to 1020 N/mm² by proper heat treatment and to 1230 N/mm² by heat treatment after cold reduction. The main use of this alloy in machinery is for springs, particularly in internal combustion engines, high-duty gears, valve sleeves, and valve seats.

The mechanical properties of the most frequently used copper alloys are given in Table 4.3.

4.12 BEARING METALS

Bearing metals may be divided into five classes. Each is named to correspond to the base, or main constituent, which makes up 50 per cent or more of the alloy. The bases are: (i) copper, (ii) tin, (iii) lead, (iv) zinc, and (v) aluminium.

Copper-base metals are stronger than those with tin, lead, or zinc bases. Also, being harder, they have a lower coefficient of friction. It is usually a misnomer to designate an alloy with a tin, lead, or zinc base as an anti-friction metal. The explanation of the use of the term lies in the fact that copper-base bearing metals are more likely to heat under abnormal operating conditions because they do not have enough plasticity. The presence of any irregularities of the journal or bearing, or of any foreign particles between them, therefore creates point of high specific pressures, which are absent in the softer alloys. The properties of some typical bearing alloys are given in Table 4.4.

Copper-base metals are used for bearings where heavy pressures and shock action occur, as in bearings for connecting rods of explosion engines. The Brinell hardness number of *Nida bronze*, a drawn phosphor bronze, ranges from 200 when hard to 80 when annealed.

Copper-lead alloys, known as plastic bronze and Allan red metal do not score or cut a journal, even if they become red-hot. Nor do they melt and run out, as do the babbitts.

Tin-base metals are commonly known as *babbitts*. Copper and antimony are added to increase the hardness of the alloy. For light bearings, lead may replace copper in order to lower the cost of the metal. Table 4.4 contains only three babbitts, but there are a great number which are adapted to various special operating conditions.

Parsons white brass is a tin-base alloy with a large amount of zinc. It is used for marine and automobile bearings. It is hard and tough; but being sluggish to pour, it cannot be cast into thin sections. After being cast it is peened to increase its resistance to wear.

Lead-base bearing metals contain antimony and a certain amount of tin, which diminishes the brittleness and increases the compressive strength of the alloy. The anti-friction metals of this class are cheapest, and are satisfactory in many services. Various widely advertised brands of babbitt come within this class.

Zinc-base metals are used when a low coefficient of friction is essential. *Lumen bronze* can be cast in sand, is easily machined, and is used for crane, motor, and pivot bearings. It is particularly suitable for high-speed bearings with low specific pressures.[11]

Aluminium alloys developed by Alcoa for bearings are of two types. One type, designated as alloy 750, is recommended for permanent-mold casting. The other, alloy XA80S, is obtainable in the form of flat-rolled sheets. The composition and mechanical properties of these alloys are given in Tables 4.4 and 4.5.[12] These aluminium alloys were originally developed for use in bearings of internal

---

[11] Data about various materials can be found in J. B. Johnson, M. O. Withey, and James Aston, *Materials of Construction*, 8th ed. (New York: John Wiley & Sons, Inc., 1939); Lionel S. Marks, ed., *Mechanical Engineers Handbook*, 5th ed. (New York;McGraw-Hill Book Company, Inc., 1951); and R. T. Kent, *Mechanical Engineers' Handbook*, 12th ed, Vol. II, Design and Production, edited by Colin Carmichael (New York: John Wiley & Sons, Inc., 1950).

[12] H. Y. Hunsicker, in *Sleeve Bearing Materials* (New York: American Society for Metals, 1949), p. 98.

Table 4.3: Mechanical properties of copper alloys

No.	Metal Name	Characteristic	Specific weight (N/m³)	Ultimate tensile strength Su (N/mm²)	Elastic limit N/mm² Tension	Compression	Shear	Modulus of elasticity N/mm² × 10³ Direct E	Transverse G	Endurance limit bending N/mm²	Elongation in 50 mm (%)	Hardness minimum Brinell (Bhn)	Rockwell B
1	Admiralty metal	Soft to hard	81530	210–620	170	–	–	–	–	–	65–14	53–157	15–83
2	Aluminum brass	Free-flowing	81000	270–350	220	–	–	100	35	–	17–52	86	49
3	Aluminum bronze	SAE 68	74100	450	100	100	80	120	44	96	20	70–190	35–91
4	Ambrac	Soft	84400	380	140	–	–	–	–	–	–	71	35
5	Ambrac	Hard	84700	550	310	–	–	130	48	–	–	160	84
6	Brass, red	SAE 40	81000	200	70	60	60	90	35	41	16	53	15
7		SAE 41, cast	81000	170	70	60	60	86	30	50	20	52	15
8	Brass, yelow	SAE 41, rolled	81300	310	170	140	100	88	41	85	–	90	54
9		SAE, 41, hard-drawn	81300	520	210	–	170	110	40	140	–	130	74
10	Bronze, bearing	SAE 62	83500	210	80	60	50	86	35	62	14	60	22
11	Beryllium copper	Soft	83400	410	50	–	–	127	52	–	–	100	60
12	Beryllium copper	Heat-treated	83400	1100	410	–	–	130	55	276	–	360	110
13	Bronze	SAE 64	82900	170	80	60	50	96	38	48	8	70	35
14	Bronze, superstrong	Sand-cast	71500	800	250	210	–	103	40	160	–	240	100
15	Copper	Soft	85000	220	20	20	15	106	41	85	–	42	5
16	Copper	Cold-drawn	85200	350	80	–	60	117	43	117	–	107	64
17	Gear bronze	SAE 65	83400	240	120	140	90	92	36	82	10	100	60
18	Gun metal	Cast	84200	270	80	85	70	180	65	–	–	60	12
19	Manganese bronze	SAE 43	74100	450	170	–	150	110	–	110	25	90	54
20	Monel metal	Cast	84200	500	170	150	160	172	62	140	–	100	60
21	Monel metal	Hot-rolled	85500	590	210	180	165	175	63	210	–	120	70
22	Monel metal	Cold-rolled	85200	830	480	410	275	175	63	345	20	145	80
23	K-Monel metal	Grade C	82100	1040	620	550	380	175	63	415	30	300	107
24	Muntz metal	Soft, cast	80200	310	150	–	–	88	35	105	30	80	45
25	Muntz metal	Hard, rolled	80200	480	310	–	–	88	35	175	30	158	54
26	Phosphor bronze	SAE 77, rolled	84500	380	120	110	105	107	43	83	–	60	22
27	Tobin bronze	SAE 73, Hard	80500	450	160	70	125	103	41	145	25	165	86
28	Tobin bronze	SAE 73 Soft	80500	370	90	50	125	103	41	–	35	90	54

Table 4.4: Properties of bearing metals

Metal	Percentages of ingredients				Compressive elastic limit N/mm²	Hardness at 28°C (Bhn)	Class of use
	Copper Cu	Lead Pb	Tin Sn	Miscellaneous			
Nida bronze Hard	91–92	–	8–9	p, 0.5	255	150–200	High shock
Annealed	91–92	–	8–9	p, 0.5	105	80–90	load
Bronze, SAE 62	86–89	>0.2	9–11	Zn, 1–3	85	60	Shaft bearings
Phosphor bronze	80–82	9–10	9–11	P, 0.7	105	55–80	Shock load
Aluminum-nickel bronze*	80	[Al, 10]	[Ni, 5]	5	310	180	High shock load
Aluminum alloy 750	1.0	[Al, 91.5]	6.5	Ni, 1.5	60	35–50	Severe loads
Aluminum alloy XA80S	1.0	[Al, 90.5]	6.5	Ni, 0.5, Si, 1.5	105	40–55	General automotive
Plastic bronze	65	30	5	–	24	30	Crank-pin
Allan red metal	50	50	–	–	17	135	bearing
Babbitt							
SAE 10	4–5	0.35	90–92	Sb, 4–5	17	17	Light loads
SAE 12	2.75–3.25	25	60–62	Sb, 10.5	9	22	Moderate loads
SAE 14	–	75	10	Sb, 15	9	23	Crank-pin bearing
Parsons' white brass	2.5–5	>0.2	64–65	Zn, 33	85	18	Light load
Lumen bronze	10	–	[Zn, 86]	Al, 4	105	116	Light load, high speed
Asarcoloy, No. 7	[Cd, 98.5]	–	–	Ni. 1.3	255	33	Severe load

*Nickel topics, Vol. 4, No. 1 (January, 1951), p. 2.

combustion engines, but they gradually found applications for other heavy-duty services, including shoes of reciprocating crossheads. Aluminum bearings have long life, withstand high pressures, resist corrosion, and have high thermal conductivity and low cost of manufacture.

Metal fluoroplastic has been developed and has very less coefficient of friction. It consists of a steel base and a layer of 0.3 mm porous bronze. The pores are impregnated with a mixture of fluoroplastic and molybdenum disulphide. This material does not require lubrication. It is light and less costly.

4.13 POWDER METALS

The commonly used powdered metals are copper, tin, and iron. They may be used singly or mixed in certain proportions. The powder, mixed with a volatile binder, is molded under high pressure. The molded piece is sintered and becomes a metal part with a desired density or porosity. A product of powdered metal can be made with such close tolerances that no machining is required, except possibly drilling or tapping of very small holes. Small and medium-size bushings made of powdered metal are used for light duty. Because of the porosity of the material, the bushing does not need an oil reservoir but is self-lubricating. Small gears, ratchets, sprockets, levers, and similar parts can be made of powdered metal. Powder-molded parts are often used as an alternative to zinc die-castings when greater strength and hardness are desired.

4.14 ALUMINIUM ALLOYS

Light alloys have wide use at present in light portable machines and particularly in various

Table 4.5: Composition and properties of Aluminum Alloys*

SAE no.	Alcoa no.	Per cent of alloy (balance aluminum and normal impurities)					Specific weight N/m³	Ultimate strength N/mm²		Yield point 0.2% Set N/mm²		Endurance limit bending N/mm²	Elongation in 50 mm (%)	Hardness 500 kG (Bhn)	Manufacturing process	General information
		Cu	Si	Mn	Mg	Ni		Tension	Shear	Tension	Compression					
38	195	4.5	1.5	0.3	–	–	26730	220	165	110	110	40	8.5	60	Sand-cast†	Structural casting requiring high strength and shock resistance
								250	210	165	170	45	5.0	75	Sand-cast‡	
								280	215	210	260	50	2.0	95	Sand-cast§	
39	142	4.0	0.7	0.3	1.5	2.0	25900	220	190	195	235	55	0.5	85	Sand-cast	Air-cooled cylinder heads, pistons in high-perf.i-c engines
								280	180	230	320	65	0.5	110	Permanent mold	
322	355	1.3	5.0	0.5	0.5	–	25200	220	210	170	200	60	2.0	80	Sand-cast	General use for high strength and pressure tightness
								260	210	190	180	60	1.5	90	Permanent mold	
35	43	0.6	5.0	0.3	–	–	25700	130	100	60	70	45	6.0	40	Sand-cast	Intricate castings with thin sections; corrosion resistance
								165	125	60	60	–	9.0	45	Permanent mold	
321	A132	0.9	12.0	0.1	1.1	2.5	25400	215	165	195	210	–	0.5	105	Permanent mold	Pistons in i-c engines low heat expansion
								280	210	150	150	65	10.0	75	Permanent mold	
380	B195	4.5	2.5	0.3	–	0.3	26700	310	220	230	230	70	5.0	90	Permanent mold†	Same as SAE 38, modified for use in permanent molds
								270	–	140	–	–	4.5	90	Permanent mold‡	
															Permanent mold§	
305	13	0.6	12.0	0.3	–	–	25400	255	–	125	–	105	1.8	–	Die-cast	Excellent casting characteristics
306	380	3.5	8.5	0.5	0.1	0.5	26200	310	–	170	–	–	2.0	–	Die-cast	Good mechanical properties
–	750	1.0	–	–	[Sn, 6.5]	1.0	27500	1140	100	60	60	60	10.0	35–50	Permanent mold	Bearing alloy
24	24S	4.5	0.5	0.6	1.5	–	26500	190	125	80	–	85	20	42	Rolled‖ and	Sheets, plates, tubes, rivets with high strength
								470	280	320	–	125	20	120	extruded#	

(Contd....)

Table 4.5: Composition and properties of aluminum alloys* (Contd...)

SAE no.	Alcoa no.	Per cent of alloy					Specific weight N/mm³	Ultimate strength N/mm²		Yield point 0.2% Set N/mm²		Endurance limit bending N/mm²	Elongation in 2 IN. (%)	Hardness 500 kG (Bhn)	Manufacturing process	General information
		Cu	Si	Mn	Mg	Ni		Tension	Shear	Tension	Compression					
260	14S	4.4	0.8	0.8	0.4	–	26730	190 430 480	125 235 290	100 280 410	– – –	75 125 125	18 25 13	45 100 135	Rolled, extruded,‡ and forged#	Highly stressed forgings, alternate for SAE 24
280	A51S	0.35	1.0	[Cr, 0.3]	0.6	–	25680	300	220	235	[Shear 150]	75	12	90	Forged#	Complicated shapes
290	32S	0.9	12.5	0.2	1.1	0.9	25800	360	260	290	–	110	5	115	Forged#	Press-forged pistons in icengines
–	XA80S	1.0	1.5	–	–	0.5	27000	45	–	105	105	–	15.0	40–55	Rolled sheet	Bearing alloy

*Modulus of elasticity; $E = 10,300,000$ psi; $G = 3,850,000$ psi
†Solution heat-treated and naturally aged
‡Solution heat-treated and artificially aged
§Solution heat-treated and stabilized
‖Annealed
#Heat-treated to stable temperature

types of transportation machinery. The three main constituents of light alloys are aluminum, which has a specific weight of 27100 N/m³; magnesium, with a weight of 17070 N/m³; and silicon, which weights 24900 N/m³.

Pure aluminum is seldom used in machinery, being too weak and not sufficiently machinable. Copper in proportions from 0.15 to 12 per cent is the main constituent of aluminum alloys, as it gives strength, hardness, and machinability. Other metals, chiefly silicon, manganese, and nickel, are also added for the same purposes.

4.15 MAGNESIUM ALLOYS

High-magnesium alloys with small amounts of aluminum, manganese, and zinc have desirable physical properties and good corrosion resistance except to marine atmospheric conditions. These alloys have two-thirds the weight of aluminum and they machine easily and smoothly. They are finding increasing use, especially in aviation, in the form of castings, forgings and rolled sheets. Magnesium pistons and connecting rods combine strength with light weight.

In Table 4.6 are given the mechanical properties of the alloys most used. The alloys AZ63, AZ61, and AZ80 are also known under the trade name *Electron*. They are used for crankcases, gear boxes, pistons and cylinder heads.

Alloy AZ63 and alloy AZ92 are those most used for sand castings.

Alloy M1 castings can be readily welded to sheets and forgings of the same alloy.

Alloy AZ92 and alloy AZ10 are used for permanent-mold castings.

Magnesium-alloy forgings are used where greater strength is required than is obtainable from castings. They are generally press-forged.

Alloy AZ61 is a general-purpose forging alloy, while AZ80 is used for forgings of highest strength and simple design. Alloy M1 may be either press-forged or hammer-forged. It is

Table 4.6: Mechanical properties of magnesium alloys*

Designation		Ultimate strength N/mm²			Yield point N/mm²		Endurance limit bending N/mm²	Elongation in 50 mm. (%)	Hardness 500 kg (Bhn)	Manufacturing process
ASTM number	American magnesium corporation	Tension	Compression	Shear	Tension	Compression				
AZ92	AM260	140	320	330	75	75	50	5	54	Permanent mold or sand-cast
AZ10	AM240	145	325	110	75	75	85	1	50	Sand-cast
		240	375	125	125	125	65	0.5	78	Sand-cast†
AZ63	AM265	185	110	85	85	70	70	6	49	Sand-cast
		270	125	130	130	125	70	5	68	Sand-cast†
M1	AM35	85	–	–	75	120	–	–	–	Cast
		195	–	100	120	50	35	5	40	Extruded
		210	–	–	125	50	–	3	44	Forged
AZ31	AM52S	255	375	125	150	85	85	12	51	Press-forged
		275	410	140	200	100	100	16	47	Extruded
AZ61	AM57S	300	475	140	210	110	120	17	54	Extruded
AZ80	AM58S	300	–	130	195	125	–	9	54	Extruded
		320	375	150	215	145	105	6	72	Forged†

*Specific weight about 17200 N/m³; $E = 45000$ N/mm²; $G = 17200$ N/mm²
†Heat-treated and aged.

used for parts of low cost with moderate strength requirements.

A wide range of extruded shapes is available in many alloys. Alloy M1 is used for low-cost extrusions with moderate strength. Extrusions made of alloys AZ31, AZ61, and AZ80 increase in strength, but also in cost, in the order named.

4.16 ZINC ALLOYS

Alloys containing 3.5 to 4.5 per cent of aluminum and small amounts of copper, iron, and magnesium, with the balance zinc, are used rather extensively for die-casting parts of various mechanisms and equipment (often of very intricate shape) in which strength is not the most important requirements. Both the American Society of Testing Materials and the Society of Automotive Engineers have established specifications for three types of alloys, the physical properties of which are given in Table 4.7.

4.17 NONMETALLIC MATERIALS

Properties of nonmetallic materials used in machines are given in Table 4.8. Most of these materials are either synthetic phenolic resin, which is widely used under the name of *Bakelite*, or laminated sheets obtained by compressing layers of paper or canvas impregnated with phenolic resin. In machinery the main use of such materials as Celoron, Formica, or Micarta is for manufacturing silent gears and bearing shells. Rawhide and vulcanized paper fiber are also used for silent gears, but these materials absorb moisture more readily.

Table 4.7: Physical properties of zinc alloys*

Trade name	ASTM specification	SAE no.	Tensile strength N/mm^2	Compressive strength N/mm^2	Shear strength	Hardness (Bhn)	Elongation in 50 mm (%)
Zamak-2	XXI	921	330	640	315	83	5.1
Zamak-3	XXIII	903	280	415	210	74	4.7
Zamak-5	XXV	925	315	600	65	76	3.0

*Specific weight of all zinc alloys is about 63530 N/mm^3.

Table 4.8: Mechanical properties of nonmetallic materials

Trade name	Material	Specific weight N/m^3	Tensile strength N/mm^2	Compressive strength N/mm^2	Shear strength N/mm^2	Modulus of elasticity E N/mm^2	Hardness, 500 kg (Bhn)
Bakelite	Phenolic resin	12900	30	220	85	7000	48–54
Celoron	Phenolic resin, paper	13000	70	165	105	7600	40
Celoron	Phenolic resin, canvas	13250	55	210	70	7600	38
Formica	Phenolic resin	13250	70	170	105	7600	65
Micarta	Phenolic resin	13250	85	215	85	7200	38
–	Rubber, hard	13250	25	–	–	–	–
–	Rubber, soft	14300	–	–	–	–	–
–	Hardwood	6350	140[†] 6[‡]	7	7	1	–
–	Leather, rawhide	9300	55	55	–	117000	–
–	Vulcanized fiber	14300	40	55	70	125	10

*Endurance limit in bending approximately 35 N/mm^2
[†]Parallel to grain
[‡]Across grain

4.18 CERAMIC MATERIALS

The advantages of ceramics are:

(a) The resistance to acids and alkalis, so useful in pipelines, containers, baths and rollers, sieves and nozzles, in heat exchangers, porcelain cladding in the chemical, sanitary and food industry;

(b) Resistance to fire and heat for thermal purposes such as fire proof stoneware and special compounds for furnaces;

(c) Their electrical strength used in insulation, coil carriers and condensers in the heavy current and light current industries, and coating in ball and roller bearings, and

(d) Their ease of forming which is convenient for making various parts.

The less known facts about them are that they can be machined to accurate dimensions using carbide tools and grinding wheels. Metal parts can be moulded into them and metallic coatings can be given over them. Also complicated and high stressed parts such as spur and worm gears, rolling and sliding bearings and even helical springs can be produced. Special ceramic compounds are also available with special properties like high impact bending strength, high thermal conductivity, high thermal fatigue resistance, extremely high hardness and less thermal expansion as in spark plugs, crucibles space applications for outside coating and chemical laboratory equipment. Properties of ceramics are given with Table 4.9.

4.19 COMPOSITE MATERIALS

Properties absent in a particular material can be achieved by intimate combination of materials of different properties. Composite materials have the following advantages:

1. Save costly or rare materials by plating the costly material on to a the cheaper one such as steel or cast iron plated with copper, bronze or carbide.

2. Impart desirable properties to a material, especially at its surface such as higher tensile strength—reinforced concrete, wire reinforced glass, fibre reinforced plastics, wear resistance as composite rail with hard head. Chemical stability— composite pipes with rust proof surface; electrical or thermal conductivity— composite contact material; low weight with high strength—aluminum carbide

Table 4.9: Mechanical properties of nonmetallic materials

	Specific weight N/mm^3	Ultimate strength N/mm^2			Impact Ncm/cm^2	Young's modulus N/mm^2 $\times 10^3$	Fire resistant up to $0°C$	Coeff off expansion $\times 10^3$	Thermal conductivity $kcal/m$ at $0°C$
		Tensile	Compression	Bending					
Hard Porcelain glazed	23,500	30	440	88	18	736	1670	4	1.35
Unglazed	–	25	390	50	–				
Steatit glazed	26,500	60	830	120	30	1030	1350	6.2	2.05
Unglazed	–	45	830	120	–	–	–	–	–
Calit glazed	27,000	65	930	140	–	1200	1350	7	2.05
Unglazed	–	45	880	140	40	–	–	–	–
Pyrodac unglazed	25,500	30	640	120	24	1000	>1750	4.6	2.4
Calpdur unglazed	23,500	10	60	15	10	–	>1750	4.2	1.5
Heschotherm	23000	15	35	30	15	–	–	3.0	5.6

composite and aluminum silicon composite, poor conductivity, good sliding properties, reflective properties, facility for joining, etc.

3. Achieve new properties such as bimetal as heat indicator and sintered carbides.

The combination is brought about by casting, welding, soldering or glueing, by sintering, diffusion, spraying, rolling or by galvanic means.

Further examples are: Armoured timber, i.e. timber with sheet metal covering, armoured copper wire-as covered steel core; sliding bearing shells, worm gears cast compositely, antivibration pads-rubber between metal plates, etc.

Selection of Material

In selecting the material, the requirements concerning the function, stressing, and life of the component are to be considered first; then those concerning the shape and manufacture and last, but not the least, the prime cost. Often the procurement question is also important.

Ordinary carbon steels are used for simple axles and shafts, keys and pins. High quality steel or special cast iron is used for crankshafts.

Gray cast iron for stands, base plates and housings. If the stresses are high then special cast iron, cast steel or welded steel are used.

Hardened steel for parts subjected to high rolling contact pressures (ball bearings, cams, heavily loaded gear teeth) cast iron, cast steel, steel with carbon 0.2% to 0.6%, heat treated steel, in special cases, plywood, plastics and non-ferrous metals are used for gear wheels.

Plastics, soft cast iron, bronze, white metal, zinc and aluminium alloys are paired for sliding surfaces.

Spring steel, rubber, spring bronze are useful for elastic spring.

Free cutting of die cast alloys are used for small mass produced components.

Heat resisting or non-scaling steel or ceramics are used for components subjected to heat or fire.

Material selection becomes a problem when the available experience is not sufficient, i.e. if new considerations, viz. new requirements, new materials, etc. arise. Then close scrutiny is needed with regard to:

1. The requirements of the component (function, loading, life).

2. The conditions of production (number of pieces, shape, method of production and cost).

3. The material properties, followed, if necessary, by tests with material in question.

In these cases it is advisable for the designer to discuss with material and production experts and with users. The decision is simple if only certain few material properties are of consequence. If several requirements are more or less fulfilled by a number of materials, the decision becomes difficult.

The problem of selecting the most suitable material for the body of a motor car—wood, plywood, plastics, light metal as aluminium alloy or sheet steel, for example, is thus to be solved by evaluating the influencing factors, viz. reliability, life, sensitive to environmental conditions, ease of forming to required shape, weight, resistance to impact and shock, maintenance, repair, cost, etc.

PROBLEMS

4.1 Define ductility, and enumerate some materials which are ductile.

4.2 Define brittleness, and enumerate five materials which are brittle.

4.3 State the reasons for the wide use of cast iron.

4.4 Compare the compositions of cast iron and carbon steel, and state the main differences between them.

4.5 State the advantages obtained through admixing nickle: (a) to cast iron, and (b) to steel.

4.6 State the main advantage of producing: (a) iron castings in an electric furnace, and (b) steel in an electric furnace.

4.7 Explain how malleable cast-iron parts are made from white-iron castings.

4.8 Point out the advantages and disadvantages of cold-rolling of steel bars.

4.9 Explain the differences in objectives, procedures, and results between annealing and normalizing.

4.10 Discuss the effects of nickel, chromium, vanadium, tungsten, and molybdenum in steel.

4.11 Discuss the effect of a small quantity of aluminum in steel, and state in what special alloy it is used in a large amount.

4.12 Explain the principal differences between a brass and a bronze.

4.13 State what elements are alloyed with aluminium to increase its strength.

4.14 Indicate the approximate ultimate tensile strength, yield point, unit elongation, and hardness of the strongest cast aluminum alloy.

4.15 (a) State the uses and advantages of phenolic resin materials, (b) compare their physical properties with those of 20C8 (SAE 1020) steel on the basis of the specific weight.

4.16 State what materials are most suitable for (a) tension, (b) compression, and (c) repeated stress action, respectively. Give reason for the selections.

4.17 Suggest suitable materials for the following parts and give reasons:
  (i) Body of screw jack
  (ii) Nut of screw jack
  (iii) Body of C-clamp
  (iv) Rigid coupling flanges
  (v) Lever of a screw-jack
  (vi) Bolts of connecting rod
  (vii) Two rods of a cotter joint
  (viii) Knuckle pin
  (ix) Nails
  (x) Screw driver
  (xi) Petrol engine block
  (xii) Piston
  (xiii) Piston rings
  (xiv) Shafts
  (xv) Levers
  (xvi) Railway wheels
  (xvii) Railway lines
  (xviii) Brake lining
  (xix) Flexible coupling bush
  (xx) Keys
  (xxi) Helical springs
  (xxii) Leaf springs
  (xxiii) Gears of watches, gadgets, cameras, etc.
  (xxiv) Gears of automobiles, machine tools, naval ships, aircraft, milk churning machines
  (xxv) Rivets
  (xxvi) Belts
  (xxvii) Chains

5 Machine Design Calculations

5.1 DEFINITIONS AND DESIGNATIONS

The elementary equations for stresses in Chapter 2 are based on the assumption that there is equilibrium between the external forces and the internal stresses. For stress in machine parts these equations give correct values only for special cases of static loading.

In most cases, in order to determine the true stress conditions in various sections, it is also necessary to consider other factors, such as the following:

(a) The type of load: constant, shock, or repeating
(b) The maximum and minimum values for repeated loads
(c) The number of cycles expected for repeated loads
(d) Discontinuities
(e) Certain properties of the material, in addition to those determined by static tests
(f) Internal stresses set up by manufac-turing or operating conditions.

Types of stresses: The actual stresses in a certain section are seldom uniform, and the determination of the maximum stress and of the point of its action is of the greatest importance for a safe and economical design.

The stress computed by one of the elementary equations is called the nominal stress and is designated s_n.

The significant stress s_{sg} is the greatest stress that exists in a section of a part. This stress determines the dimension of the section, and it is found by considering the factors mentioned above.

The term *limit stress* will be applied to the maximum stress to which a machine part can be subjected without being damaged. The limit stress S_l is a characteristic of the material, but its value and the method of determining it depend on the type of loading, the thickness of the section, the method of manufacturing and the surface conditions of the piece.

The design stress, also called the *working stress or allowable stress* or *permissible stress* is the maximum stress that should never be exceeded in a properly designed machine part. To allow for inaccuracies in assumptions, workmanship, and qualities of the material, the design stress S_d always must be lower than the limit stress S_l.

Factor of safety: The amount by which S_d is kept below S_l is expressed by their ratio, which is called the *design factor of safety* and is designated as n.

Thus,

$$n = \frac{S_l}{S_d} \qquad \ldots (5.1)$$

This definition of the safety factor differs from the older one, still sometimes used, by which the nominal stress was compared with the ultimate strength of the material. However, the present definition is more logical and has decided advantages with respect to economy of material and actual safety.

The actual safety factor is the ratio of S_l to s_{sg} and is designated by n'. Thus,

$$n' = \frac{S_l}{S_{sg}} \qquad \ldots (5.2)$$

The factor n' differs from n when the dimensions that are found by using n are changed, either by rounding them off or by using standard sizes.

Types of loads: Three main types of loads must be distinguished. They are:

1. *Steady loads*, called *static loads* or *dead loads*.
2. *Impact loads*, or shock loads.
3. *Repeated loads*, which either may vary gradually or may have the characteristics of repeated impact. So called *inertia loads* are special cases of single impact loads, or of repeated, gradually changing loads, or of repeated impact loads.

The methods of designing parts subjected to different types of loading differ greatly and must be considered separately.

5.2 STRENGTH UNDER COMBINED STRESS

A machine part is generally subjected simultaneously to several, different stresses, whose actions are combined. Therefore it is necessary to establish the combined stress that will cause failure of the part. It should be noted that a machine part may fail either by having undergone a permanent deformation or by actually breaking at the dangerous section.

Failure by yielding, when a part is subjected to steady or gradually applied loads, occurs in materials that are classified as ductile, whereas failure by fracture usually occurs in materials that are classified as brittle. However, as already mentioned in section 2.4, there is no sharp dividing line between ductile and brittle materials. Also, it was shown how a part made of ductile material may fail by fracture when subjected to repetitive loading.

Theories of failure: Four theories of failures are used at present. They are known as the maximum-normal-stress theory, the maximum-strain theory, the maximum-shear theory, and the shear-energy theory.

The maximum-normal-stress theory, also called Rankine's theory, assumes that yielding at a point begins only when the maximum principal stress on a certain plane passing through the point reaches a value equal to the elastic limit as found in a simple tension test, regardless of any other stresses that occur on other planes passing through the point. The condition of failure is

$$s' = S_e \qquad \ldots (5.3)$$

The *maximum-strain theory*, or *Saint-Venant's theory*, assumes that the elastic limit of a material is reached when the stress corresponds to maximum strain. Therefore a member is safe as long as the maximum normal stress s' found by equation 2.48 is below the elastic limit S_e of the material.

The value of Poisson's ratio μ for all materials used in machinery is close to 0.3. By inserting this value in equation 2.52 and simplifying the terms, the condition of failure for the combination of a direct stress s and a shear stress s_s is expressed by the equation

$$0.35s + 0.65\sqrt{s^2 + 4s_s^2} = S_e \qquad \ldots (5.4)$$

The *maximum-shear theory* of Guest assumes that the elastic limit of a material is reached when the greatest resultant shear stress reaches a certain limit as determined under the condition of pure shear. According to this theory a member subjected to tensile stress fails when the maximum shear stress determined by equation 2.58 exceeds $\frac{1}{2} S_e$, where S_e is the elastic limit in tension. In the case of a combined load the member will fail when its maximum shear stress s_s determined by equation 2.51 exceeds the elastic limit in shear. In general, if a member is subjected to the combination of a direct stress s, the condition of failure is expressed by the equation

$$\sqrt{s^2 + 4s_s^2} = S_e \qquad \ldots (5.5)$$

The *shear-energy theory*, or the constant-energy-of-distortion theory, also referred to as the Hencky-von Mises theory, gives the following equation[1] for the condition of failure for a combination of a direct tensile stress s and a shear stress s_s:

$$\sqrt{s^2 + 3s_s^2} = S_e \qquad \ldots (5.6)$$

Comparison of theories: For pure tension, $s_s = 0$, and all four theories give the same results. For pure shear, the maximum-normal-stress theory gives, in accordance with equation 2.46, $s_s = 0.5\,S_e$; the maximum-strain theory, equation 5.4, gives $s_s = 0.77\,S_e$; the maximum-shear theory, equation 5.5, gives $s_s = 0.5\,S_e$; and the shear-energy theory, equation 5.6, gives $s_s = 0.58\,S_e$. For combinations of tension and shear the differences are less than that for pure shear.

At present it is difficult to determine definitely which of these theories should be given preference. The available data seem to indicate that for hard, brittle materials, such as cast iron or hardened steel, the maximum-normal-stress theory or the maximum strain theory should be preferred. For ductile materials experimental data are in favour of the shear-energy theory. However, it is much simpler to use the maximum-shear theory, and the differences resulting from application of this theory rather than the shear-energy theory are very slight and on the safe side. Therefore the maximum-shear theory is often used as a basis for design calculations.[2]

However, the trend in design practice is toward the wider use of the Hencky-von Mises theory for ductile materials. The safe course to follow is to design a member so that the maximum principal stress does not exceed some selected proportion of fraction of the tensile or compressive elastic limit; and so that the maximum principal strain does not exceed the same proportion of strain corresponding to this elastic limit; and also so that the maximum shear stress does not exceed the same proportion of the elastic limit in shear.[3]

5.3 INFLUENCE OF SIZE

The values for elastic limit or yield point given in various tables of Chapter 4 were obtained with test specimens of small cross sections, about 12 mm in diameter. Tests show that, because of decreased uniformity of structure, pieces of material with larger sections have mechanical properties inferior to those with smaller sections. The influence of size upon the elastic limits in tension of various materials may be seen in Fig. 5.1, where the scale of ordinates for the three dotted curves is indicated on the right-hand side of the diagram.

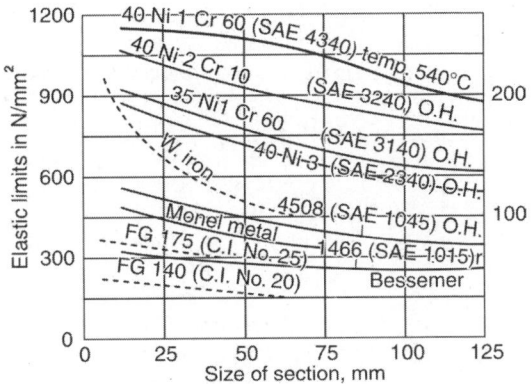

Fig. 5.1: Influence of size on elastic limits

As a first approximation the decrease of the elastic limit for all metals may be represented by an inclined straight line, as shown in Fig. 5.2. In this case, if the elastic limit given in the table is S_e, approximately corresponding to a section 12 mm thick, the elastic limit of a 75 mm section S_{e3} is known, the elastic limit S_e' for any thickness h between 12 mm and 75 mm can be determined from the relation

$$S_e' = S_e - \frac{(S_e - S_{e3})(h - 12)}{75 - 12}$$

[1] G. R. Soderbeg, "Working Stresses,"Transactions of the American Society of Mechanical Engineers, Vol. 55, No. 3 (1935), p. A-107, and Code for the Design of Transmission Shafting (New York: American Society of Mechanical Engineers 1927), p. 3.

[2] *Ibid*.

[3] F. B. Seely, Resistance of Materials, 3rd ed. (New York: John Wiley & Sons, Inc. 1947), p. 292

This equation may be written in the simple form

$$S_e' = e_{sz} S_e \qquad \ldots (5.7)$$

where, size coefficient e_{sz} may be found from Fig. 5.2 as

$$e_{sz} = 1 - \left(1 - \frac{S_{e3}}{S_e}\right)\frac{(h-12)}{63} \qquad \ldots (5.8)$$

Values of the ratio S_{e3}/S_e for a few of the more important materials are given in Table 5.1.

A further decrease of the elastic limit with the increase in the size above 75 mm in most metals is less pronounced, as may be seen from Fig. 5.1. For actual design calculations the proper data should be obtained from the manufacturer or some other reliable source.[4]

Although data are meagre, it appears that an increase in size has the same influence on the elastic limit in shear as on the elastic limit in tension. Hence, the values in Table 5.1 can be used also for shear.

It should be noted that the use of the size factor is a refinement necessary in designing light, highly stressed parts. Many ordinary

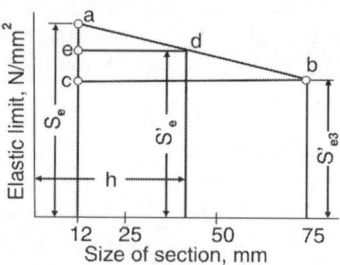

Fig. 5.2: Change of elastic limit with size of mm

parts are designed without considering this refinement.

5.4 STRESS CONCENTRATION

As explained in section 3–5, the presence of a discontinuity increases the stress at and near the discontinuity. The maximum theoretical increase in stress, compared with the average or nominal stress, is represented by the form factor K. The magnitude of K may be obtained by using the formulas and diagrams given in section 3–6, 3–7, and 3–8. The actual stress increase in a machine part is smaller than that given by the factor K, because of the elasticity

Table 5.1: Size factors for mechanical properties of materials

Material	Values of S_{e3}/S_e				
	Natural state	Annealed	Drawn at 650°C	Drawn at 540°C	Drawn at 425°C
Aluminium, strong, wrought	0.93	–	–	–	–
Tobin bronze	0.90	–	–	–	–
Monel metal, forged	0.80	–	–	–	–
Ductile iron	0.80	0.98	–	–	–
Low-carbon steel, C<0.20%	0.84	–	–	–	–
Medium-carbon steel, 0.30 to 0.50% C	–	0.85	0.72	0.59	0.53
Nickel steel, SAE 2340	–	0.86	0.80	0.74	–
Cr-Ni steel, SAE 3140	–	0.80	0.75	0.70	0.65
Cast iron, Class No. 20 FG150	0.55	–	–	–	–
Cast iron, Class No. 25 FG150	0.73	–	–	–	–
Cast iron, Class No. 35 FG260	0.60	–	–	–	–
Wrought iron	0.55	–	–	–	–

---

[4] International Nickel Company, Inc., Nickel Alloy Steel, Sec. II, Data Sheet No. 4 (New York, 1934, revised to March, 1941), pp. 64 ff.

and plasticity of the metals. In general, however, the actual increase is along the same lines as the theoretical increase.

The actual weakening effect is the effective stress increase, which is called the *stress-concentration factor* and is designated as K'. The best way to find the stress-concentration factor in a machine part is to determine the ratio of the stresses both with and without stress concentration by testing the machine part under actual or simulated service condition. Such experimental data exist, but not for all parts, materials, and types of loading.

A simpler, although less accurate, method is to find the relation between the theoretical form stress factor K and the real stress-concentration factor K' for a certain set of conditions, and to use this relation to compute K' from known values of K for the design conditions.

Index of sensitivity: The relation between K' and K may be expressed as follows:

$$K' = 1 + q(K - 1) \qquad \text{...(5.9)}$$

where, q is called the index of sensitivity of the material to abrupt changes of section. From equation 5.9,

$$q = \frac{K' - 1}{K - 1} \qquad \text{...(5.10)}$$

The magnitude of q depends on the material and on the character of loading. In general, q is higher for hard, brittle materials and decreases with an increase of ductility. Also, q is highest for repeated heavy-shock loads, decreases with a decrease of the intensity of the shocks, and has the smallest value when the load is more or less steady, as when it is static.

5.5 STATIC LOADS

The design stress for a static load can be found from the equation

$$S_d = \frac{S'_e}{n} = \frac{e_{sz} S_e}{n} \qquad \text{...(5.11)}$$

where, n is the safety factor. This factor should never be less than 1.25, and it should usually be

at least 1.5. If the material is not of the very best quality or if there is no definite information about its quality, the value of n should be taken as 2. For a brittle material it is advisable to use a value of n between 2 and 3.

Sensitivity index: Tests on various ductile metals subjected to static loads have shown that discontinuities do not affect the significant stress. *For a ductile material, therefore, the sensitivity index q is zero and the stress-concentration factor K' is 1.* Values of q for other materials may be taken as follows: for brittle materials, such as hardened steel, 0.10; for very brittle materials, such as steels that are quenched but not drawn, 0.20. *Since cast iron has so many internal discontinuities in its structure, it may be assumed that q is zero for this metal.*

In an eye bar the effect of the hole is to produce compressive stress in the outer fibers, as indicated in Fig. 3.12. Therefore, the equalizing yield cannot take out all of the localized stress; and value of K' based on the reduced normal section ck should be taken as 1.5.[5] The corresponding value of q is 0.16.

Significant stress: A discontinuity increases the stress to an effective value that is K' times as great as the nominal stress. This increase takes place only at one point or in a limited region. However, since the purpose of a correct design is to prevent the stress from exceeding the allowable design stress S_d at any place, the highest effective stress caused by stress concentration becomes the significant stress. To find this significant stress, the nominal stress s_n is first computed by applying the proper elementary stress equation and it is then multiplied by K'. Thus,

$$S_{sg} = K' s_n \qquad \text{...(5.12)}$$

Bearing stress: A special case of compressive stress encountered in machine parts is the stress caused at the surface of contact of two members that are relatively at rest, as at the surface between a rivet and a plate or between a key and a shaft. This kind of stress is called bearing stress. This term

---

[5] F. B. Seely, Advanced Mechanics of Materials (New York: John Wiley & Sons, Inc., 1932), p. 221.

should not be confused with bearing pressure, which exists at the contact surfaces of two parts in relative motion.

When there are no friction forces at the surfaces in contact, it has been found both theoretically and experimentally that the magnitude of the bearing stress at which yielding starts is[6]

$$S_d = \frac{\pi}{\sqrt{3}} S_e = 1.81 \, S_e \qquad \ldots (5.13)$$

Actually friction exists and increases the factor 1.81 materially. For practical applications the elastic limit of bearing stress may be assumed as

$$S_b = 2S_e \qquad \ldots (5.14)$$

Combined loading: Where there are combined stresses the value of the maximum significant stress S_{sg} must be determined in compliance with the theory of failure applicable to the materials used.

Factors that affect factor of safety are given below:

(i) The degree of reliability of properties of materials selected; as seamed steel, blow holes, slag in weld, shrinkage stresses affect the properties.

(ii) The change of these properties during service, e.g. stresses induced during machining, initial and residual stresses, assembly, transportation, unintentional overloading as in trains or during acceptance tests.

(iii) The extent to which the simplified assumptions made (in the analysis for nominal stresses) are true.

(iv) The degree of reliability of test results as applied to actual parts, i.e. the effect of size.

(v) The reliability of the magnitude and the kind of applied load.

(vi) The uncertainty as to the exact cause of failure.

(vii) The extent to which localized stress may be developed.

(viii) The extent to which initial stresses may be set up during processing, fabrication or assembly.

(ix) The extent to which human life and property may be endangered or cost of shut down or repairs if failure occurs.

(x) The kind of environment to which a machine may be subjected.

(xi) The reliability required of the machine.

(xii) The price class of the machine.

Safety factor: For static loading the limit stress is $S_l = S_e'$. Therefore the actual safety factor is

$$n' = \frac{S_l}{s_{sg}} = \frac{e_{sz} S_e}{K' s_n} \qquad \ldots (5.15)$$

In Fig. 5.3 the various relations between the stresses are illustrated which have to be considered for a machine part under static loading. The difference between S_d and s_{sg} is the additional safety margin Δs.

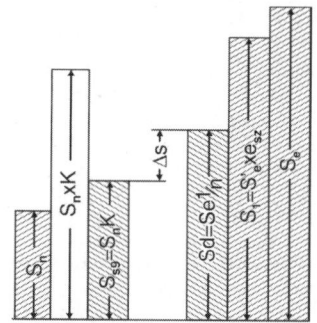

Fig. 5.3: Relation between stresses for static loading

The recommended values of the actual safety factor n' for static loads are as follows:

(a) For reliable high-grade material, when loads and stresses can be determined very accurately and a low weight is desired, n' is taken between 1.25 and 1.5.

(b) For the same conditions, when a low weight is not important, n' ranges from 1.5 to 2.

(c) For ordinary materials, when loads and stresses can be determined with sufficient accuracy, n' ranges from 2 to 2.5.

---

[6] A. Nadai and A. M. Wahl, Plasticity (Engineering Societies Monographs, New York: McGraw-Hill Book Company, Inc., 1933), p. 247.

(d) For brittle materials under ordinary conditions, n' ranges from 2.5 to 3.

(e) For ordinary materials, when loads and stresses cannot be determined accurately, n' ranges from 3 to 4.

As a general rule, the value of factor of safety is a factor of ignorance. The less we know about material, the exactness of location of load, type of load, etc., the higher is the factor of safety. *However, the factor of safety in aircraft components is the lowest 1.01 on S_e to reduce the weight of aircraft (5.1).* The failure of any component will cause disaster resulting in huge loss of life and property while in industrial machines, the factor of safety is very high even 15, when the failure will not endanger any life or property. It will disrupt work till the parts are provided with maintenance. All the parts of aircraft are subjected to accurate analysis for a specified life and tested before assembly, so factor of safety is very low. In many situations, the factor of safety is based on ultimate stresses; tensile, composure and shear as its value for most of materials is easily available.

Example 5.1: Find the safety factor for the flat tie rod shown in Fig. 5.4, if it is made of Nickel steel and is subjected to a steady pull of 1.2 MN. The dimensions are as follows: $b = 150$ mm, $h = 32$ mm, $d = 38$ mm, $h_1 = 19$ mm, $r = 19$ mm and $\alpha = 45°$.

The elastic limit for the steel is 550 MN/mm². If the value of the ratio S_{e3}/S_e is taken as that given in Table 5.1 for SAE 2340 nickel steel or 0.74, the coefficient, by equation 5.8.

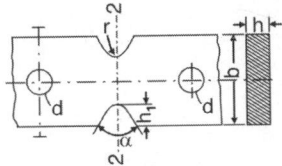

Fig. 5.4: Tie rod

$$e_{sz} = 1 - \frac{(1-0.74)(32-12)}{63} = 0.917$$

Then the corrected elastic limit, by eq. 5.7 is
$$S_e' = 0.917 \times 550 = 504.35 \text{ MN/m}^2$$

The nominal stress created at section 1.1, Fig. 5.4, by the pull of 1.2 MN is,

$$S_{n1} = \frac{1.2 \times 10^6}{(150-38) \times 32} = 334.8 \text{ MN/m}^2$$

From Fig. 3.4, $\frac{d}{b}$ is $\frac{38}{150} = 0.25$, the form factor for the hole, is $k = 2.40$. If it is assumed that the index of sensitivity is 0.10, the stress concentration factor found by eq. 5.9 is

$$k' = 1 + 0.1 (2.4 - 1) = 1.14$$

By equation 5.12, the significant stress is
$$S_{sg1} = 1.14 \times 334.8 = 381.6$$

The nominal stress in section 2.2, Fig. 5.4 is
$$S_{n2} = \frac{1.2 \times 10^6}{(150 - 2 \times 19) \times 32} = 334.8 \text{ MN/m}^2$$

From Fig. 3.15 for $r/b = \frac{19}{50} = 0.126$ and

$\frac{h}{b} = 0.126$, the form factor is $k = 1.7$, Hence stress concentration factor is

$$K' = 1 + 0.1 \times (1.7 - 1) = 1.07$$

According to Fig. 3.16, the influences of the angle α can be neglected. Therefore significant stress is
$$S_{sg2} = 1.07 \times 334.8 = 358.2 \text{ MN/m}^2$$

Since section 1.1 is subjected to the greater significant stress, that is the dangerous section. By eq. 5.2, the actual safety factor is

$$\frac{504.35}{358.2} = 1.40$$

Design procedure: The first step in the design of a machine part is to select a suitable material by considering such factors as the shape of the part, the operating conditions, and the cost. Next, the expected thickness of the sections involved is assumed, and the corrected elastic limit S_e' is estimated. The basic value of S_e may be taken from one of the tables of Chapter 4. In actual industrial design of an important part the value of S_e or S_e' may be obtained from the manufacturer of the material, and in important cases S_e' may be determined by special tests. Then the safety factor n is selected and the design stress S_d is determined.

After that the relation between the acting force F and the allowable stress S_d is expressed as a function of the dimensional characteristic of the part, such as its area A or its section modulus Z or Z_o. The equation is then solved for this characteristic, and the linear dimension are determined. Any discontinuities causing stress concentration may require an increase of the preliminary dimensions.

Example 5.2: Determine the dimension b and h of tie rod having discontinuities like those in Fig. 5.4, to sustain a dead load F of 250 kN with a safety factor of 2. The two holes on the centre line have a diameter of 25 mm and are 125 mm apart. The two notches are midway between the holes and are so shaped that $r = 6$ mm, $h_1 = 12$ mm and $\alpha = 90°$

A suitable material would be 30C8 (SAE1030) with an elastic limit S_e of 290 MN/m². The value of S_e may be decreased slightly, say by 5 per cent to take care of a thickness over 6 mm. If the steel is annealed, no stress concentration has to be taken into account from either the holes or the notches. Hence the design stress is

$$S_d = \frac{S_e'}{n} = \frac{290 \times 0.95}{2} = 137.75 \text{ MN/m}^2$$

The minimum cross-sectional area is

$$A = \frac{F}{S_d} = \frac{250}{137.75 \times 1000} = 0.001815 \text{ m}^2$$

If it is assumed that $b/h = 4$ or $A = 4h^2$, the required depth is

$$h = \sqrt{\frac{A}{4}} = \sqrt{\frac{0.001815}{4}} = 0.0213 \text{ m say 22 mm}$$

If 25 mm is added for the hole or $2 \times 12 = 24$ mm for the notches, the resulting total width becomes

$$b = \frac{0.001815}{0.022} + 0.025 = 0.1075 \text{ m say 110 mm}$$

From Table 5.1, the ratio S_{e3}/S_e may be taken as 0.85. The corrected elastic limit, by equations 5.7 and 5.8 is

$$S_e' = 290 \times \left[1 - \frac{(1 - 0.85)(22 - 12)}{63}\right]$$

$$= 290 \times 0.972 = 283.3 \text{ NN/m}^2$$

Therefore, the actual safety factor is

$$n' = \frac{283.3 \times (110 - 25) \times 22}{250 \times 1000} = 2.119$$

This is only slightly greater than that which was originally specified.

Columns: In designing a new column Ritter's formula may be used. If the column has a length l and must carry a load F, the cross-sectional

area A and the radius of gyration k are expressed by one of the dimensions of the section, the proper design stress S_d is substituted for S_c, and equation 2.41 is solved for this dimension. The same procedure is followed if the column is loaded eccentrically, but in this instance the modified equation 2.65 is used.

Example 5.3: Find the dimension of the rectangular cross section of a steel bar 500 mm long which must act as a column with the ends hinged and must support a weight of 80 kN. The width b of the bar must be twice its thickness h. The properties of the steel used are:

$S_e = 230$ MN/m² and $E = 210 \times 10^3$ MN/m². Use a safety factor $n = 2$

To find the area A needed to carry the load, it is necessary first to determine the design stress. Considering the stress as pure compression and allowing 3 per cent as a deduction for the expected influence of size, we get

$$S_d = \frac{e_{sz}S_e}{n} = 0.97 \times \frac{230}{2} = 111.55 \text{ MN/m}^2$$

Then $A = \dfrac{80}{111.55 \times 10^3} = 0.0007171 \text{ m}^2$

Since $b = 2h$, the cross-sectional area $hb = 2h^2$. Hence

$$h = \sqrt{\frac{0.0007171}{2}} = 0.0189 \text{ m}$$

According to Table 2.5, the least radius of gyration is $k = 0.289h = 0.289 \times 0.0189 = 0.00546$.

Since the ratio of slenderness is $l/k = \dfrac{0.5}{0.00546}$

$= 91.57$, the bar must be treated as short column.

According to Table 2.5 the least moment of inertia is $I = \dfrac{bh^3}{12}$. When $b = 2h$, then $I = h^4/6$. The least radius of gyration is

$$k = \sqrt{\frac{h^4}{6 \times 2h^2}} = \frac{h}{2\sqrt{3}}$$

Ritter's formula (equation 2.41), in which $n = 1$ for the specified end conditions gives

$$111.55 = \frac{80}{1000 \times 2h^2} \left[1 + \left(\frac{0.5 \times 2\sqrt{3}}{h}\right) \times \frac{230 \times 0.97}{\pi^2 \times 210 \times 10^3}\right]$$

This reduces to

$$2788.75 \, h^4 - h^2 - 0.0003228 = 0$$

from which $h = 0.02374$ m or say 24 mm and $b = 48$ mm.

To find the actual safety factor, the size influence is determined by equation 5.8

Thus,

$$e_{sz} = 1 - \frac{(1-0.85)(24-12)}{63} = 0.972$$

Equation 2.41 gives for the actual stress

$$S_e = \frac{80 \times 10^6}{1000 \times 48 \times 24} \left[1 + \left(\frac{0.5 \times 2\sqrt{3}}{0.024} \right)^2 \frac{230 \times 0.972}{\pi^2 \times 210 \times 10^3} \right]$$

$$= 108.45 \text{ MN}/\text{m}^2$$

The actual safety factor is n

$$n' = \frac{230 \times 0.972}{108.45} = 2.06$$

Long columns: In designing a long column, Euler's formula (equation 2.42) may be used. The load F that the column must carry is multiplied by a safety factor n_c, and the product $n_c F$ is substituted for the critical load F_u. The area A and the radius of gyration k of the cross section are expressed as functions of one of the dimension b of the section, the corresponding values of E and the end-condition factor n are taken, and the equation is solved for b. Since Euler's formula gives the load that corresponds to the ultimate strength S_u, which is about twice S_e, the safety factors given before must be doubled. Thus 4 to 5 should be used for ductile materials, and 6 to 8 for brittle ones. Eccentric loading does not affect a long column appreciably; but to be on the safe side, it can be taken into account by increasing the safety factor slightly.

5.6 IMPACT LOADING

When a force F acts on a machine part with impact, the magnitude of the maximum stress s' in the part is greater than the static stress s, as shown by equation 3.18. The stress s' thus becomes that effective stress and must not exceed the permissible stress, or design stress, S_d. Therefore the static stress s is the significant stress which must be used in determining the stressed cross-sectional area of the part. On the other hand, since the energy of the impact must be absorbed by the body, the resilience of the body must be at least equal to the impact energy. Therefore either of two approaches is possible. One is based on the significant static stress s, while the other is based on the resilience U of the body.

Safety factor: Since a low safety factor means higher stresses and hence greater resilience, the safety factor n is normally taken as 1.5.

Index of sensitivity: The value of q may be taken as 0.4 for those materials which are ductile and very soft, and increased up to 0.6 for other ductile materials. For hard and brittle materials, such as hardened high-carbon steel, q should be taken as 1. In this case the stress-concentration factor K' becomes equal to the theoretical form-stress factor K. Cast iron, because of its structure, can be classified among the less-sensitive materials, q having a value of about 0.5.

Design procedure: After the material has been selected, the size coefficient e_{sz} is estimated by assuming the probable thickness of the section involved and by applying equation 5.8. The corrected elastic limit S_e' can then be found by equation 5.7. The next step is to establish the value of the form-stress factor K as explained in Chapter 3 and to find the stress-concentration factor K' by equation 5.9. The design stress can then be found by the equation

$$S_d = \frac{e_{sz} S_e}{nK'} \qquad \ldots (5.16)$$

This value of S_d is substituted in equation 3.18 for s', and the significant stress is then found by the relation

$$s = \frac{S_d}{1 + \sqrt{1 + \dfrac{2h}{e}}} \qquad \ldots (5.17)$$

For tension or compression,

$$s = \frac{S_d}{2\left[\dfrac{hE}{S_d l} + 1\right]} \qquad \ldots (5.18)$$

since
$$e = \frac{S_d l}{E}$$

The dimensions of the part are now found by the usual elementary stress equations in which this value of s is taken as the maximum allowable stress.

After a part is thus designed for an impact load, its dimensions should be checked to make sure that its resilience U is sufficiently greater than the impact energy K_i. The resilience U may be determined by the general equation 3.23 or by one of the special equations 3.25, 3.30, 3.32, or 3.34.

In finding the impact energy that must be absorbed by the part, it is necessary to take into account the deformations caused by the impact. Therefore, the energy caused by the additional travel of the impact force, computed by equation 3.16, must be added to the main impact energy found by equations 3.14 or 3.16.

Because energies are proportional to the second power of stresses, the actual safety factor n' in terms of energies,

$$n' = \sqrt{\frac{U}{K_i}} \qquad \dots (5.19)$$

Example 5.4: Determine the main dimensions of a round rod of 30C8 steel (SAE 1030) stressed in tension by a weight of 1.5 kN falling from a height of 50 mm. The, free length of the rod is 1.5 m, and the ends of the rod are fastened as shown in Fig. 5.5. Take $S_e = 290$ MN/m^2.

In the preliminary calculations S_e' may be assumed as = 0.93 S_e or 0.93 × 290 = 269.7 MN/m^2 then the design stress, with $n = 1.5$ is

$$S_d = \frac{269.7}{1.5} = 179.8$$

The significant stress, found by equation 5.18 is

$$S = \frac{179.8}{2\left(\dfrac{0.05 \times 210 \times 10^3}{179.8 \times 1.5} + 1\right)} = 2.25 \text{ MN/m}^2$$

The required cross-sectional area is

$$A = \frac{1.5}{2.25 \times 1000} = 0.000666 \text{ m}^2 \text{ and the diameter is}$$

$$d_1 = \sqrt{\frac{0.000666 \times 4}{\pi}} = 0.0291 \text{ m}$$

In static loading, the actual dimensions would probably be the next larger one, which is 30 mm. For dynamic loading, in order not to decrease the deformation and resilience, it is better to keep the compound diameter.

At the ends, the diameter must be increased to take care of the weakening caused by the cross pin holes and to take care of the stress concentration due to these holes.

Assume that pins to be made of the same 30C8 steel. The elastic limit in shear for this steel is 180 MN/m^2. Since there are two pins with four shearing areas, the pins will be as strong as the rod if the cross-sectional area of the pins is $[F = A.S_e = 4 A_s \times S_s]$

$$A_3 = \frac{0.000666 \times 290}{180 \times 4} = 0.000268 \text{ m}^2$$

The diameter of pins should be

$$d_3 = \sqrt{\frac{0.000268 \times 4}{\pi}} = 0.01847 \text{ m or use 19 mm.}$$

The stress-concentration factor for the holes may be estimated at first as $K' = 1.4$, and the diameter d_2, Fig. 5.5 can be computed by equating the cross-sectional area of the end weakened by the pinhole and by stress concentration to the area of the body of the rod.

Thus,

$$\frac{\dfrac{\pi}{4} d_2^2 - 19 d_2}{1.4} = \frac{\pi}{4} \times 30^2$$

which gives $d_2 = 49.6$ or 50 mm

Now the stress concentration can be computed more accurately by using the form stress factor from Fig. 5.5.

For $d/D = 19/50 = 0.38$, the value of K is 2.02. With $q = 0.4$, the stress concentration factor is, by equation 5.9,

$$K' = 1 + 0.4 \times (2.02 - 1) = 1.408$$

This is practically the same as estimated. Thus no changes in the design are required. Naturally the

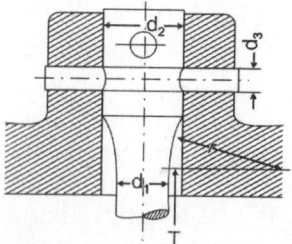

Fig. 5.5: Fastening of end of rod

change from d_2 to d_1 must be very gradual in order not to introduce a new stress concentration.

It is advisable to check the final dimensions by computing the resilience of the rod. Thus impact stress, by equation 3.20, is

$$S' = 2.25\left[1 + \sqrt{1 + \frac{2 \times 0.05 \times 210 \times 10^3}{2.25 \times 1.5}}\right]$$

$$= 179.75 \text{ MN/m}^2$$

According to equation 3.25 the absorbed energy is

$$U = \frac{179.75^2 \times 0.000666 \times 1.5}{2 \times 210 \times 10^3} = 0.0000768 \text{ MN-m}$$

The impact energy which must be absorbed is, by equation 3.16

$$K_i = \frac{1.5}{1000} \times \left[0.05 + \frac{179.75 \times 1.5}{210 \times 10^3}\right]$$

$$= 0.00007692 \text{ MN-m}$$

This check indicates that the calculations are correct.

To find the actual safety factor that corrected elastic limit must be determined. The size coefficient, by equation 5.8, is

$$e_{sz} = 1 - \frac{(1 - 0.85)(50 - 12)}{63} = 0.911$$

which gives $S_e' = 290 \times 0.911 = 264.19$. The actual safety factor is

$$n' = \frac{S_e'}{S'} = \frac{264.19}{179.75} = 1.47$$

Another method of calculating n' is by finding the maximum stress S_{max} that can be created in the main part of the rod without exceeding the elastic limit in the section weakened by the pin hole and stress concentration. Since the stresses are inversely proportional in the cross-sectional areas

$$S_{max} = \frac{264.19}{1.41} \times \frac{\frac{\pi}{4} \times 0.05^2 - 0.019 \times 0.050}{0.000666}$$

$$= 285.13 \text{ MN/m}^2$$

Since S_{max} is greater than S_e', the latter value must be used. The maximum energy that could be absorbed without exceeding the elastic limit at any point is

$$U = \frac{264.19^2 \times 0.000666 \times 1.5}{2 \times 210 \times 10^3} = 0.000166 \text{ MN-m}$$

$$n' = \sqrt{\frac{0.000166}{0.00007692}} = 1.469$$

Increase of impact strength: Equation 3.18 shows that the impact stress can be decreased by decreasing the impact travel h. If h is a clearance between two reciprocating parts, as between a wrist pin and its bearing, then the proper procedure is to reduce h to the minimum value permissible from the standpoint of a running fit.

The influence of the deformation e is in the opposite direction, that is, an increase of the deformation lowers the impact stress. Therefore machine parts subjected to impact should not be made more rigid than is necessary for proper operation.

The same conclusion may be reached by considering resilience. Equation 3.23 shows that the resilience of a member is increased if its deformation in increased; and this in turn requires an increase of stress. Whatever the nature of the stress is, the deformation of a member will be greatest when all its sections have the same maximum significant stress. Thus, to best resist shock a member must have the shape of *uniform strength*.

Increase of resilience in bending is illustrated by the shape of carriage springs in automobiles and railroad cars (Fig. 14.3). The springs have approximately the same strength in bending over the whole length. This characteristic makes them flexible and at the same time capable of absorbing heavy shocks.

In many instances the impact strength can be increased by reducing local stress concentration through a lowering of the form stress factor K.

Example 5.5: Find the impact strength in tension of the connecting-rod bolt. (Figure 5.6a), if 35 Ni1Cr60 (SAE 3140) chrome-nickel steel is used and the safety factor is 1.5.

The elastic limit in tension of this steel, drawn at 540°C is 655 MN/m². If the stock size is taken as the thickness of the head, which is assumed as 60 mm, the size of coefficient is

$$e_{sz} = 1 - \frac{(1 - 0.70)(60 - 12)}{63} = 0.772$$

The design stress is then

$$S_d = \frac{655 \times 0.772}{1.5} = 337.1 \text{ MN/m}^2$$

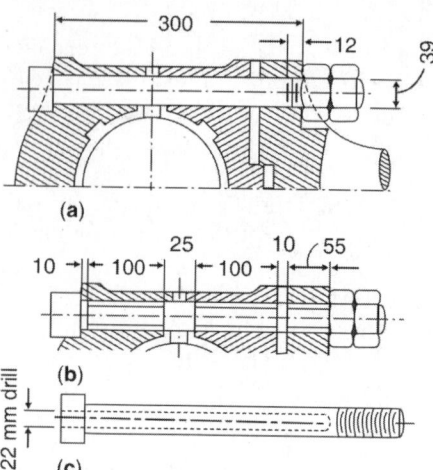

(a)

(b)

22 mm drill

(c)

Fig. 5.6: Connecting-rod bearing bolt

The form stress factor of the thread, according to section 3.6, is $K = 5.62$. The elongation of heat treated steel is 17 per cent and $q = 0.4$. Hence the stress concentration factor is

$$K' = 1 + 0.4 (5.62 - 1) = 2.85$$

Therefore, the nominal impact stress should not exceed

$$s_n = \frac{Sd}{k'} = \frac{337.1}{2.85} = 118.3 \text{ say } 120 \text{ MN/m}^2$$

Since the area of the dangerous section of the thread, or the stress, is 976 mm² (ISO pitch 4), the stress in the unthreaded part will be (39 mm)

$$s = \frac{120 \times 976}{\frac{\pi}{4} \times 39^2} = 90 \text{ MN/m}^2$$

As $l = 0.30$ m and $E = 210 \times 10^3$ MN/m², the resilience by equation 3.25 is

$$U = \frac{120^2 \times \frac{\pi}{4} \times 39^2 \times 0.30}{2 \times 210 \times 10^3 \times 10^6} = 0.0000122 \text{ MN-m}$$

In order to increase the resilience by increasing stress, the bolt can be turned down to a diameter of 30 mm, leaving 39 mm at the centre and at the ends as shown is Fig. 5.6b. At these places, the bolt must fit accurately to secure the position of the bearing. The stress in the reduced section is increased to

$$S_2 = \frac{120 \times 39^2}{30^2} = 202.8 \text{ MN/m}^2$$

The resilience now becomes

$$U = \frac{202.8^2 \times \frac{\pi}{4} \times 30^2 (0.1 + 0.1)}{2 \times 210 \times 10^3 \times 10^6} + \frac{120^2 \times \frac{\pi}{4} \times 39^2 \times (0.3 - 0.2)}{2 \times 210 \times 10^3 \times 10^6}$$

$$= 0.0000179 \text{ MN/m}^2$$

Thus the resilience or impact strength is increased without changing the maximum stress. The stress occurs in threaded part and is $S_d = 337.1$ MN/m².

Second solution: The cross sectional area of the bolt can be reduced to 814 mm² by drilling a 22 mm hole at the centre of the bolt all the way down to the threaded part, as shown in Fig. 5.6. C (The area in (b), is 707 mm²).

The stress in this part will then increase to

$$S = \frac{120 \times 39^2 \times \frac{\pi}{4}}{814} = 176 \text{ MN/m}^2$$

The resilience becomes

$$U = \frac{176^2 \times 814 \times 0.3}{2 \times 210 \times 10^3 \times 10^6}$$

$$= 0.0000180 \text{ MN-m or } 18 \text{ Nm}$$

To avoid a stress concentration at the bottom of the drilled hole, the end of the hole should have a gradual change of section, as shown in the illustration. The disadvantage of this method is that the bolt is more rigid in bending than if turned down as in Fig. 5.6b, and a high bending stress may be created if the seat is not quite normal to the bolt higher axis.

These types of bolts are called uniform strength bolts, and has the advantage of higher resilience and less stress concentration (5.4).

5.7 REPEATED LOADS

The term endurance limit is commonly applied to the highest stress to which a material can be repeatedly subjected to an infinite number of times without failing. The method of applying repeated loading when testing the material is usually by bending, and the stress changes from a maximum value in tension to the same value in compression. In a more general sense the endurance limit depends on the following factors:

(a) The amplitude of a stress variation

(b) The mean value of the stress

(c) The type of stresses invoked

(d) The method of manufacturing the material, including its heat treatment

(e) The size of the section

(f) The condition of the surface

(g) Discontinuities in the sections

Safety factor: The usual procedure in designing parts subjected to repeated loads is to use endurance limits S_{en} instead of elastic limits S_e as limit stresses and to apply the same safety factors that are given in section 5.5 for static loads.

5.8 ENDURANCE LIMITS

Methods of manufacture and of heat treatment influence the endurance limits of a metal, but this influence can be established only experimentally and is best expressed by an endurance diagram.

The influence of size on the endurance limits is as pronounced as its influence on the elastic limit, and the same size coefficient e_{sz}, determined by equation 5.8 and Table 5.1, can be used for each limit. For shafts over 75 mm in diameter, e_s may be taken equal to 0.75 until additional experimental data become available. The size correction is referred to the value of the stress amplitude S_a, and not to the whole endurance stress S'_{en}. The corrected stress amplitude is

$$S_a' = e_{sz} S_a \qquad \ldots (5.20)$$

Surface conditions: Experiments show that the nature of the surface has a very great influence upon the endurance strength of metals. The surface influences have greater effect for materials having high static ultimate strength. The highest values of endurance strength are obtained with a perfectly smooth polished surface. Slight scratches, such as produced by grinding, reduce the endurance strength by 5 to 12 per cent; a rough finish reduces it still more. A sharp, circular V groove only 1 mm deep may reduce the endurance strength of a shaft under certain circumstances

as much as 63 per cent. The curves in Fig. 5.7 show these influences for materials with various ultimate strength, the influence being expressed as a coefficient e_{sr}. This coefficient must be referred to the stress amplitude S_a and does not affect the mean stress S_m. The limit value of the stress amplitude thus becomes

$$S_a'' = e_{sz} e_{sr} S_a \qquad \ldots (5.21)$$

The values of e_{sr} from Fig. 5.7 can be used for bending, tension, and compression. For torsion the surface influence is smaller, and the corresponding value of e_{sr}' can be computed by the relation[7]

$$e_{sr}' = 0.425 + 0.575\, e_{sr} \qquad \ldots (5.22)$$

Superficial plastic distortions, such as cold-chisel marks, hammer blow marks, or centre punch marks do not affect the endurance appreciably; but any scratches, such as those that come from vise jaws or from press fits, are almost as harmful as sharp grooves, the surface coefficient for which is represented by curve d, (Fig. 5.7). Particularly detrimental is the corrosion effect, which is indicated by curves f and g.

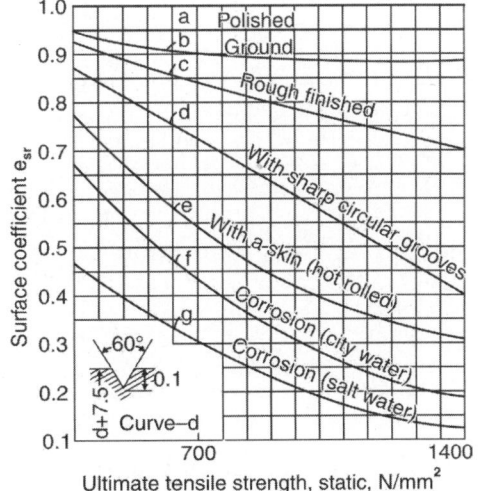

Fig. 5.7: Reciprocals of stress-concentration factors caused by surface condition

[7] Supplement to Z. VDI, Vol. 77, No. 42 (October 21, 1933), p. 4.

Surface rolling increases the endurance limit of mild annealed steel by 24 to 32 per cent.[8]

Discontinuities: Any discontinuity affects the endurance strength of a machine part. However, the effect is considerably less than indicated by the corresponding value of the form stress factor K. For repeated stresses this influence is expressed by the ratio S_{en}/S'_{en}, where S_{en} is the endurance limit without a discontinuity and S_{en}' is the endurance limit with the discontinuity. *This ratio may be called the endurance stress-concentration factor.*[9] As it has definite relation to the form stress factor K, but refers to repeated stresses, the designation K_r will be used. Thus,

$$K_r = \frac{S_{en}}{S'_{en}} \qquad \ldots (5.23)$$

5.9 STRESS AMPLITUDES

A more refined procedure may be used for parts in which the repeated stress is not completely reversed. This procedure is based on stress amplitudes.

Significant stresses: If the nominal stress amplitude S_{na} is multiplied by the stress-concentration factor K_r, the product is the significant stress amplitude; or

$$s_{sga} = K_r s_{na} \qquad \ldots (5.24)$$

Design stress: The design stress amplitude can now be determined as follows:

$$S_{da} = \frac{S''_a}{n} = \frac{e_{sz} e_{sr} S_a}{n} \qquad \ldots (5.25)$$

where the safety factor n should not be less than 1.25 for aircraft design and should be 1.5 or slightly greater in all other cases.

The actual safety factor for a part whose dimensions are established is found by considering the general definition and applying the equation

$$n' = \frac{e_{sz} e_{sr} S_a}{K_r s_{na}} \qquad \ldots (5.26)$$

The relations between the various stress amplitudes that must be considered in a machine part subjected to repeated stresses may be illustrated by Fig. 5.8. The difference Δs_a between the limit stress amplitude S_{la} and the significant stress amplitude S_{sga} represents the margin of safety of the design. This difference should be not less than $0.25\,S_{sga}$ for aircraft design and should be at least twice as much for all other cases.

Stress-load relations: For the sake of simplicity a case will be considered in which the piece is subjected to repeated direct loads, causing tension and compression. If the maximum load is F_1, the minimum load is F_2, and the cross-sectional area is A, then the nominal stresses are

$$s_{n1} = \frac{F_1}{A}, \quad s_{n2} = \frac{F_2}{A} \qquad \ldots (5.27)$$

The nominal stress amplitude is determined as follows:

$$s_{na} = \frac{s_{n1} - s_{n2}}{2} = \frac{F_1 - F_2}{2A} \qquad \ldots (5.28)$$

The maximum, or significant, stress amplitude, from equations 5.24 and 5.28, is

$$s_{sga} = K_r \frac{F_1 - F_2}{2A} \qquad \ldots (5.29)$$

If the value of s_{sga} given by equation 5.29 is equated to the value of S_{da} given by equation 5.25, the result is

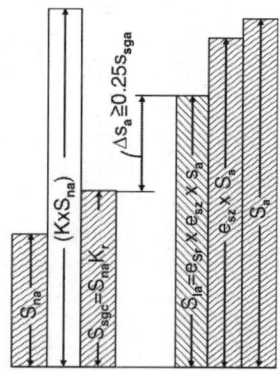

Fig. 5.8: Relations between various stress amplitudes

[8] O. J. Horger, "Increase of Endurance Strength of Steel," Journal of Applied Mechanics, Vol. 2, No. 4 (December, 1935), p. A-128.

[9] Seely, Advanced Mechanics of Materials, p. 229.

$$K_r \frac{F_1 - F_2}{2A} = \frac{e_{sz} e_{sr} S_a}{n} \qquad \ldots (5.30)$$

Hence the endurance stress amplitude is

$$S_a = nK_r \frac{F_1 - F_2}{2Ae_{sz}e_{sr}} \qquad \ldots (5.31)$$

Similarly, it can be established that the maximum mean stress, which is not affected by the surface factor, is

$$S_m = nK_r \frac{F_1 + F_2}{2Ae_{sz}} \qquad \ldots (5.32)$$

The ratio of the stress amplitude to the corresponding mean stress is

$$\frac{S_a}{S_m} = \frac{F_1 - F_2}{(F_1 + F_2)e_{sr}} \qquad \ldots (5.33)$$

Bending and torsion: In the case of bending, the stress amplitude can be expressed by the equation

$$S_a = nK_r \frac{M_1 - M_2}{2Ze_{sz}e_{sr}} \qquad \ldots (5.34)$$

where, M_1 is the moment producing the upper stress, M_2 is the moment producing the lower stress, and Z is the section modulus. Also, the maximum mean stress is

$$S_m = nK_r \frac{M_1 + M_2}{2Ze_{sz}} \qquad \ldots (5.35)$$

The ratio of the stress amplitude to the corresponding mean stress is

$$\frac{S_a}{S_m} = \frac{M_1 + M_2}{(M_1 + M_2)e_{sr}} \qquad \ldots (5.36)$$

In the same manner, for torsion,

$$S_a = nK_r \frac{T_1 + T_2}{2Z_o e_{sz}e_{sr}} \qquad \ldots (5.37)$$

where, T_1 is the torsional moment producing the upper stress, T_2 is the moment producing the lower stress, and Z_o is the polar section modulus.

Also, the maximum mean stress is

$$S_m = nK_r \frac{T_1 + T_2}{2Z_o e_{sz}} \qquad \ldots (5.38)$$

Then

$$\frac{S_a}{S_m} = \frac{T_1 - T_2}{(T_1 + T_2)e_{sr}} \qquad \ldots (5.39)$$

The values of S_a and S_m can be determined if the bending moments or torsional moment are used in place of the loads F_1 and F_2.

The endurance limit in bending is the value usually available, because the rotating beam method of testing is the simplest and most universally used. Although no real theoretical basis exists, correlation of data indicates that for polished bending specimens, there is a relationship between the endurance limit and the ultimate strength S_u such that for steel and other ferrous materials,

$$S_{en}(\text{bending}) = \left(\frac{1}{2} \text{ to } \frac{5}{8}\right)S_u$$

and for non-ferrous materials such as aluminium, brass and copper.

$$S_{en}(\text{bending}) = \left(\frac{1}{4} \text{ to } \frac{1}{3}\right)S_u$$

Since only relatively small portions of the volume and surface of a bending member are subjected to the maximum stress, where the entire volume and surface of an axially loaded member are fully stressed, it can be expected, from a statistical point of view, that the fatigue strength for repeated axial loads would be smaller than for repeated bending. Test results do not give conclusive verification, but, in general, they indicate that the following relation exists:

$$S_{en}(\text{axial}) = (0.7 \text{ to } 1.0)\,S_{en}(\text{bending})$$

It is advisable to use 0.7 to obtain endurance limit in axial loading.

In the case of repeated torsion loading, the test data are more conclusive. The relationship between endurance limit for repeated torsion and repeated bending is:

$S_{en}(\text{torsion}) = 0.55\,S_{en}(\text{bending})$ and for steel
$S_{en}(\text{torsion}) = 0.25\,S_u(\text{tension})$

5.10 ENDURANCE STRESS CONCENTRATION FACTOR

The magnitude of the stress-concentration factors K_r for endurance limits is found from tests, as shown by equation 5.23. These tests show that K_r depends on the discontinuity, on the form stress factor K, and on the material of which the specimen is made.

It will take many years to determine K_r for all materials and all possible types of discontinuities. In the meantime K_r may be determined with sufficient accuracy for practical purposes by the relation

$$K_r = 1 + q_r(K-1) \qquad \ldots (5.40)$$

The values of K and q_r are based on data already accumulated. The sensitivity index (see equation 5.10) is

$$q_r = \frac{K_r - 1}{K - 1} \qquad \ldots (5.41)$$

The index q_r is a measure of the sensitivity of the material to repeated stress in the presence of various discontinuities.

In Figs 5.9 and 5.10 values of K_r found experimentally for the bending of bars with transverse holes and for the bending of shafts with fillets are shown. These curves show the

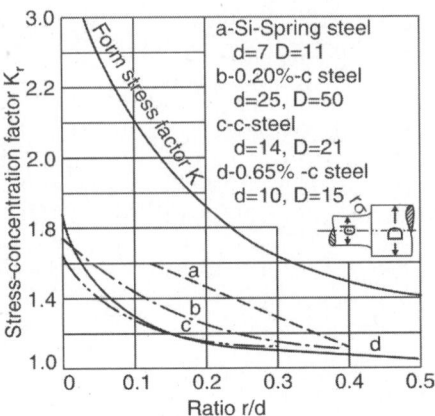

Fig. 5.10: Stress-concentration factors for shafts with fillets in bending

influence of the relative size of the discontinuity on the value of K_r.[10] The influence of the actual size of the tested bars and shafts is shown in Fig. 5.11. The curves in Fig. 5.12 show the influence upon K_r of the size of round bars tested in reversed bending and also of the fillet radius r and material. Curves a and b were obtained with specimens turned from steel having a carbon content of 0.45 per cent and normalized, whereas curve c was obtained with specimens made from Ni-Mo steel (C, 0.52; Ni, 2.96; Mo, 0.38; Mn, 0.68; Si, 0.19) heat-treated and drawn.[11]

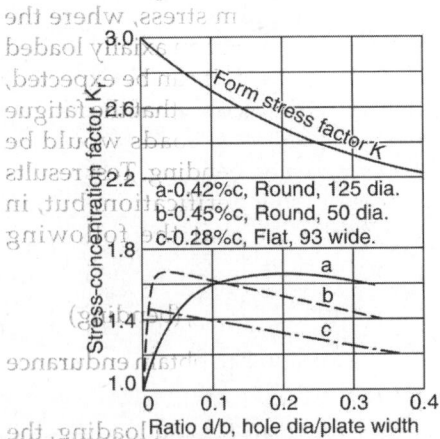

Fig. 5.9: Stress concentration factors for bars with transverse holes in bending

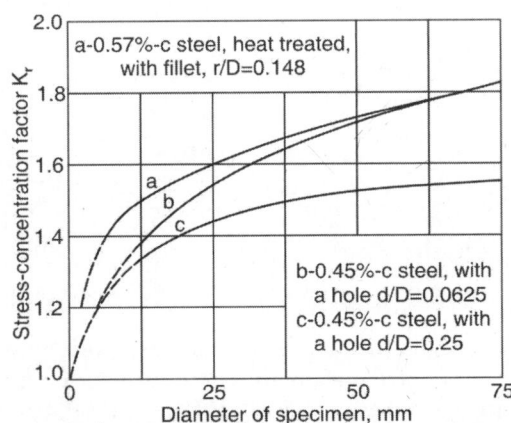

Fig. 5.11: Influence of size of hole on stress concentration

---

[10] R. E. Peterson, "Stress Concentration Phenomena in Fatigue of Metals," *Trans.* ASME, Vol. 55, APM–55–19 (1933), p. 157.

[11] R. E. Peterson and A. M. Wahl, "Fatigue of Shafts at Fitted Members with a Related Photoelastic Analysis, "JAM. Vol. 2, No. 1 (March, 1935), p. A-20.

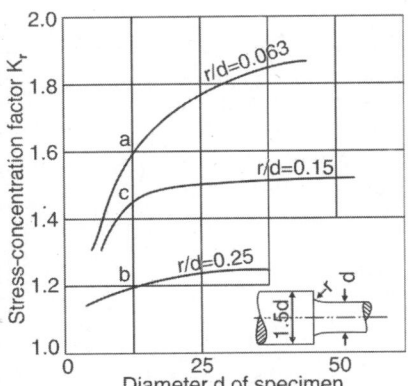

Fig. 5.12: Influence of size on stress concentration in bending

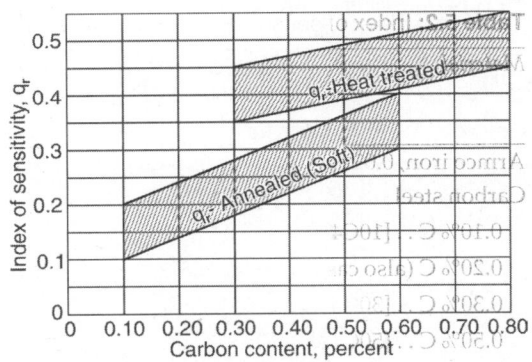

Fig. 5.14: Influence of carbon content in steel on index of sensitivity

By using data from Figs 5.12 and 3.19 and allowing a reduction of 20 per cent for bending, values of the corresponding sensitivity index q_r were computed by equation 5.41. The results are presented in Fig. 5.13. Although the curves of Fig. 5.13 should not be considered as giving accurate data for q_r because they cover a limited number of tests, they show the trends clearly and give at least the order of the values of q_r.

In Fig. 5.14 values of the sensitivity index are shown for carbon steels compiled from tests published in technical literature. The heavier, lower lines refer to average values, while the upper lines represent approximately maximum values. Finally, Table 5.2, compiled from the same sources, gives data for a wider range of

materials. These data should be considered only as a first approximation, the possible discrepancies being about 10 per cent. For this reason it is useless to segregate the influence of different types of discontinuities, such as transverse holes, grooves, or fillets, or for different types of load. Table 5.2 shows that for various material q_r ranges from zero up to 0.70. The index is smaller for ductile materials, and it increases as the ductility is decreased, as by heat treatment. Armco iron, however, is rather sensitive to stress concentration, in spite of its ductility.

Information from Figs. 5.13 and 5.14, combined with the data of Table 5.2, furnishes a fair basis for estimating q_r for any particular case.

Screw threads: The few data available for screw threads indicate high values for K_r. The following values are found by repeated tension:[12] 2.84 for an NC screw and 1.76 for a Whitworth screw, both made of 30C8 (SAE 1030) steel; 3.85 and 3.32 for the same threads, respectively, on screws made of (SAE, 2320) 20Ni3 heat-treated steel.

Keyways: Endurance tests with keyways cut with very sharp corners have shown that the stress-concentration factors are not high. For very ductile steel $K_r = 1.14$; for hard 65C8 [SAE 1065] steel, $K_r = 1.27$.[13] With a form stress factor $K = 4.8$, the corresponding values of the index

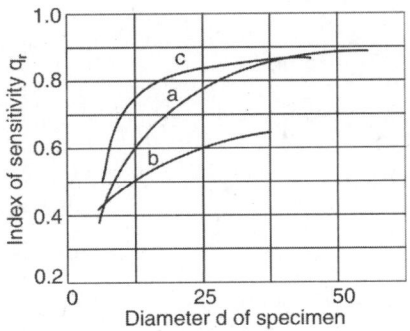

Fig. 5.13: Influence of size on index of sensitivity

[12] H. F. Moore and P. K. Henwood, The Strength of Screw Threads Under Repeated Tension, Bulletin No. 264, University of Illinois Experiment Station (1934), p. 10.
[13] Seely, Advanced Mechanics of Materials, p. 229.

Table 5.2: Index of sensitivity for repeated stresses

Material	Average index of sensitivity q_r		
	Annealed of soft	Heat-treated and drawn at 650°C	Heat-treated and drawn at 480°C
Armco iron, 0.02% C	0.15–0.20	–	–
Carbon steel			
0.10% C .. [10C4]	0.05–0.10	–	–
0.20% C (also cast steel) . [20C8]	0.10	–	–
0.30% C .. [30C8]	0.18	0.35	0.45
0.50% C .. [50C12]	0.26	0.40	0.50
0.85% C .. [85C6]	–	0.45	0.57
Spring steel, 0.56% C, 2.3 Si, rolled	–	0.38	–
SAE 3140, 0.37 C; 0.6 Cr; 1.3 Ni. [35 Ni1Cr60]	0.25	0.45	–
Cr-Ni steel, 0.8 Cr; 3.5 Ni	–	0.25	–
Stainless steel, 0.3 C; 8.3 Cr; 19.7 Ni	0.16	–	0.70
Cast iron	0–0.05	–	–
Copper, electrolitic	0.07	–	–
Duraluminium	0.05–0.13	–	–

of sensitivity q_r are 0.04 and 0.07, respectively. The width of the keyway does not affect K_r.

Other tests have shown that sled-runner keyways, cut with an ordinary milling cutter having the same width as the keyway, weaken shafts much less than profiled keyways cut by a cutter having a diameter equal to the keyway width. In these tests, failure occurred at the end of the keyway.[14]

Press and shrink fits: When a shaft is pressed or shrunk into a hub, the compressive stress in the shaft increases near the end of the hub, as indicated by curve *ab* in Fig. 5.15, because of the resistance of the fibers of the uncompressed end of the shaft. The resulting stress-concentration factor K_r may reach a considerable magnitude, as shown in Fig. 5.16.[15] With a mild-steel shaft shrunk into a hardened hub or bushing, K_r may become as high as 2.12. At the same time the surface of a highly stressed shrink fit has a tendency to oxidize. Oxidation, in turn, causes rust, and its action increases

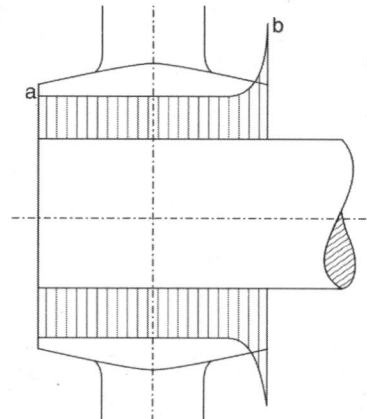

Fig. 5.15: Press fit stress-concentration

progressively with rust. *In a part subjected to repeated stresses rusting must be prevented by all available means, such as applying rust-preventing grease, making frequent inspections of exposed place, and cleaning such place and covering them with oil paint or grease (5.6, 5.7).*

---

[14] Peterson, loc. cit., p. 161.

[15] A. Thum and F. Wunderlich, "Deer Einfluss von Einspann–und Kraftagriffstellen auf die Dauerhaltbarkeit der Konstructionen:, Z. VDI, Vol. 77 (1933), p. 851.

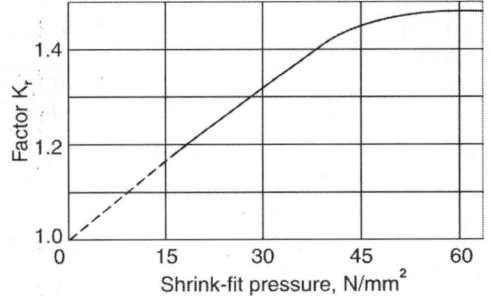

Fig. 5.16: Stress-concentration factor due to shrink-fit pressure

5.11 INCREASE OF ENDURANCE STRENGTH

Since the stress-concentration factor K_r is a function of the form stress factor K, a lowering of the latter will lower K_r and will thus increase the strength of the piece. Another way of neutralizing or diminishing the localized stresses is to set up stresses of opposite action.

Methods to increase endurance strength are given below:

1. Reduction in form-stress factor: The effect of a hole in a disk can be decreased by a symmetrical reinforcing bead, as shown in Fig. 5.17. The reduction of the form stress factor K is a function of the ratio of the cross-sectional area of the bead proper, which is $(d_2 - d)(H - h)$, to the area of the diametrical section taken out

by the hole, or hd. Designating the ratio $(d_2 - d)(H - h)/hd$ as a, and plotting the values of K against those of a, curve c is obtained. This shows that with $a = 1$ the factor K can be reduced to about one-half of its maximum value of 3.[16]

The problem of reducing the magnitude of the form stress factor is greatly simplified by using the method of force-flow lines as indicated in section 3.9.

2. Holes: Stress concentration due to a hole in a tension member, such as that in Fig. 5.18a, can be reduced by drilling additional holes which will give smoother force-flow lines, as indicated in Figs 5.18b and c. A series of holes instead of one hole or a series of grooves instead of a single groove, reduce stress concentration factor. Parts subjected to *bending can have preferably holes located near* the neutral axis.

The injurious effect of a transverse hole in a shaft subjected to bending or torsion can be reduced by filing tangential notches near the edge of the hole, as shown in Figs 5.19a at c. This spreads the stress concentration over a larger section and, by decreasing its intensity, *reduce K and K_r. Pressing in the stress-relieving grooves instead of filling them gives still better results. A similar effect is obtained by drilling holes crosswise, as in Figs 5.19b at e.*

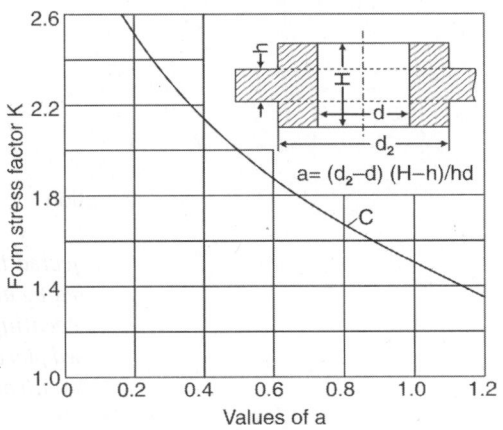

Fig. 5.17: Reinforcing bead to lower stress concentration

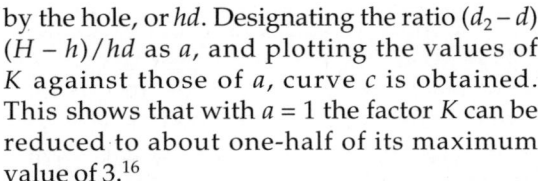

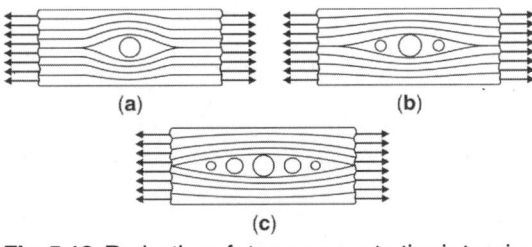

(a)   **(b)**

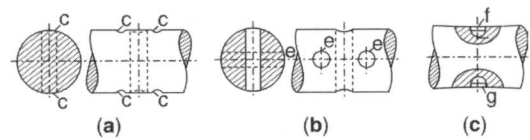

(c)

Fig. 5.18: Reduction of stress concentration in tension

(a)   **(b)**   **(c)**

Fig. 5.19: Reduction of stress concentration in bending

---

[16] S. TImoshenko and W. Dietz, "Stress Concentration Produced by Holes and Fillets," Trans ASME, Vol. 47 (1925), p. 207.

3. *Asymmetry:* Stress concentration is also caused by unsymmetrical shapes. The endurance strength of a shaft with one radial hole, as in Figs 5.19c at *f*, can be raised by drilling another hole *g* symmetrical with the hole *f*.

4. *Fillets: The best method of lowering stress concentration near a fillet is to increase the radius of the fillet.* The advantage of a larger fillet is illustrated in Figs 5.20a and b. If the fillet radius cannot be increased in the usual way, the fillet may be cut into the shoulder, as in Fig. 5.20c, or even in the side of the smaller shank, as in Fig. 5.20d. Another procedure is to smoothen the force-flow lines by cutting notches, as in Fig. 5.20e, or by drilling holes, as in Fig. 5.20f. *Fillets of hyperbolic shape at shoulders cause less stress concentration (5.9).*

In Fig. 5.21a a valve of an internal-combustion engine is shown that broke at *f* by pro-gressive fracture because of the stress concentration due to an insufficient fillet radius *r*. The trouble was overcome by increasing the fillet radius *r*, as shown in Fig. 5.21b. The material from the top is taken out to obtain a more uniform thickness and thus to reduce stress concentration. In Fig. 5.21c is shown another solution which makes it possible to bring the top of the guide *g* closer to the valve seat.

Stress concentration due to square shoulders in a shaft of 80C6 (SAE 1080) steel having the dimension shown in Fig. 5.22a lowered the endurance strength from $S_{en} = 330$ N/mm^2 to 160 N/mm^2 and thus gave a stress-concentration factor $K_r = 2.09$. A fillet with a 6 mm radius, as in Fig. 5.20b, raised S_{en} to 300 N/mm^2 and gave $K_r = 1.09$; and a 25 mm radius as in Fig. 5.22e, raised S_{en} to 327 N/mm^2 and gave $K_r = 1.01$.[17] For ingot iron the values of K_r were 1.86 for a square shoulder and 1.18 for a 6 mm radius.

If a shaft must be turned down to a smaller diameter, as in Fig. 5.23a and the radius of the fillet cannot be increased, stress concentration can be reduced in several ways.[18] A flat groove reducing the larger shaft diameter from 22 to 20 mm, as in Fig. 5.23b, increases the endurance limit by 13.5 per cent. A slightly increased fillet radius with a slight reduction of the small shaft diameter, as in Fig. 5.23c, gives an even greater

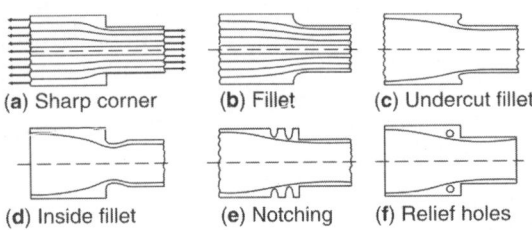

(a) Sharp corner   (b) Fillet   (c) Undercut fillet

(d) Inside fillet   (e) Notching   (f) Relief holes

Fig. 5.20: Test samples of steel of steel for determining stress concentration factors

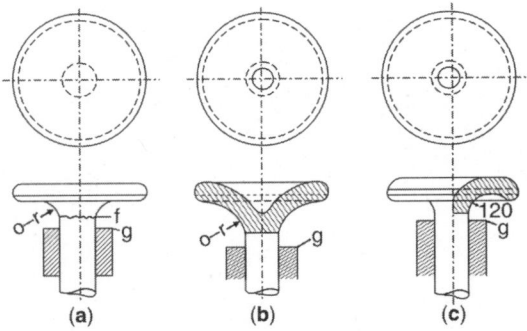

(a)   (b)   (c)

Fig. 5.21: Exhaust valve for an internal combustion engine

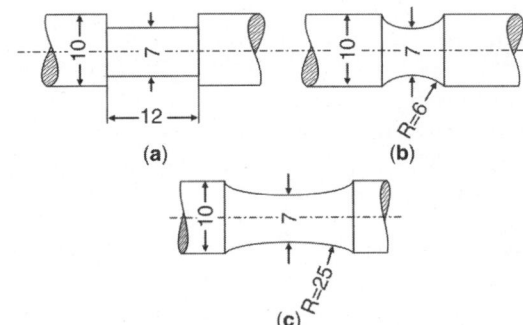

(a)   (b)

(c)

Fig. 5.22: Test samples of steel for determining stress concentration factors

[17] Battelle Memorial Institute, op. cit. p. 59.

[18] O. J. Horger and T. V. Buchwalter, "How to Increase Endurance Strength," Product Engineering, Vol. 12, No. 2 (February, 1941), p. 78.

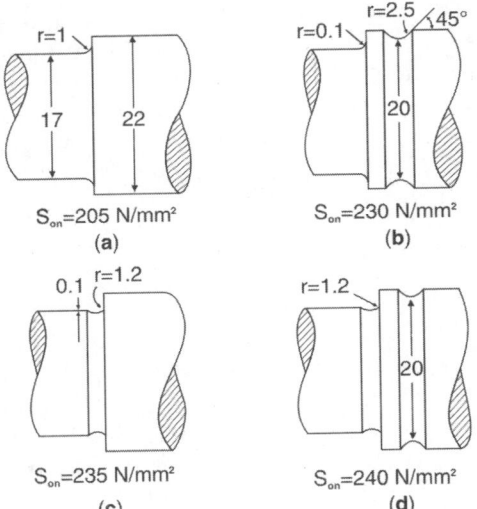

Fig. 5.23: Methods of increasing endurance of a shaft in bending

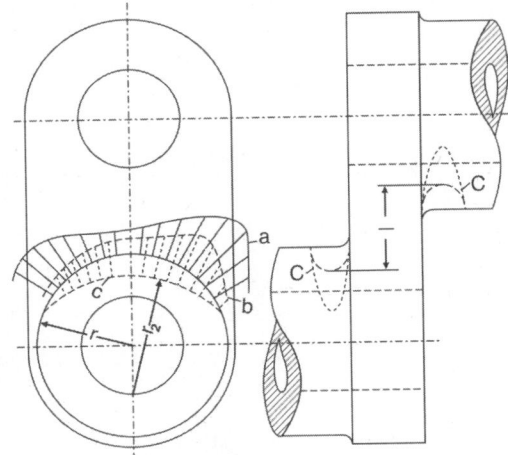

Fig. 5.24: Reducing stress concentration by removing material

endurance increase (14.8 per cent), probably because of a lower rigidity of the shaft juncture. As indicated in Fig. 5.23d, a combination of both methods shown in Figs. 5.25b and c gives only a slight additional increase (to 15.8 per cent). The explantation of the beneficial effect of the additional grooves may be found by visualizing the force-flow lines in a shaft subjected to bending or in concentric hollow shafts subjected to torsion.

In another case the stress concentration was lowered by cutting eccentric circular grooves to the depth c, Fig. 5.24.[19] This operation removed the most highly stressed material and increased the elastic length l. The torsional shear stresses that existed before the grooves were cut are shown in a polar diagram by the curve a, and those that occurred after the grooves were cut are shown by the dotted curve b. This method could be applied in this particular case because the crankshaft, originally designed for plain bearings, was used with ball bearings.

5. *Decrease of local stresses:* Parts fractured through repeated loading show that the fracture always starts at the point where the tensile stress has a maximum value.[20] Any process that can lower the tensile stress in the outer fibers will reduce the stress-concentration factor K_r and will increase the endurance strength of the piece.

The injurious effect of notches and circular grooves in a piece subjected to repeated loading can be reduced by prestretching the piece so as to produce a permanent set in the groove or notch. As a result, internal tensile stresses will be set up in the center of the piece, causing compressive stresses in the fibers next to the notch, as shown by curve a, in Fig. 5.25. The distribution of stresses due to tension is represented by curve b, and the redistribution after a small permanent set in compression is created in the outer fibers is shown by curve c. The tensile stress near the notch is reduced, and the strength of the section is thus increased. Putting a plastic deformation into the notch by local pressure accomplishes the same result. Thus the beneficial effect of the equalization grooves c, Fig. 5.19a, is materially increased if the grooves are pressed in instead of being filed.[21]

[19] O. Dietrich and E. Lehr, "Das Dehnungslinienverfahren," Z. VDI, Vol. 76 (1932), p. 981.

[20] J. O. Almen, "Shoot Peening to Increase Fatigue Resistance," SAE Journal, Vol. 51, No. 7 (July, 1943), p. 248.

[21] H. Ude, "Steigerung der Dauerhaltbareit der Konstructionen," Z. VDI, Vol. 79 (1935), p. 48

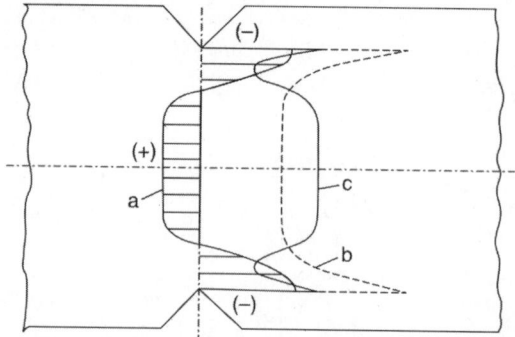

Fig. 5.25: Stress redistribution after a permanent set

6. *Cold working:* The endurance strength of a machine part can be increased if a thin layer of the part, subjected to tensile stress during its operation, is prestressed in compression by a cold-working operation such as peen hammering, swaying, shot blasting, or pressing by balls or rollers. Such an increases in endurance strength is shown by the *S-N* curves in Fig. 5.26.[22] Tests show that prestressing the surface in compression increases the endurance strength, regardless of whether the prestress is applied to highly finished parts or to parts with rough surfaces and regardless of whether the surface is soft or hard, as when it is case-hardened. The decrease of the slope of the *S-N* curves gives a measure of the increase in endurance strength.

Springs are shot preened and gear fillets are pressed with rollers to induce compression stress, thereby the endurance strength is increased.

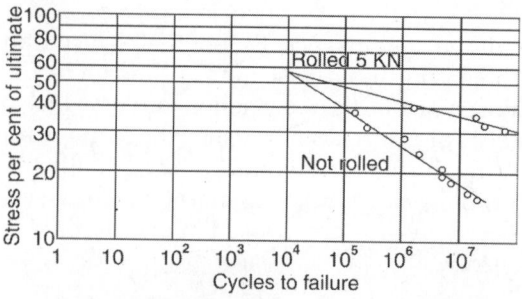

Fig. 5.26: Effect of rolling on railway axles

[22] Almen, loc. cit., pp. 249–52.

The amount of prestressing and the stress magnitude and depth depend on the size of the part, the material, and the hardness of the surface. Excessive prestressing may decrease the endurance strength, as may be seen from Fig. 5.27, which gives only a qualitative picture. The proper amount of prestressing for any particular case must be found experimentally.

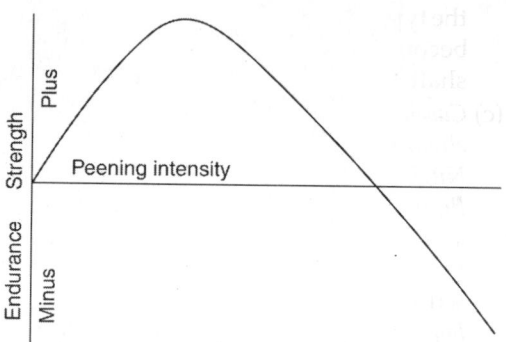

Fig. 5.27: Effect of peening

Railroad car axles are surface rolled over the region where the wheel is pressed on to induce compression stress to reduce stress concentration due to shrink fitting.

7. *Other methods of reducing stress concentration:*
(a) Shrink-fit stress concentration can be materially reduced by making the gripping part conical. The elasticity of material, even in a cast-iron hub, gives a gradually increasing shrink pressure. Such a change in design may lower K_r from 1.87 in a hub of the type shown in Fig. 5.28a to 1.38 in a hub of the type shown in Fig. 5.28b. Still better is a parabolic hub, as shown in Fig. 5.28c.
(b) Another method of lowering the stress-concentration factor is to put a compressive stress in the outer fibers of the shaft by cold rolling before shrinking the hub on. If a shaft with an ordinary hub of the type shown in Fig. 5.28a is given such a plastic deformation and is afterward ground to the proper shrink dimension, the stress concentration factor may be as

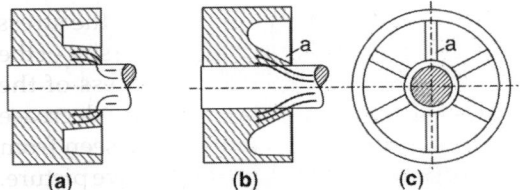

Fig. 5.28: Press-fit stress concentration

low as 1.07. A shaft with a conical hub of the type in Fig. 5.28b that is so treated may become even stronger than an untreated shaft without stress concentration.

(c) *Casehardening a mild-steel shaft practically eliminates the shrink-fit stress concentration. Nitriding does it entirely, and it also eliminates the injurious pressure oxidation that occurs in a press fit. Nitriding greatly reduces the injurious effect of notches and grooves by setting up compressive stresses in the surface layers similar to those caused by peening or some other cold-working process (5.2).*

(d) *Oxidation stress concentration can be decreased by grinding off the outer fibers affected by oxidation to a depth of 0.1 to 0.2 mm (5.6).*

(e) *Corrosion effect upon steel is reduced very materially when the piece is sherardized or cadmium-plated. Nitriding eliminates corrosion effect entirely. The reduction of endurance limits caused by corrosion effect is less for stainless steels than for other steels (5.7 and 5.8).*

5.12 VIBRATION

If an outside force changes the shape of a body without producing stresses beyond the elastic limit, it invokes restoring, internal forces which will act as soon as the outside force ceases to act. The body will return to its original shape, but because of its mass it will pass through and beyond the position of equilibrium and will create a restoring force in the opposite direction. These pendulum-like movements through the position of equilibrium will continue until the energy imparted by the outside force is absorbed by being damped out by internal friction. A series of such movements is called *vibration or oscillation.*

Vibration may be caused by longitudinal forces, by transverse forces, or by restoring torsional moments. Forces that cause vibration of certain parts of a machine or of an adjoining structure usually are of two kinds: first, those due to the inertia of reciprocating parts; second, those due to centrifugal forces created by unbalanced revolving parts (5.11).

The *natural period of vibration* is the time T, in seconds, during which a body or system, set into free vibration, will complete one cycle. The reciprocal of the natural period of vibration is called the *natural frequency.* This is designated as *f* and is expressed in vibrations per second. The natural period of a system depends on the mass of the system and on the force required to produce a unit deflection. It can be computed by the relation

$$T = 2\pi \sqrt{\frac{W}{Fg}} \qquad \ldots (5.42)$$

where, W is the weight of the oscillating system, in Newton

F is the force necessary to deflect the body in N/m

g is the acceleration of gravity, which is 9.81 m/sec/sec.

If the deflection produced by the load W is designated by y, then y = W/F and the frequency becomes

$$f = \frac{1}{T} = \frac{1}{2\pi} \sqrt{\frac{g}{y}} \qquad \ldots (5.43)$$

Resonance: If a body is set in vibration by a force recurring at regular intervals, and if these intervals happen to be equal to the natural period of the body or to some simple fraction or a multiple of that period, then the amplitude of vibration of the body may gradually increase, even though the force is small. Such conditions produce what is termed resonance, and the resulting large deflections may create dangerously high stresses in the parts involved.

Vibrations that have the same natural period of vibration and therefore the same frequency are called *synchronous vibrations.*

The critical speed of a machine with respect to a certain part is the speed at which synchronous vibrations are developed in the part by resonance.

Resonance may produce synchronous vibrations in long rods, crankshafts, springs, brackets, beams and in the whole structure supporting a piece of machinery resulting in failure of the element (5.12).

Synchronous vibrations can be developed in a body not only at the main critical speed but also at one-half, one-third, or some other simple fraction of the main critical speed. In this event each such fraction of the main critical speed also becomes a critical speed. However, such critical speeds are less dangerous than the main critical speed, because they are set up by harmonics of higher orders, which have smaller amplitudes, and the impulses are therefore smaller.

In order to avoid critical speeds in a machine resulting in resonance, the natural frequency of vibration of every part should be considerably higher than the number of impulses which the part receives during the operation of the machine (5.13).

Vibration damping: The rate of absorption of the energy of vibration of a part depends on the structure of the material, which governs its inner friction. Cast iron has a higher damping capacity than steel, which fact makes cast iron a more suitable material for parts where resonance may be expected. High-carbon steels and alloy steels have a greater damping capacity than low-carbon steels; electron, a magnesium alloy, has the smallest damping capacity known.[23]

Damping does not affect the natural frequency of a part; it affects only the amplitude of the vibration.

Engine supports: In a high-speed engine operating at variable speeds, the range of speeds may be so great that it becomes very difficult, and sometimes impossible, to avoid synchronous vibration of the supports at all speeds. Often there are several critical speeds. *In this case the transmission of vibration impulses can be stopped by inserting flexible members, such as rubber pads or springs, between the engine and the supporting structure (5.14).*

Rigidity: According to equation 5.43, the natural frequency of a body or system can be raised by decreasing the deflection *y*; that is, by making it more rigid. In the case of a structure formed by bolting together several pieces, the presence of working strips *c*, Fig. 5.29a, makes the lever arm of the resistance to the bending moment *Fl* greater than the arm with an unrelieved contact area, as in Fig. 5.29b. The contact in Fig. 5.29b can be considered, with exaggeration, as that of two cylindrical surfaces. The numerical change in the natural frequency can be computed if the conditions are known.[24]

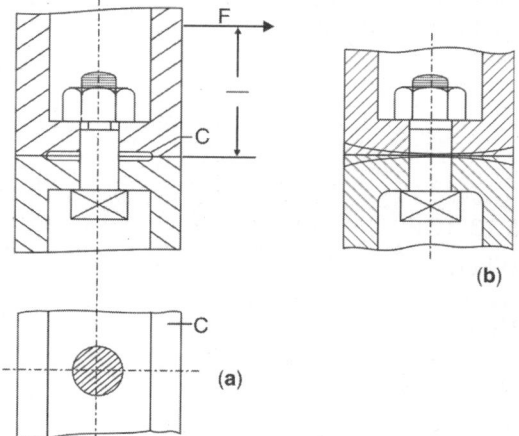

Fig. 5.29: Rigidity of junctures

Example 5.6: A small steam turbine operates at speeds from 1000 to 4800 rpm. The turbine weighs 1100 N and fastened to two standard channels 125 mm, 133.4 N laid flat side up. The distance between the supports of the channels used as simple beams is 0.66 m. Determine the possibility of resonance and the corresponding critical turbine speed.

If the channels are considered simply supported at the ends and loaded uniformly over the full length, the deflection is

$$y = \frac{5Fl^3}{384\,EI}$$

[23] A. Esan and H. Kortum, "Die Veränderlichkeit der Werkstoffdämpfung," Z. VDI, VOl. 77 (1933), p. 1133.
[24] K. Schönfelder' "Die Bedeutung der Arbeitsleisten in Teilfungen für die Steifgkeit der Konstructionen," Z. VDI, Vol. 77 (1933), p. 1070.

The load on each beam is

$$F = \frac{1}{2} \times 1100 + 0.66 \times 133.4 = 638 \text{ N}$$

$I = 26.112 \text{ cm}^4$. Therefore,

$$y = \frac{5 \times 638 \times (0.66)^3 \times (100)^4}{10^6 \times 384 \times 210 \times 10^3 \times 26.112} = 0.0000434 \text{ m}$$

The frequency of vibration found by equation 5.43 is

$$f = \frac{1}{2\pi} \sqrt{\frac{9.81}{0.0000434}} = 75.67 \text{ cycles/sec.}$$

The number of cycles or vibrations per minute = 75.67 × 60 = 4540. Therefore the critical turbine speed would be about 4540 rpm. The channels are too flexible. By turning them 90°, the moment of inertia would be increased to $I = 356.9 \text{ cm}^4$ and the frequency would be increased to the safe value

$$f = 4540 \sqrt{\frac{356.9}{26.112}} = 16785 \text{ vibr per min}$$

5.13 TORSIONAL VIBRATION

When a couple twists a shaft and then ceases to act, the internal stresses will bring the shaft sections back beyond their positions of equilibrium and will twist the shaft in the opposite direction because of the inertia of the shaft. The shaft will thus be twisted back and forth with respect to the centerline of the shaft. A series of such angular movements of one section relative to another is called torsional vibration. Torsional vibration is created in a crankshaft of a reciprocating engine because of the periodic impulses to which it is subjected; and it is created in a straight shaft when the shaft is either coupled to such a crankshaft or otherwise subjected to periodic torque variations.

Critical speeds: The natural frequency f_n of torsional vibration of a shaft is a function of its dimension, of the characteristics of its material, and of the size and arrangement of the masses attached to it. If f_n is an even multiple of the frequency f_f of the disturbing forces which come at regular time intervals, that is, if f_n / f_f is whole number, a condition of resonance will exist. The amplitude α of vibration, expressed in radians, may be found by the relation

$$\alpha = \frac{Q}{T_m \left(1 - \dfrac{f_f}{f_n} \right)} \qquad \dots (5.44)$$

where Q is the disturbing torque and T_m is the shaft constant, or the torque which will twist the shaft 1 radian.

The critical speed occurs when $f_f = f_n$. Theoretically, α could then become infinitely large, as shown in Fig. 5.30; but actually it is stopped by damping due to internal friction of the material. However, if the amplitude—even with damping—is large, torsional stresses may exceed the endurance limit and the shaft may fail by progressive fracture.

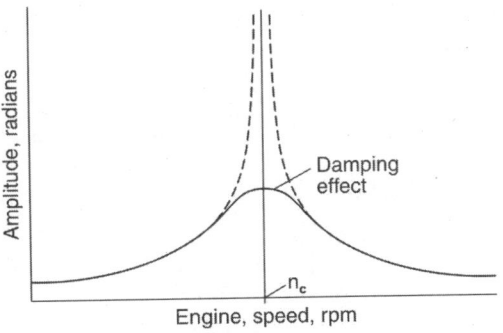

Fig. 5.30: Damping of vibration

Node: That section in which the torsional stresses reach their maximum value rotates without oscillation and is called the *nodal section* or the *node*.

The angular deflections are measured with reference to the node. The location of the node depends on the inertia masses causing torsional vibration. If one rotating mass is fastened to the shaft, the shaft will have a node at section *n-n*, Fig. 5.31, where the most distant of the tangential forces acts. If there are several inertia masses, as in Fig. 5.32, the node will be located somewhat between the first and last rotating masses. The location of the node can be found from the equation of dynamic equilibrium, the flywheel effects of the rotating masses being considered as though they were weights and the node being located at the center of gravity of the system.[25]

For three rotating masses, as a flywheel and two cranks shown schematically in Fig. 5.32, the equation of equilibrium is

$$I_1 c = I_2(l_1 - c) + I_3(l_1 + l_2 - c)$$

The distance c to the node from one end is

$$c = \frac{I_2 l_1 + I_3(l_1 + l_2)}{I_1 + I_2 + I_3} \qquad \dots (5.45)$$

Both equations 5.45 and Fig 5.32 show that a node divides a shaft into two parts, each behaving as a shaft with one oscillating mass.

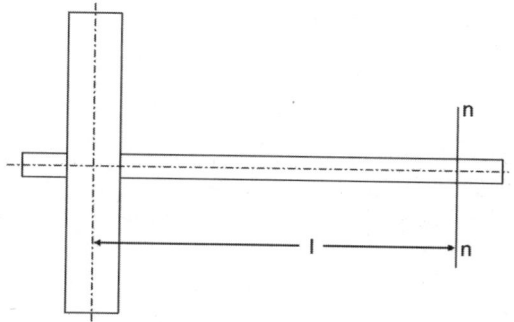

Fig. 5.31: Node location with one mass

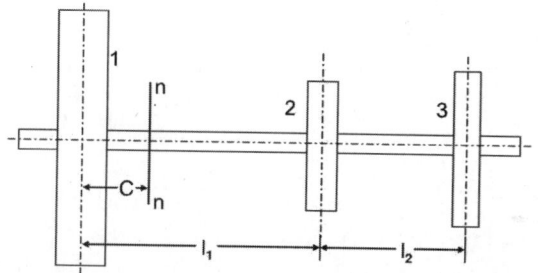

Fig. 5.32: Node location with several masses

Frequency: The natural period of torsional vibration T, in seconds, can be determined by a formula similar to equation 5.42.

Thus,

$$T = 2\pi\sqrt{\frac{I}{T_m}} = 2\pi\sqrt{\frac{Wk_o^2 l}{gGJ}} \qquad \dots (5.46)$$

where, I is the moment of inertia of the rotating mass, which is Wk_o^2/g Nm.sec^2;

T_m is the shaft constant, or the torque which produces in the oscillating shaft a twist of 1 radian in a length of 1 m in Nm.

W is the weight of the rotating mass, in N.

k_o is the radius of gyration of this mass, in m

l is the oscillating length of the shaft, in m

g is the acceleration due to gravity, which is 9.81 m/sec^2

G is the modulus of elasticity in shear, in N/m^2

J is the polar moment of inertia of the shaft section, in m^4.

When equation 5.46 is applied to a shaft with several rotating masses, the length l must be considered as the distance from the application of the inertia mass to the node for each mass. The equation for the period becomes

$$T = 2\pi\sqrt{\frac{\Sigma(Wk_o^2 l)}{gGJ}} \qquad \dots (5.47)$$

When the constant coefficients have been combined, the frequency is

$$f = 1/2\pi\sqrt{\frac{gGJ}{\Sigma(Wk_o^2 l)}} \qquad \dots (5.48)$$

Because of the approximate nature of equation 5.45, *the values of T and f determined by equations* 5.47 *to* 5.48 *are also only approximate,* being within 3 to 4 per cent of the actual values.

A shaft or a crankshaft rigidly connected to other shafts with rotating parts, such as generator rotors, pump impellers, or pulleys, cannot be considered alone; it must be considered as a part of the whole system of interconnected shafts and rotating masses. On the other hand, a flexible coupling connecting two shafts divides them into two separate systems as far as vibration is concerned.[26]

---

[25] H. F. P. Purday, Diesel Engine Design (London: Constable & Company, Ltd., 1928), p. 108. This method gives only approximate values and is used because of its simplicity. Accurate methods involve lengthy calculation and may be found in special books on vibrations.

[26] A comprehensive treatment of the subject may be found in W. K. Wilson, Practical Solution of Torsional Vibration Problem (New York: John Wiley & Sons, Inc., 1942).

The determination of the frequency of torsional vibration of a shaft systems, as well as the elimination of resonance, will be discussed in greater detail in connection with the design of crankshafts.

5.14 TEMPERATURE EFFECTS

The effects of temperature and particularly of its variation upon metals are very numerous and important for the design of certain machine parts. These effects may be divided into two groups: those produced by temperature changes during the various manufacturing processes through which a part goes; and those caused by temperature variations in a part during operation of the machine. Although the first group is very important, its discussion belongs to courses in metallurgy and technology of metals. Only the second group will be considered here.

The main effects of temperature changes in machine parts are:

(a) Expansion with the increase of temperature
(b) Stresses due to temperature differences
(c) Deterioration of surfaces
(d) Change of ultimate strength
(e) Change of elastic and endurance limits
(f) Change of the modulus of elasticity
(g) Change of hardness
(h) Change of ductility and impact strength
(i) Creep phenomenon
(j) Growth or bulk change

(a) *Expansion:* The coefficient of thermal expansion increases with the temperature; but for ordinary ranges of temperature, mean values such as those given in Table 5.3 may be used. The length l at a temperature t is

$$l = l_o [1 + \alpha t] \qquad \dots (5.49)$$

where, l_o is the length at 0°C and α is the mean linear coefficient of expansion.

At higher temperatures it is better to use more accurate data and to compute the length from the equation

$$l = l_o [1 + at \times 10^{-3} + bt^2 \times 10^{-6}] \quad \dots (5.50)$$

where values of the coefficient a and b are given in Table 5.3.

Expansion due to a temperature increase must be taken into consideration in designing parts of various heat engines, such as piston rods, crankshafts, and bedplates, and particularly in determining the diameters of the pistons of steam engines and internal combustion engines.

(b) *Temperature stresses:* The stress s set up by a change of temperature from t_1 to t_2 if the part cannot change its length is, according to equation 2.8,

$$S = \alpha(t_1 - t_2) E \qquad \dots (5.51)$$

If $t_1 < t_2$, the stress is negative, or compression; if $t_1 > t_2$, it is positive, or tension. As can be readily seen, even a moderate change of temperature may result in very high stresses.

This restriction of change of length exists in a casting with sections of varying thickness.

Table 5.3: Heat properties of metals

Metal or alloy	Mean coefficient of linear expansion between 75 and 260°C $\times 10^6$	Coefficient in equation 5.51		Temperature range (°C)	Average coefficient of heat conductivity k W/m²°C	
		$a \times 10^3$	$b \times 10^3$		At 20°C	At 100°C
Aluminium alloy	25.4	22.64	9.72	610	186	192
Brass, yellow	18.7	16.92	4.37	590	89	107
Bronze, aluminium	17.6	16.70	3.95	590	89	102
Cast iron, gray	11.3	9.79	5.67	590	49	46
Copper	17.8	16.70	4.02	680	385	380
Malleable iron	12.8	11.70	5.25	700	69	58
Monel-metal	14.9	13.86	3.95	590	26	26
Nickel	14.6	13.77	3.30	1000	60	58
Steel, mild	11.7	10.44	4.53	760	45	44
Steel, ingot	12.6	11.18	5.25	750	48	47

The smaller ones cool faster than the larger ones, and the stresses thus set up may cause a fracture even without an external load. Such temperature stresses can be eliminated, or at least reduced, by a uniform heating and cooling of the casting, and also by a careful design that avoids nonuniform and rigid sections as much as possible.

Temperature stresses of a second type are set up when the temperature at different points of a part are not the same and the part cannot change its shape, as in a cylinder liner of an oil engine with hot gases inside and cooling water outside. In such a case the stress may be determined from equation 5.51, but the meaning of the terms will be slightly different. Under these conditions t_1 and t_2 will designate the coexistent temperatures at certain points, and s will be the difference between the stresses at these points. The existence of a temperature difference $(t_1 - t_2)$ is possible only because of a continuous flow of the heat from the point with the highest temperature to those with the lowest one. The relation between the flow of heat through a wall and the temperature difference may be expressed by the equation

$$Q = A(t_1 - t_2)\tau\frac{k}{b} \qquad \ldots (5.52)$$

where, Q is the flow of heat, in kCal per hour,
A is the area through which Q flows, in square m
τ is the ratio of time of heat flow to total time
k is the heat conductivity, in Cal per hr per sq m per deg C
b is the thickness of the wall, in m

Combining equations 5.51 and 5.52, to eliminate $(t_1 - t_2)$, gives

$$s = \frac{\alpha EQb}{A\tau k} \qquad \ldots (5.53)$$

This equation shows that for given conditions fixing Q, A, and τ and for a certain material with known properties α, E, and k, the stress is directly proportional to the thickness b of the wall. Thus if stress is too high, an increase of the dimension b will only make the wall weaker because of increase of stress. If other forces acting on the wall set up a tensile or compressive stress s_2 which decreases with an increase of b, then it is necessary to find the thickness b for which the algebraic sum $(s + s_2)$ is least. If $(s + s_2)$ is greater than the allowable stress for the selected material, then the next procedure is to select a material with a higher k or a lower α. When temperature stresses are present, it is better to use ductile metals. In such metals excessive local stresses are not so dangerous as they are in brittle metals, and the excess may disappear after the metal assumes a permanent set in the overstressed section.

(c) *Deterioration of surfaces:* In parts subjected to high and fluctuating temperatures, such as the tops of the pistons in internal combustion engines, if proper heat dissipation does not take place temperature stresses near the surface may exceed the elastic limit. Then destruction will start in the shape of minute cracks, which gradually increase in size and number. As the cracks deepen, the strength of the part is decreased until its failure occurs.

Often such deterioration cannot be eliminated and can only be slowed down by using higher-grade materials and by making the piece which is subject to deterioration a separate and easily replaced part or insert.

(d) *Ultimate strength: The ultimate strength of most metals and alloys increases slightly as the temperature, however, the strength decreases under all loads—tension, compression, bending, or torsion (5.16).*

(e) *Elastic limits and endurance limits: The decrease of elastic limits with increase in temperature is more pronounced than that of strength, as may be seen from Fig. 5.33. The endurance limits decrease approximately at the same rate as the elastic limits (5.16).*

(f) *Modulus of elasticity:* The modulus of elasticity decreases with elevated temperatures, very slowly up to 260°C and thereafter progressively faster because of creep.

(g) *Hardness:* An increase of the temperature of a metal part up to 150°C does not affect its hardness. With a further temperature increase the hardness begins to decrease, at first rather slowly. The Rockwell C hardness of

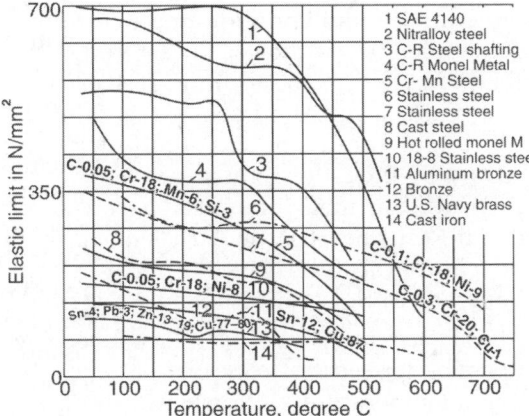

Fig. 5.33: Influence of temperature on elastic limits of metals

heat-treated steels used in ball bearings decreases from 66 to 62 at 315°C. A fully hardened 50C4 (SAE 1050) steel has a Rockwell C hardness of 55 at 200°C, 51 at 315°C, and about 47 at 425°.

(h) Impact strength: In Fig. 5.34 the change of impact strength of various steels is shown at comparatively low temperatures due to the change that takes place in the ductility.[27] Curve *a* is for high-carbon steel; *b* is for medium-carbon steel; *c* is for low-carbon steel; and *d* and *e* are for normalized nickel-chrome steel,

d being for specimens prepared for large ingots and *e* for lengthwise specimens of bars.

(i) Creep: At high temperatures, such as encountered in steam boilers, turbines, and piping, the deformation of materials ceases to be elastic and becomes plastic with a continuous increase under a constant load. The equilibrium between stress and load is not established, even after a very long time. The material under tensile stress continues to stretch, or creep. Creep is measured in terms of plastic deformation during a certain time. The limiting creep stress for a certain temperature is the maximum stress under which the material will not fail during a prescribed length of time.

Creep is a function of temperature, stress, and time. It should therefore be represented by curves in a three-dimensional system of coordinates. But in practice creep is usually of interest in connection with a certain temperature, in which case a diagram such as that shown in Fig. 5.35 serves the purpose even better.[28] The influence of temperature and stress on the total creep can be studied

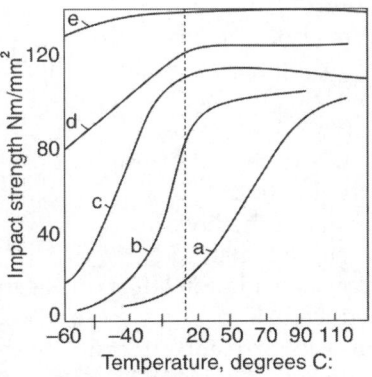

Fig. 5.34: Impact strength at low temperatures

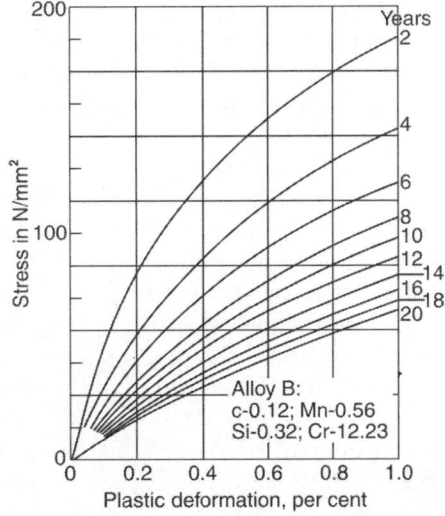

Fig. 5.35: Stress–strain curves at constant temperature of 460°C

---

[27] W. Schwining, "Die Festigkeitseigenschaften der Werkstoffe bei tiefen Temperaturen, Z. VDI, Vol. 79 (1935), p. 35.
[28] P. G. McVetty, "Working Stresses for High Temperature Service," *Mechanical Engineering*, Vol. 56, (1934), pp. 149–54; "Failure from Creep as Influence by the State of Stress," *Sulzer Technical Review*, No. 1 (1943), pp. 1–16.

by means of a diagram like that in Fig. 5.36. Chromium-molybdenum steels show great resistance to creep.

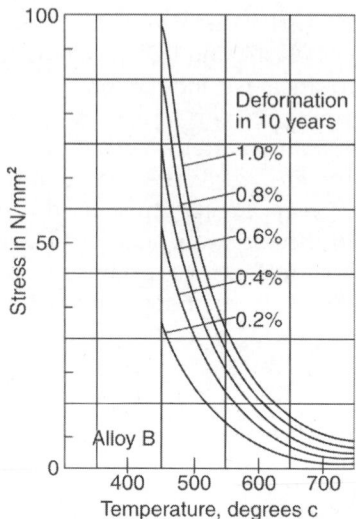

Fig. 5.36: Stress–temperature curves for various creep values

However, pure lead shows deformation or creep at room temperature (5.15).

(j) Growth of cast iron: When cast iron is heated repeatedly, its specific volume begins to increase. The dangerous temperature may vary from 300° to 540°C the actual value depending on the composition of the iron. This increase in volume, called growth, may be due to any or all of the following causes:

(a) The conversion of combined carbon to free carbon

(b) The action of the silicon and phosphorus in the cast iron

(c) The expansion of particles of gases dissolved in the metal

(d) The penetration of oil and gas under pressure into the graphite particles of cast iron, as it occurs in pistons of heat engines

This growth is dangerous in machine parts because it provides local compressive stresses similar to excessive heat stresses. It also reduces the clearance between moving parts and may result in excessive wear or seizure. The designer of a machine subjected to high temperature should provide ample clearances and should specify the use of cast iron with a minimum growth. The addition of steel to cast iron, making so called semi-steel, and the addition of about 1.5 per cent of nickel are the best methods for reducing the growth to practically harmless amounts. Ni-Tensyliron and Ni-Resist have particularly small growth and comparatively high strength at high temperatures. So has molybdenum cast iron.

Design for creep conditions: When it becomes impossible to keep the stresses in a part subjected to high temperatures within elastic limits, the allowable stresses must be determined from a diagram, such as that in Figs. 5.35 and 5.36, obtained by creep tests with the material to be used for the temperature involved. A maximum plastic deformation at the end of a certain number of years is selected to conform to the working conditions, and the corresponding stress S_{cr} is taken from the diagram. The design stress S_d is then found by dividing S_{cr} by the safety factor n, which should be at least 1.25 and preferably 1.5.

PROBLEMS

5.1 Why is the factors of safety for aircraft components very low while in industrial machines, it is very high, though the failure of aircraft results in heavy losses?

5.2 What is the effect of case hardening and nitriding on stress concentration?

5.3 What are the factors that effect factors of safety?

5.4 What is the advantage of a uniform strength bolt?

5.5 Name the methods to improve the fatigue strength of parts.

5.6 How is oxidation effect due to shrink fit reduced?

5.7 How is corrosion effect due to shrink fit reduced?

5.8 Which steel (stainless steel and other steels) has higher reduction of endurance limit due to corrosion?

5.9 Which shape of fillet causes least stress concentration?

5.10 Explain, with the help of sketches, the methods of reduction of stress concentration due to (a) notches, (b) change of cross-section, (c) grooves, (d) circular hole.

5.11 What are the causes of production of vibrations in machine members?

5.12 What is critical speed of system?

5.13 How is resonance avoided in machine members?

5.14 Describe a method to reduce the transmission of vibrations to engine supports.

5.15 Name a metal which can creep at room temperature.

5.16. What is the effect of temperature rise on elastic limit and ultimate strength of materials?

5.17 Explain the various parameters of a shaft which can be changed to avoid resonance.

5.18 A tension member in a roof truss carries an axial static load of 90 kN. The member is to be made of flat steel, 10C8, 16 mm thick, and is to be riveted to the main girder. Determine the number of 20 mm rivets required and the width of the strip.

5.19 A load F of 45 kN is suspended from a plate a, Fig. P5.1, held by bolts 1, 2, 3; the thickness of the plate is $h = 12$ mm; and

$l_2 = l_3 = 100$ mm. The bolts are made of alloy steel. (a) Determine their size with a safety factor of 1.75. (b) Find the stress in each bolt. (c) Find the stresses in bolts 1 and 3 if bolt 2 is taken out.

5.20 A spindle, Fig. P5.2, made of steel, is acted upon repeatedly by a force of F of 500N which each time travels through a clearance h of 0.8 mm. The shaft diameter is $D = 50$ mm, the spindle diameter is $d = 35$ mm and $l = 100$ m. Using a safety factor n of 1.75, determine the minimum permissible fillet radius r. Assume that the support is rigid, limit stress = 345 N/mm^2.

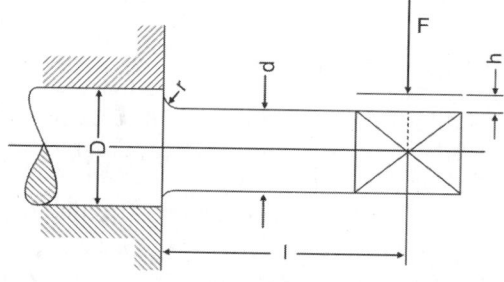

Fig. P5.2

5.21 The load on the clamp in Fig. P5.3 is $F = 18$ kN. Assuming that $h = 2b$, $l = 150$ mm and $r = 75$ mm, and using a safety factor of 1.7, determine the necessary dimensions h and b if the clamp is made (a) of cast steel with limit stress of 170 N/mm^2, and (b) of ductile iron with limit stress of 400 N/mm^2.

5.22 Determine the diameter of the shaft of problem 5.20 if the torque is not steady and may be applied suddenly.

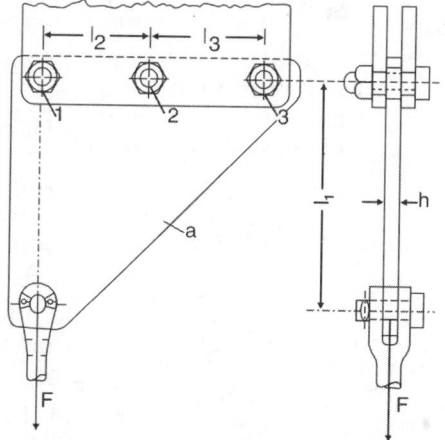

Fig. P5.1

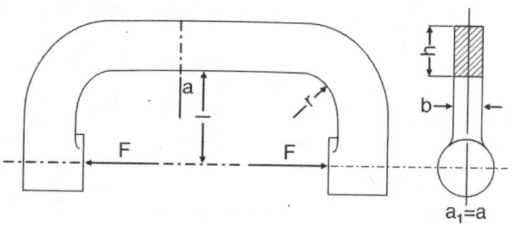

Fig. P5.3

6 General Manufacturing Considerations

6.1 MANUFACTURING METHODS

Designing a machine part means putting down on paper the dimensioned shape that the part must have to properly fulfill its functions. In order to be able to layout and draw this shape the designer must know how the shape can be produced from different raw materials. Such a knowledge requires a thorough understanding of the various manufacturing methods. These methods may be advantageously divided into two groups:

1. Preliminary shaping of machine parts, mostly, although not always, by using heat.

2. Final shaping by means of cold machining.

Under the first grouping come the processes of casting, welding, riveting, and forging. Each of these methods of forming machine parts has different possibilities, but it also has different limitations which influence the design of a part. The designer must know and keep in mind these possibilities and limitations. The main features of each method from the standpoint of machine design will be discussed in separate chapters in the order indicated above. The first group includes also such processes as rolling, drawing, extruding, and stamping. However, their purpose is not to produce machine parts as such, but to produce stock material used widely in industry, these methods can produce some finished components also.

Among the methods under the second grouping are a great number of different operations which give the piece the exact dimension required and produce the surface conditions necessary for its functioning. The main methods in this group are *turning, boring, milling, planing, shaping, drilling, reaming, spot facing, broaching, grinding, honing, and polishing.* There are additional machining processes, such as screw-cutting, tapping, and gear-cutting, which are special adaptations of the basic methods. Some of the machining operations fulfill the same object and are often interchangeable. Examples are milling, planing, and shaping. Other operations such as boring and drilling, are somewhat similar but are really different. The majority are intended for different purposes.

A good knowledge of the various operations is very helpful to a machine designer. However, such knowledge can be acquired only by working in a machine shop, and the type of machining operation does not have too much effect on the shape of a part during its design. In practice the proper operation often is selected after the designer consults the man in charge of the machine shop. Therefore no attempt will be made here to give any information about the difference between the various types of machining operations, and no suggestions will be given for selecting the best type of operation for a specific case. The information that must be given to a beginning designer is how to make simpler and easier the machining of the parts he is designing and how to determine the degree of accuracy in machining that he should prescribe on his drawings.

Production conditions: The design of a machine part depends on the facilities of the shop where the part will be built. The facilities in a small jobbing machine shop are naturally different from those in a large plant manufacturing some special machinery.

Also, the design will not be the same when the part will be produced in quantities as when only one piece, or at most a few pieces, must be made. For instance, for quantity production it may be proper to make a part as a die casting, whereas a single piece may be machined from a block. As another example, when it is necessary to replace a large sand-molded casting that was produced in quantity, the single piece may be produced more quickly and more cheaply by welding.

No rules can be given for deciding how to act in every case; the designer must be guided by this former experience and his personal judgment.

6.2 MACHINING

Removing metal by hand in order to obtain a desired dimension or to fit one part to another is much more expensive than doing the work by a metalworking machine. Therefore each machine part should be designed so that it can be finished and assembled without special fitting after it has been machined in the manner indicated on the drawing by the designer.

Limitations: The designer must know the limitations of the machine shop in which the part that he is designing will be machined. Such limitations are the biggest diameter and the greatest length or height that can go in the lathe and boring mill and the greatest length and width that can be handled by the planer. He must know for what pitches hobs are available in the shop for cutting gear teeth, as very few shops have all standard hobs in stock. He must have similar information concerning other tools, such as taps, reamers, and broaches.

The designer should know which sizes of cold-rolled and hot-rolled steel material are standard and which standard sizes are kept in stock in the shop where his design will be executed. However, if the design of his part really requires a standard size which the shop is not carrying in stock, he should not hesitate to call for such material. He should not try to use available stock sizes that involve extra machining.

Design for machining

(1) Unnecessary machining should be avoided: The area of machined surfaces should be kept to a minimum both in castings and weldments. If a part can be left rough, except for surfaces where it must be in contact with other parts, it must be provided with special strips that are raised above the unmachined surfaces. Such strips are needed along the edges of a cover and on the parts on which the cover rests. If in Fig. 6.1 the thickness of the cover a must be t_1, the thickness at the edges must be increased to t_2. Similarly, if the required thickness of the main casting b is t_3, the thickness must be increased to t_4 where the casting must be machined. Instead of machining the outside edges of the cover a in order to prepare a true surface for the nut, it is quicker and less expensive to spot-face the surface around the hole drilled for the stud to a diameter d slightly greater than that of the nut.

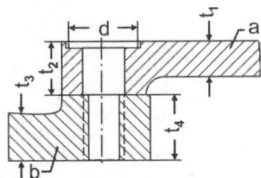

Fig. 6.1: Size of machined surfaces

(2) Spot facing is especially necessary if the surface is not normal to the axis of the bolt or stud, as shown in Fig. 6.2. This illustration also shows the spot-facing tool—a rod a with a slot into which is inserted a cutter b held by two tapered keys c. The upper end of the rod a fits into the head of a drill press, while the lower end fits into the bolt hole and acts as a guide.

In Fig. 6.3 is shown another example to indicate how the cost of machining may be reduced in several ways. The flange *a* has a larger diameter than the lower part and is left rough, since a slight difference in the overhang will not attract attention when both part *a* and part *b* are left rough. In this case the register at *r* is made short, since its length is immaterial and shortening it reduces the cost of machining of each casting. When the opening in part *b* is bored, the sharp edge at *c* is cut off at an angle of 45° to a depth of about 3 mm. This cut permits the register and the underside of the flange to be turned so as to leave a fillet at *d*. Providing such a juncture of a cylindrical surface and a flat surface is cheaper than bringing them together to a sharp corner and also gives a stronger flange. The spot facing for the stud nut is shown at *s*.

section in Fig. 6.4a, whereas it is easy to fasten a plate having the section in Fig. 6.4b; and the casting cost is the same. Special clamping devices and fixtures are permissible and advisable only if a sufficient number of pieces must be made. If only one piece or a few pieces must be made, the designer should make provision for simple and accurate clamping. He can add special bosses or extensions to support a piece during machining. In Fig. 6.5 a boss *b* for centering is shown; and in Fig. 6.6 bosses *b* on a cast compressor crankshaft are shown. It is also possible to provide openings in the casting for inserting clamping bolts. The bearing cap in Fig. 6.7 is provided with lugs to facilitate chucking during machining. In some cases the special projection must be cut off after the part has been machined, in order not to interfere with the assembling or the functioning of the part. For example, it will be

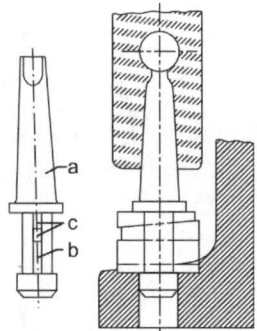

Fig. 6.2: Spot facing

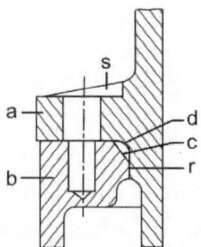

Fig. 6.3: Fitting of a flanged cylinder

(3) Provision for facility of machining: The cost of actually removing material is not always the main cost item in machining a part. Sometimes the fastening of the piece to the machine is even more expensive, requiring considerable time or special clamping devices. Thus it is difficult to fasten for machining a plate with the cross

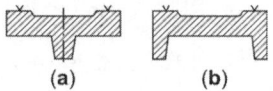

Fig. 6.4: Poor and good shapes for machining

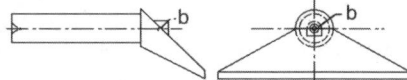

Fig. 6.5: Tool rest with boss for centering

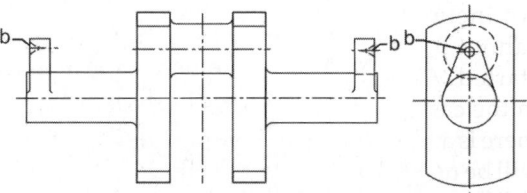

Fig. 6.6: Crankshaft with bosses for centering

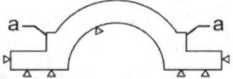

Fig. 6.7: Bearing cap with lugs for chucking

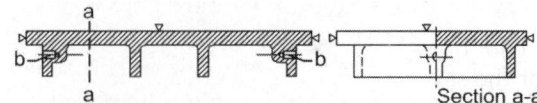

Fig. 6.8: Plate with holes for hook bolts

necessary to remove the bosses *b* in Fig. 6.6. Finally, Fig. 6.8 shows holes *b* drilled in bosses specially provided for hook bolts. These holes help to handle the heavy ribbed plate while it is being machined and when it is in service.

(4) Simple design to reduce cost: The machine part should be designed so that it can be machined without reclamping it and so that the surface to be machined will be parallel or normal to the surface to which the part is clamped. An oblique or inclined machined surface requires an inclined clamping, which always takes extra time and therefore is expensive. Arrangement of a bracket as in Fig. 6.9a requires expensive machining. The scheme in Fig. 6.9b facilitate machining and alignment in assembly. The scheme in Fig. 6.9c is still better, as it removes the danger of breaking off a bracket from a large casting.

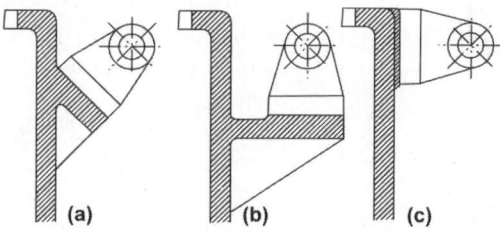

Fig. 6.9: Fastening of a bracket

(5) Provision of flat surface for ease of drilling: An inclined hole such as that in Fig. 6.10a cannot be drilled, because a drill cannot be started. A recess must therefore be provided in the casting, as shown in Fig. 6.10b. Even if there is a recess for the drilling, a special fixture will be needed to hold the piece so that the hole will be vertical or horizontal.

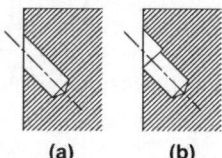

(a)   (b)

Fig. 6.10: Hole at an angle

(6a) Improved location of oil holes in castings: For the same reason a tapped hole for a drain plug should be located as in Fig. 6.11a, rather than as in Fig. 6.11b.

(6b) If a hole is to be drilled in a casting shaped as shown in Fig. 6.12a, the drill may become snagged when breaking through. The shape of the casting should be changed as shown in Fig. 6.12b.

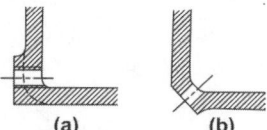

(a)   (b)

Fig. 6.11: Location of a drain hole

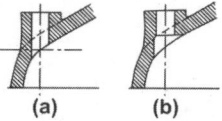

(a)   (b)

Fig. 6.12: Improving a casting for machining

(7) Provide common plane surface: Use of milled, planed, or turned surfaces in several parallel planes, as in Fig. 6.13a, is not good practice. If possible, all machined surfaces should be in one common plane, as in Fig. 6.13b. If projection *c* in Fig. 6.13a was provided to locate the part in the assembly or to resist a lateral force *F*, dowel pins *d* in Fig. 6.13b can be used instead. However, if the projection *c* is obtained by turning and must fit into a bored recess and serve as a register in assembling, this design is better than one with dowel pins.

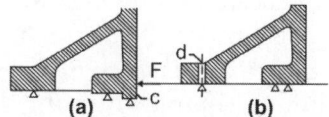

(a)   (b)

Fig. 6.13: Position of machined surfaces

(8) Provide run-out clearance for cutting tools: The designer should always provide clearance for the runout of a cutting tool. The tapping of the hole in Fig. 6.14a is impossible. However, by moving the hole a little, as shown in Fig. 6.14b the drilling operation is improved and tapping is possible. Other examples are shown in Fig. 6.15. In Fig 6.15a the clearance *c* must be provided in the rough casting; in Fig. 6.15b the clearance *c* in the comparatively small valve should be turned before turning the conical

valve seat; in Fig. 6.15c the clearance c must be turned to permit cutting of the thread; and in Fig. 6.15d clearance *c* is provided for a tapped hole. Figure 6.16a shows a runout clearance for grinding. When the keyway stops inside of the bore, as in Fig. 6.16b, a hole *l* drilled at the end of the keyway will provide the necessary clearance for the keyseater.

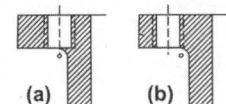

Fig. 6.14: Tapping holes in a flange

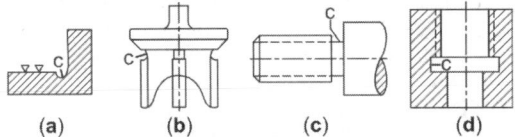

Fig. 6.15: Fastening of a bracket

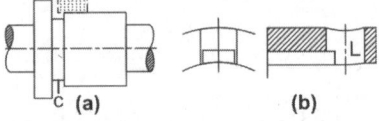

Fig. 6.16: Providing tool clearance

(9) Provide register for assembly: In order to insure correct and inexpensive assembling of concentric parts, they must have registers, as shown in Fig. 6.17 at *c*. Screw threads can never be produced accurately enough to serve as a register.

(10) Reduce the number of fitting surfaces: Simultaneous fitting on more than one surface must be avoided. If the gear hub *h* in Fig. 6.17a must be a tight fit on the tapered end of the shaft, it should not be in contact with the collar *c*, as shown in Fig. 6.17a. Actually the collar in Fig. 6.17a is not necessary, and the

correct design is shown in Fig. 6.17b. In some cases a certain design may be feasible, but it can be improved by an additional analysis. Here are a few examples: In Fig. 6.18a the upper flange is shown fastened to the lower one by a tapered dowel pin. However, drilling and reaming of the hole in the lower flange will be easier if the hole is drilled through as shown by the dotted lines. Also, a through hole will permit the pin to be knocked out when the parts are disassembled. In Fig. 6.18b is shown a register of a tubular part in a flange; and in Fig. 6.18c is shown an easier method of machining the flange without any loss of accuracy in assembly or otherwise. The design in Fig. 6.19a requires the use of a special, accurate counterboring tool, whereas the design in Fig. 6.19b is obtained by use of a simple chamfer tool and is just as good. In Fig. 6.20a is shown the conventional design of a key in a tapered shaft end. By using keyseating, as shown in Fig. 6.20b, a less expensive machine setup and easier fitting of the key are obtained.

(11) Grinding: A surface that is finished by grinding must be first machined by turning or boring. When giving the dimensions to which

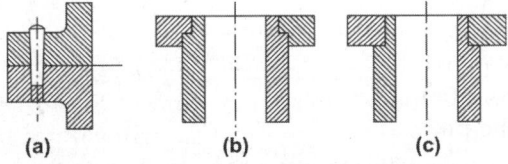

Fig. 6.18: Improved machining operations

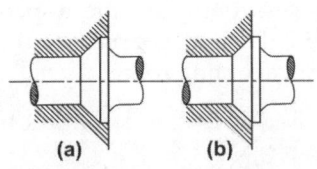

Fig. 6.19: Piston-rod fitting

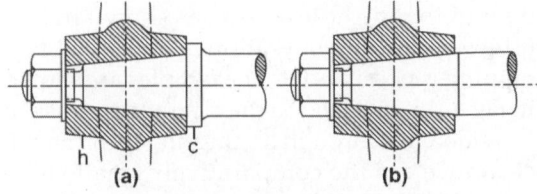

Fig. 6.17: Provision for fitting

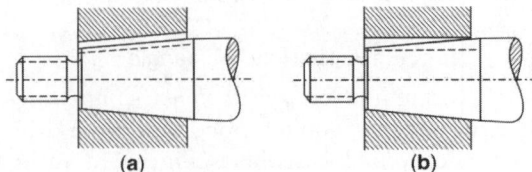

Fig. 6.20: Key in tapered shaft

the piece is machined before grinding, the designer should not forget to leave or allow enough stock to be removed by grinding.

6.3 DIMENSIONING AND TOLERANCES

Even with the best machines and workmanship it is impossible to obtain a certain dimension of a metal part exactly. The actual dimension will be a few thousandths of a mm larger or smaller than the dimension called for, the difference depending on the procedure by which the part is finished and on the skill of the machinist.

The actual dimension of a part can deviate from the nominal dimension by a certain amount. The difference between the nominal dimension and either the largest or smallest permissible dimension is called the *tolerance*. Its amount for any particular part depends on the duty of the part.

The designer must specify the tolerance for each dimension on a drawing. He should remember that the smaller the tolerance, the more accurately a machine part will be finished and the smoother will be the operations of the mechanism that he is designing—at least within certain limits. On the other hand, the smaller the tolerance, the more expensive the machining of the part will be. As a first approximation one may assume that the product of the cost of machining and the

tolerance is constant. When this statement is expressed graphically, as in Fig. 6.21, it is seen how fast the cost goes up with a decrease of tolerance. Therefore, the designer should carefully analyze what tolerance is permissible for every dimension and should not specify a tolerance smaller than that really necessary.

In the shop it is much more difficult to obtain a certain tolerance with a large dimension than with a small one. It has been found that the probable error due to both the machining operation and the measuring process is proportional to $\sqrt[3]{D}$.

Basic size: For the sake of convenience, a basic size, which is usually a whole number is given to represent the dimension of a part. This is the design size and the allowable maximum and minimum limits of dimension of the part are fixed in relation to the basic size. Basic size relates to the position of zero line.

Tolerance is the amount of permissible variation in size of a part or manufacture allowance. It is the difference between the upper limit and lower limit of a dimension.

Allowance is the minimum clearance space to be provided between two mating parts. It is an intentional prescribed difference between the hole dimension and shaft dimension of size. It is the maximum clearance or maximum interference between mating parts.

Deviation: It is the algebraic difference between the actual size and the corresponding basic size.

Upper deviation: It is the algebraic difference between the maximum limit of size and the basic size.

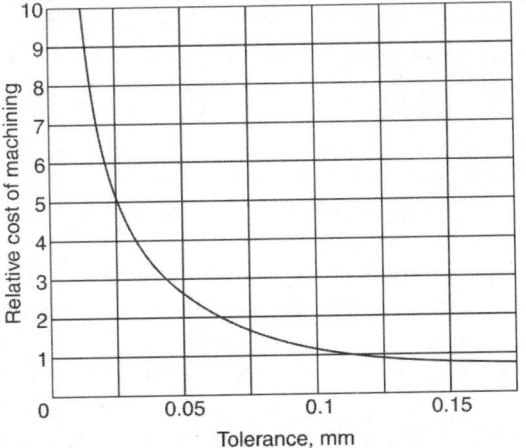

Fig. 6.21: Influence of tolerance on cost of machining

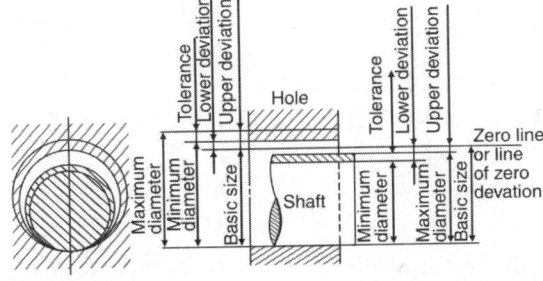

Fig. 6.22: Various terms associated with tolerances

Lower deviation: It is the algebraic difference between the minimum limit of size and the corresponding basic size.

Fundamental deviation: It is one of the two deviations which is conveniently chosen to define the position of the tolerance zone in relation to zero line.

Interference is the amount by which the dimensions of the mating parts overlap. It is also called a negative allowance as in press fit.

Fit: It is defined as the degree of tightness or looseness between two mating parts. The type of fit depends upon the size of clearance and interference.

Types of fits

Clearance fit: In this type there is always a clearance when mating parts are assembled. A shaft is always smaller than the hole into which it fits, e.g. rotating shafts in bearings, loose pulleys, etc.

Transition fit: Either clearance or interference fit may result when mating parts are assembled. It may be force fit, tight fit or push fit, e.g. keys, bushes, spigots, fasteners, pins, etc.

Zero allowance fit also called snug fit is the closest fit that can be assembled by hand and necessitates work of considerable precision. It is suitable where no perceptible shake is permissible and where moving parts are not intended to move freely under load.

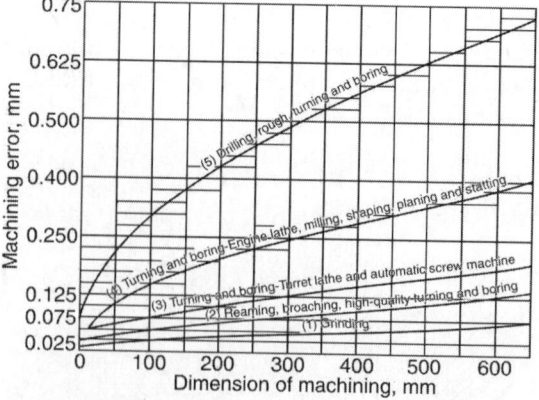

Fig. 6.23: Probable errors of different machining operations

Wringing fit: It has zero to negative allowance and is also called tuckling fit and is not interchangeable.

Interference fit: Interference always occurs between parts. The interference may be tight fit or drive fit or medium force fit or heavy drive fit or shrunk fit, e.g. pressed in bushes, crank pins in webs, car tyres, railway wheels shrunk on to axles.

Tight fit has slight negative allowance. Light pressure is required to assemble parts. These fits are used for drive fits in thin sections or on extremely long fits in other sections.

Medium force fits: Considerable pressure is required to assemble parts and the parts are considered to be permanently assembled as locomotive wheels, armatures of dynamos and motors, crank discs to shafts. This is the tightest fit recommended for cast iron external members and it stresses cast iron within safe limits.

Shrink fit has considerable negative allowance. These fits are used for steel holes where the metal can be highly stressed without exceeding elastic limit. Shrink fits are used where heavy force fits are impracticable as on locomotive wheel tyres, heavy crank discs of long engines. For heavy power transmission keys are used in addition for shaft diameter above 75 mm.

Tolerance designation: The tolerance for a dimension may be put on a drawing in several different ways. Under some conditions either the minimum permissible dimension or the maximum permissible dimension must be equal to the nominal dimension. The tolerance is then said to be *unilateral*. If the actual dimension may be either greater or less than the nominal size, the tolerance is *bilateral*.

When the tolerance is unilateral, it is good practice to indicate the nominal, or basic dimension for the maximum metal condition. The dimension of a hole or any other inside dimension indicates the minimum permissible size; and the dimension for a diameter or an outside length shows the maximum permissible size. The machinist then aims to obtain the basic dimension, and he has the tolerance as leeway for inaccuracy of his work. One method of indicating a unilateral tolerance on a drawing is to write the tolerance after the basic dimension with its proper sign and also to give the opposite sign followed by a number of ciphers. An alternate method with the same meaning is to indicate the basic dimension first and to place under it another dimension that includes the tolerance.

When the tolerance is bilateral, the nominal dimension is indicated as the average between the maximum and minimum sizes, and the tolerance is given as one-half the difference between the maximum and minimum sizes with a plus-and-minus sign or even $40^{+0.02}_{-0.03}$.

In Fig. 6.24 the three different methods of indicating tolerance for a shafts are illustrated, and Fig. 6.25 illustrates the dimensioning of a hole. In each case the first methods described for a unilateral tolerance is shown in Fig. 6.24a and Fig. 6.25a; the alternate methods is shown in Fig. 6.24b and Fig. 6.25b; and the method for a bilateral tolerance is shown in Fig. 6.24c and Fig. 6.25c.

Basic shaft and hole systems: The most common case of assembling two machine parts is when one of the parts is a rotating shaft and the other is either a hub that must fit tightly on the shaft or a bearing in which the shaft rotates.

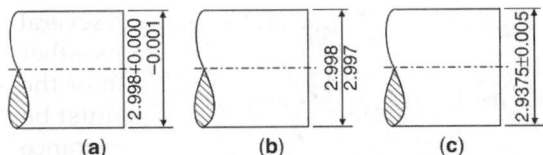

(a)   **(b)**   **(c)**

Fig. 6.24: Methods of indicating tolerance for a shaft

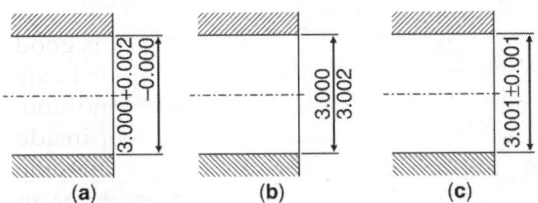

(a)   **(b)**   **(c)**

Fig. 6.25: Methods of indicating tolerance for a hole

Cold-rolled or ground shafting comes in certain standard size with close tolerance. In such a case it is logical to fit the hole to the standard or basic shaft size and to indicate the required hole dimension and tolerance on the drawing. This method of dimensioning is called the basic shaft system and is illustrated in Figs 6.24a and b.

On the other hand, when a shaft or some other male part must be machined anyway, it is convenient to use standard-size reamers for the holes and to machine the male parts to fit the reamed holes. In this case the dimensioning is said to be done on the basis of hole system. This method is illustrated in Figs 6.25a and b.

In the United States the American Standards Association and in India, BIS and International Standard Organisation (ISO) has adopted the unilateral method of expressing tolerance, with the hole as the basis of the system.[1] However, for reasons, just explained, both the bilateral method and the basic shaft system are still used in industry.

Comparison of tolerance methods: The unilateral system of dimensioning possesses distinct advantages in making interchangeable parts. Since both male and female basic dimensions are selected so as to produce the desired fit, the tolerances may be slightly changed without affecting the interchangeability of the mating parts. Also, when the unilateral system is used, fewer spoiled parts are rejected after inspection, because the machinist has the full tolerance as leeway for his accuracy. Moreover, it is easier to check drawings made with unilateral tolerances than those made with bilateral tolerances.

In certain cases, however, the bilateral method of tolerances my be better. An example is in the location of a hole center when the deviation from the basic dimension is equally critical in both directions. Also, for dimensions where large tolerances are permissible, it may be more convenient to give the mean dimension an the allowable deviations as plus and

---

[1] ASA B4A–1925.

minus values. For the same reason, welded assemblies are often dimensioned with bilateral tolerances.

Standard tolerances: Some engineering departments use drawing paper with a blanket note printed or rubber-stamped stating: "Tolerances not specified to be ±0.25 mm." A note used by a manufacturer of precision equipment reads: "All finished dimension ±0.075 mm unless otherwise specified." Another concern used a note that reads: "Unless otherwise specified, dimensions up to 75 mm to be machined with a tolerance = 0.125 mm; dimensions over 300 mm, with a tolerance = 0.25 mm."

Such a note simplifies the work of the designer to some extent. However, it does not relieve him of the responsibility of specifying smaller tolerance if the duty of the part really requires them; or of specifying larger tolerances if the duty of the part permits them, in order to cut down the machining cost.

According to IS919–1963, there are 18 grades of tolerances (according to ISO, there are 20 grades) also called grades of accuracy. These are designated as IT01, IT0, IT1 to IT16.

$$i, \text{in microns} = 0.45 \sqrt[3]{D} + 0.001D$$

where, D = geometric mean for a particular range of size [as 0 to 3, 3 to 6, 6–10, 10–18, etc.]

$$\text{IT01} = 0.3 + 0.008D$$
$$\text{IT0} = 0.5 + 0.012D$$
$$\text{IT1} = 0.8 + 0.020D$$

IT2 to IT4 have been regularly scaled between IT1 and IT5

IT5	= 7i				
IT6	= 10i	IT11	= 100i	IT16	= 1000i
IT7	= 16i	IT12	= 160i	IT17	= 1600i
IT8	= 25i	IT13	= 250i	IT18	= 2500i
IT9	= 40i	IT14	= 400i		
IT10	= 64i	IT15	= 640i		

ISO

Derivation of standard tolerance (IT) for basic sizes from 500 mm up to and including 3150 mm.

The values for standard tolerances in grades IT1 and IT18 are determined as a function of the standard tolerance factor l.

The standard tolerance factor, l, in micro-metres, is calculated from the following formula:

$$l = 0.004\,D + 2.1$$

where D is the geometric mean of the basic size step in millimetres.

The values of the standard tolerances are calculated in terms of the standard tolerance factor, l as shown in Table 6.1.

It should be noted that from IT6 upwards, the standard tolerances are multiplied by a factor of 10 at each fifth step. This rule applies to all standard tolerances and may be used to extrapolate values for IT grades above IT18.

Symbols for Tolerances, Deviation and Fits

In order to satisfy the usual requirements, both of individual parts and of fits, the system

Table 6.1: Formulae for standard tolerances in grades IT1 to IT18

Basic size		Standard tolerance grades								
mm		$IT1^{11}$	$IT2^{11}$	$IT3^{11}$	$IT4^{11}$	IT5	IT6	IT7	IT8	IT9
Above	Up to and including	Formulae for standard tolerances (Results in micrometres)								
–	500	–	–	–	–	7i	10i	16i	25i	40i
500	3150	2l	2, 7l	3, 7l	5l	7l	10l	16l	25l	40l
		IT10	IT11	IT12	IT13	IT14	IT15	IT16	IT17	IT18
		64i	100i	160i	250i	400i	640i	1000i	1600i	2500i
		64l	100l	160l	250l	400l	640l	1000l	1600l	2500l

provides for any given basic size, a range of tolerance together with a range of deviations defining the position of these tolerances with respect to the line of zero deviations. The tolerance, the value of which is a function of the size is indicated by a letter symbol (in some cases two letters), a capital letter for hole and a small letter for shaft. There are 25 fundamental deviations indicated by letter symbols for both holes and shafts: capital letters A B C D E F G H J J$_s$ K M N P R S T U V X Y Z Z$_a$ Z$_b$ Z$_c$ and corresponding small letters a to z$_c$ for shafts in diameter steps up to 500 mm.

For shafts a to h, the upper deviation is below the zero line and for shafts j to z$_c$, it is above the zero line. For holes A to H, the lower deviations is above zero line. The shafts h for which the upper deviation is zero is called *basic shaft*. Similarly, hole H for which lower deviation

is zero is called *basic hole*. A number of fits ranging from extreme interference to those of extreme clearance can be obtained by suitable combinations of fundamental tolerances. A fit is indicated by the basic size common to both components followed by symbols corresponding to each components, the hole being quoted first, e.g. 50H$_8$/g$_7$ or $50\frac{H_8}{g_7}$ which indicates—the basic size is 50 mm, the hole H has a tolerance of IT8 and the shaft g has a tolerance of IT7, the fundamental deviation for the shaft is given by g and the fit is a clearance fit. Types of fits and with symbols and applications are given in Table 6.2 and tolerance grade in various manufacturing processes are given in Table 6.3.

The values for deviations are specified in Table 6.4.

Table 6.2: The following are general recommendations for various uses

Shafts with holes				Classification
Clearance fits				
abc	H$_6$	H$_7$	H$_8$	These fits give large clearances and are not widely used
d		d$_8$	d$_8$d$_{10}$	**Loose running fits,** e.g. plummer block bearings, loose pulleys, large bearings for ball mills, roll forming, heavy bending machines, agriculture, mining, textile, rubber, candy and bread machinery.
e	e$_7$	e$_8$	e$_8$e$_9$	**Easy running fits,** e.g. for loose clearance fits and for properly lubricated bearings with high speeds and pressures $\geqslant 4$ N/mm^2 used for large, high speed heavily loaded bearings such as turbo-generators, large electric motor bearings, engines, machine tools and automotive parts.
f	f$_6$	f$_7$	f$_8$	**Running fits** used as a normal running fit, grease lubricated bearings or oil lubricated bearing viz gear box shaft bearing, bearings of small electric motor, pumps.
g	g$_5$	g$_6$		**Close running fit:** It is expensive to make as clearance is small. It is used for precision equipment for small load, e.g. bearings of accurate link work, piston and slide valves, spigot and locating fits, speed < 600 rpm, $p < 4$ N mm^2.
h	h$_5$	h$_6$	h$_7$h$_8$	**Sliding fit** (Snug) The upper shaft limit is zero, closest fit that can be assembled with hand and necessitates work of considerable precision used for normal locating fits and spigot fits on precision sliding fit.
Transition fits				
j	j$_5$	j$_6$	j$_7$	**Push fit** (Wringing fit) requires slightly less clearance, practically metal to metal fit. Light tapping with hammer is necessary to assemble parts, e.g. gear rings clamped to steel hubs, piston and piston rod.
k	k$_5$	k$_6$	k$_7$	**Easy keying fit,** e.g. inner races of bearings on shafts, belt pulleys, brake pulley or drums.

(Contd...)

Table 6.2: The following are general recommendations for various uses *(Contd...)*

Shafts with holes				Classification
m	m_5	m_6	m_7	**Tight keying fit** (Force fit), e.g. gears, belt pulley, couplings also inner races of bearings on shafts.
n	n_5	n_6	n_7	**Drive fit or light press fit,** e.g. bearing bushes, shaft and wheel with keys.
Interference fits				
p	p_5	p_6	p_7	**Light press fit,** standard press fit for steel, cast iron or brass to steel assembles
r	r_5	r_6		**Medium drive fit** for ferrous parts and light drive fit for nonferrous parts which can be dismantled, coupling on shaft ends, bearing bushes on hubs, valve seats, etc.
s	s_5	s_6	s_7	**Press fit** used for semipermanent assembly of steel and cast iron parts, provides sufficient gripping force and requires considerable pressure for assembly used for fastening locomotive wheels, armature of dynamos and motors, crank discs. Their shaft, shrink fits on medium sections, lightest fit for cast iron parts.
t	t_5	t_6	t_7	**Force fits** on ferrous parts for permanent assembly
u	u_5	u_6	u_7	**Heavy press or shrunk fit,** require considerable negative allowance. Such shrunk fits are used when heavy force fits are impractical as railway wheel tires, heavy crank discs for large engines, heavy steel rings on cast iron wheels, built up large bore guns, bronze crowns on worm wheel hubs.

Example 6.1: A sleeve bearing is to designed for a shaft of an electric motor of diameter 14 mm, find the size. From the Table 6.4, the recommended fit is H_8/f_7 or H_8/e_8.

Taking H_8/e_8 as the desired fit—easy running fit. Diameter 12 mm lies in diametral steps 10–18

$$D = \sqrt{10 \times 18} = 13.41 \text{ mm}$$

The value of tolerance IT8 = 25 i

$$i = 0.45\sqrt[3]{D} + 0.001 \cdot D \text{ micron}$$

$$IT_8 = 25\,i = 25\left[0.45\sqrt[3]{13.41} + 0.001 \times 13.41\right]$$

$$= 25 \times 1.08166 = 27 \text{ microns} = 0.027 \text{ mm}$$

Deviation e for shaft $= -11\,D^{0.41}$

$$= -11 \times 13.41^{0.41} = -31.88$$

Say = 32 micron = 0.032 mm

For shaft lower deviation = 0.027 + 0.032 = 0.059

Hole $= 14^{+0.027}_{-0.000}$

Shaft $= 14^{+0.032}_{-0.059}$

Taking running fit as H_8f_7

$$D = 13.41$$

$$IT8 = 0.027$$

$$IT7 = 16i$$

$$= 16\left[0.45\sqrt[3]{13.41} + 0.001 \times 13.41\right]$$

$$= 16 \times 1.08166 = 17.3 \text{ or } 18 \text{ microns}$$

$$= 0.018 \text{ mm}$$

$$f = -5.5\,D^{0.41}$$

$$f = -5.5 \times 13.41^{0.41} = 15.93 \text{ or } 16 \text{ microns}$$

For shaft lower deviation = 0.016 + 0.018 = 0.034

Hole size $= 14^{+0.018}_{-0.000}$

Shaft size $= 14^{+0.016}_{-0.034}$

Example 6.2: Find the size of shaft and hole for 210 mm for fixing bearing bushes as hub, from table taking medium drive fits, the recommended size is H_7/r_6

Shaft steps of 210 mm are 180–250

$$D = \sqrt{180 \times 250} = 212 \text{ mm}$$

$$IT6 = 10i = 10\left[0.45\sqrt[3]{212} + 0.001 \times 212\right]$$

$$= 10 \times 2.8952 = 28.9 \text{ say } 29 \text{ micron}$$

$$= 0.029 \text{ mm}$$

$$IT7 = 16i = 16\left[0.45\sqrt[3]{212} + 0.001 \times 212\right]$$

$$= 16 \times 2.8952 = 46.32 \text{ say } 47 \text{ micron}$$

$$= 0.047 \text{ mm}$$

Deviation r = Geom. mean of p & s

$$p = IT7 + 3$$

$$s = IT7 + 0.4D$$

$$r = \sqrt{[47+3][47+0.4 \times 212]} = \sqrt{50 \times 131}$$

$$= 80.9 \text{ say } 81 \text{ micron} = 0.081 \text{ mm}$$

Table 6.3: Tolerance grade in various manufacturing processes and tolerance in μm

Tolerance grade	Manufacturing process that can produce, examples	3 mm to 6 mm	50 mm to 80 mm	120 mm to 180 mm
16	Sand casting; flame cutting	750	1900	2500
15	Stamping (Approx)	480	1200	1600
14	Die casting or moulding; rubber moulding	300	750	1000
13	Press work, tube rolling	180	460	630
12	Light press work; tube drawing	120	300	400
11	Drilling, rough turning, boring, precision tube drawing	75	190	250
10	Milling, slotting, planning, metal rolling or extrusion	48	120	160
9	Worn capstan or automatic horizontal or vertical boring machine	30	74	100
8	Centre lathe turning and boring, reaming, capstan or automatic in good condition	18	46	63
7	High quality turning, broaching, honing	12	30	4
6	Grinding or fine honing	8	19	25
5	Machine lapping, diamond or fine boring fine grinding, ball bearing	5	13	18
4	Gauges, Fits of extreme precision produced by lapping	4	8	12
3	Good quality gauges, plug gauges	2.5	5	8
2	High quality gauges, plug gauges	1.5	3	5
1	Slip gauges, reference gauges	1	2	3.5

Notes: 1. Tolerance increases with the number
2. Tolerance increases with size of the component

For shaft upper deviation = 0.081 + 0.029 = 0.110

Hole size = $210^{+0.000}_{+0.047}$

Shaft size = $210^{+0.081}_{+0.110}$

6.4 SURFACE FINISHES

The possible types of deviations of the actual surface of a machine part from its nominal surface are shown in Fig. 6.26 in greatly magnified profile sections. These types are classified as follows:[2]

Roughness, shown in Fig. 6.26a is characterized by small irregular unevenness, such as may be felt when drawing a fingernail over a ground surface. *Waviness* shown in Fig. 6.26b, is characterised by more or less regular and repeating unevennesses that are in the nature of waves. *Surface flaws*, shown in Fig. 6.26c, comprise scratches and checks, and similar defects which occur at irregular intervals and are often caused by heat treatment. Combination of the three basic types of irregularities is often found, as shown in Fig. 6.26d.

Lay, or the direction of the prevailing surface pattern, often affects the performance of the part.

Measurement of irregularities: With the improvements in finishing operations, a demand has gradually developed for a more exact method of specifying surface finish than simply by marking on the drawing "finish," "grind," or "hone." The surface smoothness can be specified by indicating the permissible height of the irregularities above and below the mean plane.

A true magnified profile of a surface, or the contour of a section through the surface, is shown in Fig. 6.27a. In order to study the surface conditions better, Fig. 6.27b *shows the same profile when an additional magnification of 25 times is given to the vertical dimensions without any change in the horizontal dimensions. This additional magnification is obtained by preparing the specimens for viewing under the microscope by a process called taper sectioning (6.15). In this process the surface is ground at a small*

[2] Surface Roughness, Waviness, and Lay, Proposed American Standard ASA 1346.1 (July, 1947)

Table 6.4: Formulae for fundamental shaft deviations

Upper deviation (es)		Lower deviation (ei)	
Shaft designation	In microns (for D in mm)	Shaft deviation	In microns (for D in mm)
a	$= -(265 + 1.3\,D)$ for $D \leqslant 120$ $\approx -3.5\,D$ for $D > 120$	J_5 to J_8 k_4 to k_7 k for grades $\leqslant 3$ and $\geqslant 8$	No formula $= +0.6\,\sqrt[3]{D}$ $= 0$
b	$= -(140 + 0.85\,D)$ for $D \leqslant 160$ $\approx -1.8\,D$ for $d > 160$	m n p	$= +(\text{IT7} - \text{IT6})$ $= +5\,D^{0.34}$ $= +\text{IT7} + 0 \text{ to } 5$
c	$= -52\,D^{0.2}$ for $D \leqslant 40$ $= -(95 + 0.8\,D)$ for $D > 40$	r	$=$ Geometric mean of values ei for p and s
d	$= -16\,D^{0.44}$	s	$= +\text{IT8} + 1 \text{ to } 4\,D \text{ for } D \leqslant 50$
e	$= -11\,D^{0.41}$		$= +\text{IT7} + 0.4\,D \text{ for } D > 50$
f	$= -5.5\,D^{0.41}$	t	$= +\text{IT7} + 0.63\,D$
		u	$= +\text{IT7} + D$
g	$= -2.5\,D^{0.34}$	v	$= +\text{IT7} + 1.25\,D$
		x	$= +\text{IT7} + 1.6\,D$
h	$= 0$	y	$= +\text{IT7} + 2\,D$
		z	$= +\text{IT7} + 2.5\,D$
		z_a	$= +\text{IT8} + 3.15\,D$
		z_b	$= +\text{IT9} + 4\,D$
		z_c	$= +\text{IT10} + 5\,D$

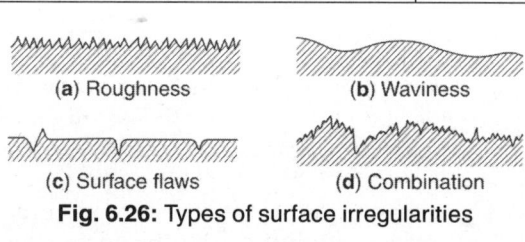

(a) Roughness   (b) Waviness

(c) Surface flaws   (d) Combination

Fig. 6.26: Types of surface irregularities

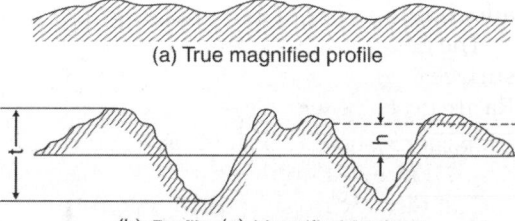

(a) True magnified profile

(b) Profile (a) Magnified 25 times in vertical direction

Fig. 6.27: Magnified cross section showing the roughness of a surface

angle with the surface, and the heights of the irregularities are thus accentuated. Evidently the magnification is equal to $1/\sin \alpha = \operatorname{cosec} \alpha$, and magnification of 25 is obtained with $\alpha = 2°17'$.

In Fig. 6.27b the distance t is the total roughness, or the distance from a peak to a valley; the mean surface is at the mean profile height; and h is a special kind of height, called the *root mean square height*, which is the average

distance above and below the mean surface. This height h is found by adding the squares of ordinates at equidistant points on the profile, dividing the sum by the number of ordinates measured, and then taking the square root of this average. The ordinates are measured from the mean surface in thousandth of a mm, written as micron, µm. The rms (root mean square) method of calculating gives more weight to the higher peaks of the surface, since high, narrow peaks seem to have considerable influence on the surface qualities but have only a small effect upon the position of the mean-surface line. The rms height is thus slightly larger than the arithmetical average of the ordinates. Sometimes the arithmetical average, or even the peak-to-valley height, is used as measure of roughness. The drawing must indicate which one is used.

Symbols used for indicating surface roughness. Fig. 6.28a shows the basic symbols used for surface roughness. It consists of two legs of unequal length inclined at 60° to the line representing the surface, wherever the removal of material by machining is required, a bar is added to the basic symbol as shown in Fig. 6.28b.

When the removal of material is not allowed, a circle is added to the basic symbol as shown in Fig. 6.28c.

If some special characteristics are to be indicated (for example, a milled surface), a line is added to the longer leg of the basic symbol as shown in Fig. 6.28d.

The value defining the surface roughness in micrometers or its corresponding grade is added to the symbol as shown in Fig. 6.28 e, f.

The grades recommended for specifying the surface roughness corresponding to roughness Ra are given below:

Roughness values Ra µm	Roughness grade symbols
50	N12
25	N11
12.5	N10
6.3	N9
3.2	N8
1.6	N7
0.8	N6
0.4	N5
0.2	N4
0.1	N3
0.05	N2
0.025	N1

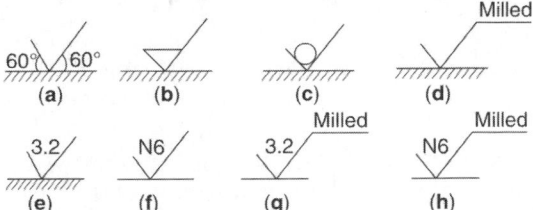

Fig. 6.28: Symbols used for surface roughness

6.5 COST

One of the most important things that the designer must keep in mind at all times is the cost of manufacturing the machinery that he is designing. The cost of any part and assembly of parts depends on the materials used; the amount of labor involved in machining and assembling; and, in any multiple reproduction, features of standardization and sales appeal.

Materials: As a result of the metallurgical developments of the last some years a great number of materials are available for every machine part. These materials vary in quality and in price. A good designer will select the least expensive material that will be satisfactory for the duty of the part. For example, he will use more expensive alloy steels only when ordinary low-carbon steels cannot give satisfactory service. When forced to use alloy steel he should compare the properties and costs of different kinds and should again follow the same principle.

In selecting materials the designer must consider not only their strength but also their rigidity and their resistance to wear. The permissible wear of a part depends on its duty and also on the length of service that it must give. If a part is subjected to severe wear when in operation but is operated intermittently and not very often, a comparatively inexpensive material may be satisfactory; whereas a similar part in continuous operation will require a more wear-resistant material. The expected life of a part of a machine must also be taken into account. The longer the intended life, the better should be the grade of the material used.

The same consideration applies to parts purchased from other manufacturers, such as bolts, setscrews, bushings, and ball bearings. For instance, if the intended life of a certain machine is 3,000 hr, its cost will be unnecessarily increased by using in it ball bearings with an expected life of 10,000 hr. Whenever possible, standard stock sizes of rolled and extruded materials should be used without changing their cross-sectional dimensions.

Machining: Any kind of machining should be specified only where it is necessary to permit the part to function properly. In former years it was considered necessary to machine the outsides of parts that were to be put together in contact, such as the top of a cylinder and the outside of the cylinder head, in order to match them accurately. This is not necessary, and the method shown in Fig. 6.3 is less expensive and just as good. Finishing cover plates on the edges and from the top is another waste of machining. All that is necessary is to spot-face around the holes for the nuts.

Where machining is necessary it should be done by the least expensive method that is consistent with the purpose of the machining. If, for example, a part must be turned for the sake of balancing, rough turning is satisfactory, and specifying finish turning would be a mistake. Similarly, if planing is satisfactory, more expensive milling should not be specified on the drawing, and reaming should not be specified if simple drilling is satisfactory.

Where fitting of parts requires tolerances, the specified tolerances should not be closer than those absolutely necessary, as explained in section 6.3. Where surface finish is considered, do not specify a smaller rms roughness than that actually necessary.

In order to reduce the cost of assembling a machine, the parts of the machine should as far as possible be so designed and built that they will place and align themselves automatically when brought together. Usually, the more complicated a machine is, the more important it is that the ease and cost of assembling be given careful consideration. If bench work and hand fitting cannot be avoided, they should be reduced to a minimum.

The number of machines to be built has an important bearing on the design of a machine and its parts. If only one machine or a few machines of a certain kind and size are to be built, the limitations of the available plant equipment should be kept in mind in the design. The capacity of the equipment of the foundry or machine shop may require a large casting to be made in two or three parts, and the available machine tools at hand may limit the methods of machining and thus influence the design. In certain cases it may be found more economical to weld a frame than to make a large, expensive pattern for it.

On the other hand, mass production or interchangeable manufacture justifies the use of special molding machines in the foundry, special jigs and fixtures, special production and inspection gages, and special tools, dies, and machines in the welding and machine shops.

Standardization: If a part is made in lots, especially if the part is manufactured on mass-production basis, it is very important to follow certain standards. Once a part is designed and developed, it should be considered standardized; and no changes should be made that would make the part not interchangeable with the original design. Standard stock parts should be used without any additional machining.

Preferred numbers: When a machine is to be made in several sizes having different powers or capacities, it is necessary to decide what capacities will cover a certain range efficiently with a minimum number of sizes. When a larger similar machine is built, its relation to the original smaller machine is complicated by the fact that lengths of parts are proportional to the first powers of the linear dimensions; areas are proportional to the second powers, or squares, of the linear dimensions; volumes and section moduli are proportional to their third powers, or cubes; and moments of inertia are proportional to their fourth powers. A certain range can be covered efficiently with a minimum number of sizes by the use of a geometrical progression with a constant ratio. With our decimal system it is convenient to select a series of numbers from 10 to 100. Then the series can be extended to 1,000, 10,000 and so on, simply by multiplying the base sizes by 10, 100, etc. or the series can be reduced in size by dividing the base sizes by 10. The numbers in such a series are called *preferred numbers*. The following series have been established:[3] the coarse series with the ratio $r = \sqrt[5]{10} = 1.6$, the next series with $r = \sqrt[10]{10} = 1.25$, the next with $r = \sqrt[20]{10} = 1.12$, and the finest with $r = \sqrt[40]{10} = 1.06$. The first series consists of the rounded numbers 1, 1.6, 2.5, 4.0, 6.3, and 10.

When preferred numbers are used, fewer stock sizes can cover certain ranges. Thus in Germany, manufacturers of power-transmission equipment formerly carried 3,600 patterns of belt pulleys. After the introduction of standards sizes selected in accordance with preferred numbers, this number was reduced to 600.[4] Catalogues of American manufacturers of power transmission machinery list 3,400 different sizes of stock belt pulleys. This number could be reduced to about 570 sizes by use of preferred numbers. Such a reduction would mean a great saving in inventory and probably in manufacturing cost too. There is a wide field for the application of preferred numbers in various fields, and machine designers can contribute a good share to our economy by using them for serial designs in proper places.

PROBLEMS

6.1 State the two main groups into which the methods of producing the shapes of various machine parts may be divided.

6.2 Enumerate the methods of preliminary, or rough, shaping of machine parts.

6.3 Enumerate (a) the various finishing operations used in machining surfaces of different machine parts, and (b) additional special machining processes necessary in the production of machine parts.

6.4 Show, by means of sketches, examples of how the cost of manufacturing can be reduced by proper design.

6.5 Give all the consideration in the design of components in machining with the help of sketches.

6.6 Show, by means of sketches, examples of how clearances may be provided for the runout of tools.

6.7 Explain what is called *grinding allowance*, and give the range of this allowance for various sizes of machine parts.

6.8 Explain what is called tolerance, and also explain why it is necessary to specify tolerances for certain machining operations.

6.9 Which letters indicate the standard deviation for clearance fits and interference fits?

6.10 What type of fit is indicated by H_7/g_6, H_7/p_6, H_7/n_6?

6.11 Compute the basic tolerances, (a) for a 55 mm shaft, (b) for a 300 mm shaft, and (c) for a 920 mm cylinder bore with IT6.

6.12 Compute the basic tolerance (a) for a 75 mm shaft, (b) for a 250 mm shaft, and (c) a 1.1 mm cylinder bore with IT7.

6.13 Explain what is termed a unilateral method of designating tolerances, and show how it would be applied to problem 6.12.

6.14 Explain what is termed a *bilateral method* of designating tolerance, and show how it would be applied to problem 6.12.

6.15 Explain how it is possible to obtain a greater magnification of the ordinates of a surface profile than of the abscissas.

---

[3] Preferred numbers, ASA Z17.1–1936 (New York: American Standards Association 1936).

[4] M. Ten Bosch, *Vorlesungen uber Maschinenelemente*, 2nd ed. (Berline: Julius Springer, 1940), p. 10.

7

Design of Castings

7.1 GENERAL CONSIDERATIONS

Both the art and the science of casting the various metals have been developed and specialized to such an extent that it is impossible for the average designer to grasp all the requirements and details which are conducive to the best results. When designing a more or less important part, the designer should always consult the foundry man and the pattern maker, whose cooperation is a prerequisite of ultimate success. However, every designer should know and observe the general rules which are given below:

Basic rules: The two main rules that should be observed in the design of castings are the following:

1. *The section thickness should preferably be uniform.*
2. *Where section uniformity is not possible, light sections should be blended into heavy ones (7.1).*

1. In the conventional design in Fig. 7.1a the section thickness is not uniform. By removing excess metal, as in Fig. 7.1b, the design is considerably improved, and the casting is made sounder, stronger, and lighter.

2. *Blending*: An abrupt change from a thin section to a heavy one, as in Fig. 7.2a, will produce a high stress concentration in the juncture and may result in shrinkage cracks or in cracking when a tensile load is applied. On the other hand, a big fillet, as in Fig. 7.2b, is apt to produce a large local accumulation of metal

which will still be liquid when the outside fibers have solidified, with the result that there will be a porous inside or even a blow hole *e*. A moderate fillet, as in Fig. 7.2c, with a radius *r* approximately equal to $0.5h$, will avoid an excessive stress concentration without producing an unsound casting. A gradual taper, as in Fig. 7.2d, will give still better results as it avoids re-entrant curve.

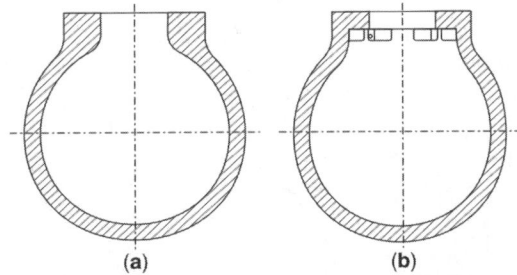

Fig. 7.1: Cylinder with opening for a pipe flange

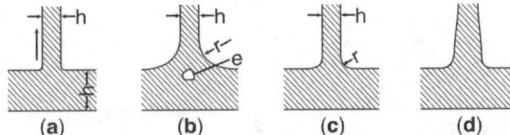

Fig. 7.2: Junctures of different sections

In Fig. 7.3a is shown a poor design that can be improved in several respects as shown in Fig. 7.3b. There is a gradual change from a thin wall to the heavy flanges; a generous radius is inserted at the juncture of the two pipes; and the inside portions of the flange faces are raised as working surfaces to decrease the amount of machining.

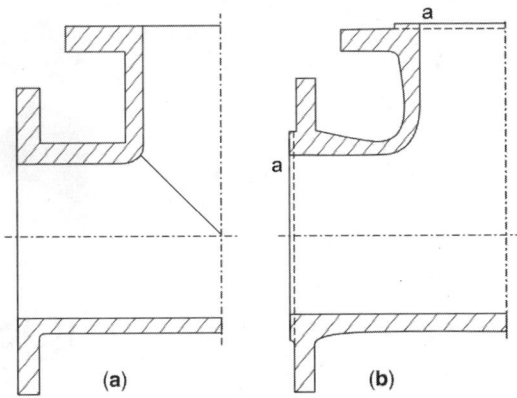

Fig. 7.3: Poor and good design of a tee-shaped casting

3. *Mass distribution*: When a heavy section or a boss is surrounded by light sections, it is important to secure proper feeding of the boss by placing ribs with a sufficiently large cross section on the light parts. This will prevent unsound casting.

The main sections should be made sufficiently heavy, or thick, to permit the addition of a gate of such size as to secure good feeding of all sections.

At a sharp corner, as in Fig. 7.4a, the casting will be unsound and there will be high stress concentration. Whenever possible a rounded corner should be used, as in Fig. 7.4b.

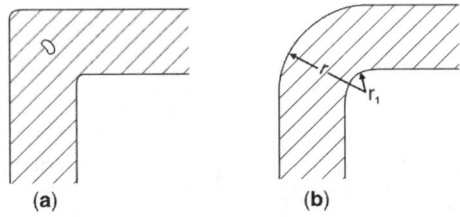

Fig. 7.4: Faulty and correct corners

Consideration must be given to possible constraining effects which may prevent complicated castings from responding freely to necessary changes of form during cooling in the mold.

Intersecting ribs of the types shown in Fig. 7.5a should be avoided, as they are apt to produce an unsound casting. Six ribs, as in Fig. 7.5b, make a poor design because of shrinkage cavities, which can be avoided by the arrangement in Fig. 7.5c.

Unsoundness due to excessive rigidity resulting from intersecting ribs on a flat surface, as in Fig. 7.6a, can be avoided by staggering the ribs as shown in Fig. 7.6b.

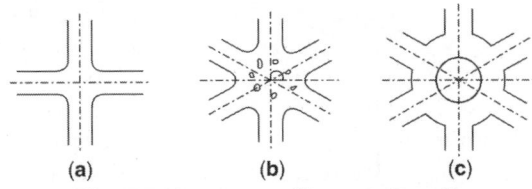

Fig. 7.5: Junctures of intersecting ribs

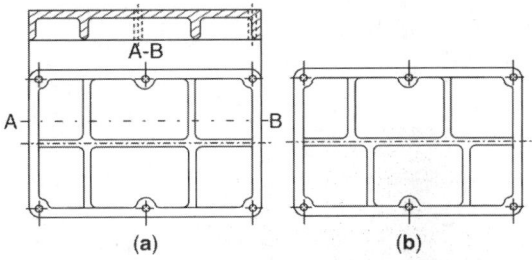

Fig. 7.6: Ribbing of a flat surface to make it more rigid

4. *Working surfaces a* should be provided, as in Fig. 7.3b, to cut down the cost of machining. Properly arranged working surfaces also contribute to the rigidity of a construction.

5. *Minimum thickness*: The required thickness of a casting determined by calculations based on strength alone is often too small to permit the production of a good casting. Other sections carry either very small loads or no loads at all. In every such case the thickness of the section must not be less than a certain practicable minimum value, which depends on the size and intricacy of the part and on the kind of metal used for the casting.

Small gray-iron castings may have sections as thin as 3 mm. However, the average minimum thickness for gray iron is about 6 mm for parts up to about 450 mm in length or diameter; and it gradually increases to 20 mm for large and heavy castings. For malleable- and ductile-iron castings, the minimum thickness may be taken as 3 mm, but the average minimum thickness of larger castings should be about 6 mm. For steel, the minimum thickness is usually regarded as 6 mm. In a small casting, however, it can be lowered to 1.5 mm. For brass and bronze the minimum thickness is 2 mm; and for aluminium it is 3 mm with 1.5 mm as an average minimum.

6. *Ribs* are added to make a construction either stronger or more rigid. While they always increase the rigidity, they may fail to increase the strength. If the addition of ribs lowers the maximum stress in a part, the ribs strengthen it. If the maximum stress in the main part of the original design is lowered, but the stress in the added rib is equal to or higher than the stress in the original design, the rib is either useless or weakens the part. *However, ribs are provided in the piston to increase the surface area for heat dissipation, it is an exception (7.5).* A crack may develop in the highly stressed rib, and it will eventually spread and cause the part to fail. It should be remembered that rigidity is measured by deflection, which is inversely proportional to the moment of inertia I of a section; whereas the stress is inversely proportional to the section modulus Z. The resulting conditions can be best illustrated by a numerical example.

Example 7.1: A cast-steel plate 0.5 m long, 0.39 m wide, and 20 mm thick is supported at the ends and carries a uniformly distributed load of 9 kN as indicated in Fig. 7.7. Find the influence of conventional ribs upon the rigidity and strength of the plate.

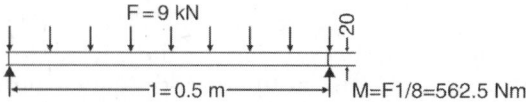

Fig. 7.7: Loading of the plate

The bending moment due to the load is

$$M = \frac{1}{8} \times 9000 \times \frac{500}{1000} = 562.5 \text{ Nm}$$

Naturally the ribs must extend in the direction of the greatest dimension, or lengthwise. Dividing the width into three strips, each 130 mm wide, and adding 6 mm ribs to double the thickness of the plate, we obtain the section in Fig. 7.8b. In Table 7.1 are given I, Z, W and s of the flat in a, and of the plate in b, after small ribs are added. Since the moment of

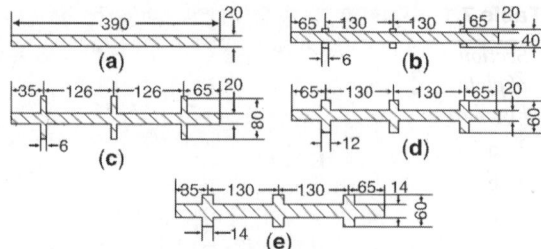

a-Strong but rigidity low
b-More rigid but stress excessive
c-Very rigid and strong but difficult to cast
d-Very rigid, very strong but heavier
e-Strong and rigid, some weight as 'a'

Fig. 7.8: Various ribs added to a plate

inertia is increased from 26 cm⁴ to 34.4 cm⁴, the rigidity is increased by about 32 per cent. However, as the maximum stress is increased from 21.63 to 32.89 N/mm², the strength is decreased by 34 per cent.

The types of ribs in Fig. 7.8c were provided in order to obtain at least the same strength as in Fig. 7.8a. By increasing the height to 80 mm, the plate became 3.9 times as rigid with a slight increase of the stress from 21.63 to 22.14 N/mm². These results are also shown in Table 7.1. However, it will actually be very difficult to cast such thin and high ribs, and they will cause additional stresses because of the abrupt change of the thickness from 20 mm to 6 mm in case (c).

A better type of rib is shown in Fig. 7.8d. The height of each rib is equal to the thickness of the plate, and the width of the ribs is twice that of the two previous designs. As shown in Table 7.1, the rigidity in Fig. 7.8d is only slightly smaller than in Fig. 7.8c, while the strength in Fig. 7.8d is about 11 per cent above that of the flat plate. The weight W is not appreciably affected by the ribs.

Finally, in Fig. 7.8e is shown a design which has still greater rigidity and greater strength, and is even slightly lighter than the flat plate. In this case the body of the plate is made thinner and the ribs are of the approximately same thickness as the plate. This plate will have only very small internal casting stresses and a little stress concentration.

A quicker and more precise method than cut-and-try for determining the proper dimension of ribs exists.[1]

[1] A. Thum and S. Berg, "Uber die Festigkeit von Rippen bei ruhender, wechselnder und Stossartigen Belastung," *Zeitschrift Verein Deutscher Ingenieure*, Vol. 77 (1933), pp. 281–81; V. L. Maleev, "Why Add Ribs That Don't Add Strength?" *Machine Design*, Vol. 8, No. 9 (September, 1936), pp. 29–32.

Table 7.1: Rigidity and strength of a plate with ribs

Section of plate (Fig. 7.8)	Moment of inertia I cm⁴	Section modulus cm³	Weight W N	Highest stress N/mm²	Relative rigidity	Relative strength	Relative weight
a	26	26	30.6	21.63	1	1	1
b	34.4	17.2	32.0	32.89	1.32	0.66	1.04
c	101.6	25.4	34.9	22.14	3.9	1.02	1.14
d	88.4	29.5	36.3	19.06	3.4	1.13	1.18
e	100	33.3	30.0	16.89	3.8	1.28	0.98

Impact loading: It has already been shown that to decrease the danger of failure of a part subjected to impact loads, the resilience of the part involved must be increased. Such a change in design, in turn, requires greater deformations. *Since ribs have an opposite effect, it is obvious that they should be avoided where impact loads are anticipated (7.6).*

General conclusions: Ribs should be used chiefly for static loads. Wide and low ribs are safer than thin and high ones.

7. *General remarks about shape*: There are many features of a good design which cannot be expressed by figures or formulas. A statement which will be found helpful in designing, perhaps more valuable to an experienced designer, *is that although a designed system which looks wrong, usually is wrong, the reverse is not always true.* The following advice is intended to be helpful in shaping machine parts.

8. *Light sections and protruding extensions should not be cast in one piece* with a heavy section or part, in order to avoid scrapping of the heavy piece through breakage of an extension. The design shown in Fig. 7.9 is incorrect because brace c is out of proportion to the rest of the bed. A correct, two piece design is shown in Fig. 7.10.

9. *Hollow or box sections should not be combined with ribbed sections* since this design complicates the casting unnecessarily. Piston is an exception.

10. *Straight surfaces* in most cases look better than curved surfaces and are cheaper to obtain in a pattern. However, the parabola is the proper shape for a heavy pedestal which resists a bending moment, such as the pedestal in Fig. 7.11, because its strength is more nearly uniform.

11. *A curved reinforcing rib a, Fig. 7.12, is better than a straight one b*, as it is less rigid and

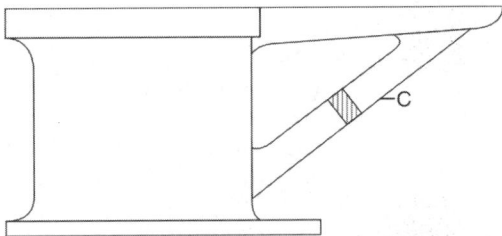

Fig. 7.9: Bed cast with an extension

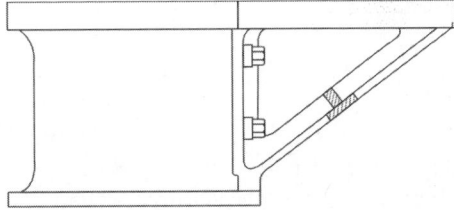

Fig. 7.10: Bed with a separate extension

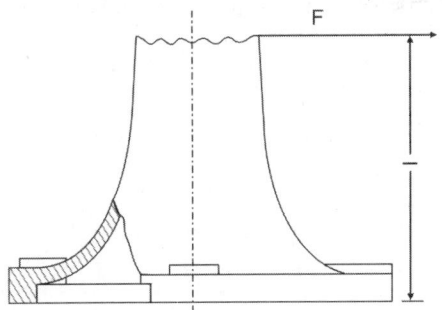

Fig. 7.11: Pedestal loaded in bending

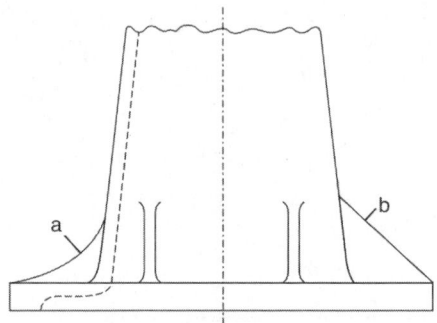

Fig. 7.12: Reinforcing ribs

causes less stress concentration. However, the maximum bending stress in rib *a* will be higher than that in rib *b*, and care must be taken to put in a sufficient number of comparatively heavy ribs.

12. *A casting without cores is preferable to one with cores:* A casting that can be molded in green sand without cores, such as that shown in Fig. 7.13, is less expensive than a casting requiring a core, such as that in Fig. 7.11 or in Fig. 7.12. If a casting must have a core, it should be designed so as to provide support for the core without the use of chaplets and so as to permit easy removal of the core from the casting. Since chaplets may cause leakage, they are especially objectionable for cast pieces that must be under fluid pressure.

13. *The old practice of hiding the reinforcing bead, as shown in Fig. 7.14 at a, is wrong.* To lower the stress concentration a symmetrical bead *b* should be used. The old advice to make a web between flanges as thin as it can be cast in order to save material, is not correct. Stress concentration at the juncture of sections that

differ widely in thickness, as in Figs 7.15a or b, is likely to cause failure. A heavy web, as in Fig. 7.15c, with a more even distribution of material, is much to be preferred.

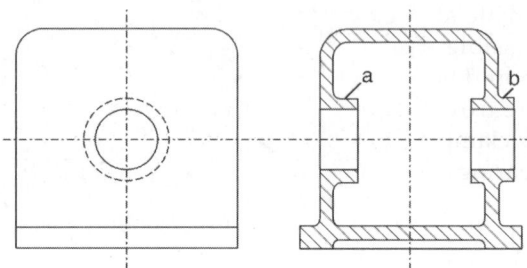

Fig. 7.14: Reinforcing an opening

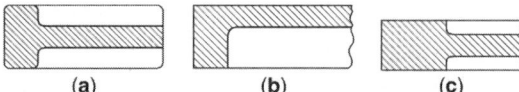

Fig. 7.15: Shape of web between flanges

14. *Pattern shaping*: Parting lines should be made as even as possible. The design should provide for ample draft for easy molding. While the pattern maker undoubtedly will put in the necessary minimum draft, even when the design does not show any, as in Fig. 7.16a, it will be considerably cheaper to mold a casting from a pattern with ample draft, as shown in Fig. 7.16b, and there will be fewer spoiled and rejected castings.

15. *Special requirements*: Because of certain distinctive properties of the various metals used for castings, different precautions should be taken with each material.

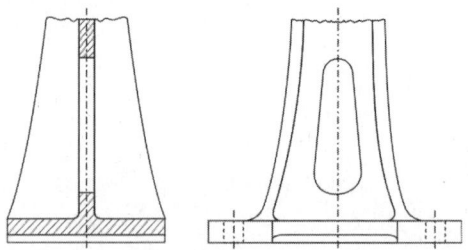

Fig. 7.13: Pedestal cast without a core

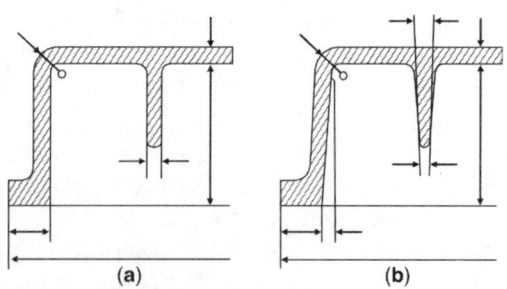

Fig. 7.16: Indication of draft

(a) *Cast iron: Tensile stresses in cast iron should be avoided*. Both the ultimate strength and the yield point of cast iron in compression are several times as great as those in tension. Cast-iron parts should therefore be so designed that tensile stresses will be avoided as much as possible. Sections subjected to tensile stress should be either reinforced or relieved of the stress by appropriate tie rods, as in Fig. 7.17, or by clamp-shaped devices such as forged-steel bearing caps *b* in Fig. 7.18.

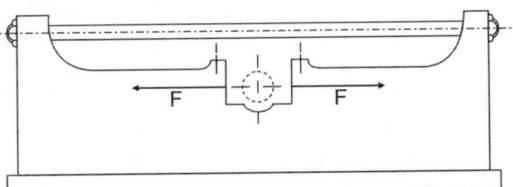

Fig. 7.17: Bedplate with a tie rod

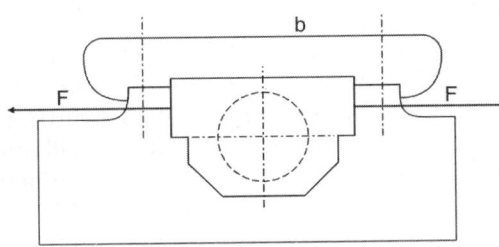

Fig. 7.18: Steel bearing cap

For the same reason, sections of parts subjected to bending in one direction should not be made symmetrical with respect to the neutral plane. In Fig. 7.19 is given a typical section for a cast-iron beam.

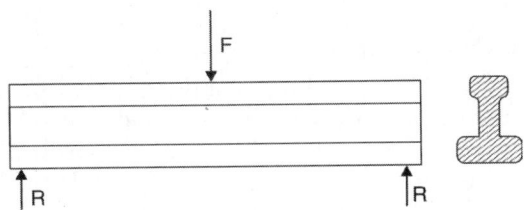

Fig. 7.19: Cast-iron beam

In the case of a heavy load, as in the punch frame in Fig. 7.20, the box section is made with a heavy wall in tension. If the shape at section *c-c* is as shown in Fig. 7.20a, stress concentration may cause the frame to break; whereas the shape shown in Fig. 7.20b, because of the gradual change of wall thickness, gives a satisfactory frame. However, in this case a welded steel construction is a still better solution, being stronger, more rigid, and less expensive.

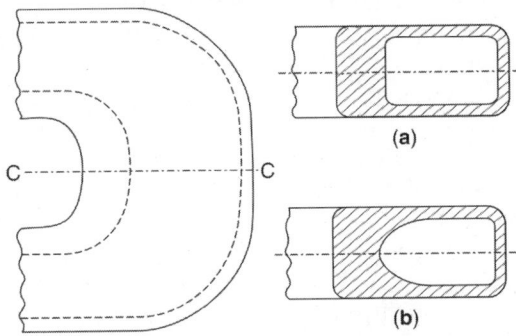

Fig. 7.20: Punch-press frame

(b) *Malleable iron*: Dry-sand cores for malleable iron castings should be avoided wherever possible, especially where the casting, in cooling, contracts around a core.[2]

(c) *Steel casting: Steel is not as fluid as cast iron. Therefore, complicated shapes, sharp corners, and thin sections cannot be obtained (7.2).* If a cast-iron rocker arm with the cross section shown in Fig. 7.21a is to be cast in steel for greater

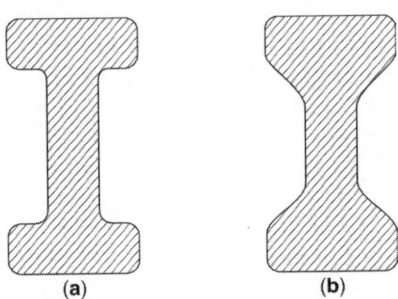

Fig. 7.21: Beam sections of cast iron and steel

---

[2] F. A. Halsey, Handbook for Machine Designers, 2nd ed. (New York: McGraw-Hill Book Company, Inc., 1916), p. 480.

strength, the pattern must be changed to produce the shape shown in Fig. 7.21b. If made of phosphor bronze, which is almost as strong as cast steel and is more fluid, the section need not be changed.

All steel castings must be annealed to relieve them of internal stresses. Their strength can be increased by subsequent heat treatment. This fact should be borne in mind when designing a piece.

Light alloys: It should be remembered that the mechanical properties given in Tables 4.5 and 4.6 refer to small test sections. In castings the values are 10 to 20 per cent lower.

The comparatively low modulus of elasticity E of light alloys, particularly magnesium alloys, requires the use of sections with large section moduli if a certain degree of rigidity is desired. Thus the engine bed in Fig. 7.22a, which is designed along the lines of a cast-iron casting, would be strong but would not be rigid enough to take the load F. The change to the design in Fig. 7.22c increased the rigidity, but not sufficiently. Finally the box-shaped construction in Fig. 7.22d proved to be entirely satisfactory. The same result could be attained by using a heavy T-section, as in Fig. 7.23b. However, this construction is heavier and presents the probability of a porous center e because of an excessive accumulation of material. The use of lower allowable stresses is therefore necessary.

In Fig. 7.23a is shown the proper section for a light-alloy beam bent in one direction, as

compared with the section of a cast-iron beam in Fig. 7.23b.

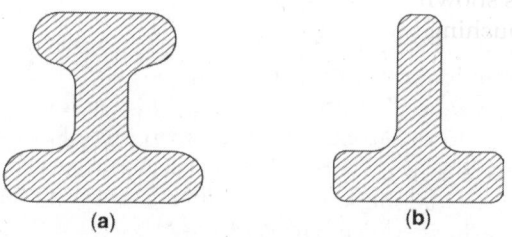

(a)   (b)

Fig. 7.23: Beam sections of aluminium and cast iron

Light alloys are more sensitive to stress concentration produced by fillets and notches than is cast iron. Fillets should be made larger; notches should be avoided; and as much uniform wall thickness as possible should be provided.

The strength of aluminium sections in tension can be increased by cast-in steel reinforcing rings or anchors.

The high coefficient of linear expansion of a light alloy with a change in temperature should be kept in mind. Also the designer must not overlook the fact that light alloys lose a considerable part of their strength at temperatures above 260°C and that they should not be used at all for parts whose operating temperature is over 260°C.

With proper distribution of material an aluminium casting will show a saving of weight up to 50 per cent.

A screw joint should be designed so as to divide the load over a comparatively large number of small screws. *Large washers should be used under boltheads and nuts to lower the specific bearing pressures, which should not exceed 1.4 MN/m² for static loads and should never be more than 2 MN/m² (7.4).* The bosses under such washers must be proportionately large. Holes can be tapped in the alloy if the screws are not unscrewed frequently. The length of thread should be 2.5d. If the screws must be unscrewed often, the holes should be drilled and tapped in cast-in buttons of a harder material, such as brass or Monel metal, with which aluminium castings form a good bond.

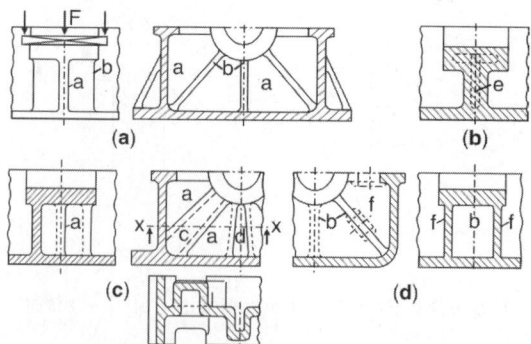

Fig. 7.22: Bearing bridge in a bedplate cast of aluminium alloy

In Fig. 7.24 the use of cast-in, transverse steel buttons *a* for distributing the load of a stud *b* is shown. Another example of a cast-in tapped bushing is given in Fig. 11.30.

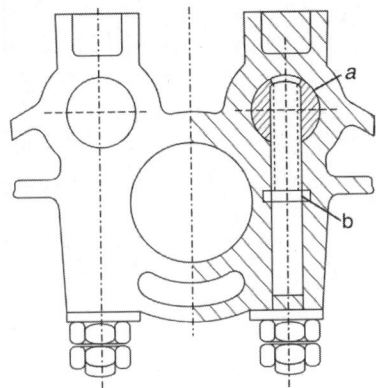

Fig. 7.24: Cast-in steel button for a stud

Silumin: A heat-treated aluminium-silicon alloy, known as silumin-gamma, has such high strength characteristics that parts cast of it can be made with the same wall thickness as when cast iron is used.[3] However, the other design requirements must be observed. Working stresses in tension for silumin-gamma can be up to 10 per cent higher than those given in Table 8.3 for shielded-arc steel welds. The exact increase depends on the character of the loading. However, the allowable crushing stress is 120 MN/m² for a static load and 60 MN/m² for variable loads.

PROBLEMS

7.1 What are the two basic rules for design of castings?

7.2 Describe all the consideration for design of castings.

7.3 Why is it necessary to have larger thickness of steel castings?

7.4 Why is it desirable to have large washer below nuts and bolt heads?

7.5 Why are ribs provided in the piston through the design gets more complicated and rigid?

7.6 What is the effect of ribs in parts subjected to impact loads?

7.7 Determine suitable cross sections x-x, y-y, and z-z for a cast-iron rocker arm, Fig. P7.1, assuming that $F = 5$ kN, $l_1 = 44$ mm, $l_2 = 150$ mm, $l_3 = 225$ mm and $l = 175$ mm, $S_e = 80$ N/mm².

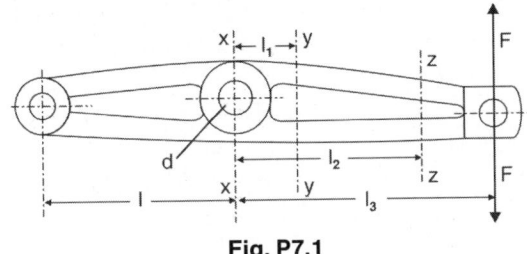

Fig. P7.1

7.8 Using data of problem 7.7 but assuming that the rocker arm is cast of gun metal, determine suitable cross sections, $S_e = 85$ N/mm².

7.9 Using the data of problem 7.7 but assuming that the rocker arm is of 0.20% C cast steel, determine suitable cross sections. $S_e = 170$ N/mm².

7.10 Using the data of problem 7.7 but assuming that the rocker arm is of ordinary malleable iron, determine suitable cross sections. $S_e = 200$ N/mm².

7.11 Using the data of problem 7.7 but assuming that the rocker arm is made of ductile iron, determine suitable cross sections. $S_e = 400$ N/mm².

7.12 Determine the proper cross sections and weight W of a cast-iron beam, Fig. P7.2, to carry a dead load F of 45 kN, if $l_1 = 1.11$ m

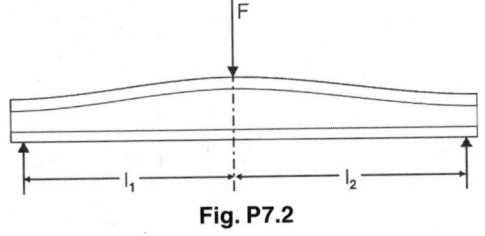

Fig. P7.2

---

[3] J. Dornant, Silumin-Gamma (Frankfurt am Main: Metallgesellschaft A. G., 1938).

and $l_2 = 1.3$ m. Use cast iron, and design the beam so as to obtain a low weight.

7.13 (a) Determine the proper cross section and the weight per linear meter of a cast-iron beam of constant section, Fig. P7.3, to carry a load F of 50 kN. The spans are $l_1 = l_2 = 1.2$ m, (b) Determine the number, size and spacing of the bolts which connect the two halves of the beam,

(c) Give the necessary sketches with all dimensions of the centre joints, showing in detail the necessary machining surfaces.

7.14 Determine the width b and the height h of a hollow beam of class FG150 cast iron, Fig. P7.4, assuming that $h = 2 b$. The span is $l = 3$ m, $l_1 = 1.5$ m and $l_2 = 0.9$ m and the loads are $F_1 = 10$ kN and $F_2 = 15$ kN.

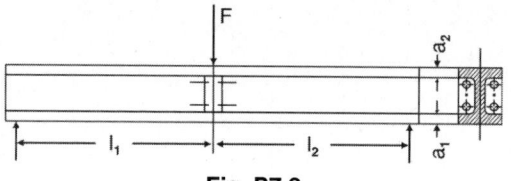

Fig. P7.3

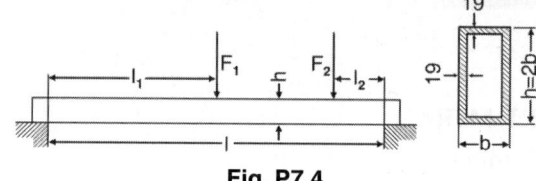

Fig. P7.4

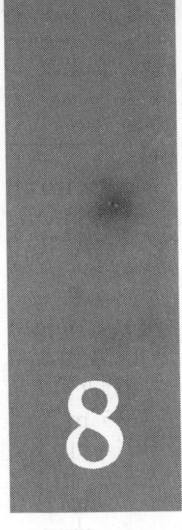

8 Design of Weldments

8.1 METHODS OF WELDING

A machine part or structure whose component parts are joined by welding is called a *weldment*.

Welding is defined as the localized intimate union of metal parts in the plastic and molten states with the application of blows or mechanical pressure, or the union of parts in the molten state without any pressure. There exist three main methods of welding: Forge welding, pressure welding, and fusion welding.

Forge welding is the oldest process known. It is accomplished either by hand hammering, as practiced by blacksmiths, or by machines, as in the manufacturing of wrought-iron pipes and pressure vessels. Forge welding can be applied only to wrought iron or low-carbon steel.

Pressure welding, or *resistance welding*, utilizes the heat created by an electric current which passes through the two pieces to be welded. The current meets with a much higher resistance at the surfaces of contact than in the body of the metal, and it raises their temperature to a molten state; and the applied pressure upsets the edges, which become united. This welding, if properly done, results in a joint which is practically as strong as the body of the metals joined. Because it is difficult to obtain a uniform area contact and to secure a uniform heat distribution, the method is not suitable for large sections—larger than about 25 × 350 mm.

Spot welding is another form of resistance welding. In this process, two sheets are put together as for a lap joint, the overlapped portions being gripped between two electrodes. Pressure is applied to the electrodes and a heavy low voltage current is passed from one electrode to the other through overlapping plates. Heat is developed at the spot of contact until the necessary welding temperature is reached, when the pressure is increased. As a result, the pieces are tacked together. Spot welding is done on thin plates not over 10 mm thick, since it is difficult to develop pressures required for thicker plates.

Seam welding is a development of spot welding. In this process the electrodes are in the form of two rollers, between which the overlapping edges of the thin plates are passed. A continuous strip of welded surfaces is thus produced.

Fusion welding does not require any pressure to form the weld. The places of contact of the two metal pieces to be joined are heated to the fusion temperature of the metal; additional metal is usually applied in the corner of the joint by melting a filler rod of suitable composition, and the joint is allowed to cool. The heating of the metals is produced in one of the following ways: by a burning gas, by an electric arc, in an electric furnace, or by thermit.

Gas welding is often called *autogenous welding*. The *oxyhydrogen process*, the older of the gas-welding processes, uses a blowpipe flame produced by the combustion of oxygen and hydrogen, a temperature of about 2200°C being obtained. It is employed chiefly for welding nonferrous metals of low fusibility.

The *oxyacetylene process* uses the flame produced by the combustion of oxygen and acetylene, and a temperature up to 3500°C is attainable. This process is used for welding steel, steel castings, wrought iron, cast iron, copper and its alloys (including Monel), aluminium, and other commercial alloys. It is also used to weld a metal to a dissimilar metal, such as steel to cast iron or copper to steel.

Arc welding was developed about the end of the previous century but has been employed extensively in production only during the last sixty years. The necessary temperature is produced by an electric arc formed between the pieces to be welded and an electrode held by the operator or moved by an automatic machine. The electrode may be either a carbon rod with a separate filler rod furnishing the additional molten metal or a metal rod which acts as a filler rod.

In *metallic-arc welding* globules of metal from the tip of the electrode are carried across the arc deposited in the joint, where they solidify. The globules can be deposited upward against the force of gravity; this property is a great advantage in structural and bridge work. In carbon-arc welding the heat of the arc forms a small pool of molten metal in the joint to which metal is being added from the filler rod. This process cannot be used for vertical or overhead work.

Deoxidizing and shielding: When the weld metal from the bare filler rod is being deposited, it absorbs oxygen and nitrogen from the atmosphere. These form oxides and nitrides in the completed weld that lower its strength, ductility, and resistance to corrosion. By introducing into the arc stream a *deoxidizing agent*, which has a greater affinity for oxygen and nitrogen than has the weld metal, or by *shielding* the arc and the molten metal from contact with the atmosphere, a weld can be produced which is free from impurities and is

therefore much stronger than when the metal is deposited by an ordinary arc.

One of the deoxidizing methods uses a jet of hydrogen which is projected through an arc formed between two tungsten electrodes and produces a reducing atmosphere. The shielding of the arc is accomplished by employing electrodes consisting of a metal core with a coating of a material which fluxes out the impurities as it burns away. Shielded-arc welds are free from slag inclusion and blow holes and have a finer crystalline structure and a greater integration of weld and base metal in the fusion zone. At present all arc welds where strength is important are made with a shielded arc.

In *electric-furnace welding* the parts to be welded are assembled to give a snug fit by inserting one part into the other, by spot welding, or by pining the parts together. Copper in the form of wire or paste is applied at the joints, and the assembled piece is put into the hydrogen-filled heating zone of an electric furnace. The parts are gradually brought up to a temperature of 1150°C. This process melts the copper, which is drawn by capillary attraction both downward and upward into the seams. The hydrogen atmosphere completely reduces all oxides, scale, and other foreign matter on the surface of the parts and secures chemically clean surfaces for the welds. After the welding is complete, the piece is gradually cooled in the same reducing atmosphere. This procedure insures a complete absence of internal stresses, and surfaces that are clean and free from scale.

Although resembling brazing, this process should nevertheless be classified as welding because the iron and copper form an alloy. Some of the copper goes into a solid solution in the steel, and some of the iron is dissolved by the copper, to produce an integral copper-iron alloy bond between the parts. The weld thus obtained is actually stronger than the steel itself, as proved by a number of tensile, compressive, shearing, and bursting tests.

The main field of application of electric-furnace welding is for manufacturing machine parts of intricate shape, such as diesel cylinder heads. Such parts produced by copper-hydrogen electric welding not only are stronger, lighter, and of better appearance but with proper design can be produced at a lower cost.

Thermit welding is based on the generation of heat by the chemical combination of iron oxide and powdered aluminium. When the mixture is ignited, the aluminium reduces the oxide to molten steel, which unites with the parts to be joined. The temperature of the reaction reaches almost 2800°C. This process is of great value in repairing heavy cast-iron and steel parts but is not practical for production manufacturing.

8.2 FIELD OF APPLICATION

Of the many types of welding, the one most used in the manufacture of machinery is arc welding; next comes gas welding. Only these two methods will be discussed in connection with design of weldments.

Welding of steel: Most weldments are made of low-carbon steel. There are two general fields of application for welding steel: (a) when welding is substituted for riveting and (b) when welding is used as an alternate method for casting or forging.

The main advantages of welded joints over riveted ones are:

(a) Lighter weight, due to elimination of straps, gusset plates, clip angles, etc. (the weight of welded joints is about 2–3% less than those of rivelded joints).

(b) Greater strength. With proper care the joint can be made with an efficiency of 100 per cent, and with the development of X-ray and Gamma ray and other methods of inspection the strength of a welded joint can be determined with out destroying the piece.

(c) Lower cost, chiefly because of the lighter weight.

(d) Noiselessness. Since it has been established that noise is not only disagreeable but harmful to both the efficiency and health of workers, noiselessness has become a rather important factor.

An important development is the arc-welding of low-creep Cr-Mo-alloy steels used for making boilers and superheater tubes subjected to pressures of 7.3 MN/m^2 and temperatures of 500°C.[1]

The main advantages of welded machine parts over cast parts are:

(a) Lighter weight, because the metal is utilized better.

(b) Greater strength. The metal is put exactly where it is needed and the danger of stress concentration due to discontinuities can be eliminated.

(c) Lower cost, due to elimination of patterns and to lighter weight.

Table 8.1 in connection with Fig. 8.1 gives some idea of how different fabrication methods

Table 8.1: Comparison of fabrication methods for a bearing support (Fig. 8.1)

Basis for comparison	(a) Cast iron (per cent)	(b) Cast steel (per cent)	(c) Forged steel (per cent)	(d) Welded steel (per cent)	(e) Welded steel (per cent)
Static strength	100	100	100	100	100
Thickness	100	60	54	54	74
Deflection	100	191	29	259	100
Vibration damping	100	28	26	26	36
Weight	100	76	72	71	85
Cost	100	110	300	45	54

[1] H. Schottky, "Das Schweissen der warmfesten und hitzbestandigen Stahllegierungen,"*Zeitschrift Verein Deutscher Ingenieure*, Vol. 79 (1935), pp. 41–46.

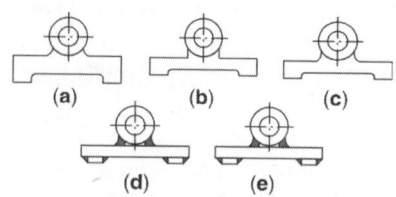

Fig. 8.1: Shaft support fabricated by different methods

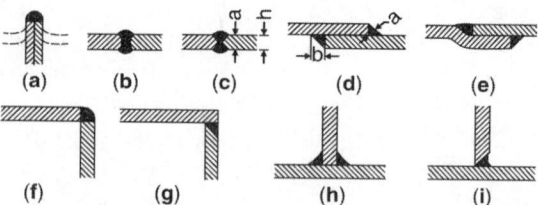

Fig. 8.2: Forms of welded joints

affect the characteristics of a certain machine part. The five alternative designs of a bearing support in Figs 8.1a, b, c, d and e have the same static strength but differ in rigidity, vibration-damping capacity, weight, and cost. If rigidity and damping capacity are not important, the welded design in Fig. 8.1d will result in a large saving in cost. If rigidity is essential, the welded design in Fig. 8.1e will again give a substantial saving in cost with the same rigidity as a casting and will give a greater strength. However, if vibration damping is important, the best results are obtained by using a gray-iron casting.

Welding of aluminium: Aluminium and its alloys can be welded by any of the various methods used in welding steel plates. Since many aluminium shapes and castings are heat-treated, it should be remembered that welding anneals the joint and thus decreases its strength almost to the untreated values. The weld itself has a cast structure, and to increase its strength it must be reinforced in the same manner as a steel weld. More detailed information may be found in special trade literature.[2]

8.3 WELDED JOINTS

There are five basic forms of welded joints: *edge, butt, lap, corner,* and *tee.* These forms are illustrated in Fig. 8.2, where the weld is shown in black. The joint shown by full lines in Fig. 8.2a is a plain edge joint; and the dotted lines show a so-called *leaf edge joint* obtained by bending the plates after the weld has been made. In Fig. 8.2b a single-welded butt joint is shown; that in Fig. 8.2c is a double-welded butt joint; those in Fig. 8.2d and e are lap joints; those

in Figs 8.2f and g are corner joints; and those in Figs 8.2h and i are tee joints.

A joint in which the base metal plates are brought close together, as in Figs 8.2e or h, is called a *closed joint*. A joint with a gap between the plates prior to the welding, as in Fig. 8.2b, is called an *open joint*.

In an edge weld, the weld metal is deposited along the edges of the plates, as in Fig. 8.2a. In a *butt weld,* the weld metal is deposited between the edges to be joined, as in Figs 8.2b, c, g, or i. In a fillet weld, the weld metal is deposited in a corner formed by the two surfaces to be joined, as in Figs 8.2d, e, f, or h.

Spot-welded joints are formed by so-called *tack welds*.

From the standpoint of strength, welds are classified as reinforced, flush, or concave. The welds in Figs 8.2a, b, c, and f are reinforced; those in Figs 8.2d, e, g, and h are flush; and that in Fig. 8.2i is concave. The usual amount of reinforcement is one-fourth the depth of a flush weld. A *calk weld* has for its main purpose the tightness of the joint. The main requirement of the majority of joints is tensile strength, but tightness is also important in some cases. The dimension *a*, in Fig. 8.2, is called the throat, and *b* is the leg of a wed.

The edges of the plates to be welded may have various forms some of which are shown in Fig. 8.3. The selection of the edge form depends upon the welding procedure, the plate thickness, and whether both sides of the joint or only one can be welded.

Designations on drawings: In Fig. 8.4 are shown samples of the designations used for welds on drawings. According to American

---

[2] Aluminium Company of America, *Welding and Brazing Alcoa Aluminium* (Pittsburgh: 1944).

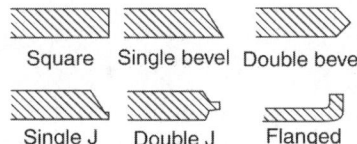

Square   Single bevel   Double bevel

Single J   Double J   Flanged

Fig. 8.3: Forms of edges for welding

practice, a weld that appears in cross section is shown filled in and is given its true shape as it is to be made—reinforced, flush, or concave. A weld that is visible in a side view, or on the near side, is represented by a row of crosses; and a weld that is invisible, or on the far side, is represented by a row of short lines inclined at an angle of 45°. A butt joint with double welds is indicated by a zigzag line when reinforced and by a coil when flush.

The International Committee on Standardizations recommends the use of the following symbols in a cross section: a full quadrant for a reinforced fillet weld, a triangle for a flush weld, and an angle with an inscribed quadrant for a concave weld. In a side view the type of weld, the throat dimensions t and the length l are given over the leader line when the weld is visible, and under the leader line when the weld is not visible. The American Welding Society advocates the use of this system with a few additional refinements.

8.4 STRENGTH OF WELDED STEEL JOINTS

Welded joints subjected to tension may be divided into three classes with respect to the internal stresses set up by the external load. These are shown in Fig. 8.5. In class 1 the weld is subjected only to longitudinal shear as in Fig. 8.6a; in class 2 the weld is in transverse shear and in tension and bending as shown in Fig. 8.6b. It is quite complex particularly since the distribution of these stresses may not be uniform due to variations in workmanship and metallurgical conditions. *It is conservative to assume that the failure occurs along the throat section in both the cases as in Fig. 8.6c due to shear stress (8.1)*

$$S_s = \frac{F}{tl} \text{ or } t = \frac{F}{S_s \times l}$$

and size s or h is $\sqrt{2t}$. The value of S_s is taken as 95 N/mm². In class 3 the weld is in pure tension.

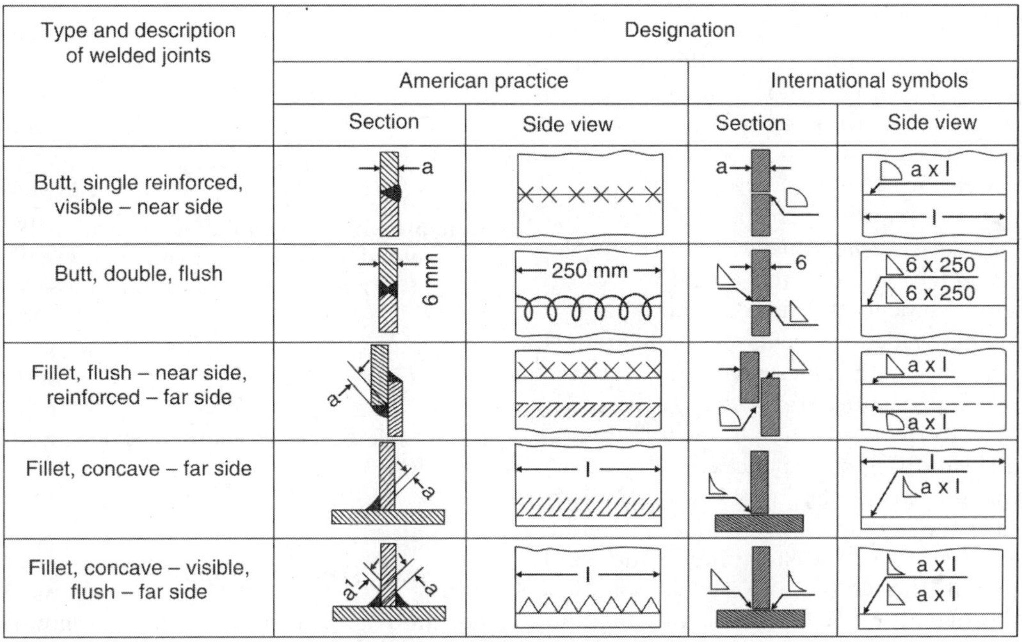

Type and description of welded joints	Designation			
	American practice		International symbols	
	Section	Side view	Section	Side view
Butt, single reinforced, visible – near side	←a	××××××	a→	a x l
Butt, double, flush	6 mm	← 250 mm →	6	6 x 250
Fillet, flush – near side, reinforced – far side	a	××××××		a x l
Fillet, concave – far side	a	← l →		l
Fillet, concave – visible, flush – far side	a, a	← l →		a x l

Fig. 8.4: Designations of welds on drawings

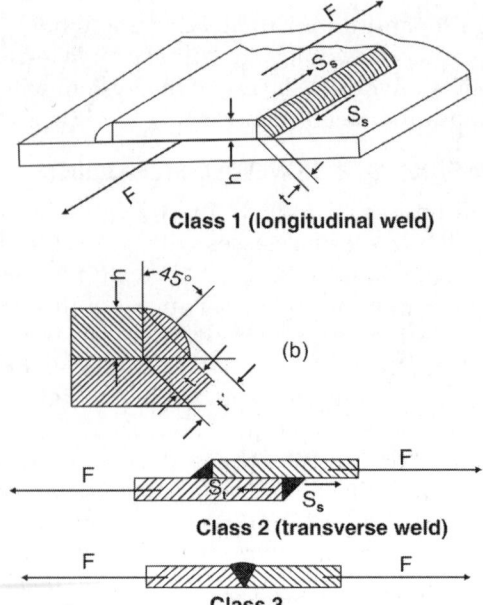

Class 1 (longitudinal weld)

(b)

Class 2 (transverse weld)

Class 3

Fig. 8.5: Force action in various classes of welds

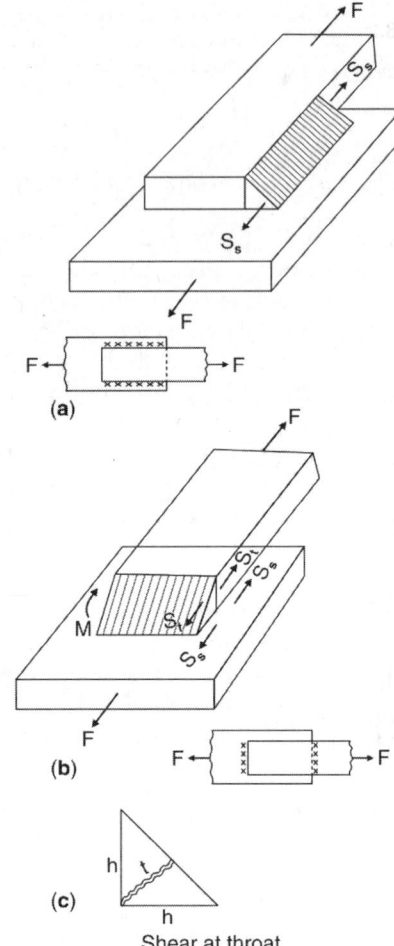

(a)

(b)

(c)

Shear at throat

Fig. 8.6: Stresses in welds

It should be noted that in a weld of class 1, the shear area is $2tl$ where l is the effective length of each weld. In a weld of class 2 the shear area is $2tl$, where t is the throat thickness of the weld and is often made equal to the thickness of the plate. In a flush weld, $t = 0.707\,h$. Repeated tests have shown that the length of weld and all other factors being equal, weld of class 2, in which the linear dimension is normal to the lines of stress are about 30 per cent stronger than welds of class 1, in which the linear dimension is parallel to the lines of stress.

Approximate strength calculations: In the design of new machine parts and structures under moderate static loads, the required lengths of joints with shielded-arc welds may be determined in accordance with values found satisfactory in practice. These values are given in Table 8.2. For bare-electrode welds the values in Table 8.2 should be reduced by 20 per cent.

In arc-welded joints, about 12 mm should be added to the calculated length of each weld to allow for starting and stopping of the weld, because of the weakening by the magnetic blowing of the arc. For approximate calculations 12 mm is omitted.

Example 8.1: Determine the total load that a shielded-arc-welded joint of class 1 (Fig. 8.5) can carry if it joins 10 mm plates and each fillet is 120 mm long.

The effective length of the two 10 mm welds is

$$l = 2 \times (120{-}12) = 216 \text{ mm}$$

From Table 8.2 the safe load per mm length is 440 N. Therefore,

$$F = 216 \times 440 = 95040 \text{ N}$$

8.5 DESIGN OF STEEL WELDS

For more accurate design the values of design stresses for shielded-arc flush welds are given in

Table 8.2: Allowable loads on mild steel welds

Size of weld mm	Allowable static load per linear mm of weld N			
	Bare welded rod		Shielded arc	
	Normal weld	Parallel weld	Normal weld	Parallel weld
3	170	135	210	170
5	275	220	345	275
8	440	350	550	440
10	550	440	700	550
12	660	530	840	660
15	840	660	1050	840
20	1100	900	1400	1100

Table 8.3.[3] These values apply for painstaking workmanship, proper supervision, and careful inspection; with less skill or less care, the values should be lowered. For bare-electrode welds the allowable stresses in Table 8.3 should be multiplied by 0.8. For gas welds these values should be multiplied by 0.8 to 0.85, the factor depending on the skill of the welder.

In computing the area of the cross section of a reinforced fillet, the throat dimension t' Fig. 8.5b, should be taken as $1.2t$.[4] For a concave weld it is safe to take t' as $0.5t$.

The *efficiency* of a welded joint is the ratio of the resistance of the joint to the resistance of the weakest section of the base metal. Since the lengths of the welds in many cases are not limited, an efficiency of 100 per cent or even higher can be easily obtained.

Initial stresses: Initial stresses are unavoidable in welds, but they are not very dangerous for static loading because if the yield point is passed at the most stressed point, the metal starts to creep at this point and the load then redistributes itself more evenly over the cross section.[5] After several local overloadings the stress distribution in the structure approaches that which would exist without initial stresses. *However, this is true only for ductile materials. Therefore the most important requirement for a weld material is high ductility.*

Table 8.3: Strength of shielded-arc flush-steel welds

Kind of stress	Limit stress N/mm²			Recommended design stress N/mm²		
	Base metal Elastic limit	Deposited metal		Safety factor 2 static load	Safety factor 2 load varies from 0 to F	Safety factor 2.75 load varies from +F to −F
		Elastic limit	Endurance limit			
Tension	220	275	145	110	100	55
Compression	240	300	–	125	110	55
Bending	240	300	180	125	110	65
Shear	140	165	–	95	70	35
Shear and tension	–	–	–	95	70	35

[3] O. Graf, "Dauerfestigkeit von Schweissverbindungen," Z.VDI, Vol. 78 (1934), pp. 1423–27; G. D. Fish, *Arc-Welded Steel-Frame Structures* (New York; McGraw-Hill Book Company, Inc., 1933), p. 89.
[4] Graf Poc. cit.
[5] J. Mather, "Determination of Initial Stresses by Measuring Deformations around Drilled Holes," *Transactions of the American Society of Mechanical Engineers*, Vol. 56 (April 1934), IS-56-2, p. 254.

With variable loads initial stresses become dangerous and must be relieved by annealing. Especially dangerous is the presence of initial stresses with impact loads. The weldment must be carefully stress-relieved. During the process of welding, each layer of deposited metal should be peened; and the finished weldment should be heat-treated.

Stress concentration: Since abrupt changes in cross sections at welds are unavoidable, stress concentration will develop at these places. For static or steady loads, stress-concentration effects may be neglected; for variable loads, and particularly for shock loads, they must be taken into account in the usual way. Values of the stress-concentration factor K' are given in Table 8.4.

Table 8.4: Stress concentration factor K' for welds

Type of weld	K'
Reinforced butt weld	1.2
Toe of transverse fillet weld	1.5
End of longitudinal weld	2.7
T-butt joint with sharp corners	2.0

In a weld, stress concentration depends to a great extent on workmanship. Thus an insufficient fusion at e in Fig. 8.7a or a hump at f would create serious stress concentration. In the welds of Figs 8.7b and c, point h is called the *heel* and t is the *toe*. Imperfections in the weld at either the heel or the toe cause stress concentration.

When a welded joint is subjected to bending or reversed stresses, the weld should be so proportioned as to be stiffer than the adjacent material, in order to throw the deflection into the base-metal parts.

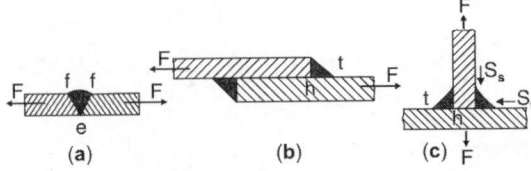

(a)   **(b)**   **(c)**

Fig. 8.7: Stress concentration in welds

Example 8.2: Determine the load which an arc welded joint of class 1, Fig. 8.5, can safely carry if it joins 10 mm plates and each fillet is 120 mm long and reinforced. The load varies from zero to a certain maximum values.

The cross-section area of a flush fillet weld resisting shear would be $t(l-12)$ and the area of a fully reinforced fillet may be taken as $1.2t \ (l-12)$. Thus the total cross-sectional area for both fillets is

$$A_1 = 0.707t \times 1.2 \ (l-12) \times 2 = 0.707 \times 10 \times 1.2 \times (120-12) \times 2$$
$$= 1832.5 \ \text{mm}^2$$

The working stress being taken from Table 8.3 as $S_e = 70 \ \text{MN/m}^2$, the safe load is

$$F_1 = A_1 S_e = \frac{1832.5 \times 70}{1000} = 128 \ \text{kN}$$

Example 8.3: Determine the load which welded joints of class 2 and class 3, Fig. 8.5 can safely carry, using the conditions and data of the previous example, and compare them with the strength of the strip.

In the joint of class 2 the section area in tension is equal to that in shear. The area of the weld is

$$A_2 = 10 \times 0.707 \times 1.2 \times (120 - 12) \times 2 = 1832.5 \ \text{mm}^2$$

and the safe load, with $s = 70 \ \text{MN/m}^2$ is

$$F_2 = A_2 S_2 = \frac{1832.5}{10^3} \times 70 = 128 \ \text{kN}$$

Thus one may say that a class 2 joint is considerably stronger than a class 1 joint, although they are quite comparable. In the joint of class 3, the area of the dangerous section with a reinforced joint is

$$A_3 = 10 \times 1.2 \times (120-12) = 1296 \ \text{mm}^2$$

From Table 8.4, the allowable stress in tension $S = 100 \ \text{MN/m}^2$, and the safe load is

$$F_3 = A_3 S = \frac{1296}{1000} \times 100 = 129.6 \ \text{kN}$$

Assume that the base metal is 10C8 steel. For a force amplitude $F_a = \frac{1}{2} F_{max}$, the stress amplitude is $S_a = 100 \ \text{MN/m}^2$. The safe load amplitude for a strip 10×120 mm, with a safety factor $n = 2$, is then

$$F_a = \frac{10 \times 120 \times 100}{10^3 \times 2} = 60 \ \text{kN}$$

Hence, $F_{max} = 2 F_a = 120 \ \text{kN}$

Thus all are welded joints theoretically are stronger than the strip.

Example 8.4: (a) Compare the strength of the joints in the preceding two examples with class 1 and class 2 joints made with flush welds and class 3 joint with a reinforced weld ground flush afterwards. All the welds are to be made by oxyacetylene process by a first class weld. (b) Find the efficiencies of the joints.

(a) In a class 1 joint with a flush weld the area of the dangerous cross section is

$$A_1 = 0.707 \times 10 \times 2 \times 120 = 1697 \text{ mm}^2$$

With the stress coefficient 0.85 for gas welds, the safe load is

$$F_1 = A_1 S_a \times 0.85 = \frac{1697}{10^3} \times 70 \times 0.85 = 101 \text{ kN}$$

In a class 2 joint, the area of the dangerous cross-section is $A_2 = 0.707 \times 10 \times 2 \times 120 = 1697 \text{ mm}^2$ and the safe load for a gas weld is

$$F_2 = A_2 S_a \times 0.85 = \frac{1697}{10^3} \times 70 \times 0.85 = 101 \text{ kN}$$

In a class 3 joint the area of the dangerous section is $A_3 = 10 \times 120 = 1200 \text{ mm}^2$
and the safe load with the coefficient 0.85 is

$$F_3 = \frac{1200}{10^3} \times 100 \times 0.85 = 102 \text{ kN}$$

The efficiencies of the three types of gas welds are

$$e_1 = \frac{0.101}{0.12} = 0.841, \text{ or } 84.1 \text{ per cent}$$

$$e_2 = \frac{0.101}{0.12} = 0.841, \text{ or } 84.1 \text{ per cent}$$

$$e_3 = \frac{0.102}{0.12} = 0.85, \text{ or } 85 \text{ per cent}$$

with flush gas weld, the joints of class 1 and class 2 have a strength lower than that of the strip, and the strength of class 3 joint is higher. However, it must be remembered that Table 8.4 assumes the best workmanship, which is not always available.

The leg size is the basis of specifying a weld. The size of a fillet weld is specified by the leg length of the largest inscribed isosceles right-angled triangle or the leg lengths of the largest inscribed right triangle. The leg length of a filled weld with equal legs is given by w or s or h and the leg lengths of a fillet weld with unequal welds, both the legs sizes are given.

Bending loads: When the load acts on the welded joint which makes the beam a cantilever as shown in Fig. 8.8. The weld is subjected to a

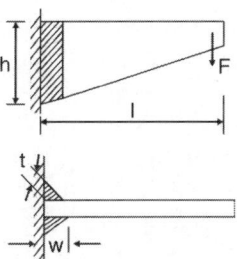

Fig. 8.8: Eccentric load on a welded beam

combination of direct load causing shear stress and bending stress.

Maximum bending moment is $M = Fl$

Primary shear stress $S = \dfrac{F}{2ht} = \dfrac{F}{\sqrt{2}hw}$

Secondary shear stress

$$S_b = \frac{M}{Z} = \frac{Fl}{2th^2/6} = -\frac{3\sqrt{2}Fl}{wh^2} \qquad \dots (8.1)$$

Maximum shear stress =

$$= \sqrt{\left(\frac{3\sqrt{2}Fl}{wh^2}\right)^2 + \left(\frac{F}{\sqrt{2}wh}\right)^2}$$

$$= \frac{F}{wh}\sqrt{\frac{18l^2}{h^2} + \frac{1}{2}} \qquad \dots (8.2)$$

In terms of t,

Maximum shear stress $= \dfrac{F}{th}\sqrt{\dfrac{9l^2}{h^2} + \dfrac{1}{4}}$

Annular fillet weld subjected to a bending moment. There are many instances of circular beams welded and forming centilevers subjected to a moment as shown in Fig. 8.9. In this case there is a bending stress only. Considering an

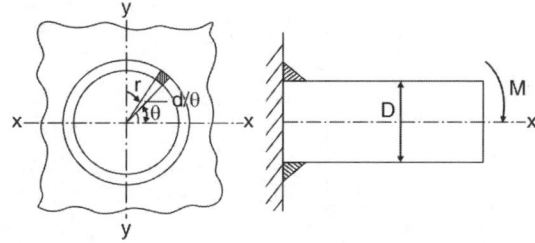

Fig. 8.9: Circular fillet weld subjected to bending

element of area = $rd\theta t$, if s is the stress acting at this area, the force acting on this area = $rd\theta\ ts$

The intensity of stress may be considered proportional to the distance from the neutral axis xx or if s_m is the maximum stress at the topmost point at a distance r, the stress at any point at a distance $r\sin\theta$ in,

$$\frac{s_m}{r} = \frac{s}{r\sin\theta}$$

Then $s = s_m\sin\theta$

Moment of the force about the central axis, is

$$M = \int_0^{2\pi} trd\theta s.\ r\sin\theta$$

$$M = s_m tr^2 \int_0^{2\pi} \sin^2\theta d\theta = \pi r^2 ts_m$$

$$s_m = \frac{M}{\pi r^2 t} = \frac{4M\sqrt{2}}{\pi D^2 w} = \frac{5.66M}{\pi D^2 w} \quad \dots (8.3)$$

or $\quad w = \dfrac{5.66M}{\pi D^2 s_m}$ or $t = \dfrac{M}{\pi r^2}$

Annular weld subjected to torque: A shaft may be welded to a pulley or hub may be welded to a coupling or pulley subjected to a torque as shown in Fig. 8.10. The stress set up in the weld is due to torque only. In this case also, by considering an element of area $rd\theta t$.

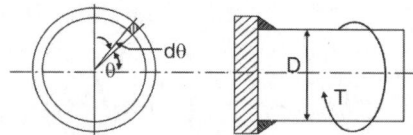

Fig. 8.10: Welded circular bar subjected to torsion

The stress acting on whole of the annular area is s, the resisting torque is given by

$$T = \int_0^{2\pi} srd\theta t.r = sr^2 t.2\pi \text{ or } t = \frac{T}{2\pi r^2}$$

$$s = \frac{T}{2t\pi r^2} = \frac{2\sqrt{2}T}{\pi D^2 w} = \frac{2.83T}{\pi D^2 w} \quad \dots(8.4)$$

Eccentric loads: When the load on a welded joint is applied eccentrically, the welds will be subjected to a combination of shear caused by the direct load and shear caused by torque. The state of stress in such a joint is complicated; and in order to determine the value of the significant stress even approximately, it is necessary to assume that the torsional shear stress at any point is proportional to its distance from the centroid of all weld areas.

If the weld shown in Fig. 8.11 is one of several forming a joint with the centroid of the weld areas at O, the torsional shear stress s on an element dA of the weld will be perpendicular to r and can be presented as

$$s = nr$$

where, n is a constant of proportionality and r is the distance from the element to O. If s_m is the maximum shear stress at a farthest distance r_1, then

$$s = \frac{s_m r}{r_1}$$

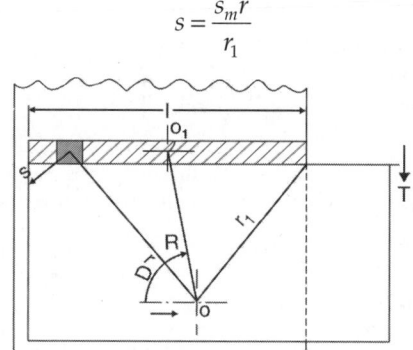

Fig. 8.11: Element of an eccentrically loaded weld

The external torque T is equal to the torque resistance of the element, or $sdAr$, integrated overall the welds in the joint. Thus,

$$T = \sum sdAr = \frac{s_m}{r_1}\sum r^2 dA$$

or $\qquad T = \dfrac{s_m . J}{r_1}$

where, J is the polar moment of inertia with respect to O for all elements of the weld.

The stress in any element can then be found by the relation

$$\frac{T}{J} = \frac{s_m}{r_1} = \frac{s}{r} \quad \dots (8.5)$$

or $\qquad s = \dfrac{Tr}{J}$

The maximum stress will be found by using for r the distance to the point farthest away from the center of gravity O.

The significant stress will be found by adding geometrically the maximum torsional stress and the direct shear stress. For static loads it is standard practice to assume that the direct stress in a weld is uniformly distributed throughout the area.

The procedure for finding the polar moment of inertia J of all welds in a joint is as follows: First the polar moment of inertia of each weld about its own center of gravity, as O_1 in Fig. 8.11, is computed from the relation

$$J_1 = \frac{tl^3}{12} + \frac{t^3l}{12} \simeq \frac{tl^3}{12} \qquad \dots (8.6)$$

After this the polar moment of each weld J_0 with respect to the common center of gravity O is found by using the relation for parallel axes, which is

$$J_0 = J_1 + AR_1^2 \qquad \dots (8.7)$$

The sum of all these moments of inertia J_0 will then give the numerical value of J to be used in equation 8.5.

The method described for finding the significant stress is only a simplified approach, but it gives satisfactory results if a proper safety factor—$n = 1.75$ or higher is used.

Example 8.5: Determine the maximum stress in the reinforced weld of the bracket plate in Fig. 8.12. Assume that the load varies from zero to the maximum value.

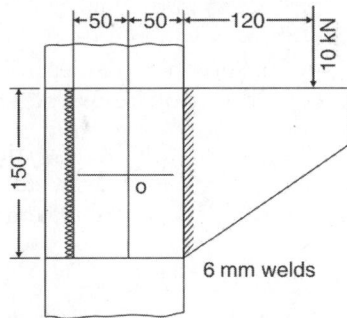

Fig. 8.12: Bracket plate welded to a column

The throat area of each weld is

$$A = 0.707 \times 6 \times (150\text{--}12) = 585.4 \text{ mm}^2$$

For each weld, $r_1 = 50$ mm, and the total polar moment of inertia is

$$j = 2A\left(\frac{l^2}{12} + r_1^2\right) = 2 \times 585.4 \times \left(\frac{138^2}{12} + 50^2\right)$$

$$= 4785 \times 10^3 \text{ mm}^4$$

The torque is

$$T = 10 \times \frac{170}{10^3} = 1.700 \text{ kNm}$$

The distance R in equation 8.5 is

$$R = \sqrt{69^2 + 50^2} = 85.2 \text{ mm}$$

Hence the maximum torsional stress, by equation 8.5, is

$$s = \frac{1.7 \times 10^3 \times 85.2 \times 10^{12}}{4785 \times 10^3 \times 10^3} = 0.0302 \times 10^9 \text{ N/m}^2$$

$$= 30.2 \text{ MN/m}^2$$

This stress is resolved into a vertical component

$$S_v = 30.2 \times \frac{50}{85.2} = 17.72 \text{ MN/m}^2$$

and a horizontal component

$$S_h = 30.2 \times \frac{69}{85.2} = 24.46 \text{ MN/m}^2$$

The primary shear stress has only a vertical component which acts downwards and is

$$S_v = \frac{F}{2A} = \frac{10,000}{2 \times 585.4} = 8.54 \text{ N/mm}^2$$

The total vertical component is

$$S_v = 17.72 + 8.54 = 26.26 \text{ MN/m}^2$$

Therefore the resultant stress is

$$S = \sqrt{26.26^2 + 24.46^2} = 35.88 \text{ MN/m}^2$$

With a stress concentration factor $K' = 2.7$ from Table 8.4, the significant stress is

$$S_{max} = 35.88 \times 2.7 = 96.9 \text{ MN/m}^2$$

8.6 REPEATED STRESSES

Experiments with repeated tension stresses have shown that the endurance limit of butt-welded joints, Figs 8.13a and b, is $S_{en}' = 0.85\, S_{en}$, where S_{en} is the endurance limit of a whole plate.[6] For the joint in Fig. 8.13c, with a ground-off

[6] A. Thum and W. Schick, "Dauertestigkeit von Schweissverbindungen bei vershiedener Formgebung," Z.VDI, Vol. 77 (1933), p. 493.

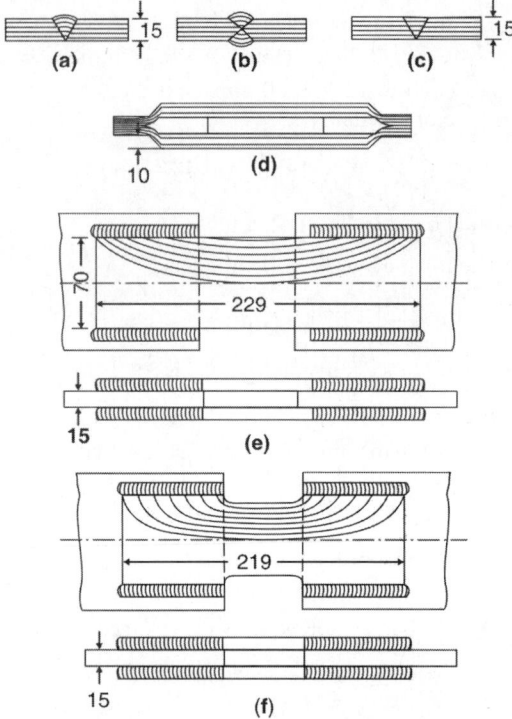

Fig. 8.13: Force-flow lines in welded joints

(SAE 1010) steel. As shown, changing the blunt edge in Fig. 8.14a to a sharper one, as in Fig. 8.14b, raised the endurance limit from 93 to 108 MN/m². By making the unwelded edge still smaller, as in Fig. 8.14c, the endurance limit is raised to 146 MN/m².

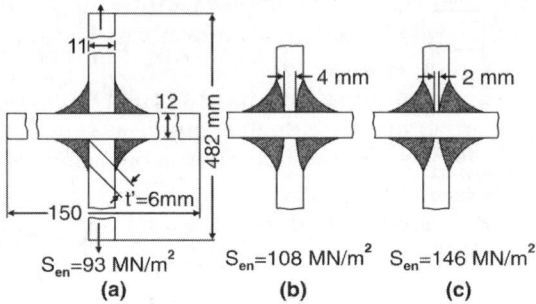

Fig. 8.14: Endurance of various types of welds

weld, $S_{en}' = 0.87 S_{en}$. Joints with cover plates, if of class 2 as in Fig. 8.13d, have $S_{en}' = 0.74 \, S_{en}$; those of class 1, as in Fig. 8.13f, have $S_{en}' = 0.81 S_{en}$. The cause of such a variation is the difference in the force, or stress flow, which is shown diagrammatically in these sketches.

The more uniform the flow is, the higher can be the magnitude of S_{en}'. This consideration should guide the designer in laying out the connections.

Additional data can be obtained from the results of test conducted with various types of connections.[7] The connections were subjected to tensile loads repeated 2 million times for each specimen. The stress amplitude ranged from 4.8 N/mm² to the indicated endurance limit. The material corresponded to 10C8

In Fig. 8.15 it is shown that with the same cross-sectional area the thicker and narrower straps have a higher endurance limit than the thinner and wider ones, regardless of their location.

As shown in Fig. 8.16 the endurance limit of channel-iron straps can be raised from 90 to 120 MN/m² by cutting the ends in a V shape and welding the ends as well as the sides. This increase of S_{en}' is due to a more direct force flow.

The grinding off the porous welds, and welds with cracks, increases the endurance limit, since these imperfections act as notches and produce stress concentration.

Design stress for structures and machine parts subjected to repeated loads may be based on the values in the last two columns of Table 8.3 or may be taken from Table 8.5,[8] which is more conservative. Values in Table 8.3 take into consideration repeated stresses, according to Fig. 3.43 and actual tests.[9] All statements made previously in connection with the use of Table 8.3 for static loads apply also for repeated loads.

---

[7] Graf, loc. cit., and *Dauerfestigkeit mit Schweissverbindungen* (Berlin: VDI Verlag, 1935).

[8] Based on *Standard Specifications for Welded Highway and Railway Bridges* (D2.0–47) and *Standard Code for Arc and gas Welding in Building Construction* (D1.0–46) (New York: American Welding Society, 1946)

[9] G. E. Thorton, *Study of Welded Metals*, Bulletin No. 34, Washington State Colleage Engineering Experiment Station (September 1930); C. H. Jennings, "Welding Design," *Trans.* ASME, Vol. 58 (October, 1936) MSP-58-1, p. 504).

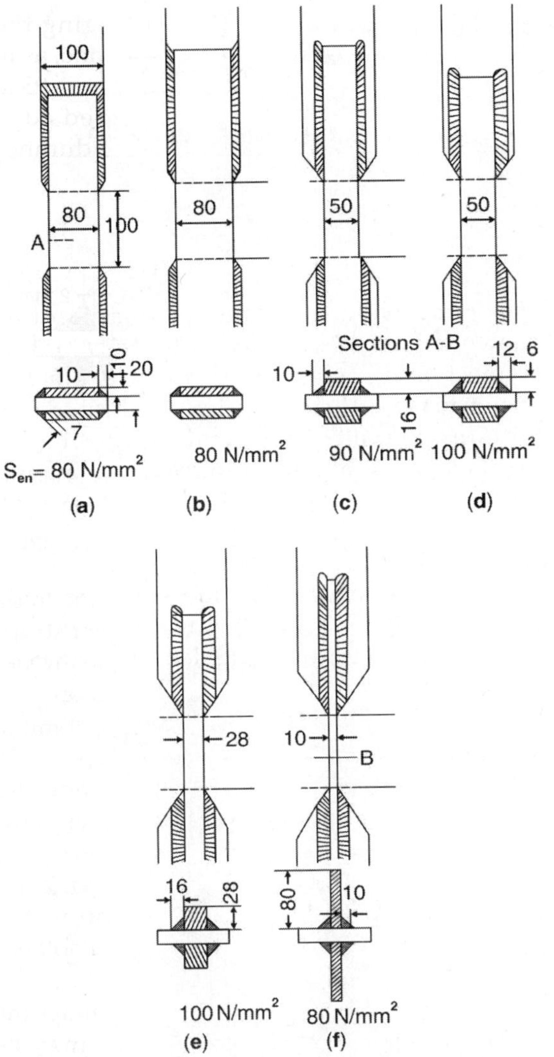

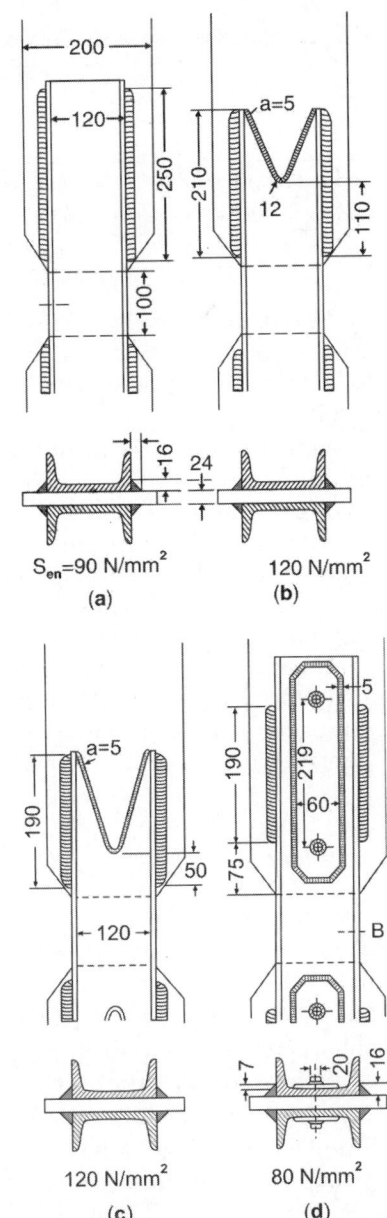

Fig. 8.15: Endurance of various connections stressed in tension

If single-V, backed-up welds are used, the values in Table 8.5 must be reduced by 15 per cent, and the stress-concentration factors given in Table 8.4 should be used.

8.7 APPLICATION OF WELDING

Welding is becoming more and more popular for various kinds of construction.

Pressure vessels: The latest editions of the ASME Boiler Construction Code and its Section VIII on

Fig. 8.16: Endurance of class 1 welds connecting channels to a plate

Unfired Pressure Vessels devote considerable attention to welding.[10] For boilers the maximum design stress is prescribed as 40.0 MN/m² for fusion welding and 55 MN/m² for forge welding.

---

[10] ASME *Code for Unfired Vessels* (New York: American Society of Mechanical Engineers, 1950).

Table 8.5: Strength of shielded-arc steel welds subjected to repeated loads

Type of joint	Kind of stress	Recommended stress N/mm²	
		Load varies from 0 to F	Load varies from +F to –F
Butt-welded from both sides	Tension or compression	90	60
Butt-welded from both sides	Shear	60	40
Filet-welded	Tension, compression, or shear	50	35

For unfired pressure vessels the maximum design stresses for fusion welding may vary from 40 to 55 MN/m², the value depending on the type of joint.

Fusion-welding of boiler joints may be produced either by one of the autogenous processes or by shielded-arc welding. The physical properties of the welding material used in autogenous welding correspond more closely to those of the plate than in the case of arc welding, which gives a weld that is stronger than the boiler plate. If it is desired to have the finished work as uniform throughout as possible, autogenous welding should be used. If a high tensile strength or high elastic limit of the weld is desired, arc welding is to be preferred.[11] Fusion welding should not be used where bending stresses may occur. Longitudinal seams of high-pressure boiler drums are forge-welded.

Fusion welding is extensively used as calk weld for riveted joints, screwed-in or riveted nipples, and other connections.

Tests on sample welds and full-size welded boiler drums have shown that metallic arc welding, properly done, has an efficiency of almost 100 per cent.[12] The ASME Code allows a maximum efficiency of 80 per cent that can be increased by 10 per cent if the joint is radiographed and can be increased by additional 5 per cent if the joint is thermally stress-relieved. Thermal stress-relieving by annealing is beneficial also for other parts of the structure, such as flanged or cold-bent plates.

A good weld for a longitudinal joint is shown in Fig. 8.17. Welded girth joints for the connection of a tank bottom to the drum are shown in Fig. 8.18. That in Fig. 8.18a is good but requires accurate fitting; that in Fig. 8.18b is much superior but is feasible only for a tank with a manhole;[13] that in Fig. 8.18c is a very common method; that in Fig. 8.18d is another common method which, however, reduces somewhat the tank capacity and is not advisable for high inside pressures because of its convex shape; that in Fig. 8.18e is excellent but is limited to a tank with a manhole.

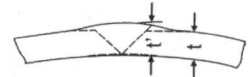

Fig. 8.17: Reinforcement of a welded joint

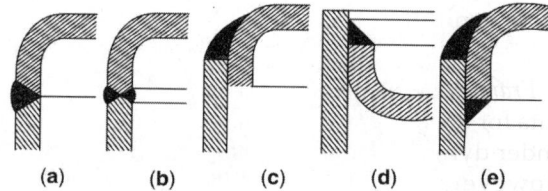

(a)   (b)   (c)   (d)   (e)

Fig. 8.18: Welding of tank heads to drum

Figure 8.19 shows a seamless press-forged steel nozzle (produced commercially), which is welded to a pressure-vessel shell by a butt weld a.[14] The reinforcing b counteracts the weakening of the shell by the large hole.

Steel structures under static loads: The design of welded steel structures subjected to static loads does not present any peculiarities.

[11] *Sulzer Technical Review*, No. 1. (1928), pp. 1–6.

[12] Linder Air Products, Engineering and Management Phases of Oxyweld Connections (New York: 1935), p. 63.

[13] A. J. Moses, "Tests on Welded Boiler Drums," *Combustion*, Vol. 2, No. 5 (November, 1930), p. 23.

[14] ASME *Rules for Construction of Unfired Pressure Vessels (New York: American Society of Mechanical Engineers, 1950).*

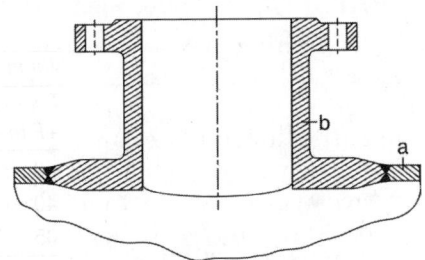

Fig. 8.19: Forged nozzle welded to a shell

Solid beams under dynamic loads: In the design of a solid beam the following requirements should be observed:[15]

(a) Only continuous welds should be used.

(b) Cross welds in a shelf working in tension should be avoided; where they cannot be avoided, they should be made as small as possible.

(c) An accumulation of welds should be avoided by all means.

(d) Shelf plates superimposed on other plates should be made narrower toward the ends in order to obtain a gradual change from a heavier cross section to a lighter one.

(e) Welds used at the intersection of shelf and web plates should always be butt welds.

Frame beams under dynamic loads: Complete data for the economical design of frame beams under dynamic loads are not easily available. However, many valuable pointers may be obtained from the test results discussed in connection with Figs 8.13 to 8.16.

Allowable stresses: For building construction the code of the American Welding Society recommends the following stresses in a section through the throat of a steel weld for bare or lightly coated electrodes: in shear, 80 MN/m², in tension, 95 MN/m², in compression, 125 MN/m². When a shielded arc is used, the allowable shearing and tensile stresses may be increased by 20 per cent, or to 95 and 110 MN/m², respectively.

Typical connections: In Figs 8.20a, b and c and Fig. 8.21c are shown some typical connections of *symmetrical structural shapes* in which the load is applied halfway between the two edges. Regardless of whether the end is cut off at a right angle, as in Figs 8.20a or c and Fig. 8.21c, or on a bias, as in Fig. 8.20b, the length of each fillet weld must be the same, or $l_1 = l_2$. This length may be found from the equation

$$F = 2t\,(l_1 - 12)S_s \qquad \ldots (8.8)$$

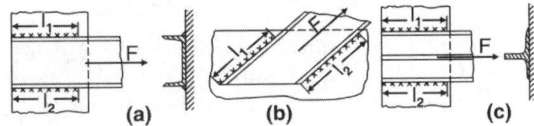

Fig. 8.20: Typical channel and tee connections to a plate

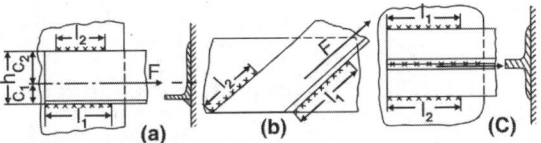

Fig. 8.21: Typical single-angle and double-angle connections to a plate

The T shape in Fig. 8.20c may be obtained by splitting an H beam along its center line by a gas- or arc- welding flame.

In the case of an *unsymmetrical member*, as in Figs 8.20a or b, the load F is applied at the center of gravity; and the lengths l_1 and l_2 should be selected so that they give equal moments with respect to the center of gravity. Thus they can be found, when the 6 mm imperfect ends are taken into account, from the following simultaneous equations:

$$F = (l_1 + l_2 - 2 \times 12)S_s t \qquad \ldots (8.9)$$

and

$$(l_1 - l_2)c_1 = (l_2 - 12)c_2 \qquad \ldots (8.10)$$

Without considering the end imperfection in the welds, the above equations change to the following:

$$F = F_1 + F_2 = t(l_1 + l_2)\,S_s$$

[15] A. Sceaper, "Die Dauerfestigkeit der Schweissverbindungen," Z.VDI, Vol. 77 (1933), p. 560.

For unequal welds

$$F_1 \times c_1 = F_2 \times c_2$$
$$F_1 = tl_1 S_s$$
$$F_2 = tl_2 S_s$$
$$\frac{F_1}{F_2} = \frac{l_1}{l_2} = \frac{c_2}{c_1}$$
$$F = F_1 + F_2$$

Therefore

$$F_1 = \frac{Fc_2}{c_1 + c_2} \text{ and } F_2 = \frac{Fc_1}{c_1 + c_2}$$

Example 8.6: Determine (a) the size of the single angle iron in Fig. 8.21a, loaded in tension by a static load $F = 135$ kN and (b) the dimension of the welds.

(a) For structural steel, which is approximately 20C8 (SAE 1020) $S_e = 234$ MN/m^2. With a safety factor $n = 2$, the allowable stress is $S_d = 117$ MN/m^2, and the necessary cross-sectional area

$$A = \frac{F}{S_d} = \frac{135}{10^3 \times 117} = 0.001154 \text{ m}^2$$

From a structural steel handbook, the nearest size that is not too thick is an angle $100 \times 75 \times 8$ mm for which $A = 13.36$ and $c_1 = 31$ mm.

(b) The thickness of the throat of the fillet weld is

$$t = 0.707s = 0.707 \times 8 = 5.656 \text{ mm}$$

According to Table 8.3, $S_s = 75$ MN/m^2. From equation 8.9

$$l_1 + l_2 = \frac{F}{S_s} + 24 = \frac{135 \times 10^6}{75 \times 10^3 \times 5.656} + 24$$

$$= 342.2 \text{ mm} \quad \text{(a)}$$

Substituting the corresponding values in equation 8.10 gives

$$31(l_1 - 12) = (100 - 31)(l_2 - 12) \quad \text{(b)}$$

from which $l_2 = \dfrac{31l_1 - 372}{69} + 12 = 0.45l_1 + 6.6 \quad \text{(c)}$

By substituting the value of l_2 from equation (a) and solving for l_1, we get

$$l_1 = 342.2 - 0.45l_1 - 6.6$$

$$l_1 = \frac{335.6}{1.45} = 232$$

then, $l_2 = 110$ mm

8.8 DESIGN OF WELDED MACHINE PARTS

Welding is coming into wide use in the manufacture of frames, bases, crankcases, flywheels, gears, various jigs, and many other machine parts. Rolled plates and structural shapes are then used instead of cast iron and cast steel, with a considerable saving of weight and cost. Various welding processes are employed, the best one depending on the thickness of the metal. The component parts are usually cut to size with shears or cutting torch and are assembled in a preliminary way be tacking in a few spots, and the welds are completed when all parts are accurately assembled. Some component parts can be prepared advantageously of weldable cast steel.

Welding of machine parts is particularly economical when only a limited number of them is to be made and also when comparatively heavy pieces have thin or medium-thick walls, or when the pieces have unusually heavy walls, as in the frames of large presses.

Welded parts can be fabricated in large quantities more economically if the component parts are stamped or die-punched. Before a designer decides whether a piece should be cast or welded, or welded of cast component parts, he should prepare careful designs of the parts by the different methods. He should also take into account not only the cost but also various other considerations, such as rigidity, resistance to corrosion, and even pleasing appearance.[16] There is much literature on the design of welded parts.[17]

In making a layout of a weldment the designer should remember that the cost of deposited metal in the welds is to be included in the cost of the mild steel from which the component parts are made. Therefore it is usually advisable to shape the component parts by bending wherever possible, and thus to reduce the amount of welding.

---

[16] J. L. Brown, "Casting or Welding in Machine Design," Trans. *ASME*, Vol. 58 (October, 1936), MSP-58–9, pp. 553 ff.
[17] E. J. Charlton, "Trends in the Use of Welded Machinery Parts," *Mechanical Engineering*, Vol. 67 (1945), pp. 109 ff; Lincoln Electric Company, *Procedure Handbook of Arc Welding—Design and Practice*, 9th ed. (Cleveland: 1950); also articles in *Machine Design, Product Engineering, Steel, Welding Journal*, and other periodicals.

Also, in designing a piece to be made by welding, it is advisable to forget how this piece looks when it is cast or riveted; to keep in mind only its functions; to obtain these functions with the simplest combination of available stock material—plates, angle irons, and channels, and occasionally to use special steel castings or forgings.

In Fig. 8.22 is shown a pedestal which was designed to replace a casting and which looks like a casting. In Fig. 8.22b is shown a design in which better advantage is taken of the possibilities presented by welding. The top, front, and back are made of one sheared piece bent to conform with the sides, and the sides have flame-cut openings which facilitate the handling of the piece and also add to its appearance. The design in Fig. 8.22b permits a considerable cost saving over the design in Fig. 8.22a because of a reduction of 30 per cent in the length of welds and about 30 per cent in weight. This design costs much less than a casting shaped like the design in Fig. 8.22a

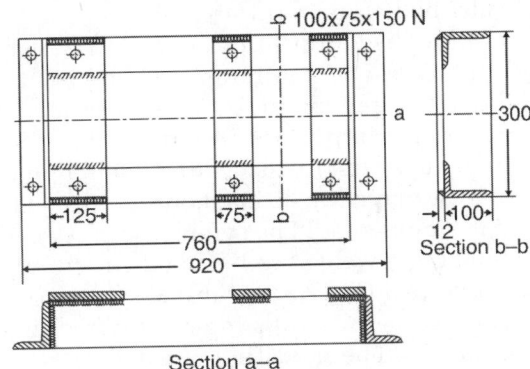

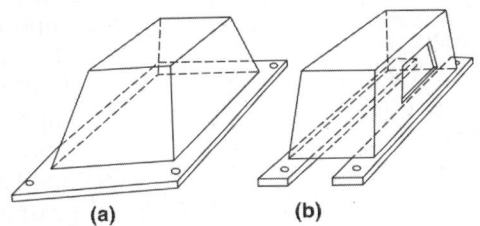

Fig. 8.23: Welded bedplate

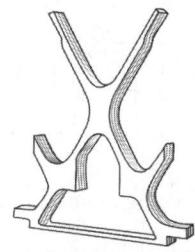

Fig. 8.24: Flame-cut part of a crankcase

(a)   **(b)**

Fig. 8.22: Welded pedestal

A bedplate for a motor-driven pump appropriately made from three sheared plates and four angle irons is shown in Fig. 8.23. The weight of this weldment is only 45 per cent of that of a casting, and its cost is about one-half that of a cast bed because it does not require expensive machining of the supporting surfaces.

The drawings in Figs 8.24, 8.25 and 8.26 illustrate the construction of a crankcase for a 12-cylinder, lightweight, 735 kW diesel engine.[18]

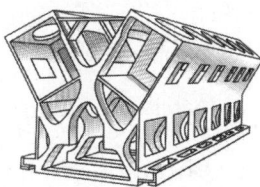

Fig. 8.25: Crankcase in process of fabrication by welding

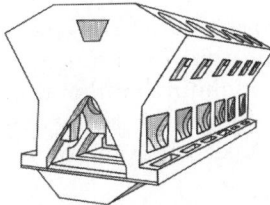

Fig. 8.26: Welded crankcase

Stress distribution: In designing a welded machine part the chief aim should be to secure a proper distribution of material in order to

[18] E. Chapman, "Welded-Steel Diesel Structures," *Motorship*, Vol. 18 (1933), pp. 404–9.

obtain a body of uniform strength as well as to avoid the discontinuities of section which result in stress concentration and endurance failure. Arrangement of the material is particularly important in parts subjected to repeated high loads, as in diesel-engine crankcases. Eccentric action should be avoided, and the joints should be symmetrical. A gear hub that was welded to I-shaped structural-steel arms, as in Fig. 8.27a, with one-sided welds at *c*, failed because of progressive fracture cracks at *b*. The modified construction in Fig. 8.27b, with symmetrical welds, *c* from both sides of the flange, proved satisfactory.

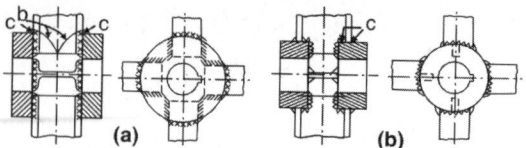

Fig. 8.27: Welded gear hub with arms

An effective method of determining points of maximum stress in a three-dimensional structure is by painting it with a fast-drying varnish, called *stress-coat*, which possesses a low modulus of elasticity and a low yield, point.

When a static load is applied to the structure, the varnish cracks at the points of maximum strain while the structure is only lightly loaded. The varnish will crack first at those points where the fillet radius is not large enough, at the contours of an improper weld, or at the root of an undercut. These cracks will give valuable information as to how to improve the structure in order to give it more uniform strength throughout. Similar although not as accurate results may be obtained by using lime wash, which peels off from points subjected to strain.

Rigidity: In machine construction the rigidity or flexibility is often of great importance. Proper arrangement of weld joints gives a control in this respect. In Fig. 8.28a is shown a

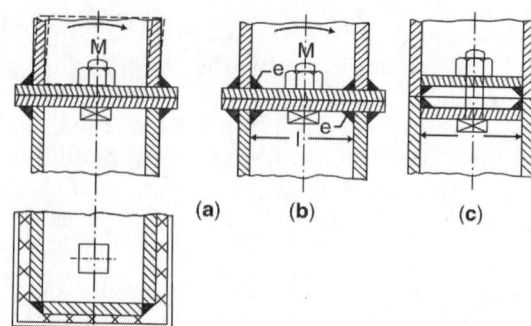

Fig. 8.28: Welded joints with different rigidity

connection of two columns that is conveniently made but is not very rigid. The rigidity, and the strength also, can be materially increased by adding the inside welds *e*, as in Fig. 8.28b, if the distance *l* allows such construction, but this is seldom the case. In Fig. 8.28c is shown a rigid connection which is easy to make.

Other considerations: Welding also presents the advantage of employing different materials, each most suitable for the particular service, to form an integral part. In Fig. 8.29, for example, a horizontal planer support with hardened cast-steel guides *g* welded to a ductile-steel U-shaped box b is shown.

Two methods of welding used with flange couplings to replace tapered keys if the flanges do not have to be removed are shown in Fig. 8.30.[19] The simultaneous use of both methods, as in Fig. 8.31, reduces the throat dimensions *a*; this is desirable with a large torque.

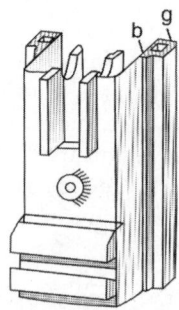

Fig. 8.29: Welded frame of plates and castings

[19] L. v. Roessler, "Schweissungen statt Keilbefestigungen," *Electroschweissungen*, Vol. 7 (1936), p. 209; also *Z.VDI*, Vol. 81 (1937), p. 305.

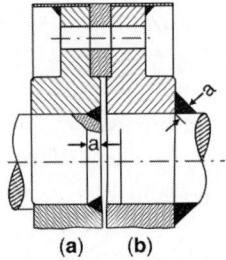

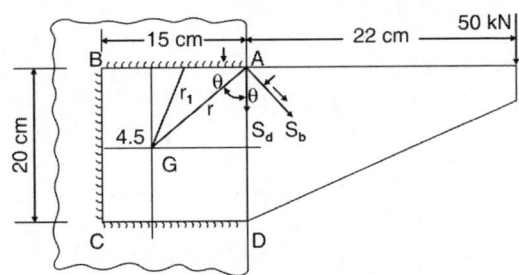

Fig. 8.30: Flange coupling

Fig. 8.32: Unsymmetrical weldment

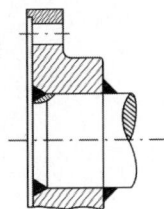

Fig. 8.31: Stub-shaft flange

Example 8.7: Find the size of weld for the joint shown in Fig. 8.32. The allowable shear stress in the weld is 80 N/mm².

For a throat thick t, the position of centroid of weld G is found as

$$\bar{x} = \frac{2 \times 15 \times t \times \frac{15}{2}}{2 \times 15 \times t + 20 \times t} = 4.5 \text{ cm}$$

$$r_1 = \sqrt{10^2 + 3^2} = 10.44 \text{ cm and}$$

$$r = \sqrt{10.5^2 + 10^2} = 14.5 \text{ cm}$$

The polar moment of inertia for AB and CD about G is

$$I_a = \frac{2 \times t \times 15^3}{12} + 2 \times t \times 15 \times 10.44^2 = 3832.35\, t$$

The polar moment of inertia for BC about G is

$$I_b = \frac{t \times 20^3}{12} + t \times 20 \times 4.5^2 = 1071.6\, t$$

Total $I_G = I_a + I_b = 3832.3\, t + 1071.6\, t = 4904\, t$

Total Area of weld $= 2 \times 15\, t + 20\, t = 50\, t$

Primary shear stress $S_d = \dfrac{50{,}000}{50\, t} = \dfrac{100}{t}$

Secondary shear stress

$$S_s = \frac{50{,}000 \times 32.5 \times 14.5}{4904\, t} = \frac{4804}{t}$$

at A, $\cos\theta = \dfrac{10.5}{14.5} = 0.7241$

The resultant shear stress

$$= \sqrt{\left(\frac{1000}{t}\right)^2 + \left(\frac{4804}{t}\right)^2 + 2 \times \left(\frac{1000}{t}\right)\left(\frac{4804}{t}\right) \times 0.7241}$$

$$= \frac{5570}{t}$$

Equating to the allowable shear stress

$$8000 = \frac{5570}{t}$$

$t = 0.696$ cm say 7 mm

Size of weld $s = 0.7 \times \sqrt{2} = 10$ mm

Example 8.8: Find the size of the weld when the circular cantilver of diameter 150 mm is 300 mm long and carries a concentrated load of 25 kN at its end and also when the same cantilever is subjected to a twisting moment of 6000 Nm. The allowable shear stress is 60 N/mm².

1. Size of weld using by eq. 8.3,

$$w = \frac{5.66M}{\pi D^2 S_s}$$

$$= \frac{5.66 \times 25 \times 1000 \times 300}{\pi \times 150^2 \times 60}$$

$$= 5.6 \text{ mm say 6 mm}$$

2. Size of weld using equation 8.4

$$w = \frac{2.83 \times 6000 \times 1000}{\pi \times 150^2 \times 60}$$

$$= 4.04 \text{ say 4 mm}$$

PROBLEMS

8.1 Which is the weakest area of a fillet weld?

8.2 Which type of welded joint is preferable for steel structures and why?

8.3 (a) Determine the approximate static load which a welded class 1 joint, Fig. 8.5, can safely carry if it connects two 6 mm plates and each fillet is 125 mm long and flush. The upper plate is of 20C8 (SAE 1020) steel and is 125 mm wide, and a shielded arc is used. (b) Find the efficiency of the joint. (c) Compare with results obtained by using the more accurate data of Table 8.3.

8.4 (a) Determine the safe static load for a class 2 welded joint. Fig. 8.5, if it connects two 6 mm plates and each fillet is 75 m long and concave. The plates are of the 20C8 (SAE 1020) steel and 75 mm wide. (b) Find the efficiency of the joint.

8.5 (a) Determine the safe load for a class 3 welded joint, Fig. 8.5, if it connects two strips 6 mm thick and 75 mm wide. The strips are of 7C4 (SAE 1010) steel and the weld is ground flush. (b) Determine the efficiency of the joint.

8.6 Determine the necessary minimum distance between the welds in Fig. P8.1 for a static load F of 36 kN if l = 100 mm and a = 150 mm. Use 5 mm shielded arc fillet welds, and assume that the direct shear stress is distributed uniformly over the throat area.

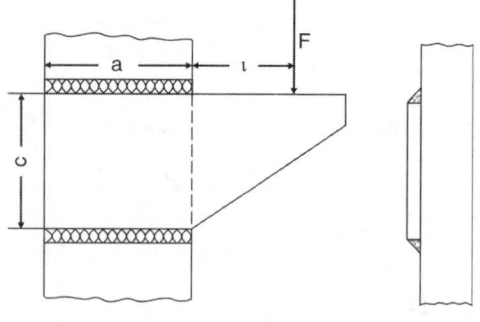

Fig. P8.1

8.7 Work problem 8.6, assuming that the load F varies continually from 0 to 36 kN. Increase the weld size to 10 mm.

8.8 Work problem 8.6 for the following conditions: F = 40 kN, l = 240 mm, a = 180 mm and the size of the welds is 10 mm.

8.9 Work problem 8.8, assuming that the load varies continuously from 0 to 40 kN. Increase the weld size to 12 mm.

8.10 A clip angle is fastened to a column by two 10 mm fillet welds loaded in transverse shear. (a) Determine the length of each weld to support a steady load F of 30 kN if a bare electrode is used. (b) Determine the length if a coated electrode is used.

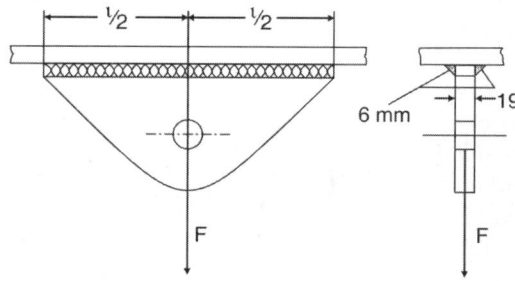

Fig. P8.2

8.11 Work problem 8.10 for the following conditions: F = 22 kN and the size of the welds is 8 mm.

8.12 A load of 130 kN is suspended from a vertical 10 mm plate fastened to a beam by two fillet welds so as to form a tee joint. Determine the necessary size and length of each weld, using (a) bare electrodes, (b) coasted electrodes, and (c) oxyacetylene-gas welding.

8.13 Work problems 8.12 for a load of 100 kN.

8.14 Determine the size of the steel channel in Fig. 8.20b, and determine the dimensions of the welds, for a tensile load F that fluctuates from zero to 180 kN using (a) flush fillet welds and (b) reinforced fillet welds. Assume first-class gas welding.

8.15 Find the permissible static load F for the connection of Fig. P 8.2 if l = 300 mm and shielded arc weld is used.

9 Design of Riveted Constructions

9.1 GENERAL REMARKS

Riveting was the standard method of joining plates and structural parts before welding began to replace it with increasing rapidity.

Rivets: A rivet is a round bar consisting of an upset end called the *head* and a long part called the *shank*. The rivet blank is heated to a red glow and inserted into one of the holes; and while the head is held firmly against the plate by a heavy sledge, the projecting end is formed into a second head, called the *point*, by means of a hand hammer and a forming tool called a *set*, or by a press. In Fig. 9.1 are shown various shapes of heads and points, with the length of shank necessary to form the corresponding point in each case. The head *a* is a button head or snap head; *b* is a double-radius button, or conoidal head; *c* is a pan head; and *d* is a countersunk head. The point *e* is a cone point; *f* is a steeple point; and *g* and *h* are countersunk points. Button heads are used for small rivets up to 8 mm in diameter, which are driven cold; pan heads are used chiefly in ship work; countersunk heads are used only in special cases, chiefly in structural work and below the water line in ships. The countersunk points weaken the plate so much that they should be used only when unavoidable; the others, including points similar to button heads, are used in boiler and structural work.

Rivet diameter in common use increase from 4 to 10 mm in steps of 1 mm; 10 mm to 24 mm in steps of 2 mm and 24 to 36 in steps of 3 mm.

Material: Rivets are made of tough and ductile low-carbon steel or nickel steel. According to the ASME Boiler Construction Code, the rivet material must have a tensile strength between 300 to 380 MN/m^2 and a minimum yield point equal to one-half the tensile strength. (Almost same according to BIS.) *It should be bent through 180° without cracking (9.1).* Brass rivets are used only cold and in small sizes.

Rivet holes: Holes for rivets are either punched or drilled. Punching injures the metal around the holes. Therefore they should be punched at least 3 mm undersize, and then be reamed. Simultaneous reaming of the matching holes through the plates to be connected by a rivet straightens the holes and eliminates offsets, which weaken the rivet. Holes of a diameter smaller than the thickness of the plate cannot be punched, since the punch is likely to crush.

The maximum force F needed to punch a hole in a plate may be estimated conservatively as follows:[1]

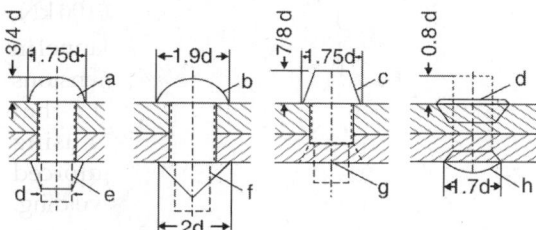

Fig. 9.1: Rivet heads and points

---

[1] C. D. Albert, Machine Design Drawing Room Problems, 4th ed. (New York: John Wiley & Sons, Inc., 1949), p. 212

177

$$F = 1.25\,\pi dt S_s \qquad \ldots (9.1)$$

where, d is the diameter of the hole, mm

t is the thickness of the plate, mm

S_s is the ultimate strength in shear of the plate material, in N/mm^2 and is about 0.75 times the tensile strength S_u.

The work of punching a hole, Nm may be taken as

$$W = Ft \qquad \ldots (9.2)$$

Drilling the holes is the best method. With the present improved machinery it can be accomplished almost as cheaply as punching and reaming.

Rivet holes are made 1.5 mm larger than the shank of the cold rivet. When the red-hot rivet is driven and set, it fills out the full size of the hole. In order to reduce the stress concentration at the junctures of the stem with the head and the point, the holes should be countersunk about 1.5 mm deep.

9.2 RIVETED JOINTS

Structures in which riveted joints or connections are used may be grouped as follows:

   (a) Tanks and pressure vessels.

   (b) Bridges, buildings, cranes, and machinery in general.

   (c) Hulls of ships.

In structures in the first group, the joints must be leakproof. In structures in the second group the connections must resist given outside loads and have sufficient rigidity. In structures in the third group, it is important to consider strength, rigidity, and durability, as well as security against leakage.

Types of joints: Two arrangements used in joining plates by means of rivets are equally well adapted to all three groups of structures, namely *lap joints* and *butt joints*.

Lap joints consist of overlapping plates held together by one or more rows of rivets. A single-riveted lap joint is shown in Fig. 9.2. A double-riveted lap joint can have rivets arranged either staggered, as in Fig. 9.3a., or in a chain form, as in Fig. 9.3b. In triple-riveted lap joints

the rivets are staggered. The eccentric force action in a lap joint produces bending; and the resulting distortion, shown exaggerated in Fig. 9.4, *tends to produce a concentrated crushing load on the corners of the plates, a tensile stress in the rivet shank, and a shear stress in the heads (9.2).* This action thus tends to decrease the frictional resistance. However, tests have shown that a lap joint has a 30 per cent greater frictional resistance than a butt joint with two straps, [2] like that shown in Fig. 9.5. The explanation seems to be in the pinching of the lap joint at the edges and in an unavoidable difference in the thickness of the two plates in a butt joint. Also, it is more difficult to keep lap joints tight, because of the bending that occurs.

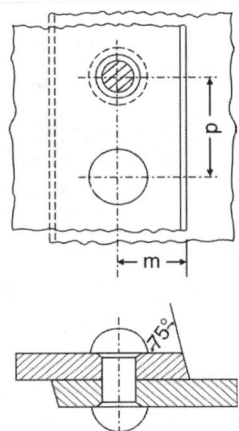

Fig. 9.2: Single riveted lap joint

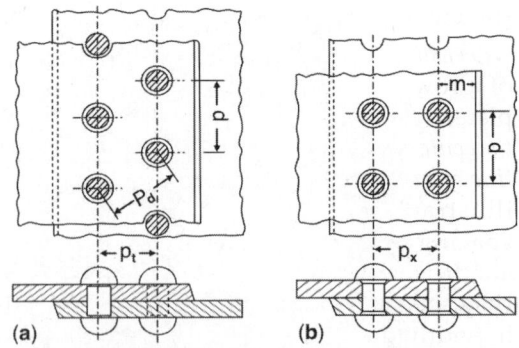

Fig. 9.3: Double-riveted lap joints

[2] C. A. Norman, *Principles of Machine Design* (New York: The Macmillan Company, 1925) p. 57.

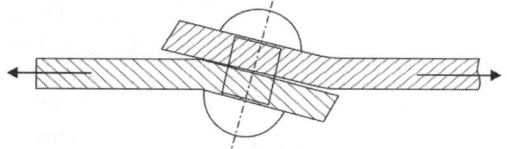

Fig. 9.4: Bending in a lap joint

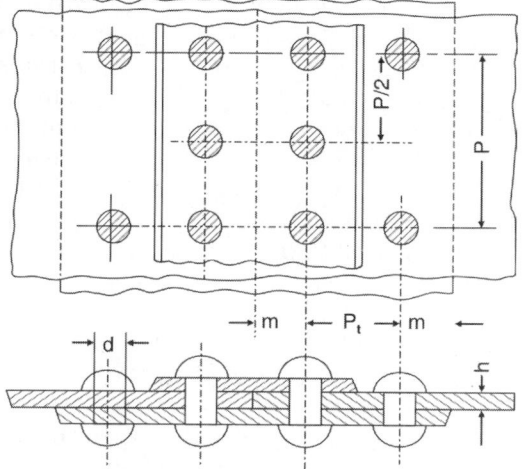

Fig. 9.5: Double-riveted chain butt joint

Butt joints are formed by butting the plates against each other and joining them by overlapping straps, or cover plates. A butt joint may have one strap on the outside, but most butt joints have one strap on the outside and another on the inside. The cover plates may be of the same width, or the outside plate may be narrower, as in Fig. 9.5. Butt joints are made single-double-, triple-, quadruple-, and sometimes even quintuple-riveted.

Terminology: The line *g* through the centers of a row of rivets, parallel to the edge of the plate, is called the gage line. *The distance between the centers of adjacent rivets measured on the gage line is called the pitch*; it is denoted by *p* in the illustrations and tables of this book. If the spacing on one gage line is greater than on another, as in Fig. 9.5, the pitch is the maximum rivet spacing. A *unit strip* or *unit length* is equal in width to the pitch. *The distance between two adjacent gage lines in the same plate is called the*

back pitch, or *transverse pitch*; it is denoted by p_t, as in Figs 9.3 and 9.5. In staggered riveting the distance between the centers of adjacent rivets on adjacent gage lines is called the *diagonal pitch*; it is denoted by p_d, as in Fig. 9.3. The distance from the gage line to the edge of the plate is called the *lap distance, marginal distance*, or *margin*; it is denoted by *m*, as in Figs 9.2, 9.3, and 9.5.

9.3 STRESSES IN JOINTS

Rivets should always be placed at right angles to the acting forces, and the maximum stress induced in them should be either shear or crushing. In a long rivet the initial stress set up at the junction of the shank and the point, when the rivet cools, is dangerous. This initial stress increases with the relative length of the rivet, and in a very long rivet it may cause the head to snap off without any load. For this reason the length of the rivet between the heads should not exceed four or five times its diameter. Longer rivets can be used if the head end is cooled before the rivet is placed in its hole.[3]

Calking was developed primarily for the purpose of making riveted joints leakproof. It is good practice to calk not only the plate but also around the rivet heads, as shown in Fig. 9.6.

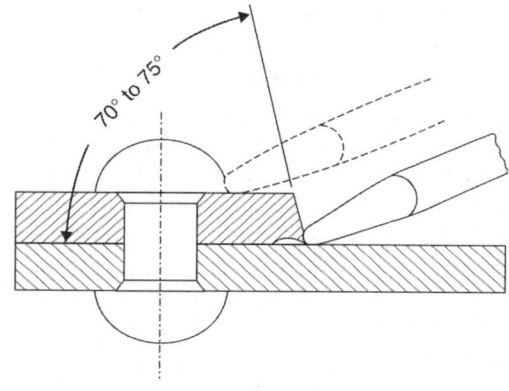

Fig. 9.6: Calking

Failure of a riveted joints: The conventional assumption is that the load is distributed

---

[3] H. J. Spooner, *Machine Design*, 6th ed. (London: Longmans, Green & Company, 1925), p. 127.

equally among all rivets. This assumption does not hold approximate even roughly the actual conditions, particularly for multiple-riveted joints. However, the load distribution upon the rivets depends on so many factors and is so indefinite that no better theory exists. According to the conventional theory, the failures of joints may occurs in one of the following ways:

(a) Shearing of the rivets, as in Fig. 9.7a.
(b) Rupturing of the plate by tension, as in Fig. 9.7b.
(c) Crushing of the plate or of the rivets, as in Fig. 9.7c.
(d) Shearing of the margin, as in Fig. 9.7d.
(e) Tearing of the margin, as in Fig. 9.7e.
(f) Rupturing of the plate by tension in a zigzag line passing diagonally between the rivet holes in staggered riveting.

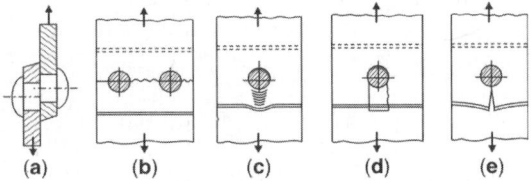

Fig. 9.7: Ways of failure of a riveted joint

In a multiple-riveted joint ultimate failure may be due to one or more of the causes just cited. However in all cases, the failure will be preceded by overcoming the frictional force between the surfaces of two plates or plates and the cover plates.

Strength of rivets: It should be noted that in designing riveted joints of pressure vessels, the diameter of the hole is used in calculating the resistance of the rivets both in shear and in crushing. In *structural joints* the rivets sometimes are driven cold and sometimes are replaced by bolts; it is therefore customary to use the nominal rivet diameter in calculating the resistance of rivets.

Efficiency: The ratio of the strength of the weakest section of a unit length of the joint to the tensile strength of an equally wide unperforated plate is called the *efficiency* of the joint. A well-designed joint should have the same efficiency

in regard to every possibility of failure. If this equality does not exist, the governing lowest efficiency of the joint can be raised by strengthening the corresponding weaker element at the expense of the stronger one.

Margin: The width of the margin m is independent of other elements. *With the usual proportions, however, if the margin is equal to 1.5d, it will be safe against both shearing and tearing by the rivet pressure (9.3).* For the sake of additional safety, in single-riveted joints m is given the width $2d$. On the other hand, in boiler work the margin should not be made larger than necessary, as it is then more difficult to calk the joint and to make it steam-tight.

Transverse pitch: In chain riveting the transverse pitch p_t, in Fig. 9.3b, may be determined by the requirement of room for the heading tool. It should not be less than $2d$ and preferably should be $2.5d$.

In staggered riveting, as in Fig. 9.8, the minimum value of p_t is determined by the conditions that the plate will not rupture along the diagonal pitches p_d. Then

$$2(p_d - d) \geq p - d$$

The elimination of p_d by means of the geometric relation in this diagram gives

$$p_t \geq \sqrt{0.5p_d + 0.25d^2} \qquad \dots (9.3)$$

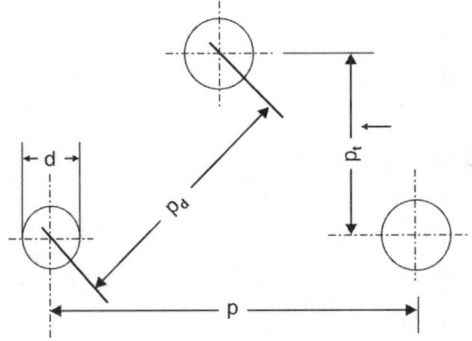

Fig. 9.8: Pitch relations

With the usual proportions of joints, p_t is greater than or equal to 1.7d, and a safe value of p_t is 2d, for convenience in head formation usual minimum pitch p is 2.25 to 2.5d or even 3d.

If one of the cover plates is narrower, as in Fig. 9.5, the transverse pitch must be made longer to give room for the riveting tool. A practical value for the transverse pitch in this case is 2.75d.

9.4 THEORETICAL STRENGTH ANALYSIS

In a theoretical analysis of strength the problem is to find the general relation between the thickness of the plates, the rivet diameter, and the pitch when a riveted joint is equally strong against failure in any of the six possible ways. Since the margin and the transverse pitch do not affect the longitudinal pitch p and can be either determined independently or assigned values based on the diameter of the rivet, it follows that the last three ways of failure, designated d, e, and f in section 9.3, may be omitted from the general analysis. Hence the problem narrows down to the analysis of three types of failure, designated a, b, and c.

Design steps: To design a lap joint or a butt joint, pitch will be considered as a unit.

(i) Calculate thickness of plate as if it is solid considering a reasonable value of efficiency depending on the type of joint

(ii) Find the diameter of the rivet by using the equation

$$d = 6\sqrt{h} \text{ to } 7\sqrt{h}$$

(iii) Find the resistance of the perforated plate in tearing,

$$F_t = (p - d) hS$$

(iv) Find the resistance of the rivet in shearing

$$F_s = \frac{\pi}{4} d^2 S_s$$

or

$$F_s = n \times 1.875 \frac{\pi}{4} d^2 S_s$$

if in double shear and n is the number of rivets per pitch.

(v) Find the resistance of rivet/plate in crushing
$$F_c = nhd\, S_c$$

(vi) By equating the minimum of the resistance in shearing or crushing to the resistance in tearing, find pitch p and take the minimum $p = 2.5d$ or $3d$

(vii) Find the efficiency of the joint

$$= \frac{\text{Minimum resistance}}{\text{Resistance of solid plate } (phS)} \times 100$$

Minimum resistance is the minimum value of tearing or shearing or crushing resistance.

If the efficiency is less than the expected value, repeat calculations by taking higher value of $d = 7\sqrt{h}$

(viii) The margin failures is avoided by taking $m = 1.5d$.

For a complex design analysis Fig. 9.9 will be used only as an illustration of a complex joint with various rivet spacings, some of the rivets working in double shear and some in single shear. In accordance with the ASME Boiler Code, the shear stress S_s will be considered the same, whether induced in single shear or in double shear. Also, elasticity will be disregarded and the crushing stress, S_c will be considered to be the same in all rivet rows. In addition to the designations in Fig. 9.9, n_2 will designate the number of rivets working in double shear in a unit length, whose width is p, and n_1 will designate the number of rivets in single shear. For the joint in Fig. 9.9, $n_2 = 8$ and $n_1 = 3$.

The tensile strength of the solid plate is

$$F_p = phS \qquad \qquad \dots (9.4)$$

The rivet diameter $d = 6\sqrt{t}$ to $d = 7\sqrt{t}$

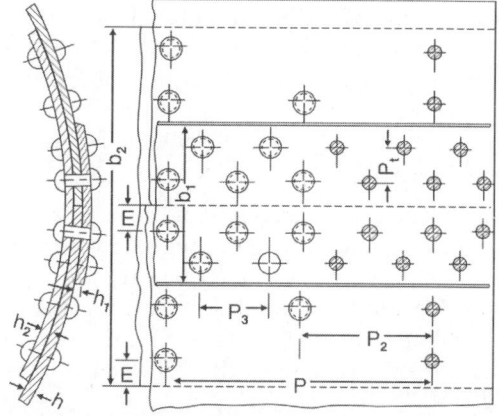

Fig. 9.9: Quadruple-riveted double-strap butt joint

The tensile strength of the perforated strip along the outer gage line is

$$F_t = (p - d)hS \qquad \dots (9.5)$$

The general expression for the resistance to shear of all the rivets in a unit strip is

$$F_s = (2n_2 + n_1)\tfrac{1}{4}\pi d^2 S_s \qquad \dots (9.6)$$

The general expression for the resistance to crushing of the rivets is

$$F_c = (n_2 h + n_1 h_2)dS_c \qquad \dots (9.7)$$

where h_2 is the thickness of the wider strap.

If the joint is designed for maximum efficiency, $F_t = F_s = F_c$ and each of these three quantities divided by F_p gives the theoretical efficiency e of the joint. Thus, dividing equation 9.6 by equation 9.4 gives

$$e = \frac{(2n_2 + n_1)\pi d^2 S_s}{4phS} \qquad \dots (9.8)$$

The value of F_t is equated to be value of F_s or F_c whichever is less and the equation is solved for p. For $F_s < F_c$, the value of p is given by $F_t = F_s$

$$p = \frac{(2n_2 + n_1)\pi d^2 S_s}{4hS} + d \qquad \dots (9.9)$$

The minimum value of p is 2.5 d or 3d.

If this expression of p is substituted in equation 9.8, the result is

$$e = \frac{(2n_2 + n_1)\pi d^2 S_s}{(2n_2 + n_1)\pi d^2 S_s + 4dhS} \qquad \dots (9.10)$$

Equation 9.10 applies to any form of riveted joint and gives the limit value of efficiency for any form and material. The actual proportions adopted may give a lower efficiency, but they can never give a higher than the theoretical efficiency. However a slight increase in efficiency is possible by increasing the rivet diameter to the higher value of $7\sqrt{h}$.

In a multirow joint, if one of the resistances F_t, F_s, or F_c is not sufficiently high, failure may occur through combined action, instead of in one of the simple ways. For example, tearing of the plate in Fig. 9.9 between the rivet holes in the second row with the pitch p_2 may be combined with failure of the rivets in the first row. These rivets may fail either by shearing or by crushing. For the first case, the permissible load is

$$F_{s1} = (p - 2d)\, hS + \frac{\pi}{4} d^2 S_s \qquad \dots (9.11)$$

For the second case the permissible load is

$$F_{c1} = (p - 2d)\, hS + dhS_c \qquad \dots (9.12)$$

Another way of failing can be through the shearing of the rivets in the outer row and the crushing of the rivets in the two inner rows. If their number in a unit is designated by n, the allowable load is

$$F_{sc} = \frac{\pi}{4} d^2 S_s + ndhS_c \qquad \dots (9.13)$$

The efficiency of the joint will be the smallest of the values F_t, F_s, F_c, F_{s1}, F_{c1}, or F_{sc} divided by F_p. If the efficiency thus determined happens to be lower than the values assumed before, it should be raised by changing the pitch or the size of the rivets or by changing their distribution.

Table 9.1 gives the ranges of values of the efficiency for the various types of joints used in boiler work and may serve as a guide in making the necessary assumptions when starting to design a certain joint.

Actual stresses: It is most difficult to evaluate the straining actions to which the different elements of a riveted joint are subjected, because of the elasticity of the parts through which the loads are transferred and the unavoidable occurrence of bending stresses. This is particularly true for multiriveted joints. Tests show[4] that while actual efficiencies of lap joints

Table 9.1: Efficiency of riveted joints

Type of joint		Efficiency (per cent)	
		Normal range	Maximum
Lap joints	Single-riveted	50–60	63
	Double-riveted	60–72	77
Butt joints with two cover plates	Double-riveted	76–84	87
	Triple-riveted	80–88	95
	Quadruple-riveted	86–94	98

[4] J. Considere, *European Experiments on Riveted Joints*, Bulletin No. 62, American Railway Engineering Association (Chicago: April, 1905), p. 149.

are 1 to 3 per cent higher than the theoretical values, the actual efficiencies of double-strap butt joints are from 2 to 12 per cent lower than the theoretical values.[5] Furthermore, the method of riveting—by hand or by a machine—and the workmanship have a great influence upon the actual efficiency of a riveted joint. To take care of the unknown factors a sufficiently large factor of safety, as 2.5 to 3.0 referred to the elastic limit, is used.

9.5 BOILER AND TANK JOINTS

The stress set up by the pressures on the ends of a circumferential section or joint of a boiler shell is only half as great as the stress set up by the radial pressure in a longitudinal joint. Therefore a circumferential, or girth, joint does not need to be more than half as strong as a longitudinal joint. Even single-riveted lap joints, when properly designed, have an efficiency over 50 per cent. *Therefore single-riveted or double-riveted lap joints, being simpler and cheaper to make than butt joints, should be used for girth joints. For longitudinal joints the more efficient although more expensive double-strap butt joints are more suitable (9.6).*

Rivet diameter: The diameter of the rivets in a boiler shell or a tank should be between $6\sqrt{h}$ and $7\sqrt{h}$, where h is the thickness of the plate.

Pitch: Equating the values of F_t and F_s given by equations 9.5 and 9.6, and solving for p, gives

$$p = \frac{0.7854(2n_2 + n_1)d^2 S_s}{hS} + d \quad \dots (9.14)$$

However, it is preferable that the pitch be found by equating the tensile strength to the minimum of shearing and crushing strength.

In a lap joint the pitch is taken as a multiple of 6 mm. In a quadruple-riveted butt joint the long pitch p, Fig. 9.9, is taken as a multiple of 12 mm so that the middle pitch p_2 will be a multiple of 6 mm and the short pitch p_3 will be a multiple of 3 mm. In order to obtain a tight joint that is suitable for calking, the pitch p_2 should not be more than $8h$.

Cover plates: When only one cover plate, or strap, is used, its thickness should be at least equal to that of the main plates. A butt joint with a single strap is equivalent to two lap joints and is used mainly where a smooth surface is desired. When double straps are used, each should be slightly thicker than half thickness of the main plate, about $\frac{5}{8}h$ to $\frac{3}{4}h$. Where the outer strap is narrower than the inner one, as in Fig. 9.10, the narrower strap is often made of the same thickness as the main plates, and the thickness of the inner one should be from $\frac{5}{8}h$ to $\frac{3}{4}h$.

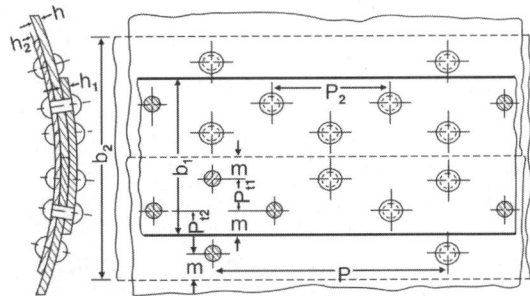

Fig. 9.10: Triple-riveted double-strap butt joint

Calking: Each joint is made tight against leakage by calking—hammering the edge of the plate and rivet into contact with the plate underneath it by blows on the calking tool. The edge of the plate must be beveled as shown in Fig. 9.6. Calking cannot be applied to plates less than 5 mm thick.

Joining of plates: The form of the joint at the intersection of a longitudinal joint and a circumferential joint may be as shown in Fig. 9.11.

The methods of joining plates at right angles are shown in Fig. 9.12. The types in Fig. 9.12a, b, c are used mostly for cylindrical vessels; that in Fig. 9.12d is used for connecting flat plates.

Materials: The ASME Boiler Construction Code specifies definite physical properties for *the materials used and a safety factor $n_u = 5$, based on ultimate strength.* Wrought iron being

[5] W. M. Wilson, J. Mather, and C. O. Harris, *Tests of Joints in Wide Engineering Plates*, Bulletin No. 239, University of Illinois Engineering Experiment Station (November, 1931), p. 44.

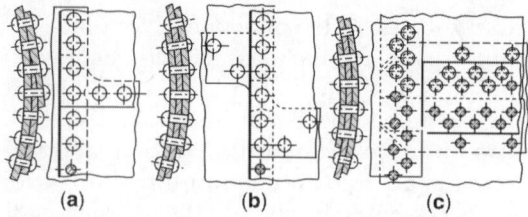

Fig. 9.11: Forms of riveted joints at junction of three or four plates

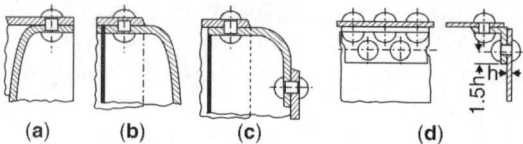

Fig. 9.12: Joining plates at right angles

practically no longer used, the allowable stresses are commonly assumed as 75 N/mm² in tension, 130 N/mm² in crushing, and 60 N/mm² in shear, for both steel plates and rivets. These values correspond approximately to data of Table 4.2 when the factor of safety n is taken as 3 and when SAE 1010 steel (10C8) is used for the rivets and SAE 1020 (20C8) steel is used for the plates. With such low allowable stresses, since the materials of the rivets and plates are very ductile and the loads are fairly constant, the influence of stress concentration from the holes can be neglected. For grade A alloy steel, the ASME Code allows the use of a crushing strength of 800 N/mm² and a shear strength of 400 N/mm² which correspond to allowable stresses of 162 N/mm² and 80 N/mm² respectively.

9.6 DESIGN PROCEDURE FOR VESSELS

The first step in designing a cylindrical vessel is to select the type of longitudinal joint to be used, the choice depending on the diameter D of the vessel and the internal, or fluid pressure p_1. The ASME Boiler Code permits lap joints for power boilers 900 mm or less in diameter provided the steam pressure does not exceed 700 kN/m²; for unfired pressure vessels not over 600 mm in diameter with a maximum pressure of 1.4 MN/m² and for unfired pressure vessels not over 1.2 m with a maximum pressure

of 1 MN/m². In all other cases butt joints must be used.

The next step is to determine the number of rivet rows. A greater number of rows gives a higher efficiency and permits the use of thinner plates, but it increases the cost of manufacturing. The materials selected for the plates and rivets determine the allowable stresses S, S_c and S_s.

Next, the thickness h of the plates can be determined by equating the fluid pressure upon a ring 25 mm wide, or $p_f D$, to the tensile stress of the material in two sections of the ring on opposite ends of a diameter and solving the equation for h. Thus

$$h = \frac{p_f D}{2Se} \qquad \ldots (9.15)$$

With the type of joint and the number of rivet rows selected, an expected efficiency e is assumed, based on the data in Table 9.1.

Plate thicknesses commonly used vary from 5 mm upwards by 1 mm up to 10 mm and in steps 2 mm up to 24 mm.

Thereafter the size of the rivets is selected so that it will be between $6\sqrt{h}$ and $7\sqrt{h}$.

Then the main pitch is determined by equation 9.16, or by equating F_t to F_c or F_s whichever is lower. The value of pitch is increased to 2.5d if it is lower. The efficiency of the joints is computed, all possible cases of failure being considered. This efficiency must be equal to or greater than the one assumed in computing h by equation 9.15. The minimum values of pitch is taken as 2.25 to 2.5d or even 3d and maximum value of pitch $\leq 6d$ for leakproof joint.

Finally, the margins and transverse pitches are determined chiefly from practical considerations, as explained in section 9.3.

An example will illustrate the method of procedure.

Example 9.1: Design the longitudinal joint of a boiler 1.5 m diameter for a working pressure of 1.4 MN/m².

It is obvious that only a double strap butt joint can be used. In view of the pressure and diameter no smaller than a triple-riveted joint should be selected. Suppose that the upper strap goes over only two rows of rivets on each side and every other rivet is omitted in the third row as in Fig. 9.10, to increase the efficiency. From Table 9.1 an average efficiency $e = 0.84$ is selected. The design stresses may be taken as prescribed by the ASME Boiler Code:

In shear $S_s = 60$ MN/m^2, tension $S = 75$ MN/m^2, crushing $S_c = 130$ MN/m^2

By using thin cylinder's equation, the thickness of the plate is

$$h = \frac{1.4 \times 1.5}{2 \times 75 \times 0.84} = 0.0166 \text{ m}$$

The nearest larger size is 18 mm, which will be used.

The diameter of the rivets should be between $6\sqrt{18} = 26$ mm to $7\sqrt{18} = 30$ mm. Selecting the higher size for higher efficiency, i.e. 30 mm. After the rivet is driven in, the size will be equal to the hole diameter. Calculations are based on rivet diameter.

By equating the shearing and tearing loads

$$(p - d)hS = \frac{\pi}{4}(2n_2 + n_1)d^2 S_s$$

$$p = \frac{\frac{\pi}{4}(2n_2 + n_1)d^2 S_s}{hS} + d$$

$$= \frac{\frac{\pi}{4}(2 \times 4 + 1)30^2 \times 60}{0.018 \times 75 \times 10^6} + 0.030 = 0.313 \text{ m}$$

As stated in section in 9.5, the pitch p_2 should not be greater than $8h$ or $8 \times 18 = 144$ mm and the corresponding value of p is $2 \times 144 = 288$ mm. This pitch will be used in order to insure a tight joint. The thickness h_2 of the narrow cover plate may be taken equal to 18 mm, with these values, the strength of solid plate by equation 9.4 is

$$F_p = \frac{288 \times 18 \times 75}{10^6} = 0.3888 \text{ MN}$$

By equation 9.5 strength of the plate along the outer gage line is

$$F_t = \frac{(288 - 30) \times 18 \times 75}{10^6} = 0.3483 \text{ MN}$$

The resistance to shear, by equation 9.6, is

$$F_s = (2 \times 4 + 1) \times \frac{\pi}{4} \times 30^2 \times \frac{60}{10^6} = 0.387 \text{ MN}$$

The resistance to crushing by equation 9.7, is

$$F_c = \frac{(4 \times 18 + 1 \times 12) \times 30 \times 130}{10^6} = 0.3275 \text{ MN}$$

The resistance against failure of the plate through the second row and simultaneous shearing of the rivet in the first row by equation 9.13 is

$$F_{s1} = \frac{(288 - 2 \times 30) \times 18 \times 75}{10^6} + \frac{\frac{\pi}{4} \times 30^2 \times 60}{10^6}$$

$$= 0.3078 + 0.04241 = 0.3502 \text{ MN}$$

Similarly, the resistance of the plate through the second row and simultaneous crushing of the rivets in the first row by equation 9.14, is

$$F_{c1} = 0.3078 + \frac{30 \times 18 \times 130}{10^6}$$

$$= 0.3078 + 0.0702 = 0.378 \text{ MN}$$

The resistance of the rivets in shearing in the outer row and the crushing of the rivets in the two inner rows, by equation 9.15, is

$$F_{sc} = \frac{4 \times 30 \times 18 \times 130}{10^6} + 0.04241$$

$$= 0.2808 + 0.04241 = 0.3232 \text{ MN}$$

Evidently the deciding minimum efficiency is

$$e = \frac{F_{sc}}{F_b} = \frac{0.3232}{0.3888} = 0.8313$$

For a triple riveted joint the efficiency is good and is slightly lower than the assumed value of 0.84. However, it can be improved. The simplest way to increase the efficiency is by decreasing pitch. For instance, making $p = 275$ mm gives

$$F'_p = \frac{275 \times 18 \times 75}{10^6} = 0.37125$$

The other resistances are also affected by the change of p. Their new values are:

$$F'_t = \frac{(275 - 30) \times 18 \times 75}{10^6} = 0.3307$$

$$F'_{s1} = \frac{(275 - 2 \times 30) \times 18 \times 75}{10^6} + 0.04241$$

$$= 0.29025 + 0.04241 = 0.3327$$

$$F'_{c1} = 0.29025 + 0.0702 = 0.36045$$

Thus the new efficiency $= \dfrac{F_{sc}}{F'_p} = \dfrac{0.3232}{0.37125} = 0.8705$

The margin for a multiriveted joint is

$m = 1.5\,d = 1.5 \times 30 = 45$ mm

The inner transverse pitch is taken as

$p_{t1} = 2 \times 30 = 60$ mm

The outer transverse pitch p_{t2} must be wider, and as stated in section 9.3, it is made equal to $2.75d$ or

$p_{t2} = 2.75 \times 30 = 82.5$ or 83 mm

From Fig. 9.10, the width of the upper cover plate is

$b_1 = 4m + 2p_{t1} = 4 \times 45 + 2 \times 60 = 300$ mm

and width of the lower cover plate is

$b_2 = b_1 + 2p_{t2} = 300 + 2 \times 83 = 416$ mm

Rivets: Structural rivets are available in sizes from 8 to 24 mm in 2 mm increments and 24 to 36 mm in 3 mm increments. The size of the rivet depends chiefly on the thickness of the connected members, but the commonly used sizes are: 12 or 16 mm for channels or built-up sections up to 150 mm, and for 50 mm angles, 16 mm; for 200 mm sections, and for 65 mm angles; 20 mm; for sections up to 300 mm and for 76 mm angles, and 20 or 22 mm for larger sections. Rivets are driven either by machines or by power-operated hand tools—only seldom by hand hammers. The maximum size of rivets to be used with a certain structural shape is given in handbooks.[6]

Rivet holes: Holes for rivets are either drilled or punched 1.5 mm larger than the nominal rivet diameter for $d \leq 20$ mm and are made 2 mm larger for $d \geq 22$ mm. However, in all cases, the computed resistance of rivets is based on their nominal size.

Spacing: Practical rules give for the pitch the limits $16h \geq p \geq 3d$, where h is the thickness of the thinnest plate used in the joint. An excessively large pitch prevents intimate contact between the members. Water may then collect in the crack and deteriorate the joint by rusting.

If the force is acting parallel to the edge, the distance from the center of a rivet to the edge should be $m_1 \geq 1.5d$. If the tensile force is acting normal to the edge, the edge distance should be $m_2 \geq 2d$.

Types of joints: The lap and butt joints used in structural work are similar to those used in pressure vessels. In addition the structural shapes are fastened to each other directly or by means of intermediate members, such as connection angles, clip angles, splice plates, or gusset plates. In Fig. 9.13 two connection angles are shown in a typical construction—the fastening of a beam to another beam or to a column. The size of connection angles have been standardized, the handbooks giving their detailed dimensions and the maximum safe load to be borne by the rivets in the parallel legs *a*, Fig. 9.13, as well as by the rivets in the outstanding legs *b*. If the standard connection angles for a given beam are not sufficiently strong, a special connection must be designed with a greater number of rivets. This may be accomplished either by increasing the size of the angles or by adding a seat angle *c*, Fig. 9.13.

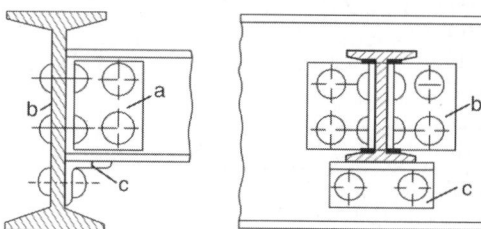

Fig. 9.13: Connection angles

Splice plates are used to connect two members placed end to end in a straight line, as *a* in Fig. 9.14.

Gusset plates, as *g* in Fig. 9.14, are used to connect members that intersect or meet at an angle.

Filler plates, as *c,* are used to line up members of different thicknesses (Fig. 9.14).

Clip angles, as *i* in Fig. 9.15b, are intended to secure a central loading of angle-shaped members.

Stress determination: In computing the stress in a tension member it is customary to

[6] Lionel S. Marks, ed., *Mechanical Engineers' Handbook*, 5th ed. (New York: McGraw-Hill Book Company, Inc., 1951), pp. 1558–76; *Steel Construction Manual*, 5th ed. (New York: American Institute of Steel Construction, 1951); and handbooks published by the major steel companies. Design Data Handbook–K Mahadevan and K Balaveera Reddy.

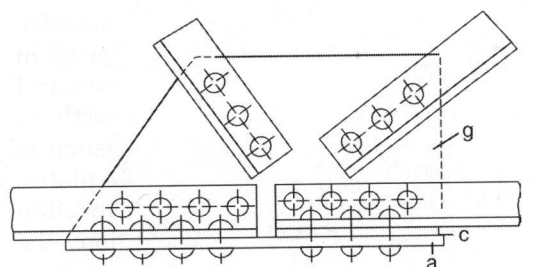

Fig. 9.14: Gusset, splice, and filler plates

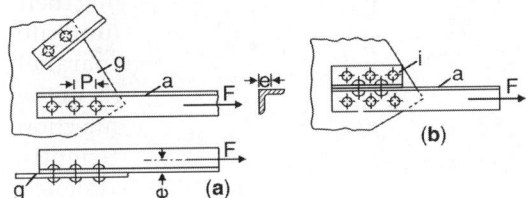

Fig. 9.15: Riveting of an angle to a gusset plate

determine the net area, by considering the size of the rivet holes as 3 mm larger than the rivet diameter. For compression members the area of the rivet hole is not deducted from the total cross section.

The connection of structural shapes to other members is often accompanied by eccentric loading, as shown in Fig. 9.14. This eccentricity increases the stress considerably over that due to the central loading. The determination of additional stress due to the bending moment does not complicate the problem a great deal and the analysis should be made as complete as possible. Thus, for the conditions in Fig. 9.15a, the total stress is

$$s = \frac{F}{A} + \frac{Fe}{Z} \qquad \ldots (9.16)$$

where, A is the cross-sectional area of the angle iron a, in square mm;

e is the eccentricity of the force, which may be considered to be applied at the center of gravity of the section, in mm;

Z is the section modulus of the angle iron, taken from a reference book, in mm^3

The allowable stresses are selected in the manner described for pressure vessels, but the safety factor is taken lower—about 2.0. The allowable stresses also depend on the method of driving the rivets. Recommended values are given in Table 9.2.

Aluminium structures: The design of aluminium riveted structures is similar to that of steel structures, the main difference being in the greater variety of rivet alloys available. The safe working stresses for the stock alloys are given in Table 9.3. In selecting the rivet material the general rule is to use the rivet with about the same properties as those of the structure to be riveted. Aluminium rivets can be obtained with the same heads as steel rivets and in sizes from 6 mm to 24 mm in 2 mm increments. Additional information may be found in special trade literature.[7]

9.7 DESIGN PROCEDURE FOR STRUCTURAL JOINTS

The following sequence of steps applies to the calculations for structural joints:

(a) The load transmitted by each member of the structure is determined analytically or graphically.

(b) The shape and size of each member is determined from the magnitude of the load that it takes.

Table 9.2: Allowable stresses in structural riveting MN/m^2

Load-carrying member	Type of stress	Rivet-driving method	Rivets acting in single shear	Rivets acting in double shear
Rolled steel, SAE 1020 (20C8)	Tension	–	125	125
Rivets, SAE 1010 (10C8)	Shear	Power	95	95
	Shear	Hand	70	70
	Crushing	Power	165	210
	Crushing	Hand	110	140

[7] Aluminium Company of America. Riveting of Alcoa Aluminium and Its Alloys (Pittsburgh: 1950) or BIS or ISO standards.

Table 9.3: Allowable stresses for aluminium rivets

Rivet allow	Procedure of driving	Allowable stress*	
		Shear MN/m²	Bearing MN/m²
2S (pure aluminium)	Cold, as received	20	45
17S	Cold, immediately after quenching	70	180
17S	Hot, 500°C to 510°C	60	180
61S–T6	Cold, as received	55	100
53S	Hot, 515°C to 525°C	40	100

* Factor of safety, 1.5.

(c) The diameter of the rivets is determined by the thickness of the structural shapes.

(d) The number of the rivets required in each member is based upon the shearing or crushing stress, whichever determines the cause of failure (i.e. lower).

(e) The rivets in the joint are spaced and arranged in such a manner as to utilize the material in the most economical way, avoiding eccentric loading as far as possible.

An example will illustrate the sequence of calculations.

Example 9.2: Determine the size of the angle and the size and number of rivets required in the connection shown in Fig. 9.15a, if the force F acting on the member a is 52 kN.

Step a. This step is omitted in this case, since the load F is given.

Step b. From Table 9.2, the stress S for the angle is 125 MN/m² and for power driven rivets $S_s = 95$ MN/m² and $S_c = 165$ MN/m². Hence the necessary area of the cross section for axial loading by taking the value of 100 MN/m²

$$A = \frac{52}{10^3 \times 100} = 0.00052 \text{ m}^2$$

However because of the comparatively large eccentricity of the load, a section somewhat more than twice as large will be assumed, and a 90 × 60 × 12 mm angle will be tried. After deduction is made for the hole for one 22 mm rivet, the net area of this (with area = 16.57 cm²) angle is $16.57 - \frac{\pi}{4} \times 2.2^2 = 12.77$ cm². It can be assumed that the force F is applied at the centre of gravity of the angle and that the reaction acts in the plane where the longer leg is in contact with the gusset plate. Therefore the eccentricity e in equation 9.19 is equal to the distance

from the centre of gravity of the angle to the outer surface of the longer leg. This is the distance designated as $\bar{x}$ in the handbook table and is 1.63 cm. Also, the section modulus is $Z = 10.3$ cm³.

By equation 9.19

$$S = \frac{52}{12.77} + \frac{52 \times 1.63}{10.3} = 4.072 + 8.229$$

$$= 12.3 \text{ kN/cm}^2 = 123 \text{ MN/m}^2$$

The stress is slightly lower than the permissible value, but the stress in a 90 × 60 × 10 or 100 × 75 × 10 angles will be excessively high.

The thickness of the gusset plate may be taken slightly less than the thickness h of the angle. So 10 mm would be a suitable value.

Step c. The rivet diameter is

$$d = 6\sqrt{12} = 20.78 \text{ or } 22 \text{ mm}$$

which is the same already assumed in determining the net area of the angle.

Step d. To determine the number of rivets, it is necessary first to find out whether the rivet is stronger in shear or crushing. In shear the resistance is

$$F_s = \frac{\pi}{4} \times \frac{2.2^2}{10^4} \times 95 = 0.0361 \text{ MN}$$

In crushing, the resistance is

$$F_c = \frac{1.0 \times 2.2}{10^4} \times 165 = 0.0363 \text{ MN}$$

Using the lower of the two values, the number of rivets required is

$$n = \frac{F}{F_c} = \frac{52}{10^3 \times 0.0361} = 1.44 \text{ or } 2$$

Step e. The arrangement of the rivets is simple—in a straight line and as close as possible to the vertical leg of the angle, that is, to the centre of gravity of the section. In accordance with the recommendations given previously, the minimum and maximum values for the spacing p are

$p_{min} = 3 \times 22 = 66$ mm, $p_{max} = 16 \times 10 = 160$ mm

A spacing of 70 mm will be satisfactory.

Theoretically it would seem to be useful to connect both legs of the angle to the gusset plate by means of a clip angle, as in Fig. 9.15b, in order to obtain a more central loading of the angle a. However, tests made with such a clip-angle connection indicate that very little is gained by it, evidently because of the elastic deflection of the connection.[8]

PROBLEMS

9.1 What is the most important property of rivet materials?

9.2 Why is a lap joint not preferable to a butt joint?

9.3 What is the value of margin in a riveted joint?

9.4 What type of failures are possible in the margin of c riveted joints?

9.5 Which type of riveted joint is stronger?

9.6 What type of riveted joint is suitable for boiler joints?

9.7 The girth seam of a horizontal cylindrical boiler is of the single-riveted lap-joint types. The thickness of the shell is 10 mm, the diameter of the rivets is 22 mm, the pitch is 50 mm and the margin is 36 mm. Using the allowable stresses recommended by the Indian Boiler Code/ASME Boiler Code, determine the manner in which this joint can fail, and its efficiency.

9.8 The longitudinal seam of an air-pressure tank is of the double-riveted lap-joint type. The shell is 12 mm thick, the nominal diameter of the rivets is 24 mm, the pitch is 86 mm, the transverse pitch is 54 mm, and the margin is 42 mm. Determine the manner in which this joint can fail, and its efficiency, using the allowable stresses recommended by the ASME Boiler Code/Indian Boiler Code.

9.9 The longitudinal seam of a cylindrical boiler drum is of the double-riveted butt joint type. The shell is 12 mm thick, the nominal rivet diameter is 22 mm, the pitch is 125 mm, the short pitch is 65 mm, the transverse pitch is 70 mm, and all margins are 36 mm. The thickness of the inner, wide cover plate is 10 mm, and that of the narrow cover plate is 12 mm. Determine the manner of possible failure and the efficiency of this joint, using the ASME Boiler Code/Indian Boiler Code allowable stresses.

9.10 Determine the manner of the possible failure and the efficiency of a triple-riveted butt joint, Fig. 9.10 in the text, having the following dimensions: $h = 20$ mm, the thickness of both cover plates is $h_1 = h_2 = 12$ mm, $d = 27$ mm, $p = 205$ mm, the inner transverse pitch is $p_{t1} = 60$ mm, and the outer pitch is $p_{t2} = 82$ mm and each margin is $m = 46$ mm. Use allowable stresses as given by the ASME Boiler Code/Indian Boiler Code.

9.11 The beam or jib supporting the head sheave of a crane consists of two ISMC200 channels, Fig. P9.1. Determine the thickness h of the square reinforcing steel plates supporting the stationary sheave pin, the number of rivets to fasten them, and the dimension b. Use the maximum diameter of the rivets recommended by the steel manufactures. The load is $Q = F = 90$ kN and the sheave-pin diameter is 90 mm. Use a safety factor of 2.5.

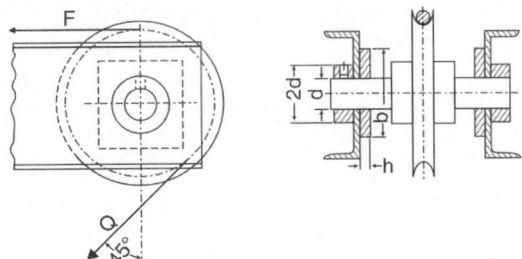

Fig. P9.1

---

[8] O. A. Leutwiler, Elements of Machine Design (New York: McGraw-Hill Book Company, Inc., 1917), p. 66.

9.12 As angle of a, with equal legs, is riveted to the gusset plate b, which in turn is riveted between two angles c. A load F of 45 kN is acting on a free end of the beam formed by the angles c. A load F of 45 kN is acting on a free end of the beam formed by the angles c. Assuming a safety factor of 2.5, determine (a) the size of the angle a if the relative positions of the rivets are as shown in Fig. P9.2, $\alpha = 30°$ and $l = 76$ mm, and (b) the size, number, and spacing of the rivets in both angles a and c. Make the gusset-plate thickness equal to the leg thickness of the angle a.

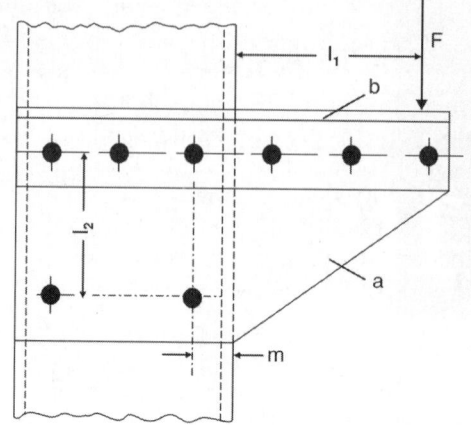

Fig. P9.3

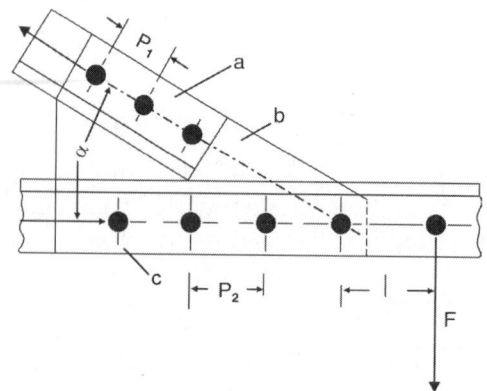

Fig. P9.2

9.13 Determine the safe load F which can be carried by the crane bracket in Fig. P9.3, using the following data. The steel plate a is 8 mm thick; the size of the angle b is $90 \times 60 \times 6$; the 200 mm channel weights 92 N/m all rivets are 16 mm in diameter; and $l_1 = 220$ mm, $l_2 = 125$ mm and $m = 40$ mm. Assume a safety factor n of 2.25.

9.14 Design a guide-sheave bracket similar to that in Fig. P9.4, for a load F of 45 kN and a height h of 150 mm. Use 12 mm plates c and a safety factor n of 1.75.

9.15 A small crane has a riveted connection similar to that shown in Fig. P9.5. The

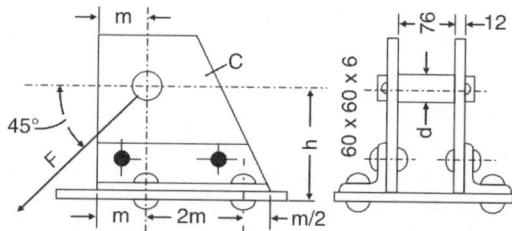

Fig. P9.4

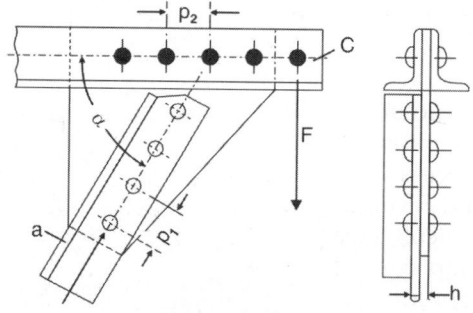

Fig. P9.5

strat is made up of a $150 \times 90 \times 7$ angle and has a free length of 2.4 m. Determine the shear and crushing stresses in each of the eight 22 mm rivets if the load F is 60 kN, the upper pitch is $p_2 = 100$ mm, the thickness of gusset plate is $h = 13$ mm and $\alpha = 60°$.h

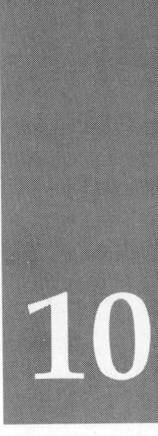

10

Design of Forgings

10.1 GENERAL CONSIDERATION

Manufacturing machine parts by forging is more expensive than by most other methods. Nevertheless the process is used quite extensively because forged pieces have great strength and resistance to shock.

Raw materials for forgings is usually in the shape of bars, plates, or billets. Before being forged the material is heated to a plastic state and is then subjected to several of the following operations: *bending, slitting, cutting out, drawing down (thinning), swaging, upsetting (thickening), punching and drifting, welding, and swaying (10.1).*

By a combination of these primary operations it is possible to produce parts of comparatively complicated shape. Since forging is an expensive process the designer should know what shapes and forms can be produced without undue difficulty, and should always strive to use the simplest forms in order to keep down the cost of production.

In spite of the wide use of forgings, comparatively a little specific information is available for the designer. The main reason for this is probably the fact that with a certain skill the blacksmith shop can reproduce without great trouble any reasonable shape worked out by the designer.

Forging methods may be divided into the following groups: (a) hand forging, (b) machine forging, (c) die forging, or drop forging, and (d) press forging (10.2).

Fiber flow: A great advantage of producing a part by forging, rather than by some other method, *is the possibility that the fiber lines in a piece may be kept unbroken (10.3)*. This statement may be illustrated by Fig. 10.1. In Fig. 10.1a, a piece made of steel by casting is shown, the casting having a granular structure; in Fig. 10.1b is shown a similar piece produced by cutting from bar stock, in which case the fiber lines are cut; and in Fig. 10.1c is shown another similar piece produced by forging, the bar being bent hot and the fiber lines remaining unbroken. If the pieces are made of about the same metal but by using the three different methods, then (as indicated in Fig. 10.1c) the grain slip planes in the forging are arranged in the most favorable way and the forging can withstand greater stresses, particularly those due to impact or repeated loads. In addition the ultimate strength and yield strength increase in cold forging. This important fact must be kept in mind in designing every forging, regardless of whether it is made by hand or by any type of machinery.

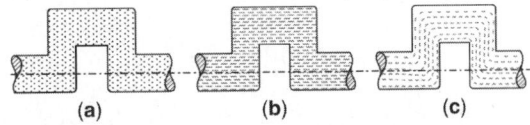

(a)   (b)   (c)

Fig. 10.1: Structure of material in a piece produced by different methods

10.2 HAND FORGING

By using proper combinations of the primary operations, a skilled blacksmith can shape a piece of metal into very intricate forms. An example of hand forging is the agitator propeller shown in Fig. 10.2. First one blade is drawn out from the side of a cylinder 0.18 m long and about 0.36 m in diameter. Then the opposite side is split, and the other two blades are formed by spreading and drawing them out from the split parts. Finally the hub is formed and a hole is punched in the center.

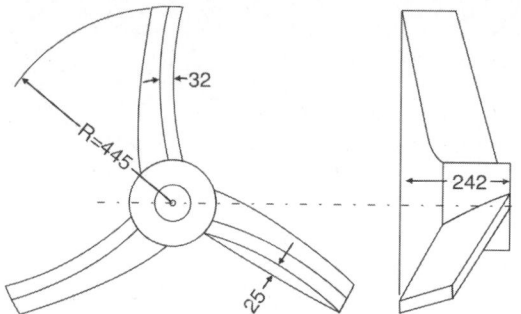

Fig. 10.2: Forged agitator propeller

It may be interesting to know that a propeller like that in Fig. 10.2 can be produced by a less-skilled worker, and probably more cheaply, by forging the three blades separately to conform to appropriate templates and then welding them to the hub. However, the direct forging of the whole piece should be preferred if a high-carbon alloy steel is to be used, as welding gives best results only with low-carbon steel.

10.3 MACHINE FORGING

By machine forging we mean forging of large, heavy pieces by using in general the same primary operations as in hand forging, but by striking the metal with some kind of power hammer such as a steam or air hammer or a friction-operated board-drop hammer. These hammers are used for such forgings as heavy crankshafts, connecting rods, and gear blanks.

The designer of such heavy forged pieces should strive to place as much material as possible in one plane and to avoid undercut surfaces and deep recesses. Deep recesses can

be obtained, if actually necessary, but they increase the cost of production very much.

10.4 DIE FORGING

Die forging is often called drop forging, the selection of the name depending on whether the speaker desires to emphasize the use of dies or the methods of applying the energy that forces the metal to assume the form determined by the shape of the dies (10.6).

Before deciding to use die forging the designer should take into account the fact that because the dies must be made of steel or steel castings, and also must be heat-treated, they are very expensive. Moreover, most die forgings require a series of dies, usually three sets, for a gradual changes of the formless piece of steel into an accurately shaped machine part; therefore die forging is economical only in mass production, that is, when the cost of dies is absorbed by a large number of forged pieces.

Die forgings should be so designed as to (a) permit the use of cheaper, longer-lasting dies; (b) obtain stronger forgings; and (c) permit subsequent machining operations to be performed efficiently.

10.5 DIES

The main points to be considered on a die are the draft angles, the partings, the recesses, and the fillets.

Draft angles: The two dies shown in Fig. 10.3a and b will have practically the same weight and the same strength if the dimensions a and b are made alike. However, a die should be provided with sufficient draft angles α and β, Fig. 10.3b, as specified in Table 10.1. The draft angle β for

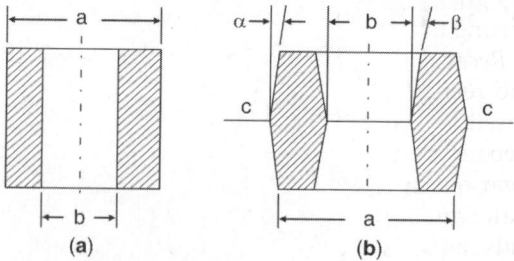

Fig. 10.3: Modification of a part for die forging

Table 10.1: Forging draft angles

Forging machine to be used	External surfaces		Internal surfaces		Remarks
	Normal angle (deg)	Commercial limits (deg)	Normal angle (deg)	Commercial limits (deg)	
Board drop hammer	7	0–10	10	0–13	Modern machines permit reduction to 5 deg
Steam drop hammer	$5\frac{1}{2}$	0–8	7	0–10	Modern machines permit reduction to 2 deg
Upsetter	3	0–4	5	0–6	Draft required only on deep dies
Press	0–3	–	0–3	–	No draft needed if knockout pins eject part

internal surfaces must be made larger than the angle α for external surfaces because the forging, which shrinks while cooling, tends to grip the projecting part of the die.

Parting: The parting line should be a straight line; all parting surfaces preferably should be in one plane, as plane *c–c* in Fig. 10.3b; and the parting surfaces should divide the volume of metal into approximately two equal amounts. In this respect Fig. 10.4a shows a poor design and 10.4b shows a good design.

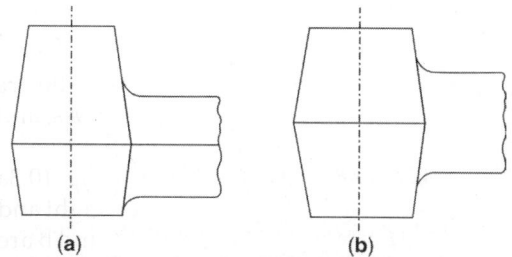

Fig. 10.4: Incorrect and correct parting of a die

If a one-plane parting is not feasible, the designer should select the simplest irregular parting that will produce the required results.

Recesses: Deep recesses should be avoided, and recesses should be as simple in shape as possible. The relatively sharp corners in the web recess as *c*, in Fig. 10.5a, will result in rapid wear of the die and in difficulty in removing scale; and simple recess with well-rounded ends, as in Fig. 10.5b, gives a cheaper and a longer-lasting die.

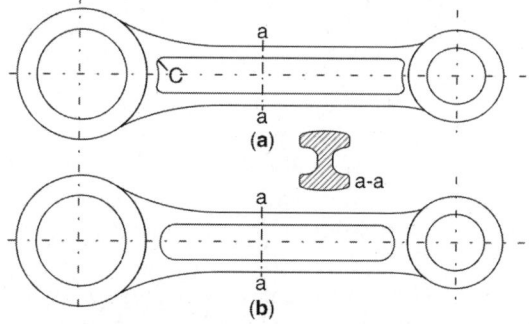

Fig. 10.5: Poor and good designs of a connecting rod

Fillets should be provided to avoid sharp inside corners and sharp exterior edges and corners. Fillets that are too small result in die cracks and reduce the life of a die. The minimum fillet runs from 1 mm for a small part to 3 mm for a die forging weighing 450 N.

However, if the forging must be machined, an excessively large fillet is also undesirable, as illustrated in Fig. 10.6. If there is a very large fillet as in Fig. 10.6a, the machined surfaces will cut many more fiber lines than when the fillet in the forging is of the same order as the fillet after machining, as shown in Fig. 10.6b.

Practical advice: After having completed the preliminary design of a die forging, a designer who did not specialize in this field should consult a manufacturer of die forgings, who may have valuable suggestions with respect to possible improvements. The suggestions may pertain to possible changes in the design of the dies in order to increase their life or to lower

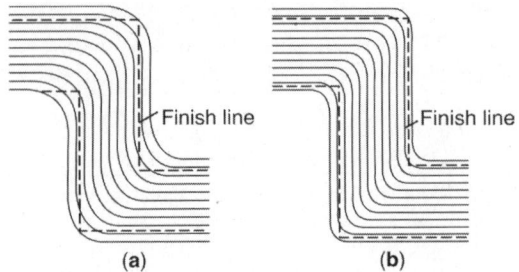

Fig. 10.6: Effect of fillet size in a machined part

their cost or to produce a stronger or less expensive forging.

10.6 STRENGTH OF FORGINGS

The important factors affecting the strength of a forging are its thickness and freedom from scale on its surface.

Thickness: Abrupt changes in thickness should be avoided. Since thin sections cool quicker than thicker ones, heavy shrinkage stresses are likely to be created at the places of abrupt changes. A poor design is illustrated in Fig. 10.7a, and the correct design, with a gradual change of thickness, is shown in Fig. 10.7b.

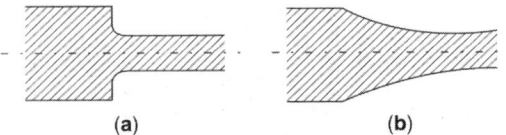

Fig. 10.7: Blending of different thicknesses

Very thin sections should be avoided in forgings because it is more difficult to forge them and to remove them from the die without damaging them. Also, quench cracks are likely to occur in thin sections during heat treatment. The minimum thickness recommended for moderate-size drop forgings is 3 mm.

Cleaning: The drawing for a drop forging should specify that the forging must be delivered free from scale. Forgings for parts that will be subjected to impact loads, and particularly to repeated loads must be cleaned by blasting. The action of blasting is similar to

that of peening and improves the endurance strength of steel parts. Forgings not subjected to impact loads may be cleaned by pickling.

10.7 MACHINING CONSIDERATION

To allow for subsequent machining, the locating points for machining should be carefully selected, sufficient material should be provided for finishing, and drill holes should be spotted.

Locating points: The designer should give careful consideration in selecting the location points at which a part is held during machining and should indicate them on the drawing. Locating points should be kept away from the parting line because the unavoidable wear of the edges of the dies gradually increases the thickness of the flash, which is the metal that must be removed—usually by a shear die. The dotted lines in Fig. 10.8 show how die wear occurs at the parting line.

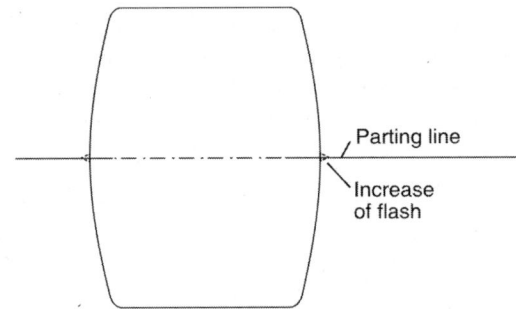

Fig. 10.8: Influence of die wear

Machining stock: Sufficient material should be allowed for machining at all places where finishing to accurate dimensions is necessary. The allowance for machine finish ranges from 1 mm for small hammer die forgings to 8 mm for large drop forgings. This is not an overall allowance but indicates the amount of metal to be removed from each machined surface. For press die forgings the allowance for machining can be much smaller, the maximum allowance being 1.5 mm.

Drilling: When large holes parallel to the die motion are to be drilled in forgings, it is economical to specify that the holes be spotted by the forging dies. This reduces the amount of layout work and helps to start the drill at the proper point. Spotting by the die is used only for holes 12 mm in diameter and larger.

Even though drilling is to be done from only one side of the forging, spots are placed on both sides of the piece, as shown in Fig. 10.9. This is done to a certain extent to obtain a more favorable fiber flow and to bring the fiber flow lines closer together. Where accurate center distances between two holes are required, only one of the holes should be die-spotted, the other being drilled from a template.

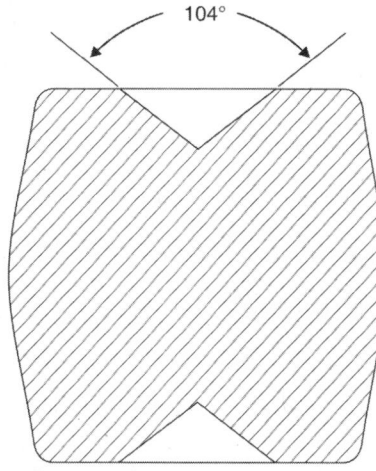

Fig. 10.9: Spotting hole in forging

10.8 TOLERANCES

The general rule, explained in section 6.3, that a tolerance should never be unnecessarily close must be observed in drop forging drawings even more strictly than in machining operations. The Drop Forging Association has set up tolerances for thickness, length and width, draft angle, fillets, and corners. The designer should never specify tolerances closer than 'close' figures given in the tables of the DFA.[1]

10.9 PRESS FORGING

In press forging the piece is shaped by the applications of steady pressure rather than by repeated blows. This procedure requires special machines and is used only for mass production. The dies used are similar to those used for drop forging. However, a much smaller draft may be specified, for both external and internal surfaces. A hole whose depth is less than its diameter can be produced practically without any draft.

Press forging is better than drop forging for pieces with heavy sections because it refines the grain of the entire piece, whereas the effect of hammering does not penetrate so deep below the surface. For the shaping of pieces made up of thin sections, drop forging is good.

Press forging is a relatively new process, but its use is increasing rapidly, especially in the production of nonferrous parts. By the use of several steps and retractable split-type dies, it permits the production of such complicated forgings as automobile-engine pistons made of aluminium alloys.

10.10 GENERAL CONCLUSION

The points discussed above are only the main points that a designer should know and observe. There are a great number of minor points which are also important and which may be found in special and general literature.[2]

PROBLEMS

10.1 Enumerate the primary operations which are used in forging.

10.2 Enumerate the four main groups of forging methods.

10.3 State one of the main advantages of forgings and the requirements to be followed in producing forgings.

---

[1] Erik Oberg and Franklin D. Jones, *Machinery's Handbook*, 14th ed. (New York: The Industrial Press, 1953), pp. 1176–77.

[2] *Forging Handbook* (Cleveland: American Society for Metals, 1950); Herbert Chase, *Handbook on Designing for Quantity Production*, 2nd ed. (New York: McGraw-Hill Book Company, Inc., 1944); *SAE Handbook*; articles in *Machine Design*, *Product Engineering*, and other periodicals; and Aluminium Company of America, *Designing for Alcoa Forgings* (Pittsburgh: 1950).

10.4 Show by a sketch an example of a hand-forged machine part.

10.5 Explain the process called machine forging.

10.6 What is the difference between a die forging and a drop forging?

10.7 What are main points to be considered for die forgings?

10.8 What are the important factors affecting the strength of a forgings?

10.9 Enumerate some constructive criticisms that may be expected from a practical designer who is working in the field of die forgings.

10.10 Enumerate the requirements that a designer of a drop forging must observe in order that a superior product may be obtained.

10.11 State what requirements a designer of a drop forging must keep in mind in regard to the subsequent machining.

10.12 State in what respect the design of press forgings differs from that of drop forgings.

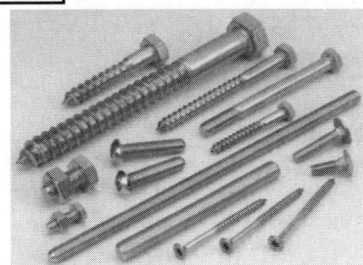

Screw Fastenings

11.1 UNIFIED AND AMERICAN SCREW THREADS

Screw fastenings are used for holding two or more machine parts together or for adjusting one part with relation to another. In screw fastenings the threads are made in several forms but are always of the triangular types, single-thread.

Screw threads are made right-hand and left-hand. Unless otherwise stated, it is always understood that the thread is right-hand.

After prolonged efforts of special committees, screw-thread standards in the United States, Canada, and the Great Britain were unified in 1948. The American standard screw thread is shown in Fig. 11.1. The Unified and American screw threads standard,[1] shown in Fig. 11.2, combines the good features of the old standards of the respective countries. The thread angle of 60°, the depth of thread h, the truncation f, and the number of thread n are identical with the old American National standard, and threads made to the new and old standards are interchangeable.

The Unified standard recommends the rounded root contour for the external screw thread, as was used in the British Whitworth standard, but permits a flat root when existing tools are used.

Designations used in formulas for threads are shown in Fig. 11.1.

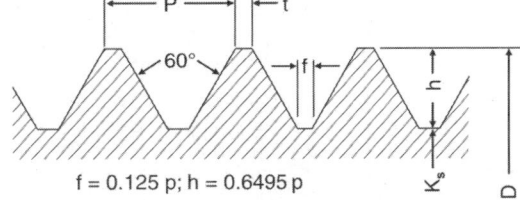

f = 0.125 p; h = 0.6495 p

Fig. 11.1: American standard screw thread

p = *pitch*, or the axial distance between two consecutive threads, in mm

h = depth of the threads, in mm

n = number of threads per unit length.

D = *major diameter* of the thread on a screw or nut—the largest diameter of the thread, in mm.

K = *minor diameter* of the thread on a screw (D) or nut (D_1)—the smallest diameter of the thread, in mm

E = *pitch diameter* of an imaginary cylinder whose surface passes through the thread profiles at such points as to make the width of the groove equal to one-half the pitch.

By definition there exists the general relation

$$p = \frac{1}{n} \qquad \ldots (11.1)$$

The following equations apply to the unified form of thread:

The basic depth of thread engagement is

$$h_e = 0.5417p = \frac{0.5417}{n} \qquad \ldots (11.2)$$

---

[1] ASA B1.1–1949.

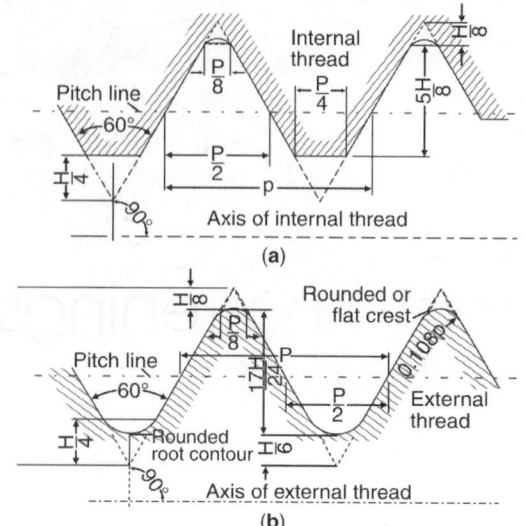

Fig. 11.2: Unified and American screw-thread forms [ISO Metric Screw Thread-Profile]

The minor diameter of the screw thread is given by the formula

$$K_s = D - 1.22687p \qquad \ldots (11.3)$$

The minor diameter of the nut thread is given by the formula

$$K_n = D - 1.08253p \qquad \ldots (11.4)$$

The basic pitch diameter is given by the formula

$$E = D - 0.64952p \qquad \ldots (11.5)$$

The stress area A_s is the assumed area of a screw and is used for the purpose of computing tensile strength. It is given by the formula

$$A_s = \frac{\pi}{4}\left(\frac{E_m + K_n}{2}\right)^2 \qquad \ldots (11.6)$$

This area is somewhat greater than the area corresponding to the minor diameter and takes into account the strengthening effect of the helix of the thread. In computing the stress in a screw in bending or torsion it is safer to base the calculations on the minor diameter.

Two Unified and six American standard thread series are in use.[2]

The UNC coarse thread series, is for general use and for all materials, including materials of lower strength than steel, such as cast iron, bronze, brass, and aluminium.

The UNF fine thread series, is for automotive and aircraft use or for use where maximum strength is required and the nut is also of steel.

The *American extra-fine thread series, NEF*, is not a unified standard. It covers threads from No. 12 to 2 in. and has four to six threads per inch more than the UNF series.

The American eight-thread series, 8N, in which $p = \frac{1}{8}$ in., is for threads from 1 in. to 6 in. It is used for fastenings where an initial tension must be set up by elastic deformation of the bolt and of the parts that it holds together. Examples are bolts for high-pressure pipe flanges, and studs for engine cylinder heads.

The *American twelve-thread series*, 12N, which $p = \frac{1}{12}$ in., is for threads from $\frac{1}{2}$ in. to 6 in. It is used for thin nuts on shafts and sleeves. In diameters larger than $1\frac{1}{2}$ in. it serves as a continuation of the fine thread series, UNF.

The *American sixteen-thread series*, 16N, in which $p = \frac{1}{16}$ in., is for threads from $\frac{3}{4}$ in. to 6 in. It is used chiefly for details where very fine close adjustment is required, such as threaded adjusting collars and retaining nuts.

ISO Metric Thread

Bureau of Indian Standards has also decided to adopt ISO (International Organization for Standardization) metric threads. These are being followed by a number of countries including United States, Canada, and Great Britain. The basic profile of the ISO metric screw threads is shown in Fig. 11.2a. The design profiles of external and internal threads are shown in Fig. 11.2b. ISO recommends the rounded root contour for the external threads

---

[2] Complete information is given in ASA B1.4–1949l; Lionel S. Marks, ed., *Mechanical Engineers' Handbook*, 5th ed. (New York: McGraw-Hill Book Company, Inc., 1951); R. T. Kent, Mechanical Engineers' Handbook, 12th ed., Volume 2, Design and Production, ed. by Colin Carmichael (New York: John Wiley & Sons, Inc., 1950); and Erik Oberg and Franklin D. Jones, Machinery's Handbook, 14th ed. (New York: The Industrial Press, 1949).

Table 11.1: Basic dimensions for design profiles of ISO metric screw threads

Basic dia mm $D_1 d$	Pitch mm p	Major diameter mm d	Pitch diameter, threads d_2	Minor diameter, mm		Lead angle at Basic pitch Dia		Tensile stress area, A_s mm^2
				External threads d_3	Internal D_1	deg.	min.	
1	0.25	1.0	0.83762	0.693283	0.729367	5	27	0.46
	0.20	1.0	0.870096	0.754626	0.783494	4	11	0.53
2	0.40	2.0	1.740192	1.509252	1.566987	4	11	2.07
	0.25	2.0	1.837620	1.693283	1.729367	2	29	2.45
2.5	0.45	2.5	2.207716	1.947909	2.012861	3	43	3.39
	0.35	2.5	2.272668	2.070596	2.121114	2	20	3.70
3.0	0.50	3.0	2.675240	2.386565	2.458734	3	24	5.03
	0.35	3.0	2.772668	2.570596	2.621114	2	18	5.61
4.0	0.70	4.0	3.545337	3.141191	3.242228	3	36	8.78
	0.50	4.0	3.675240	3.386565	3.458734	2	29	9.79
5.0	0.80	5.0	4.480385	4.018505	4.133975	3	15	14.2
	0.50	5.0	4.675240	4.386565	4.458734	1	57	16.1
6.0	1.00	6.0	5.350481	4.773131	4.917468	3	24	20.1
	0.75	6.0	5.512861	5.079848	5.188101	2	29	22.0
7.0	1.00	7.0	6.350481	5.773131	5.917408	2	52	28.9
	0.75	7.0	6.512861	6.079848	6.188101	2	6	31.3
8.0	1.25	8.0	7.188101	6.466413	6.646835	3	10	36.6
	1.00	8.0	7.350481	6.773131	6.917468	2	29	39.2
10	1.50	10.0	9.025721	8.159696	8.376202	3	2	58.0
	1.25	10.0	9.188101	8.466413	8.646835	2	29	61.2
	1.00	10.0	9.350481	8.773131	8.917468	1	57	64.5
12	1.75	12	10.863342	9.852979	10.105569	2	56	84.3
	1.50	12	11.025721	10.159696	10.376202	2	29	88.1
	1.25	12	11.188101	10.466413	10.646835	2	2	92.1
	1.00	12	11.350481	10.773131	10.917468	1	36	96.1
14	2.00	14	12.700962	11.546261	11.834936	2	52	115
	1.50	14	13.025721	12.159696	12.376202	2	6	125
	1.25	14	13.188101	12.466413	12.646835	1	44	129
16	2.00	16	14.700962	13.546261	13.834936	2	29	157
	1.50	16	15.025721	14.159696	14.376202	1	49	167
18	2.50	18	16.376202	14.932827	15.293671	2	47	192
	2.00	18	16.700962	15.546261	15.834936	2	11	204
	1.50	18	17.025721	16.159696	16.376202	1	36	216
20	2.50	20	18.376202	16.932827	17.293671	2	29	245
	2.00	20	18.700962	17.516261	17.834936	1	57	258
	1.50	20	19.025721	18.159696	18.376202	1	26	272
22	2.50	22	20.376202	18.932827	19.293671	2	14	303
	2.00	22	20.700962	19.546261	19.834936	1	46	318
	1.50	22	21.025721	20.159696	20.376202	1	18	333
24	3.00	24	22.051443	20.319392	20.752405	2	49	353
	2.00	24	22.700962	21.546261	21.844936	1	39	384
	1.50	24	23.025721	22.159696	22.376202	1	11	401

(Contd…)

Table 11.1: Basic dimensions for design profiles of ISO metric screw threads *(Contd...)*

| Basic dia mm $D_1 d$ | Pitch mm p | Major diameter mm d | Pitch diameter, threads d_2 | Minor diameter, mm | | Lead angle at Basic pitch Dia | | Tensile stress area, A_s mm² |
				External threads d_3	Internal D_1	deg.	min.	
25	3.00	25	23.051443	21.319392	21.752405	2	36	385
30	3.50	30	27.726683	25.705957	26.211139	2	18	561
	3.00	30	28.051443	26.319392	26.752405	1	57	581
	2.00	30	28.700962	27.546261	27.834936	1	16	621
	1.50	30	29.025721	28.159696	28.376202	0	57	642
35	1.50	35	34.025721	33.159696	33.376202	0	48	886
42	4.5	42	39.077164	36.479088	37.128607	2	6	1120
	4.0	42	39.401924	37.092523	37.669873	1	51	1150
	3.0	42	40.051443	38.319392	38.752405	1	22	1210
	2.0	42	40.700962	39.546261	39.834936	0	52	1260
	1.5	42	41.025721	40.159696	40.376202	0	40	1290
45	4.5	45	42.077164	39.479088	40.128607	1	57	1300
	4.0	45	42.401924	40.092523	40.669873	1	43	1340
	3.0	45	43.051443	41.319392	41.752405	1	16	1400
	2.0	45	43.700962	42.546261	42.834936	0	50	1460
	1.5	45	44.025721	43.159696	43.376202	0	37	1490
52	5.0	52	48.752405	45.865653	46.587341	1	52	1760
	4.0	52	49.401924	47.092523	47.669873	1	29	1830
	3.0	52	50.700962	48.319392	48.752405	1	6	1900
	2.0	52	50.700962	49.546261	49.834936	0	43	1970
	1.5	52	51.025721	50.159696	50.376202	0	32	2010
60	5.5	60	56.427645	53.252219	54.046075	1	47	2360
	4.0	60	57.401924	55.092523	55.669873	1	16	2490
	3.0	60	58.051443	56.319392	56.752405	0	57	2570
	2.0	60	58.700962	57.546261	57.834936	0	37	2650
	1.5	60	59.025721	58.159696	58.376202	0	28	2700

at the minor diameter to a theoretical radius equal to $0.14434\,p$ below the flat width $p/4$. In the case of internal threads also the roots are rounded and cleared beyond a pitch $p/8$ at the major diameter to avoid sharp corners.

Designation used in formulas for threads are shown in Fig. 11.2a and Fig. 11.2b.

D = major diameter of internal thread.
D_1 = minor diameter of internal thread.
D_2 = pitch diameter of internal thread.
d = major diameter of external thread.
d_2 = pitch diameter of external thread.
d_3 = minor diameter of external thread.
H = height of the fundamental triangle.
H_1 = maximum depth of engagement.
h_3 = basic depth of external thread.

p = pitch
v_r = root radius of external thread.
r_c = crest radius of external thread.

The following equation apply to ISO metric threads:

Internal Threads
$H = 0.86603\,p$
$D = d$ = major diameter
$D_2 = d_2 = d - 0.64952\,p$

External Threads
$D_1 = d - 1.08253\,p$
$d_3 = d - 1.22687\,p$
$H_1 = 0.54127\,p$
$h_3 = 0.61343\,p$
$r = 0.14434\,p$
$r_c = 0.108215\,p$

The basic design profiles of external thread show the crest as flat but modern methods of manufacture result in large quantities of external thread with crests partially or even completely rounded. The departure from flat crests is not detrimental. The following advantages are associated with external threads with rounded crests:

(a) External thread with rounded crests are less susceptible to damage by burring in handling and transport than those having flat crests, which result in sharp or semi-sharp edges at the major diameter of external threads.

(b) Troubles associated with plating are far less serious if the crests of external threads are rounded. In the plating of external threads by the usual barrel plating process, the burring of flat crested threads may be quite serious and in still-vat process, the plating tends to build up round the two edges at the major diameter and encroaches upon the flanks.

(c) The threads on thread rolling dies are stronger, less subject to fatigue failure and easier to grind if their roots are rounder rather than sharp cornered.

As per BIS recommendations (ISO), a thread series shall consist of a graduated series of diameters associated with suitable pitches. The recognized thread series shall be as follows:

(a) Course series
(b) Fine series
(c) Constant pitch series

The largest of the pitches associated with a thread diameter is called the course pitch for that particular diameter, the rest of the pitches being always finer than the course pitch. Of these one is graded as standard fine pitch for the specific diameter involved, the remaining pitch or pitches being graded as constant pitch or pitches. No standard fine pitch has been established for thread size below 8 mm diameter, while constant pitch series start from 1 mm thread diameter.

11.2 FITS

The following three main classes of screw-thread fits are provided by the standards:

Easy running fit is recommended only for screw-thread work where a substantial clearance between the screw and nut is required for rapid assembly and where shake or play is not objectionable. Tolerances and allowances for external threads and tolerances and allowances for internal threads are established for UNC and UNF thread series corresponding to tolerance grades 7, 8 and 9.

Normal running fit is recommended for a high quality of commercial screw-thread work, and tolerances are established for UNC, UNF, 8N, and 12N thread series corresponding to tolerance grade 6.

Close running fit is recommended for an exceptionally high grade of commercially threaded product and is recommended only in cases where the high cost of precision tools and continual checking of tools and products is justified. Tolerances are established with tolerance grades 3, 4, and 5.

ISO specifies tolerances for ISO metric screw threads for diameter range 1 to 300 mm. The tolerancing system provides for tolerances defined by tolerance grades, tolerance position and also a selection of grades and positions constituting tolerance classes.

Tolerance grades. For each of the two main elements, pitch diameter and crest diameter, a number of tolerance grades have been established as follows:

	Tolerance grade
Minor diameter of nut threads	4, 5, 6, 7, 8
Major diameter of bolt threads	4, 6, 8
Pitch diameter of nut threads	4, 5, 6, 7, 8
Pitch diameter of bolt threads	3, 4, 5, 6, 7, 8, 9

The grades 3, 4 and 5 correspond to precision threads where a little variation of fit is required as well as to short lengths of thread engagement.

The grade 6 should be used for medium quality corresponding to the general run of commercial bolts and nuts.

The grades 7, 8 and 9 correspond to thread produced by less precise methods, such as cutting screw threads directly on hot rolled bars and long blind holes. These grades are intended for coarse quality as well as for long lengths of thread engagement.

The tolerance position shall be so as to meet the current quoted thicknesses and the requirements of easy assembly. The combination of tolerance grade and tolerance position shall constitute a tolerance class. However in order to guarantee sufficient overlap, the finished components should preferably be made to form fits H/g, H/h or G/h.

Metric threads are specified by writing the diameter and pitch in millimeters and shall be designated by the letter M, e.g. M8 × 1 indicates a thread size or a major diameter of 8 mm and pitch of 1 mm. If there is no indication of pitch, it implies course pitch. M8 × 1 – 5H/4h also indicates the type of fit between nut and bolt.

Screw-thread designation: For brevity on drawings, a designation for a screw thread includes the nominal size, the diameter in inches (or the number, for sizes under $\frac{1}{4}$ in.), the number of threads per inch, the initial letters of the thread series, and the screw-fit class. For example, the designation $\frac{1}{2}$–13 UNC–2A indicates a $\frac{1}{2}$-in. screw with 13 threads per inch which is of the UNC coarse-thread series and which is made to conform to class 2A tolerances and allowances. Unless the designation just described is followed by the letters LH, which stand for left-hand, the screw is understood to be right-hand.

11.3 PIPE THREADS

American standards taper pipe-threads are used in the United States for pipes and fittings. The thread is formed with a taper of 1 in 16 in order to make a tight joint. The threads have slightly flattened roots and crests for a distance L_2, Fig. 11.3, which includes two threads with slightly imperfect crests. Following these threads are three or four that are imperfect because of the leads of the die. The number n of threads, the outside diameter D, and the length L_1 of normal engagement when the part is turned by hand. The pitch p is determined by equation 11.1, and the other dimensions are as follows:

The depth of thread is
$$h = 0.8\,p \qquad \ldots (11.7)$$
The pitch diameter at the end is
$$D_o = D - (0.05D + 1.1)\,p \qquad \ldots (11.8)$$
The length of the effective thread is
$$L_2 = (0.8D + 6.8)p \qquad \ldots (11.9)$$
American standard straight pipe threads are cut with a pitch diameter given by the relation
$$D_1 = D - (0.05D + 1.1)p + \tfrac{1}{16}\,L_1 \quad \ldots (11.10)$$
where the length L_1 is the length of normal engagement.

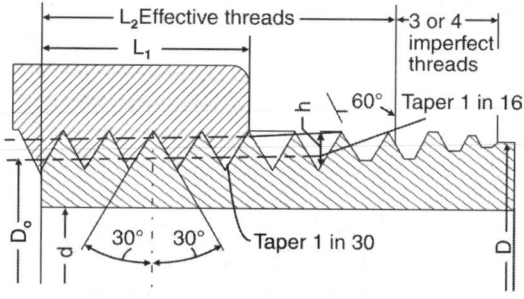

Fig. 11.3: ISO standard taper pipe thread

11.4 TYPES OF SCREW FASTENING

Screw fastenings may be classified as follows: (a) through bolts, (b) stud bolts, (c) tap bolts and cap screws, (d) machine screws, (e) set screws, (f) special screws, and (g) special details.

Through bolts: A through bolt, also called simply a *bolt*, is a round bar one end of which is threaded and fitted with a nut while the other end is upset to form a head. Nuts are made either hexagonal or square, as shown in Figs 11.4a and b, which show plain through bolts. The heads also are made either hexagonal or square. Square heads and nuts are used mostly with rough bolts, as in construction work.

Through bolts are the best form of screw fastening because when the nut has been tightened the shank of the bolt is subjected only to tension. If a through bolt is loaded in shear, it carries the load on the shank area.

Through bolts should be used whenever the available space and other conditions permit their use. If conditions are such that a through bolt cannot be inserted, as in a split pulley, a rod threaded from both ends with two nuts may be used, as shown in Figs 11.5 and 11.37.

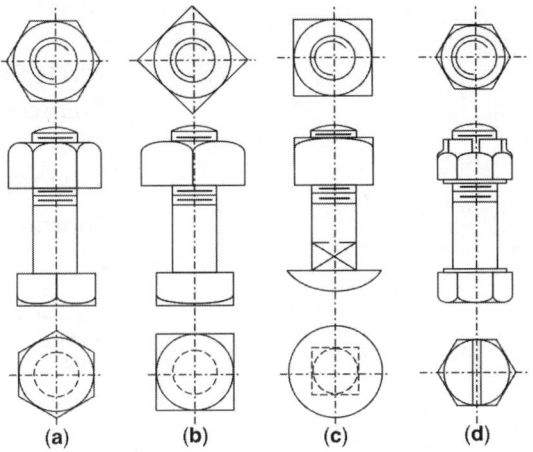

Fig. 11.4: Types of bolts

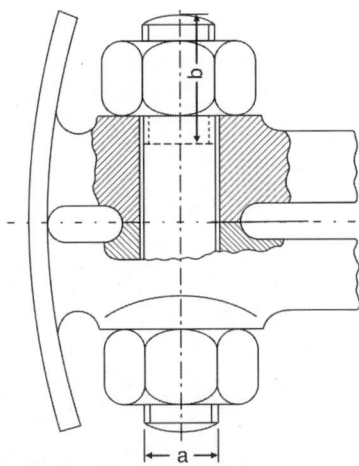

Fig. 11.5: Bolt with two nuts

A *Machine bolt* has a rough shank, but the head and the nut may be rough or finished, as desired. Commercial machine bolts are made and carried in stock in sizes from $d = 1.6$ mm to 150 mm.

A *coupling bolt* is a machine bolt that is finished all over and is made to be fitted into a reamed hole having the same diameter as the bolt.

A *carriage bolt*, Fig. 11.4c, is used when the head must rest against wood. The part of the

shank at the head is square to prevent the bolt from turning when the *nut is tightened*.

An *automotive bolt* is fitted with ISO or UNF threads, is finished all over, and has a somewhat smaller hexagonal head and nut. Such bolts are made in sizes from 6 mm up to 39 mm and in various lengths up to 150 mm. The head is often slotted for a screwdriver. Nuts of the ordinary type and of the castellated type are available. A castellated nut is shown in Fig. 11.4d.

Stud bolts: A stud bolt, often called simply a *stud*, is shown in Fig. 11.6. A stud is used when it is desired to fasten two parts *c* and *d* together, but when it is not possible or not desirable to drill a hole entirely through the second part *d*. To prevent unscrewing of the inner end, its thread is made to give a tight fit, while the thread at the nut end is of standard size.

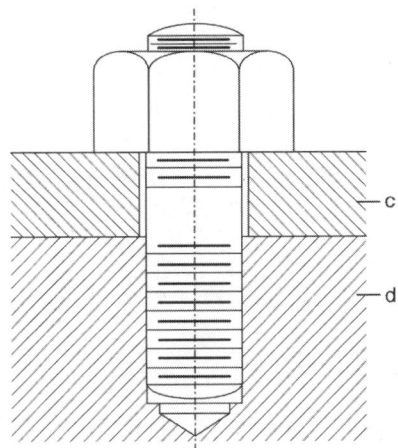

Fig. 11.6: Stud bolt

There is no standard for the length of the thread on the end of a stud. *However, it is advisable to make the depth of the tapped hole equal to the length of the threaded end, which should be at least equal to the diameter, in steel, and $1\frac{1}{2}$ diameters in cast iron and brass (11.1). This prevents stripping of the thread and provides enough frictional resistance against turning when the nut is unscrewed.*

A stud is the next-best screw fastening if a through bolt cannot be used. Studs are particularly convenient for positioning the covers of a cylinder head.

Wrenches: Wrenches for tightening nuts and screws are usually made with two ends, the openings in which fit two adjoining sizes. If the handle is straight or inclined at 30°, then the tightening of a hexagonal nut requires a 60° swing of the wrench, as indicated in Fig. 11.7. If the handle is inclined at 15° or 45°, the nut can be tightened even where the swing is limited to 30°. The nut is first turned 30°, as shown in Fig. 11.8a. Then the wrenches taken off the nut, is reversed so that the other face is up, and is put back on the nut as in Fig. 11.8b and turned another 30°. *So a very large number of bolt heads and nuts are hexagonal shaped as compared to square shaped as these can be tightened in narrow spaces (11.6).*

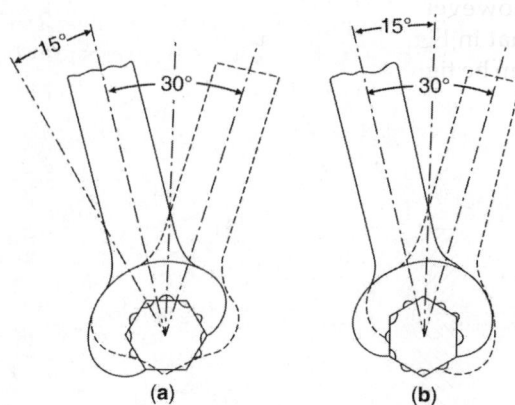

Fig. 11.8: Use of 15-deg wrench to reduce the swing

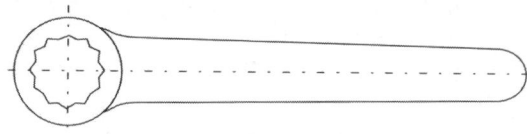

Fig. 11.9: Closed wrench

A *hook wrench*, Fig. 11.10, is used in places where there is no room for a standard open-end wrench and where a closed or socket wrench also cannot be used, such as for locknuts for ball bearings or stuffing-box glands.

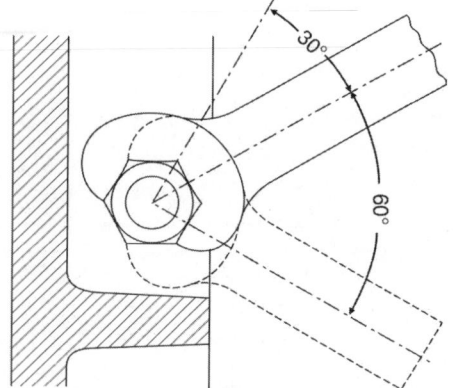

Fig. 11.7: Tightening a nut in a confined space

Steel nuts that are to be frequently unscrewed should be casehardened on the outside in order to protect them from wear by the wrench.

In locating bolts and studs in a confined place, as indicated in Fig. 11.7, the designer must remember to allow enough room for the jaws of the wrench and a sufficient angle of swing more than 30° for the wrench.

By using a *closed wrench* with twelve cut-in corners, like that in Fig. 11.9, the nut can be brought somewhat closer to a wall or rib parallel to which the centre line of the bolt or stud lies, and the necessary angle of swing can be reduced considerably. The use of a *socket wrench* helps in the same respect.

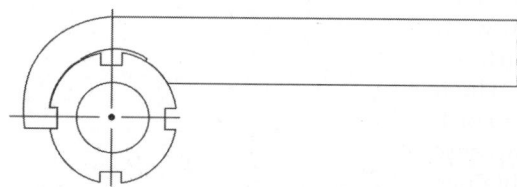

Fig. 11.10: Hook wrench

The usual length of a wrench is made equal to 12D, where D is the nominal diameter of the screw thread (11.7).

Tap bolts and cap screws: As shown in Fig. 11.11, a tap bolt is a through bolt without a nut.

Cap screws are made with various heads: hexagonal, square, socket, fillister, round, flat countersunk, and oval countersunk. Commercial cap screws are available in sizes from 6 mm to 30 mm and up to 150 mm long; cap screws with screwdriver slots are made in sizes from 3 mm up to 24 mm and in lengths up to 125 mm.

However, cap screws with socket heads like that in Fig. 11.15b are preferred because they can be tightened better.

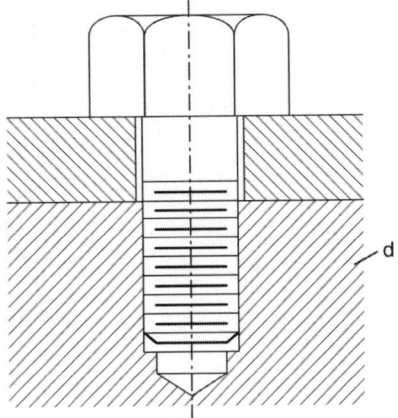

Fig. 11.11: Tap bolt

Tap bolts and cap screws are cheaper than studs and hence are often used instead of them. However, if a tap bolt or cap screw is loaded in shear, it carries the load on the root area, whereas a through bolt carries a shear load on the larger shank area. Long cap screws and tap bolts should not be used, because they twist when they are tightened; and later, when they untwist, this reduces the pressure between the surfaces they hold together.

The majority of cap screws are made with ISO or UNC threads. For screwing into steel, cap screws with ISO, UNF threads may be obtained.

Commercial cap screws are made of ordinary machine steel and various alloy steels, some having a tensile strength up to 1750 MN/m² Brass and Monel-metal cap screws are also obtainable.

Properly fitted cap screws are good for parts that are seldom removed, but their use should be avoided where they have to be unscrewed frequently, unless the materials of the part *d*, Fig. 11.11, is steel. To secure a good fastening by means of cap screws, the depth of the tapped hole should not be less than one and one-half times the diameter, in steel, and two times the diameter in cast iron.

Machine screws: Small cap screws with various types of heads, most of which have a slot for a screwdriver, are called machine screws. They are made with ISO UNC and UNF threads. The sizes are designated by numbers from 0 to 12. The diameter of a No. 0 machine screws is $D = 1.5$ mm, and the diameter increases 0.33 mm per number. Thus the diameter of any machine screw can be found from its number by the relation

$$D = 1.5 + 0.33\,N \qquad \ldots (11.11)$$

where, N is the size number. The diameter of a No. 12 screw is $D = 5.5$ mm. For larger sizes regular cap screws are used.

In addition to the number, a machine screw is specified by the number of threads and the length or pitch and length.

Set screws: *A set screw is a screw whose point presses against a piece and thus holds two parts together by friction (11.24).* A set screw may have a square head, Figs 11.12a or b; it may be headless, with a screwdriver slot, Fig. 11.12c; or it may have a hexagonal or fluted socket, Figs 11.12d, e or f. The point may be cup-shaped, Figs 11.12a or f; cone-shaped, Fig. 11.12e; oval, Fig. 11.12b; flat, Fig. 11.12c; or dog-shaped, Fig. 11.12d. The point is generally hardened.

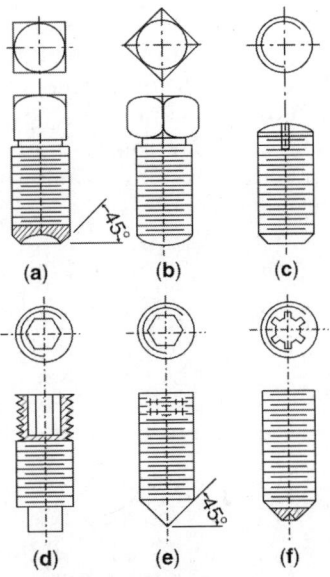

Fig. 11.12: Types of set screws

The cone tip or dog tip is subjected to shear for holding the shaft as these get inserted into corresponding grooves for transmission of small torque as in fans (11.24).

Special screws: Screw fastenings of special shapes are used freely in various design problems. Only a few typical examples will be given here.

Anchor bolts, or *T-head bolts*, shown in Fig. 11.13, are used with slots instead of holes in the connected parts to facilitate quick removal.

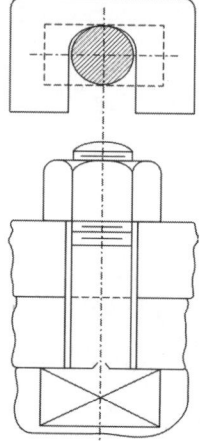

Fig. 11.13: Anchor bolt

Eyebolts are used for fastening parts requiring frequent removal when misplacing of the bolt must be prevented. The construction and method of application of an eyebolt are shown in Fig. 11.14. The pin *a* is made to give a drive fit in the cast-iron lugs *b* and goes freely through the bolt eye.

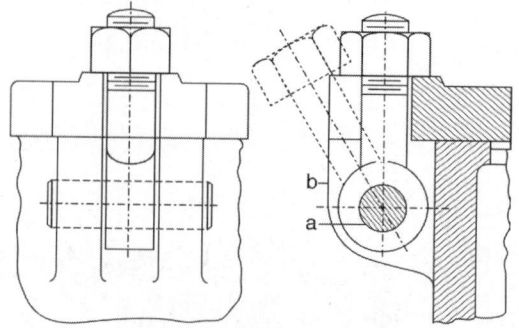

Fig. 11.14: Swinging eyebolt

Shoulder screws, Figs 11.15a and b, are chiefly used as fulcrums for rotating or rocking parts, since they prevent binding or clamping of the moving parts.

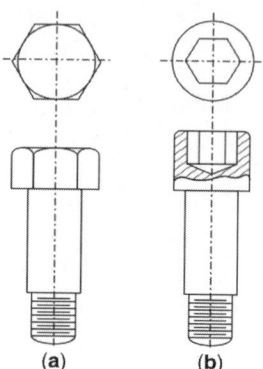

(a)   (b)

Fig. 11.15: Shoulder screws

Tie rods, also called *threaded rods*, are rods with threaded ends which are either screwed into other details or have nuts on both sides. A tie rod may be fitted with a right-hand thread on one end and a left-hand thread on the other, as shown in Fig. 11.16. This makes it possible to adjust the distance *l* between the connected parts by turning the rod, without disturbing other parts.

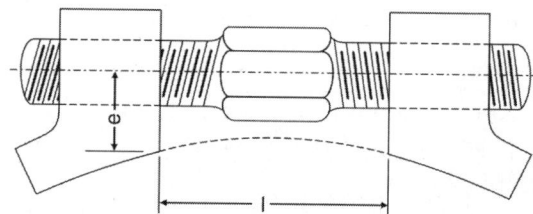

Fig. 11.16: Tie rod

Various details are fastened to rods by means of threads. The attachment of a piston to the piston rod by means of a thread and nut is shown in Figs 11.17 and 11.34; and the connection between a piston rod and a crosshead by means of a tapped hole and locknut is shown in Fig. 11.18.

Special details: A few special screw devices are shown in Fig. 11.19. The *turnbuckle*, Fig. 11.19a, connects two rods, one of which has a right-hand thread and the other a left-hand thread.

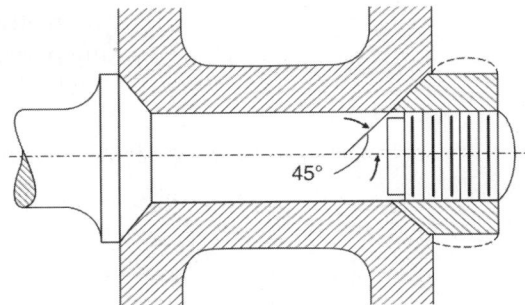

Fig. 11.17: Fastening of piston to rod

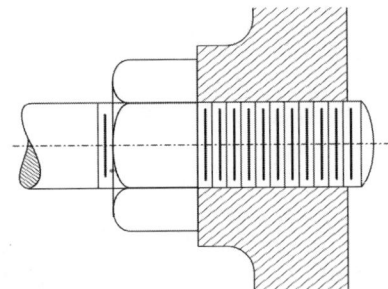

Fig. 11.18: Fastening of piston rod to crosshead

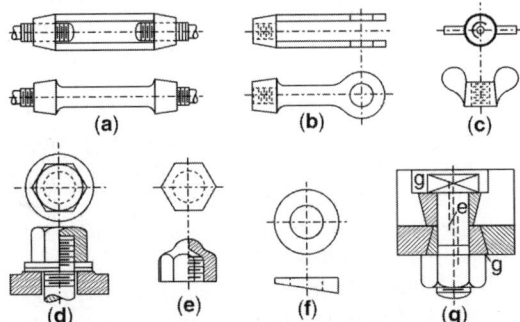

Fig. 11.19: Special screw-fastening details

The *clevis*, Fig. 11.19b, is used to fasten a rod to a plate or to a fulcrum pin.

The *wing nut*, Fig. 11.19c, is used where the holding device must be screwed down and unscrewed frequently.

The *cap nut*, Fig. 11.19d, is used to prevent leakage past the threads of studs. A cap nut of the type shown in Fig. 11.19e is sometimes used for the same purpose but often for the sake of appearance.

Washers: A thin ring, or washer, of steel is placed under the nut, when the surface of the part is not finished, in order to decrease the friction that occurs when the nut is tightened (11.8). A special wedge-shaped washer, Fig. 11.19f, is used to prevent bending of a bolt if the surface under the nut is not normal to the axis of the bolt, as in an I beam (11.9).

Conical washers, as shown in Fig. 11.19g, with eccentrically drilled holes are used to assemble parts in which the center lines of the holes do not coincide. For large sizes this method is cheaper than reaming the holes in place. The washers are made of soft steel or, when there is no impact action or vibration, of a special lead alloy.

11.5 FRICTION LOCKING DEVICES

Nuts on bolts and studs, when subjected to vibration, tend to work loose. There are many kinds of locking devices. Of the two main groups, those using friction and those based on a positive engagement, the first group will be discussed first.

Lock nuts: One of the most used devices is a *second nut*. When the upper nut is tightened while the lower one is held with a wrench, the bolt stretches and the pressure on the threads of the lower nut is relieved, with the result that the latter acts more like a spacer. For this reason the lower nut can be made shorter. Its height may be only $\frac{1}{2}D$, or half that of a regular nut. However, the average mechanic has a tendency to put the thinner lock nut on top. To prevent this mistake it is a better practice, if a saving in height is desired, to make both nuts of the same height, which may be about $\frac{3}{4}D$.

The elastic stop nut (Fig. 11.20a) uses a fiber collar *c* which is set in a recess in the nut and grips the thread of the bolt as the nut is screwed home. Elastic stop nuts are made of various materials—steel, brass, and duralumin–in sizes up to 50 mm, and the either UNC, UNF or metric threads.

The spring washer, or lock washer, Fig. 11.20b, tilts the nut slightly so as to produce a pinching action, but its use may dent the lower surface (11.11).

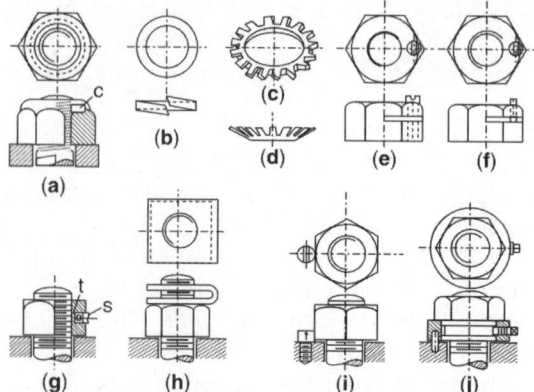

Fig. 11.20: Nut-locking devices based on friction

A shakeproof lock washer, Figs 11.20c and d, depends on twisted hardened-steel teeth to form a multiple lock of sharp edges which cut into the nut and the machine part. The conical lock washer in Fig. 11.20d is the only locking device for flat, cone-shaped machine-screw heads.

A slotted nut has a saw-cut on the sides. A small screw pulls the slot together, Fig. 11.20e, or spreads it, Fig. 11.20f, and thus cause the nut to pinch the thread.

The *lock screw*, Fig. 11.20g, is used successfully in a special joint or with a heavy nut. The set screw *s* presses against a piece of copper or fiber *t* to protect the screw thread and to create additional friction. The fine threads of the small set screw prevent it from unscrewing when the assembly is subjected to vibration which would cause the big nut to unscrew. The boxer lock, Fig. 11.20h, is made of a spring steel plate which is tapped and which has one side slightly twisted. Screwed on top of a regular nut, it produces a pinching action which makes it one of the safest locking devices. The arrangements of lock screws shown in Figs 11.20i and j have almost positive engagements but can be applied only in special cases.

The wrench-type lock *a*, Fig. 11.21, is also almost positive because the flat-head machine screw *b* will not unscrew under severe vibration.

The *Dardelet thread lock*, Fig. 11.22, consists of a self-locking bolt and nut. The effectiveness of

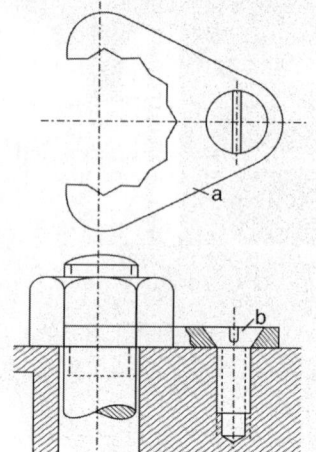

Fig. 11.21: Wrench type nut lock

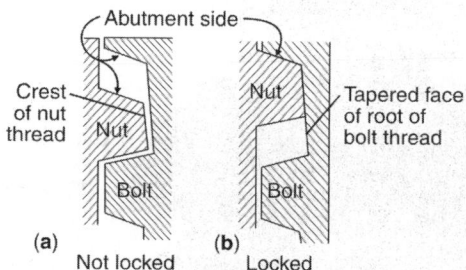

Fig. 11.22: Dardelet thread lock

the grip depends on the tightness with which the spiral sloping surfaces of the root of the bolt and the crest of the nut thread jam against each other when the nut is screwed home (Fig. 11.22b). The difficulty of cutting the special thread restricts the use of this otherwise very effective device. Also it is often difficult to apply a sufficient initial tension in the bolt, because the crest of the nut thread may jam against the face of the bolt root before the bolt is stretched sufficiently.

A nut having a conical bottom with a cone angle of 45° or smaller, Fig. 11.17, is a very dependable lock and is used rather extensively in special screw joints.

11.6 POSITIVE LOCKING DEVICES

The most positive lock is the cotter pin, Fig. 11.23a, which is inserted in a hole drilled through the bolt after the nut has been screwed home.

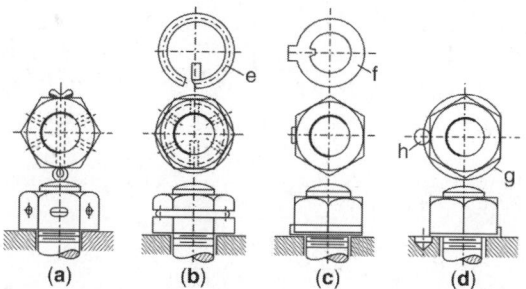

Fig. 11.23: Nut-locking device with a positive engagement

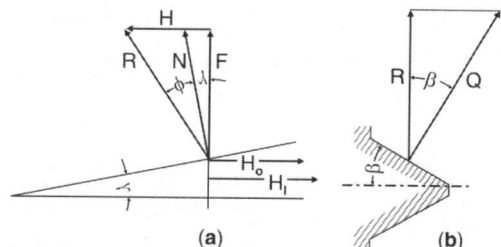

Fig. 11.24: Forces acting on a triangular thread

Where the nut is to be removed frequently or pulled up in service, the use of a *castellated nut* (Fig. 11.4d) with a finer UNF or ISO thread is a satisfactory solution. Insertion of the cotter pin above the nut makes its removal easier but has the disadvantage that the nut can back up before it is stopped by the pin.

The wiring together of cross-drilled screw heads or tap-bolt heads, or of bolt ends protruding through the nuts, is used when a slight turning of the bolts and nuts is not objectionable.

The *spring wire lock*, Fig. 11.23b, requires a groove to be turned in the nut and six small holes to be drilled through it. The hole in the screw is drilled after the nut is screwed home.

The *tongued washer*, Fig. 11.23c, requires the nut to be in a certain position to be really positive and also requires milling of a groove in the bolt end in order to accommodate the inner tongue.

The *upturned washer*, Fig. 11.23d, is used with large nuts, such as those which hold pistons, and is made of soft steel. If the surface under the nut is flat, it is necessary to form (drill) a recess h into which part of the washer is hammered. This lock has two advantages: The bolt is not weakened by having holes or grooves cut in it, and the nut may be stopped in any position.

11.7 EFFICIENCY OF TRIANGULAR THREADS

Friction in a screw thread creates a torsional stress in the screw. The turning moment acts on the screw in a plane normal to the screw axis. The useful force F, Fig. 11.24a, is exerted by the

thread in the direction of the axis. If there were no friction, this force would exert a pressure N normal to the surface of the thread.

The angle between the forces F and N is evidently equal to the helix angle λ. In a V thread this angle changes from the outside to the inside of the thread. Its mean value is at the pitch diameter E and may be found from Fig. 11.25, which shows a development of the pitch cylinder. Thus,

$$\tan \lambda = \frac{p}{\pi E} = \frac{1}{\pi En} \qquad \dots (11.12)$$

where, with sufficient accuracy, $E = \frac{1}{2}(D + K)$.

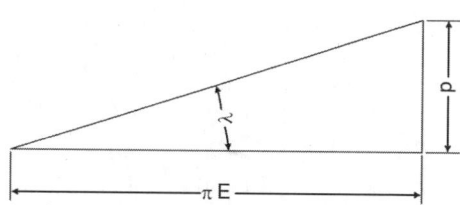

Fig. 11.25: Helix angle of a V thread

Because of friction the resultant force R, Fig. 11.24a, will be inclined from the normal by the amount of the angle of friction ϕ, and will have a component H in the plane normal to the screw axis. To do its work, the reaction R must have a component in the direction of F equal to the load but oppositely directed, and the component H must be overcome and balanced by an opposite force H_0. From Fig. 11.24a

$$H_0 = F \tan(\lambda + \phi) \qquad \dots (11.13)$$

If $\tan(\lambda + \phi)$ is expressed in terms of functions of the component angles λ and ϕ, and $\tan \phi$ is replaced by the coefficient of friction f_1, the resulting equation is

$$H_0 = F_1 \frac{\tan\lambda + f_1}{1 - f_1 \tan\lambda} \qquad \ldots (11.14)$$

In order to take into account the triangularity of the thread, the acting reaction Q, Fig. 11.24b, can be taken with sufficient accuracy as

$$Q = \frac{R}{\cos\beta} \qquad \ldots (11.15)$$

Since the friction force is directly proportional to the thread pressure, the friction terms in equation 11.14 must be divided by $\cos\beta$. The corrected tangential resistance is

$$H_1 = F \frac{\tan\lambda + f_1 \sec\beta}{1 - f_1 \tan\lambda \sec\beta} \qquad \ldots (11.16)$$

Values of the coefficient of friction compiled from various sources are given in Table 11.2.[3] The lower values apply to high-grade workmanship; the higher ones apply to average or poor workmanship.

Efficiency: The efficiency of a screw is the ratio of the force that has to be applied for a given output if there is no friction in the threads, to the force actually required in the presence of friction. If there is no friction, $f_1 = 0$ and equation 11.14 becomes

$$H_0' = F \tan\lambda \qquad \ldots (11.17)$$

The efficiency e is equal to H_0'/H_1. Substituting values from equations 11.17 and 11.16, and simplifying the result, gives

$$e = \frac{\cos\beta - f_1 \tan\lambda}{\cos\beta + f_1 \cot\lambda} \qquad \ldots (11.18)$$

Equation 11.18 allows for thread friction only. To find the overall efficiency when a nut or a screw with a head is being turned, it is necessary to consider the additional friction against the surface on which the nut or head bear. If the outside diameter of the bolt is D, the standard width across the flats of a hexagonal head or nut is $b = 1.5D$, and the diameter at which this friction is applied may be considered to be

$$D' = \tfrac{1}{2}(D + b) = 1.25D \qquad \ldots (11.19)$$

The tangential friction force at this diameter is Ff_2, where f_2 is the coefficient of friction against the outside surface. Referred to the same mean thread diameter E as the force H_1, the additional friction force is

$$H_2 = Ff_2 \times \frac{1.25D}{E} \qquad \ldots (11.20)$$

Thus the total friction if $H_1 + H_2$, and the overall efficiency is

$$e_2 = \frac{H_0'}{H_1 + H_2} \qquad \ldots (11.21)$$

Substituting values of H_0', H_1, and H_2 in equations 11.17, 11.16, and 11.20, and simplifying the result, gives

$$e_2 = \frac{\tan\lambda}{\dfrac{\tan\lambda + f_1 \sec\beta}{1 - f_1 \tan\lambda \sec\beta} + 1.25 f_2 \dfrac{D}{E}} \qquad \ldots (11.22)$$

For the UNC thread the Lewis empirical formula may be used. This is

$$e_2 = \frac{p}{p + D} \qquad \ldots (11.23)$$

Equation 11.23 is only approximate, but it checks well for $f_1 = 0.10$ and $f_2 = 0.15$.

Table 11.2: Coefficient of friction f for screws and nuts

Material of screw	Lubrication	Material of nut		
		Steel	Brass	Cast iron
Steel, mild or	Dry	0.15–0.25	0.15–0.23	0.15–0.25
casehardened, or	Machinery oil	0.11–0.17	0.10–0.16	0.11–0.17
bronze	Oil with graphite	0.08–0.12	0.04–0.06	0.06–0.09

---

[3] C. W. Ham and D. G. Ryan, *An Experimental Investigation of the Friction in Screw Threads*, Bulletin No. 247, University of Illinois Engineering Experiment Station (June, 1932), pp. 9 and 47.

11.8 STATIC STRESSES IN SCREW FASTENINGS

The following stresses must be considered in screw fastenings with a static loading:

(a) Initial stresses due to screwing-up forces.
(b) Stresses due to external forces.
(c) Stresses due to a combination of forces of types a and b.

Stress concentration: In computing the tensile and torsional stresses in the root section of a screw, the stress-concentration effect should be considered. With a ductile material, however, there will be a localized yielding with a consequent readjustment of the internal stresses. A certain permanent set may occur in the outer fibers at the root. Under a steady load, however, the bolt will not fail if the highest stress is higher than the elastic limit of the material but lower than its ultimate strength.

11.9 INITIAL STRESSES

The stresses in a bolt, screw, or stud when it is screwed up tightly are: (a) a tensile stress due to the stretching of the bolt; (b) a torsional stress caused by the frictional resistance of the thread during its tightening; and (c) a bending stress if the surfaces under the head or nut are not perfectly normal to the bolt axis.

Tensile stress: The stretching of the bolt depends on the force F, Fig. 11.24, which can be computed theoretically from the proportions of the thread. The following analysis will explain the procedure. The axial tension F may be treated as a function of the force $H = H_1 + H_2$ which must be applied when screwing down the nut and which may be found from equations 11.16 and 11.20. By taking as two more or less extreme cases a 12 mm screw and 76 mm screw with ISO or UNC threads, assuming as average friction coefficients $f_1 = 0.15$ and $f_2 = 0.20$, and determining the values for the other terms from the corresponding geometrical relations, we find that $H = 0.507F$ for the 12 mm screw and $H = 0.473F$ for the 75 mm screw. The values

of the coefficients of friction, Table 11.2, vary much more than these values of H. Therefore an average value of $0.49F$ can be assumed. This gives for the tension in the screw,

$$F = 2.04H \qquad \ldots (11.24)$$

The tensile stress is then

$$s = \frac{F}{A_s} \qquad \ldots (11.25)$$

where the stress area A_s is given in Table 11.1.

Torsion: The torsional stress s_s can be computed by the relation

$$s_s = \frac{16T}{\pi K^3} \qquad \ldots (11.26)$$

In this case the torque is

$$T = \tfrac{1}{2} H_1 E \qquad \ldots (11.27)$$

where the force H_1 can be found by equation 11.16, E is the pitch diameter and K is the minor diameter of the thread.

The magnitude of the resultant stress can be found by combining s and s_s as in equation 2.50. For UNC and ISO standard screws with sizes from 12 mm to 76 mm and for $f_1 = 0.15$ and $f_2 = 0.20$, the resultant stress is found to be from 28 per cent to 22 per cent greater than the tensile stress alone. However, for a screw with a diameter of 20 mm or less, the initial stress depends to such an extent upon the judgment and experience of the mechanic that an attempt to calculate it is practically useless.

Repeated tightening: According to accurate measurements, repeated unscrewing and tightening of steel nuts increases the torsional moment that must be applied, because there is a gradual scoring of the threads. *After 50 tightenings the torsional resistance is about 100 per cent greater than the resistance at the first tightening, and after 200 tightenings the increase is about 150 per cent.*[4]

Initial tension: According to experiments made at Sibley College[5] the load in Newtons

[4] A. Thum, "Vorspannung und Dauerhaltborkeit von Schraubenverbindungen," *Mitteilungen der MPA der T. H. Darmstadt*, Heft 7 (Berlin: VDI-Verlag, 1936), pp. 1–72.
[5] D. S. Kimball and J. H. Barr, *Elements of Machine Design*, 3rd ed. (New York: John Wiley & Sons, Inc., 1935), p. 277.

produced by screwing a threaded member up tight may be estimated as

$$F' = 2800 \, D \qquad \ldots (11.28)$$

where, D is in mm.

Equation 11.28 was confirmed surprisingly well by German experiments[6] with screws having diameters of $\frac{1}{2}$ in., $\frac{3}{4}$ in., and 1 in. (12, 20, 25 mm).

Substituting the value of F' from equation 11.28 in equation 11.25 and using the values of D and K from Table 11.1, we get for screws with sizes from 12 mm to 75 mm an average stress in MN/m² (N/mm²) with D mm.

$$S = \frac{4900}{D} \qquad \ldots (11.29)$$

This expression shows that even 24 mm screws are subjected to an excessively high initial stress. Smaller screws are easily stretched above the elastic limit and even broken. For this reason, bolts that must carry external loads preferably should not be under 12 mm and sizes under 10 mm should not be used. *Usual breaking of a bolt takes place at the first thread inside the nut due to stress concentration caused by high pressure (11.12).*

Equations 11.28 and 11.29 apply to cases where the nuts have metal-to-metal contact. If a bolt fastens two parts with a flexible gasket between them, a smaller effort is applied in order not to crush the gasket, and the initial load may be taken as one-half that given by equation 11.28. Thus

$$F' = 1400 \, D \qquad \ldots (11.30)$$

Bending: If the outside surfaces of the parts that are bolted together are not parallel, the bolt will be subjected to bending by a moment M that is constant over the whole free length l, Fig. 11.26. The stress induced by the bending may be found by considering the deformation of an element of the bolt, Fig. 11.27. The unit elongation of the outer fibers is

$$\epsilon = \frac{\Delta dx}{dx} = \frac{r \, d\alpha}{R \, d\alpha} = \frac{r}{R} \qquad \ldots (11.31)$$

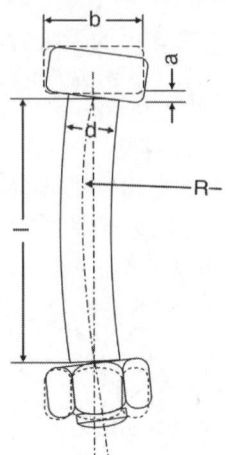

Fig. 11.26: Bending of a bolt

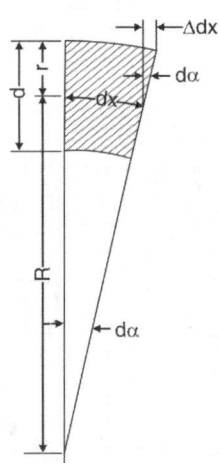

Fig. 11.27: Element of a bolt subjected to bending

Expressing the stress by equation 2.4 and substituting for ϵ its value from equation 11.31 gives

$$s = \epsilon E = \frac{rE}{R} \qquad \ldots (11.32)$$

From Fig. 11.26

$$\frac{1}{R} = \frac{2a}{b} \qquad \ldots (11.33)$$

Substituting the value of R from equation 11.33 in equation 11.32 and noticing that

[6] E. Bock, "Das Verhalten der Schraubenverbindung beim Anziehen und Losen," *Zeitschrift Verein Deutscher Ingenieure*, Vol. 78 (1934), p. 780.

$r = d/2$ and $b = 2d$, as a limit value, we finally get

$$s = \frac{aE}{2l} \qquad \dots (11.34)$$

Equation 11.34 shows that for a given material the stress depends only on the ratio a/l.

Example 11.1: Determine the bending stress induced in a 24 mm bolt that must hold together two flanges which have a combined thickness of 60 mm, and are so placed that the gap a, Fig. 11.26 is 0.2 mm.

By equation 11.34, the stress is

$$s = \frac{0.2 \times 210 \times 10^3 \times 10^3}{10^3 \times 2 \times 60} = 350 \text{ MN/m}^2$$

This small deviations of the surface from being parallel, which corresponds to about 0.3° between two surfaces, the stress exceeds the elastic limit.

This example shows how important it is to have the outside surfaces true. If they are not machined, then the surfaces around the holes must be spot-faced (11.14). With structural shapes, wedge-like washers, Fig. 11.19f, must be used.

The sum of the bending stress and the initial stress from tightening the nut can be very easily exceed not only the elastic limit but also the ultimate strength of the bolt material. However, because of the ductility of steel the bolt will be bent before breaking. This bending will relieve some of the stress, but it is not permissible in a machine part.

11.10 STRESSES DUE TO EXTERNAL FORCES

Generally the external load applied to a bolt tends to separate the connected machine parts in the direction of the bolt axis, and this action sets up a tensile stress in the bolt. Sometimes when bolts are used to prevent the relative movement of two or more parts, a shear stress is induced in the bolt. When the action of the load is at an angle to the axis, the stress in the bolt becomes tension and shear combined.

Tension: In a bolt subjected to a load F which induces tension, the stress is found by equation,

$$S = \frac{F}{A_s}$$

Shear: Bolts should preferably not be subjected to shear. When shear stresses cannot be avoided

in the design, the bolt shank should be accurately fitted to the holes in the connected parts, at least in the portion near the joint. A better arrangement is to use dowel pins fitted accurately into reamed holes after the bolts have been inserted and tightened.

(a) *Eccentric load on rectangular base:* The bolts holding a bracket as in Fig. 11.28 are not efficient. The load F tends to cause the bracket to rotate clockwise about the edge a, thus stretching each bolt to a degree that depends on its distance from the edge a. Since stress is a function of elongation, the loads on the bolts are different. However, for convenience and economy all six bolts are made of the same size.

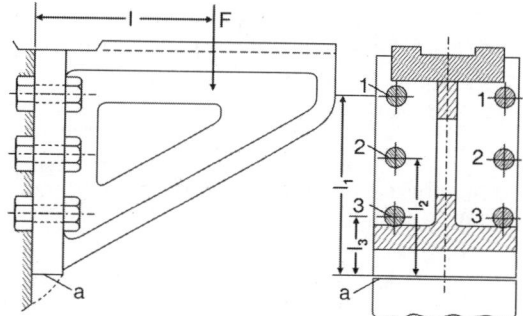

Fig. 11.28: Fastening of a bracket

The moment Fl of the load F must be balanced by the sum of the moments of the bolt loads about the edge a. If the loads on the bolts are designated by F_1, F_2 and F_3 and their moment arms are designated as in Fig. 11.28.

$$Fl = 2(F_1 l_1 + F_2 l_2 + F_3 l_3) \qquad \dots (11.35)$$

Since the stresses induced in the bolts are directly proportional to the elongations produced, and the elongations are proportional to the distances from the edge a, the maximum load comes on the bolts marked 1 and the following relations can be written:

$$\frac{F_1}{l_1} = \frac{F_2}{l_2} = \frac{F_3}{l_3} = k$$

$$F_1 = l_1 k, \ F_2 = l_2 k, \ F_3 = l_3 k$$

Substituting these values in equation 11.35

$$Fl = 2\left[l_1^2 k + l_2^2 k + l_3^2 k \right] = 2k \sum l^2$$

$$k = \frac{Fl}{2\Sigma l^2}$$

So values of F_1, F_2, F_3 are found as:

$$F_1 = \frac{Fll_1}{2\left(l_1^2 + l_2^2 + l_3^2\right)} \qquad \ldots(11.36)$$

and similarly F_2 and F_3.

(b) *Eccentric loading on circular base:* A machine member is frequently made with a circular base fastened by bolts located on a circle. To simplify the discussion a round-flange bearing with four bolts, as shown in Fig. 11.29, will be analyzed. By proceeding in the same way as with a rectangular base, we find the magnitude of the load on bolt 1 to be

$$F_1 = \frac{Fll_1}{l_1^2 + l_2^2 + l_3^2 + l_4^2} \qquad \ldots(11.37)$$

From Fig. 11.29a

$l_1 = a - b \cos \alpha \quad l_3 = a + b \cos \alpha$

$l_2 = a + b \sin \alpha \quad l_4 = a - b \sin \alpha$

When these values are substituted in equation 11.37, the result is

$$F_1 = F\frac{l(a - b \cos \alpha)}{4a^2 + 2b^2} \qquad \ldots(11.38)$$

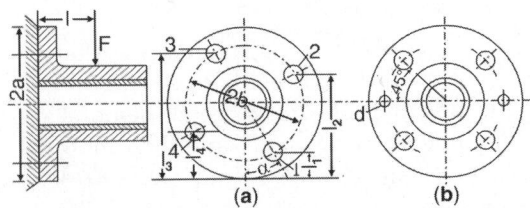

Fig. 11.29

By repeating the discussion for i bolts, we can write the general expression for the load as

$$F_i = F\frac{2l(a - b\cos\alpha)}{(2a^2 + 2b^2)i} \qquad \ldots(11.39)$$

The force F_i evidently has its maximum value when $\cos \alpha$ is a minimum, or when $\cos \alpha = -1$. Then $\alpha = 180°$, and

$$F_{max} = F\frac{2l(a + b)}{(2a^2 + 2b^2)i} \qquad \ldots(11.40)$$

Also on visual observation when a bolt is in the top position, l_i will be maximum and force in that bolt will be maximum, i.e. $F = k\,l_i$.

Equation 11.40 gives the absolute maximum value and should be used if the direction of the load F can change with relation to the bolts, as in the case of the base of a pillar crane. If the direction of F is fixed, the maximum loads on the bolts can be reduced by locating the bolts so that two of them will be equally stressed, as shown in Fig. 11.29b. In this case the angle α in equation 11.39 becomes $180° - 360°/2i$, the equation for the maximum load is

$$F'_{max} = F\frac{2l\left[a + b\cos\left(\dfrac{180°}{i}\right)\right]}{(2a^2 + 2b^2)i} \qquad \ldots(11.41)$$

The bolts should be relieved of shear stresses by using dowel pins d, Fig. 11.29b.

Alternately, primary shear force $= \dfrac{F}{i}$ where i is the number of bolts

Maximum shear force

$$= \frac{1}{2}\left[4\left(\frac{F}{A}\right)^2 + \left(\frac{F_{max}}{A}\right)^2\right]^{\frac{1}{2}}$$

This is equated to allowable shear stress to find A and then diameter or

Maximum tensile stress

$$= \frac{1}{2}\left[\frac{F}{A} + \sqrt{4\left(\frac{F}{A}\right)^2 + \left(\frac{F_{max}}{A}\right)^2}\right]$$

and equated to allowable tensile stress to calculate A and then diameter of bolt.

(c) *Eccentric load on rectangular plates:* Another example of eccentric loads is shown in Fig. 11.30 wherein a beam is fixed on to a member or another beam or it may be a part of a structure. G represents the C. G. of the group of bolts and it is assumed that all bolts are of the same diameter and bolts are numbered as 1, 2, 3, ... The total load on the bolts is calculated in three steps

(a) Vertical shear is equally divided among the bolts so that each bolt takes P/n

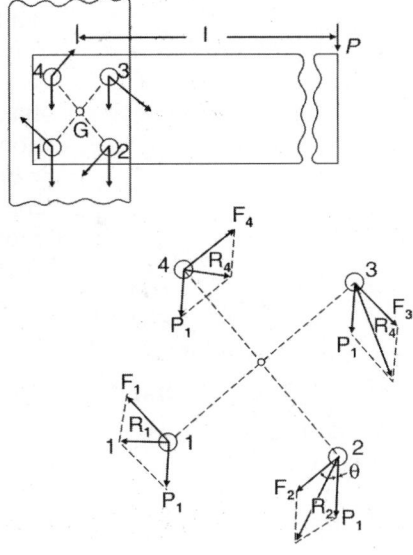

Fig. 11.30

$$\therefore \quad M = kr_1^2 + kr_2^2 + \dots = k \sum r^2$$

$$k = \frac{M}{\sum r^2}$$

$$F_n = kr_n = \frac{Mr_n}{\sum r^2} \qquad \dots (11.41a)$$

(c) Direct and moment loads are added vectorially to obtain the resulting load or each bolt. The diameter is found from the heaviest loaded bolt. In Fig. 11.30b bolts 2 and 3 are having the greatest load shown by the resultant.

Resultant for bolt 2, $R = \sqrt{P_1^2 + F_2^2 + 2P_1F_2 \cos\theta}$

The resultant is equated to resistance of bolt in shear to find diameter of bolt d as

$$R = \frac{\pi}{4}d^2 S_s$$

If instead of bolts rivets are used, the treatment is exactly similar.

Nuts: The computation of the stresses in bending, shear, and compression that are set up in the threads of a nut may be omitted if the effective height of a nut or of a tapped hole is made at least equal to the values given in Table 11.3. The lower figures refer to first-class workmanship, and the higher ones to average workmanship.

Table 11.3, in connection with Fig. 11.31, gives the dimension b and l for nuts made of different materials. If a nut on a rod takes only part of the load that the rod can stand, the nut may be treated as a tubular connection and the minimum length l may be determined as a function of the thickness h, as shown in the right-hand column in Table 11.3.

load = P_1 considering the members to be absolutely rigid.

(b) Moment load or secondary shear is the additional load on each bolt due to moment $M = Pl$. If $r_1, r_2, r_3, \dots$ are the radial distances from the C. G. to the centre of the bolt, then the moment loads $F_1, F_2, \dots$ on bolts 1, 2,…are related as $M = F_1 r_1 + F_2 r_2 + F_2 r_3 + \dots$

The force taken by each bolt depends upon its radius (proportional to the strain produced), i.e. the bolt farthest from C. G. takes the greatest load, while the nearest bolt, takes the smallest as

$$\frac{F_1}{r_1} = \frac{F_2}{r_2} = \dots = k$$

Table 11.3: Dimensions for angular threads (Fig. 11.30)

Nut			Tubular connection	
Material*	$\dfrac{l}{d}$	$\dfrac{b}{d}$	Material	$\dfrac{l}{h}$
Steel	0.8–1.0	1.4–1.5	Steel	2.5–3.0
Bronze	1.0–1.2	1.5–1.6	Bronze	2.5–3.0
Cast iron	1.3–1.5	1.6–2.0	Cast iron	2.0–2.5
Aluminium alloy	2.2–2.6	2.0–2.5	Aluminium alloy	2.5–3.0

* Material of screw or rod: SA 1120; 20C12S14

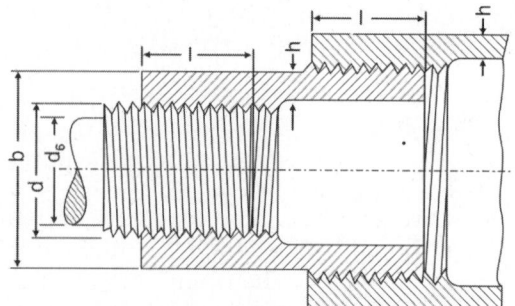

Fig. 11.31: Thread fastenings

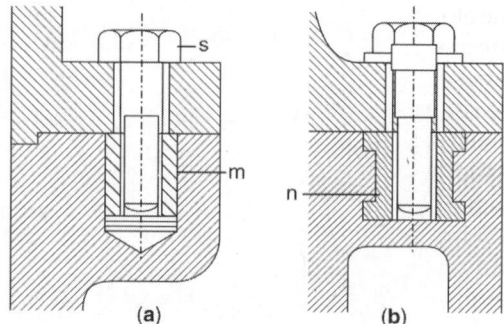

Fig. 11.32: Screw bushing

Example 11.2: An angular thread on a 90 mm steel shaft takes a steady tensile load of 220 kN. Determine the minimum length and outside width of a suitable steel nut.

The elastic limit of 20C8 (SAE 1020) steel according to Table 4.2 is $S_e = 240$ MN/m². If a size factor $e_{sz} = 1.30$ and a factor of safety $n = 2$ are used, the design stress is $S_d = 92$ MN/m². The corresponding outside diameter, Fig. 11.31 may be found from the equation

$$\frac{220}{10^3} = \frac{\frac{\pi}{4}(b^2 - 90^2)}{10^6} \times 92$$

gives $b = 106$ mm. This corresponds to

$$h = \frac{1}{2} \times (106 - 90) = 8 \text{ mm}$$

By the ratio in the right hand column of Table 11.3,

$$l = 8 \times 3 = 24 \text{ mm}$$

The dimensions of the nut will probably be $\left[\dfrac{b}{d} = 1.4\right]$, $b = 126$ mm and $l = 24$ mm $\left[\dfrac{l}{n} = 3\right]$.

Where cast iron or aluminium is used, angular threads are permissible only for permanent fastenings because the threads in these materials are easily damaged by repeated unscrewing and tightening. When the bolts are to be screwed and unscrewed repeatedly, a screwed-in steel bushing m, Fig. 11.32a, should be used in the case of cast iron, and for aluminium a cast-in insert n, Fig. 11.32b, of bronze or Monel metal should be used and should be drilled and tapped in place.

11.11 STRESSES DUE TO COMBINED LOADS

The stress resulting in a bolt from the combined action of the initial load put on when the thread is screwed tight and of the external load may be found by the following analysis.

Assume that no flexible gasket is used in a bolted joint like that shown in Fig. 11.33. For one unit, or one bolt with the corresponding parts of the flanges, the following designations will be used:

F_i is the initial load due to tightening of bolt

F is the external load applied

F_b is the part of external load on bolt

F_m is the part of external load on the members

l is the length of bolt = total thickness of both members

k_b is the stiffness of bolt

k_m is the stiffness of member

K_g is stiffness of gasket

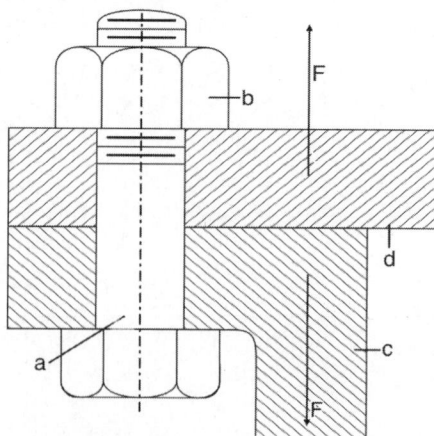

Fig. 11.33: Metal-to-metal joint

Δb is the elongation of bolt due to external load F

Δm is decrease in thickness of members due to F

C, Stiffness coefficient

A_g, Area of gasket

E_b, Elastic modulus of bolt

E_g, Elastic modulus of gasket

When the bolt is tightened but before the external load is applied, the bolt stretches and the flanges and the gasket are compressed. However, the modulus of elasticity of the metal of the flanges is so much higher than the modulus of the gasket that the compressive deformation of the flanges can be neglected.

However first by neglecting the presence of gasket, the force compressing the members will be equal to the force stretching the bolt, i.e. initial compression of the members will be equal to the elongation of the bolt. When external force is applied, the increase in length of bolt will permit the members increase in thickness by the same amount. $F = F_b + F_m$

$$E = \frac{F/A}{\delta l / l} \qquad \ldots (11.42)$$

or stiffness $k = \dfrac{F}{\delta l} = \dfrac{EA}{l}$ or $\delta l = \dfrac{F}{k}$ $\ldots (11.43)$

As $\Delta b = \Delta m$

$$\frac{F_b}{k_b} = \frac{F_m}{k_m} = \frac{F - F_b}{k_m} \qquad \ldots (11.44)$$

$$F_b = \frac{F\,k_b}{k_m + k_b} \qquad \ldots (11.45)$$

and $$F_m = \frac{F\,k_m}{k_m + k_b} \qquad \ldots (11.46)$$

Total force in bolt $= F_i + F_b = F_i - \dfrac{F\,k_b}{k_m + k_b}$

$$\ldots (11.47)$$

Total force in members $F_i + F_m = F_i - \dfrac{F\,k_m}{k_m + k_b}$

$$\ldots (11.48)$$

To decrease the load on the bolt, the fraction

$c = \dfrac{k_b}{k_m + k_b}$ should be minimum.

Now considering a gasket of thickness t is placed on the member and bolt length is l, the fraction of load on the gasket is,

$$c = \frac{k_b}{k_b + k_g} \qquad \ldots (11.49)$$

$$= \frac{A_b E_b / l}{A_b E_b / l + A_g E_g / t} = \frac{t A_b E_b}{t A_b E_b + l A_g E_g} \qquad \ldots (11.50)$$

Example 11.3: Estimate the value of the factor C in equation 11.50 for a joint with 24 mm bolts and a rawhide gasket 1.5 mm thick.

Equation 11.49 must be used. In this case $t = 1.5$ mm

and $A_b = \dfrac{\pi}{4} \times 24^2 = 452.4$ mm^2; and from Table 4.2, for

20C8 $E_b = 210 \times 10^3$ MN/m^2.

The thickness of each flange may be estimated as 32 mm. Therefore $l = 32 \times 2 + 1.5 = 65.5$ mm. The bolt spacing may be estimated as $4d = 4 \times 24 = 96$ mm, the width of the gasket is approximately $2d = 48$ mm; the diameter of the hole punched for the bolt will be about 27 mm; and therefore consider the gasket to

be 48 mm $\times$ 96 mm for one bolt, $A_g = 48 \times 96 - \dfrac{\pi}{4} \times 27^2$

$= 4035$ mm^2. From Table 4.8, $E_g = 125$ MN/m^2. In equation 11.49

$$t A_b E_b = \frac{1.5 \times 452.4 \times 210 \times 10^3}{10^3 \times 10^6} = 0.1425 \text{ MNm}$$

$$l A_g E_g = \frac{65.5 \times 4.035 \times 125}{10^3 \times 10^6} = 0.033 \text{ MNm}$$

$$C = \frac{0.1425}{0.1425 + 0.033} = 0.81$$

Metal-to-metal joint: For the case the factor C becomes 0, and the external load does not affect the stress in the bolt so long as F is less than F_i. The load F_i may be computed by equation 11.28. For a tight joint, F must be smaller than F_i by a safe margin, usually at least 20 per cent. If F is greater then F_i, the joint will separate and the stress will be due to the external load alone.

If the joint is to be kept tight, F_i must be greater than F, and the contact surfaces of the parts c and d, Fig. 11.33, must be true and smooth. Usually they are ground. The presence of a thin, inelastic gasket, such as one made of a copper or lead sheet or a very thin asbestos sheet, does not change the conditions materially and helps to obtain a tight joint when the surfaces are machine-finished and not ground.

In parts subjected to fatigue loads, the component of load shared by the bolt can be reduced in comparison with the initial tightening load. This will result in a small external fluctuating stress (of the total stress) to which the bolt is subjected in comparison with the initial tightening stress. *The following are the obvious advantages of using hard metal gasket.*

1. *Improvement in fatigue strength of bolt.*
2. *Improvement in the locking effect. If the loosening of the nut is due to the variation of stress within a fastener, preloading reduces the magnitude of external stress within the fastener. For such effects it is advisable to avoid soft gaskets or washers as these will reduce the stiffness of the members with respect to the bolts (11.10).*

Example 11.4: For a bolted joint, the initial tightening load is 12 kN. If the stiffness of member is three times the stiffness of the bolts, find the load in the bolt and members if the external load is 12 kN. At what external load the separation of the members will occur?

$$F_i = 12 \text{ kN}$$
$$K_m = 3 \text{ kb}$$
$$F_b = F_e \times \frac{K_b}{K_b + K_m} = 12 \times \frac{K_b}{K_b + 3K_b} = 3\text{kN}$$

Total bolt load = 12 + 3 = 15 kN
Total load in members = 12 – 3 = 9 kN

For separation to occur = $F_e \times \dfrac{K_m}{K_b + K_m} - F_i = 0$

$$F_e = 12 \times \frac{4}{3} = 16 \text{ kN}$$

Gasket joint: With a gasket, the bolt carries, in addition to the initial load, a certain fraction of the external load, as shown by equation 11.49. In this case F' may be computed by equation 11.30. Equation 11.49 shows that the additional load depends on the rigidity of the gasket and on the rigidity of the bolt itself. The value of the factor C may be computed by equation 11.49. However, the value of E_g is not always known for certain, and Table 4.8 may serve as a guide.

Table 11.4: Values of C in equation 11.45

Type of joint	Factor C
Soft, elastic gasket with studs	1.00
Soft gasket with through bolts	0.90
Copper-asbestos gasket	0.60
Soft-copper corrugated gasket	0.40
Lead gasket with studs	0.10
Narrow copper ring	0.01
Metal-to-metal joint	0.00

Bolt spacing: To obtain a tight joint when two surfaces are held together by a row of bolts, the distance c between the centers of each two bolts, called the *spacing*, or *pitch*, must not be too great. Bolt spacing is discussed in section 15.6.

11.12 DYNAMIC STRESSES IN SCREW FASTENINGS

The stress-concentration factor K' for ISI and American Standard threads subjected to impact loads may be taken equal to 2.85 for soft steel, such as (SAE 1120) or (SAE 1030). For heat-treated SAE 2320 steel, $K' = 3.9$. (Same for ISO metric threads).

Shock loads: As already discussed in 5.6 the shock absorbing capacity of a bolt can be improved by making the bolt shape of uniform strength (Figs 5.6b and c) which has a reduced shank diameter equal to the core diameter of threads (11.3, 11.4, 11.5).

Repetitive loads: For ductile materials the stress-concentration factor k_r is small and increases with a decrease of ductility. Regardless of the nature of external loading cycle, it has been discussed that the total load on bolt = $F_i + CF$. By decreasing the external component with respect to initial load, the resistance of bolt can be increased.

Example 11.5: Determine the safe useful load for a 39 mm bolt 115 mm long with ISO metric thread and made of 20C8 (SAE 1020) steel for three conditions: (a) static load; (b) sudden change of load from zero to a maximum; (c) impact action resulting from a variable load when the nut is unscrewed one sixteenth turn.

Static loading: The elastic limit in tension of 20C8 steel is 245 MN/m². If the bolt was turned from a

71 mm hexagonal bar, the size coefficient calculated from equation 5.7 and Table 5.1, is

$$e_{sz} = 1 - \frac{(1-0.85)(71-12)}{63} = 0.8596$$

The factor of safety may be taken as $n = 1.5$. The design stress is then

$$S_d = \frac{245 \times 0.8596}{1.5} = 140.4 \text{ MN/mm}^2$$

The stress area from Table 11.1 is 976 mm², and the maximum safe load is

$$F = 976 \times \frac{140.4}{10^6} = 0.137 \text{ MN} = 137 \text{ kN}$$

According to equation 11.28, the initial load produced by screwing the nut tight is of the order of

$$F = 2800 \times 39 = 0.109 \times 10^6 \text{ N}$$

If a separation of the bolted surfaces must be avoided, the useful load should not exceed.

$$F_1 = 0.8 \times 0.109 = 0.0872 \text{ MN}$$

Sudden load: When the load is applied suddenly, the stress concentration must be taken into account by using a factor $K = 2.85$. However, this factor does not apply to the static stress created by screwing down the nut. This stress can be estimated as

$$S_1 = \frac{F}{A} = \frac{0.109 \times 10^6}{976} = 111.6 \text{ MN/m}^2$$

The relation between the stress S_1 and the stress S_2 set up by outside force may be expressed as

$$S_1 + KS_2 = S_d$$

From the equation

$$S_2 = \frac{S_d - S_1}{K} = \frac{140.4 - 111.6}{2.85} = 10.10 \text{ MN/m}^2$$

According to equation 3.21, the stress created by a sudden load is twice as high as that due to static load. The nominal allowable static stress is therefore lowered to

$$S_2 = \frac{1}{2} \times (111.6 + 10.10) = 59.6 \text{ MN/m}^2$$

and the external sudden load must not exceed

$$F = 59.5 \times \frac{976}{10^6} = 0.6085 \text{ MN} = 60.85 \text{ kN}$$

or only about 43.2 per cent of the static load of 140.4 kN

Impact loading: The allowable nominal stress with an impact load can be computed by equation 3.21. The travel of the impact force is given as $\frac{1}{16}p$ where p is the pitch of the screw.

From Table 11.1, $p = 4$

$$h = \frac{4}{16} = \frac{1}{4} \text{ mm}$$

The unthreaded length of the screw, which is 115 mm, takes the shock. The stresses in equation 3.21 should therefore be referred to the unweakened area, which is

$$A' = \frac{\pi}{4} \times 39^2 = 1195 \text{ mm}^2$$

The allowable static load, taking into account stress concentrations is

$$F_s = \frac{F}{K} = \frac{0.137}{2.85} = 0.048 \text{ MN}$$

and the maximum stress in the unweakened section is

$$S = \frac{0.048 \times 10^6}{1195} = 40.16 \text{ MN/m}^2$$

If these values of h and s are substituted in equations 3.20, the result is

$$40.16 = S \left[1 + \sqrt{1 + \frac{2 \times 0.25 \times 210 \times 10^3 \times 10^3}{10^3 \times 115 S}} \right]$$

Solving for S gives $S = 1.623$ MN/m². Thus the allowable load is

$$F_s' = 1.623 \times \frac{1195}{10^6} = 0.000194 \text{ MN} = 1.94 \text{ kN}$$

which is only 1.38 per cent of the maximum static load of 140.4 kN.

Example 11.6: Determine the diameter d_1 of the end of the piston rod, Fig. 11.34 of a double acting steam engine. The cylinder bore is 450 mm; the maximum steam pressure is 1.20 MN/m²; and the steam is exhausted to atmosphere.

A suitable material is 45C8 (SAE 1045) steel,

The force acting on the full face of the piston is

$$F = \frac{\pi}{4} \times \frac{450^2}{10^6} \times 1.2 = 0.191 \text{ MN}$$

On the crank side, because of the presence of the piston rod, the area of the piston face will be about 1.5 per cent smaller than that on the headside, so that $F = 0.191 \times 0.985 = 0.188$ MN.

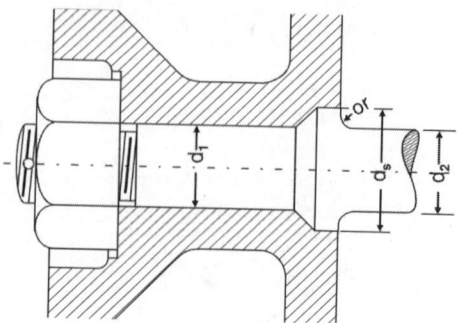

Fig. 11.34: Piston rod end

In order to prevent the piston from being lifted from the conical seat, the stress set up by tightening the nut must be about 25 per cent higher than the coming from the steam pressure. This stress corresponds to an additional load on the rod end.

$$F'' = 0.25 \times 0.188 = 0.047 \text{ MN}$$

With a static load and with $S_e = 350 \text{ MN/m}^2$, and with a safety factor of 2, the design stress $S = 175$ MN/m^2. Thus the minimum stress area of the thread is

$$A_s = \frac{0.188 + 0.047}{175} = 0.0013428 \text{ m}^2$$

The next larger stress area from Table 11.1 is 1470 mm^2 which corresponds to a major diameter of 48 mm. Anticipating influence of size factor, assume a size of 56 mm. The rod diameter will be about $d_3 = 80$ mm and the size factor from Table 5.1 will be $e_{sz} = 0.85$.

The corrected stress area is

$$A_s = \frac{0.0013428}{0.85} = 0.00158$$

which correspond to a size of 52 mm or say 56 mm.

With a nut tightened as here assumed, the stress in the end section of the rod will not change with the fluctuation of the steam pressure, and static load conditions will exist. So it is not necessary to take into consideration the stress concentration in the threads.

Example 11.7: Find the size of the thread necessary for the case of example 11.6 when the nut is tightened only enough to induce a stress equal to one half that caused by the steam pressure.

In this case the varying loads on the rod are

$$F_1 = 0.188 \text{ MN}$$

and $F_1 = \dfrac{1}{2} \times 0.188 = 0.094$ MN

$$\frac{S_a}{S_m} = \frac{0.188 - 0.094}{0.188 + 0.094} = 0.333$$

using the same size factor of 0.85 gives $S_a' = 72.25$ MN/m^2 with $S_a = 85$ MN/m^2.

The stress-concentration effect of the threads, taken into account by using curve d, Fig. 5.7 and assuming that $S_a = 640$ MN/m^2 is $e_{sa} = 0.74$. With $n = 2$ and $K_r = 1.25$ since, the effect of the threads is taken into account as a surface effect, the required area is

$$A_s = \frac{(0.188 - 0.094) \times 2}{2 \times 0.74 \times 72.25} = 0.0017581 \text{ m}^2$$

instead of $A_s = 0.001343$ m^2 for a static load even though the maximum load is reduced from 0.235 to 0.188 MN.

Force flow: The endurance limits of screw joints depend to a great extent on the flow of forces in the material. The difference in force flow in various cases is shown in Fig. 11.35. In a stud, Fig. 11.35a, the flow is deflected but keeps the same direction; in a standard bolt and nut, Fig. 11.35b, the force flow is turned 180° and the lower threads carry a much greater load than the upper ones. By removing part of the threads with a conical reamer, Fig. 11.35c, the force flow is deflected upward, relieving the stress in the lower threads and increasing it in the upper ones, and thus gives a more uniform stress distribution. The same result is obtained, but to a greater degree, by undercutting the bottom of the nut, Fig, 11.35d, to about one-half its height. If the endurance of an ordinary bolt and nut, Fig. 11.35b, is taken as unity, the

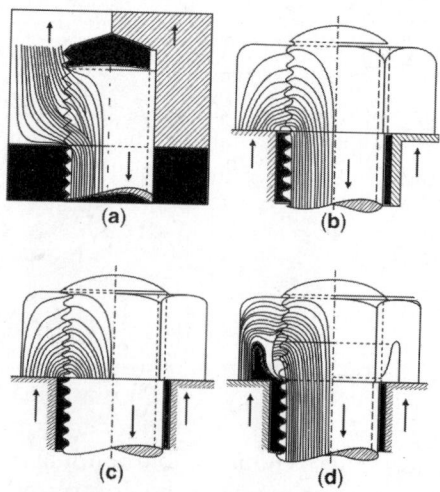

Fig. 11.35: Force flow in different types of bolts and nuts

conical boring of the nut, Fig. 11.35c, increases the limit about 20 per cent,[7] and the undercutting, Fig. 11.35d, increase it about 30 per cent. The strength of a screwed-in stud end, Fig. 11.35a is about 95 per cent greater.

Increase of endurance: The chief methods by which the endurance limit of screw stock can be raised, and the danger of failure of a screw through progressive fracture decreased, are as follows:

(a) Nitriding of the bolt, which practically eliminates the stress-concentration effect of the threads

(b) Creating a compressive stress at the root of the threads by rolling them instead of cutting them,[8] and thus decreasing stress concentration.

(c) Blending the thread and stock by a large-radius fillet, as shown in Fig. 11.36, and counterboring the hole for a stud so as to remove bending stresses from the last thread.

(d) Boring the nut conically or under cutting it.

(e) Using a lock nut or any other device that prevents the nut from unscrewing and maintains a certain initial tension in the bolt, thereby eliminating rattling and impact effect.

A stud may break as a result of progressive fracture at the end of the thread, as in Fig. 11.36. Cutting away the stock diameter to less than the minor thread diameter is helpful if the large-radius fillet extends beyond the first female thread. Counterboring the hole removes the bending stress from the dangerous section.

11.13 DESIGN OF SCREW FASTENINGS

Screw fastenings include all types of threaded parts.

Bolts: It is not possible to determine the initial stresses produced in bolts when screwing them up. This is particularly true of sizes up to

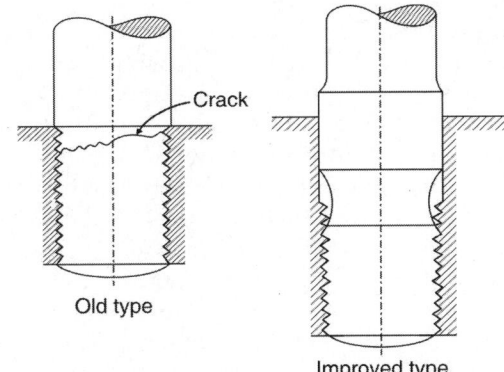

Old type

Improved type

Fig. 11.36: Stress-relieving of studs

and including 20 mm. To overcome this handicap it is general practice to take the external force alone as the total load on a bolt, and to neglect the initial stress and allow a low working stress instead. Thus Unwin[9] recommends the following formulas for the allowable stresses in bolts of ordinary steel used to make a fluid-tight joint: For rough joints,

$$S_d = 11.1\left(\frac{d^2}{310} + 1\right) \qquad \ldots (11.51)$$

and for faced joints,

$$S_d = 17.3\left(\frac{d^2}{540} + 1\right) \qquad \ldots (11.52)$$

These values seem to be rather low, but they are fairly close to those given by Carl Bach for the stresses in bolts of 12–24 mm sizes. Refigured for the same material, the allowable stress for average workmanship (rough joints) given by Bach is 16 MN/m² for 12 mm bolt. It increases with the size of the bolt up to 32 MN/m² for 24 mm bolt and over. For high-grade workmanship (faced joints), Bach raises the allowable stress for average workmanship by 25 per cent.

Factors of safety: It is customary to compute the allowable design stress by taking the factor

---

[7] *Ibid.*

[8] H. F. Moore and P. E. Henwood, *The Strength of Screw Threads under Repeated Tension*, Bulletin No. 264, University of Illinois Engineering Experimental Station (1934), pp. 12 and 15.

[9] W. C. Unwin and A. N. Mellanby, *The Elements of Machine Design*, rev. ed., Part I (London: Longmans, Green & Company, 1927), pp. 198–204.

of safety n as 2 for low-carbon steels and copper-base alloys, and as 1.5 for high-strength alloy steels.[10] These figures apply to good workmanship and to screws with d not less than 24 mm. For inferior workmanship the allowable stresses must be multiplied by 0.85; and for poor workmanship, by 0.7. For screws under 24 mm the design stresses should be lowered. For 12 mm screws, they should be about one-half the values just given.

When a number of bolts work together, especially if conditions may cause a great difference in their tightnesses, still lower stresses should be used.

Safety stop: Sometimes it may be desirable to make one of the bolts the weakest part of the machine, so that when the machine is overloaded the bolt will break and machine will stop. In such a case breaking load should be equal to the load that causes the weakest member of the machine connected by this bolt to be stressed close to the elastic limit.

Materials: Table 11.5 shows the main properties of various steels and noncorrosive copper alloys suitable for making bolts, studs, screws, and other threaded parts.

High temperature: Mild steels with ultimate tensile strengths of less than 450 MN/m² should

not be used for bolts whose temperatures may exceed 300°C.[11] Harder steels can be used for temperatures up to 450°C. Special steels should be used for temperatures above 450°C.

The influence of temperatures above 200°C on the elastic limit of steel may be taken into account by decreasing the elastic limit at room temperature by 1 per cent for every 5° above 280°C, up to 500°C.

Design stresses prescribed by the ASME Boiler Construction Code for alloy-steel bolts for temperatures from 370 to 510°C are given in Table 15.2.

Sizes: Machines should be designed so as to require the smallest possible number of sizes of bolts and screws, in order to reduce the number of drills and taps needed in manufacturing them and the number of wrenches required to service them.

Location of bolts: Screw fastenings are generally subjected to very high stresses. Therefore a good design should avoid any additional stresses that may arise from an inappropriate location of the fastening. Thus the adjusting screw in Fig. 11.16 is subjected to an additional bending stress due to a bending moment equal to the product of the force and the lever arm e. By

Table 11.5: Properties of some bolt materials

Material	Ultimate strength MN/m²	Yield point MN/m²	Elastic limit MN/m²	Endurance limit MN/m²
Low Carbon steel (hot rolled) C 0.10 to 0.20	410	270	210	170
Low Carbon steel (hot rolled) C 0.20 to 0.30	456	295	240	260
Medium Carbon steel (annealed) C 0.30 to 0.40	480	320	290	240
Ni steel (heat treated drawn at 540°C) Ni 3.25 to 3.75, C 0.35 to 0.45	865	710	650	340
Ni Cr. Steel (Heat treated drawn at 540°C) Ni 1.00 to 1.5, C 0.35 to 0.45, Cr 0.45 to 0.75	1030	960	650	550
Cr. Va Steel (Heat treated drawn at 425°C) Cr 0.80 to 1.10, C 0.45 to 0.55, Va 0.18	1570	1440	1170	550

---

[10] C. Höhner, "Richtlinien fur Schrauben und Verschraubungen," *Z. VDI*, Vol. 78, (1933), p. 299.
[11] *Ibid.*

decreasing the distance e the additional stress can be lowered. When a threaded rod holds together the parts of a split flywheel or pulley rim, as in Fig. 11.37, centrifugal force creates a similar unavoidable bending stress in the rod under the nuts. If the rod fits the holes in the lungs tightly, the bending action will become greater, but it will be confined to the shank in the middle, where the bolt is not weakened by threads. In all such cases it is very difficult to determine the additional load. The only practical way to cope with the situation is to lower the design stress by a certain amount. A case of additional shear stress is illustrated in Fig. 11.28. If the bolts alone are depended upon to resist the downward force F, they must be carefully fitted to insure that each bolt receives its full share of the shearing load. Only through bolts can be used in such a case, because tap bolts or studs cannot be fitted accurately. A better design is to have either dowel pins or a projecting lip a take the shear. In this case the bolts need not fit the holes closely, as tap bolts or studs can be used.

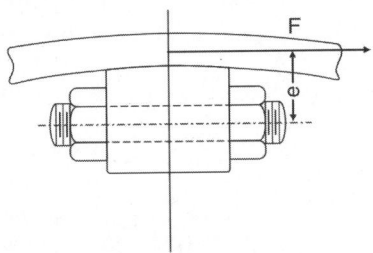

Fig. 11.37: Split-pulley rim

Assembling: The designer of a screw fastening should make sure that all heads and nuts are accessible for tightening, and that room is provided for the insertion of standard wrenches or screwdrivers. When through bolts or threaded rods are used, provision should be made to prevent them from turning when the nuts are tightened.

11.14 SET SCREW

Table 11.6 gives the engineering data for cup-point set screws with hexagonal socket heads made of molybdenum alloy steel having a tensile strength of 1730 MN/m². In computing the size and number of set screw to be used instead of a shaft key, the holding power listed in Table 11.6 must be divided by a safety factor of 1.5 to 2.0. The holding power of a set screw made of steel of a lower grade is usually given as about one-sixth that shown in Table 11.6. A headless set screw has a still smaller holding power.

The disadvantage of the cup point is that it raises a burr on the shaft, thus making the removal of the piece more difficult. Filing a flat spot under the point overcomes this difficulty. If a conical point is used, a conical hole should be drilled in the shaft.

The proper size of set screws may be found from the following equation,

$$d = \frac{D}{8} + 8$$

d = diameter of set screw mm
D = Shaft diameter mm

If F_t is frictional resistance set up (or the holding power)

$$\text{Torque set up} = F_t \times \frac{D}{2}$$

$$\text{Power transmitted} = \frac{2\pi n F_t D}{60 \times 2 \times 1000} \text{kW}$$

Table 11.6: Holding power of set screws

Diameter of set screw mm	Holding power N	Diameter of set screw mm	Holding power N	Diameter of set screw mm	Holding power N
6	300	14	3000	22	10,000
8	700	16	4000	24	12,000
10	1000	18	5500	27	15,000
12	2000	20	8000		

Example 11.8: Design the bolts for the cover plate of a cylinder in which the pressure fluctuates from zero to 2N/mm². The cylinder bore is 300 mm. The bolts are made of alloy steel for which the yield strength is 700 N/mm² and ultimate strength is 900 N/mm². The factor of safety can be taken as 1.5. The initial load on the bolts to be applied is 30% larger than the maximum tensile load. The stress concentration factor is 2.5 and stiffness coefficient is 0.2.

Assume the gasket diameter of 400 mm

Maximum tensile load,

$$F_t = pA = 2 \times \frac{\pi}{4} \times 400^2 = 251327 \text{ N}$$

The initial tightening load

$F_i = 1.3 \times 251327 = 326725 \text{ N}$

Maximum bolt load

$F_b = 0.2 \times 251327 + 326725 = 376990 \text{ N}$

Average bolt load

$$F_{av} = \frac{F_b + F_i}{2} = \frac{376990 + 326725}{2} = 351858 \text{ N}$$

Fluctuating load $F_r = 251327$

Applying the stress concentration factor of 2.5,

$$F_r' = 2.5 \times 251327 = 628317 \text{ N}$$

Maximum load $= 351858 + 628317 \text{ N} = 980175 \text{ N}$

Now $\dfrac{S_r}{S_{av}} = \dfrac{F_r'}{F_{ar}} = \dfrac{628317}{381858} = 1.7857 = \tan^{-1}\theta$

$\theta = 60.75\,°$

Taking the endurance stress S_e of the given alloy steel $\cong 0.45 \times 900 = 400 \text{ N/mm}^2$

$S_{av} \tan\theta = 400 - S_{av} \tan\theta$

$S_{av}(1.7857 + 0.5714) = 400$

$S_{av} = 170 \text{ N/mm}^2$

$S_{rv} = 170 \times 1.7857 = 303 \text{ N/mm}^2$

Therefore maximum stress

$$= S_{av} + S_v = 170 + 303 = 473 \text{ N/mm}^2$$

Design stress $= \dfrac{S_{max}}{F.S} = \dfrac{473}{1.5} = 315 \text{ N/mm}^2$

Total area of bolts

$$= \frac{\text{Maximum load}}{\text{Design stress}} = \frac{980175}{315} = 3112 \text{ mm}^2$$

Considering 8 bolts, area per bolt

$$= \frac{3112}{8} = 389 \text{ mm}^2$$

Diameter of a bolt $= \left[\dfrac{389 \times 4}{\pi}\right]^{\frac{1}{2}} = 22.2$

Specifications may be 8M22 × 2.5 or 8M24 × 2.5

Example 11.9: Design a bolted joint for a bracket shown in Fig. 11.38. The allowable shear stress = 75 N/mm². F = 20 kN acting vertically downwards.

Direct shear load $= \dfrac{20,000}{4} = 5000 \text{ N}$

Equating the external moment to the resisting moment of the bolts about the CG of the bolts of the bracket G.

$$20,000 \times 250 = \Sigma Fr = k\,\Sigma r^2 = 2k(25^2 + 75^2)$$

$$k = \frac{20,000 \times 250}{2[25^2 + 75^2]} = 400$$

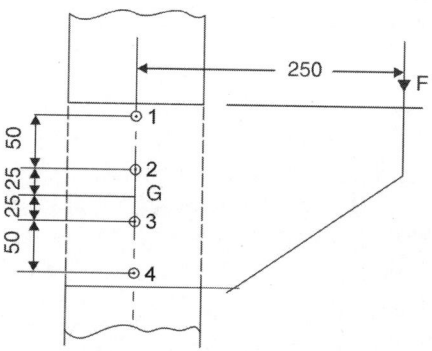

Fig. 11.38

Load on bolts 1 and 4 $= 400 \times 75 = 30,000 \text{ N}$

Total shear load on bolt

$$= \sqrt{5000^2 + 30,000^2} = 30.4 \times 1000 \text{ N}$$

Diameter of the bolt

$$= \left[\frac{30.4 \times 1000}{\dfrac{\pi}{4} \times 75}\right]^{\frac{1}{2}} = 22.7 \text{ mm}$$

Adopting M24 as the bolt diameter for all the bolts for uniformity. Instead these can be substituted for rivets.

Example 11.10: Find the rivet/bolt diameter for fixing a bracket as shown in Fig. 11.39. There are five rivets/bolts having allowable shear stress 80 N/mm². If the bracket plate is 12 mm thick, find the bearing stress in the plate and rivet/bolt.

The centre of gravity of the bolts is at the centre o. The load can be resolved into components.

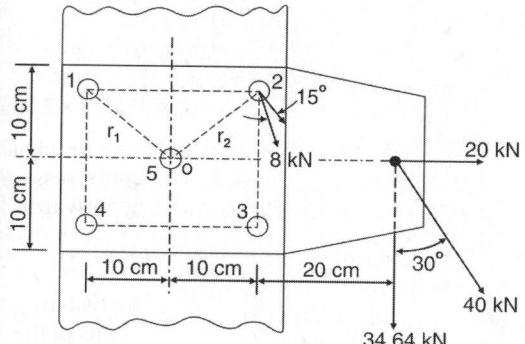

Fig. 11.39

The vertical component = 40 cos 30° = 34.64 kN
The horizontal component = 40 sin 30° = 20 kN
The vertical component causes direct shear force and rotation while the horizontal component cause direct shear stress. The direct shear force due to the total load of 40 kN is considered instead of components and the direction of shear force is parallel to 40 kN force.

Primary or direct shear force on each bolt

$$= \frac{40}{5} = 8 \text{ kN}$$

Distance of centre of gravity from each bolt

$$= 10\sqrt{2} = 14.14 \text{ cm}$$

Considering force $P_1, P_2 \ldots$ in bolt 1, 2 ......
Sum of moments of forces on all the bolts = Moment of external force.

$$P_1 r_1 + P_2 r_2 + ... = \Sigma r^2 = 34.64 \times (20 + 10)$$

$$4 \times 14.14^2 \, k = 1039.2$$

$$= 1039.2 \text{ kN cm}$$

$$k = \frac{1039.2}{4 \times 14.14^2} = 1.3$$

Secondary force in rivets 1, 2, 3 and 4 is $F_n = 1.3 \times 14.14 = 18.38$ kN and acts perpendicular to the radii.
The maximum resultant force will be acting on bolt 2.

Resultant shear force on bolt 2 is

$$R = \sqrt{8^2 + 18.38^2 + 2 \times 8 \times 18.38 \times \cos 15°} = 26.19 \text{ kN}$$

The diameter d of the bolt is found by equating it to resistance of bolt

$$\frac{\pi}{4} d^2 S_s = R$$

$$d = \left[\frac{26.19 \times 1000 \times 4}{\pi \times 80} \right]^{1/2} = 20.4 \text{ say 22 mm}$$

The bearing stress in the plate or bolt

$$= \frac{26.19 \times 1000}{22 \times 12} = 99.2 \text{ N/mm}^2$$

which is quite safe.

Example 11.11: Find the bolt diameter to fix a circular bar of 200 mm to a pillar. The free end of the bar is to carry a load of 20 kN and the bar is 300 mm long. The allowable shear stress in bolts is limited to 65 N/mm², Fig 11.29.

Assume a flange diameter of 350 mm and pitch circle diameter of 270 mm.

l_1, l_2, l_3, l_4 be the distances of bolt 1, 2, 3, 4 from the axis of rotation at 0.

$2a = 350, a = 175,$ $\quad 2b = 270, b = 135$
$l_1 = a - b \cos\alpha$ $\quad\quad l_2 = a + b \sin\alpha$
$l_3 = a + b \cos\alpha$ $\quad\quad l_4 = a - b \sin\alpha$

The best position of bolts is when $\alpha = 45°$

$l_1 = l_4 = a - b \cos\alpha = 175 - 135 \cos45°$
$\quad\quad\quad\quad = 175 - 95.45 = 79.55$
$l_2 = l_3 = a + b \sin\alpha = 175 + 135 \sin45°$
$\quad\quad\quad\quad = 175 + 95.45 = 270.45$

$$k = \frac{F \times l}{l_1^2 + l_2^2 + l_3^2 + l_4^2} = \frac{20,000 \times 300}{2 \times 79.55^2 + 2 \times 270.45^2} = 37.75$$

$F_4 = F_1 = 37.75 \times 79.55 = 3002.96 \text{ N}$
$F_3 = F_2 = 37.75 \times 270.45 = 10209.49 \text{ N}$

As the maximum force is 10209.49 N and is shared by two bolts.

$$\text{Force/bolt} = \frac{10209.49}{2} = 5104.74 \text{ N}$$

$$\text{Direct shear stress} = \frac{20,000}{4} = 5000 \text{ N}$$

Considering tensile force of 5105 N, the core diameter of bolt is

$$d_c = \left[\frac{5105 \times 4}{\pi \times 65} \right]^{1/2} = 10, \quad d = 1.2 \times 10 = 12 \text{ mm}$$

Taking next higher size of 14 mm with core diameter of 11.5

$$\text{Shear stress} = \frac{5105 \times 4}{\pi \times 14^2} = 32.5 \text{ N/mm}^2$$

$$\text{Tensile stress} = \frac{5105 \times 4}{\pi \times 11.5^2} = 48.7 \text{ N/mm}^2$$

Principal stress

$$= \frac{1}{2} \left[48.79 + \sqrt{48.79^2 + 4 \times 32.5^2} \right] = 65 \text{ N/mm}^2$$

So M14 is safe.

PROBLEMS

11.1 To what length should studs, tap bolts and cap screw be screwed into cast iron and steel? Explain.

11.2 To what depth should blind holes to be taped into cast iron be drilled; to what depth in steel? Explain.

11.3 What is the advantage of a uniform strength bolt and what is the value of shank diameter?

11.4 Why connecting rod bolts have a reduced diameter along the shank of bolt with small width of larger diameter at three places?

11.5 How can the shock absorbing property of a bolt be increased?

11.6 Why is it preferable to use a hexagonal headed bolt or nut in comparison with a square headed one?

11.7 What is the desirable length of a wrench?

11.8 Two plates are clamped by means of a bolt and nut with initial tension F_i. A separating force F is applied to the plates. What is the bolt tension?

11.9 Why is wedge type of washer used in certain locations?

11.10 Why is the use of soft metal gaskets avoided in assemblies subjected to fatigue loads?

11.11 Why are spring washers or gaskets used in bolted assemblies? Why is their use avoided when the assemblies are used for fatigue loads?

11.12 Which is the most dangerous section in the length of a bolt tightened up under load?

11.13 What are the common types of locking devices? Why is the thick nut usually placed on the top of another thinner nut?

11.14 What is the purpose of spot facing for a bolted joint? How does the stress in the bolt affected if spot-facing is not done?

11.15 When a nut is tightened what is the type of stress set up in a bolt.

11.16 Name various methods for increasing the endurance strength of bolts.

11.17 (a) Find the pull that must be applied to a 380 mm long wrench when tightening a M 33 standard bolt to produce a maximum tensile stress of 40 N/mm^2 with no lubrication. (b) Find the maximum true normal stress.

11.18 Determine the efficiency when screwing up a M50, ISO metric bolt (a) by using the theoretical expression and assuming oil-graphite lubrication and (b) by using the empirical formula.

11.19 Determine the torque to which a torque wrench must be set when tightening a M22, ISO bolt, in order not to exceed a tensile stress of 42 N/mm^2. Assume that coefficient of friction is 0.12 in the threads and 0.15 on the bottom of the nut.

11.20 Determine the relative strength in tension and torsion of a M36 coarse thread as compared to the fine standard thread.

11.21 Why is sometimes a small hole drilled in the bolt shank? Why is it preferable?

11.22 Why are washers used along with nut?

11.23 How does a set screw transmit torque?

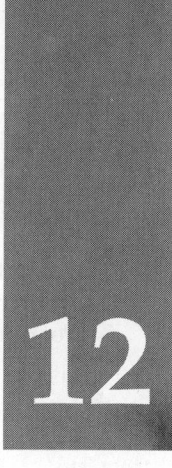

12
Keys, Pins, Cotters and Levers

12.1 TYPES OF KEYS

The main function of a key is to transmit torque between a shaft and a machine part assembled on it. In most cases keys prevent relative motion, both rotary and axial (12.1). In some constructions they allow an axial motion between the shaft and the hub; such keys are called feather keys, or spline keys. In spite of the tendency to standardize, there are over a dozen key types in use by various manufacturers.

According to various characteristics keys may be distinguished as straight and tapered; rectangular, dovetailed, chamfered, and disk-shaped; radial and tangential; and (according to their use) light-duty and heavy-duty.

However, the distinguishing features so overlap that a single, all-embracing classification is impossible. Keys of various types will be briefly described and illustrated, but they will be classified only according to their intended duty.

Light-duty and medium-duty keys: Keys used for light duty and medium duty are shown in Fig. 12.1. A straight *square key* (Fig. 12.1a) is a standard type used extensively for light duty. A modification of this key is the *rectangular key* (Fig. 12.1b) which has a standard depth in the shaft but is shallow above the shaft for use with a thin hub. Another modification is the key shown in Fig. 12.1c, which is shallow in both ways for a thin hub and a hollow shaft or sleeve.

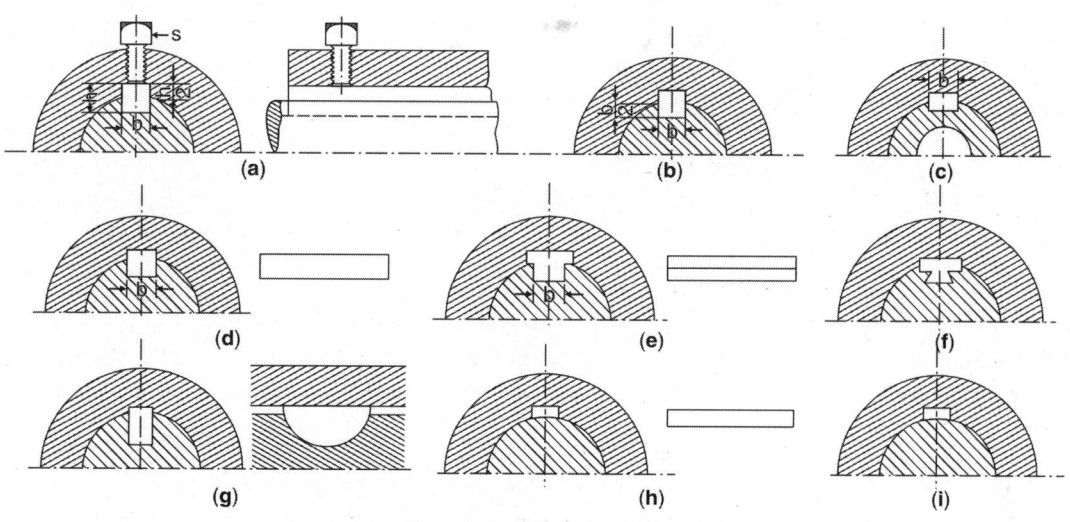

Fig. 12.1: Shaft keys for light and medium duty

227

The standard taper key (Fig. 12.1d) is made also with a gib head to make the removal of the key easier. A taper key can be used for medium duty or a variable torque.

The two-width key (Fig. 12.1e) which fits two sizes of keyseats, is used with a thin bushing and standard-size keyseats in the hub and shaft.

The dovetailed key (Fig. 12.1f) is fitted in the shaft to prevent its working loose, and is used as a feather key.

The Woodruff key (Fig. 12.1g) is used for small torques to avoid troublesome fitting (12.2). The keyseat in the shaft is milled with a special cutter. The keyseat in the hub is usually straight.

The flat key (Fig. 12.1h) is used for light duty when it is not desired to cut a keyseat in the shaft. Usually the key has a taper of 1 in 100. A straight key can be used but it requires a setscrew to hold it in place, as in the type of key shown in Fig. 12.1a.

The saddle key (Fig. 12.1i) depends upon friction alone to transmit the torque. It is used only with very light loads or for temporary service, as in setting eccentrics or cams. Usually it has a taper of 1 in 100. If it is made straight, it requires a setscrew.

Heavy-duty keys: Keys used for heavy duty are shown in Fig. 12.2.

The *round key*, or *pin key* (Fig. 12.2a) also called the *Nordberg key*, was originally used for light and small work. When made tapered, however, it proved satisfactory for heavy duty, because it can easily be fitted accurately.

The *Barth key* (Fig. 12.2b) is a rectangular key with two corners beveled off. This key does not require a tight fit, because the torque itself tends to force the key into its seat and, producing a compressive stress instead of shear, does not tend to turn the key in its seat. Another form of this key is turned 180°, so that the beveled sides are in the shaft instead of in the hub. The Barth key is used mostly as a feather key, and it is also used to replace a rectangular feather key that has given trouble.

The *Kennedy key* (Fig. 12.2c) consists of two rectangular taper keys driven from opposite ends of the hub. Diagonals through the keys to the shaft center form an angle of 90°. Like the Barth key, this key works in compression; two keys are required to transmit the torque in both direction. It is used for heavy duty, as in rolling mills.

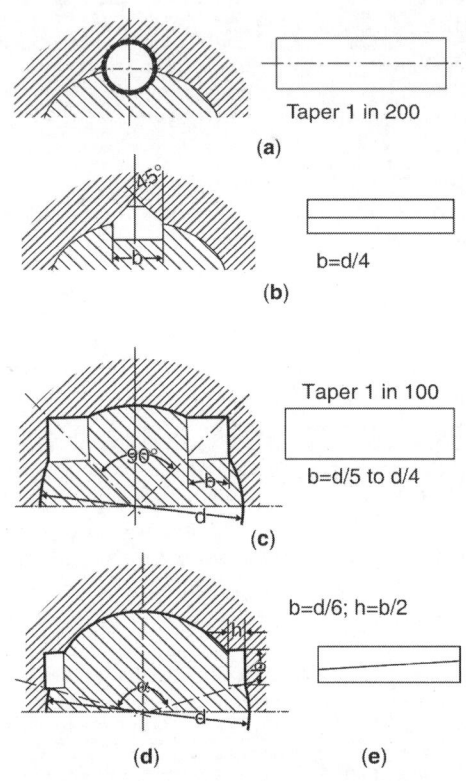

Fig. 12.2: Shaft keys for heavy duty

The *Lewis key* (Fig. 12.2d) consists of two tangential taper keys. This key is similar in characteristics to the Kennedy key but is much easier to fit and is used for medium and heavy service. The key is subjected to practically pure compression in the direction of its width *b* and therefore does not loosen in operation under the heaviest torque and shock action.

The *double-taper tangential key* (Fig. 12.2e) is a modification of the Lewis key and is used for heavy duty in place of the Kennedy key.

12.2 KEY DATA

As already pointed out, keys are made either straight or tapered. A straight key transmits torque by means of shear and compression. It does not disturb the alignment of the keyed parts, but this advantage is offset by the requirement of an accurate fit on the sides. Moreover, it cannot transmit a heavy torque. A tapered key transmits torque chiefly through friction. By taking up any play between the shaft and the hub, it has a tendency to throw the pulleys or gears out of alignment. It can transmit a higher torque, however, and it eliminates any axial motion between the hub and the shaft.

Square keys: In Table 12.1 are shown the dimensions of standard straight keys.[1] Standards have been established for shafts up to 500 mm; the data in Table 12.1 correspond to average practice. The keys are made square up to $b = 6$ mm; in larger sizes h is often made smaller than b. In all sizes, b is made about equal to $\frac{1}{4}D$, where D is the shaft diameter.

The depths of the keyseats in both the shaft and the hub are made equal to one-half the key thickness, except for the type shown in Figs 12.1(g, h, i).

Every straight key is fitted on the sides of the keyseats and has a slight clearance at the top. To prevent any axial motion of the hub a setscrew s, Fig. 12.1a, or some other holding means, should be provided. A setscrew also reduces the wear of the whole connection if the torque varies.

Keys made as in Fig. 12.1b have a width and thickness approximately as given below

$$w = \frac{D}{4} \text{ to } \frac{D+13}{4}$$

$$h = \frac{D}{6} \text{ to } \frac{D+13}{6}$$

while the actual values are shown in Table 12.1.

For calculations the width and depth of a square key

$$w = h = \frac{D}{4} \text{ to } \frac{D+13}{4}$$

Keyseats: In the hub the keyseat is cut by a special keyseater with a reciprocating motion of the cutter. In the shaft the keyseat is cut by a milling cutter that is either of the disk type, Fig. 12.3a, or of the end type Fig. 12.3b. The keyseat in Fig. 12.3a is cheaper to make. It is also better, because it does not weaken the shaft by stress concentration as much as the keyseat

Table 12.1: Standard dimensions of straight keys

Diameter of shaft inclusive mm	Key dimensions mm		Diameter of shaft inclusive mm	Key dimensions mm	
	Width b	Thickness h		Width b	Thickness h
6–8	2	2	85–95	25	14
8–10	3	3	95–110	28	16
10–12	4	4	110–130	32	18
12–17	5	5	130–150	36	20
17–22	6	6	150–170	40	22
22–30	8	7	170–200	45	25
30–38	10	8	200–230	50	28
38–44	12	8	230–260	50	32
44–50	14	9	260–290	62	32
50–58	16	10	290–330	70	36
58–65	18	11	330–380	80	40
65–75	20	12	380–400	90	45
75–85	22	14	400–500	100	50

[1] Leaflets AESC B17b and AESC B17d, prepared by the American Society of Mechanical Engineers

in Fig. 12.3b, though the latter does hold the key in place axially.

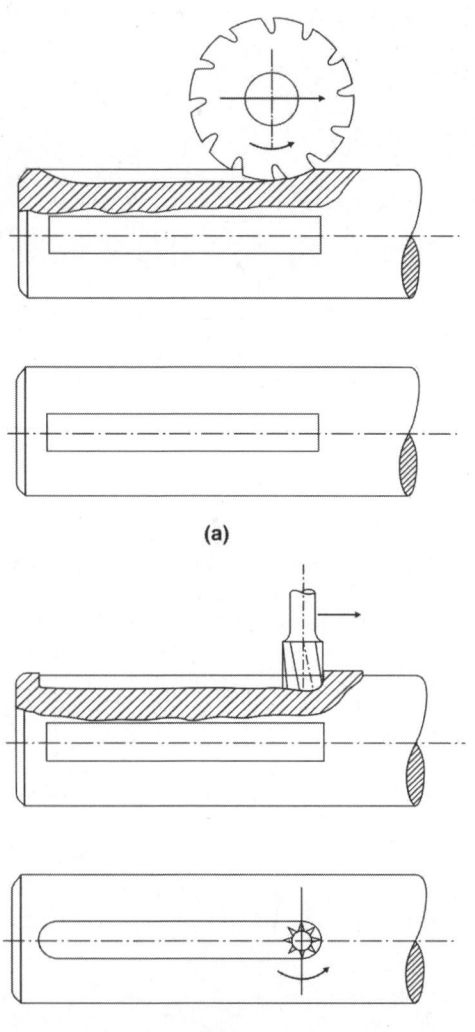

(a)

(b)

Fig. 12.3: Methods of cutting keyways

Feather keys: A feather key is used when the hub must slide along the shaft. The key is fastened to the shaft by countersunk machine screws, as shown in Fig. 12.4a, or by through-pins riveted over, Fig. 12.4b. If the keyseat extends to the end of the shaft, a key with end lips, Fig. 12.4c, may be used; it slides along the seat with the hub. When the key is fastened to the shaft, the keyseat may be cut either by an

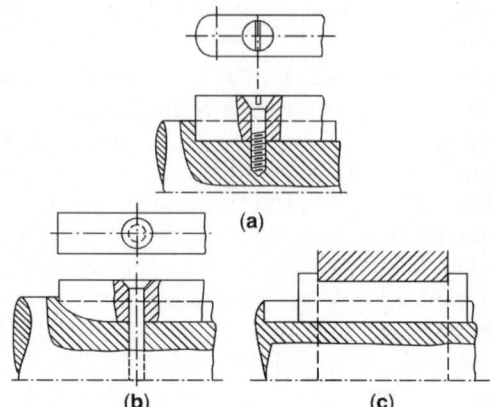

(a)

(b)   **(c)**

Fig. 12.4: Methods of fastening feather keys

end miller, resulting in a keyway like that shown in Fig. 12.3a, or by a circumferential cutter, as in Fig. 12.3b.

Standard taper key: The thick end of a standard taper key has the dimensions given in Table 12.1 for straight keys. The standard taper is 1 in 100 and starts at a distance b from the end, Fig. 12.5. The keyseat in the shaft has the same depth as that used with a standards square or rectangular key, and only the keyseat in the hub is tapered. The deep end of the keyseat in the hub is cut slightly shallower than the corresponding projection of the key, to allow for fitting. The allowance is about 0.5 mm for shafts up to 50 mm increasing slightly with the size of the shaft. The key is fitted at the top and bottom and is often relieved on the sides to make fitting easier. However, both theory and practical experience indicate that a taper key holds better when it is fitted accurately on the sides as well as on the top and bottom.

The gib-head dimensions, Fig. 12.5, are approximately $c = 1.2b$ and $H = 1.7b$. The gib

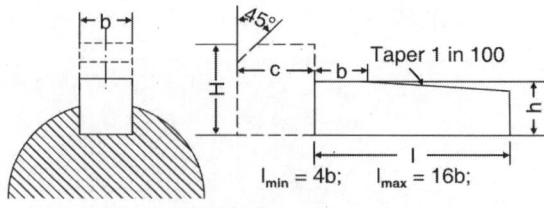

Fig. 12.5: Taper key

head is helpful for removing a key, but it is a dangerous protrusion on a rotating shaft and should be covered by a stationary guard or one fastened to the shaft. At present gib heads are not often used.

Woodruff keys: Woodruff keys, also known as Whitney stock-size keys, are used in shafts up to 70 mm in diameter. *The extra depth of the keyseat, Fig. 12.6, weakens the shaft, but at the same time it precludes all possibility of the key tipping. Another advantage of this key is its tendency to adjust itself to a tapered keyseat in the hub (12.3).* When a long hub must be secured,

the depth of the keyway in the shaft can be decreased by using two or more Woodruff keys having the same thickness but a smaller diameter.

The dotted circle in Fig. 12.6 illustrates how two keys are cut from a disk, which itself is cut from round bar stock. Table 12.2 gives dimensions of Woodruff keys of standard sizes.

The depth of the keyway in the hub is $c = \frac{1}{2} b$, and the length is practically equal to the stock diameter, or $l = a$. The size of key to be used in a shaft with a certain diameter D is given by the manufacturers in a special table or may be determined by selecting a key having a thickness equal to or slightly greater than $0.17D$.

Others keys: The width b and the height above the shaft of other keys shown in Fig. 12.1 are made equal to the standard dimensions given in Table 12.1, in order to use hubs with standard keyseats.

Round, or pin, keys: Taper-pin keys are fitted halfway into the shaft and hub, as shown in Fig. 12.2a. If a taper pin is used for light duty, it is advisable to use commercial standard

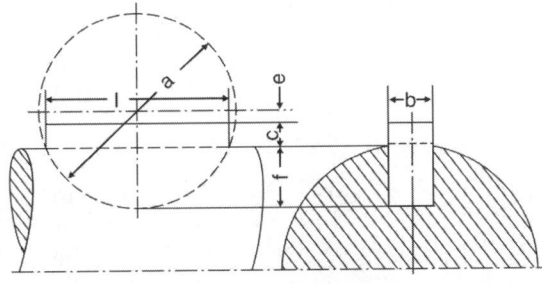

Fig. 12.6: Woodruff key

Table 12.2: Dimensions of Standard Woodruff Key

Shaft diameter d				Key size
Series 1		Series 2		
Over	Including	Over	Including	$b \times f \times a$
3	4	3	4	$1.0 \times 1.4 \times 4.0$
4	5	4	6	$1.5 \times 2.6 \times 7.0$
5	6	6	8	$2.0 \times 2.6 \times 7.0$
6	7	8	10	$2.0 \times 3.7 \times 10.0$
7	8	10	12	$2.5 \times 3.7 \times 10.0$
8	10	12	15	$3.0 \times 5.0 \times 13.0$
10	12	15	18	$3.0 \times 6.5 \times 16.0$
12	14	18	20	$4.0 \times 6.5 \times 16.0$
14	16	20	22	$4.0 \times 7.5 \times 19.0$
16	18	22	25	$5.0 \times 6.5 \times 19.0$
18	20	25	28	$5.0 \times 7.5 \times 19.0$
20	22	28	32	$5.0 \times 9.0 \times 22.0$
22	25	32	36	$6.0 \times 9.0 \times 22.0$
25	28	36	40	$6.0 \times 10.0 \times 25.0$
28	32	40	–	$8.0 \times 11.0 \times 28.0$
32	38	–	–	$10.0 \times 13.0 \times 32.0$

taper pins. The taper of these pins is 1 in 50 and the corresponding data may be taken from Table 12.4. When such a pin is used as a key, its large diameter d should comply with the relation

$$d = 2.8\sqrt{D} \text{ to } 3.5\sqrt{D} \qquad \ldots (12.1)$$

where, D is the shaft diameter in mm.

Heavy-duty keys: The dimensions of a taper-pin key, as used by the Nordberg Manufacturing Company for heavy duty, are given in Table 12.3. The total taper of the reamer is 1 in 200.

The width of a Barth key is given in Table 12.1, and the thickness is halfway between that of a square key and that of a shallow key.

A Kennedy key is formed by using standard taper keys chosen according to Table 12.1. To facilitate erection the hub is first bored for a press fit with the shaft and is then rebored, the center being offset about 1 mm to produce the clearance shown in Fig. 12.2c.

A Lewis key, Fig. 12.2d, gives excellent service in heavy work. However, since it is rather expensive to fit, it is not used extensively. The angle α is made equal to 150° or slightly smaller.

The keyseating and fitting are considerably simplified if each key is composed of two tapered keys with the tapers turned in opposite directions so as to make the outside edges parallel, as shown in Fig. 12.2e.

12.3 STRENGTH OF KEYS

In regard to stress analysis and wear, keys of all the various types can be classified in four main groups, as illustrated diagrammatically in Fig. 12.7.

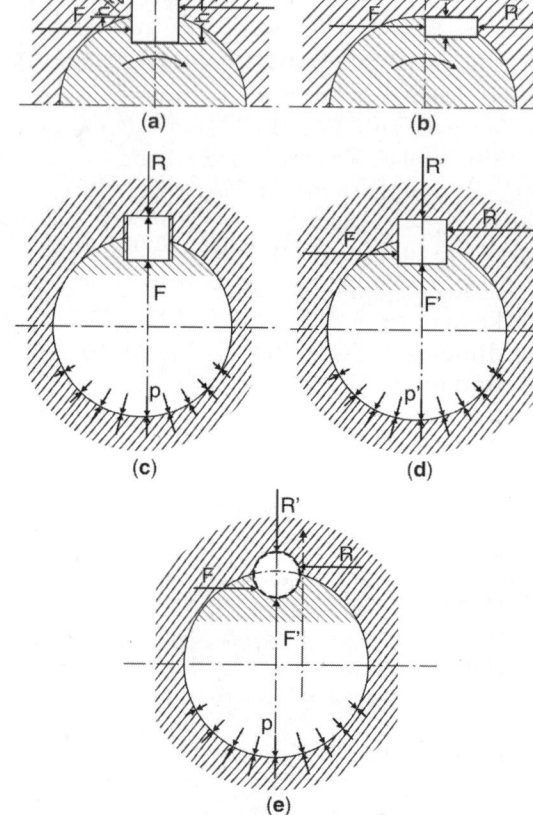

Fig. 12.7: Diagrams of key action

(a) Rectangular fitted keys, Fig. 12.7a, in which the torque is transmitted by means of compressive and shear stresses.

(b) Tangential keys, Fig. 12.7b, in which the torque is transmitted by means of compressive stresses alone.

Table 12.3: Standard taper pins

Small diameter ds	Lengths available	Small diameter ds	Lengths available	Small diameter ds	Lengths available
1.5	8–25	5	20–60	16	40–180
2	10–35	6	25–90	20	45–200
2.5	10–35	8	25–130	25	50–200
3	12–45	10	30–160	30	55–200
4	14–55	12	35–180	40	60–200

(c) Tapered keys, Fig. 12.7c, in which the torque is transmitted by means of friction induced by compressive stresses.

(d) Tapered keys fitted on the sides, Fig. 12.7d, in which torque is transmitted by a simultaneous action of compressive and shear stresses and friction. Round tapered pins, Fig. 12.7e, fall in this group.

Rectangular fitted key: Rectangular stock keys of machine steel are made to standard dimensions b and h, and the designer has to find only the length l of the key necessary to transmit the given torque T. The stresses produced in a fitted key are crushing and shear, and both should be investigated.

Crushing strength: Since a hub is always much more rigid than a shaft, the shaft will be twisted by the torque, whereas the hub will remain practically undistorted. The torque has its maximum value at the point where the shaft enters the hub, and it gradually decreases toward the other end of the hub. The tangential pressure p per unit length of the key, (Fig. 12.8), is proportional to the torque. If its maximum value where the shaft enters the hub is p_1, it decreases to p_2 at the end of the key, the length of which is l_2. The tangential pressure p at any intermediate distance l from the hub edge can be computed from the relation.

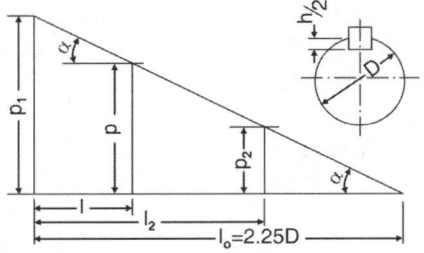

Fig. 12.8: Pressure between key and keyseat

$$p = p_1 - l \tan \alpha \qquad \ldots (12.2)$$

where, $\tan \alpha = (p_1 - p_2)/l_2 = p_1/l_0$.

The torque T transmitted by the key can be determined from the elementary torque formula

$$dT = p \times dl \times \tfrac{1}{2} D \qquad \ldots (12.3)$$

where, D is the shaft diameter. By substituting the value of p from equation 12.2 in equation 12.3 and integrating it from $l = 0$ to $l = l_2$, the torque can be expressed by the equation

$$T = \tfrac{1}{2} p_1 D l_2 - \tfrac{1}{4} D l_2{}^2 \tan \alpha \quad \ldots (12.4)$$

Since the pressure per unit length of the key is equal to the product of the bearing stress S_b and the area $0.5h$ of the key side 1 unit long, it follows the $p_1 = 0.5 s_{b1} h$. According to practical experience, a length of key over $2.25D$ is useless. If the tangential pressure p_2 is considered zero when

$$l_2 = l_0 = 2.25D,$$

$$\tan \alpha = \frac{p_1}{l_0} = \frac{s_{b1} h}{4.5 D} \qquad \ldots (12.5)$$

In general the torque transmitted by the key is

$$T = \tfrac{1}{4} s_{b1} h D l_2 - \tfrac{1}{18} s_{b1} h l_2^2 \quad \ldots (12.6)$$

where, s_{b1} is the nominal stress at the dangerous point.

The significant stress, according to equation 5.12, is the nominal stress times K', where the stress-concentration factor, K' may be determined by equation 5.9. The theoretical stress factor K due to a concentrated load may be taken as 4, as found for key fitted at the sides.[2] For impact action q may be taken as 0.4. For a well-fitted key, the impact is reduced to sudden load and q may be taken as 0.2 which gives a stress concentration factor $k' = 1.6$. For a steady load, $K' = 1$. The safety factor n may be taken as 1.5 for a steady torque, and it should be increased up to 2.5 for a strongly fluctuating torque.

To determine the length l_2 of a square key required to transmit a certain torque by applying equation 12.6, the design stress S_d must be used instead of the stress s_{b1}. If solving equation 12.6 gives a negative quantity under the square root, this condition will indicate that one key is not enough to transmit the required torque T. If solving equation 12.6 gives a length l_2 less than D, a practical rule is to make l_2 equal to D.

Strength in shear: The resistance of a key to shear, when the flexibility of the shaft is taken into consideration, may be represented by the same diagram as that shown in Fig. 12.8, with the maximum unit resistance of $p_1 = s_{s1}b$, where s_{s1} is the maximum shear stress at the end of the key. The torque can also be determined by an equation similar to equation 12.6. If it is assumed that the maximum useful length of a key is 2.25D,

$$\tan \alpha = \frac{p_1}{l_0} = \frac{s_{s1}b}{2.25D} \qquad \text{... (12.7)}$$

The torque becomes

$$T = \tfrac{1}{2}s_{s1}bDl_2 - \tfrac{1}{9}s_{s1}bl_2^2 \qquad \text{... (12.8)}$$

where, s_{s1} is the nominal shear stress at the dangerous point. From equation 12.8,

$$s_{s1} = \frac{T}{l_2b(0.5D - 0.11l_2)} \qquad \text{... (12.9)}$$

The safety factor n may be taken as 1.5 to 2.5, the exact value depending on the character of the torque, but stress concentration may be neglected because of the low value of K'.

For simplification, the width of a rectangular key

$$w = \frac{D}{4} \text{ to } \frac{D+13}{4}$$

and depth h or $t = \frac{D}{6}$ to $\frac{D+13}{16}$

Length of key required to transmit torque to the shaft. If the key has the same material as that of the shaft, the strengths will be equal so

$$T = lwS_s\frac{d}{2} = \frac{\pi}{16}S_sd^3$$

$$wl = \frac{\pi}{8}\frac{d^2}{w} = \frac{\pi}{8}\frac{d^2}{d/4} = \frac{\pi}{2}d = 1.57d$$

i.e. key length = 1.57d to have same strength as shaft (12.8).

Usually length is taken as 1.25d to 1.5d or 2d.

Example 12.1: Find the dimension of a square fitted key for a 80 mm steel shaft to transit 70 kW at 200 rpm; the torque fluctuates and the key is well fitted.

Torsional moment, or torque is

$$T = \frac{10^3 P}{2\pi n} = \frac{10^3 \times 70 \times 60}{2\pi \times 200} = 3342.5 \text{ Nm}$$

Considering square key, the width of the key is $b = 22$ mm and its height is $h = 22$ mm. For 10C8 (SAE 1010), the elastic limit compression is $S_e = 206$ N/mm^2 and from equation 5.14, $S_b = 2S_e = 412$ N/mm^2. The size effect according to equation 5.8 is

$$e_{se} = 1 - \frac{(1-0.84)(22-12)}{63} = 0.975$$

Because the torque fluctuates, a safety factor n of 2.5 should be used. Hence the allowable bearing stress is

$$S_d = \frac{412 \times 0.975}{2.5} = 161 \text{ N/mm}^2$$

With a well fitted key, $K' = 1.6$, whence

$$S_d' = \frac{161}{1.6} = 100 \text{ N/mm}^2$$

Substituting the value of S_{b1} and other data in equation 12.6 gives,

$$3342500 = \frac{1}{4} \times 100 \times 22 \times 80 \times l_2 - \frac{1}{18} \times 100 \times 22 \times l_2^2$$

Solving this equation for l_2 gives two roots, 108 and 252 mm. As the second one is extraneous, as it contradicts the condition that $l_2 < 2.25D$, the first one is the answer. Therefore, the proper key length is 110 mm.

Check for shear. The maximum shear stress by equation 12.9

$$S_{S_1} = \frac{3342500}{110 \times 22[0.5 \times 80 - 0.11 \times 110]} = 47.8 \text{ N/mm}^2$$

Since the elastic limit in shear is $S_{es} = 103$ N/mm^3 $S_{es}' = 0.975 \times 103 = 100.5$, the safety factor is

$$n = \frac{S_{es}'}{S_{Si}} = \frac{100.5}{47.8} = 2.1$$

This result indicates that the shear stress, with $n = 2.1$ is more dangerous. Approximate calculation with uniform pressure is given below.

$$3342500 = 100 \times 22 \times l_2 \times \frac{80}{2} \text{ gives}$$

$l_2 = 76$ mm say 80 mm

For this length the shear stress is,

$$S_{S1} = \frac{3343500}{80 \times 22 \times \dfrac{80}{2}} = 47.49$$

for this result also the shear stress is more dangerous.

Length of key is usually 1.25 to 1.5d, i.e. $l_2 = 100$ to 120 mm which will be somewhat safer.

12.4 TAPER KEY

The relation involving the circumferential force F_t, Fig. 12.9, and the pressure F between the shaft and the hub is

$$F_t = f_1 F \qquad \dots (12.10)$$

where, f_1 is the coefficient of friction between the shaft and the hub and can be taken as 0.25. But

$$F = blp \qquad \dots (12.11)$$

where, p is the pressure, or compressive stress, in the key. Therefore the allowable pressure is only one-half that for a straight key. When the key is driven home, p may attain a high value.

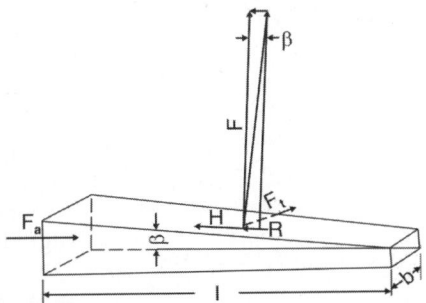

Fig. 12.9: Diagram of forces acting on a taper key

Also, the relation between the torque T and the force F_t is

$$T = \tfrac{1}{2} F_t D \qquad \dots (12.12)$$

Hence

$$T = \tfrac{1}{2} f_1 blp D \qquad \dots (12.13)$$

The necessary key length is then

$$l = \frac{2T}{f_1 bpD} \qquad \dots (12.14)$$

The axial effort F_a, Fig. 12.9, necessary to drive the key home is

$$F_a = H + R = 2Ff_2 + F \tan\beta \qquad \dots (12.15)$$

It is a good practice to have the key greased. The coefficient of friction f_2 may then be taken as 0.10 on both the top and bottom surfaces of the key. If the taper is 1 to 100, $\tan\beta = 0.01$. From equation 12.11.

$$F_a = 2 \times 0.1\, pbl + 0.01\, pbl = 0.21 pbl \quad \dots (12.16)$$

Example 12.2: Find the dimensions of a tapered key for the same conditions as those in example 12.1.

Width of key, $b = 22$ mm, with a drive fit, a safety factor n of 1.5 is sufficient. The allowable pressure is

$$p = \frac{206 \times 0.975}{1.5} = 133.9 \text{ MN/m}^2$$

From equation 12.14, the required length is, taking $f = 0.25$

$$l = \frac{2 \times 0.003343 \times 10^3 \times 10^3}{0.25 \times 22 \times 1.339 \times 80}$$

$$= 0.11345 \text{ m say } 114 \text{ mm}$$

This is almost the same length as that of a fitted key. The effort required to drive the key home, by equation 12.16, is

$$F_a = 0.21 \times 133.9 \times \frac{22}{10^3} \times \frac{110}{10^3} = 0.068 \text{ MN}$$

Effect of elasticity: With a tapered key, as in the case of a fitted key, the shaft has a tendency to twist in the hub, especially if the torque fluctuates. The torque distribution is approximately the same as with a fitted key. As a result, where the torque is a maximum the shaft will move slightly in the hub. This motion produces a gradual wear of the surfaces pressed together. The wear extends the motion further into the hub, and so on until the shaft becomes loose. A good fit on the sides limits the torsional motion of the shaft and lengthens the life of this key connection. *If the torque fluctuates, however, the key will ultimately loosen and produce eccentricity. Therefore a tapered rectangular key is not satisfactory for a heavy and fluctuating torque (12.6); key of other special types should be used.*

12.5 FRICTION OF FEATHER KEYS

When a hub connected to a shaft by a feather key and subjected to a torque T is moved along the shaft, it presents a resistance which can be found from Fig. 12.10a. The torque T produces a circumferential force, which is

$$F_t = \frac{T}{a} \qquad \dots (12.17)$$

Two opposite forces $F' = F'' = F_t$ applied in the center plane do not change the existing equilibrium but give a couple formed by the force F_t and F'' and a single force F'. The couple, whose moment is $F_t a = T$, tends to rotate the hub about the center O. The force F' presses the

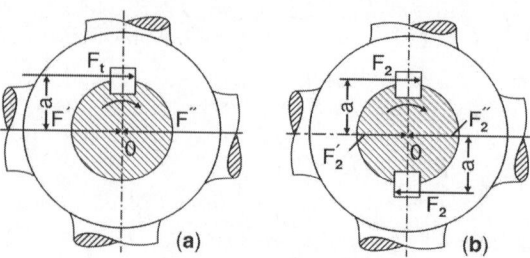

Fig. 12.10: Action of feather keys

hub against the shaft. When the hub is shifted lengthwise, a resistance on the key and on the shaft must be overcome. This resistance is

$$R = fF_t + f_2F' \qquad \dots (12.18)$$

Since f and f_2 are approximately equal,

$$R = 2fF_t \qquad \dots (12.19)$$

If two feather keys are used, as in Fig. 12.10b,

$$T = F_2a + F_2a = 2F_2a \qquad \dots (12.20)$$

A comparison of equations 12.17 and 12.20 shows that

$$F_2 = \tfrac{1}{2}F_t \qquad \dots (12.21)$$

Two opposite forces $F_2' = F_2'' = F_2$ applied in the center plane give two couples, both tending to rotate the hub clockwise. When the hub is being shifted, the only resistance to be overcome is that on the keys:

$$R_2 = 2fF_2 \qquad \dots (12.22)$$

Substituting values of F_2 from equation 12.21 and F_t from equation 12.19 results in

$$R_2 = \tfrac{1}{2}R \qquad \dots (12.23)$$

The addition of the second key halves the shifting resistance by eliminating the eccentric application of forces. Naturally, a hub with two feather keys requires very accurate fitting in order to give the advantage disclosed by the preceding analysis.

12.6 PARALLEL-SIDE SPLINES

The development of the automobile required a connection between circular shafts and hubs that would be strong, light, and suitable for mass production. This demand led to the development of the integral spline shaft. The first standard spline connections were developed and adopted by the SAE in 1914. *These splines consist of multiple integral keys with parallel sides milled on the outside surface of the shaft (12.4).* The hub is bored to the small diameter of the shaft and has keyseats broached in it to receive the projecting ribs on the shaft, as shown in Fig. 12.11. Evidently this construction can be used only with hubs made of a material soft enough to be broached. The SAE has standardized four types of spline fittings, namely the four-, six-, ten-, and sixteen-spline types. All dimensions are given as functions of the shaft diameter D, (Fig. 12.11). The bore of the hub is $d = D - 2h$. The height h of the spline for each type depends on the operating conditions. Press fit, which is recommended for a permanent fit, has the smallest height h. For close fit, which is recommended where the hub is to slide when

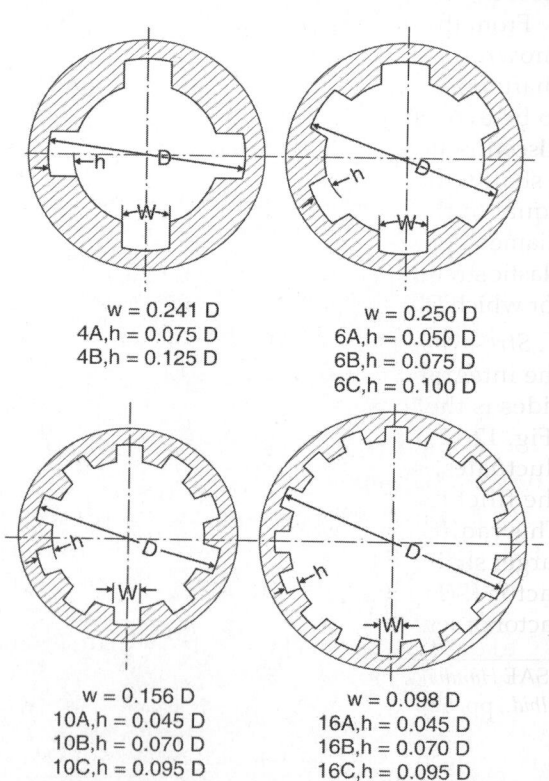

w = 0.241 D
4A,h = 0.075 D
4B,h = 0.125 D

w = 0.250 D
6A,h = 0.050 D
6B,h = 0.075 D
6C,h = 0.100 D

w = 0.156 D
10A,h = 0.045 D
10B,h = 0.070 D
10C,h = 0.095 D

w = 0.098 D
16A,h = 0.045 D
16B,h = 0.070 D
16C,h = 0.095 D

Fig. 12.11: SAE standards parallel-side splines

not under load, the specified height is about 50 per cent greater than that for press fit. For sliding fit, which is recommended where the hub is to slide when under load, the specified height is about 100 per cent greater than that of press fit.

The torque which an integral multispline shaft can transmit safely by pressure between the sides of the spline can be expressed as

$$T = \tfrac{1}{2} phli(D - h) \qquad \ldots (12.24)$$

where, p is the allowable pressure, in N per mm^2

i is the number of splines

l is the engaged length of each spline, in mm.

The length l required for a given torque T and a selected pressure p may also be found by equation 12.24.

The pressure p must be much lower than the elastic limit S_e in compression, to prevent wear of the sliding surfaces. Since p must be considered a bearing pressure, the very low pressure of 7 N/mm^2 is recommended[2] for sliding fit. The value 14 N/mm^2 may be used for close fit; and 21 N/mm^2 is suitable for press fit.

From the standpoint of strength, tests have shown that the spline grooves weaken the shaft more in torsion than in bending, as was to be expected in view of the nature of the discontinuity. The elastic strength in torsion of a shaft with spline grooves is approximately equal to the strength of a straight shaft the diameter of which is $D' = D - 2h$. In bending the elastic strength is equal to the strength of a shaft for which $D' = D - h$.

Stress concentration: The main drawback of the integral spline connection with straight sides is the forming of cracks in the fillets a (Fig. 12.12) of the splined shaft if the torque fluctuates. According to the SAE standards, the fillet radius should not exceed 0.4 mm. This radius is very small, especially for the larger shafts, and results in a high form-stress factor K (Fig. 3.31). In ductile materials the factor of sensitivity q is low, giving a moderate

Fig. 12.12: Failure of a spline shaft

stress-concentration factor K'. In brittle materials, such as tempered steel, q is high, ranging up to 0.6, and the factor K' is high.

12.7 INVOLUTE SPLINES

Lately there has been an increasing tendency to use involute stub splines instead of straightside splines, both for automotive and general machinery applications. These splines are produced by the methods and tools used for cutting involute gears. A standard pressure angle of 30° is used and half the depth as compared to standard spur gear teeth (12.4). Involute splines are considerably stronger than straight-side splines, and no stress concentrations or cracks, such as those shown in Fig. 12.12, ever occur (12.5).

The SAE has established fifteen standard pitches, with diametral pitches from $\tfrac{1}{2}$ to $\tfrac{48}{96}$ and with 6 to 50 teeth. The pitch designation is the diametral pitch p_d, which determines the pitch diameter of a spline for a given number of teeth, and the second figure designates the diametral pitch p_d', which determines the size of the addendum a and dedendum d. For a flat root, a and d are equal and

$$a = d = 0.5 \text{ m} \qquad \ldots (12.25)$$

As per standards, splines have modules 2 and 2.5 and number of teeth from 6 to 50. The size of addendum a and dedendum d is given below

$$a = d = 0.5 \text{ m}$$

All data for manufacturing and checking the accuracy of the obtained splines are given in elaborate table in the SAE Handbook.[3]

Fits: The SAE standards establish three methods of fitting spline members together. In one method the fit is controlled by varying the major, or outside, diameter D_0 of the external spline, Fig. 12.13a. In a second method the fit is

[2] SAE *Handbook* (New York: Society of Automotive Engineers, 1953), pp. 605–7.

[3] *Ibid.*, pp. 521–79; also ASA B5.15–1950 (New York: American Standard's Association 1950).

controlled by varying the tooth thickness t. This fit is used for splines of the full-fillet type, Fig. 12.13b. In the third method the fit is controlled by varying the minor, or root, diameter D_r of the internal splines, Fig. 12.13c.

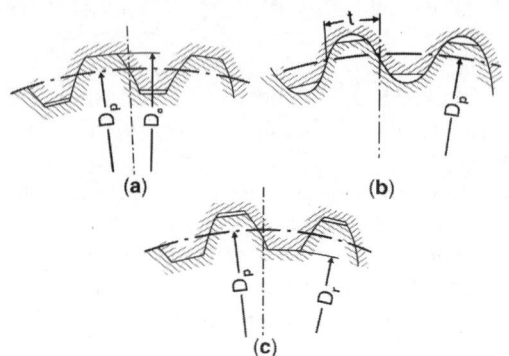

Fig. 12.13: Ways of fitting involute splines

For each of these three methods there are three classes of fits—sliding, close, and press fits. Sliding fits must have clearance at all points; close fits must be close at one point (dimension) of the tooth profile; and press fits must have an interference at only one point (dimension) of the tooth profile.

The SAE Handbook gives the limits for clearances and interferences for the three classes of fits. These limits vary with the diametral pitch.

The involute splines are made either with a so-called flat root, Fig. 12.13a and c, or with a full-fillet root, Fig. 12.13b.

The advantages of involute splines with a 30° pressure angle are:

(a) The teeth have the maximum strength through the minor diameter, where it is needed.

(b) Involute splines are self-centering and therefore tend to equalize the bearing and shear stresses among all teeth.

(c) The tooth surface is smooth, being obtained by the generating action of the hobbing operation and are easier to manufacture as compared to square teeth splines.

(d) The stress concentration is less.

Length of spline: The stresses created in the spline by the torque transmitted by the shaft are shear in the teeth, and bearing stress at the contacts between the teeth.

The area resisting shear is that of the pitch cylinder with the diameter D_p, which is the same for both the external and internal splines. This area is

$$A_s = \frac{\pi D_p L}{2} \qquad \text{...(12.26)}$$

where, L is the length of the spline. If the design stress in shear is designated as S_{ds}, the torque capacity of the teeth in shear is

$$T_s = \left(\frac{\pi D_p L}{2}\right)\left(\frac{D_p}{2}\right) S_{ds} = \frac{\pi}{4} D_p^2 L S_{ds} \quad \text{...(12.27)}$$

The minimum height of contact on one tooth is

$$h = \frac{0.8}{p_d} = \frac{0.8 D_p}{i} \qquad \text{...(12.28)}$$

or 0.8 m where m is module

The corresponding area of contact of all i teeth is

$$A = \left(\frac{0.8 D_p}{i}\right) Li = 0.8 D_p L \quad \text{...(12.29)}$$

The torque capacity of the spline in bearing stress, with $S_b = 2S_{dc}$, is

$$T_b = (0.8 D_p L)\left(\frac{D_p}{2}\right) 2 S_{dc} = 0.8 D_p^2 L S_{dc} \quad \text{...(12.30)}$$

or $= 0.8 \ miL \ D_p S_{dc}$ \qquad ...(12.31)

The maximum height can be half of the normal spur gear teeth.

For steel, the ratio of the elastic limit in shear to the elastic limit in compression is about 0.6. If the safety factor is the same, $S_{ds} = 0.6 S_d$ and dividing equation 12.27 by equation 12.30 gives $T_s = 0.59 \ T_b$. Therefore, the design of splines is critical in shear.

The necessary length of a spline may be obtained by equating the torque capacity of the spline to that of the shaft. The effective shaft diameter may be considered equal to the pitch diameter D_p of the spline. Then

$$T = \frac{\pi D_p^3 S_{ds}}{16} \qquad \ldots (12.32)$$

Experience shows that because of inaccuracies in spacing and **tooth** form, only about 25 per cent of the teeth **are** in actual contact. Therefore, introducing the **factor** 0.25 in the second member of equation 12.27, equating the resulting expression to the second member of equation 12.31, and solving for L, we get $L = D_p$. This value is in accordance with standard practice. It is also sufficient to obtain stability of the part with the internal spline.

Involute serrations: In addition to the splines previously described, the Society of Automotive Engineers and the American Standards Association have adopted standards for splines with much finer pitches and a pressure angle of 45°. These splines have been given the name *involute serrations*. The shape of an involute serration is shown in Fig. 12.14. The pitches range from 10/20 to 128/256. The three coarser pitches may use 6 to 100 teeth; the six finer pitches use from 6 to 50 teeth. The pitch diameters range from 2.5 mm to 250 mm.

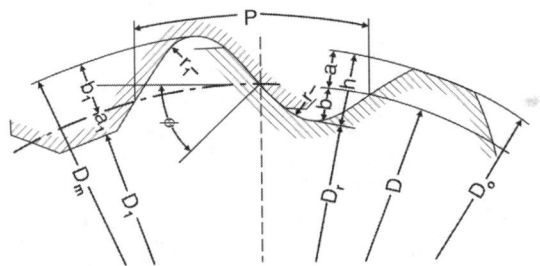

Fig. 12.14: Involute serration

Involute serrations are used for permanent connections of shafts with parts mounted on them. They are made with three classes of fits. A class fit is loose; a class B fit is close; and a class C fit is a press fit. Involute serrations are fitted on the sides of the teeth. All values necessary for manufacturing and checking the accuracy are given in Tables.[4]

----

[4] SAE *Handbook*, pp. 580–604; also ASA B5.26–1950.

12.8 PINS

Geometrically, pins may be divided into cylindrical pins, called *straight pins*, and conical pins, or *taper pins*. Dynamically, pins may be classified as those used only to locate the relative position of two parts when there is a little or no force acting upon the pin, and those that fasten two or more parts together and are subjected to considerable stresses, which are mostly shear stresses but sometimes bending stresses. Locating pins are called *dowel pins, or* simply *dowels*. A connecting pin, like a dowel, may be used either as a permanent connection or as a fulcrum for a movable joint.

The various uses of dowel pins are illustrated in Fig. 12.15. To insure accuracy of assembly the dowel is sometimes made tapered; Fig. 12.15d.

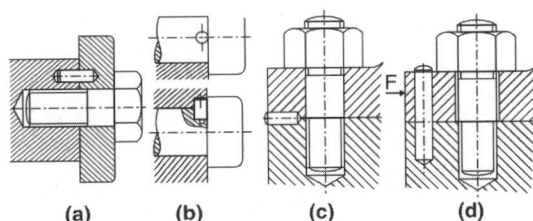

(a) (b) (c) (d)

Fig. 12.15: Dowel pins

Connecting pins are made tapered. When a transverse force F acts on the flange, as shown in Fig. 12.15d, the dowel pin becomes a connecting pin which must be properly dimensioned to resist the force F. To remove a taper pin when there is no through hole, its large end may be threaded, Fig. 12.16, and a nut

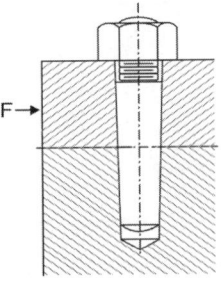

Fig. 12.16: Taper dowel

may be used as a puller. The round key, Fig. 12.2a, is another example of a connecting pin. To secure a part on a shaft in regard to both rotation and axial movement, the pin is inserted as shown in Fig. 12.17a. However, if the torque transmitted to the hub is not heavy and it is desired to avoid unnecessary weakening of the shaft, the pin may be inserted as shown in Fig. 12.17b. All connecting pins must be fitted well, using standard pins and reamers, and they must be driven home to prevent them from working loose.

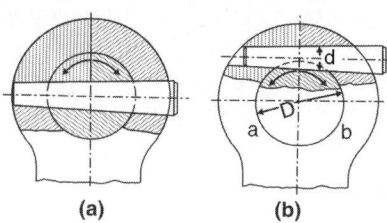

Fig. 12.17: Taper pins

Several types of pins used as fulcrums in knuckle joints are shown in Fig. 12.18. Shoulder screws, Fig. 11.15, are used for the same purpose. In a knuckle joint with all holes of the same diameter, like that shown in Fig. 12.18a, a simple cap screw with a nut and lock nut may be used.

Grooved pins are now finding wide application because they do not require reaming of the drilled holes. These pins, as shown in Fig. 12.19, have three tapered rolled grooves with protruding edges. The edges deform elastically and prevent loosening of the pin under the action of variable loading and even under severe vibration. These pins are used in place of

straight or conical pins in the manner shown in Fig. 12.20. They can be reused up to twenty-five times.

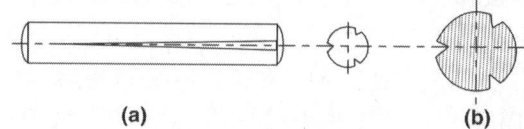

Fig. 12.19: Grooved pin

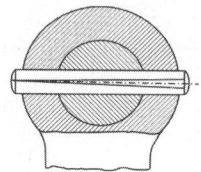

Fig. 12.20: Grooved pin as a key

Roll pins are also as recent innovation, designed by the Elastic Stop Nut Corporation to replace various dowel, pivot, and tapered pins. The roll pin is made of 45C8 (SAE 1045) steel or type 420 stainless steel, and is formed by rolling a plate into the shape of a cylinder with an axial gap. Both ends of the pin are chamfered, which permits it to be driven into a hole with a diameter slightly smaller than that of the pin. The spring action of the compressed pin holds it securely against any vibration or shock.

At present, roll pins are made in thirteen diameters to fit holes with diameters ranging from 2.5 mm to 12 mm and in a wide range of stock lengths. The use of a roll pin is illustrated in Fig. 12.21. Figure 12.21a shows how easily

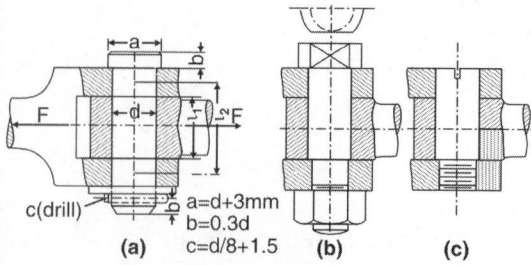

Fig. 12.18: Knuckle-joint pins

$$a = d + 3\text{mm}$$
$$b = 0.3d$$
$$c = d/8 + 1.5$$

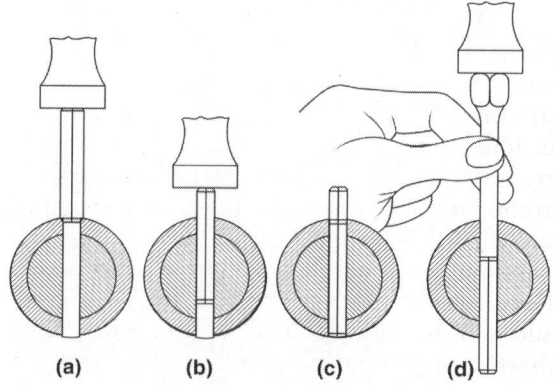

Fig. 12.21: Joint with a roll pin

the pin, with its chamfered ends, can be driven into a predrilled hole; Fig. 12.21b shows the compression of the pin as it is driven; Fig. 12.21c shows how it locks permanently in place; and Fig. 12.21d shows how easy it is to remove the pin when necessary.

Taper pins are made of ordinary machine steel, or of 0.10 carbon steel. The diameter of a taper pin is selected so that the force acting upon the pin does not induce an excessive shear stress. A pin may work in single shear, like the one shown in Fig. 12.16, or in double shear, like the one shown in Fig. 12.17a. A tangential pin, shown in Fig. 12.17b, produces friction between the shaft and the hub on the arc *ab*. To ensure sufficient pressure the hole is first reamed to a taper with the hub on the shaft, and the hub must then be taken off and the hole in it reamed slightly larger. In a pin of this type, satisfactory operation depends on the skill of the mechanic to such an extent that any theoretical calculations are futile.

The mean diameter of the pin is made

$$d_m = 0.20D \text{ to } 0.25D \qquad \ldots (12.33)$$

where, D is the shaft diameter.

A *knuckle pin* should be designed to have sufficient bearing area, sufficient strength in bending, and sufficient strength in double shear.

With the designations of Fig. 12.18, the bearing pressure is

$$p = \frac{F}{dl_1} \qquad \ldots (12.34)$$

For ordinary machine steel, p should not exceed 20 MN/m² if the rocking motion is small and if a small amount of wear is not objectionable. Otherwise, p should not be more than 14 MN/m².

In computing the bending stress, one should regard the load as applied at the center of the eye, and the points of support should be taken at the center of each fork shank. The designer should determine the required diameter d for all three stress conditions. The largest of the three values for d will be the correct size to use.

Snap rings are used to prevent axial motion of two concentric parts. They are of two types: external, Fig. 12.22a, and internal, Fig. 12.22b. The rings can be easily slipped in place, and the holes in the ends serve to permit their removal. They are made of high-carbon heat-treated steel in sizes from 12 to 210 mm in diameter, and in thicknesses from 1 mm to 3 mm. Figure 12.23 illustrates the use of an external snap ring e for locating the inner race of a ball bearing, and the use of an internal ring i to act as a stop in the housing. The grooves for snap rings must be made with a slight clearance for ease of insertion, and the rings cannot be inserted while the parts are under axial load.[5]

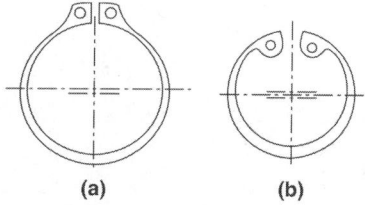

(a)   **(b)**

Fig. 12.22: Snap rings

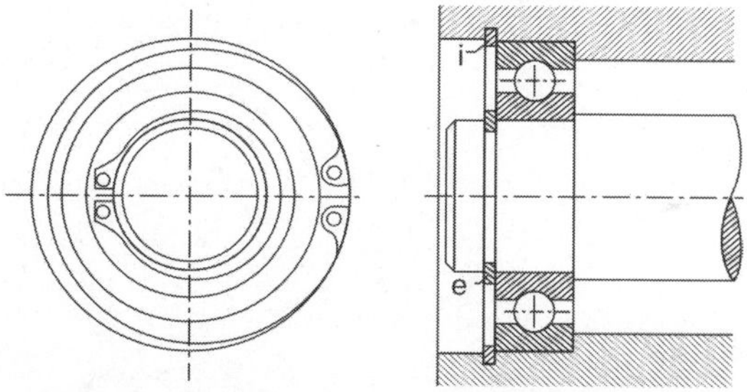

Fig. 12.23: Ball bearing installation with snap rings

12.9 COTTER JOINTS

A cotter is a cross key used for fastening a rod to some part having a socket, the rod being subjected to tension or compression.

Analysis of a cotter joint: An analysis of stresses in a cotter joint, which can serve as a guide in the determination of proportions for such joints in general, will be made with reference to the type of joint shown in Fig. 12.24b. Because of the elasticity of materials, a cotter joint—like

[5] P. F. Rossman, "Designing Snap Ring Fastenings." *Machine Design*. Vol. 13 (May, 1941), pp. 49–51, 106.

a screw fastening—can work satisfactorily only if the cotter is previously tightened. When the joint is in tension, the total force on some parts is equal to the sum of the initial tighting force F set up due to wedge action and the external force Q. The maximum force at any instant is taken as 1.25 Q in tension. The force H, Fig. 12.24c, necessary to produce F can evidently be found from the relation

$$H = F \tan (\alpha + \Phi) \qquad \ldots (12.35)$$

where, α is the angle of the cotter slope and φ is the angle of friction.

The joint may fail if the stress induced by the force $F + Q$ or Q, whichever is acting at a

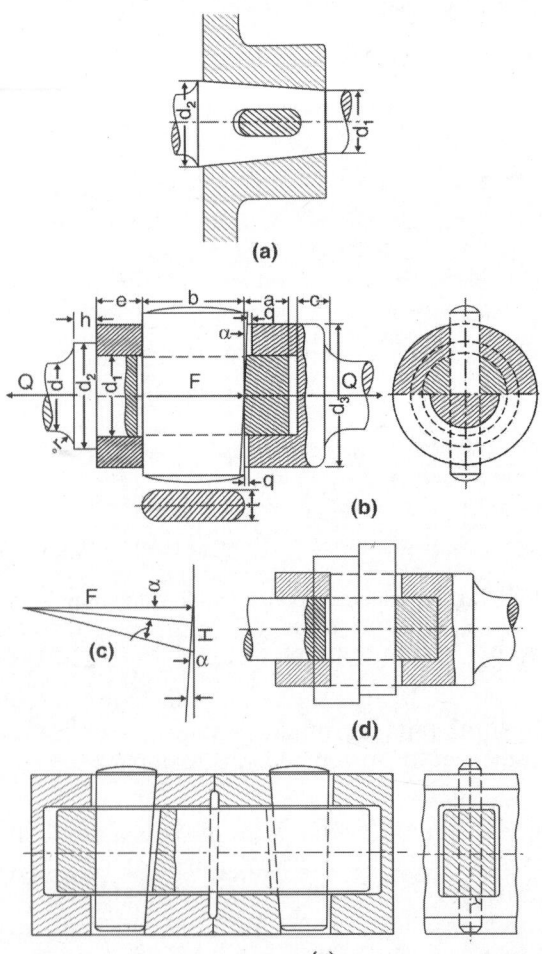

(a)

(b)

(c)

(d)

(e)

Fig. 12.24: Types of cotter joints

certain place, exceeds the safe value in any one of twelve places.

(a) *Rod in tension*: The tearing resistance of the solid rod is

$$Q = \frac{\pi}{4} d^2 S \qquad \ldots (12.36)$$

the diameter of rod d can be calculated and rounded off to next mm standard size; the stress concentration due to the change from d to d_3 for the socket and d to d_2 for the spigot side with a fillet radius r must be taken into consideration.

The force F set up by the initial tightening force on the cotter affects the following places, whether load Q acts or not. However, the analysis has been done with force Q ignoring F by taking it in the factor of safety.

(b) *Rod across the slot*: The rod may fail in tension at this section. Stress concentration at the ends of the slot in the rod exists but may be neglected in view of constant load conditions. In the case of changing load, the consideration may be done in the factor of safety. The resistance of the rod at the slot is

$$Q = \left(\frac{\pi}{4} d_1^2 - d_1 t \right) S \qquad \ldots (12.37)$$

In order to start the design, the cotter dimension may be taken as:

$$t = 0.4\, d \text{ or } 0.25\, d_1 \qquad \ldots (12.38)$$

It is preferable to take $t = 0.25\, d_1$ to calculate d_1 to facilitate the following check.

(c) *Crushing between cotter and spigot:* Crushing may occur at the surface of contact between cotter and rod at the slot because of excessive compressive or bearing stresses if the dimension are too small, i.e. checking for compressive stress.

$$Q = d_1 t\, S_c \qquad \ldots (12.39)$$

The dimensions d_1 and t can be suitably proportionately increased to satisfy this condition and final dimensions of d_1 and t are to be used in all the calculations.

(d) *Socket across the slot (Tensile failure)*: This section is stressed in tension and stress concentration at the ends of the slot may again be neglected.

$$Q = \left[\frac{\pi}{4}(d_3^2 - d_1^2) - (d_3 - d_1)t\right] \times S \quad \dots (12.40)$$

This equation gives the value of d_3

(e) *Rod end:* The end of the spigot rod may fail in double **sheer** as

$$Q = 2 d_1 a S_s \quad \dots (12.41)$$

which gives the dimension a.

(f) *Socket and cotter (Crushing failure):* Crushing may occur between cotter and socket contact area. For saving material dimension of the socket may be increased to d_4 to satisfy the following relation or the whole length of the socket may be made of large diameter equal to d_4 instead of d_3.

$$Q = (d_4 - d_1) t S_c \text{ or } (d_3 - d_1) t S_c \quad \dots (12.42)$$

The design may be either as shown in Figs 12.24b or 12.25.

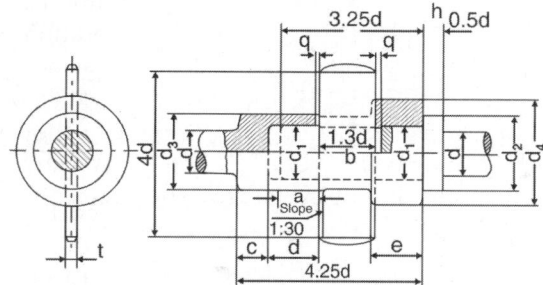

Fig. 12.25: Socket and spigot joint

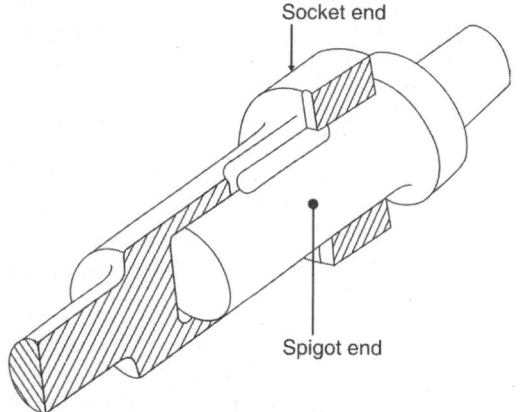

Socket end

Spigot end

Fig. 12.25a

(g) *Socket end:* The socket may fail in double shear in the dimension e, which is found as:

$$Q = 2(d_4 - d_1) e S_s \quad \dots (12.43)$$

(h) *Collar in contact with socket (Crushing of collar):* There is crushing of the surfaces contact between collar diameter and socket due to initial tightening load or due to compressive force given by the following equation to get the diameter d_2,

$$Q = \frac{\pi}{4}(d_2^2 - d_1^2) S_c \quad \dots (12.44)$$

(i) *Shearing of collar:* The collar may fail due to shearing if its thickness h is too small as,

$$Q = \pi d_1 h S_s \quad \dots (12.45)$$

to find the dimension h.

(j) *Shear of cotter:* The cotter may fail in double shear. To find the width of cotter b, the following equation is used

$$Q = 2 b t S_s \quad \dots (12.46)$$

(k) *Rod socket connection:* The rod can fail the socket by shearing if the dimension c is not large enough, and c is given by:

$$Q = \pi d c S_s \quad \dots (12.47)$$

(l) *Bending of cotter:* The cotter may also fail in bending. For the sake of safety, it may be assumed that the load is concentrated at the centre of the cotter and the support offers a varying reaction, having maximum value at the inner side of the socket and decreasing to zero at the outer surface of the socket.

$$Q = \frac{2}{3}\frac{t b^2 S_b}{d_3} \text{ or } \frac{2t b^2 S_b}{3d_4} \quad \dots (12.48)$$

to find the value of S_b. If $S_b > S$, it may be assumed that the cotter is weak and might fail earlier keeping the other two parts safe.

In actual practice, bending is not likely to take place unless there is clearance between spigot and socket to allow for bending. The value of bending stress is likely to be higher than the allowable tensile stress which indicates that cotter is weaker than other elements. *If failure occurs in the cotter due to Q, the high load will not damage spigot and socket (12.12).*

The clearance and the taper in the cotter help to draw in of the cotter and also gives adjustment for wear between cotter and spigot and cotter and socket (12.10).

By tapering cotter on both sides, the cost of cotter and cutting of slots will increase (12.11).

The following empirical proportions are used in practice based on experience.

$d_1 = 1.2d$
$d_2 = 1.5d$
$d_3 = 1.75d$ to $2d$
$d_4 = 2.4d$
$b = 1.6d = 4t$
$a = e = 0.75d$
$c = h = 0.5d$

The clearance $q = 2$ to 3 mm

Sufficient clearances q must be provided for taking-up of wear. In order to decrease the stress concentration in the slots of the rod and of the socket, the cross section of the cotter and slot should not be rectangular, but should be semicircular on the short sides, as shown in Figs 12.24a and b.

Fillets at the junctions of the collar and the socket with the rods must be large enough to avoid dangerous localized stresses.

Bearing stresses: To avoid large and clumsy proportions, rather high bearing stresses S_c must be allowed. An examination of successful practice shows that actual bearing stresses computed from the initial force F are as high as 140 N/mm^2 for machine steel 30C8 (about

SAE 1030). A conical rod end, (Fig. 12.24a), should be preferred to one with a collar (Fig. 12.24b), since it can be fitted more easily and it is not likely to loosen. The bearing stress must be computed from the projected ring area of the cone. Thus the same dimensions d_1 and d_2 are needed regardless, whether a cone or a collar is used.

For ease of calculation the external force is used in calculation and the tightening force effect is taken in the factor of safety.

Jib and cotter joints: These types of joints (Fig. 12.26) are used mostly in square rods. One end of the rod is shaped in the form of a fork or straps. *When the joint is made with a cotter without jib, there is a tendency of the fork ends to open. To prevent the opening of fork, jib is used along with cotter (12.13).* The thickness of jib and cotter is same, the total width is divided equally into the jib and cotter.

1. Tensile failure of square rod. The tearing resistance of solid rod of side a is equated to the external load to find a, i.e.

$$Q = a^2 S \qquad \text{... (12.49)}$$

2. The rod is increased in size to a_1 to allow for weakening due to slot the resistance of tearing of the slotted rod is equated to Q to find a_1 by taking $t = 0.3a_1$.

$$Q = a_1 [a_1 - t]S \qquad \text{... (12.50)}$$

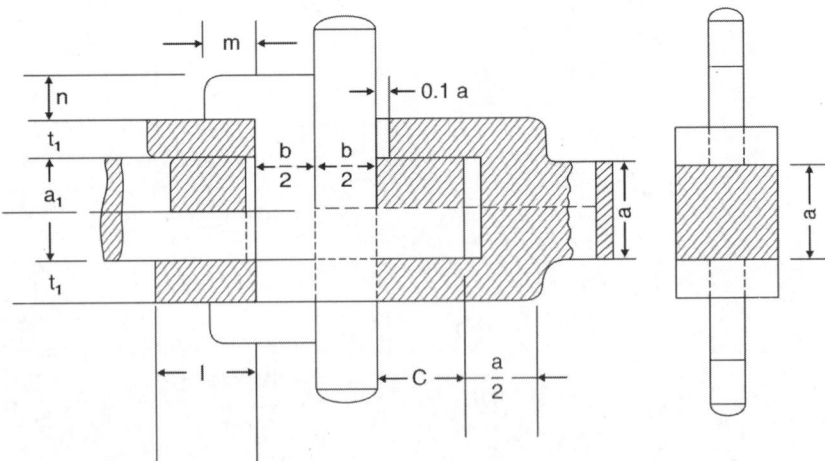

Fig. 12.26: Jib and cotter joints

$$Q = [a_1^2 - a_1 t]S$$

By taking $t = 0.3a_1$, calculate a_1 and t.

3. Crushing or bearing strength of the cotter with rods is used for checking the calculated dimensions

$$a_1 \times t \times S_c \geqslant Q \qquad \dots (12.51)$$

If dimensions are not suitable, a_1 and t need to be increased.

4. Shear of cotter: The total width b of the cotter is calculated from shearing strength of the cotter

$$Q = 2b \times t \times S_s \qquad \dots (12.52)$$

The width b divided equally, i.e. jib and cotter each is $= \dfrac{b}{2}$

5. Shear of the rod end: The end of the rod beyond the slot has a length c and is subjected to double shear, i.e.

$$Q = 2\,ca_1 t \qquad \dots (12.53)$$

Maximum value of $c = a$

6. Tensile failure of strap containing slot having a thickness t_1 which can be found from the equation

$$Q = 2\,(a_1 - t)\,t_1 S \qquad \dots (12.54)$$

7. Shear failure of fork and having length l beyond the slot is given by the following equation.

$$Q = 2 \times l \times 2\,t_1\,S_s$$

usually $t_1 = \dfrac{a_1}{2}$

8. Crushing failure of fork and jib can be used as a check for t_1

$$2t_1 t S_c = Q$$

9. Other dimensions of jib. The jib width m and thickness n can be taken as $m = n = 0.4a$.

10. Check in bending. As in the case of cotter, the bending stress of combined jib and cotter is

$$S_b = \frac{3Ql}{2bt^2}$$

where, $\qquad l = \dfrac{a + 2t_1}{3}$

Example 12.3: Design a cotter joint for transmitting a tensile and a compressive load of 20 kN. The allowable stresses for the material 20C8 (SAE 1020) of the three parts may be taken as: tensile 60 N/mm^2; compressive 90 N/mm^2; shear 45 N/mm^2.

(a) The diameter of the rod d is given by equation 12.36

$$d = \left[\frac{4Q}{\pi S}\right]^{1/2} = \left[\frac{4 \times 20{,}000}{\pi \times 60}\right]^{1/2} = 20.6 \text{ mm}$$

adopt the next standard size as 22 mm

(b) The rod diameter d_1 across the slot by equation 12.37 is by considering 25% additional load for initial tightening

$$Q = 1.25 \times 20{,}000 = \left(\frac{\pi}{4}d_1^2 - d_1 t\right)S$$

Taking $t = \dfrac{d_1}{4}$

$$d_1 = \left[\frac{1.25 \times 20{,}000}{(\frac{\pi}{4} - 0.25)60}\right]^{1/2} = 28 \text{ mm}$$

The empirical value of $d_1 = 1.2 \times 22 = 26.4$.

Therefore 28 mm is safe and $t = \dfrac{28}{4} = 7$ mm

(c) Checking for crushing resistance by equation 12.39

$$28 \times 7 \times 90 = 17640 \text{ N}$$

which shows that the dimensions are not safe. Taking the higher value of d_1 as 30 and t as 7.5, the crushing resistance $= 30 \times 7.5 \times 90 = 20250$ N which is ok. As it is more than 20,000, so

$$d_1 = 30 \text{ mm}$$
$$t = 7.5 \text{ m}$$

(d) The diameter of the socket d_3 is found from the equation 12.40

$$1.25 \times 20{,}000 = \left[\frac{\pi}{4}(d_3^2 - 30^2) - (d_3 - 30) \times 7.5\right]60$$

Solving for d_3 gives $d_3 = 39$ mm. The empirical value is $1.75 \times 22 = 38.5$ to 39 mm is safe.

(e) Rod end length a beyond the slot is found by equation 12.41

$$a = \frac{Q}{2d_1 S_s} = \frac{1.25 \times 20{,}000}{2 \times 30 \times 45} = 9.25$$

The empirical value $= 0.75 \times 22 = 17$ mm or $a = 17$ mm

(f) The outer diameter of the socket d_4 may be found from equation 12.40

$$d_4 = \frac{Q}{tS_c} + d_1 = \frac{1.25 \times 20,000}{7.5 \times 90} + 30 = 57 \text{ mm}$$

The empirical value is $2.4 \times 22 = 53$ mm. So 57 mm is safe.

(g) The length of the flange of the socket beyond the slot is found from equation 12.43

$$e = \frac{Q}{2(d_4 - d_1)S_s} = \frac{1.25 \times 20,000}{2(57 - 30)45} = 10.28 \text{ mm}$$

The empirical value of $e = 0.75 \times 22 = 17$ mm which is taken.

(h) The outer diameter of collar d_2 is found from equation 12.44

$$d_2 = \left[\frac{4Q}{\pi S_c} + d_1^2 \right]^{\frac{1}{2}} = \left[\frac{4 \times 1.25 \times 20,000}{\pi \times 90} + 30^2 \right]^{\frac{1}{2}}$$

$= 35$ mm which is safe as empirical value is $1.5 \times 22 = 33$ mm

(i) The thickness of the collar is found from equation 12.45

$$h = \frac{Q}{\pi d_1 S_s} = \frac{1.25 \times 20,000}{\pi \times 30 \times 45} = 5.9 \text{ mm}$$

The empirical value $= 0.5 \times 22 = 11$ mm is adopted.

(j) The width of the cotter b is given by equation 12.46

$$b = \frac{Q}{2tS_s} = \frac{1.25 \times 20,000}{2 \times 7.5 \times 45} = 37 \text{ mm}$$

The empirical value is $1.6d = 1.6d = 1.6 \times 22 = 35.2$, so the higher value of 37 mm is adopted.

(k) The thickness of the socket beyond the hole, i.e. c is given by equation 12.47

$$c = \frac{Q}{\pi d \times S_s} = \frac{20,000}{\pi \times 22 \times 45} = 6.4 \text{ mm}$$

Which is also taken as $0.5 \times 22 = 11$ mm.

(l) Checking the bending stress in the cotter by equation 12.48

$$S_b = \frac{3 \times 20,000 \times 57}{2 \times 37^2 \times 7.5} = 166.5 \text{ N/mm}^2$$

This stress is higher than 60 N/mm². This results in the cotter being weaker than other parts. However, the cotter is properly fitted and this higher stress may be ignored. The load is taken as concentrated and actually it is distributed.

Example 12.4: Design a jib and cotter joint for connecting two square rods subjected to a tensile/

compressive load of 15 kN. The allowable stresses are: tensile 75 N/mm², compression 120 N/mm² and shear 60 N/mm².

1. Tensile load on the square rod of sides a_1.
   $15 \times 1000 = a_1^2 \times 75$, $a_1 = 14.14$ mm, taking the standard size as 16 mm.
2. However, this needs to be calculated with a slot

$$15 \times 100 = 75 \left(a_1^2 - a_1^2 \times 0.3 \right)$$

$$a_1 = \left[\frac{15000}{75 \times 0.7} \right]^{\frac{1}{2}} = 16.9 \text{ say 18 mm}$$

$$t = 0.3 \times 18 = 5.4 \text{ say } 6$$

3. Checking in crushing strength
   $18 \times 6 \times 120 = 12960$ which is not safe.

$$\text{Increasing } a_1 = \left[\frac{15000}{120 \times 0.3} \right]^{\frac{1}{2}} = 20.4 \text{ say 21 mm}$$

$$t = 0.3 \times 21 = 6.3 \text{ say 7 mm}$$

4. Width of cotter $= \dfrac{15000}{2 \times 7 \times 60} = 17.8$ say 18 mm

   Empirically $= 21$ mm
   Width of gib $=$ width of cotter $= 11$ mm
5. Length of rod end beyond slot

$$c = \frac{15000}{2 \times 21 \times 60} = 5.9 \text{ mm}$$

Empirical value of $c = a = 21$ mm

6. Thickness of strap $t_1 = \dfrac{15000}{2(21 - 7)75} = 7.1$ mm

   Minimum value of $t_1 = \dfrac{a}{2} = 11$ mm

7. Length of fork beyond slot

$$e = \frac{15000}{4 \times 11 \times 60} = 5.6 \text{ mm}$$

Minimum value of $c = a = 21$ mm

8. Checking for crushing failure of fork and cotter
   $(11 + 11)7 \times 120 = 18480$ N which is quite safe.
9. Other dimensions $n = 0.4a_1 = 0.4 \times 21 = 8.4$ say 9 mm

   $m = 0.4 a_1 = 0.4 \times 21 = 8.4$ say 9 mm
10. Check for bending stress of cotter

$$l = 21 + \frac{2 \times 11}{3} = 28.3$$

$$S_b = \frac{3 \times 15000 \times 28.3}{2 \times 22^2 \times 7} = 187.9 \text{ N/mm}$$

which is high. In case the bending takes place, cotter is likely to fail keeping other parts safe. The calculation is based on concentrated load which is actually distributed and bending of cotter is not likely to take place.

12.10 KNUCKLE JOINT

A knuckle joint is used to connect two rods or bars loaded preferably in tension as shown in Fig. 12.27. An excessive load Q may cause the joint to fail due to any of the following induced stresses and this analysis is used to find the dimensions.

1. *Rod in tension*: The main rod of diameter d is subjected to tensile stress given by the following equation.

$$F = \frac{\pi}{4} d^2 S_s \qquad \dots (12.55)$$

This gives diameter d, which is to be standard size.

2. *Shear of pin:* The pin is subjected to double shear and the diameter of pin d_1 is found from the equation given below:

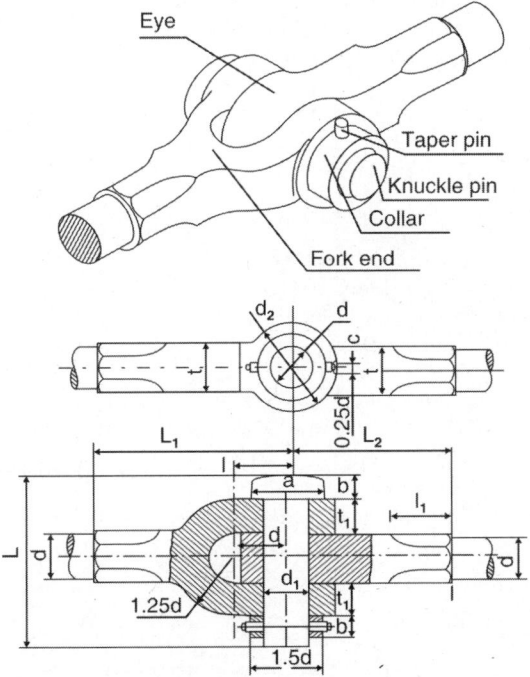

Fig. 12.27: Knuckle joint

$$F = 2 \frac{\pi}{4} d_1^2 S_s \qquad \dots (12.56)$$

The bearing pressure should be checked and also for bending stress. However, the diameter of pin is taken equal to the main rod, i.e. $d_1 = d$.

3. *Compression in the eye*: The above dimensions are again checked for bearing stress, i.e.

$$F = d_1 t S_c \qquad \dots (12.57)$$

This also gives the bearing stress in the pin.

4. *Eye is tension*: The outer diameter of the eye d_2 is found by considering the tensile failure of eye, i.e.

$$F = (d_2 - d_1) t S \qquad \dots (12.58)$$

By taking $t = 1.2d$, d_2 known. However, d_2 is taken $= 2d_1$.

5. *Eye in shear*: These dimensions can be checked by considering double shear of the eye. The resisting area is approximately taken

$$= 2 \times \frac{(d_2 - d_1)t}{2} = (d_2 - d_1)t \quad \dots (12.59)$$

Therefore $F = (d_2 - d_1) t S_s$.

6. *Fork in tension*: The thickness of the limbs of fork or clevis $t_1 = 0.75d$. The tensile stress across the fork is checked by:

$$F = (d_2 - d_1) 2t_1 S \qquad \dots (12.60)$$

7. *Fork in shear*: The fork is checked for failure in double shear and the approximate equation is given by:

$$F = 2 \frac{d_2 - d_1}{2} 2t_1 S_s \qquad \dots (12.61)$$

8. *Fork in compression*: The bearing stress is checked because of contact between fork and the pin i.e.

$$F = 2 d_1 t_1 S_s \qquad \dots (12.62)$$

This also gives the bearing stress in the pin.

9. *Pin in bending*: Considering the pin to be supported by the fork and loaded through the eye as the bearing stress gives a uniformly distributed load. The maximum bending moment

$$M = \frac{Fl}{8}; \quad I = \frac{\pi d_1^4}{64}; \quad y = \frac{d_1}{2}, \quad l = t + \frac{2t_1}{3}$$

Bending stress

$$= \frac{My}{I} = \frac{4Fl}{\pi d_1^3} \qquad \ldots (12.63)$$

The bending stress in the knuckle pins is likely to be higher than allowable stress so the pin will be the weakest element in the joint so that in case of failure, the other elements will not get damaged. In general, bending does not take place till sufficient wear has occurred (12.14).

The following empirical proportions are used in practice.

$$d_1 = d$$
$$t = 1.2 d_1$$
$$t_1 = 0.75 d$$
$$d_2 = 2 d_1 = 2 d$$
$$a = 1.5 d$$
$$b = 0.3 d \text{ to } 0.5 d$$
$$c = d/8 + 1.5 \text{ or } 6 \text{ mm}$$
$$l = 1.2 d$$
$$l_1 = 1.5 d$$
$$L = 4 d$$
$$L_1 = 4.5 d$$
$$L_2 = 4 d$$

Example 12.5: Design a knuckle joint subjected to a load of 35 kN. The permissible stresses are: tensile 70 N/mm², shear 60 N/mm² and compression 110 N/mm².

1. Tensile failure of rod

$$F = \frac{\pi}{4} d^2 S$$

$$d = \left[\frac{35 \times 1000 \times 4}{\pi \times 70} \right]^{1/2} = 25.23$$

Take the next standard value 27 mm.
Take $t = 1.2 d = 32.4$ say 33, $t_1 = 0.75 d = 20.25$ say 21.

2. Shear of pin

$$F = 2 \times \frac{\pi}{4} d_1^2 S_s$$

$$d_1 = \left[\frac{35 \times 1000 \times 4}{2 \times \pi \times 60} \right]^{1/2} = 19.27 \text{ mm say 20 mm}$$

The empirical value of $d_1 = d = 27$ mm
Outer diameter $d_2 = 2 d_1 = 2 d = 54$ mm
Diameter of head of pin = diameter of bush = $1.5 d$ = $1.5 \times 27 = 40.5$ say 41 mm.
Thickness of head/bush = $0.3 d = 0.3 \times 27 = 8.1$ say 8 mm. Taper pin diameter c

$$= \frac{d}{8} + 1.5 = \frac{27}{8} + 1.5 = 3.4 + 1.5 = 5 \text{ mm (say)}$$

3. Outer diameter of eye d_2

$$F = 35 \times 10^3 = (d_2 - d_1) t S = (d_2 - 27) 33 \times 70$$
$$d_2 = 42.1 \text{ say 43 mm}$$

Empirical value of $d_2 = 2 d_1 = 54$ mm, so $d_2 = 54$ mm
4. Considering shear of eye
Shear stress in eye

$$S_s = \frac{2F}{2(d_2 - d_1) t} = \frac{35000}{(54 - 27)33} = 39.28 \text{ N/mm}^2$$

This is o.k. as it is safe.
5. Compression stress in the eye and pin
Compression stress

$$S_c = \frac{F}{d_1 \times t} = \frac{35 \times 10^3}{27 \times 33} = 39.28 \text{ N/mm}^2$$

This is safe
6. Tensile stress is the fork. Tensile stress in the fork which has also $d_2 = 54$ mm is

$$S = \frac{F}{(d_2 - d_1) t_1 \times 2} = \frac{35 \times 10^3}{(54 - 27) \times 21 \times 2}$$

$$= 30.86 \text{ N/mm}^2 \text{ which is safe.}$$

7. Shear stress in fork is given by:

$$S_s = \frac{F}{2(d_2 - d_1) 2 t_1} = \frac{35 \times 10^3}{(54 - 27) \times 21 \times 2}$$

$$= 30.86 \text{ N/mm}^2 \text{ which is safe.}$$

8. Compression stress in fork or pin
Compression stress in the fork or pin is given by

$$S_c = \frac{F}{2 \times d_1 \times t_1} = \frac{35 \times 10^3}{2 \times 27 \times 21} = 30.86 \text{ N/mm}^2$$

9. Checking the pin for bending stress which is given by

$$S_b = \frac{4F \times t}{\pi d_1^3} = \frac{4 \times 35 \times 10^3 \times 33}{2 \times 27^3} = 74.7 \text{ N/mm}^2$$

Bending is likely to take place if there is clearance between pin and hole. However, bending stress is higher than the allowable stress in tension. *So the pin can be considered to be the weakest part. In case the stress is high and pin fails, it will be replaced and costlier parts will not be damaged. The other dimensions are:*

$$L = 4 d = 4 \times 27 = 108 \text{ mm}$$
$$L_1 = 4.5 d = 4.5 \times 27 = 121.5 \text{ or 122 mm.}$$
$$l = 1.2 d = 33 \text{ mm}$$

Length of octagonal shape $l_1 = 1.5 d = 1.5 \times 27 = 41$ mm.

12.11 LEVER

A lever is simply a rigid bar or rod capable of turning about a fixed point, called the fulcrum.

It may be straight or curved and the forces exerted on or by it may be parallel or inclined to one another. A lever is the basic machine to get a desired mechanical advantage, i.e. to displace or raise a load with a smaller effort. A lever is supported by a pin at the point of fulcrum and the forces, i.e. load and effort are applied at suitable distances to get a required mechanical advantage (ratio of load to effort). The levers are classified as: Simple and compound levers; hand operated or foot operated levers; straight and cranked levers. There are three classes of levers depending on the position of each of F, W, P in between the other two, the force diagrams are similar as shown in Fig. 12.28 a, b, c. In all these three cases, the bending moment diagram is similar to Fig. 12.28d. It may be positive or negative. The bending moment at the ends is zero and maximum at B and the mechanical advantage W/P can be increased by changing the two arms.

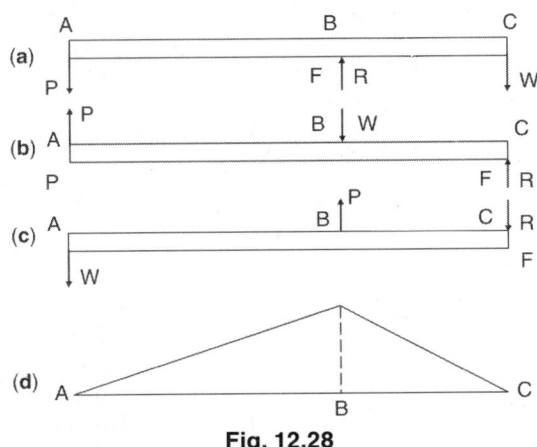

Fig. 12.28

Design Procedure

1. Depending upon the load W and effort P, the lengths of the two arms are assumed.
2. The reaction at the fulcrum is found with given load W and effort P and the direction found depending on the direction of W and P

$$R = \sqrt{W^2 + P^2 + 2WP \cos\theta}$$

3. The cross-section of the lever, usually assumed as rectangular, with width w

and depth t is estimated from maximum bending moment at B by considering $\dfrac{t}{w} = 2$ *to 5 to make economical use of material as this ratio gives a larger value of moment of inertia for the same weight (12.19). The depth is tapered to lower values at the ends depending upon the end connections and reducing bending moment (12.18).*

4. The fulcrum pin diameter d and length l are designed from bearing pressure, i.e. $R = d\, l\, f_b$ taking $l/d = 1$ to 2.5 and checked for shear $R = \dfrac{\pi}{4} d^2 \times 2 \times S_s$ and sometimes bending and deflection also.

5. *The loss in strength at the fulcrum is compensated by providing bosses which also increases the length of bearing (12.16).*

6. The other two pins are similarly designed and usually same dimensions are taken for them.

7. The dimensions of the fork are similar to design of knuckle eye.

8. As there is relative motion of pin in bearing, the wear of pin is reduced by providing soft material bushes in the eyes which can be replaced when wear is more (12.17).

Example 12.6: Design a bell-crank lever to apply a load $F_2 = 20$ kN with arm $L_1 = 400$ mm as shown in Fig. 12.29. The end B and the fulcrum C are to be fixed with the help of pins inside forked shaped supports and the end A is forked itself. Determine the dimensions of the arms and pins to be used at A, B and C. The lever is to have a mechanical advantage of 4. The ultimate stresses is tension and shear for the lever and pin materials are 500 and 400 N/mm² respectively and allowable bearing pressure for the pins is 12 N/mm².

For mechanical advantage to be 4, $F_1 = 5$ kN

$L_2 = 100$ mm

For fulcrum pin, C, resultant reaction

$$R = \sqrt{20^2 + 5^2} = 20.6 \text{ kN}$$

Assuming length to diameter ratio of 2 for the bearing, i.e. $l = 2d$

$2060 = f_b l d$

$$d^2 = \frac{20.6}{12 \times 2}, \quad d = 29.3 \text{ say } 30 \text{ mm}$$

and $l = 60$ mm

Checking for shear failure of pin

$$S_s = \frac{20600}{2 \times \frac{\pi}{4} \times 30^2} = 14.57 \text{ N/mm}^2$$

which is quite safe

Checking for bending stress:

$$\text{Max. bending moment} = \frac{Wl}{6} = \frac{20600 \times 60}{6}$$
$$= 206000 \text{ Nmm}$$

Max. bending stress:

$$\frac{2006000}{\frac{\pi}{32} \times 30^3} = 77.7 \text{ N/mm}^2$$

O.K. as it gives a F.O.S of more than 6.

The bending of pin will take place after sufficient wear of pin and bush.

A bush of about 3 mm thickness will be used into the boss as a bearing so that it can be replaced when it wears off. So the hole diameter in the lever is 30 + 2 × 3 = 36 mm.

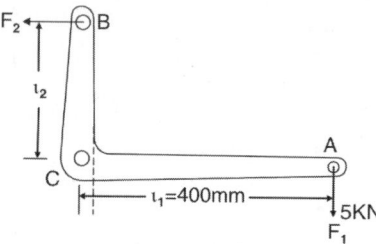

Fig. 12.29: Lever

Section of the lever arms:

Assuming section of lever with depth $t = 2.5$ times the thickness b

$$\text{Distance AX} = 400 - \frac{\text{Boss diameter}}{2}$$

Assume a boss diameter of about twice the hole diameter, say 70 mm

$$\text{AX} = 400 - \frac{70}{2} = 365 \text{ mm}$$

Maximum bending moment on the lever arm
= 5 × 365 = 1825 kNmm

Taking a factor of safety of 6, the allowable tensile stress

$$= \frac{500}{6} \text{ say } 80 \text{ N/mm}^2$$

$$\frac{b \times (2.5b)^2}{6} \times 80 = 1825 \times 1000$$

$$b = \left[\frac{1825 \times 1000 \times 6}{(2.5)^2 \times 80} \right]^{\frac{1}{3}}$$
$$= 27.97 \text{ say } 28 \text{ mm}$$

Depth $t = 28 \times 2.5 = 70$ mm

Section of lever at fulcrum is shown in Fig. 12.30, lever thickness = 28 and is increased to 60 mm thickness by providing bosses on both sides of 16 mm thickness to suit the length of pin of 60 mm.

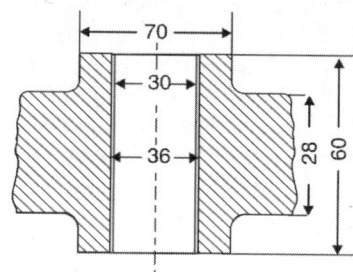

Fig. 12.30: Section of lever at fulcrum

Actual section modulus inclusive bosses

$$z = \frac{I}{y} = \frac{60 \left[70^3 - 36^3 \right]}{12 \times 70 / 2} = 42334.8 \text{ mm}^3$$

Bending stress

$$= \frac{1825 \times 1000}{42334.8} = 43.1 \text{ mm}^3 \text{ O.K.}$$

Pin at A

Assume $\frac{l}{d} = 2$

From bearing pressure, $12 \times l \times d = 5000$

$$d = \left[\frac{5000}{12 \times 2} \right]^{\frac{1}{2}} = 14.43 \text{ say } 15 \text{ mm}$$

and $l = 15 \times 2 = 30$ mm.

As A is a forked end, each side of fork will be at least half of this length, say 16 mm each with phosphor bronze bush 3 mm thickness.

Outer diameter of eye = 2[15 + 2 × 3] = 42 mm

Checking the pin for bending

$$\text{Maximum bending moment} = \frac{Wl}{6} = \frac{5000 \times 30}{6}$$
$$= 25,000 \text{ Nmm.}$$

$$\text{Bending stress} = \frac{25000 \times 32}{\pi \times 15^3} = 75.45 \text{ N/mm}^2 \text{ O.K.}$$

The eye will be of 30 mm width

Pin B

Pin B has a load which is not much different from pin C so same dimensions will be adopted, i.e. $l = 60$ mm and $d = 30$ mm.

PROBLEMS

12.1 What is the primary purpose for which a key is used?

12.2 In what class of work, Woodruff keys are used and why?

12.3 Name two advantages of Woodruff key over rectangular key.

12.4 What is meant by a splined shafting and under what conditions of service is it used? How are splined shafts and hubs made? What are involute splines? What pressure angle is usually used?

12.5 Why is involute spline usually preferable to a straight spline?

12.6 Why is a tapered rectangular key not satisfactory for a heavy and fluctuating torque?

12.7 Determine the dimension of a standard square key for a gear bored for a 80 mm shaft and designed to transmit 45 kW at 225 rpm. The outside load is steady.

12.8 What length of key of width $d/4$ will have strength equal to strength of shaft?

12.9 Why is a cotter tapered?

12.10 Why is a clearance provided between cotter and spigot and cotter and socket?

12.11 What is the disadvantage of providing taper on both sides of cotter?

12.12 Why is the cotter weakest part in the cotter joint?

12.13 Why is jib provided in the cotter joint of square rods?

12.14 Why is knuckle pin designed for least factor of safety?

12.15 Name the stresses produced in the cotter when the joint is subjected to tensile loads.

12.16 Why is a boss generally needed at the fulcrum of the levers?

12.17 Why are bushes of softer materials fitted in the eyes of the levers?

12.18 Why are the arms of the levers usually tapered?

12.19 Why is it preferable to have a rectangular section for the levers?

12.20 Determine the dimensions of a standard square key for a gear bored for a 80 mm shaft and designed to transmit 45 kW at 225 rpm. The outside load is steady.

12.21 Find (a) the force necessary to drive home the taper key of problem 12.12, (b) the rate of speed which must be used with a 45 N sledge hammer to drive this key home, assuming that the last blow will move the key 0.4 mm and (c) the number of hammer blows of the same intensity required to drive the key home. (Refer to section 3.3).

12.22 Determine the size and the number, if more than one key is required, to transmit 3.7 kW at 200 rpm by a pulley to be fastened by Woodruff keys to a 36 mm shaft. Assume that the torque is steady.

12.23 Determine all the dimensions of a permanent 4-spline fitting for a 56 mm shaft to transmit 75 kW at 800 rpm.

12.24 Determine all the dimensions of a 6-spline fitting for a 56 mm shaft that will permit sliding when not under load, to transmit 75 kW at 800 rpm.

12.25 Determine all the dimensions of a 10-spline fitting for a 56 mm shaft that will permit sliding when under load, to transmit 70 kW at 800 rpm.

12.26 (a) Determine all the dimensions of a knuckle joint which carries a load of 5 kN and has a rocking motion of 20°, (b) Find the stress in the pin, (c) Give a sketch of the joint.

12.27 Determine all the dimensions of a cotter joint similar to that in Fig. 12.24a, if the external tensile load F is 55 kN and it is applied alternatively to the rod ends.

12.28 Determine all the dimensions of a cotter joint similar to that in Fig. 12.24b in the text. For an external load of 90 kN which changes continuously from tension to compression.

12.29 Determine the main dimensions of a cotter joint for a flywheel rim similar to that in Fig. 12.24e in the text, assuming that a centrifugal force of 140 kN tends to separate the two halves.

13 Press, Shrink and Friction Joints

13.1 GENERAL EXPLANATIONS

A *press joint*, also called, *force joint*, is obtained by forcing a shaft into a hole that is slightly smaller than the shaft. This is possible because of the elasticity of the materials. The tendency of the materials to return to their original dimensions produces the grip that holds the hub and shaft together.

A *shrink joint* differs from a press joint chiefly in the method of assembling it. The hub is heated to expand its bore in order to slip it on the shaft. When the hub cools down to the temperature of the shaft, the grip is produced in the same way as in the force joint. The shrink joint is also used to connect machine parts by means of special rings, anchors, and tie rods.

In a *friction joint* the holding grip is produced either by the conical shape of the shaft end and the hub bore and by the pull of a nut, or by a slotted hub whose bore is smaller than the shaft and which is spread by a wedge when the joint is being assembled.

Comparison of joints: The assembling of a shaft and a hub by means of a press joint is simpler than with a shrink joint, especially if a hydraulic press of sufficient capacity is available. Shrink joints are used mainly in places where it is difficult or impossible to assemble a press joint, as in the case of rings or anchors. However, tests have shown[1] that shrink joints assembled with the same interference as press joints give more than three times the holding power against both torsion and axial pull. *This superior effectiveness is due to the absence of abrasion between the surfaces of the shaft and the hub during assembly (13.8).*

Press joints and shrink joints are permanent connections, while friction joints can be dismantled. Any one of these joints can be used when machine parts must be connected more securely than can be accomplished with a key or screw joint, especially when they are subjected to shock or vibration.

13.2 PRESS FITS

Medium force fits, are the tightest recommended for holes in cast iron since they stress cast iron to its elastic limit. Heavy force fits are used for holes in steel, whose elastic limit is considerably higher.

Stress due to a force fit: If e is the elongation or contraction of any radius r, the elongation or contraction per unit of length is e/r, in both the radial and circumferential directions. If s is the stress that accompanies this strain, then from equation 2.4.

$$e = \frac{sr}{E} \qquad \qquad \dots (13.1)$$

If r_2, (Fig. 13.1a), is the radius of the contact surface of the hollow shaft a pressed into the hub b, then before assembling, the outer radius of the shaft was $(r_2 + c)$ and the inner radius of

---

[1] J. J. Wilmore, "Shrink and Force Fits." *American Machinist*, Vol. 22 (February 16, 1899) p, 120.

the hub was $(r_2 - c')$. The shaft a is in the condition of a thick solid cylinder with open ends subjected to an external pressure, which induces a compressive stress in it. The hub b is in the condition of an open-end thick hollow cylinder subjected to an internal pressure which induces a tensile stress in the tangential direction. Because of the elasticity of metals, the highest compressive stress s_c will be at the inner surface of the shaft and the highest tensile stress s will be in the inner fibers of the hub.

The radial interference before the parts are assembled, designated by i is obviously

$$i = (r_2 + c) - (r_2 - c') = c + c' \qquad \ldots (13.2)$$

The common radius r_2 can be used both for the original outside shaft radius and for the original inside hub radius, without too great inaccuracy. From equation 13.1, remembering that the compressive stress is negative

$$i = \left(\frac{s_c}{E_s} + \frac{s}{E_h} \right) r_2 \qquad \ldots (13.3)$$

where, E_s and E_h are the moduli of elasticity of the shaft and hub, respectively.

The relation between the stresses s_c and s, the radii of the shaft and hub and the pressure between them may be found by applying equations 2.70 and 2.71. If equation 2.70 is applied to the surface of the hollow shaft with the designations of Fig. 13.1, where $p_i = 0$, $p_o = p_2$, $r_i = r_1$, and $r_o = r_2$, and also $s_r = s_c$, the result is

$$s_c = -p_2 \qquad \ldots (13.4)$$

The tangential stress at the inner surface of the hub may be found from equation 2.71, substituting $p_i = p_2, p_o = 0, r = r_i = r_2,$ and $r_o = r_3$. Thus

$$s = \frac{p_2(r_3^2 + r_2^2)}{r_3^2 - r_2^2} \qquad \ldots (13.5)$$

The variation of the tangential stresses in the shaft and hub is shown diagrammatically in Fig. 13.1b.

If the ratio of the absolute values of the stresses is designated by n,

$$\frac{s_c}{s} = -n \qquad \ldots (13.6)$$

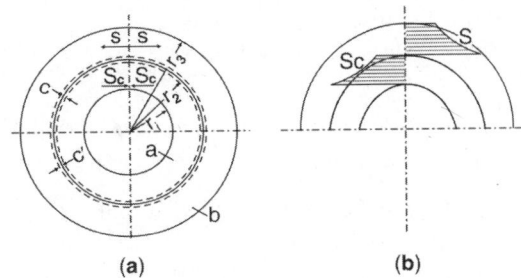

Fig. 13.1: Hub shrunk on hollow shaft

Substituting the values of the stresses from equations 13.4 and 13.5 gives

$$n = \frac{r_3^2 - r_2^2}{r_3^2 + r_2^2} \qquad \ldots (13.7)$$

From equations 13.3 and 13.6

$$s = \frac{i}{\left(\dfrac{n}{E_s} + \dfrac{1}{E_h} \right) r_2} \qquad \ldots (13.8)$$

and

$$s_c = \frac{-in}{\left(\dfrac{n}{E_s} + \dfrac{1}{E_h} \right) r_2} \qquad \ldots (13.9)$$

Example 13.1: A steel crank is to be pressed upon the end of a hollow steel shaft with an outside diameter of 225 mm and an inside diameter of 76 mm; the outside diameter of the crank hole is 400 mm. Using a recommended fit, determine the maximum stress in the hub and the pressure between the hub and shaft. Also find the corresponding stresses at the outer and inner surfaces of the shaft.

With the steel on steel, interference fit may be used. The maximum interference corresponding to force fit is about 0.226 mm. The interference referred to the radius = 0.113 mm.

The modulus of elasticity is the same for both the shaft and hub: $E_s = E_h = 210 \times 10^3 \, \text{MN/m}^2$.

From equation 13.7,

$$n = \frac{200^2 - 112.5^2}{200^2 + 112.5^2} = 0.519$$

From equation 13.8, the maximum tangential stress in the hub is

$$S = \frac{0.113 \times 210 \times 10^3}{(0.519 + 1) \times 112.5} = 138.9 \, \text{MN/m}^2$$

From equation 13.6, the compressive stress on the outer surface of the shaft is

$$S_c = 138.9 \times (-0.519) = -72 \, \text{MN/m}^2$$

From equation 13.4, the pressure on the contact surface is

$$p_2 = 72 \, \text{MN/m}^2$$

The stress at the surface of the shaft can be found from equation 2.71. Since $p_i = 0$, $r = r_1 = 38$ mm

$$S = \frac{-112.5^2 \times 72 \times 112.5^2 \times 72}{112.5^2 - 38^2} = -162.5 \, \text{MN/m}^2$$

The fact that the maximum stresses in the hub and the shaft, both of which are made of steel, have approximately the same values, show that the relations r_3/r_2, r_1/r_2, and i were selected properly. If these stresses should differ considerably, the value of i should be slightly altered.

This discussion assumes that the materials are elastic and the Hooke's law applies. For cast iron, Hooke's law applies only approximately. For the combination of a cast-iron hub on a steel shaft, there is a semirational formula based on Lame's theory and the records of several manufactures.[2] This formula, slightly modified to conform to the designations of this text, gives the stress s at the inner surface of the cast-iron hub. It is

$$s = \frac{Ei}{r_2 + 0.14r_3} \qquad \ldots (13.10)$$

The following simple formula for a steel hub on a steel shaft, also modified, gives the stress at the inner surface of a hub:

$$s = Ei \frac{r_3^2 + r_2^2}{2r_3^2 r_2} \qquad \ldots (13.11)$$

Equation 13.11 may be used for preliminary design, but the final figures should be checked by proceeding as has been explained.

It should be remembered that the maximum stress in the hub should always be well below

the elastic limit. Otherwise a permanent set will occur, with the result that the pressure between the shaft and the hub will be lost and the joint will become loose.[3]

Forcing pressures: The force required to assemble a force-fit joint depends to a great extent on the material, the finish of the surfaces, and the lubricant applied. Although experimental data relating to this feature are rather incomplete, they indicate that the pressure required, in addition to the influence mentioned, is directly proportional to the interference i, the shaft diameter d, and the hub length l.

If the radial pressure p_2 is computed from equation 13.4 or equation 13.5 for a given value of i, then the axial force, in N theoretically necessary to press the shaft into the hub is

$$F = \pi dl f p_2 \qquad \ldots (13.12)$$

where, f is the friction coefficient, which may vary from 0.085 to 0.125 for unlubricated surfaces but with special lubricants can be lowered to about 0.05.[4]

The value of F, in kN for a steel shaft in a cast-iron hub may be computed by the formula[5]

$$F = 93.3 \frac{(r_3 + 0.3r_2)li}{r_3 + 6.33r_2} \qquad \ldots (13.13)$$

And for a steel shaft in a steel hub,

$$F = 64 \frac{(r_3^2 - r_2^2)li}{r_3} \qquad \ldots (13.14)$$

Example 13.2: Find the force that a press must produce to assemble the shaft and hub discussed in the example 13.1, assuming the hub length is 380 mm.

By equation 13.12

$$F = \frac{\pi \times 225 \times 380 \times 0.10 \times 72}{10^3} = 1934 \, \text{kN}$$

[2] A. L. Jenkins, "Formulas for Forced and Shrink Fits," *American Machinist*, Vol. 42 (March 4, 1915), pp. 377–84.
[3] "Design of Press and Shrink Fit Assemblies," *Journal of Applied Mechanics*, Vol. 5, No, 1 (March, 1938), p. A-32; Joseph Marin, "Designing Shrink Fit Assemblies." *Machine Design*, Vol. 14 (June and July, 1942), pp. 68–73 and pp. 72–75.
[4] H. L. Guy, "Factors Affecting the Grip in Force, Shrink, and Extension Fits," *Mechanical Engineering*, Vol. 56 (1934), p. 235.
[5] Jenkins, *inc. cit.*

By equation 13.14

$$F = \frac{64 \times (200^2 - 112.5^2) \times 380 \times 0.113}{200^2} = 1878.6 \text{ kN}$$

The agreement between the two results is satisfactory.

Transmission of torque: Experiments by the Westing House Electric Corporation indicate that the elastic torque of a press joint does not depend on the length of the joint, if the latter is longer than half of the diameter. The term *elastic torque* means the torque at which slip begins at one end of the joint because of twisting of the shaft. The *ultimate torque* is the torque at which the joint slips throughout its full length. A press fit should be designed to prevent elastic torque, especially if an alternating torque is applied. In the presence of an elastic torque, wear and abrasion take place and cause a final loosening of the joint. A sideways-fitted key inserted in a press fit prevents the starting of wear and abrasion. The ultimate torque may be used as a basis of the design only in the case of a steady torque application.

The ultimate torque T of a press fit or a shrink fit may be found by the relation

$$T = \tfrac{1}{2}\pi d^2 l f p_2 \qquad \dots (13.15)$$

where, f may be taken as 0.10 for press fits and 0.125 for shrink fits.[6]

The torque that can be safely transmitted by a press fit or a shrink fit may be found by dividing the ultimate torque from equation 13.15 by a safety factor n which may be between 1.5 and 2.5, the value depending on the uniformity of the transmitted torque.

Disk with hub: All calculations may be made by assuming that the disk and the hub are two separate pieces placed side by side.

Stress concentration: As pointed out in section 12.3, a shaft that transmits a torque always twists. As pointed out in section 5.10, when a hub is pressed or shrunk on a shaft, the radial compressive stress (designated p_2 in equations 13.4 and 13.5) increases near the end where the shaft protrudes from the hub as shown in Fig. 5.15. This stress concentration may increase p_2 to a value above the elastic limit, and the twisting of the overstressed shaft entering the rigid hub will gradually loosen it, regardless of the accuracy with which it was assembled. The lengthening of the fit will increase the time for which the fit will hold. The rate at which the shaft will work loose can be materially decreased by making the end of the hub less rigid as shown in Fig. 5.28b. A hyperbolic slope as shown in Fig. 13.2, is still better. The ribs a support the rim without increasing the torsional rigidity of the hub. Naturally, if the torque is transmitted to both shaft ends, the hub must be made symmetrical.

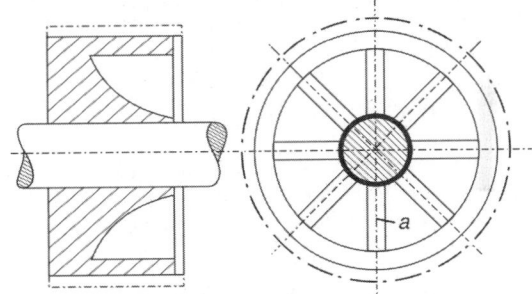

Fig. 13.2: Gear with a flexible hub

13.3 SHRINK JOINTS

Since the holding power of a shrink joint is considerably greater than that of a press joint with the same interference, there is no reason to use a greater interference in a shrink joint. However, a rather common practice is to make the interference in a shrink joint almost twice as large, even though this is often detrimental to the connection because a large interference may produce stresses above the elastic limit.

Types: Besides being used instead of press joints to connect shaft to hubs, shrink joints are used to connect machine parts by means of rings and anchors. Figure 13.3a shows one of several hoops used to connect heavy frames; Fig. 13.3b shows a ring shrunk on the hub end of a split pulley or flywheel; and Fig. 13.3c shows one of the four anchors used to connect two halves of a split flywheel. Another example

---

[6] *Design Work Sheets* (New York: Product Engineering, 1932), p. 99.

of a shrink connection is the heating of a long tie rod having nuts on each end, with a blowtorch, before tightening the nuts. Finally, shrink joints are used to fasten rims of a certain material on wheels of another material, as in the case of gears and rail-car wheels. Circular hoops should be preferred to oval hoops because of the greater ease of machining circular hoops and the greater accuracy in determining the actual interference and the resulting stresses.

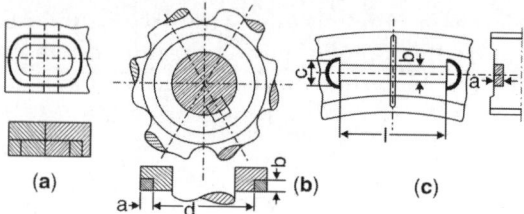

Fig. 13.3: Types of shrink joints

Stresses: The stresses in hubs shrunk on shafts are exactly the same as those that occur in the case of press joints. If a hoop which is shrunk on to hold two parts together is radially thin compared to its diameter, the assumption can be made without appreciable error that the stress is uniform throughout the cross section of the hoop. The contraction in the part on which the hoop is placed is so small in such a case as to be negligible, and the stress in the hoop may be considered as due to stretching it over an incompressible body. This assumption is in favor of strength. The method of computation can be best illustrated by the following example.

Example 13.3: Two circular steel rings are to be shrunk on the ends of a hub of a split flywheel Fig. 13.3b. The outside diameter of the hub is $d = 400.51$ mm, and the centrifugal force that must be resisted by the hoops is 800 kN, determine the inside diameter of the hoops and their cross-section, provided the maximum tensile stress does not exceed 210 MN/m^2.

If it is assumed that $E = 210 \times 10^3$ MN/m^2 and that the hub is incompressible, equation 13.3 gives the magnitude of the radial interference as

$$i = \frac{s r_z}{E} = \frac{210 \times 400.51}{2 \times 210 \times 10^3} = 0.200255 \text{ mm}$$

The inside diameter of the rings must be

$$d_o = 400.51 - 2 \times 0.200255 = 400.1095 \text{ mm}$$

The interference per mm of diameter is $\frac{2 \times 0.200255}{400} = 0.001$ mm which corresponds to medium force fit and the tolerance is 0.423 mm. The initial tension set up by the shrinkage must be greater than the centrifugal force in order that the hub halves shall be held together. If a safety margin of 50 per cent assumed, the force that each ring must resist is

$$F = \frac{1}{2} \times 800 \times 1.5 = 600 \text{ kN}$$

Since this force is taken up by two cross-sections $a \times b$

$$ab = \frac{600 \times 10^3}{210} \times 2 = 1429 \text{ mm}^2$$

The section dimensions may be $a = b = 38$ mm

Anchors: The necessary linear interference i for shrink anchors may be computed by the equation

$$i = \frac{S_d l}{E} \qquad \ldots (13.16)$$

where, l is the effective length, Fig. 13.3c, and the stress S_d may be taken with a safety factor n of 1.25.

The force that is exerted by such an anchor is

$$F = abS_d \qquad \ldots (13.17)$$

where it is convenient to make the ratio b/a between 2 and 3 and to select the dimension a as the thickness of an available stock plate in order to avoid unnecessary machining.

Shrink-on temperatures: The temperature t_2 to which a piece ready to be shrunk on must be heated to go on the larger part having a temperature t_1 depends on the radial interference i and may be computed from the equation

$$\alpha' d(t_2 - t_1) \geqq 2i \qquad \ldots (13.18)$$

where, α' is the coefficient of linear expansion and d is the diameter or length of the shrink link.

Example 13.4: Find the temperature to which the hoops of example 13.3 must be heated for assembling.

The coefficient of linear expansion for steel from Table 5.3 is $\alpha = 11.7 \times 10^{-6}$ per K.

From equation 13.18, the temperature should be

$$t_2 \geq \frac{2 \times 0.200255}{11.7 \times 10^{-6} \times 400} + t_1 \geq 85.6 + t_1$$

If we assume a safety margin of about 20% and a room temperature of $t_1 = 21°C$, the temperature to which the hoops must be heated will be

$$t = 85.6 \times 1.2 + 21 = 123.7°C$$

Shrink fits are also assembled by using dry ice to cool the piece that goes into a hole, such as a pin, to a temperature t_3, thus creating the required temperature difference $(t_1 - t_3)$.

13.4 FRICTION JOINTS

The connection of a shaft to another rotating part by means of a friction joint is attained either by a cylindrical fit or by a conical fit.

For a cylindrical friction joint a light keying fit is used. The split hub, Fig. 13.4a, slides on the shaft freely when a steel wedge is driven into the slot. After the wedge has been removed, an additional grip on the shaft is obtained by tightening the bolts b. The diameter of the bolts could be figured from the torque transmitted by the pulley, but usually it is made equal to $\frac{1}{6}d$. A square, side-fitted key may be used. If a taper key is used, it should be placed so as not to act against the bolts. In a large pulley or flywheel the hub may be split into two pieces, as shown in Fig. 13.4b. In this case the initial gripping force is set up by the rim tension through the arms.

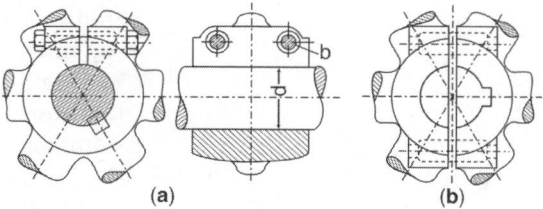

Fig. 13.4: Split-pulley hubs

In a conical fit the force of friction is created by the pull of a nut, as shown in Fig. 13.5. The shank is turned down to the minor-diameter size in order to increase the elasticity of the

fastening and thus to insure more constant pressure and friction between the conical surfaces. Still better results may be obtained by inserting a spring-type washer instead of the slid washer a.

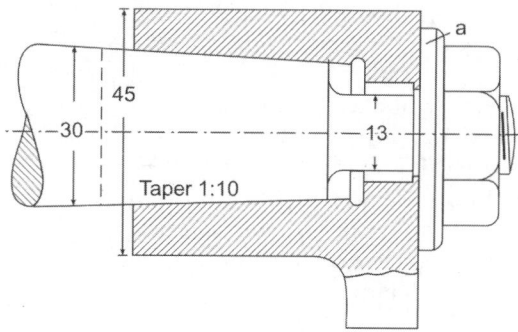

Fig. 13.5: Hand crank

The static coefficient of friction f between the conical surfaces may be taken from Table 13.1.[7]

Table 13.1: Coefficient of friction between conical surfaces

Metals in contact	Condition of surfaces	Coefficient f
Any	Greased with tallow	≤ 0.10
Any	Oil-lubricated	≈ 0.15
Steel on cast iron	Dry	0.16
Steel on steel	Dry	0.22
Steel on cast iron or on steel	Shrink joint, "cold-welded"	≈ 0.33

The diametrical taper is made from 120 mm/m, or 1:8, down to 60 mm/m or 1:16, and even smaller.

The torque transmission in a conical fit is often secured by a key that may be either rectangular or round. A round key is driven into a hole drilled half-and-half in the connected parts.

Figures 13.6a shows a built-up crankshaft of a motorcycle engine, and Fig. 13.6b shows the center part of a two-piece crankshaft of a radial airplane engine. In the latter, the two halves are connected by a lapped-in conical fit with a

---

[7] J. Bach, "Kegelreibungsverbindungen," *Zeitschrift Verein Deutscher Ingenieure*, Vol. 79 (1935), pp. 1570–71.

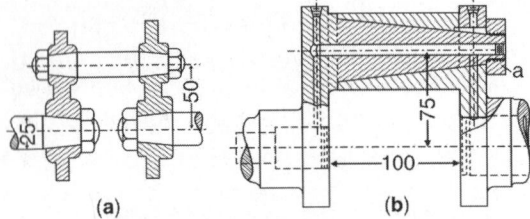

(a)   (b)

Fig. 13.6: Small built-up crankshafts

taper of 1:8 and a round-pin key. The propeller hub is also fastened with a conical fit. The taper of this hub is given by the SAE standards. The hub is secured additionally by either a rectangular key or an integral multiple spline.

The proper size of the pulling nut in a conical fit may be determined by an analysis similar to that applied to taper keys.

PROBLEMS

13.1 What is the advantage of surface rolling of railroad car axles over the region where wheels are press fitted?

13.2 Compare advantages and disadvantages of shrink joints, press joints and friction joints.

13.3 Rims of rail road car wheels are usually shrunk joints. Explain.

13.4 Describe the process of manufacture of a built up crank.

13.5 A thick walled steel hub of 300 mm outside diameter and 250 mm length is to be press fitted over a 150 mm diameter steel shaft. The maximum tangential stress is to be 35 N/mm². $E = 200$ kN/mm². Poissons ratio is 0.3 and coefficient of friction is 0.12. Find the maximum diametral interference and the axial force required to press the hub. What is the maximum torque which can be transmitted by the assembly?

13.6 Determine the dimensions for the hole of hub and a shaft to be shrunk fitted with the shaft diameter of 75 mm.

13.7 A gun barrel is made by assembling two steel cylinders having nominal diameters of 25 mm inside and 50 mm outside diameter and 50 mm inside diameter and 75 mm outside diameter if the tangential stress is not to exceed 80 N/mm². Determine the required interference and stresses at the inner and outer surfaces of two members.

13.8 Why are shrink joints superior to press joints for the same interference?

Springs

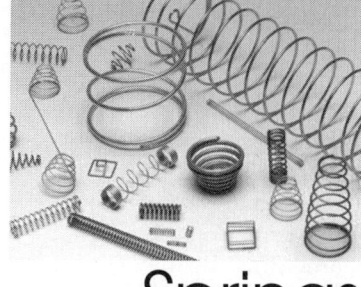

14.1 GENERAL CONSIDERATIONS

Springs are used to connect two parts or bodies by a flexible joint. Their functions are:

(a) To cushion, absorb, or control energy either due to shock, as in car springs or railway buffers, or to vibration, as in spring-supports and vibration dampers.

(b) To control motion—by maintaining contact between two elements, such as a cam and its follower; by restoring a machine part to its normal position when the disturbing force is removed, as in a governor or valve; or by producing the necessary pressure in a friction device, as in a brake or clutch.

(c) To store energy, as in clockworks or starters.

(d) To measure forces, as in spring balances, gages, or engine indicators.

Classification: Springs can be classified according to shape, as indicated in Fig. 14.1. In Fig. 14.1a *plate spring*, or *leaf spring* is shown; and in Fig. 14.1b a helical conical spring is shown. If the radius of the coils of a helical spring is constant, as in Fig. 14.1c, it becomes a *cylindrical helical spring*. If the angle of the helix is zero, as shown in Fig. 14.1d, it is a *spiral spring*.

Special springs for large loads and deflections are made in the form of a series of flat-tapered or dish-shaped disks.[1]

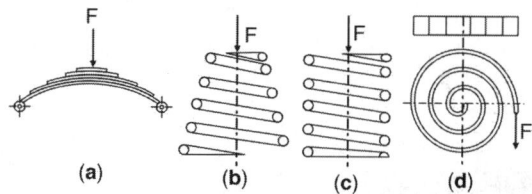

Fig. 14.1: Shapes of springs

In leaf, spiral, and disk springs the stress induced by the load F is *bending*. In a helical spring the main stress induced by an axial load F is *torsion*. If a helical spring works in torsion similarly to a spiral spring, the main stress is *bending*.

A confined quantity of gas or air as in the wheels of bicycle, car or bus is an excellent spring to reduce vibrations effects (14.8).

Solid rubber and cushioning devices in the form of solid block or hollow cylindrical

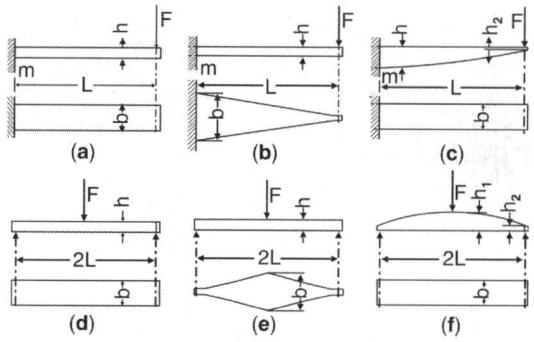

Fig. 14.2: Beams with a rectangular section

---

[1] F. W. Brecht and A. M. Wahl, "The Radially Tapered Disk Spring," *Transactions of the American Society of Mechanical Engineers*, Vol. 52 (1930), APM–52–4, p. 45.

shapes are finding an increasing range of applications.

14.2 LEAF SPRINGS

A *flat spring* may have the form of a cantilever beam, as in Figs 14.2a, b, or c; or it may have the form of a simple beam, as in Figs 14.2d, e, or f, which can be considered as a double cantilever beam. Some beams shown in Fig. 14.2 are of constant cross section, and others are of uniform strength obtained by keeping either a constant thickness h or a constant width b. The main stresses are tension on one side of the neutral axis and compression on the other. The additional transverse shear may be neglected as far as strength is concerned.

The maximum stress in each of the six cases in Fig. 14.2 is $s = M/Z$, and it may be expressed by the one general equation.

$$s = \frac{c_1 Fl}{bh^2} \qquad \ldots (14.1)$$

Similarly, the maximum deflection is

$$y = \frac{c_2 Fl^3}{Ebh^3} \qquad \ldots (14.2)$$

The constants c_1 and c_2 can be taken from Table 14.1, which also contains data on the resilience for the various types of springs.

By dividing equation 14.1 by equation 14.2, the terms F and b are eliminated and the following convenient relation for the design of a spring is obtained:

$$h = \frac{c_2 sl^2}{c_1 Ey} \qquad \ldots (14.3)$$

where, l and y are given; s is the allowable stress which must be selected for the material; and E is the modulus of elasticity. After that, the width for a given load F can be found from equation 14.1. Thus,

$$b = \frac{c_1 Fl}{sh^2} \qquad \ldots (14.4)$$

In Fig. 14.2 it is shown that beams of uniform strength permit a considerable saving in material. At the same time they have a greater deflection, as shown in Table 14.1. This means that their resilience and capacity for absorbing impact energy are also greater in the same proportion, as was shown in section 3.4. The shear stress, which can be neglected in the section with the maximum width b, becomes important toward the end of the beam. Therefore, the beam cannot come to a point, but it must have either a certain minimum height h_2 or a certain minimum width b_2, to resist this shear.

Example 14.1: Determine the thickness and width of a flat steel spring to carry a load of 3.4 kN with a deflection of 32 mm. The spring must be supported at the ends, the distance between the supports being 0.66 m, and is loaded at the centre. Allow a maximum stress of 350 MN/m^2 and make the spring with a constant thickness and varying width.

The steel for the spring must have a modulus of elasticity 210×10^3 MN/m^2. The conditions correspond to those in Fig. 14.2e. From Table 14.1, the constants are $c_1 = 3, c_2 = 3$. According to Fig. 14.2e.

$$l = \frac{0.66}{2} = 0.33 \text{ m}$$

By equation 14.3,

$$h = \frac{3 \times 350 \times 330^2}{2 \times 210 \times 10^3 \times 32} = 5.61 \text{ mm or 6 mm}$$

By equation 14.4,

$$b = \frac{3 \times 3400 \times 330}{350 \times 6^2} = 267 \text{ mm or 270 mm}$$

Table 14.1: Constants in beam equations

Constant	Cantilever beam (Fig. 14.2)			Simple beam (Fig. 14.2)		
	a	b	c	d	e	f
c_1–for the stress	6	6	6	3	3	3
c_2–for the deflection	4	6	8	2	3	4
Unit resilience, Nm/m^3	$\dfrac{s^2}{18E}$	$\dfrac{s^2}{6E}$	$\dfrac{s^2}{6E}$	$\dfrac{s^2}{18E}$	$\dfrac{s^2}{6E}$	$\dfrac{s^2}{6E}$

14.3 LAMINATED SPRINGS

In order to decrease the width b, if it becomes too large, as in example 14.1, the lozenge-shaped plate can be assumed to be cut into narrow strips, as indicated in Fig. 14.3, and assembled with a clamp c, as in Fig. 14.3b. Evidently

$$b' = \frac{b}{i} \qquad \dots (14.5)$$

where, i is the number of strips or leaves.

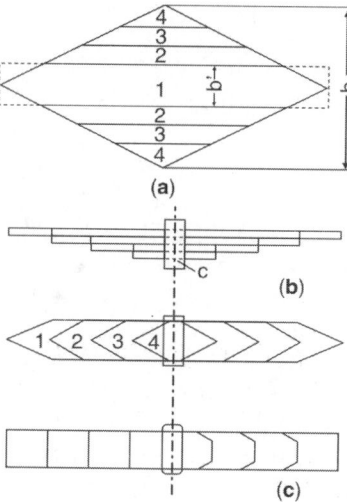

(a)

(b)

(c)

Fig. 14.3: Laminated leaf springs

This laminated spring is a beam of approximately uniform strength. The unit stress and the maximum deflection, if friction between the leaves is neglected, will be the same as the values for the original plate.

Equations 14.1 to 14.4 apply to laminated springs with the substitution of ib' for b. Thus the load, from equation 14.1 is

$$F = \frac{sib'h^2}{c_1 l} \qquad \dots (14.6)$$

and the maximum deflection is

$$y = \frac{c_2 F l^3}{Eib'h^3} \qquad \dots (14.7)$$

In an actual laminated spring the full-length leaves must have square ends by which they are fastened to the supports. Sometimes the shorter, or *graduate*, leaves are simply cut

square at the ends, as shown in Fig. 14.3c at the left. To make the gradual transition, from one leaf to the next, it is better to taper the ends in width or in thickness, or in both dimensions, as shown on the right side in Fig. 14.3c. This change does not affect the stress found by equation 14.6, but it may influence the deflection found by equation 14.7.

The resilience of a laminated spring is equal to that of an equivalent lozenge-shaped spring.

Example 14.2: Change the spring of example 14.1 to a laminated leaf spring and find (a) the stress which will be induced if the load comes down with a shock deflecting the spring 76 mm; (b) the magnitude of the impact energy which the spring will absorb in this case; (c) the height from which the load must drop; and (d) the corresponding impact force.

(a) If the spring is made of 6 leaves, the width b' is $\dfrac{270}{6} = 45$ mm

The stress under the increased deflection can be found by noticing that according to equation 14.3, the stress is proportional to the deflection. Hence,

$$S = \frac{350 \times 76}{32} = 831 \text{ MN/m}^2$$

(b) The impact is equal to the resilience of the spring which, according to Table 14.1, is

$$U = V \times \frac{S^2}{6E} = \frac{660 \times 270 \times 6}{2} \times \frac{831^2}{6 \times 210 \times 10^3}$$

$$= 0.293 \text{ kNm}$$

(c) Since the total impact energy $K_i = W(h + e)$, where, e is the deflection or 76 mm, the height of drop must be

$$h = \frac{293}{3.4} - 76 = 10.2 \text{ mm}$$

Checking h by equation 3.18 gives

$$h = \left[\left(\frac{831}{350} - 1 \right)^2 - 1 \right] \frac{32}{2} = 14.2 \text{ mm}$$

(d) The force of the impact is found by equation 3.23 as

$$F = \frac{2U}{e} = \frac{2 \times 293 \times 10^3}{76} = 7710 \text{ kN}$$

This force F will act upon the end supports of the springs.

Other factors: This analysis did not take into consideration several factors which influence the design of actual springs. First comes the shape of the leaves. The most common type of leaf spring is the semielliptic, Fig. 14.4. The unloaded spring is curved, or cambered, the magnitude of the camber being such that the spring is approximately straight under the full static load. Under the maximum dynamic load, the camber will be negative.

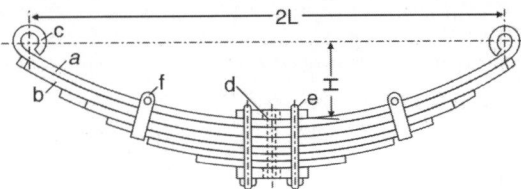

Fig. 14.4: Laminated semielliptic spring

The long leaf *a* fastened to the supports is called the main leaf or *master leaf*. Its ends are bent to form an eye *c*. The center bolt *d* holds the spring leaves together. The hole drilled for this bolt weakens the spring. Fortunately, the pressure exerted by the *U* clips *e*, which hold the spring to the seat, materially reduces the bending stresses in the center part with the bolt. However, the bending action is not reduced enough to allow the omission of this part from the length *l* of the spring in computations relating to deflections and stresses. Also, the *U* clips *e* create a stress concentration at the edge of the spring seat. A soft pad placed between the leaf and the seat will reduce the stress concentration. The *rebound clips f* serve to distribute to the shorter leaves some of the load of the rebound which otherwise would be taken by the master leaf alone.

Full-length leaves: Laminated springs for heavy loads usually have under the master leaf one or more additional full-length leaves, as *b* in Fig. 14.4. With such an arrangement the spring ceases to be of uniform strength. However, the relation between the load *F* and the stress *s* remains the same, being given by

equation 14.6. The deflection can be calculated by introducing a correction factor k_4. Thus,

$$y = \frac{c_2 F l^3 k_4}{E i b' h^3} \quad \dots (14.8)$$

This factor k_4 may be determined by the formula[2]

$$k_4 = \frac{1 - 4r + 2r^2(1.5 - \log_e r)}{(1-r)^3} \quad \dots (14.9)$$

where, $r = i'/i$, with i' the number of full-length blunt-ended leaves and i the total number of leaves.

Or alternatively $k_4 = 0.7r^{0.1}$ for r from 2 to 20 and $k_4 = 1$ for $r > 20$.

Support influence: The ends of the master leaf of a laminated spring are fastened to the supports by means of hinges. This arrangement produces longitudinal loads and additional stresses. To these may be added stresses from transverse forces and from twisting. One method of taking care of this increase of stresses in the master leaf is to make the master leaf of a stronger material than the rest of the spring.

Another way is to reduce the combined stress in the master leaf by making it thinner than the rest,[3] as can be seen from equation 14.3. *The simplest way to reduce the stress is to make the radius of curvature of the master leaf a, Fig. 14.4, greater than that of the next one. When the spring is assembled, the master leaf is given an initial bend by the center bolt and is thus given an initial stress in a direction opposite to that which the load will create. Therefore, when the load is applied, the stress in the master leaf is first relieved of the initial stress and is then only stressed in the opposite direction. The other leaves receive initial stresses in the same direction as those coming from the load. Hence, the stress from the vertical load in the master leaf will always be lower than that in the rest of the leaves. Such prestressing caused by a difference in the radii of curvature is termed nipping (14.15). In automobile springs all leaves are nipped to a certain extent*

[2] J. B. Peddle, "Chart for Full and Semielliptic Leaf Springs," *American Machinist*, Vol. 38 (April 17, 1913), p. 645; also F. A. Halsey, *Handbook for Machine Designers*, 2nd ed. (New York; McGraw-Hill Book Company, Inc., 1916), p. 201.
[3] F. Franz, "Remedy for Spring Breakage," *Automotive Industries*, Vol. 45 (1921), p. 924.

to reduce the maximum stress (14.13, 14.15). This helps to keep them in contact on the rebound, and also keeps out dirt and reduce maximum stress (4.15).[4]

14.4 LEAF-SPRING DESIGN

Carbon steel 98 C4 (SAE 1095) that is properly heat-treated is commonly used for laminated springs. Silicon-manganese steels, 50C9 (SAE 9250) and 60C9 (SAE 9260), the latter having a slightly higher carbon content and higher mechanical properties, are both standardized by the Society of Automotive Engineers for leaf springs. Chrome-vanadium steel has a still greater strength, as well as greater toughness and resistance to failure through repeated stresses.

Sizes: Spring plates are made in accordance with the Birmingham wire gage, Table 14.2, and also in thicknesses varying in 1 mm up to 8 mm, by 2 mm from 10 to 16 mm, 5 mm

steps from 40 to 90 mm and 100, 120, 150 mm widths.

Limit stresses: All springs must be heat-treated. The mechanical properties of spring materials that are properly tempered are given in Table 14.3. The influence of the thickness h is considerable in spring leaves and can be taken into account by multiplying the values in Table 14.3a by a size coefficient which is computed by the formula[5]

$$e_{sz} = 0.8 + \frac{2.5}{h} \qquad \dots (14.10)$$

Safety factor: For static loads, as in governor springs, a safety factor n of 1.5 is sufficient. If the calculations are based on elastic limits, in railroad service the values of n can range from 2 to 2.25; in automobile design n is taken as 2.25 to 2.5 for rear springs and from 3.25 to 3.5 for front springs. If the design takes into account the shock action and is based on

Table 14.2: Standard dimensions of wire in mm

Cold drawn steel wire unalloyed	Hardened and tempered spring steel wire and valve spring wire	Stainless steel wire for normal corrosion resistance
Increment	Increment	Increment
0.07 to 0.12–0.01	1.00 to 1.10–0.05	0.10, 0.11, 0.125
0.14 to 0.22–0.02	1.2, 1.25	0.14 to 0.22–0.02
0.25	1.30 to 2.10–0.10	0.25
0.28 to 0.40–0.22	2.25	0.28 to 0.40–0.02
0.43, 0.45, 0.48	2.40 to 2.60–0.10	0.43, 0.45, 0.48, 0.50
0.50, 0.53, 0.56	2.80 to 4.00–0.20	0.53, 0.56, 0.60, 0.63
0.60, 0.63	4.25 to 5.00–0.25	0.65 to 1.30–0.05
0.65 to 1.30–0.05	5.30, 5.60, 6.00, 6.30	1.40 to 2.10–0.10
1.40 to 2.10–0.10	6.50 to 11.00–0.50	2.25, 2.40, 2.50, 2.60
2.25, 2.40, 2.50	12.0, 12.5, 13.0	2.80, 3.00, 3.15
2.60, 280, 3.00	14.0	3.20 to 4.00–0.20
3.2 to 4.0–0.20		4.25 to 5.00–0.25
4.25 to 5.00–0.25		5.30, 5.60, 6.00, 6.30
5.30, 5.60, 6.00, 6.30		6.50 to 10.00–0.50
6.50 to 11.00–0.50		
12.0, 12.50		
13.00 to 17.00–1.00		

[4] P. M. Heldt, *Motor Vehicles and Tractors* (Nyack, N. Y.: published by the author, 1929), pp. 468 ff.
[5] Derived from data given in D. S. Kimball and J. H. Barr, *Elements of Machine Design*, 3rd ed. (New York: John Willey & Sons, Inc., 1935), p. 215.

Table 14.3: Properties of spring materials

Material	Ultimate strength in tension S_u N/mm²	Elasticity limits		Endurance limits		Modulus of elasticity	
		Tension S_e N/mm²	Shear S_e N/mm²	Bending S_{en} N/mm²	Torsion S_{en} N/mm²	Direct E kN/mm²	Transfer G kN/mm²
Music wire	2100	1720	1050	900	550	206	78
Cr-V steel, SAE 6150 (50 Cr6V23)	1650	1300	760	620	380	210	83
Si-Mn Steel, SAE 9250 (50 Si2Mn90)	1520	1100	660	550	330	206	79
Carbon steel, SAE 1095, (98C6)	1400	1000	590	480	290	204	79
Carbon steel, SAE 1050, (50C4)	850	520	310	390	240	204	77
Monel metal, cold-rolled	850	480	290	350	210	175	65
Phosphor bronze, SAE 77 hard and SAE 81	680	350	210	320	140	105	43
Brass wire, SAE 80 wire grade A	650	325	195	140	85	100	38
Brass, SAE 70, hard-drawn	620	310	190	100	70	100	38
Duraluminum C17 ST, SAE 26	450	230	140	90	55	70	27

Table 14.3a: Properties of spring materials

Material	UTS N/mm²	Ultimate torsion strength N/mm²	Elastic limit N/mm²		Modulus of N/mm² × 10⁵	
			Tension	Torsion	Elasticity	Torsion
Watch spring steel C 1.10–1.19, Mn 0.15–0.25	2260–2350	–	2060–2275	–	2.21	–
High carbon steel C 0.85–0.95, Mn 0.25–0.60	1380–1725	1100–1450	1100–1380	760–1035	2.06	0.7845
Chrome-Venadium steel C 0.45–0.55, Mn 0.50–0.80 Cr 0.80–1.1, V 0.15–0.18	1380–1725	965–1205	1240–1590	690–900	2.06	0.7845
18–8 stainless steel	1100–2275	830–1650	415–1795	300–965	1.93	0.6865
Spring brass Cu 64–72 Zn 28–36	690–900	310–620	275–414	206–414	1.03	0.3825
Nickel silver Cu 56 Zn 25 Ni 18	900–1035	590–685	550–760	414–480	1.1	0.3825
Phosphor bronze Cu 91–93 Sn 7–9	690–1035	520–725	414–760	345–590	1.03	0.4315
Monel Ni 64 Cu 26 Mn 2.5 Fe 2.25	690–965	550–760	550–830	310–480	1.8	0.671
K. Monel Ni 66 Cu 29 Al 2.75 Fe 0.90	1100–1240	725–860	795–1000	450–590	1.8	0.6571
Music wire C 0.7–1.0 Mn 0.3–0.6	1725–3790	1035–2060	1035–2410	620–640	2.06	0.7845
Inconel Ni 80 Ce 14 Fe 6	965–1210	660–830	660–830	380–550	2.1	0.7551
Beryllium copper Cu 98 Be 2	1100–1380	690–900	685–1035	450–660	1.28	0.48

endurance values, a factor of safety of 1.5 or even 1.25 is sufficient.

Design procedure: So many variable factors enter into the design calculations of a leaf spring that no definite rules can be given and cut-and-try methods have to be applied.

The first step is to decide what material will be used and to assume its approximate thickness h. With the data of Table 14.3 and equation 14.10 the limit stress S_l can be determined. Having selected a suitable safety factor n, the design stress $S_d = S_l/n$ is determined.

The value of ib' can be found from equation 14.6 by substituting the selected value of S_d for S, using the given load F, and using a length l that is given or is determined by other considerations. The number of leaves i usually should be from 5 to 10, but in railroad cars it may be increased up to 14. If this does not give a strip width b' within the recommended limits of 50 to 150 mm, the thickness h must be changed correspondingly and the whole procedure repeated.

In a particular case other factors, such as nipping and variable leaf thickness, must be taken into consideration in finding both the maximum stress and the deflection.

Test check: Because of the involved calculations and the influence of the friction, the actual performance of a laminated spring may differ somewhat from the indicated by the design figures. Before the final adoption of the design for quantity production, the spring performance should be checked under expected working conditions.

14.5 DISK SPRINGS

Disk springs, also called Belleville springs, are used where high-capacity compression springs must fit into small spaces (14.7). Each spring consists of several annular disks that are dished to a conical shape, as in Fig. 14.5. They are stacked up one on top of another, as in Fig. 14.6, in order to increase the deflection.

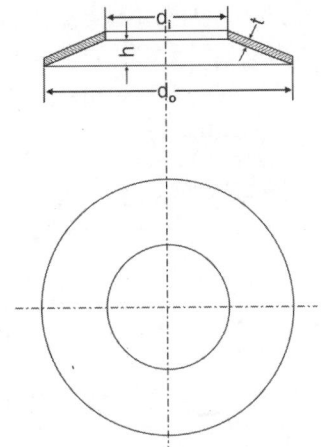

Fig. 14.5: Disk spring

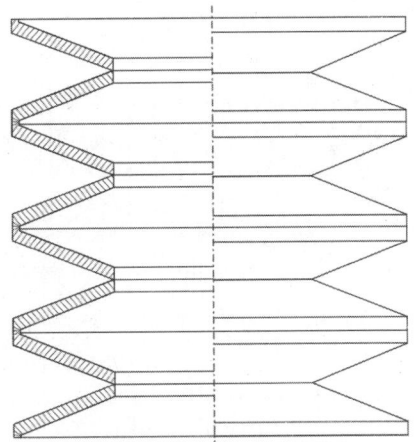

Fig. 14.6: Nest of disk springs

The load is applied uniformly around the edge, and the relation between the load F and the axial deflection y of each disk is given by the equation

$$F = \frac{4Ey}{(1-\mu^2)Md_o^2}\left[(h-y)\left(h-\frac{y}{2}\right)t+t^3\right] \quad \ldots(14.11)$$

where, μ is Poisson's ratio; M is a constant which depends on the ratio d_o/d_i, is indicated in Fig. 14.7; and d_o, d_i, h, and t are spring dimensions.

The maximum stresses induced at the edges are given by the equation[6]

---

[6] J. O. Almen and A. Lazzlo, "Disk Spring Facilitate Compactness," *Machine Design*, Vol. 8 (June, 1936), p. 80; W. W. Boyd, "Defection and Capacity of Belleville Springs," *Product Engineering*, Vol. 3 (September, 1932), and Vol. 4 (February, 1933), p. 63; and Brecht and Wahl, *loc. cit.*

$$s = \frac{4Ey}{(1-\mu^2)d_o^2}\left[C_1\left(h - \frac{y}{2}\right) \pm C_2 t\right] \quad \ldots (14.12)$$

where the plus sign is used for the stress at the inner edge, the minus sign is used for the stress at the outer edge, and C_1 and C_2 are constants depending on the ratio d_o/d_i, as indicated in Fig. 14.7.

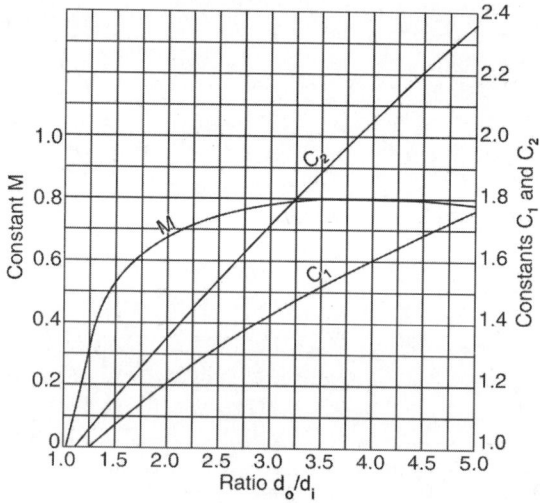

Fig. 14.7: Diagram of disk constants

Experience with springs of this type shows that under static conditions the stresses computed by equation 14.12 may reach 1500 MN/m² for steel having an elastic limit in tension of 830 MN/m². Endurance tests show that the maximum stresses computed by equation 14.12 may reach 1240 MN/m². Actual stresses are unknown. The endurance life of a disk spring is considerably increased if the corners of the disk edges are rounded off. The ratio d_o/d_i should be between 1.5 and 5.

14.6 HELICAL SPRINGS

Helical springs are used to take up forces which tend to shorten, lengthen, or twist them. However, most springs work in compression. The advantages of compression springs are that they are cheaper to make and that they continue to function to a certain extent even if

a coil breaks. In Fig. 14.8 two compression springs are shown that differ in the type of coiling and in the method of finishing the ends. The best end finish, which gives a uniform compression without additional distortion, is by bending the ends, closing them, and grinding them square to the axis (Fig. 14.8b). In such a spring practically a whole coil on each end is inactive, and this fact must be taken into consideration in figuring the spring deflection. Helical springs are usually made of round wire.

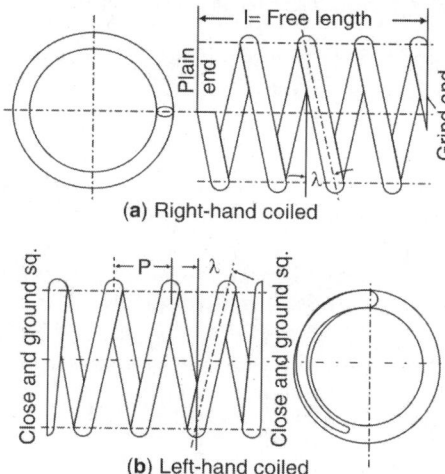

Fig. 14.8: Compression springs

14.7 CYLINDRICAL COMPRESSION SPRINGS

Compression springs may be made of round wire, or wire with a rectangular cross section. The first part of this discussion will deal with round-wire springs.

Stress analysis: A quadrant of a coil of a round-wire spring, Fig. 14.9, will be considered as a free body. The load F acting along the axis of the spring, which has a mean diameter, or pitch diameter, D, produces a torsional moment determined by the formula

$$T = \frac{1}{2}FD \quad \ldots (14.13)$$

The same torsional moment acts upon all sections of the loaded spring. The internal resisting moment, by equations 2.17 and 2.18, is

$$s_s Z = \tfrac{1}{16} s_s \pi d^3 \quad \ldots (14.14)$$

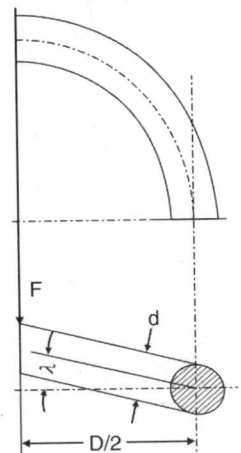

Fig. 14.9: A quadrant of a coil

where, d is the wire diameter, in mm. Equating the external moment to the internal one and solving for s_s results in

$$s_s = \frac{8FD}{\pi d^3} \qquad \ldots(14.15)$$

In addition to the shear stress s_s, there will be a compressive stress due to the component F' acting along the coil, and also a direct shear stress $s_{s2} = F/A$, where $A = \frac{\pi}{4}d^2$. Unless the helix angle or pitch angle λ is large, the compressive stress is negligible. The direct shear stress s_{s2} is usually neglected. However, if the spring diameter d is comparatively small, s_{s2} may become too great to be neglected.

By a theoretical analysis, which has been confirmed by actual tests, a more accurate formula for the stress has been established.[7] This formula is

$$S_s = \frac{8FDk}{\pi d_3} \qquad \ldots(14.16)$$

where the stress factor k is a function of the diameter ratio $c = D/d$, which is called the spring index. For closely coiled springs, the Wahl factor k is

$$k = \frac{4c-1}{4c-4} + \frac{0.615}{c} \qquad \ldots(14.17)$$

For the sake of convenience a curve for finding k is given in Fig. 14.10. *This curve shows that the stress factor increases very rapidly as the spring index c decreases due to the fact that inner fibres have a smaller length as compared to outer fibres (14.11).* It is advisable to have c at least 3.

Deflection: The angular deflection θ, found by using the general equation 2.19 and substituting for T_s its value from equation 14.15, is

$$\theta = \frac{16FDl}{\pi d^4 G} \qquad \ldots(14.18)$$

The length l of the bar is approximately equal to πDi, where i is the number of active coils, which is greater by one-half coil than the number of free coils.[8] Since the axial deflection y of the whole spring is equal to the angular deflection times the mean radius of the coil.

$$y = \frac{8FD^3 i}{d^4 G} \qquad \ldots(14.19)$$

Values of y calculated by equation 14.19 check well with those measured in actual springs.

Substituting for F its expression from equation 14.16 gives

$$y = \frac{\pi i s_s D^2}{k d G} \qquad \ldots(14.20)$$

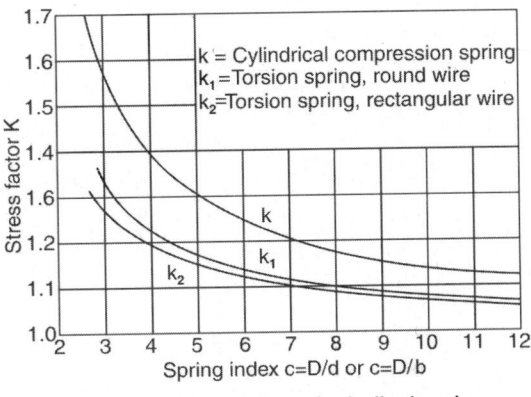

Fig. 14.10: Stress factors for helical springs

[7] A. M. Wahl, "Stresses in Heavily Coiled Helical Springs," *Trans. ASME*, Vol. 51 (1929), APM–51–17, p. 186, and Vol. 52 (1930), APM–52–18, p. 217; A. M. Wahl, "Helical Compression and Tension Springs," *Journal of Applied Mechanics*, Vol. 2 (March, 1935), PA-35.

[8] R. F. Fogt, "Number of Active Coils in Helical Springs," *Trans. ASME*, Vol. 56 (1934), RP–56–4, pp. 469–72.

This equation allows finding the maximum deflection permissible for a selected stress limit. Solving equation 14.20 for s_s gives

$$s_s = \frac{kydG}{\pi i D^2} \quad \dots (14.21)$$

This equation is convenient for checking the stress in a spring in operation.

Spring scale: A useful characteristic of a spring is the force F_o necessary to compress it by a unit length. This force, which is termed the *spring scale*, or stiffness is

$$F_o = \frac{F}{y} \quad \dots (14.22)$$

Combining equations 14.22 and 14.19 gives

$$F_o = \frac{d^4G}{8iD^3} \quad \dots (14.23)$$

From equation 14.22, the deflection is

$$y = \frac{F}{F_o} \quad \dots (14.24)$$

By equation 14.19 the deflection is directly proportional to the load. If the force F_1 exerted by a spring under certain conditions is given, and the force F_2 exerted with an additional compression y' of the spring is also given, then

$$F_o = \frac{F_2 - F_1}{y'} \quad \dots (14.25)$$

Conversely, the total deflection y_2, by equation 14.24, is

$$y_2 = \frac{F_2}{F_o} = \frac{y'F_2}{F_2 - F_1} \quad \dots (14.26)$$

This relation is illustrated by Fig. 14.11.

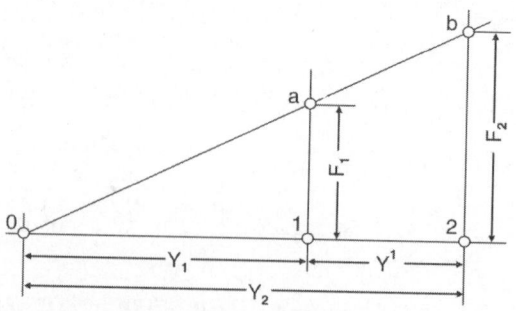

Fig. 14.11: Relation between loads and deflection in a helical springs

Resilience: The resilience U of a spring is equal to the energy absorbed. By the general equation 3.23, in which the value of y is given by equation 14.19,

$$U = \frac{Fy}{2} = \frac{4F^2D^3i}{d^4G} \quad \dots (14.27)$$

Substituting for F its value from equation 14.16 gives

$$U = \frac{\pi d^2 Dis_s^2}{16k^2G} \quad \dots (14.28)$$

Since the volume V of the spring is $\pi Di \times \frac{1}{4}\pi d^2$,

$$U = \frac{Vs_s^2}{4k^2G} \quad \dots (14.29)$$

This equation could be written directly by taking the expression for resilience from Table 3.3 and substituting the corrected stress s_s/k for the elastic limit S_s. Equation 14.29 shows that, in order to have a greater resilience and a better utilization of the material, k should be low. In turn, the spring index c should be large.

Finally, another equation for U may be obtained by substituting in equation 14.28 the value of s_s from equation 14.21. The result is

$$U = \frac{y^2 d^4 G}{16iD^3} \quad \dots (14.30)$$

This equation is very convenient for checking the resilience of a given spring directly from its dimensions and deflection.

Rectangular-section springs: The equation for the stress produced in the spring shown in Fig. 14.12 can be derived by proceeding as just explained for equation 14.15. Taking into consideration the additional direct shear stress

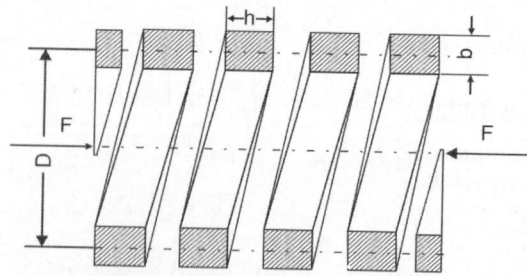

Fig. 14.12: Spring with rectangular section

and the influence of the curvature of the coils, the resultant equation is

$$s'_s = \frac{kFD(1.5h + 0.9b)}{b^2 h^2} \qquad \ldots (14.31)$$

where, k is determined by equation 14.17 or Fig. 14.10 with $c = D/b$.

The deflection may be found from the equation

$$y = \frac{2.83iFD^3(b^2 + h^2)}{b^3 h^3 G} \qquad \ldots (14.32)$$

For design purposes it is convenient to let $b/h = m$. Then

$$s'_s = \frac{kFD(1.5 + 0.9m)}{m^2 h^3} \qquad \ldots (14.33)$$

and

$$y = \frac{2.83iFD^3(1 + m^2)}{m^3 h^4 G} \qquad \ldots (14.34)$$

For a spring made of square wire, $m = 1$. In this case

$$s'_s = \frac{2.4kFD}{h^3} \qquad \ldots (14.35)$$

and

$$y = \frac{5.66iFD^3}{h^3 G} \qquad \ldots (14.36)$$

As may be found by using equation 3.29, the resilience of a spring with a rectangular cross section, generally speaking, is smaller than that of a round-wire spring. However, it increases as the ratio b/h decreases.

The main advantage of a rectangular-section spring, as compared with a round-section spring, is the possibly of obtaining a stronger spring within given space limitations.

If any helical spring breaks into two halves, The stiffness of each half $K = \dfrac{F}{y/2} = \dfrac{2F}{y}$, *i.e. twice the stiffness of the whole spring (14.12).*

14.8 DESIGN DATA

The material most commonly used for cylindrical compression springs is carbon steel with a carbon content from 0.50 to 1.0 per cent. Alloy steels are used to increase the life of springs subjected to high stresses and repeated loads. Spring brass and phosphor bronze are used when corrosion must be prevented. Monel-metal is used for springs exposed to corrosion and also to high temperatures.

Safe stresses: Mechanical properties of all spring materials are given in Table 14.3 for 12.5 mm section. *As thinner sections are obtained by drawing and successive drawing operations, the strength of wire increases with each successive drawing operation due to strain hardening (14.9).* For other sections having a different diameter d, the values for elastic and endurance limits must be multiplied by a size factor e_{sz}. For steel e_{sz} can be determined by the relation

$$e_{sz} = 0.86 + \frac{1.78}{d} \qquad \ldots (14.37)$$

If the diameter of the wire is less than 2.2 mm, this value should be used for d. For rectangular sections the value of b or h, whichever is smaller, is used instead of d in equation 14.37.

For Monel-metal wire the influence of the diameter is much less and can be taken into account by applying the equation[9]

$$e_{sz} = 0.986 + \frac{0.11}{d} \qquad \ldots (14.38)$$

For spring in infrequent or intermittent service the calculations may be based on elastic limits in shear with a safety factor from 1.5 to 2, the value depending on the service. For continuous service and rapid regular applications of the load, spring manufacturers advise lowering of the stress by 20 to 50 per cent, which means using a safety factor from 1.8 to 2.5. However, in the case of repeated loading the correct method of designing the spring is to use the endurance diagram. With such a procedure a safety factor n of 1.5 is again sufficient.

Properly done, shot peening increases the endurance life of helical springs more than

[9] *Inco*, Vol. 11, No. 1 (1931), p. 21.3 (Published by the International Nickel Company).

200 per cent and permits the use of a nominal safety factor n of 1.25. Shot peening must be specified on the drawing (14.2).

14.9 DESIGN PROCEDURE

As in the design of leaf springs, the design of a coil spring involves a cut-and-try methods, and the result should be checked by actual testing of the spring. Furthermore, a slight variation in the spring scale found by equation 14.22 must be expected in springs that seem to have identical dimensions.

The data usually given are the load on the spring, or the force which it must exert, and the desired force increase with a certain deflection. Often certain space limitations are given, such as the minimum inside diameter or the maximum outside diameter and the length of the spring under working conditions.

The values to be found usually are: the pitch or mean coil diameter D, the size of the wire d, the number of coils i, and the free length l of the spring. From these basic values, the spring scale F_o can be found.

The usual design procedure is as follows. First the pitch diameter D is selected to conform to given space limitations or to other conditions. The material and safety factor n are selected next, and the approximate value for the safe stress S_d is then determined by taking c between 5 and 8. With these data the wire diameter d can be found from equation 14.15. Thus,

$$d = \sqrt[3]{\frac{8FDk}{\pi S_d}} \qquad \ldots (14.39)$$

The standard wire diameters are given in Table 14.2.

Based on this preliminary value of d, the size factor e_{sz} is found by equation 14.37 or equation 14.38. This value can be used to obtain a more accurate value for S_d. Also, the stress factor k is determined from Fig. 14.10 or by equation 14.17. Now the more accurate value of d is found from equation 14.16. The resulting relation is

$$d = \sqrt[3]{\frac{8FDk}{\pi e_{sz} S_d}} \qquad \ldots (14.40)$$

If there are no space limitations, it is convenient to set $D = cd$ and to select a suitable value for c which will determine the value of k. In this case equation 14.16, when solved for d, gives

$$d = \sqrt[3]{\frac{8Fck}{\pi S_d}} \qquad \ldots (14.41)$$

Final dimension: The necessary minimum diameter d having been determined, the actual d is taken as the nearest size from Table 14.2. A larger size decreases the working stress and gives a slightly stiffer spring, while a smaller size increases the stress and gives a softer spring.

With the final value d, space limitation can be checked either $D + d$ if the limitation is outer diameter and $D - d$ if the limitation is inside diameter.

High-grade steel wire, particularly the kind used for internal-combustion engine valve springs, may be obtained from 3 to 6 mm in diameter. The exact sizes should be obtained from the wire manufactures.

With d determined and the total deflection y given, the number of active coils is found from equation 14.19. Thus,

$$i = \frac{yd^4G}{8FD^3} \qquad \ldots (14.42)$$

If y is not known, it must be determined by equations 14.24 or 14.26 as the case may be.

If i becomes too small, the spring is too soft. According to equation 14.42, the pitch diameter D should be decreased. This change will also slightly decrease d, and the calculations must be repeated. If i is too great, D must be increased.

When i is finally determined, the minimum free length l_o of the spring can be found. If the ends are bent before grinding, there will be 2 additional end turns

$$l_o \geq (i + 2)d + y \qquad \ldots (14.43)$$

To this a certain length, a mm, should be added to prevent the spring from being compressed solid with the application of the maximum load to avoid damage due to repeated contact called clash allowance. This is approximately 10 per cent of wire diameter, i.e.

$$l_o = (i + 2)\,d + y + 0.1\,(i + 2)d \quad \ldots(14.43a)$$

Buckling: A helical compression spring having a great length in proportion to its pitch diameter D may buckle at a comparatively low load. Such a spring is a very flexible column, and the critical axial load F_{cr} that can cause buckling may be found by the formula[10]

$$F_{cr} = F_o K_l l_o \quad \ldots(14.44)$$

where, F_o is the spring scale, in kN/m

K_l is a factor depending on the ratio l_o/D, as shown in Fig. 14.13;

l_o is the free length of the spring, m

In using Fig. 14.13, plain ends as in Fig. 14.8a may be considered hinged ends, and closed and ground square ends as in Fig. 14.8b may be considered built-in ends.

However, buckling can be prevented by enclosing it in a housing or inserting a cylindrical rod or pipe of length slightly less than solid length (14.10).

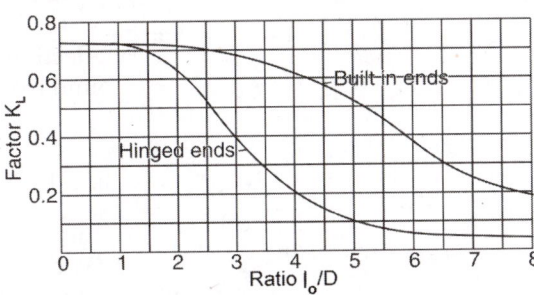

Fig. 14.13: Buckling factor for helical compression springs

Example 14.3: Design a valve spring for an automobile motor. When closed, the spring should produce a force of about 45 N, the spring must fit over the valve bushing, which has an outside diameter of 20 mm, and must go inside a space 36 mm in diameter; the valve lift is 6 mm. In designing the spring, take into account the repeated stresses only indirectly, by the usual simplified method.

Si-Mn steel may be selected as suitable material. As a small size wire and hence larger e_{sz} must be anticipated, a preliminary value of $S_d = 0.50\,S_e$ may

be assumed. Since S_e is 690 N/mm², $S_d = 0.5 \times 690 = 345$ N/mm².

The load with the valve open may be assumed as

$$F_2 = 1.2\,F_1 = 1.2 \times 45 = 54\ \text{N}$$

The mean diameter can be assumed as a mean value between two space limitations. Thus

$$D = \frac{1}{2}(20 + 36) = 28\ \text{mm}$$

By equation 14.39

$$d = \left[\frac{8 \times 54 \times 28}{\pi \times 345}\right]^{\frac{1}{3}} = 2.235\ \text{mm}$$

By equation 14.37

$$e_{sz} = 0.86 + \frac{1.78}{2.235} = 1.65$$

Spring index $c = \dfrac{28}{2.235} = 12.5$ and by equation 14.7 the stress factors

$$k = \frac{4 \times 12.5 - 1}{4 \times 12.5 - 4} + \frac{0.615}{12.5} = 1.1144$$

The safety factor for continuous service and rapid stress change should be taken as $n = 2 \times 1.5 = 3$. Then

$$S_d = \frac{690}{3} = 230\ \text{N/mm}^2, \text{by equation 14.40}$$

$$d = \left[\frac{8 \times 1.1144 \times 54 \times 28}{\pi \times 1.65 \times 230}\right]^{\frac{1}{3}} = 2.244$$

The nearest size in Table 14.2 is $d = 2.5$ mm. The total deflection by equation 14.26 is

$$y_2 = \frac{6 \times 54}{54 - 45} = 36\ \text{mm}$$

With these data the number of active coils, by equation 14.42 is

$$i = \frac{36 \times 2.5^4 \times 0.7845 \times 10^5}{8 \times 54 \times 28^3} = 11.6$$

If the number of coils seems to be too great, a slightly large mean coil diameter D should be selected without ignoring the space limitations, i.e. D may be 31 and $i = 8.5$.

The free length by equation 14.43 is

$$l_o = (8.5 + 2) \times 2.5 + 36 + 0.1 \times 10.5 \times 2.5$$

$$= 64.875\ \text{say } 65\ \text{mm}$$

i.e. 65 mm free length fully compressed to $65 - 36 = 29$ mm would be used.

---

[10] A. M. Wahl, "When Helical Springs Buckle," *Machine Design*, Vol. 15 (May, 1943).

Impact loading: Depending on the limitations imposed on the design, the dimensions of a spring needed to sustain impact loading are based on either equation 14.28 or equation 14.30. In either case it is best to select a value of $c = D/d$ and a number of coils i, and to determine the wire diameter by equating the spring resilience to the impact energy that the spring must absorb. If only the strength of the spring must be considered, equation 14.28 should be used. If a certain deflection y is prescribed, equation 14.30 should be used. However, in the latter case, after the wire diameter d is found, the stress must be checked by equation 14.21 or equation 14.28. If this stress exceeds the permissible design value S_d, the spring resilience must be increased by increasing either the number of coils i or the pitch diameter D, or by increasing both. It should be noted that in this case equations 14.16 and 14.19 cannot be used unless a fictitious force F is found which will produce the same deflection and stress as when the impact energy is absorbed by the spring.

Example 14.4: Design the springs for a mechanical sieve. The weight of the dropping frame and the material being sifted is 900 N; the height of drop is 12 mm; no stops except the springs themselves are used; eight springs are taking the full impact, and each time the frame is lifted by the mechanism they open up to their free length; the load may be assumed as distributed evenly among all eight springs.

The energy which each spring must absorb is

$$K_i = \frac{900 \times (12 + y)}{8}$$

where, y is the deflection of the spring. From preliminary calculations it may be assumed that $y = 48$ mm, and the estimated energy will be

$$K_i = \frac{900 \times 60}{8} = 6750 \text{ Nmm}$$

Selected $c = \dfrac{D}{d} = 10$. Then $D = 10d$, and from Fig. 14.10, $k = 1.14$. Next select the material. Choosing Si–Mn Steel, the endurance limit in torsion may be $0.5 \times 690 = 345$ N/mm². Also take the safety factor n as 1.7 because repeated stress fluctuation is not accompanied by a reversal of the stress. If a wire

diameter d of 6 mm is assumed, the size factor found from equation 14.37 is

$$e_{sz} = 0.86 + \frac{1.78}{6} = 1.15$$

The design stress is then

$$S_d = \frac{345 \times 1.15}{1.17} = 233 \text{ N/mm}^2$$

Try using $i = 10$ coils. Substituting these values in equation 14.28 and solving it for d, result in

$$d = \left[\frac{6750 \times 16 \times 1.14^2 \times 0.7845 \times 10^5}{\pi^2 \times 10 \times 10 \times 233^2} \right]^{1/3} = 5.9 \text{ mm}$$

The nearest standard size is $d = 6$ mm. Therefore in accordance with the preceding assumption

$$D = 10 \times 6 = 60 \text{ mm}$$

The increase of d from 5.9 to 6.0 mm will lower the stress to about 222 N/mm². The corresponding actual deflection y found by equation 14.20 is

$$y = \frac{\pi \times 10 \times 222 \times 60^2}{1.14 \times 6 \times 0.7845 \times 10^5} = 46.8 \text{ mm}$$

Therefore, the energy to be absorbed will be a little smaller than the value assumed.

$$K_i = \frac{900(12 + 46.8)}{8} = 6615 \text{ Nmm}$$

The free length l_o of the springs is found by assuming the total number of coils as $10 + 2 = 12$ and allowing a clearance between the coils, i.e.

$$l_o = 12 \times 6 + 46.8 + 11 \times 0.1 \times 6$$
$$= 125.4 \text{ mm or } 126 \text{ mm}$$

The scale of the spring F_o can be found from equation 14.19. Thus,

$$F_o = \frac{6^4 \times 0.7845 \times 10^5}{8 \times 60^3 \times 10} = 5.88 \text{ N/mm}$$

The actual safety factor can be found by finding the energy

$$U = \frac{\pi^2 \times 6^2 \times 60 \times 10 \times (345 \times 1.14)^2}{16 \times 1.14^2 \times 0.7845 \times 10^5} = 20215 \text{ Nmm}$$

Then safety factor by equation 5.19 is

$$n = \sqrt{\frac{20215}{6615}} = 1.748$$

Repeated loading: In the case of repeated loading, encountered particularly in valve springs of internal-combustion engines or plunger return springs of oil engines, the calculations should be based on the endurance diagram, the stress amplitude S_a, from equation 5.38, and the mean stress S_m, from equation 5.39. The corresponding torque amplitude T_a, based on equation 14.13, is

$$T_a = \tfrac{1}{2}(T_1 - T_2) = \tfrac{1}{2}(F_1 - F_2) \times \tfrac{1}{2}D$$
$$= \tfrac{1}{4}(F_1 - F_2)D \qquad \ldots (14.45)$$

Substituting this value of T_a in equation 5.38, and k instead of K_r, gives

$$S_a = nk\frac{(F_1 - F_2)D}{4Z_o e_{sz} e'_{sr}} \qquad \ldots (14.46)$$

Similarly, the mean torque is

$$T_m = \tfrac{1}{2}(T_1 + T_2) = \tfrac{1}{4}(F_1 + F_2)D \quad \ldots (14.47)$$

and the mean stress, found by applying equation 5.39, is

$$S_m = nk\frac{(F_1 \times F_2)D}{4Z_o e_{sz}} \qquad \ldots (14.48)$$

For a round bar, $Z_o = \tfrac{1}{16}\pi d^3$. Therefore the necessary wire diameter d, determined from equation 14.46, is

$$d = \sqrt[3]{4nk\frac{(F_1 - F_2)D}{\pi e_{sz} e'_{sr} S_a}} \qquad \ldots (14.49)$$

In equations 14.46, 14.48, and 14.49 the stress factor k determined from equation 14.17 must be used for the stress-concentration factor K_r. The surface coefficient e'_{sr} should be determined by equation 5.23 and Fig. 5.7. Curve b will correspond to a usual clean surface, and curve e will apply for a spring with scale as is often found after quenching. The removal of this scale and the addition of a smooth finish allows an increase of the stress and therefore the use of a small diameter d.

Example 14.5: Design the springs of example 14.4, making a more accurate allowance for the danger of failure through repeated stresses.

Considering force $F_2 = 0$ and $F_1 = F_o y = 5.88 \times 46.8 = 275$ N, the stress amplitude is found to be $S_a = 268$ MN/m². The various factors in equation 14.49 can at first be assumed to be based on the value of $d = 6.3$ mm found by static method. Thus, with the spring index $C = 10$, the stress factor from Fig. 14.10

is $k = 1.14$. Since the size factor remains the same, $e_{sz} = 1.15$. The surface factor, taken from curve b in Fig. 5.8 is $e_{sr} = 0.88$. By equation 5.23

$$e'_{sr} = 0.425 + 0.57 \times 0.88 = 0.931$$

A safety factor of 2.5 is sufficient. With these data, equation 14.49 gives

$$d = \left[\frac{4 \times 2.50 \times 1.14 \times 275 \times 63}{\pi \times 1.15 \times 0.931 \times 268}\right]^{1/3} = 6.02$$

This corresponds to the wire size of 6.0 mm. Thus the accurate method gave the same size of wire but a greater assurance that the spring would not break in operation.

Rectangular section: In designing a spring with a rectangular cross section, the procedure remains the same but the ratio m must be assumed. The smaller the value of m, the stiffer the spring will be according to equation 14.34. Usually m is made from 1 down to about 0.1, although m is occasionally made greater than 1.

14.10 CONCENTRIC SPRINGS

Concentric springs are used for one of the following three purposes:

(a) To obtain a greater spring force in a given space.

(b) To insure the operation of a mechanism in the event that one spring breaks.

(c) To obtain a spring force which does not increase in a direct relation to the deflection, but increases faster.

In cases a and b two or more concentric springs have the same free lengths and are compressed equally. In case c the springs are made of different lengths, as l_1 and l_2 in Fig. 14.14, the shorter

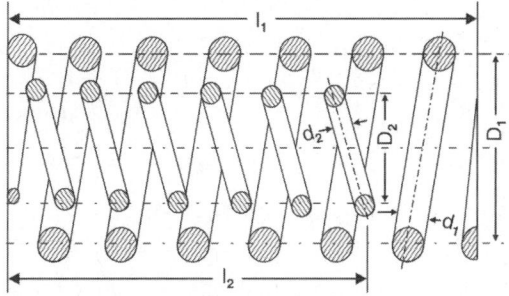

Fig. 14.14: Concentric springs

spring begins to act only after the longer spring is compressed a certain amount.

To prevent any tendency to bind, the adjacent coils are wound in opposite directions. In the design of concentric springs, if the same material is used, the springs should be designed with approximately the same safety factor. In order to have approximately the same stress factor k, the same spring index c is desirable. This requirement in turn calls for a smaller wire diameter d with the decrease of the pitch diameter D. Because the size factors are different, the design stresses will be different.

The approximate relation between the sizes of two concentric springs wound from round wire of the same material can be obtained by solving equation 14.40 for S_d, substituting for e_{sz} a simpler though less accurate expression $e_{sz} = (0.5/d)^{0.15}$, and canceling k. This gives the relation.

$$\left(\frac{d_1}{d_2}\right)^{2.85} = \frac{F_1 D_1}{F_2 D_2} \qquad \dots (14.50)$$

The diameters D_1 and D_2 are selected from geometrical considerations. Also, F_2 is expressed in terms of F_1 by the relation $F_2 = mF_1$, where $m \leq 1$. This gives $F_1 = F/(1 + m)$, where F is the total maximum spring load. The wire diameter d_1 of the outer spring is found in the usual way, and the wire diameter d_2 is then found by equation 14.50.

Variable force–deflection ratio: Where the force–deflection ratio is variable, the springs have different free lengths and different deflections. In Fig. 14.15 is shown a case involving three springs. Their free lengths are l_1, l_2, and l_3; their maximum deflections are y_1, y_2, and y_3

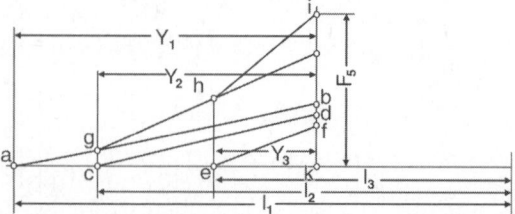

Fig. 14.15: Multiple spring with a variable force–deflection ratio

respectively; the inclined lines *a–b*, *c–d*, and *e–f* represent the forces of the springs; and the curve *aghi* represents the gradual increase of the spring force of the combination up to $F_s = i–k$.

Such multiple springs are used in governors of variable-speed engines to take care of the variable centrifugal force, which is proportional to the square of the speed.

14.11 VIBRATION OF CYLINDRICAL SPRINGS

If the natural frequency of a spring is a small multiple of the frequency of the outside force which acts upon it, the spring will have a tendency to vibrate with its own natural frequency. Under these conditions the middle coils will begin to surge back and forth. This surging changes the spring force and interferes with the proper operation of the spring. Furthermore, surging of the coils increases the maximum deflection between adjoining coils and therefore raises the stress above that figured by equation 14.21. This excessive stress may cause a failure of the spring through progressive fracture. The farther the natural frequency of the spring is above the frequency of the external force, the smaller is the danger of surge interference. Therefore, it is desirable to use springs with a high natural frequency.

The natural frequency of the springs may be found from equation 5.43. In determining the values of F and W, two cases must be considered: first, when one end of a spring is at rest; and second, when both ends of the spring are fixed. In the first case, the amplitude of the coils moving back and forth gradually decreases toward the coil which is at rest. The mass in motion may be considered as one-half the whole spring mass, or $W/2g$. Therefore, from equation 5.43, the frequency is

$$f = \frac{1}{T} = \frac{1}{2\pi}\sqrt{\frac{2F_o g}{W}} \qquad \dots (14.51)$$

where, F_o is the scale of the spring, in Newton per mm.

With both ends fixed, the maximum vibration or surge will take place at the middle coil. The force required to move the middle coil the same distance y will be four times as large as in the former case, because the coils on one side must be compressed while those on the other side are stretched and because the same deflection y is produced in one-half the coils. The active weight is again about one-half the spring weight. Thus,

$$f = \frac{1}{\pi}\sqrt{\frac{2F_o g}{W}} \qquad ...(14.52)$$

Some authors[11] recommend formulas which give values about 10 per cent higher.

Example 14.6: Determine the dangerous engine speed for a valve spring of a four stroke engine. The outside diameter of the spring is 76 mm, the wire diameter is 6 mm, and the spring has 8 coils, with the valve closed the spring is compressed 64 mm and has a force of 650 N.

The scale of a spring is

$$F_o = 650/64 = 10.15 \text{ N/mm}$$

The weight of the spring is

$$W = \frac{\pi}{4} \times \frac{6^2 \times \pi(76-6) \times 8 \times 78600}{10^9} = 3.82 \text{ N}$$

By equation 14.51, the frequency is

$$f = \frac{1}{2\pi}\sqrt{\frac{2 \times 10.15 \times 9.81 \times 1000}{3.82}}$$

= 36.33 vibr per sec or 2180 vibr per min.

Since the cam shaft runs at half of the speed of the engine, the most dangerous engine speed would be $2180 \times 2 = 4360$ rpm. Other dangerous speeds would be $4360/2 = 2180$ rpm, $4360/3 = 1453$ rpm and $4360/4 = 1090$ rpm. Lower engine speeds are not likely to set up surge action, because of the damping effect of the spring material between the impulses. The surge frequency of the middle coil will be twice as high, or still less dangerous.

Surge elimination: The computation of the stress caused by surging is rather involved and does not give reliable results because of the uncertainty of the damping effect of the spring material. Insofar as the design is concerned, the best policy is to have the natural frequency considerably above the number of impulses of the disturbing force. To find the frequency, substitute in equation 14.51 the value of F_o given by equation 14.22, and replace W by $\frac{1}{4}\pi^2 d^2 Diw$, where i is the number of active coils, and w is the specific weight of the spring material, in Newtons per cubic metre. Then, reducing the terms gives

$$f = \frac{d}{iD^2}\sqrt{\frac{G}{w}} \qquad ...(14.53)$$

Examination of equation 14.53 in connection with equation 14.21 shows that it is not easy to raise f without increasing the stress. Tests show that the spring deflection due to surge depends on the initial spring load. This load permits reduction of the ill effects of surging but requires a careful study of the spring behavior under operating conditions.[12] Means for adjusting the initial load by changing the working length of a spring should be provided in the design if the theoretical analysis indicates a possibility of spring surge.

14.12 EXTENSION SPRINGS

The equations for compression springs apply as well to extension springs. A compression spring is usually wound with the coils touching, as in Fig. 14.16, and often with some initial tension. Extensions springs are used in machinery considerably less than compression springs. The main reasons for this are:

(a) They are more expensive to manufacture.
(b) They require a more elaborate fastening of the ends.

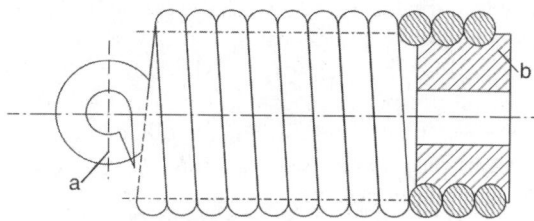

Fig. 14.16: Tension springs

[11] C. H. Kent, "Don't Overlook Surge in Designing Spring," *Machine Design*, Vol. 7 (October, 1935), p. 38; A. Swan, "Valve Springs," *Automobile Engineer*, Vol. 16 (1926), pp. 218, 290.
[12] Ricardo, *op. cit.*, p. 233

(c) They are more likely to be stressed beyond the elastic limit.

(d) In case of failure, the kinematic constraint is lost.

(e) *For forming hooks at the ends, there will be bends sometimes sharp producing stress concentration responsible for reducing the life of springs as compared to compression springs (14.5).*

In Fig. 14.16 are shown two types of spring ends—a hook or eye *a* bent from the spring wire itself, and a cast-iron or steel plug *b* which is screwed into the spring and has a tapped hole for a threaded connecting detail.

To overcome the last two disadvantages listed, the arrangement shown in Fig. 14.17 may be used instead of an extension spring.

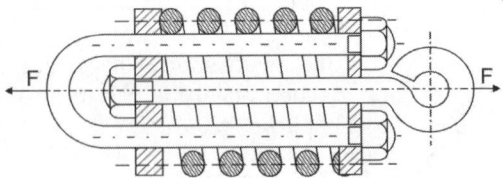

Fig. 14.17: Compression spring for tension purposes

14.13 CONICAL SPRINGS

Conical helical springs usually work in compression and are made of round or rectangular stock. A conical spring is used either where space limitations prohibit the use of a cylindrical spring or where a single spring with a variable stiffness is desired.

From Fig. 14.18 and equation 14.16 it can be seen that the largest coil with the diameter D_2 is subject to the highest stress, and that the stress gradually decreases toward the end with the smallest diameter D_1. Naturally, in designing the spring the highest stress must be taken into account. For a round stock, this stress will be found by equation 14.16. For a square stock, it will be found by equations 14.31 or 14.33, with the substitution of D_2 for D.

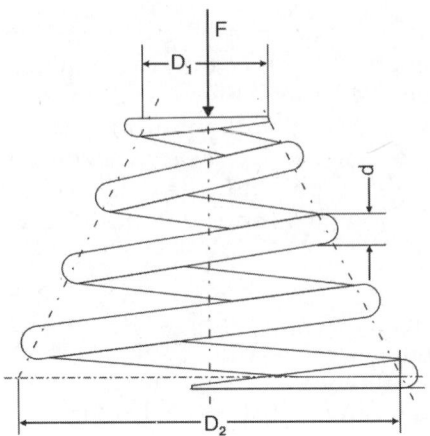

Fig. 14.18: Conical spring

If the spring is so designed that with an increase of the load the coils having the large diameters will gradually compress solidly, the smallest coil must be designed for the maximum load.

The axial deflection y for i coils of round stock may be computed by the relation[13]

$$y = \frac{2iF(D_2^3 + D_2^2 D_1 + D_2 D_1^2 + D_1^3)}{d^4 G} \quad \ldots (14.54)$$

By substituting for F its value from equation 14.16, the deflection can be expressed in terms of the maximum stress. Thus

$$y = \frac{\pi i S_s (D_2^3 + D_2^2 D_1 + D_2 D_1^2 + D_1^3)}{4 d D_2 k G} \quad \ldots (14.55)$$

The axial deflection of a conical spring made of rectangular stock with a radial thickness b and an axial dimension h, as in Fig. 14.12, is

$$y = \frac{0.71 i F(b^2 + h^2)(D_2^3 + D_2^2 D_1 + D_2 D_1^2 + D_1^3)}{b^3 h^3 G}$$

$$\ldots (14.56)$$

14.14 TORSION SPRINGS

The torsional moment T on a spring produces a bending stress in the wire, the bending moment on the wire being numerically equal to T. In addition to this stress, there is a direct tensile

---

[13] O. A. Leutwiler, *Elements of Machine Design* (New York: McGraw-Hill Book Company, Inc., 1917), p. 141.

or compressive stress due to the force F that is tangential to the coil. Therefore, the maximum stress is

$$s = \frac{T}{Z} + \frac{F}{A} \qquad \ldots(14.57)$$

where, Z is the section modulus of the wire and $F = 2T/D$. By reason of the curvature of the coil the actual bending stress will be greater than that given by equation 14.57. The correction can be made as described for compression springs by introducing a factor k', giving the equation

$$s = \frac{k'T}{Z} + \frac{2T}{DA} \qquad \ldots(14.58)$$

The deflection, measured by the distance traveled by a point on the pitch diameter of the end coil to which the pull is applied, is approximately

$$y = \frac{TLD}{2EI} \qquad \ldots(14.59)$$

where, L is the length of the coil part of the spring and is equal to $i\pi D$, and I is the moment of inertia of the wire.

Round-wire spring: In a round-wire spring, Fig. 14.19, $Z = \frac{1}{32}\pi d^3$; $A = \frac{1}{4}\pi d^2$; and $k' = k_1$ is a function of the spring index $c = D/d$ and may be found from the curve k_1 in Fig. 14.10, therefore

$$s = \frac{8T(4k_1 D + d)}{\pi d^3 D} \qquad \ldots(14.60)$$

Rectangular-wire spring: For a rectangular section, as in Fig. 14.12, $Z = \frac{1}{6}hb^2$, $A = bh$, and $k' = k_2$. The value of k_2 can be found from Fig. 14.10, where $c = D/b$. The stress is then

$$s = \frac{6k_2 T}{b^2 h} + \frac{2T}{Dbh} \qquad \ldots(14.61)$$

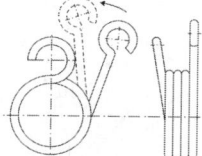

Fig. 14.19: Torsion spring

Additional data, such as resilience for the springs discussed and formulas for other special types of springs, can be found in handbooks.[14]

Example 14.7: Design a spring to operate the valve of a petrol engine having the following particulars:

Spring load when valve is open = 400 N
Spring load when valve is closed = 260 N
Length of the spring when the valve is open = 40 mm
Length of the spring when the valve is closed = 48 mm
Approximate inside diameter of the spring = 25 mm

Take the maximum allowable shear stress in the spring wire as 400 N/mm² and modulus of rigidity as 85000 N/mm². Find the free length of the spring.

The compression of the spring = 48 – 40 = 8 mm and is caused by a load = 400 – 260 = 140 N.

The maximum torque on the spring is due to a load of 400 N. If D is mean coil diameter, then taking $k = 1.15$, $D = 29$ slightly more than approximate diameter of 25 mm, the wire diameter is given by

$$d = \left[\frac{8FDK}{\pi S_s}\right]^{1/3} = \left[\frac{8 \times 400 \times 29 \times 1.15}{\pi \times 400}\right]^{1/3}$$

$$= 4.4 \text{ say } 4.5 \text{ mm}$$

Now $D = 25 + 4.5 = 29.5$ mm

and $\quad c = \dfrac{29.5}{4.5} = 6.65$

$\therefore \quad K = \dfrac{4c - 1}{4c - 4} + \dfrac{0.615}{c}$

$$= \frac{4 \times 6.65 - 1}{4 \times 6.65 - 4} + \frac{0.615}{6.65} = 1.234$$

By substituting $k = 1.234$ instead of 1.15, the value of d will still be 4.5 mm

Now number of turns, $i = \dfrac{yGd^4}{8FD^3}$

$\therefore \quad i = \dfrac{8 \times 85000 \times (4.5)^4}{8 \times 140 \times (29.5)^3} = 9.8$

Taking 2 end coils, total no. of coils = 2 + 9.8 = 11.8 turns say 12 turns

Maximum deflection under 400 N load

$$= 8 \times \frac{400}{140} = 22.8 \text{ mm}$$

keeping $\dfrac{d}{10} = 0.45$ mm say 0.5 mm as clearance between coils at maximum load

---

[14] Lionel S. Marks, ed., *Mechanical Engineers' Handbook*, 5th ed. (New York: McGraw-Hill Book Company, Inc., 1951) pp. 455 ff; R. T. Kent, *op. cit.*, pp. 10–21 ff.

Free length of the spring $= nd + \delta + \dfrac{d}{10} \times n$

$= 12 \times 4.5 + 22.8 + 0.5 \times 12$

$= 82.8$ mm

Pitch of coils $= \dfrac{82.8}{12} = 6.9$ mm

Example 14.8: Design a spring for four buffers of a railway wagon weighting 220 kN. It is moving at a speed of 1.5 km/hr and is to be brought to rest with the help of buffers. The springs are to be housed in a pipe of 250 mm internal diameter. Modulus of rigidity is 86000 N/mm². The allowable shear stress is 500 N/mm². The springs undergo deflection of 120 mm.

Velocity $v = 1.5$ km/hr $= \dfrac{1.5 \times 1000}{60 \times 60} = \dfrac{5}{12}$ m/s

The kinetic energy of the wagon $= \dfrac{Wv^2}{2g}$

$= \dfrac{22000 \times (5)^2}{2 \times 9.81 \times (12)^2} = 1946.7$ Nm

As there are four such springs. This energy is equivalent to work done by a force F deflecting the springs by 120 mm.

Therefore, $4 \times \frac{1}{2} F\delta = 1946.7 \times 1000$

$F = \dfrac{1946.7 \times 1000}{2 \times 120} = 8111$ N

Assuming a mean coil diameter of 250 mm and $k = 1.01$

$d = \left[\dfrac{8 \times 1.01 \times 8111 \times 200}{\pi \times 500}\right]^{\frac{1}{3}} = 20.28$ say 21 mm

The value of $k = 1.013$ o.k.
Outside diameter of coil $= 200 + 21 = 221$ mm o.k. as pipe has a diameter of 250 mm

Number of turns of the spring $i = \dfrac{yGd^4}{8FD^3}$

$= \dfrac{120 \times 86000 \times 22^4}{8 \times 8111 \times 200^3} = 3.86$ say 4

Assuming 2 end turns
Free lengths of springs $= nd + 0.1nd + y$
$= 6 \times 21 + 0.1 \times 6 \times 21 + 120 = 258.6$ say 259 mm
i.e. the pipe length is 259 mm to accommodate compression spring which is satisfactory.

Example 14.9: Design a spring balance to weigh 1000 N. The allowable shear stress for this service may be taken as 580 N/mm². Assume the scale length as 40 mm. The tube in which the spring is to be placed has internal diameter of 34 mm.

Assume mean coil diameter of 28 mm, and $k = 1.3$
Wire diameter

$d = \left[\dfrac{8 \times 1.3 \times 1000 \times 28}{\pi \times 580}\right]^{\frac{1}{3}} = 5.4$ say 5.6 mm

The standard gauge size is 5.6 mm.

The outside diameter of coil $= 28 + 5.6 = 33.6$ mm o.k. as it can be accommodated in 34 mm.

As the scale length is 40, this is the maximum deflection

Calculate number of turns

$i = \dfrac{40 \times 86000 \times 5.6^4}{8 \times 1000 \times 28^3} = 19.2$ say

Taking 2 end turns for a tensile spring, one hook will be outside for attaching to the load. Taking 22 turns.

Free length $= 22 \times 5.6 + 0.1 \times 22 \times 5.6 = 136$ mm and it can extend to $136 + 40 = 173$ mm.

Example 14.10: Design springs for a five passenger lift moving from the first floor to the ground floor considering failure of all controls and cage has a free fall of 3.5 m. Take the average weight of passenger as 700 N and the empty cage weighs 1.5 times the weight of passengers. Sixteen spring are placed on the ground pit so that the top surface of springs are in level with the ground floor. The allowable shear stress in the springs is 500 N/mm² and the maximum deflection of springs is 100 mm.

Take the spring coil diameter $D = 200$ mm
Weight of 5 passengers $= 5 \times 700 = 3500$ N
Weight of empty cage $= 1.5 \times 3500 = 5250$ N
Total weight $= 3500 + 5250 = 8750$ N
Velocity v of loaded cage after free fall is
$$v^2 = 2gh = 2 \times 9.81 \times 3.5$$
$$= 68.67$$

The kinetic energy of cage is destroyed by the deflection of springs, therefore the equivalent force F is

$$16 \times \frac{1}{2} F \times y = \dfrac{Wv^2}{2g}$$

$$F = \dfrac{8750 \times 68.67 \times 2}{16 \times 2 \times 9.81 \times 0.1} = 38281.25 \text{ N}$$

Diameter of spring wire is calculated by taking $k = 1.1$

$$d = \left[\dfrac{8 \times 1.1 \times 38281.25 \times 200}{\pi \times 500}\right]^{\frac{1}{3}} = 35 \text{ mm}$$

$$c = \frac{200}{35} = 5.71$$

Now
$$k = \frac{4 \times 5.71 - 1}{4 \times 5.71 - 4} + \frac{0.615}{5.7} = 1.26$$

As $k = 1.26$ will slightly change the wire diameter, the wire diamter of 35 mm is retained.

The number of turns

$$i = \frac{100 \times 84000 \times 35^4}{8 \times 38281.25 \times 200^4} = 5.14$$

say 6 turns. Consider two end turns for square and ground ends, total turns are 8.

Free length of springs = $35 \times 8 + 100 + 0.1 \times 35 \times 8$ = 408 mm

The pit will be 408 mm deep from the ground level.

PROBLEMS

14.1 Name four uses of springs giving example.

14.2 Name the processes for increasing the fatigue strength of springs.

14.3 Sketch an arrangement for the use of compression springs for tensile loads.

14.4 Which spring will have more fatigue life, (a) tensile spring, (b) compression spring?

14.5 What is the reason for a shorter life of a tensile spring?

14.6 If two spring have stiffness k_1 and k_2, what will be the combined stiffness when used in series and in parallel?

$$\left[\frac{k_1 k_2}{k_1 + k_2}, k_1 + k_2 \right]$$

14.7 In which situations disk springs are preferable?

14.8 Give an example of a confined quantity of gas performing the function of a spring.

14.9 Why is the strength of thinner wire greater than that of thicker wires of the same material?

14.10 How can a long compression spring be prevented from buckling?

14.11 What type of stress is taken care of by using Wahl's factor for spring?

14.12 What is the effect on the spring stiffness of the helical spring length is made half?

14.13 Why are laminated springs given an initial curvature?

14.14 In what way a solid block of the same material and shape be inferior to the one made up to laminations?

14.15 What is the purpose of 'Nip' in the leaf springs?

14.16 A quarter-elliptic (cantilever) laminated spring 680 mm long has, in addition to the master leaf, five graduated leaves, all made of BWG No. 4. SAE 9250 (60C9) steel, 45 mm wide. For a safety factor n of 2.25, find the maximum load, the deflection, and the energy absorbed. $S_e = 1035$ N/mm^2.

14.17 (a) Design semielliptic spring to carry a steady load of 4 kN with a deflection of 50 mm. The distance between the hinges is $2l = 760$ mm, (b) Find the necessary camber of the unloaded spring for the condition that if the load is applied suddenly the camber becomes 25 mm.

14.18 A semielliptic street-car spring has a length of 1.1 m and carries a load of 45 kN. It is made up to 18 leaves, 76 mm wide, two of which are full length. Use SAE 6150 (50 Cr 1 V23) steel and a safety factor n of 2.25. Determine (a) the necessary thickness of the leaves and (b) the deflection of the spring, $S_e = 1170$ N/mm^2.

14.19 The top of a shaking table is supported on four flat spring leaves S, Fig. P14.1, with the ends rigidly fixed both in the floor and to the table. The free length l is 660 mm and spring strips are of SAE 1050 (50C4) BWG No. 6 steel 40 mm wide. Find (a) the force F necessary to move the top 40 mm from its middle positions and (b) the maximum bending stress in the strips.

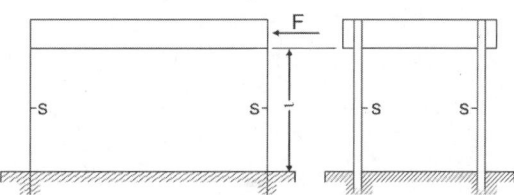

Fig. P14.1

14.20 A round-wire cylindrical compression spring has an outside diameter of 76 mm and is made of 12 mm steel wire. The spring supports an axial load of 5 kN. Determine (a) the maximum shear stress and (b) the total deflection, if it has 8 coils with closed and ground square ends and is made of SAE 9250 (60C9) steel. $S_e = 690 \text{ N/mm}^2$. ⋅

14.21 Determine the continuously repeated load, varying from the maximum value down to 25 per cent of it, which the spring of problem 14.20 can carry if made of SAE 9250 steel (a) with a safety factor of 3, when using the simplified method of calculation; (b) with the same safety factor of 3, when using the endurance limit method; and (c) with a safety factor of 1.5, when using the endurance-limit method.

14.22 Design a helical spring for the front-wheel suspension of an automobile, using silicon-manganese steel. The spring must have a scale of 35 N/mm, an inside diameter of 105 mm, and a free length of 356 mm with closed and ground ends. The spring must be able to carry a static load of 3.5 kN and when compressed solid it should not receive a permanent deformation. $S = 520 \text{ N/mm}^2$.

14.23 Design a cylindrical, round-wire spring for an overhead valve for a vertical four-stroke oil engine. When the valve is closed, the spring force must be 600 N. The valve lift is 20 mm, the valve weights 30 N, the weight of other parts which must be accelerated is 34 N and the engine runs at 300 rpm. The valve must close during 95 deg of the crank travel. The outside diameter of the spring must not be over 80 mm. Use the simplified method first, and then check the design by the endurance diagram.

14.24 A helical spring of round wire must support a load 900 N. The inside diameter must be not smaller than 50 mm. When the load is applied, the spring must compress 50 mm. Use SAE 1095 (98C6) steel wire and a safety factor n of 2. Find (a) the wire size, (b) the number of active coils, and (c) the maximum shear stress, taking into account also the direct stress. Elastic limit stress in shear = 270 N/mm².

15 Cylinders, Heads, Cover Plates and Pipe Joints

15.1 THIN-WALL CYLINDERS

Cylindrical vessels may be divided into two general classes: (a) those having thin walls, such as boiler shells, pressure tanks, and pipes; and (b) those having comparatively thick walls.

The analysis of stresses induced in the walls of a thin-wall cylinder by an internal pressure will be made by assuming that the stresses are distributed uniformly over the cross section of the cylinder and by neglecting the restraining action of the heads at the ends of the cylinder. In a cylinder having a diameter d, with the ends closed by heads, the internal pressure p against the heads produces an axial force $\frac{1}{4}\pi d^2 p$. If h is the thickness of the walls, the area of any circumferential cross section is πdh. Hence, the axial force on the ends causes in any circumferential section a longitudinal tensile stress s_1 found by the equation

$$s_1 = \frac{\frac{1}{4}\pi d^2 p}{\pi dh} = 0.25\frac{pd}{h} \qquad \ldots (15.1)$$

The magnitude of the hoop stress s_2 is found by taking a ring section of the cylinder 1 unit long. If it is imagined that this ring is cut in two by an axial plane, the force acting on each half ring is pd and it induces a reaction $2hs_2$. Equating these forces gives

$$s_2 = 0.5\frac{pd}{h} \qquad \ldots (15.2)$$

Thus the simple longitudinal stress is one-half the hoop stress. When the biaxial loading

is taken into account by equation 2.44, the resulting longitudinal stress becomes

$$s_1' = 0.25\frac{pd}{h} - 0.5\frac{\mu pd}{h} \qquad \ldots (15.3)$$

In the same way, the resulting hoop stress is found to be

$$s_2' = 0.5\frac{pd}{h} - 0.25\frac{\mu pd}{h} \qquad \ldots (15.4)$$

For steel, for which $\mu = 0.303$, the governing stress found by equation 15.4 is

$$s_2 = 0.424\frac{pd}{h} \qquad \ldots (15.5)$$

Since the difference is not very large and it is on the safe side to do so, the simpler equation 15.2 is commonly used in practice rather that equation 15.4. Therefore the necessary thickness may be determined by the equation

$$h = \frac{0.5pd}{s_2} \qquad \ldots (15.6)$$

15.2 THICK-WALL CYLINDERS

It the ratio of diameter to wall thickness, i.e. $\frac{d}{h} \geq 20$, the cylinder will be considered as thick walled. In a cylinder with thick walls, the actual hoop stress varies along the wall thickness. It has the highest magnitude at the inner surface of the cylinder and gradually decreases toward the outer surface, the distribution being similar to that of the stress in the outer ring of Fig. 13.1b. The magnitude of the stress depends on the cylinder-end condition; that is, on whether the cylinder is closed or open at the ends.

At the same time, the design procedure depends on whether the material is ductile or brittle. This characteristic indicates which theory of failure should be applied for the given case, as explained in section 5.2, and hence what equation must be used.

As the shape of engine and pump cylinders is complicated, it is preferably manufactured by casting from cast iron (15.2).

Cylinder with close ends: If the cylinder is closed at the ends, the hoop stress induced by the inner pressure governs the design. For a cylinder of ductile material having an inside diameter d_1 and an outside diameter d_2 and subjected to an inner pressure p, the governing stress may be computed by Clavarino's formula, which is based on the maximum-strain-energy theory. Thus,

$$s = \frac{p\left[(1-2\mu)d_1^2 + (1+\mu)d_2^2\right]}{d_2^2 - d_1^2} \qquad \dots (15.7)$$

Substituting the wall thickness h for $(d_2 - d_1)/2$ in equation 15.7 and solving the resulting equation for h, results in

$$h = \frac{d_1}{2}\left[\sqrt{\frac{s+(1-2\mu)p}{s-(1+\mu)p}} - 1\right] \qquad \dots (15.8)$$

where, Poisson's ratio μ may be taken from Table 2.1.

For a *brittle* material the governing stress should be based on the maximum-normal-stress theory. From equation 2.71 (Lami's equation),

$$s = \frac{p\left(d_2^2 + d_1^2\right)}{d_2^2 - d_1^2} \qquad \dots (15.9)$$

The necessary wall thickness h may then be computed from the Lami's equation

$$h = \frac{d_1}{2}\left(\sqrt{\frac{s+p}{s-p}} - 1\right) \qquad \dots (15.10)$$

Cylinder with open ends: In a cylinder with open ends the pressure seal is obtained by a separate piston. In a cylinder of this type the longitudinal stress is zero.

For a *ductile* material the stress may be determined in accordance with the distortion–

energy theory. The equation is

$$s = \frac{p\sqrt{d_1^4 + d_2^4/\mu}}{d_2^2 - d_1^2} \qquad \dots (15.11)$$

The necessary wall thickness h is then

$$h = \frac{d_1}{2}\left[\sqrt{\frac{s^2 + p\sqrt{4s^2 - p^2/\mu}}{s^2 - p^2/\mu}} - 1\right] \qquad \dots (15.12)$$

Equation 15.12 may be made more convenient for design purposes by dividing the numerator and the denominator under the radical sign by p^2 and designating the ratio s/p by a. Then

$$h = \frac{d_1}{2}\left[\sqrt{\frac{a^2 + \sqrt{4a^2 - 1/\mu}}{a^2 - 1/\mu}} - 1\right] \qquad \dots (15.13)$$

Other equations for calculating thickness are: Grashof's equation for thick cylinders

$$h = \frac{d_1}{2}\left[\sqrt{\frac{3s+2p}{3s-4p}} - 1\right] \qquad \dots (15.14)$$

Patterson's equation is

$$h = \frac{pd_1}{2(s-p)} \qquad \dots (15.15)$$

Birnie's equation for open cylinder

$$h = \frac{d_1}{2}\left[\sqrt{\frac{s+(1-\mu)p}{s-(1+\mu)p}} - 1\right] \qquad \dots (15.16)$$

From these equation, it is clear that Birnie's equation will give greater values of h. Birnie's equation can be used for open or closed cylinders.

For a *brittle* material the maximum-normal-stress theory should be applied again. In this case there is no difference between open ends and closed ends, and equations 15.9 and 15.10 should be used.

It should be noted that for a ductile material and the same values of d_1, p, and s, the thickness h found by equation 15.8 is smaller than the thickness found by equations 15.12 or 15.13.

In determining the necessary wall thickness, the value of the design stress S_d is used for s in all of the foregoing equations.

The endurance stress and pressure capacity of the cylinder can be increased by auto frettage and is commonly used in gun barrels. In this process the cylinder is subjected a large internal pressure so that, the inner surface is subjected plastic range and the external surface is still in the elastic range. When the pressure is released, the outer surface contracts and residual compressive stress is induced in the inner surface.

Relative thicknesses of walls: To determine whether the simpler formulas given for thin-wall cylinders may be used or the more accurate formulas given for thick-wall cylinders must be used, it is necessary to ascertain when the latter will give larger values. On the basis of a mean value 0.3 for μ, it will be found that for closed ends and ductile materials, when $p \leq S_d/6$, the thin-wall formula can be used; when $p \geq S_d/6$, the thick-wall formula must be used. For open ends the thick-wall formulas give larger values.

15.3 CIRCULAR HEADS, COVERS AND PLATES

The conditions of stress distribution in cylinder heads and covers are rather involved, and more or less empirical formulas must be used.

Dished head: The thickness of a dished head (Fig. 15.1) that is riveted or welded to a cylindrical shell, should be[1]

$$h = \frac{3pr}{S_e} \qquad \dots (15.17)$$

where, S_e is the elastic limit of the material from Table 4.2. The radius r should not be greater than the diameter d of the shell. If there is a manhole in the head, the coefficient should be multiplied by 1.2.

Flat heads: The minimum required thickness h of an unstayed flat head or cover plate should be determined by the relation[2]

$$h = d\sqrt{\frac{cp}{S_d}} \qquad \dots (15.18)$$

where, d is the diameter, or shortest span, as indicted in Fig. 15.2, in mm.
c is an empirical coefficient, given in Table 15.1;
p is the maximum inside pressure, in N/mm²
S_d is the allowable design stress in N/mm² be taken from Table 15.2.

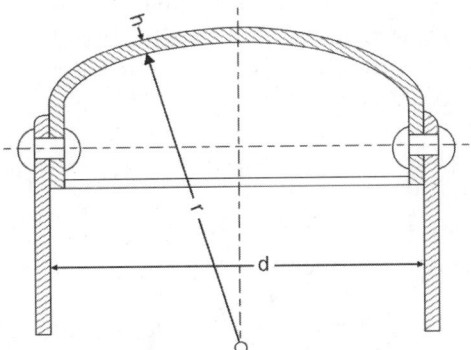

Fig. 15.1: Dished boiler head

Table 15.1: Coefficients for determining head thickness

Type of head in Fig. 15.2	Coefficient c	Remarks
A or A′	0.162	Plate rigidly riveted or bolted to the shell flange.
B	0.162	Integral flat head; $d \leq 60$ cm; $h \geq 0.05d$.
C	0.30	Flanged plate attached by a lap joint; $r \geq 3h$.
D or E	0.25	Plate butt-welded or forged integral; $r \geq 3h_f$.
F	0.50	Plate fusion-welded with fillet weld; throat h_1 1.25h_s.
G or H	0.30 + K	Bolts tend to dish the plate; K is found by the relation $K = 1.4Wh_G/Hd$, where, W = total bolt load in newton; H = total pressure on area bounded by the outside diameter of the gasket, in newton; and h_G and d are as shown in Fig. 15.2.

[1] *ASME Boiler and Pressure Vessel Code*, 1952 Edition, Section I, *Power Boilers* (New York: American Society of Mechanical Engineers, 1952).
[2] "Revisions and Addenda to the Boiler Construction Code," *Mechanical Engineering*, Vol. 56 (1934), p. 310.

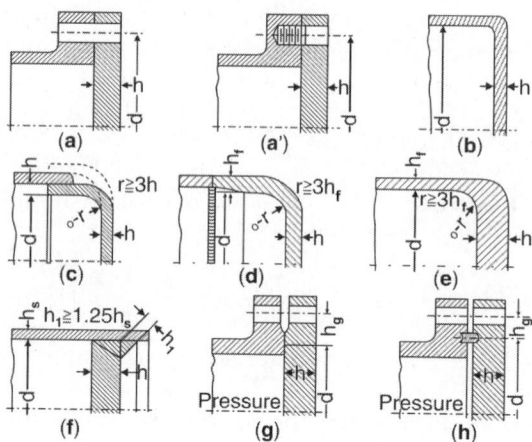

(a) **(a')** **(b)**
(c) **(d)** **(e)**
(f) **(g)** **(h)**

Fig. 15.2: Types of cylinder heads and covers

Plate uniformly loaded: The thickness h of a plate with a diameter d supported at the circumference and subjected to a pressure p distributed uniformly over the total area may be calculated by the equation

$$h = k_1 d \sqrt{\frac{p}{S_d}} \qquad \dots (15.19)$$

The maximum deflection in this case is given by the equation

$$y = k_2 d^4 \frac{p}{Eh^3} \qquad \dots (15.20)$$

The coefficients k_1 and k_2 depend on the method of holding the edges and the material of the plate and are given in Table 15.3.

Plate loaded centrally: The thickness h of a flat cast-iron plate supported freely at the circumference with a diameter d and subjected to a load F distributed uniformly over an outside area $\frac{1}{4}\pi d_o^2$ may be calculated by the equation

$$h = 1.2 \sqrt{\left(1 - \frac{0.67 d_o}{d}\right) \frac{F}{S_d}} \qquad \dots (15.21)$$

The deflection is given in this case by the equation

$$y = \frac{0.12 d^2 F}{Eh^3} \qquad \dots (15.22)$$

If the plate with the above given type of loading is fixed rigidly around the circumference, its thickness may be calculated by Grashof's formula, which is

$$h = 0.65 \sqrt{\frac{F}{S_d} \log_e \left(\frac{d}{d_o}\right)} \qquad \dots (15.23)$$

Table 15.2: Design stress for bolted flanged heads

Maximum temperature °C	Minimum of specified range of tensile strength of flange material at room temperature N/mm²					Alloy bolt steel N/mm²
	310	345	380	415	485	
370	74	82	90	98	115	98
398	65	72	80	87	101	87
426	56	62	68	75	87	75
454	47	52	57	62	73	62
482	37	42	46	50	58	50
510	28	32	34	37	44	37

Table 15.3: Coefficients in formulas for cover plates

Material of cover plate	Method of holding edges	Circular plate		Rectangular plate		Elliptical plate
		k_1	k_2	k_3	k_4	k_5
Cast iron	Supported, free	0.54	0.038	0.75	1.73	1.5
	Fixed	0.44	0.010	0.62	1.4; 1.6*	1.2
Mild steel	Supported, free	0.42	–	0.60	1.38	1.2
	Fixed	0.35	–	0.49	1.12; 1.28	0.9

* With gasket

The deflection for this case may be found by equation 15.22, with the numerical coefficient 0.12 changed to 0.055.

Example 15.1: Determine the thickness of a cover of nickel cast iron grade II for a cylinder having an inside diameter of 300 mm and carrying a pressure of 2 MN/m².

From Table 4.1, the elastic limit in tension of grade II nickel cast iron is 120 N/mm². If we first assume a safety factor of 2 and a size coefficient e_{sz} of 0.95, the allowable stress is

$$S_d = \frac{120 \times 0.95}{2} = 57 \text{ N/mm}^2$$

By equation 15.15, in which C is taken as 0.162 and diameter d is estimated to be 370 mm for inside diameter of 300 mm.

$$h = 370 \sqrt{\frac{0.162 \times 2}{57}} = 27.9 \text{ mm or say 28}$$

This is the thickness required at the edge, the centre thickness may be made 26 mm. The edge of the plate should be checked in bending. The moment per mm of circumference at the inner diameter d_o = 300 mm is

$$m = \frac{\frac{\pi}{4} \times d_o^2 \times p \times h}{\pi \times d} = \frac{\pi \times 300^2 \times 2 \times 28}{4 \times \pi \times 370} = 3405.4 \text{ Nmm}$$

The section modulus for a width of 1 mm is

$$Z = \frac{1}{6} \times 1 \times 28^2 = 130.7 \text{ mm}^3$$

$$S = \frac{3405.4}{130.7} = 26 \text{ N/mm}^2$$

This is below the allowable stress of 57 N/mm².

ASME design data: The Applied Mechanics Division of the American Society of Mechanical Engineers has worked out, for various types of loading of circular plates, simple formulas that are convenient for use and involve numerical factors which may be taken from curves.[3]

15.4 RECTANGULAR PLATES

In deriving a formula for the strength of a rectangular plate it is assumed that the critical section passes through the center of the plate. A square cast-iron plate fixed at the edges and subjected to a uniformly distributed load or a load concentrated at the center fails as shown in Fig. 15.3. It fractures first along the diagonal lines from c to d, e, f, and g and then breaks along the fixed edges. When loaded uniformly and clamped between rigid edges, a thin plate may fail by shearing off along the edges between d, e, f, and g. A square plate freely supported at the edges fails by breaking along the diagonal lines only.

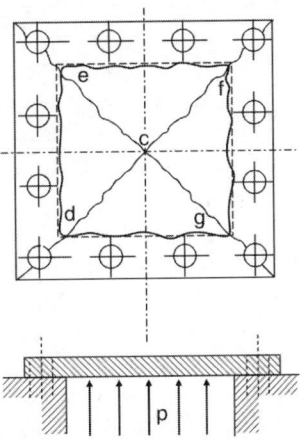

Fig. 15.3: Failure of a square plate

The probable manner of failure of a rectangular plate is shown in Fig. 15.4. A plate freely supported along the edges will fail by breaking along the center line c–c and the diagonals to d, e, f, and g. A plate held rigidly at the edges presents a greater resistance to failure than does a plate freely supported. This is evident if the plates are considered as beams, in which case the bending moments depend on the method of fastening, as shown in Table 2.6. This table also shows that in a plate with rigidly held edges the sections at the edges are stressed at least as heavily as if not more heavily than that at the center.

The existing formulas for strength of plates are based on certain assumptions which are not fully corroborated by tests. For design purposes these formulas are used with empirical

---

[3] A. M. Wahl and Stewart way, "Stress and Deflection of Circular Plates," *Journal of Applied Mechanics*, Vol. 3 (March, 1936), p. A-28.

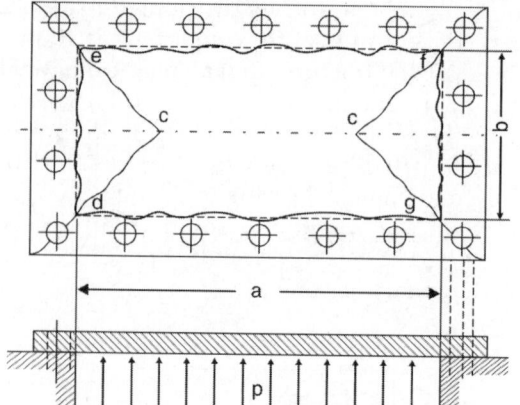

Fig. 15.4: Failure of a rectangular plate

coefficient and give only approximate results. Therefore sufficiently large safety factors must be used.

Uniform load: According to Grashof and Bach, the thickness h of a rectangular plate subjected to a pressure p, as in Fig. 15.4, can be found from the equation

$$h = abk_3\sqrt{\frac{p}{S_d(a^2+b^2)}} \qquad ...(15.24)$$

where, a is the length of the plate, in mm;

b is its breadth, in mm;

k_3 is a coefficient

Values of k_3 are given in Table 15.3. However, k_3 depends also on the conditions of the contact surface of the plate, the material of the gasket, and the initial force required to make a tight joint. The values in Table 15.3 are the low-limit values for the most favorable conditions.

Concentrated load: If a rectangular plate similar to that already discussed carries a load F at the intersection of the diagonals, the relation between the thickness h and the stress S_d is determined by the equation

$$h = k_4\sqrt{\frac{abF}{S_d(a^2+b^2)}} \qquad ...(15.25)$$

where, the values of k_4 are given in Table 15.3. If the edges are not fixed rigidly, but only held more or less securely, as with a gasket, k_4 should be taken as 1.6 for cast iron and 1.28 for steel.

ASME design data: Simple formulas that are convenient and involve numerical factors determined by curves have been worked out by the American Society of Mechanical Engineers.[4]

15.5 OTHER PLATES

The relation between the thickness h and the stress for a uniformly loaded elliptical plate can be computed by equation 15.24, in which the value of k_5 taken from Table 15.3 is substituted for k_3 and in which the dimensions a and b are the major and minor axes, respectively.

Data for computing the strength and deflection of plates of other shapes for various types of loading may be found in handbooks.

15.6 BOLTED JOINTS

Flanges of vessels under internal pressure are subjected to bending, as shown to an exaggerated degree in Fig. 15.5a. When a joint is closed by means of through bolts, the arm of the bending moment is comparatively large. The stress reaches its maximum at the corner where the flange joins the wall. Here the stress is still increased by a stress concentration, the magnitude of which can be estimated from Figs. 3.24 and 3.25. It is desirable to use a large fillet radius r. In order to keep l as small as possible, it is advantageous to spot-face the surfaces on which the bolt head or nut rests.

The flange thickness h_2, Fig. 15.5a, according to common practice, is made equal to $1.25d$ to $1.5d$, where d is the bolt diameter; but it should

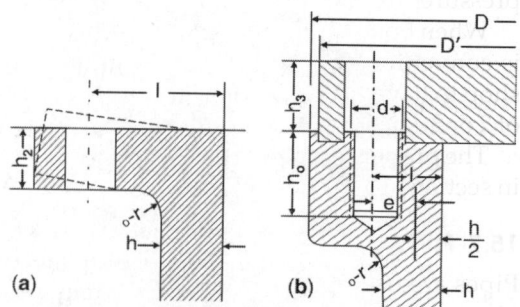

Fig. 15.5: Flanges on vessels

---

[4] I. A. Wojtaszak, "Stress and Deflection of Rectangular Plates," *JAM*, Vol. 3 (June, 1936), p. A-71.

not be less than $1.1h$ to $1.25h$. After h_2 has been thus selected, the thickness should be checked in bending by considering the flange or the cover edge of thickness h_3, Fig. 15.5b, as a cantilever beam.

Studs: The use of studs, as in Fig. 15.5b, has many advantages for heavy pressures. It decreases the bending stress at the flange root since the moment arm e can be made very small. The distance l can be made equal to $1.25d$ to $1.5d$. The depth h_o of the tapped hole is made equal to $1.25d$ in a steel casting, $1.5d$ to $1.75d$ in cast iron, and $1.75d$ to $2d$ in aluminium to take care of shear strengths of these materials. If the flange is not turned on the outside, it is advisable to make the cover diameter D', 3 or 6 mm smaller than the flange diameter D. Rough castings never match, and if the diameters are supposed to be equal, the joint looks worse than when the two pieces are purposely made of slightly different sizes.

Bolt spacing: To insure a tight joint the distance c between two bolts, often called the pitch, must not be greater than a certain limit, which depends on the rigidity of the flange or cover and the inner pressure p (15.9). On the other hand, this distance should not be made smaller than $3d$ in order to leave room for the wrench when tightening the nuts. The United States Navy has standards for the pitch c for water and steam joints which may serve as a guide in the design. According to these standards, $c = 7d$ for low pressures from 0.345 kN/m^2. The pitch c gradually decreases with the increase of pressure, and $c = 3.5d$ for pressures from 1.2 to 1.4 MN/m^2.

When bolts are spaced equally on a circle, it is convenient for laying out the holes in the shop to have the number of bolts and multiple of 2 or 3, such as 4, 6, 8, or 12.

The proper size of bolts to use was discussed in sections 11.9, 11.11 and 11.13.

15.7 PIPES

Pipes are used for transporting fluids. These are made of cast iron, wrought iron, steel, copper, rubber and plastics. *For low pressure of fluids like raw water, rain water, soil waste including*

underground, cast iron pipes are mainly used (15.1). The ends are usually flanged and are spigot and socket type. Wrought iron and steel is very common for transporting water, gas and steam. Steel pipes are manufactured in three variants; butt welded, lap welded and seamless as solid drawn. The welded construction pipes are cheaper than solid drawn. They have screwed ends and provided with one socket at the end for joining with other pipes. The bore and thickness is standardized. Pipes with plain ends are also available. The lap welded pipes are stronger than butt welded pipes. Both of these are available from 3 mm to 150 mm bore. For various types of layout, sockets, bends, tee and barrel nipples are available in various sizes.

The terms gas (light), water (medium) and steam (heavy) denote three standards for pressure. The water (medium) quality and steam (heavy) quality pipes are one and two gauge thicker respectively than the gas (light) quality, both welded. For distinguishing the different qualities of pipes/tubes, each screwed and socket pipe is marked as follows:

Gas (light) pipe is available in natural colour grey or black. Water (medium) pipe is provided with blue band. Steam (heavy) pipe is provided with red band (15.1). All these pipes are available as galvanised and unglavanised.

The gas and water types are specified by 3.5 and 4.5 N/mm^2 respectively hydraulic test pressure while the heavy type can have a test pressure of 4.5 N/mm^2 and ever higher.

Pipe joints: There are three types of joints commonly used for joining pipes. These are: (a) Screwed joint, (b) socket and spigot joints, and (c) flanged joints.

(a) *Screwed joints*: In this case pipes are joined together by using a coupler or socket having half its length inserted in each pipe. The joint is made leak proof with threads wound round threads emulsified with lead oxide and then socket is inserted over the ends. Sometimes locking nuts are provided at the end of the socket. In certain cases, the pipes are joined by screwed unions.

(b) *Socket and spigot pipe joint*: The socket and spigot pipe joints are used for cast iron pipes for water and gas purposes. These are used for low pressure fluids. The joint is effected by inserting a gasket or by inserting jute thread and then filling with lead.

(c) *Flanged pipe joints*: These joints are very common for every type of pipe. As compared to other joints, these resist the longitudinal forces similar to the pipes. The flanges also help in connecting with the flanges of stop valves also. *The flanges are not shrunk fitted to avoid shrinkage stresses (15.4).*

To avoid leakage, packing is inserted between flanges. When bolts are tightened the packing is subjected to a larger pressure as compared to internal fluid pressure so that there is no leakage (15.5).

Design procedure of flanged pipe joints: Pipes used for transporting fluids are provided with pipe bore d, or alternatively the discharge is specified to calculate the size of pipe.

1. The thickness of pipe is calculated by using Lami's equation or other equations

$$t = \frac{d_1}{2}\left[\sqrt{\frac{S_d + p}{S_d - p}} - 1\right] \qquad \ldots(15.26)$$

The thickness can also be checked by using Birnies equation (5.16).

The outer diameter of pipe $d_2 = (d_1 + 2t)$

To take care of thread depth and corrosion effect, this value is increased by a factor 1.1,

so $\qquad d_2 = (d_1 + 2t) \times 1.1$

2. To prevent leakage a packing needs to be used between the flanges. *Though it is convenient to place the packing of the same diameter as the flange diameter, the force required for tightening will increase (15.6).* So a lower packing diameter is selected so that width w of packing is 10 to 20 mm. In order to make a leak proof joint, the pressure on the packing should be equal to or greater than the inside fluid pressure. For ease of calculation it is desirable to have the pressure p on the packing created

by tightening of bolts. An alternate design of the flanges to prevent leakage is shown in Fig. 15.6.

Taking the outside diameter of packing

$$d_p = d_1 + 2w$$

Tightening force of bolts or separating force on the pipes,

$$F = \frac{\pi}{4}d_1^2 p + \frac{\pi}{4}(d_p^2 - d_1^2)p$$

$$= \frac{\pi}{4}d_p^2 p \qquad \ldots(15.27)$$

This force is to be resisted by n number of bolts.

Force per bolt $F_1 = \dfrac{F}{n}$

Taking,

n = 2 for oval shaped flange

= 4 for square or circular flange

= 6 or higher for circular flange

First trial should be with minimum number of bolts if d_o is the core diameter of bolts

$$\frac{\pi}{4}d_o^2 S_d = F_1$$

or $d_o = \left[\dfrac{F_1 \times 4}{\pi S_d}\right]^{1/2} \qquad \ldots(15.28)$

Nominal bolt diameter $d \simeq 1.2\, d_o$ which is to be of standard size.

Flange size and shape: If the diameter of bolts is suitable with respect to pipe diameter by comparing head of bolt or nut size $\simeq 2d$, then two bolts are o.k., the shape of the flange will be oval. If diameter of bolt appears to be large, increase the number of bolts to four. If four bolts are not suitable, increase the number of bolts from 4 to 6 or to 8. If four bolts are suitable, the shape of the flange can be circular or square. If the number of bolts is more than four, the shape will be circular.

Bolt centre distance or diameter d_3, diagonal D_1 = outer diameter of pipe + clearance + 2 diameter of bolt (or distance across corners). This distance D_1 will apply to oval flange, circular flange or square flange.

Now major axis of oval flange or diameter of a circular flange or diagonal of square flange $D_2 = D_1 + 2d +$ clearance.

Clearance is provided for providing space for tightening with wrench.

(a) *For oval flange*: Figure 15.6 for easy manufacture. The oval flange is made of circular arcs and straight lines. The circular profile is drawn at the tip of minor axis, with radius = outer radius of pipe + clearance. The arcs at major tip have a radius = bolt diameter + clearance and arcs having centre as centre of bolt. The arcs are joined by drawing tangents to complete the oval shape.

Thickness of flange t_1. When the bolts are tightened, the flanges will be subjected to bending about the tangent to the outer diameter of pipe in the case of oval flange as shown in Fig. 15.6b.

The arm a = distance of bolt center from the tangent = $\dfrac{D_1}{2} - \dfrac{d_2}{2}$

Length of tangent = b, it is found from the drawing.

Bending moment = $F_1 \times a$

Resistance in bending of flange $= \dfrac{bt_1^2}{6} S_d$

$$t_1 = \left[\frac{6F_1 a}{bS_d}\right]^{1/2} \qquad \dots (15.29)$$

(b) *For square flange*: Figure 15.6c D_2 is the approximate diagonal as corners are replaced by circular arcs similar to oval flanges. Flange drawing is developed on the sheet. As in the case of oval flange, the thickness t_1, is to be calculated. There are possibly three planes about which bending is possible.

(i) Considering bending about AA.

Bending moment = $F_1 \times a_1$

Length b of AA is the length of tangent, but the effective width about which fracture may take place is along the arc of circle with radius a, i.e. the length

$$b'b' \simeq \frac{2}{3} \text{ of } b.$$

Therefore, $F_1 a_1 = \dfrac{2}{3} b \dfrac{t_1^2}{6} S_d.$

$$t_1 = \left[\frac{9F_1 a_1}{bS_d}\right]^{\frac{1}{2}} \qquad \dots (15.30)$$

(ii) Second possibility of bending is about axis BB.

There are two bolt forces which will be considered for bending

Distance of bolt centres from axis BB is

$$a_2 = \frac{D_1}{2} \sin 45°$$

or it can be scaled from the drawing also. Maximum bending moment about BB = Moment due to bolt – Moment due to pressure of fluid.

$$= 2F_1 a_2 - \frac{F_1}{2} \times 0.6366 \frac{d_1}{2} \qquad \dots (15.31)$$

$0.6366 \dfrac{d_1}{2}$ is the distance of centre of pressure from the axis BB.

Effective width of flange b_2 = side of square $- d_2$

Bending resistance of flange $= \dfrac{b_2 t_2^2}{6} S_d$

$$t_2 = \left[\frac{6\left[2F_1 a_2 - \dfrac{0.6366}{4} F_1 d_1\right]}{b_2 S_d}\right]^{\frac{1}{2}} \qquad \dots (15.32)$$

(iii) Third possibility is bending about CC. In this case only one bolt force is responsible for bending.

Distance of bolt center from axis CC

$$a_3 = \frac{D_1}{2}$$

Effective width $b_3 = D_2 - d_2 - 2d.$

Therefore, $\quad t_3 = \left[\dfrac{6 F_1 a_3}{b_3 S_d}\right]^{\frac{1}{2}} \qquad \dots (15.33)$

It is found that case (b) gives the maximum thickness.

(c) *For circular flanges*: In Fig. 15.6d, D_2 is the outside diameter and D_1 is the bolt circle

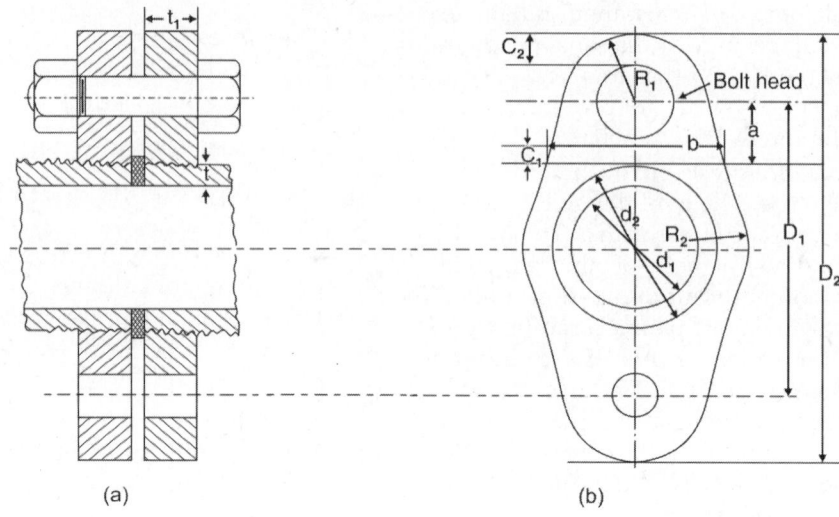

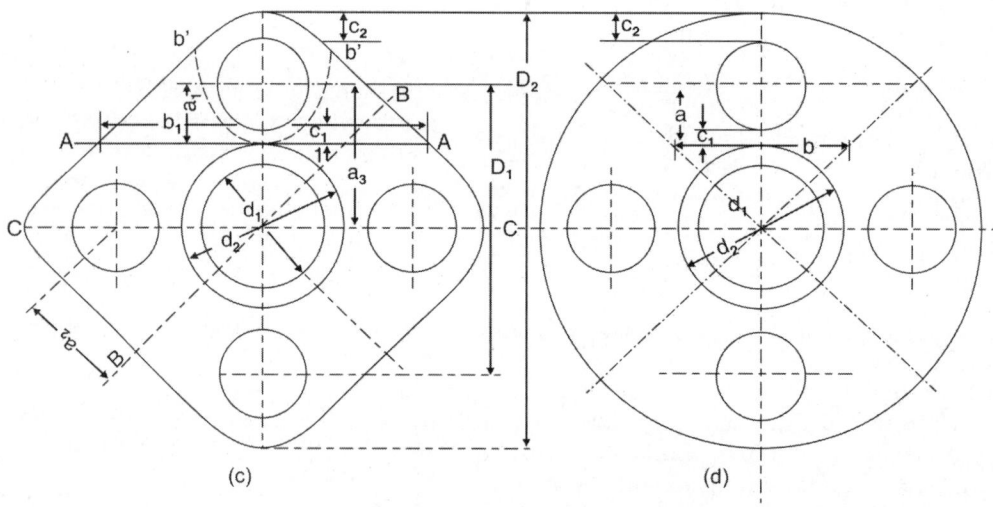

Fig. 15.6: Various shapes of pipe joints

diameter. The influence of bolt is a sector of a circle/n. In the case of 4 bolts, the distance b is given by

$$\frac{b/2}{d_2/2} = \tan\theta, \quad \theta = \frac{\pi}{n}$$

or $b = d_2 \tan\theta$

while distance $a = \dfrac{D_1}{2} - \dfrac{d_2}{2}$

Therefore $\dfrac{bt^2}{6} S_d = F_1 \times a$

$$t = \left[\frac{6 F_1\, a}{b\, S_d}\right]^{\frac{1}{2}} \qquad \ldots (15.34)$$

Example 15.2: Find the number and size of stud bolts to hold the cover of example 15.1 with a metal-to-metal joint.

Since h should be between $1.25d$ and $1.5d$, the bolt diameter d should be approximately equal to $0.67h$ to $0.8h$. Their number may be determined by the least distance between two bolts for insuring a tight joint. The cover not being very rigid, for $p = 2$ MN/m², according to the BIS, $c = 3.5d$. If 24 mm bolts are used, $c = 80$ mm and the diameter D of the bolt circle can be made equal to $300 + 2 \times 24 = 348$ or say 350 mm. Thus, the required number of bolts is

$$i = \frac{\pi \times 350}{80} = 13.7 \text{ say } 14$$

The stress in the bolts must be checked. Use 20C8 with elastic limit 240 MN/m² with a safety factor n of 2, and take the size coefficient as

$$e_{sz} = 1 - 0.4 \times (1 - 0.85) \times (1 - 0.5) = 0.97$$

The safe stress is then

$$S_d = \frac{240 \times 0.97}{2} = 116.9 \text{ MN/m}^2$$

The safe load on one bolt, based on the stress area in Table 11.1, is

$$F = 353 \times 116.9 = 41089 \text{ N}$$

However, according to equation 11.28, the initial load from tightening the nuts is $F_o = 2800 \times 24 = 67200$ N. Thus the actual safety factor n' of a tightened 24 mm bolt is only $67200/41089 = 1.64$.

According to section 11.11, for a metal-to-metal contact the required number of bolts based on the consideration of strength alone is

$$i = \frac{\pi}{4} \times \frac{350^2 \times 2}{41089} = \frac{192422}{41089} = 4.68 \text{ say } 5$$

For comparison the allowable stress given by Unwin's formula (equation 11.52) may be used, although the formula applies to a thin gasket rather than to a metal-to-metal joint. By this formula,

$$S_d = 17.3\left(\frac{24^2}{540} + 1\right) = 35.75 \text{ N/mm}^2$$

The safe load on one bolts is

$$F = 353 \times 35.75 = 12619 \text{ N}$$

and the number of bolts should be

$$i = \frac{192422}{12619} = 11.2 \text{ say } 12$$

According to Bach (section 11.13), the allowable stress for this case of average workmanship is $32 \times 1.25 = 40$ N/mm² and

$$i = \frac{192422}{353 \times 40} = 13.6 \text{ say } 14$$

Since Unwin's and Bach's stresses are very low, fourteen 24 mm bolts may be used safely.

Example 15.3: Find the number and size of stud bolts to hold the cover of example 15.1, using a gasket joint.

For a joint with a flexible gasket the load on the bolts, according to section 11.11, is equal to the sum of the initial load F_o and the fluid pressure. Evidently, the size of the bolts must be increased if the same convenient number $i = 14$ is to be used. Increasing the major diameter to $d = 27$ mm gives, by Unwin's formula (equation 11.52),

$$S_d = 17.3\left(\frac{27^2}{540} + 1\right) = 40.655 \text{ N/mm}$$

The safe load on one bolt is

$$F = 459 \times 40.655 = 18661 \text{ N}$$

The diameter of the bolt circle must be increased to $300 + 27 \times 2 = 354$ say 360 mm and the number of bolts should be

$$i = \frac{\dfrac{\pi}{4} \times 360^2 \times 2}{18661} = 10.9$$

However, 12 bolts will be used.

The safety factor can now be found as follows:

By equation 11.30, with a factor of 0.5 for a gasket joint, the initial load is approximately $1400 \times 27 = 37800$ N. The load from the fluid pressure is $F = 192422/12 = 16035$ N, and the stress in each bolt is

$$S = \frac{37800 + 16035}{353} = 152.5 \text{ N/mm}^2$$

The size coefficient for 27 mm, steel is 0.96, and the safety factor is

$$n = \frac{240 \times 0.96}{152.5} = 1.51$$

Example 15.4: Design an oval-flanged pipe joint for a C.I. pipe 50 mm bore, subjected to an internal fluid pressure, of 7 N/mm², gauge. The maximum tensile stress in the pipe and flange material is not to exceed 20 N/mm² and in the bolt 80 N/mm².

Using Lami's equation for thickness of the pipe, t

$$t = \frac{d}{2}\left[\sqrt{\frac{S_t + p}{S_t - p}} - 1\right]$$

$$= \frac{50}{2}\left[\sqrt{\frac{20+7}{20-7}} - 1\right] = 11.1 \text{ say } 12 \text{ mm}$$

Outside diameter of the pipe $d_2 = (50 + 2 \times 12) 1.1 = 81.4$ say 82.

Assuming a width of the packing = 15 mm, so the outside diameter of the packing = 50 + 2 × 15 = 80 mm.

Separating force on the pipe

$$= \frac{\pi}{4} 50^2 \times 7 + \frac{\pi}{4}\left(80^2 - 50^2\right) \times 7 = 35185.92 \text{ N}$$

As there can be two bolts,

Force per bolt, $P = 35185.92 / 2 = 17592.96$ N say 17593.

The core diameter, d_c of the bolt is given by

$$P = \frac{\pi}{4} d_c^2 s_t$$

where, s_t is the allowable tensile stress for the bolts.

$$d_c = \left[\frac{17593}{\pi/4 \times 80}\right]^{1/2} = 16.73 \text{ mm}$$

The nominal diameter of the bolt

$$d_o = 16.73 \times 1.2 = 20.08$$

The standard diameter of the bolt can be taken as 20 mm or a bolt M 20 × 2.5 will be used.

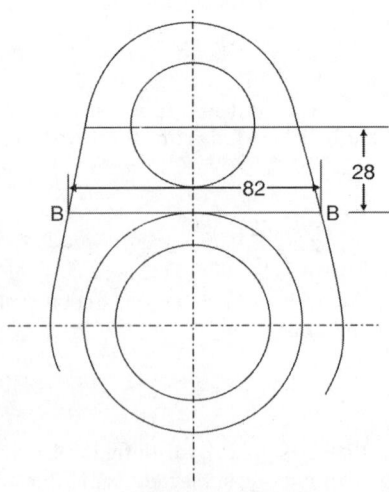

Fig. 15.7: Oval flange joint

The center line of the bolts should have sufficient distance to accommodate the fillets of the flange and nuts (= outside dia of pipe + 2 × fillet size + $2 \times \frac{1}{2}$ of corner to corner distance of nut).

$$D_1 = 82 + 2 \times 8 + 2 \times \tfrac{1}{2} \times 40 = 138 \text{ mm}$$

Major axis $= D_1 + 2d_o + 2$ clearance

$$= 138 + 2 \times 20 + 2 \times 10 = 198 \text{ mm}$$

Minor axis $= d_2 + 2 \times 10 = 82 + 20 = 102 \text{ mm}$

After drawing the elevation of the flange and measuring BB = b = 82 mm and a = 28 mm.

Max. B. M. = Force/bolt × a

$$= 17593 \times 28$$

$$= \text{Resistance of flange in bending}$$

$$= S_t \frac{1}{6} b t_1^2$$

∴ $$t_1 = \left[\frac{17593 \times 28 \times 6}{82 \times 20}\right]^{1/2}$$

$$= 42.45 \text{ say } 43 \text{ cm}$$

i.e. the flange thickness is 43 mm

Example 15.5: Design a pipe joint for pipes of 50 mm bore subjected to an internal pressure of 16 N/mm². The allowable stress in the pipe is 50 N/mm² and in bolts 70 N/mm².

1. Thickness of pipe using Lami's equation

$$t = \frac{d_1}{2}\left[\sqrt{\frac{S_d + p}{S_d - p}} - 1\right]$$

$$= \frac{50}{2}\left[\sqrt{\frac{50+16}{50-16}} - 1\right]$$

$$= 9.83 \text{ mm say } 10 \text{ mm}$$

2. Outside diameter of pipe $d_2 = 1.1(d + 2t)$ $= 1.1[50 + 20] = 77$ mm.

Considering Clavarino's formulas, equation 15.8

$$h = \frac{50}{2}\left[\sqrt{\frac{50 + (1 - 2 \times 0.3)16}{50 - (1 - 2 \times 0.3)16}} - 1\right]$$

$$= 25\left[\sqrt{\frac{50 + 6.4}{50 - 6.4}} - 1\right] = 3.43 \text{ say } 4$$

$$d_3 = 1.1 [50 + 2(4)]$$

$$= 63.8 \text{ say or taking wall thickness as } 10, \text{ the higher value}$$

$$d_o = 77 \text{ mm}$$

Assume an outside diameter of packing d_p

$$= d + 2 \times 15 = 80 \text{ mm}.$$

3. Separating force

$$F = \frac{\pi}{4} d_p^2 \times 16 = \frac{\pi}{4} \times 80^2 \times 16 = 80424.7 \text{ N}$$

4. Using two bolts, force/bolt

$$F = \frac{80424.7}{2} = 40212.35 \text{ N}$$

5. Core diameter of bolt is given by

$$\frac{\pi}{4} d_c^2 \times 70 = 40212.35$$

$$d_c = \left[\frac{40212.35 \times 4}{\pi \times 70} \right]^{\frac{1}{2}} = 27 \text{ mm}$$

Nominal diameter $d_o \simeq 1.2 \times 27 = 32.45$ say 33 mm

This size is very large so we go for 4 bolts, core diameter is given by

$$d_c = \left[\frac{80424.7 \times 4}{4 \times \pi \times 70} \right]^{\frac{1}{2}} = 19.12$$

Nominal diameter $d_o \simeq 19.12 \times 1.2 = 22.9$ say 24 mm
Even this is large, 6 bolts are taken and core diameter is

$$d_c = \left[\frac{80424.7 \times 4}{6 \times 70 \times \pi} \right]^{\frac{1}{2}} = 15.6$$

Nominal diameter $d_o \simeq 18.73$, we may take 18 mm.

The flange shape will be circular

Diameter of the circular flange = d_2 + clearance $(2 \times 10) + 4 \times d_b$ + clearance (2×10)

$$= 70 + 20 + 72 + 20 = 182 \text{ mm}$$

The width of the tangent to the outer diameter of pipe is b.

$$\frac{b}{2} = \frac{70}{2} \tan 30° \qquad b = 41.41 \text{ or } 41 \text{ mm}$$

$$a = d_o + 10 = 18 + 10 = 28$$

Maximum bending moment of bolt force about the tangent

$$= 13404 \times 28 = 375312 \text{ Nmm}$$

$$= \frac{bt_1^2}{6} \times 50$$

$$t_1 = \left[\frac{375312 \times 6}{41 \times 70} \right]^{\frac{1}{2}} = 28.01 \text{ say } 30 \text{ mm}.$$

PROBLEMS

15.1 For what purpose do we find cast iron pipes commonly used? Why are they not used for steam lines?

15.2 Why is cast iron so much used for engine and pump cylinders?

15.3 In what way, very thick cast iron cylinders be found defective and unreliable?

15.4 In general why is it poor practice to shrink flanges on to a pipe?

15.5 What purpose is served by a packing in a pipe joint?

15.6 It is not desirable to have a packing diameter of very large size. Why? Explain.

15.7 In many applications, the diameter of packing is kept equal to the diameter of the flange, though it is not required. Explain.

15.8 Why is the depth of tapped hole is more in cast iron as compared to that of steel?

15.9 Why is the pitch of bolts on the cylinders cover restricted to $3.5d$?

15.10 (a) Find the thickness of the shell of a 1.5 m boiler drum for a steam pressure of 1.2 N/mm^2. The longitudinal seam is a triple-riveted butt joint having an efficiency of 85 per cent. Use an allowable stress of 76 N/mm^2.

(b) Determine the true hoop stress in this shell induced during a hydrostatic test when the pressure is raised to $1\frac{1}{2}$ times the working pressure.

15.11 A standard lap-welded 150 mm pipe has an inside diameter of 152 mm and an outside diameter of 166 mm. Determine the stress induced in the pipe when it is tested with a pressure of 4 N/mm^2.

15.12 Determine the wall thickness of a removable cast iron oil-engine liner if the engine bore is 362 mm. Assume a maximum gas pressure of 4 N/mm^2.

16

Packings and Seals

16.1 GASKETS

If there is no relative motion between the parts forming a joint, the joint can be made tight by means of a gasket. When a gasket is used, the surfaces in contact need not be as accurately finished as in the case of a metal-to-metal joint. The material of the gasket depends on the pressure, temperature, and chemical properties of the fluid against which the joint must be sealed. *Rubber* is a good material for cold water and cold gases under moderate pressure. *Paper* and *cork-composition* sheets are used to seal oil, gasoline, or water. *Asbestos* gaskets can be used for water, steam, or hot gases without regard to the temperature. For pressures over $0.7 \, \text{N/mm}^2$ it is better to use special gasket materials made of asbestos fiber with the addition of binders for increased tensile strength. For high temperatures and pressure, such as encountered in automobile engines, special copper-asbestos gaskets are made in which a thin copper sheet gives the necessary strength and prevents blowouts. For high and sudden pressures, gaskets are confined in a ring space, as in Fig. 16.1. The width b is made so small that the gasket is compressed above the elastic limit. For ordinary temperatures the gasket g may be made of lead; for higher temperatures it is a ring of soft copper.

Thickness: A fibrous gasket should be thin in order to prevent a blowout. Gaskets are made with thicknesses from 0.25 to 1.5 mm and not over 3 mm, the value depending on the surface. Asbestos gaskets are made 0.8 to 1.6 mm. This thickness t_g of a copper gasket, Fig. 16.1, need not be over 0.4 mm, provided the surfaces are finished. However, since the danger of a blowout is eliminated, t_g can be made considerably greater.

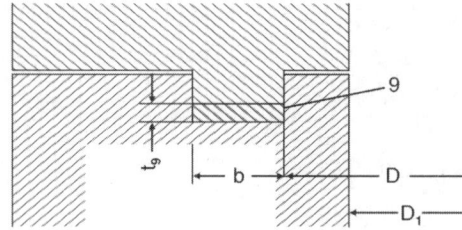

Fig. 16.1: Sealing of high pressures

Ground joints without any gaskets are used for very high pressures. Both the cylinder and the head must be very rigid, so as not to be distorted by tightening of the bolts or by the inside pressure.

16.2 PACKINGS

To seal the joint between two parts when one of them is in motion with respect to the other, *stuffing boxes* are used. In most instances the sealing is produced by a pressure exerted against the moving part. This pressure may be produced either externally or by the fluid itself.

Packings for reciprocating motion: The most commonly used soft-packing stuffing box is shown in Fig. 16.2. The packing material is hemp or cotton rope soaked in grease, for cold water or air, and graphited asbestos rope, for steam or hot water. The bottom of the box and the end of the gland are usually conical, to produce a radial pressure on the rod. However, a flat bottom and a double-cone end of the gland, as shown in the lower half of Fig. 16.2 and the upper half of Fig. 16.3, give a seal at the box wall and at the rod without creating excessive friction.[1] The gland flange, which is elliptic for two stud bolts, is made round for three bolts and square for four bolts. A screwed-on cap, Fig. 16.3, is used for smaller rod diameters. In the lower half is shown another type of gland end.

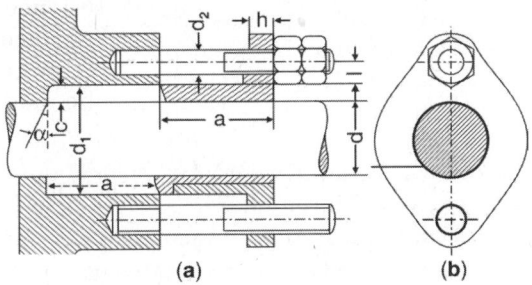

(a)　　　　　**(b)**

Fig. 16.2: Stuffing box with bolted gland

Main proportions

$$c = 0.2d + 5 \text{ if } d \le 100$$

$$c = 2.5\sqrt{d} \text{ if } d > 100$$

$$h = \frac{d}{8} + 12$$

$$a = d + 2c$$

$$\alpha = 10 \text{ to } 15°$$

$$d_2 = \frac{0.2(d + 100)}{\sqrt{n}}$$

where, n is the number of bolts.

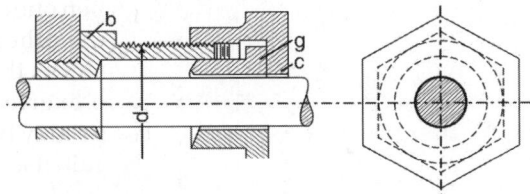

Fig. 16.3: Stuffing box with screwed cap

In Fig. 16.4 is shown a stuffing box with metal packing rings. The rings are split into two or more parts to obtain the radial pressure from the 45° cones. They are made of babbitt or brass, or in larger sizes of soft cast iron. A layer of soft packing p is necessary to give the packing a certain flexibility. This packing can be used only with a perfectly cylindrical rod where there is no side play.

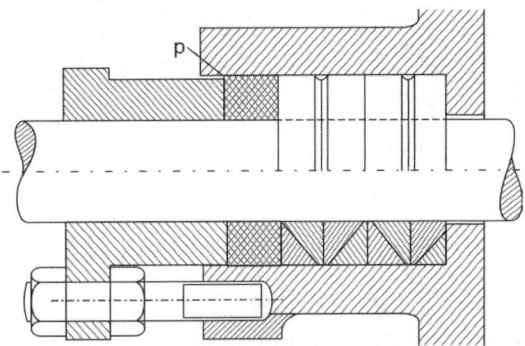

Fig. 16.4: Stuffing box with metallic packing

A flexible metal packing for larger rod diameters is shown in Fig. 16.5. Each ring is split into three parts which are pressed against the rod by garter coil springs s, and pins a prevent the gaps of a pair of rings from aligning. The split rings are made of cast iron. A copper-asbestos gasket g prevents any leakage along the wall of the box. This packing is built in units or sections, two of which are shown in Fig. 16.5. For higher pressures the number of sections is increased up to ten or more.

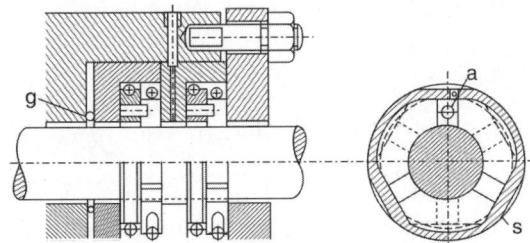

Fig. 16.5: Flexible metal packing

---

[1] E. D. Waters, D. B. Wesstrom and Frank S. G. Williams, "Design of Bolted Flanged Connections," *Mechanical Engineering*, Vol. 56 (1934), p. 736; F. A. Halsey, *Handbook for Machine Designers*, 2nd ed. (New York: McGraw-Hill Book Company, Inc., 1916), p. 435.

In Fig. 16.6 are shown packings in which the fluid pressure helps to keep the joint tight. In Fig. 16.6a a U-shaped leather collar used either alone or with a soft packing ring *f* is shown. In Fig. 16.6b a U-shaped leather collar *l* is used without any gland; to facilitate its insertion, it may be cut diagonally without impairing its sealing effect. In Fig. 16.6c there is a flanged collar *l*, and the sealing along the wall is obtained by the pressure of the gland *g*. Finally, Fig. 16.6d shows a so-called *chevron packing*— a series of V-shaped collars *c* molded of a cotton duck and rubber compound for water, and of an asbestos cloth and rubber cement for high temperatures; the collars *c* are split at an angle, as are also the top and bottom collars *t* and *b*, which are molded of the same composition. These packings act automatically, the pressure with which they are held against the sliding rod and their sealing effect varying with the fluid pressure itself.

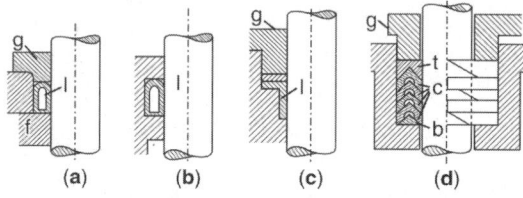

Fig. 16.6: Self-sealing packings

Figure 16.7 shows methods of sealing the water jacket in cylinders of internal-combustion engines. In Fig. 16.7a each of the two rubber rings is made of a round rubber band with *d* greater than *a*. When confined in the space having a cross-sectional area *a* × *b* they do not quite fill it, allowing a slight expansion of the liner *l* in the direction indicated by the arrow. In Fig. 16.7b the rubber ring slides along with the end of the liner. The diameter *d* of the ring is made equal to 6 mm for small cylinders and is gradually increased up to 9 to 11 mm for the largest sizes; the depth *c* may be made equal to 0.8*d*.

A packingless leakproof joint can be obtained by using a very close fit. The advantage of such a joint is the absence of friction, except that

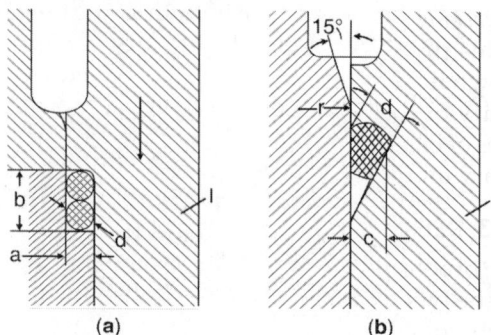

Fig. 16.7: Joints with rubber packing rings

produced in the film of the fluid which is sealed. The joint is made by grinding and then lapping the plunger in the bushing. In order to have a leakproof joint the clearance must be of the order of 0.0025 mm. This requires precision and expensive work. Figure 16.8 shows a packingless steam and guide of a steam-engine poppet valve. The circular grooves *a* increase the resistance to leakage of steam or gas; hence the name *labyrinth packing* is applied. With liquids such grooves are useless and only decrease the effective length of the seal. Packingless fuel-oil pump plungers of oil engines give good service for pressures up to 28 N/mm² and more, if made of very hard material such as heat-treated Nitralloy. Because of the exceedingly small clearance, special care must be exercised to avoid distortion through uneven clamping in assembling.

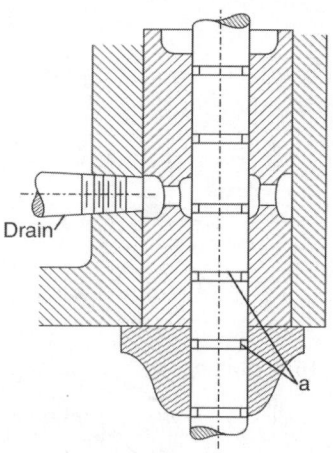

Fig. 16.8: Labyrinth packing

Packing for rotary motion: Many of the stuffing boxes previously described, such as those shown in Figs 16.2 to 16.5, can also be used for rotary or rocking motion. Packings along the lines of Fig. 16.6 wear out too fast when used for rotary motion.

16.3 DESIGN REMARKS

The procedure in designing a packing depends on the type of packing.

Elastic packings: The friction force F_r exerted by a soft packing upon the reciprocating rod may be computed from the relation

$$F_r = kdp \qquad \ldots (16.1)$$

where, k is an empirical coefficient;

d is the diameter of the sliding member, in mm.

p is the fluid pressure to be sealed, in N/mm^2.
If p is less than $0.35 \, N/mm^2$ use

$$p = 0.35 \, N/mm^2$$

When the nuts holding the gland are tightened only enough to prevent leakage, the coefficient k may be taken equal to 0.2.[2]

Gland: The flange thickness h, found from data of Fig. 16.2, should be checked by considering the flange as a cantilever beam having a length l and carrying a load equal to the working strength of the bolt.

The gland is made of bronze in small sizes, and of cast iron with a bronze bushing for larger sizes.

To reduce friction the gland should be bored slightly larger, by about $0.03d$, than the diameter d of the sliding member. The outside diameter of the gland should be smaller than the bore of the box by the same amount.

Bolts: There is no rule governing the selection of the number of bolts, n. A survey of existing machinery shows that two bolts are used in small boxes, with d up to approximately 60 or 75 mm; three bolts are used up to 100 mm; and four bolts are used on all larger sizes. Six bolts are seldom found, except on size over 250 mm.

The bolt diameter d_2 is found by equating the working strength of the bolts to the pressure p exerted by the fluid upon the gland and the frictional force F_r. With the notations of Fig. 16.2,

$$\frac{\pi}{4} d_o^2 n S_d = \frac{\pi}{4}(d_1^2 - d^2)p + F_r \qquad \ldots (16.2)$$

where, d_o is the minor bolt diameter. The allowable stress S_d can be taken as about $70 \, MN/m^2$ and should not be over $85 \, MN/m^2$ because the bolts are subject to an additional load due to friction between the packing and the reciprocating rod, and also because the bolts may be tightened under load.

For reciprocating motion the stud bolts should be made long enough to take two nuts when the gland barely enters into the packing space.

To prevent uneven tightening of the bolts and cocking of the gland in large boxes, the nuts are sometimes equipped with gear-shaped collars which are engaged by a gear concentric with the box. By turning this gear all nuts are tightened equally.

Box body: In a box with a threaded cap, as in Fig. 16.3, the root diameter of the threads must be equal to or greater than the minimum outside diameter of the box. Thus

$$d_3 \geq d + 2c + 2h \qquad \ldots (16.3)$$

where, c is the gland thickness and h is the wall thickness determined to withstand the fluid pressure p.

The thread used is either straight pipe or according to Table 11.2.

Lubrication of the moving member should be provided. In small and low-speed boxes lubrication may be by gravity. With a metal packing, Fig. 16.5, pressure lubrication should be used.

Self-sealing packings: The thickness h, Fig. 16.9, of a U-shaped collar for great pressures can be determined by combining data of Houghton,[3] Welch, and Jenkins.[4] The approximate thickness should be[5]

[2] Halsey, *op. cit.*, p. 244.

[3] *Hydraulic Engineering* (Philadelphia: E. F. Houghton & Company, 1926), p. 35.

[4] H. J. Spooner, *Machine Design*, 6th ed. (London: Longmans, Green & Company, 1930), p. 66, and Halsey, *op. cit.*, p. 242.

[5] Hutte, *Des Ingenieurs Taschenbuch*, 26th ed., Vol. II (Berlin: Wilhem Ernst and Son, 1931), p. 66.

$$h = 1.6 \, d^{0.2} \qquad \ldots (16.4)$$

The width b is made as small as possible, about equal to $4h$, and the depth l can be made from $1.2b$ to $1.8b$.

Speeds: Leather collars should be used only for moderate speeds, not over $1 \, \text{m/s}$. With high speeds the wear is excessive and a soft packing with a gland should be used.

Friction: The friction resistance F_r can be computed from the relation[6]

$$F_r = F_o + fAp \qquad \ldots (16.5)$$

where, F_o is the friction force of the stuffing box when there is no fluid pressure, in N

f is the coefficient of friction;

A is the area of the leather collar in contact with the sliding member, in square mm;

p is the pressure of the fluid, in N/mm^2.

The coefficient f may be taken as 0.01 for rubber and soft lubricated leather;[7] for hard leather, f is many times as great, ranging up to 0.15.

Lubrication: Leather packings must have provision for lubrication by pressure oil, to reduce friction and to prolong the life of the packing.

Rotary motion friction: The tangential friction force F_r for rotary motion can be determined from equation 16.1, as for reciprocating motion. The torsional resistance therefore becomes

$$T = \frac{F_r d}{2} = \frac{k d^2 p}{2} \qquad \ldots (16.6)$$

16.4 PACKINGLESS SEALS

Leakage of the fluid past a rod, as in Fig. 16.8, can be computed with fair accuracy by the formula[8]

$$V = 260(100c)^3 (p_1 - p_2) \frac{d}{lZ} \qquad \ldots (16.7)$$

where, V is the discharge, in cubic mm per second;

c is radial clearance between the rod and the bushing, in mm; p_1, p_2 are pressures on each end of the joint, in N/mm^2;

d is the rod diameter, in mm;

l is the length of joint, in mm

Z is the absolute viscosity of the fluid, in centriposes.

Table 16.1 gives the viscosities of fluids commonly encountered in machines. Viscosities at other temperatures may be determined by interpolation.

Table 16.1: Absolute viscosities Z

Fluid	Temperature °C	Absolute viscosity (centipoises)	Temperature °C	Absolute viscosity
Steam	20	0.0097	260	0.018
Air	20	0.018	93	0.022
Water	0	1.79	39	0.69
Water	20	1.0	71	0.40
Gasoline	20	0.6	62	0.3
Kerosene	20	2.7	82	1.3
Fuel oil, 30° Baumé	20	5	82	1.6
Fuel oil, 24° Baumé	20	40	82	4
Spindle oil	20	20–35	82	3–4
Machine oil	20	200–500	99	5.5–16
Castor oil	20	1,000	43	200

[6] *Ibid.*, p. 66.

[7] K. Kutzbach, "Versuche uber Stopfbuchsen mit hohen Flussigkeitsdruck," *Zeitschrift Verein Deutscher Ingenieure*, Vol. 80 (1936), p. 609.

[8] A. M. Rothrick and E. T. Marsh, *Effect of Viscosity on Fuel Leakage between Lapped Plungers and Sleeves on the Discharge from a Pump Injection*, Report No. 477, National Advisory Committee for Aeronautics (1934), p. 9.

Lubrication: When a packingless joint is used for sealing gas, provision should be made for lubrication, preferably under pressure.

16.5 COMPARISON OF PACKINGS

Simple stuffing boxes with a soft packing are inexpensive but can be used only for low pressures, not over $1.04 \, MN/m^2$ or $1.04 \, N/mm^2$.

Self-sealing leather collars have small friction and long life and can be used up to the highest pressures, but they are suitable only for cold water or oil.

Metallic packings are more expensive and take up more space, but they have small friction and long life and can be made for any pressure that may occur, and for any fluid.

PROBLEMS

16.1 Describe with sketches various types of packings and gaskets.

16.2 Describe a packingless leak proof joint.

16.3 What is a gland? Why is it used?

16.4 How is self-sealing packing designed?

16.5 Compare a stuffing box, self-sealing leather collars, and metallic packages.

16.6 (a) Determine the main dimensions for a simple stuffing box for a water-pump piston rod. The rod diameter is 50 mm, and the pump delivers water against a head of 96 m. (b) Give a sketch with dimension of the box assembly.

16.7 (a) Determine the main dimension for a stuffing box with a soft packing to be used on a steam engine. The piston-rod diameter is 65 mm and the steam pressure is $0.8 \, N/mm^2$ gage. (b) Give a sketch with dimension of the box assembly.

16.8 Determine the main dimension of a stuffing box with a threaded cap, Fig. 16.3, for a 22 mm rod. Make both threads of the same size. Assume that the box is made of cast bronze and that a steam pressure of $1.6 \, N/mm^2$.

16.9 Determine the main dimensions of a self-sealing leather packing. Fig. 16.6a, for the piston rod of problem 16.6.

16.10 Determine the friction force which must be overcome by the piston rod of problem 16.6, using the packing box of that problem, and compare them with those of problem 16.8. Assume that the friction force without fluid pressure is 18 N.

16.11 (a) Determine the main dimensions of a leather cup packing for the plunger of a hydraulic elevator, similar to that of Fig. 16.6b. The plunger diameter is 100 mm, and the pressure is $10 \, N/mm^2$ (b) Find the friction force and compare it with that of a soft packing gland. Without pressure the gland friction is 90 N.

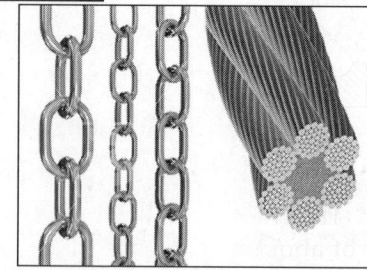

17

Chains and Wires Ropes

17.1 CHAINS

Chains are of two main classes, known as coil chains and stud-link chains.

Coil chains: The type of chain used on hoists, cranes, and dredges is shown in Fig. 17.1 and is known as a coil chain. Such a chain is designated by the diameter d of the link stock, but no rigid standards exist for the link length l or the width b. These dimensions as well as the breaking load F_u or proof load F may be obtained from catalogs of chain manufacturers.[1] The usual sizes run from 6 mm to 65 mm.

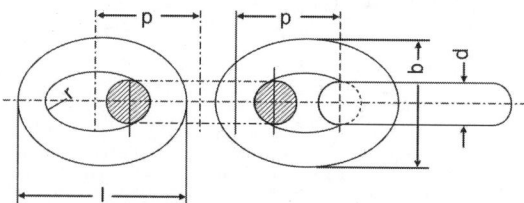

Fig. 17.1: Coil chain

The material used for crane chains is basic open-hearth steel for the smaller sizes and high-grade wrought iron for the larger sizes.

Load capacity: Chains are tested with a proof load equal to approximately one-half the breaking load, or practically equal to the load at the elastic limit. The safe load is often taken as one-half the proof load. This practice corresponds to using a safety factor n of 2.

However, since chains are likely to be subjected to shock action, it is better to take n as 3 for intermittent machine operation, and as 3.5 for continuous machine operation.

In chain, the weakest link will break first so the strength of chain is the strength of the weakest link (17.2).

Stud-link chains: A stud-link chain is shown in Fig. 17.2. Such chains are used for heavy hoisting and in marine work for anchors and moorings. Within the elastic limit a stud-link chain will carry a load of 20 to 25 per cent greater than that which is safe for open-link chains.

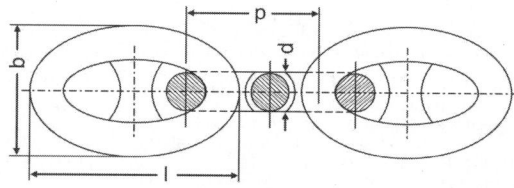

Fig. 17.2: Stud-link chain

An additional advantage of stud-link chains is that they do not kink or tangle as readily as open-link chains.

17.2 CHAIN DRUMS AND SHEAVES

A drum used for winding up a chain should be provided with machined helical grooves. Two forms of such grooves are used, the more common form being shown in Fig. 17.3a.

---

[1] Lionel S. Marks, ed., Mechanical Engineers' Handbook, 5th ed. (New York: McGraw-Hill Book Company, Inc., 1951), p. 931.

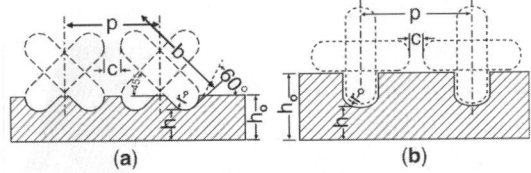

Fig. 17.3: Grooves in chain drums

The grooves are cut so as to leave a clearance c of about 5 mm. From the geometry of the sketch, the pitch p is

$$p = 0.707b + 0.293d + 5 \quad \dots (17.1)$$

where, all values are expressed in mm.

The radius of the groove should be

$$r = 0.5d + 1.5 \quad \dots (17.2)$$

Deeper grooves, like those in Fig. 17.3b, require a heavier drum wall and also, with the same clearance, c, a greater drum length to take the same length of chain, since their pitch is greater. This pitch is

$$p' = b + 5 \quad \dots (17.3)$$

The diameter D of the drum depends on the size of the chain, its speed, and its desired life. For short-link chains the drum diameter should be

$$D \geq 20d \quad \dots (17.4)$$

and it should preferably be about $30d$. If a chain is wound on a relatively small drum, the bending of the links is increased and the life of the chain is reduced.

The length of the drum should always be such that the required chain length will go on the drum in a single layer. In order to reduce the load coming upon the chain anchor, one or two coils of the chain should remain on the drum when the load is in its lowest position.

Anchors: In Fig. 17.4a is shown an anchor type which is easily assembled but has the disadvantage of subjecting the tongue a to bending. In case the chain should assume the position shown by the dotted lines, it imposes a much greater load on the cap screw c than on the screw b. The design shown in Fig. 17.4b is cheaper to make and gives a good load distribution, but it may not be as convenient for assembling.

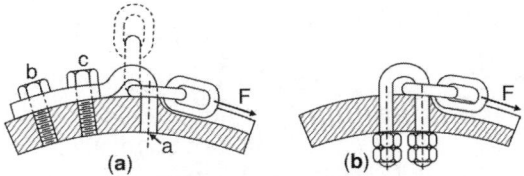

Fig. 17.4: Chain anchors

Design of drums: A crane drum may be made with bushings and may revolve on a stationary shaft, or the drum may be fastened to a revolving shaft.

An exact stress analysis for a hoist drum is a rather difficult problem, but an approximate procedure consisting of the following steps may be applied:

(a) The minimum thickness of the metal, as h in Fig. 17.3, is determined from considerations relative to casting the drum and machining the grooves.

(b) The bending stress s_1 is determined by considering the drum as a hollow round beam loaded at the middle and supported at the ends. The stress s_1 should not exceed 21 MN/m^2 for cast iron[2] and 35 MN/m^2 for cast steel.

(c) The tangential compressive stress s_2 due to the tension in the chain when it is wound on the drum may be determined by the relation

$$s_2 = \frac{F}{p'h} \quad \dots (17.5)$$

where, p' is the pitch of the grooves, or the distance between the centers of two adjacent grooves. The stress s_2 should not exceed 15 MN/m^2 for ordinary cast iron, 25 MN/m^2 for the best cast iron, or 80 to 115 MN/m^2 for cast steel.

(d) The shear stress due to the torsional moment created by the load may be calculated. Usually, however, this stress is negligibly small.

---

[2] Erik Oberg and Franklin D. Jones, *Machinery's Handbook,* 14th ed. (New York: The Industrial Press), p. 445.

By considering the wall thickness equal to *h* and neglecting the metal above the groove, a strong yet not excessively heavy drum is obtained.

Sheaves: Chain sheaves are of two types: plain sheaves which only guide the chain in changing its directions, and sheaves with pockets which serve to pull the chain.

Figures 17.5a and b show sheave designs with grooves similar to those used on drums. The center webs may be made plain, but round holes are helpful for handling the sheave and for decreasing the weight. Figure 17.5c shows a simpler but satisfactory groove. The sheave diameter *D* should not be less than 20*d*.

Pocket sheaves are used in place of drums, particularly for anchor chains. A sheave for this type has a rim similar to that in Fig. 17.5a, but there are cross ribs to catch the horizontal links, the vertical links going into the central groove.

17.3 WIRE ROPE

Wire ropes are built up of strands made of wires twisted together. Ordinarily the strands are twisted into rope in the opposite direction to the twist of wires in the strands, as shown in Fig. 17.6. The rope is then known as *regular lay rope*. When wires and strands are twisted in the same direction, as in Fig. 17.7, the rope is known as *lang lay rope*. Tests show that lang lay rope has a life several times as long as regular lay rope.[3] However, regular lay rope is more generally used than *lang lay rope* because it

has a less tendency to twist and spin. Ropes with strands twisted in the right direction, as in Figs 17.6a or 17.7a, are known as *right lay ropes*; those twisted in the left direction, as in Figs 17.6b, or 17.7b, are *left lay ropes*.

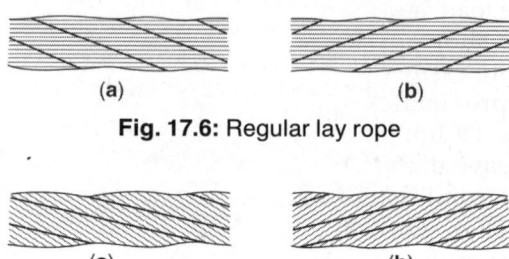

Fig. 17.6: Regular lay rope

Fig.17.7: Lang lay rope

Standard wire ropes are made of six strands with a hemp core saturated with a lubricant. The number of wires in a strand may be 7, 19, or 37. *The rope is designated accordingly as* 6 × 7, 6 × 19, or 6 × 37, *i.e. there are 6 strands and each strand is consisting of 7, 19 or 37 wires (17.5).* *Extra-flexible hoisting* rope is made of 8 strands of 19 wires, or 8 × 19 construction. 6 × 12/6/1 means there are six strands. In each strand, there are 12 wires in the outer circle, six wire in the inner circle and one wire in the center. 6 × 12 × 6 +6 F/1 means there are 6 strands. In each strand there are 12 wires in the outer circle and six wires in the inner circle, in addition there are 6 thinner (fine) wires and one thinner wire in the centre. Hoisting ropes also differ in

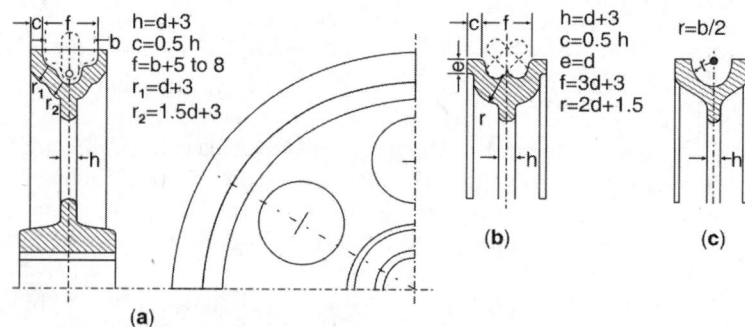

Fig. 17.5: Chain sheaves

---

[3] [Symposium,] "Drantseilforschung," *Zeitschrift Verein Deutscher Ingenieure*, Vol. 75 (1931), p. 1485.

the metal of the wires, as indicated in Table 17.1. *The breaking strength of wire ropes as found by tests is about 83 per cent of the combined strength of all wires. This reduction is due to the difficulty of getting a perfect grip on the rope.* As a result, not all the wires carry their full share of the load; and since the inner wires in a strand are shorter than the outer wires, they are more easily overloaded. Table 17.1 gives the approximate breaking strength of standard 6 × 19 hoisting ropes, and the minimum sheave diameters that are recommended are given in Table 17.2.

High-grade steel increases the strength of the rope but reduces its flexibility. When using one of the stronger ropes, it is therefore good practice to make the sheave or drum diameter equal to that required by a rope of mild plow steel of the same strength.

The strength of extra-flexible 8 × 19 or special flexible 6 × 37 hoisting rope is about 10 to 12 per cent lower than that of standard 6 × 19 rope of the same material (17.8). Also, the more-flexible ropes are made of thinner wire and are more subject to outside abrasion.

The life of rope depends to a great extent on the number of bendings it undergoes at every operation. If bendings are more, life is less; the life also decreases with increase of speed (17.6). A bending over a guide sheave, as in Fig. 17.8b, is equivalent to two bendings, as it is a reversed bending; that is, it requires bending the rope first in one direction and then in the opposite direction. However, it makes no difference whether the angle of contact with the sheave is 90°, 180°, or 360°. Both the stress and the life of the rope will remain the same in each case. Thus, if all conditions are equal and the life of a rope in

Table 17.1: Specifications for steel wire ropes

Construction	Diameter of rope, mm	Approximate weight N/m	Nominal breaking strength of rope kN	
			Tensile strength of wire 1600 to 1750 MN/m²	Tensile strength of wire 1750 to 1900 MN/m²
Group 6 × 19	8	2.4	34.0	37.0
	10	4.4	66.0	72.0
6 × 12/6/1[a]	12	5.4	86.0	94.0
	14	7.6	109.0	119.0
6 × 12/6+6F/1	16	9.4	134.0	147.0
6 × 9/9/1	18	12.5	193.0	211.0
6 × 10/5+5F/1	20	14.7	226.0	246.0
	22	18.4	259.0	284.0
	24	21.0	300.0	330.0
	25	24.1	340.0	376.0
	29	30.5	432.0	472.0
	32	37.5	533.0	582.0
	35	45.5	645.0	706.0
	38	54.3	767.0	843.0
	41	63.7	904.0	991.0
	44	73.8	1047.0	1148.0
	48	84.8	1199.0	1321.0
	51	96.4	1372.0	1504.0
	54	108.9	1544.0	1697.0

Table 17.2: Ratio of drum and sheave diameter to rope diameter

Purpose	Construction	Minimum ratio		
Mining installations	All	100		
		Class 1	Class 2 and 3	Class 4
Cranes and allied	6 × 37			
hoisting equipment	8 × 19 Filler wire	15	17	22
	8 × 19			
	8 × 19 Warrington	17	18	24
	8 × 19 Seale			
	34 × 7 Non-rotating			
	6 × 24	18	19	25
	6 × 19 Filler wire	18	20	23
	6 × 19	19	23	27
	6 × 19 Warrington			
	17 × 7 Non-rotating			
	18 × 7 Non-rotating			
	6 × 19 Seale	24	28	35

These ratios are valid for rope speeds up to 50 m/m. These need to be increased by 5% for each additional 50 m/m of rope speed.

Fig. 17.8a is taken as unity, the life of the rope in Fig. 17.8b will be about one-fourth as long, and that in Fig. 17.8c will be only one-sixth or one-seventh as long. A design that eliminates reversed and unnecessary bendings will lengthen the life of the rope. The life of the rope decreases also with the speed.

Working loads: The working load of a rope is based not on the elastic limit but on its breaking strength. Therefore, to avoid confusion with the safety factor the ratio of the breaking strength to the effective load (defined in section 17.4) will be designated as the working factor n_0. Average practice is to take n_0 as 6. However, n_0 may be taken either lower or higher, as shown in Table 17.3, the actual value depending on the service. For transportation of persons, state law requires a minimum value of 8 for n_0. In mine hoists lower values of n_0 are used for hoisting materials from considerable depths. For hauling persons, n_0 is increased to 8 by decreasing the maximum useful load. For passenger service the maximum rope speed usually should not exceed 300 mpm. Only for very tall buildings and great depths should it be increased up to 375 mpm (6.25 m/s).

For wrought-iron ropes it is difficult to obtain a value of n_0 greater than 5, and this restriction automatically limits their use.

Friction and efficiency: When a rope is wound on a sheave, the friction between the fibers offers resistance to bending, which must be overcome by the pull applied to the running-off side of the rope.

In hoisting practice, for average conditions, a friction loss of 5 per cent for each bending, or a rope efficiency e_r of 0.95, is considered as safe values; and a friction loss of 3 per cent for each bending, or a rope efficiency, e_r of 0.97, is used for very flexible ropes with extra-large pulleys.

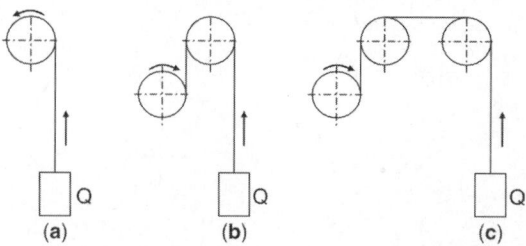

Fig. 17.8: Schemes of rope direction

Table 17.3: Hoisting rope data

Type of service	Working factors n_o		Rope speeds (m/s)	
	Minimum	Maximum	Normal	Maximum
Cranes				
Hand-operated	3	5	0.5	–
Motor-operated	4	7	1.0	–
Hot-ladle cranes, slings	8	–	1.0	–
Derricks, haulage rope	3	–	–	10.0
Small electric and air hoists	7	–	1.0	–
Mine hoists (according to shaft depths)				
Up to 150 m	8	10	6	8.0
150 to 300 m	8	10	8	10.0
300 to 600 m	7	9	10	12.5
600 to 900 m	6	8	12.5	15.0
Over 900 m	5	8	15.0	15.0
Miscellaneous hoisting	5	–	1	
Passenger elevators	8	12	1.25–1.5	6.25

17.4 STRESSES IN HOISTING ROPE

The following main stresses are created in a hoisting rope under load:

  (a) Direct stress due to the load hoisted and the weight of the rope.

  (b) Stress due to the bending of the rope about the sheave.

  (c) Stress during starting.

  (d) Stress due to acceleration.

  (e) Stress due to change of rope speed, including stops.

  (a) *Direct stress*: If Q is the load hoisted, W_r is the weight of wire rope. Direct load = $Q + W_r$.

  (b) *Bending stress*: The commonly used formula proposed by Reuleaux for the bending stress in the wire is

$$s = \frac{E_r d_w}{D} \qquad \ldots (17.6)$$

where, E_r is the modulus of elasticity of the rope as a whole, in N/mm².

d_w is the diameter of the wire, in mm
D is the sheave diameter, in mm.

According to experiments, E_r can be taken equal to 76 GN/m² for wrought-iron ropes and to 83 GN/m² for steel ropes.[4] The modulus of elasticity E_r is less than one-half that of the wire material because the rope is composed of twisted wires, which act as helical springs.

The wire diameter d_w may be taken as in Table 17.4, where d is the nominal diameter of the rope.

If s is the bending stress in each wire, the load on the whole rope due to bending can be taken as

$$Q_b = \frac{\pi}{4} d_w^2 \times n \times s \qquad \ldots (17.7)$$

where, n is the total number of wires in the rope section.

Table 17.4: Wire diameter of hoisting rope

Rope	d_w	D
6×7	$0.106d$	$75d$
6×19	$0.063d$	$45d$
6×37	$0.045d$	$27d$
8×19	$0.050d$	$31d$

[4] J. F. Howe, "Determination of Stresses in Wire Rope as Applied to Modern Engineering Problems," *Transactions of the American Society of Mechanical Engineers*, Vol. 40 (1918), p. 1043.

(c) *Stress during starting*: The general case of starting is when the rope has a slack h which must be taken out before the rope is taut and starts to exert a pull on the load. The stress due to this slack depends on the acceleration a which is necessary in order to impart to the drum and the rope velocity v_s, at the instant when the rope is taut. If this acceleration is considered constant, which is a natural assumption, then

$$v_s = \sqrt{2ah} \qquad \ldots (17.8)$$

where, v_s is in m/s, a is in m/s^2 and h is in m.

Taking up the slack is an impact action, and the corresponding stress may be found by using equation 3.20. However, this equation was derived for a weight falling under the influence of the force of gravity, with an acceleration g. Since the acceleration of the hoist rope is a, the distance h in equation 3.20 must be multiplied by the ratio of the accelerations. The stress due to starting and slack may therefore be determined by the relation

$$s' = s\left(1 + \sqrt{1 + \frac{2ahE_r}{slg}}\right) \qquad \ldots (17.9)$$

Equation 17.9 shows that the length of rope l decreases the impact stress. If the comparatively small resistance of the hemp core is disregarded, the static stress s may be found from the equation

$$s = \frac{Q + W_r}{\dfrac{\pi}{4} d_w^2 n} \qquad \ldots (17.10)$$

If loads are substituted for stresses, the starting load in the rope near the drum becomes

$$Q_{st} = (Q + W_r)\left(1 + \sqrt{1 + \frac{2ahE_r}{slg}}\right) \qquad \ldots (17.11)$$

This equation, in conjunction with equation 17.8, gives

$$Q_{st} = (Q + W_r)\left(1 + \sqrt{1 + \frac{v_s^2 E_r}{slg}}\right) \qquad \ldots (17.12)$$

If there is no slack, $h = 0$, $v_s = 0$, and

$$Q_{st} = 2(Q + W_r) \qquad \ldots (17.13)$$

i.e. there is a sudden load.

(d) *Stress due to acceleration*: Stress due to acceleration is directly proportional to the acceleration a given to the load. The magnitude of the corresponding load is equal to the mass multiplied by the acceleration:

$$Q_a = (Q + W_r)\frac{a}{g} \qquad \ldots (17.14)$$

Usually it is not the acceleration a but the time t necessary to attain a velocity v that is given. Then if v is in m/s and t is in second, a may be found by the relation

$$a = \frac{v}{t} \qquad \ldots (17.15)$$

(e) *Stress due to change in speed*: The additional load created by a change of speed may be found by equation 17.12, and the acceleration may be found by equation 17.15 by using for v, the change of velocity $(v_2 - v_1)$.

Sudden stopping of the hoist drum when lowering the load produces a stress several times as great as the static stress because of the kinetic energy of the moving masses. This kinetic energy is absorbed by the rope, and the resulting stress may be computed by equating the kinetic energy to the resilience of the rope and solving for this stress. If during stopping the load moves down a certain distance, the corresponding change of potential energy must be added to the kinetic energy, and it is necessary to add also the work of stretching the rope during stopping, which may be computed from the impact stress.

Effective load: The sum of the useful load Q, the weight W_r of the rope, and the load Q_b equivalent to bending is called the effective load. During starting, the starting load Q_{st} takes the place $Q + W_r$; and during acceleration of the load the effective load is increased by Q_a.

If the wire rope moves with constant velocity, the effective stresses are due to Q, W_r and Q_b.

i.e. total load $= Q + W_r + Q_b$

or $\qquad = Q + W_r + Q_b + Q_a$

or $\qquad = Q_{st} + Q_b$

Example 17.1: Determine the size of wire rope necessary for a mine hoist carrying a load Q of 84 kN to be lifted from a depth of 225 m. A rope speed of 8 m/s must be attained in 10 sec.

The simplest method is to assume a certain size of rope from preliminary calculations and to find the actual working factor n_o by taking all stresses into consideration. The breaking strength of the rope should be about six times the maximum load. At starting the stress is double. Also, to this stress is added the stress due to bending, which with the minimum permissible drum diameter is about equal to the static stress. Therefore a nominal working factor of about 15 may be assumed. The minimum breaking strength must then be 84 × 15 = 1260 kN.

The commonly used type is the 6 × 19 rope. Since the cost of various ropes for a given strength is about the same, it is desirable to use a high-quality steel giving a smaller size and a small weight for the same strength. Thus according to Table 17.1, a 48 mm plow-steel rope seems to be the proper choice.

The diameter of the drum may be assumed 50d = 2400 mm say to 2.5 m.

The wire diameter, by Table 17.4, is

$$d_w = 0.063 \times 48 = 3.024 \text{ mm}$$

The bending stress, according to equation 17.6, will be

$$s = 83000 \times \frac{3}{2500} = 99.6 \text{ N/mm}^2$$

The load equivalent to bending, by equation 17.7, is

$$Q_b = \frac{\pi}{4} \times 3^2 \times 6 \times 19 \times 99.6 = 80.26 \text{ kN}$$

The weight of the rope, from Table 17.1, is

$$W_r = 83.16 \times 225 = 18.7 \text{ kN}$$

The starting load without slack, from equation 17.13, is

$$Q_{st} = 2(84 + 18.7) = 205.4 \text{ kN}$$

The acceleration, found by equation 17.15, is

$$a = \frac{8}{10} = 0.8 \text{ m/s}^2$$

The corresponding additional load, by equation 17.14, is

$$Q_a = \frac{(84 + 18.7) \times 0.8}{9.81} = 8.375 \text{ kN}$$

Thus, the effective load during starting is

$$Q_{max} = Q_b + Q_{st} = 80.26 + 205.4 = 285.66 \text{ kN}$$

The working factor during starting is

$$n_0 = \frac{1260}{285.66} = 4.41$$

During the first 10 sec after starting, the effective load $= Q + W_r + Q_b + Q_a$, the working factor is,

$$n_0 = \frac{1260}{84 + 18.7 + 80.26 + 8.37} = 6.58$$

During the uniform lifting or lowering of the load, effective load $= Q + W_r + Q_b$, the working factor is,

$$n_0 = \frac{1260}{84 + 18.7 + 80.26} = 6.88$$

This shows that the 48 mm rope is satisfactory. However, if a higher working factor n_0 is desired, the drum diameter must be increased and all calculations must be repeated.

Example 17.2: Determine the influence of a rope slack of only 300 mm on the hoist rope discussed in Example 17.1.

The acceleration of the rope while it is made taut is $a = 0.8$ m/s as found in Example 17.1. By equation 17.8, the rope velocity, when the rope becomes taut, is

$$v_s = \sqrt{2 \times 0.8 \times 0.3} = 0.693 \text{ m/s}$$

The static stress in the rope, by equation 17.10, is

$$s = \frac{(84 + 18.7) \times 1000}{\frac{\pi}{4} \times 3^2 \times 6 \times 19} = 127.4 \text{ N/mm}^2$$

By equation 17.12, the starting load is

$$Q_{st} = (84 + 18.7) \times \left(1 + \sqrt{1 + \frac{0.693^2 \times 83 \times 10^3}{127.4 \times 225 \times 9.81}}\right)$$

$$= 212.44 \text{ kN}$$

Thus the slack increases the static load by 3.4 per cent. If the length l of the rope is smaller, the influence of slack will be greater. However, in this case the amount of slack and the velocity v_s at the moment when the rope is made taut will also be smaller.

An alternative method of solving the problem is given below:

Example 17.3: Determine the size of a suitable wire rope (6 × 19) to lift a cage of a vertical mine hoist 800 m deep. The cage weighs 10 kN and it has to lift 30 kN of iron ore at a speed of 12 m/s which is to be attained in 10 s. The sheave diameter may be assumed 80 times the diameter of wire rope. Assume a working factor of 6 and $E = 8 \times 10^4 \, \text{N/mm}^2$. Weight of rope = $0.036 \, d^2 \, \text{N/m}$, d is in mm. The ultimate stress of wire rope is 1800 N/mm². Area of rope $A = 0.38 \, d^2$ and wire diameter $d_w = 0.063 \, d$.

Acceleration $a = \dfrac{v}{t} = \dfrac{12}{10} = 1.2 \, \text{m/s}^2$

Sheave diameter $D = 80 \, d_w$

Area of wire rope = $0.38 d^2$

Weight of wire rope
$$W_r = 800 \times 0.036 d^2 = 28.8 d^2$$

Acceleration load
$$Q_a = \frac{(Q + W_r)a}{g} = \frac{(10,000 + 3000 + 28.8 d^2)}{9.81} \times 1.2$$
$$= 4892.96 + 3.52 d^2$$

Bending load
$$Q_b = \frac{A E d_w}{D} = \frac{0.38 d^2 \times 8 \times 10^4 \times 0.063 d}{80 d}$$
$$= 23.94 d^2$$

Effective load = $10,000 + 30,000 + 28.8 d^2 + 4892.96 + 3.52 d^2 + 23.94 d^2$
$$= 44892.96 + 56.26 d^2$$

Breaking load $= \dfrac{1800 \times 0.38 d^2}{6} = 114 d^2$

$$114 d^2 = 44892.96 + 56.26 \, d^2$$

$$d = \sqrt{\frac{44892.96}{57.74}} = 27.88 \, \text{mm}$$

The nearest standard size is 29 mm.

17.5 BEARING STRESS

As a result of recent research work on failure of wire ropes due to repeated bending, a new design criterion has been offered.[5] This criterion is based on the bearing pressure exerted on the wire of a rope and is a dimensionless ratio B which can be calculated from the relation

$$B = \frac{2 Q_c}{S_u d D} \qquad \ldots (17.16)$$

where, Q_c is the total load in tension on the wire, in Newtons

S_u is the ultimate tensile strength of the wire rope, in N/mm²

d is the rope diameter, in mm;

D is the sheave or drum pitch diameter, in mm;

In Fig. 17.9 average curves showing the relation between the bearing-pressure ratio B and the number of bends causing failures for various constructions of rope are given. All curves have similar shape and are of the same order of magnitude. From Fig. 17.9 it appears that a value of B below 0.0015 will ensure long life for a stiff 6 × 12 rope; below 0.0017, for a standard 6 × 19 rope; below 0.0023, for extra-flexible 6 × 37 hoisting rope; and below 0.0017, for 6 × 24 and 8 × 19 ropes.

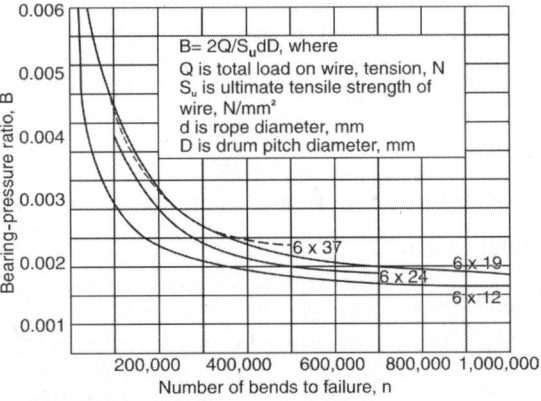

Fig. 17.9: Rope life as a function of the bearing-pressure ratio

There is not enough practical data to discard the conventional design of a rope that takes into account bending stresses by equation 17.6, and instead to base the selection of the rope size and the drum diameter exclusively on equation 17.16. However, it seems appropriate to check d and D by this equation.

[5] D. C. Drucker and H. Tachau, "A New Design Criterion for Wire Rope," *Trans. ASME*, Vol. 66 (1945), p. A-43.

Example 17.4: Check by the bearing-pressure ratio, the rope size and the drum diameter found in Example 17.1.

The total load in equation 17.16 is the sum of the useful load, or 84 kN, and the weight of the rope, or 18.7 kN. Thus

$$Q_c = 84 + 18.7 = 102.7 \text{ kN}$$

The ultimate strength S_u may be found from the breaking strength of 1260 kN. It is

$$S_u = \frac{1260 \times 1000}{\frac{\pi}{4} \times 3^2 \times 6 \times 19} = 1563.6 \text{ N/mm}^2$$

By equation 17.16, the bearing-pressure ratio is

$$B = \frac{2 \times 102.7 \times 1000}{1563.6 \times 48 \times 2500} = 0.00109$$

This value is lower than the high limit of 0.0017 based on Fig. 17.9.

According to equation 17.16, the value of d is taken 44 mm and $D = 2400$, weight of wire rope

$$= \frac{72.376 \times 225}{1000} = 16.283 \text{ kN}$$

and gives the total load $Q_c = 84 + 16.283 = 100.283$ kN.

Therefore $B = \dfrac{2 \times 100.283 \times 1000}{1563.6 \times 44 \times 2400} = 0.00121$

If $d = 41$, $D = 2050$, weight of will rope

$$= \frac{62.468 \times 225}{1000} = 14.05 \text{ kN and}$$

total weight $= 84 + 14.05 = 98.05$

and $B = \dfrac{2 \times 98.05 \times 1000}{1563.6 \times 41 \times 2050} = 0.0015$

This is more close to the limit value of 0.0017.

17.6 ROPE SHEAVES AND DRUMS

Table 17.4 gives the minimum diameters of sheaves and drums recommended by rope manufactures. As a general rule the sheave diameter D should not be much less than $1,000 \times d_w$, where d_w is the diameter of the heaviest wire in the rope. However, D is often made equal to $500\, d_w$, or even to $300\, d_w$, naturally with a material decrease of the life of the rope. With respect to nominal diameter d of wire rope, sheave size varies between $18d$ and $75d$

depending on the rope construction. With 6×7 rope, D lies between $42d$ and $75d$. Drum diameters can be made about 10 per cent smaller than sheave diameters with the same length of rope life.

Sheaves for rope are made similar to those for chains (see Fig. 17.3), the main difference being in the rim groove, as shown in Fig. 17.10. For light and medium service the sheaves are made of cast iron, but for heavy crane service they are often made of steel castings. To prevent wear of the individual wires, the grooves are finished smooth. The life of the rope can be increased materially by use of an insert i, Fig. 17.10b, at the bottom. This may be made of leather, hard rubber, wood, or an artificial Bakelite-like material with a fibrous insert such as used for noiseless gear. *The radius r of the bottom of the groove should be made slightly larger than the radius of the rope so as to prevent wedging of the rope in the groove (17.9).* A good relation is

$$r = 0.53d \qquad \qquad \dots (17.17)$$

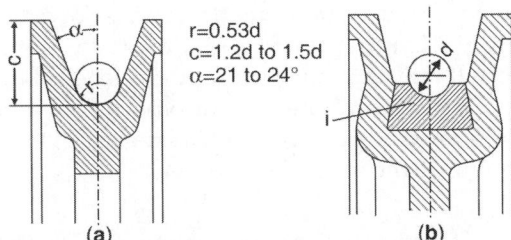

Fig. 17.10: Rope sheave rims

A still-larger radius does not give sufficient bearing area and decreases the life of the rope.

Small drums in hand hoists are made plain. A hoist operated by a motor or an engine has a drum with helical grooves, as shown in Fig. 17.11. The pitch p of the grooves must be made slightly larger than the rope diameter, to avoid friction and wear between the coils. A satisfactory relation is

$$p = 1.15d \qquad \qquad \dots (17.18)$$

The radius of curvature of a groove may be made the same as that for a sheave, or $r = 0.53\text{d}$.

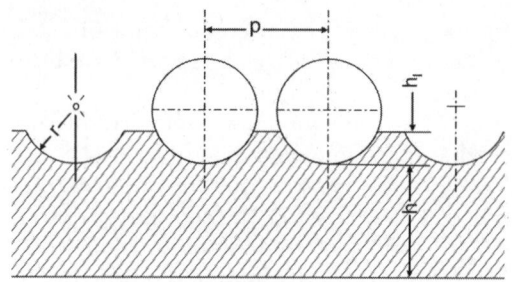

Fig. 17.11: Rope drum grooving

The height h_1 of the groove ribs may be found by the relation

$$h_1 = 0.25d \qquad \dots (17.19)$$

The drum thickness h is determined by the procedure outlined for a chain drum. When the bending stress s_1 is computed, the total weight of the rope is added to the useful load, although actually one-half the rope weight is evenly distributed over one side of the drum. A practical rule is to make h equal to d and then to check h for strength.

Rope anchors: The rope may be anchored to the drum with a steel clamp (shown in Fig. 17.12a) that catches a whole turn of the rope. If only the end of the rope is anchored, as in Fig. 17.12b, a longer clamp with four tap bolts must be used in order to take the full load applied to the rope when necessary.

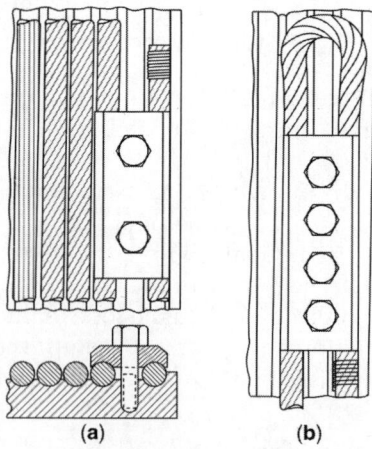

(a)   (b)

Fig. 17.12: Rope anchors

PROBLEMS

17.1 What is the load capacity of chains?

17.2 The strength of a chain is the strength of the weakest link. Explain.

17.3 Sketch chain anchors.

17.4 Explain with the help of sketches regular lay and lang-lay rope.

17.5 How are wire ropes specified?

17.6 On what factors does the life of a wire rope depend? Explain with sketches.

17.7 What stresses cause the effective stresses is a wire rope running with constant velocity?

17.8 Is the flexible rope stronger or weaker than standard wire rope and why?

17.9 Why should the radius of groove at bottom of a drum or sheave be made slightly larger than the radius of the rope?

17.10 (a) Determine the size, and find from a handbook all other dimensions and weight per meter of length, of a link crane chain to withstand a pull to 10 kN for continuous machine operation but without shock action. (b) Find the number of links in a 30 m length.

17.11 Determine the size and all other dimensions of a wrought-iron link chain for a working load of 100 kN for intermittent machine operation with heavy shocks. Also find the number of links and the weight of a 30 m length.

17.12 (a) Determine the size of a 6 × 19 standard plow-steel rope to be used with a drum hoist to lift 25 kN from a depth of 60 m. Assume a rope speed of 5 m/s and an acceleration of 2 m/s² when starting, with no slack. (b) Find the influence of a slack of 0.6 m when starting. (c) Find the necessary drum diameter and length. (d) Check the rope size and drum diameter by the bearing-pressure ratio.

17.13 (a) Determine the size of a 6 × 19 wrought-iron rope used with a drum hoist to lift 10 kN from a depth of 150 m. Assume a rope speed of 2.5 m/s and an acceleration of 1.5 m/s² when starting, with a slack of 0.6 m. (b) Determine the desirable diameter of the drum.

17.14 The pulley system shown in Fig. P17.1 is used to raise a load Q of 25 kN with a maximum acceleration of 2.4 m/s². Assume a 6 × 19 standard-plow-steel rope, sheaves 42 rope diameters in diameter, and a loss of 5 per cent with each pulley, (a) determine the diameter d of the rope required for a working factor n_o of 5. (b) Determine the effort F. (c) Check the rope size by the bearing-pressure ratio.

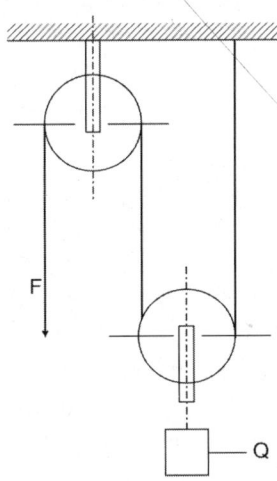

Fig. P17.1

Weight = 0.0343 d^2 N/m, Area = 0.38d^2
d_w = 0.045d, UTS = 1800 N/m², F.O.S = 8,
$E = 8 \times 10^4$ N/mm² [22 mm, 880 N].

17.15 (a) Determine the size of special flexible 6 × 37 wire rope and the drum diameter

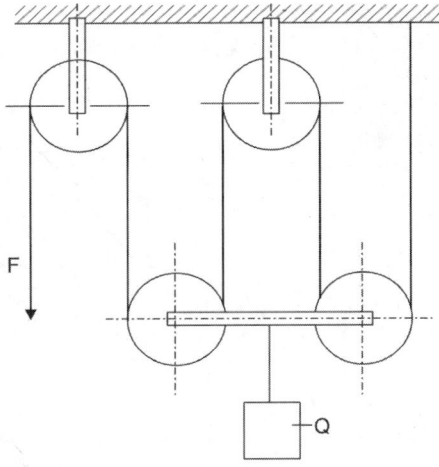

Fig. P17.2

required for an elevator in a building 135 m tall for a load of 20 kN. Assume a maximum rope speed of 5 m/s and an acceleration of 1.8 m/s² when starting with no slack. According to the state law, not less than four ropes must be used. (b) Check the rope size and the sheave diameter by the bearing-pressure ratio.

17.16 An oil well is drilled to a depth of 1500 m with the use of 110 mm drill pipe which has the same dimensions and weight as 100 standard pipe. Assume a weight of 220 N for pipe joints in every 12 m of pipe length. The pipe must be raised by using 24 mm special improved plow-steel 6 × 19 rope with 0.9 m sheaves and an acceleration of 3 m/s². (a) Determine the number of ropes required, using a working factor n_o of 3. (b) Check the rope size by the bearing-pressure ratio.

Brakes

18.1 GENERAL CONSIDERATIONS

The functions of a brake is to regulate the speed of a mechanism by transforming, through friction, the energy of a moving body into heat, and dissipating it.

Classification: Only these brakes will be discussed here which are applied to rotating sheaves, drums, or wheels.

A general classification of brakes and clutches is shown in Fig. 18.1.

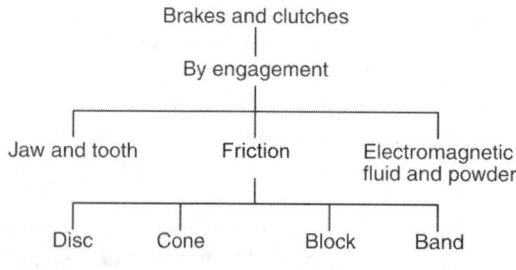

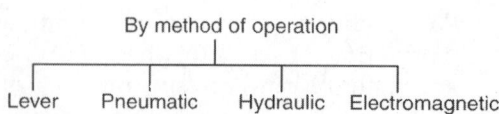

Fig. 18.1: Classification of brakes and clutches

Energy equations: In general the energy absorbed by a brake is equal to the sum of the energy given up by the live load and the energy of all moving parts that are being retarded. To deduce an expression for the tangential force F_t which must be applied to the brake sheave in order to decrease the velocity of a load Q,

the case of a hoisting drum lowering a load will be considered. The following additional designations will be used:

v_1, v_2 = speed of the live load before and after the brake is applied, respectively, in metres per second;

ω_1, ω_2 = angular velocity of the rotating parts, in radians per second;

n_1, n_2 = speed of the brake sheave, in revolutions per seconds;

t = duration of the brake application, in seconds;

D = diameter of the brake sheave, in metres.

The change of the speed of the live load from v_1 to v_2 represents a decrease of kinetic energy by the amount

$$E_k = \frac{Q(v_1^2 - v_2^2)}{2g} \qquad \ldots (18.1)$$

This must be absorbed by the brake. During the same time t the brake must absorb the change of the potential energy, which is equal to the load times the mean velocity $\frac{1}{2}(v_1 + v_2)$ times t. Thus

$$E_p = \frac{1}{2}Q(v_1 + v_2)t \qquad \ldots (18.2)$$

Finally, the brake must also absorb the change of the kinetic energy of all rotating parts, such as the hoist drum and various gears and sheaves. This is,

$$E_r = \Sigma \frac{Wk_o^2(\omega_1^2 - \omega_2^2)}{2g} \qquad \ldots (18.3)$$

312

where, W and k_o designate the weight and the radius of gyration, respectively, of each of these parts. Total change of energy $E = E_k + K_b + E_r$

The work to be done by the tangential force F_t at the brake sheave surface, in t sec, is

$$W_k = \frac{F_t \pi D(n_1 + n_2)t}{2} \qquad \ldots (18.4)$$

Equating this work to the sum $(E_k + E_p + E_r)$, and solving for F_t, gives

$$F_t = \frac{2(E_k + E_p + E_r)}{\pi D(n_1 + n_2)t} \qquad \ldots (18.5)$$

The magnitude of F_t depends on the final velocities v_2 and n_2 and on the braking time t. It attains a maximum value if $v_2 = 0$ and $n_2 = 0$, when the load is stopped.

The torque which the brake must absorb is

$$T = \tfrac{1}{2}F_t D \qquad \ldots (18.6)$$

Heat dissipation: The energy E absorbed by the brake and transformed into heat must be dissipated to the surrounding air in order to prevent an excessive temperature rise of the brake. The temperature rise depends on the amount of energy which the brake is required to absorb per unit of time and on the weight of the heated parts, chiefly of the sheave rim. The highest permissible temperature t_2 depends on the material of the friction surfaces. For leather, fiber, and wood facing, t_2 should not exceed 65° to 70°C to prevent charring. For asbestos and metal surfaces that are slightly lubricated, t_2 should not exceed 90° to 100°C, to prevent burning of the oil film. In automobile brakes with asbestos block lining, t_2 goes up to 200°C, and even to 260°C.

Instead of computing the temperature rise, it is more practical to establish the relation between the energy to be absorbed and the factors influencing its absorption and dissipation. The chief factors are the size and character of the friction surface and of the surface dissipating the heat. The energy absorbed by a brake per second can be computed by the equation

$$E = fpA_f v \qquad \ldots (18.7)$$

where, f is friction coefficient

p is the specific pressure, in N/m^2

A_f is the contact area of the friction surfaces, in square m;

v is the relative velocity of the friction surfaces, in m/s.

This energy must not be greater than the capacity of the brake to dissipate the heat of friction. This capacity can be considered as the product of the area A_d that dissipates the heat and a factor k which is a function of the service conditions. Thus

$$fpA_f v \leq kA_d \qquad \ldots (18.8)$$

Values of k taken from actual brake performance are[1]

0.04 for continuous operation, lowering brakes, wood on cast iron

0.08 for intermittent operation, stopping brakes, wood on cast iron

0.12 for good heat dissipation, metal on cast iron in an oil bath

0.08 for continuous operation, woven asbestos on steel

0.12 for intermittent operation, woven asbestos on steel

The magnitudes of f and the limiting values for p in equation 18.8 are given in Table 18.1.

The coefficient of friction depends on the nature of the friction surfaces, the specific pressure, and the rubbing velocity. The results of tests for friction between steel wheels and cast-iron block,[2] for rubbing velocities v from 0 to 25 mps, may be expressed by the equation

$$f = \frac{0.4}{\sqrt[3]{v}} \qquad \ldots (18.9)$$

The influences of velocity and specific pressure on the coefficients of friction between

---

[1] Hütte, *op. cit.*, p. 159

[2] Lionel S. Marks, ed., Mechanical Engineers' Handbook, 5th ed. (New York: McGraw-Hill Book Company Inc., 1951), p. 221.

Table 18.1: Friction coefficients and allowable pressures

Materials in contact	Friction coefficient f			Allowable pressure kN/m²
	Dry	Greasy	Lubricated	
Cast iron on cast iron	0.2–0.15	0.10–0.06	0.10–0.05	1000–1750
Bronze on cast iron	–	0.10–0.05	0.10–0.05	550–850
Steel on cast iron	0.30–0.20	0.12–0.07	0.10–0.06	800–1400
Wood on cast iron	0.25–0.20	0.12–0.08	–	400–620
Fiber on metal	–	0.20–0.10	–	70–200
Cork on metal	0.35	0.30–0.25	0.25–0.22	50–100
Leather on metal	0.5–0.3	0.20–0.15	0.15–0.12	70–200
Wire asbestos on metal	0.5–0.35	0.30–0.25	0.25–0.20	280–550
Asbestos blocks on metal	0.48–0.40	0.30–0.25	–	280–1100
Asbestos on metal, short action	–	–	0.25–0.20	1400–2100
Metal on cast iron, short action	–	–	0.10–0.05	1400–2100

steel drums and different brake linings in automobiles are shown in Fig. 18.2.[3]

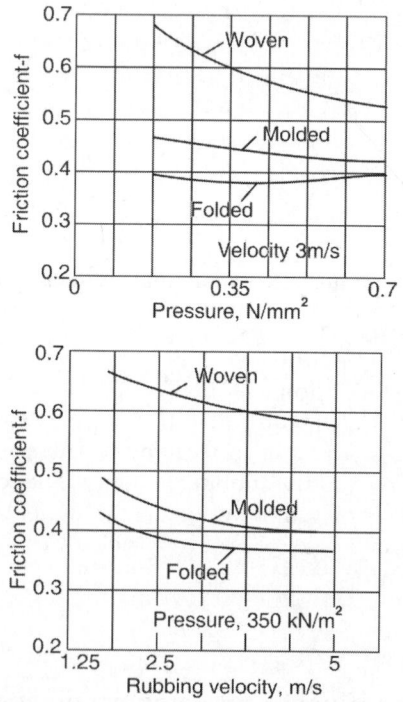

Fig. 18.2: Friction coefficient of brake linings

Finally, it may be readily seen that the tangential force F_t in equation 18.6 may be computed by the relations

$$F_t = f p A_f \qquad \text{... (18.10)}$$

From equation 18.8,

$$A_d \geq \frac{F_t v}{k} \qquad \text{... (18.11)}$$

Rating of brakes: The natural and frequently used way of rating a brake is by the torque T (in N-mm) which it can absorb. Substituting the value for F_t from equation 18.10 in equation 18.6 gives

$$T = \tfrac{1}{2} f p A_f D \qquad \text{... (18.12)}$$

Another way of rating, also often used, is in terms of power. The power P of a brake is usually considered to be equal to the power of the motor used for hoisting the load. Actually the load may be lowered at a higher rate of speed than it is raised. This increases the energy or power which the brake has to absorb, but the safety margin in the rating of the brake must anticipate such an overload.

[3] A. Vallance and V. L. Doughtie, Design of Machine Members, 2nd ed. (New York: McGraw-Hill Book Company, Inc., 1943), p. 216.

Classification: Brakes may be divided into two groups, according to the direction of the acting force: (a) *radial brakes* and (b) *axial brakes*.

Radial brakes, in turn, may be subdivided into *external brakes* and *internal brakes*. Another basis of classification may be the shape of the friction detail; thus there may be distinguished *block brakes* and *band brakes*.

Axial brakes may be subdivided into *cone brakes* and *disk brakes*.

18.2 BLOCK BRAKES

When the brake is applied by operating a single lever, the operating lever with the friction block can be considered as a free body in equilibrium under the action of the applied force F, Fig. 18.3a, the normal reaction N between the sheave and block, the friction force fN between them, and the pin reaction R, the last being unknown in magnitude and direction. For a body in equilibrium the moments with respect to any point must balance. Therefore, if moments are taken with respect to the axis of the fulcrum pin, and clockwise rotation is considered

$$Fa + fNc = Nb, \quad F = \frac{Nb - fNC}{a} \quad \text{...(18.13)}$$

For counterclockwise rotation

$$Fa = Nb + fNc \quad \text{...(18.14)}$$

When the frictional force is responsible for additional force requirement, i.e. when the force

$$F = \frac{Nb + fNc}{a},$$

the shoe is called a lagging shoe.

From these two equations the force needed to operate the brake, in general, is

$$F = \frac{Nb + fNc}{a} \quad \text{...(18.15)}$$

But

$$fN = F_t \quad \text{...(18.16)}$$

where, F_t is given by equation 18.5. Hence equation 18.15 may be written as follows:

$$F = \frac{N(b \pm fc)}{a}$$

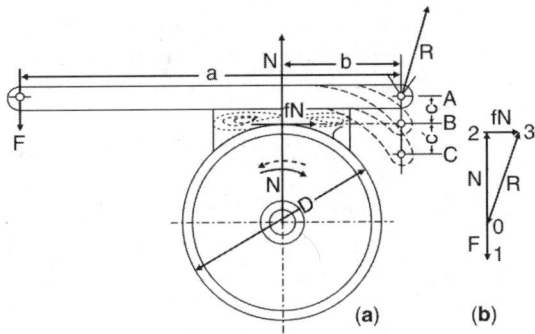

Fig. 18.3: Single-lever block brake

or
$$F = F_t \frac{(b \pm fc)}{fa} \quad \text{...(18.17)}$$

The magnitude of F depends on the direction of rotation. It is smaller for clockwise rotation. This condition is known as self energizing brake *as friction force is helping in application of brake (18.1). This is also called leading shoe*. In fact, it may be zero if $b = fc$; and it may be negative if $b < f_c$, *i.e. this condition is called self-acting brake or grabbing brake. The drum does not move when the block is touching it.* The magnitude of the friction coefficient f is rather uncertain and varies with the condition of the surfaces and with the velocity and specific pressure. Therefore, to avoid grabbing or self actuating of the brake, b must be kept sufficiently larger than fc. On the other hand, a negative value of F means that the sheave can turn only if the lever is slightly lifted. This is also known as self locking section. This gives an automatic brake used for lowering loads. To be certain that the automatic brake will not begin to slip as the result of a change of f, the lever ratio b/c must be smaller than the smallest value of f for the existing conditions.

If $c = 0$ in Fig. 18.3a, i.e. fulcrum is at B, F does not depend on the direction of rotation. This is an advantage if the brake is operated in both directions. If, on the other hand, the fulcrum point is lowered by an additional amount c', to position C the influence of the direction of rotation is reversed.

The magnitude of the force F determines the dimensions of the operating lever. To determine the size of the fulcrum pin and its bearings, the magnitude of the force R coming upon it must be known. This force is easily found graphically, as shown in Fig. 18.3b. It is the geometrical resultant of the forces F, N, and fN drawn to scale, with their respective directions of action.

Long piveted shoe: In this analysis it is assumed that the pressure between the shoe and drum is uniform. This is valid for a short shoe. If the angle subtended by shoe is more than 60°, the pressure is not uniform as it can be seen that when the angle is 180°, the shoe is pressed by the lever, the pressure will be zero at the end of the shoe. For exact analysis, consider an element of shoe Fig. 18.4a having width b.

Area of the element of shoe = $brd\phi$

On this small area, pressure, p is uniform.

The normal reaction will be opposite to the force applied, i.e.

$$dN = pbrd\phi$$

The tangential force of friction $dF_t = fpbrd\phi$. Frictional torque $dT = fpbr^2d\phi$

Considering the pressure distribution between shoe and the drum, Fig. 18.4b, the pressure at the centre C is maximum p_m and the pressure at any other point A is given by $p = p_m \cos \phi$.

Force exerted by the shoe along the vertical direction is given by

$$F = \int_{-\theta}^{\theta} dN \cos\phi = \int_{-\theta}^{\theta} p_m br \cos^2 \phi \, d\phi$$

$$= \frac{p_m br}{2} (2\theta + \sin 2\theta)$$

Frictional torque

$$T = \int_{-\theta}^{\theta} dT = \int_{-\theta}^{\theta} dF_t \times r = \int_{-\theta}^{\theta} fp_m br^2 \cos\phi \, d\phi$$

$$= 2fp_m \, br^2 \sin\theta$$

Torque $T = f' Fr$

By substituting value of F and T,

$$f' = \frac{4f \sin\theta}{2\theta + \sin 2\theta} \qquad \dots (18.18)$$

is called equivalent coefficient of friction for simplifying calculation, f will be substituted by f' and pressure is considered as uniform.

Although it is simple and reliable, the single-lever block brake is not used much, because the normal force N exerts a heavy pressure on the shaft bearings and produces bending of the shaft. This is particularly objectionable in large brakes.

Double-block brakes, with two brake blocks located diametrically opposite to each other, are used to overcome the drawbacks of single-lever brakes. A double-block brake used on cranes is shown in Fig. 18.5. The brake is set by a spring s which pulls the upper ends of the brake levers together. When a force F is applied to the bell crank lever b, which has its middle fulcrum i on the end of the brake lever, the

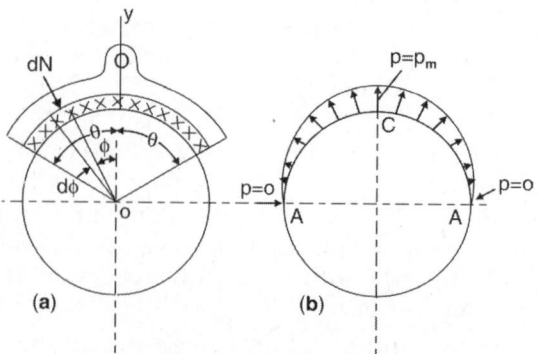

Fig. 18.4: Long paroled shoe

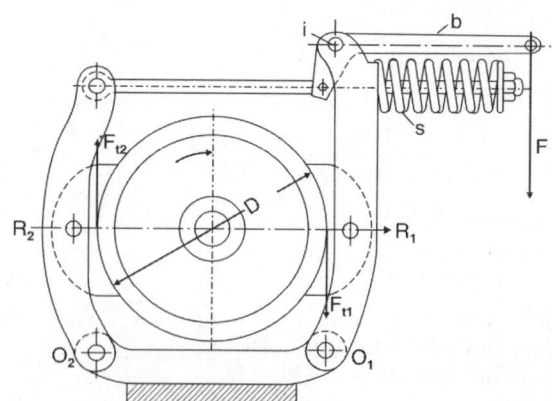

Fig. 18.5: Double-block crane brake

spring s is compressed and the brake is released. This type of brake is often used on electric cranes, and the effort F is produced by an electromagnet or solenoid. When the current is off, there is no pull F on the bell crank, and the brake is set automatically, thus preventing the load from moving down. This type of brake is called holding brake.

Design procedure: It is first necessary to select a sheave or drum diameter D and width b in accordance with the given load, Table 18.2 the given rotative speed, and the other data by equation 18.5. The second step is to determine the tangential effort F_t by equation 18.5 and to compute the torque capacity T by equation 18.6. The next step is to decide what materials to use on the friction surfaces, to select the corresponding friction coefficient from Table 18.1, and to find the normal force N from equation 18.16 by using equivalent friction coefficient by equation 15.15.

Table 18.2: Size of wheels of brakes

Diameter mm	Width mm	Radial clearance mm
160	50	
200	65	
250	80	1 to 1.5
320	100	
400	125	
500	160	1.5 to 2.0
630	200	
800	250	2 to 2.5

After this the necessary area A_{sf} of the brake shoes is found from the relation.

$$A_{sf} = \frac{N}{p} \qquad \ldots (18.19)$$

where, A_{sf} is the projected area normal to the direction of N, in square mm. In regard to p, there is another limiting factor—the capacity of the brake to dissipate the generated heat of friction. According to experience, for wooden or molded asbestos blocks, if v is the rubbing velocity, in m per second, the following limits should be assumed;

$pv \leq 1$ for continuous operations, as in lowering a load,

$pv \leq 2$ for intermittent operations with comparatively long periods of rests; and

$pv \leq 3.0$ for continuous application of load and good dissipation of heat as in oil bath.

The actual value of p for calculating A_{sf} should be less than found from equation 18.19. The actual width of the brake lining will be less than the width of the drum from Table 18.2 by 5 to 10 mm. Then the projected length L of the lining is,

$$L = \frac{A_{sf}}{b}$$

For this value of projected length L, find the angle made by it at the centre $\theta = 2 \sin^{-1} \dfrac{L}{D}$. *This angle θ should be $\leq 120°$ (preferably between $60°$ and $120°$) as it will not be effective if $\theta > 120°$, (18.4), if not repeat with a new drum diameter.*

Distance between the fulcrums of the two levers on the lower side $\leq D$

Distance of fulcrum from the centre of drum $= 0.6D$

Height of levers from the centre of drum $= 0.6D$

Length of shorter arm of bell crank lever $= 0.1D$ to $0.2D$.

Example 18.1: Determine (a) the capacity and (b) the main dimensions of a double block brake for the following conditions: The brake sheave is mounted on the drum shaft; the hoist with its load weighs 27 kN and moves downwards with a velocity of 1.2 m/s, the pitch diameter of the hoist drum is $D = 1.2$ m; the hoist must be stopped in a distance of 3 m, the KE, of the drum may be neglected.

(a) the shaft speed is $n = \dfrac{1.2 \times 60}{\pi \times 1.2} = 19.1$ rpm

as $\left[v = \dfrac{\pi D n}{60} \right]$

Select a brake sheave diameter $D = 800$ mm and width 250 mm. The time t for stopping in 3m with $v_1 = 1.2$ m/s and $v_2 = 0$.

$$t = \frac{3}{0.5(1.2 + 0)} = 5s$$

Change of K.E. of load

$$E_k = \frac{27 \times 1.2^2 \times 1000}{2 \times 9.81} = 1981.6 \text{ Nm}$$

Change of P.E. of load

$$E_p = \frac{27000 \times 1.2 \times 5}{2} = 81000 \text{ Nm}$$

Total change of energy = 1981.6 + 81000 = 82981.6 Nm

The tangential effort by equation 18.5 is

$$F_t = \frac{82918.6 \times 2 \times 1000 \times 60}{\pi \times 800 \times 19.1 \times 5} = 41487 \text{ N}$$

Considering $\frac{F_t}{2}$ as frictional force

at each lever $\frac{41487}{2} = 20743.5 \text{ N}$

The torque capacity by equation 18.6 is

$$T = \frac{1}{2} \times 41487 \times \frac{800}{1000} = 16595 \text{ Nm}$$

and the power is, therefore

$$P = \frac{2\pi \times 19.11 \times 16595}{1000 \times 60} = 33.19 \text{ kW}$$

(b) A suitable brakeshoe material is asbestos block. A cast iron sheave will be used and from Table 18.1 a safe value of coefficient of friction is $f = 0.4$. The normal force, by equation 18.16, is

$$N = \frac{20743.5}{0.4} = 51858.75 \text{ N}$$

For continuous operation, with $v = 1.2$ m/s and $pv \le 1$.

$p = \frac{1}{1.2} = 0.833 \text{ N/mm}^2$. The maximum value is 2.1 N/mm². Using $p = 0.8$ N/mm². The brakeshoe area, by equation 18.18 is

$$A_{sf} = \frac{51858.75}{0.8} = 64823.4 \text{ mm}^2$$

For the drum chosen width = 250 mm. Taking the effective width $b = 240$ mm, the projected length L of each shoe

$$L = \frac{64823.4}{240} = 270 \text{ mm}$$

Angle θ at the centre is $2 \sin^{-1} \frac{270}{800} = 40° $ O.K.

Example 18.2: Design a pivoted double shoe spring operated holding brake for a crane hoisting loads up to 30 kN. Maximum lifting speed is 25 m/s. Moment of inertia of rotating parts is 10 Nm²s. Motor speed is 760 RPM. Time of application of brake is 1 second.

Lifting speed $v = 25$ m/m $= \frac{25}{60} = 0.42$ m/s

Design load $Q = 1.5 \times 30 = 45$ kN

Change of kinetic energy $E_k = \dfrac{Q(v_1^2 - v_2^2)}{2g}$

$$= \frac{45000 \times (0.42)^2}{2 \times 9.81} = 404.58 \text{ Nm}$$

Change of potential energy $E_p = \dfrac{Q(v_1 + v_2)t}{2}$

$$= \frac{45000 \times 0.42 \times 1}{2} = 9450 \text{ Nm}$$

Change of kinetic energy of rotating parts $E_r = \dfrac{1}{2} I w^2$

$$w = \frac{2\pi N}{60} = \frac{2\pi \times 760}{60} = 79.59 \text{ rad/s}$$

$$E_r = \frac{1}{2} \times 10 \times (79.59)^2 = 31672.84 \text{ Nm}$$

Total change of energy $E_T = E_k + E_p + E_r$

$$= 404.58 + 9450 + 31672.84$$

$$= 41527.42 \text{ Nm}$$

Taking a standard drum diameter D of 320 mm and width $b = 100$ mm, radial clearance 1.5 mm W_k frictional work done on the drum

$$= \frac{F_t \pi D(n_1 + n_2)t}{60 \times 2}$$

$$= \frac{F_t \pi \times 0.32 \times 760 \times 1}{60 \times 2} = 6.367 \ F_t$$

Equating it to total change of energy

$$F_t = \frac{41527.42}{6.367} = 6522.2 \text{ N}$$

Considering asbestos brake lining with μ = 0.4, taking average pressure of 0.2 to 0.6 N/mm² it as 0.4 N/mm² and can sustain a maximum temperature of 350°C.

Taking equal reactions on the two sides, i.e.

$$R_1 = R_2 = R \ [2F_t = F_{t1} + F_{t2}]$$

Then $R = \dfrac{F_t}{2\mu} = \dfrac{6522.2}{2 \times 0.4} = 8152.75 \text{ N}$

Width of lining b_1 = width of drum −10

$$= 100 - 10 = 90 \text{ mm}$$

$R = p\, l_p\, b_1$

$$l_p = \frac{R}{p\, b_1} = \frac{8152.75}{0.4 \times 90} = 226.46 \text{ say } 227 \text{ mm}$$

$$\theta = \sin^{-1}\frac{l_p/2}{D/2} = \sin^{-1}\frac{227}{320} = 45.18°$$

$$2\theta = 90.36° \text{ o.k.} < 120°$$

Checking for heat generation and heat description as product pv

$$pv = 0.4 \times 0.42 = 0.168 \text{ as it is } < 1$$

So suitable, i.e. continuous service with short rest periods and poor heat dissipation.

Assuming various dimensions of levers, etc. as referred to Fig. 18.4.

Distance between hinges = 0.8D

Height of hinges from the center of drum = 0.6D height of top pull rod from the centre of drum = 0.6D

Length of short arm of bell crank lever = 0.2D

0.2D = 64 mm, 0.6D = 192 mm; 0.8D = 256 mm

Calculate equivalent $f' = \dfrac{4f \sin\theta}{2\theta + \sin 2\theta}$

$$= \frac{4 \times 0.4 \times 0.7093}{1.577 + 0.9949}$$

$$= 0.44$$

Taking moments about the right hinge O_1.

Subscript 1 is for right side and 2 for left side

$R_1 \times 0.6D + F_{t1} \times 0.1D = S \times 1.2D$

$0.6R_1 + 0.1\, F_{t1} = 1.2S$

$F_{t1} = 12S - 6R_1$

Taking moments about O_2

$R_2 \times 0.6D - F_{t2} \times 0.1D = S \times 1.2D$

$F_{t2} = 6R_2 - 12S$

$F_{t1} + F_{t2} = F_t = 6\,R_2 - 6R_1$

$$R_2 - R_1 = \frac{6522.2}{6} = 1037 \text{ N} \qquad \text{... (i)}$$

Also $F_{t1} + F_{t2} = \mu'(R_1 + R_2) = 6522.2$

$$R_1 + R_2 = 14823 \qquad \text{... (ii)}$$

From (i) and (ii)

$R_2 = 7930$ N

$R_1 = 6893$ N

Maximum value of $S = \dfrac{F_{t1} + 6R_1}{12}$

$$= \frac{0.44 \times 6893 + 6 \times 6893}{12} = 3699 \text{ N}$$

In the opposite rotation, $S = \dfrac{F_{t2} + 6R_2}{12}$

$$= \frac{0.44 \times 7930 + 6 \times 7930}{12} = 4256 \text{ N}$$

Spring design force = 1.25 × 4256 = 5320 N

Taking mean diameter = 100

$C = D/d = 6$

$$K = \frac{4 \times 6 - 1}{4 \times 6 - 4} + \frac{0.615}{6} = 1.25$$

Wire diameter $d = \left[\dfrac{8 \times 1.25 \times 5320 \times 100}{\pi \times 400}\right]^{\frac{1}{3}}$

$$= 16.17 \text{ say } 18 \text{ mm}.$$

Taking a larger deflection of 20 mm, the number of turns is given by

$$n = \frac{\delta G d^4}{\delta F D^3} = \frac{20 \times 84000 \times 18^4}{8 \times 5320 \times 100^3} = 4.15 \text{ say } 5$$

Considering two (2) end turns, total no. of turns = 7

Free length of spring = $nd + 0.1nd + y$

$$= 7 \times 18 + 0.1 \times 7 \times 18 + 20$$

$$= 126 + 12.6 + 20 = 158.6 \text{ say } 160 \text{ mm}$$

Levers and pins

When the brake is engaged, forces are shown in Fig. 18.4.

With increased value of $S = 5320$, $R_2 = 9913$ and $Ft_2 = 4361$. When each lever moves backwards by 1.5 mm., the spring force will increase to

$$5320 + \frac{5320}{160} \times 3 = 5420 \text{ N}$$

Maximum B.M at $B = 5320 \times 192 = 1021440$ Nmm

When brake is away from drum

Maximum bending moment at $C = 5420 \times 1.2 \times 320 = 2081280$ Nmm.

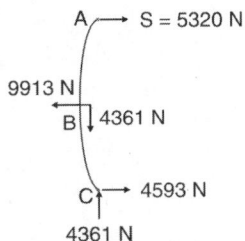

Fig. 18.5: Free body diagram of lever

Taking the larger value, and b the width, the depth of the lever $d = 3b$, $S_b = 70$ N/mm^2

$$\frac{bd^2}{6} \times S_b = M$$

$$b = \left[\frac{2081280 \times 6}{9 \times 70}\right]^{\frac{1}{3}} = 27 \text{ mm}$$

depth $d = 3 \times 27 = 81$ mm and increasing it to 90 mm.

For easy operation and for locating spring, one lever is replaced by two half levers of width 0.65 × 27 ≈ 20 mm each. Both levers will have the same dimensions. Taking width = 30 mm

Bottom hinges are subjected to a reaction

$$= \sqrt{4593^2 + 4361^2}$$

$$= 6333 \text{ N}$$

Pins will be subjected to double shear having diameter d_1 and core diameter d_c

$$2\frac{\pi}{4}d_c^2 \times 45 = 6333$$

$$d_c = \left[\frac{6333 \times 4}{2\pi \times 45}\right]^{\frac{1}{2}} = 9.4 \text{ mm}$$

The nominal diameter = 12 mm.

This diameter can be checked for bearing pressure in the lever

$$f_b = \frac{6333}{12 \times 2 \times 15} = 17.59 \text{ N/mm}^2 \text{ o.k.}$$

The shoe will also be fixed by pins. The resultant force acting on pins of the shoe

$$= \sqrt{9913^2 + 4361^2} = 10829.8 \text{ N}$$

This pin is also subjected to double shear having core diameter d_2

$$2 \times \frac{\pi}{4} \times d_2^2 \times 45 = 10829.8$$

$$d_2 = \left[\frac{10829.8 \times 4}{2\pi \times 45}\right]^{\frac{1}{2}} = 12.3$$

Nominal diameter can be taken as 14 mm.

The tie rod for supporting the spring has a diameter d_3

$$\frac{\pi}{4}d_3^2 \times s = 5420$$

$$d_3 = \left[\frac{5420 \times 4}{\pi \times 70}\right]^{\frac{1}{2}} = 9.92$$

The nominal diameter = 12 mm.

The diameter is further increased to 14 mm for providing pin to connect with shorter arm of bell crank lever. The bell crank lever has a shorter arm = 64 mm and longer arm greater than compressed spring, i.e. 160–20 = 140 mm. To accommodate a solenoid or a cylinder for actuating the lever.

The longer lever length is taken

$$= 300 \text{ mm}$$

Maximum bending moment = 5420 × 64

$$= 346880 \text{ Nm}$$

Force to be applied for actuating brake

$$= \frac{5420 \times 64}{300} = 1156 \text{ N}$$

Taking width of bell crank lever b_1 and depth t_1 as $t_1 = 3b_1$

$$\frac{b_1 t_1^2}{6} s_t = M$$

or $\quad b_1 = \left[\frac{346880 \times 6}{9 \times 70}\right]^{\frac{1}{3}} = 14.89 \text{ say } 15 \text{ mm}$

and $\quad t_1 = 45$ mm

Maximum reaction at the fulcrum of bell crank lever

$$= \sqrt{5420^2 + 1156^2} = 5541.9 \text{ N}$$

Diameter of pin to connect bell crank lever and right side lever d_1 is found as:

$$2 \times \frac{\pi}{4} d_1^2 \times 45 = 5541.9$$

$$d_1 = \left[\frac{5541.9 \times 4}{2 \times \pi \times 45}\right]^{\frac{1}{2}} = 8.85 \text{ say } 9 \text{ mm}$$

The boss diameter at the pin may be taken as 50 mm diameter. The modulus of section of lever at the pin section (having width 15 mm and a hole of 9 mm)

$$= \frac{1}{25}\left[\frac{15 \times 50^3}{12} - \frac{15 \times 9^3}{12}\right] = 6213.5 \text{ mm}^3$$

as compared to the requirement of

$$\frac{15 \times 45^2}{6} = 5062.5 \text{ mm}^3 \text{ so it is o.k. but it is}$$

desirable to take $b_1 = 20$ mm and $t_1 = 60$ mm.

Grooved sheaves: Grooved sheaves, as in Fig. 18.6a, are used to increase the force F_t without increasing the normal force N. Summing up the forces along the vertical axis in Fig. 18.6b results in

$$N = N' (\sin \alpha + f \cos \alpha) \qquad \dots (18.20)$$

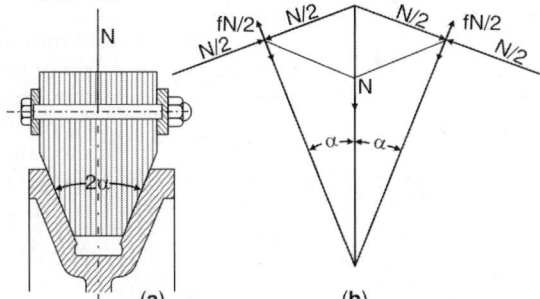

Fig. 18.6: Grooved brake sheave

For the case of a brake operating in both directions, apply equation 18.15, in which $c = 0$, the value of N is taken from equation 18.19, and N' is then eliminated by using the relation $fN' = F_t$. The result is

$$F = \frac{F_t b(\sin \alpha + f \cos \alpha)}{fa} \quad \ldots (18.21)$$

The expression $f/(\sin \alpha + f \cos \alpha)$ is called the apparent *coefficient of friction*. The groove half-angle α should not be made smaller than 20°, to prevent grabbing and locking, if the friction coefficient should increase. On the other hand, it is seldom made greater than 30°, as then the advantage of the grooves is decreased.

Internal block brakes: Internal block brakes are very compact and are particularly suitable if rotation occurs in both directions. Figure 18.7a shows a widely used automobile brake arrangement with metal-asbestos lining. The brake is applied by turning the elliptical cam c clockwise. When the cam is turned back, the springs s pull the blocks away from the drum surface, releasing the brake. In order to prevent grabbing and locking of the brake, the smallest angle α, formed by the lines going through the end of the lining and the centers o and o_1, must be greater than the angle of friction ϕ. For $f = 0.5$, the maximum center angle of the lining is $\beta = 72°$; and for $f = 0.4$, the angle β may be 90°.

Force analysis: For the purpose of analysis, Fig. 18.7b it is assumed that,

(i) The normal pressure at the point of contact between brake lining and brake cylinder is proportional to its vertical distance from the fulcrum

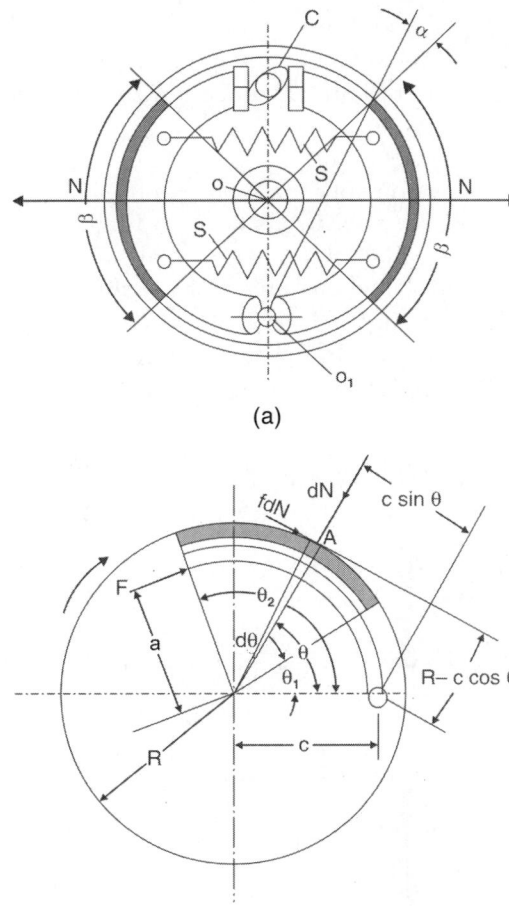

(a)

(b)

Fig. 18.7: Internal block brake

(ii) There is no contact between lining and cylinder near the fulcrum

(iii) The coefficient of friction is constant and is not affected by pressure and velocity.

The pressure p at any point of contact $= p_m \sin\theta$, p_m is the intensity of maximum pressure.

θ, angle of the contact point from the axis passing through the fulcrum.

θ_1, angle from the fulcrum to heal of shoe at the centre

θ_2, angle from the fulcrum to the toe of shoe at the centre

p_m, maximum pressure on the right shoe

p'_m, maximum pressure on the left shoe

R, radius of drum

c, distance of the centre to the fulcrum

a, perpendicular distance of the applied force F from the fulcrum

b, width of lining

At any point of contact $p = p_m \sin\theta$

At $\theta = 0$, pressure is zero and maximum pressure is at $\theta = 90°$

For an elementary area at $A = bR\,d\theta$

Normal force $dN = pb\,R\,d\theta$

Frictional force $dF_t = f\,dN = fpb\,R\,d\theta$

Frictional torque $dT = RdF_t = fpb\,R^2\,d\theta$

Total torque

$$T = \int_{\theta_1}^{\theta_2} f\,p_m\,b\,R^2 \sin\theta\,d\theta$$

$$= fp_m\,b\,R^2\,(\cos\theta_1 - \cos\theta_2) \qquad \ldots (18.22)$$

Moment of normal force about fulcrum

$$M_n = \int dN\,c\sin\theta = \int_{\theta_1}^{\theta_2} p\,b\,Rc\sin\theta\,d\theta$$

$$= p_m\,b\,R\,c\int_{\theta_1}^{\theta_2} \sin^2\theta\,d\theta$$

$$\frac{1}{2}p_m bRc\left[(\theta_2 - \theta_1) + \frac{1}{2}(\sin 2\theta_1 - \sin 2\theta_2)\right]$$
$$\ldots (18.23)$$

Moment of the frictional force about the fulcrum

$$M_f = \int_{\theta_1}^{\theta_2} f\,p\,b\,Rd\theta(R - c\cos\theta)$$

or $$M_f = \int_{\theta_1}^{\theta_2} f\,p_m\,b\,R\sin\theta(R - \cos\theta)\,d\theta$$

$$= fp_m bR\big[R(\cos\theta_1 - \cos\theta_2) + \frac{c}{4}(\cos 2\theta_2 - \cos 2\theta_1)\big] \qquad \ldots (18.24)$$

To find the actuating force F, summation of moments of all forces about fulcrum is zero

$$F = \frac{M_n - M_f}{a} \qquad \ldots (18.25)$$

and for anticlockwise rotation

$$F = \frac{M_n + M_f}{a} \qquad \ldots (18.26)$$

For the left hand shoe $M_n' = \dfrac{M_n p_m'}{p_m}$ $\ldots (18.27)$

and $$M_f' = \frac{M_f p_m'}{p_m} \qquad \ldots (18.28)$$

In case the maximum pressure is same, the expressions for M_n and M_f will be same as for right shoe.

The brake shoe will be self energizing, if $M_n > M_f$ and it is self locking, if $M_n \le M_f$.

Instead of cam, a convenient method of applying forces and displacement to the ends of levers is by fluid pressure through rams. There will be one ram on each side of the cylinder to operate both levers. When the brake pedal is pressed, the movement and force on the levers is proportional to increase of pressure.

Example 18.3: Design an internal shoe brake for an automobile to overcome a maximum torque of 136 Nm at 1500 rpm. The drum diameter is 250 mm. Coefficient of friction between steel drum and lining is 0.25. For speed of automobile up to 100 km/hr. maximum pressure is limited to 0.28 N/mm². Find the force required to operate the brake and spring force (Fig. 18.8).

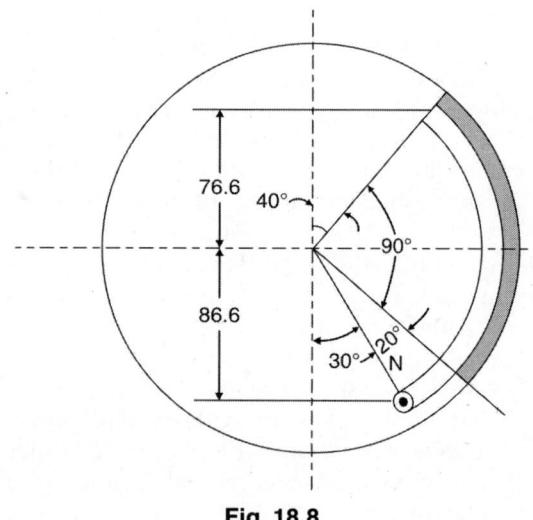

Fig. 18.8

Assume 60% of the torque to be overcome by front wheels, torque for front wheels = 0.6 × 136 = 81.6 Nm. Considering overload of 1.5, torque to be resisted = 1.5 × 81.6 = 122.4 Nm.

$$\text{Torque per wheel} = \frac{122.4}{2} = 61.7 \text{ Nm}$$

$$T = f p_m b R^2 (\cos\theta_1 - \cos\theta_2)$$

Assume the hing is located at a radial distance of 100 mm from the centre and makes an angle of 30° to vertical. The lining is started making 20° at the centre from fulcrum and included angle of lining is taken as 90°

$$\theta_1 = 20°, \theta_2 = 110°$$

Vertical distance of fulcrum from the centre = 100 cos 30° = 86.6 mm

Distance of toe from center = 100 cos 40° = 76.6 mm

$a = 86.6 + 76.6 = 163.2$ mm

$c = 100$

Assuming larger value of torque on the right hand lining as 50 Nm.

$$b = \frac{50,000}{f p_m R^2 (\cos\theta_1 - \cos\theta_2)}$$

$$= \frac{50,000}{0.25 \times 0.28 \times 125^2}(\cos 20° - \cos 110°)$$

$$= 35.66 \text{ say } 40 \text{ mm}.$$

Calculating torque due to normal force and frictional force

$$M_n = \frac{1}{2} p_m b R c \left[(\theta_2 - \theta_1) + \frac{1}{2}(\sin 2\theta_1 - \sin 2\theta_2) \right]$$

$$= \frac{1}{2} \times 0.28 \times 40 \times 125 \times 100$$

$$\left[(110 - 20) + \frac{1}{2}(\sin 40° - \sin 220°) \right]$$

$$= \frac{1}{2} \times 0.28 \times 40 \times 125 \times 100$$

$$\left[\frac{90 \times \pi}{180} + \frac{1}{2} \times 1.2854 \right]$$

$$= 84889 \text{ Nmm}$$

$$M_f = f p_m b R \left[R(\cos\theta_1 - \cos\theta_2) + \frac{C}{4}(\cos 2\theta_2 - \cos 2\theta_1) \right]$$

$$= 0.25 \times 0.28 \times 40 \times 125$$

$$\left[125(\cos 20° - \cos 110°) + \frac{C}{4}(\cos 220° - \cos 40°) \right]$$

$$= 0.25 \times 0.28 \times 40 \times 125$$

$$\left[125 \times 1.2816 + \frac{100}{2} \times (-1.5320) \right]$$

$$= 29260 \text{ Nmm}$$

$$\text{Force } F = \frac{M_n - M_f}{a}$$

$$= \frac{84889 - 29260}{163.2} = 340.86 \text{ N say 341 N}$$

$$p'_m = \frac{Fa \, p_m}{M_n + M_f}$$

$$= \frac{341 \times 100 \times 0.28}{84889 + 29260} = 0.0836$$

Torque available on the left lining

$$= \frac{50 \times 0.0836}{0.28} = 14.93 \text{ Nm}$$

Total torque which can be overcome by one brake = 50 + 14.93 = 64.93 Nm o.k. as it is more than 61.7 Nm.

Maximum spring force = N

$$\text{Normal force } N = \int_{\theta_1}^{\theta_2} p \, b \, R \, d\theta$$

$$= \int_{\theta_1}^{\theta_2} p_m \, b \, R \sin\theta$$

$$= p_m \, b \, R \, (\cos\theta_1 - \cos\theta_2)$$

$$= 0.28 \times 40 \times 125 \, (\cos 20° - \cos 110°)$$

$$= 1794.25 \text{ say } 1800 \text{ N}.$$

Spring has to be designed for a force 1.25 × 1800 = 2250 N.

Force to be generated by application of brake by fluid pressure = 341 N.

Assuming a ram diameter of 20 mm.

Fluid pressure p is given by

$$\frac{\pi}{4} \times 20^2 \times p = 341$$

$$p = \frac{341 \times 4}{\pi \times 20^2} = 1.08 \text{ N/mm}^2 \text{ say } 1.5 \text{ N/mm}^2$$

Reactions at the hing are calculated similarly.

18.3 BAND BRAKES

A general arrangement of band brakes is shown in Fig. 18.9. Before a relation is deduced between the tangential force F_t and the pull F at the lever end, it should be noted that owing to friction between the sheave and the band tensions, F_1 and F_2 in the two band ends are different. Such brakes are called *differential brakes*.

Differential brakes: The desired relation for differential brakes may be found by using Fig. 18.9a. An infinitesimal length of the band is subtending an angle $d\theta$. The tension at one end is F, and at the other end is $(F + dF)$; each of these tensions makes an angle $(\frac{1}{2}\pi - \frac{1}{2}d\theta)$ with the vertical center line. The pressure between the band and sheave rim is designated by N, and with a friction coefficient f, the friction force is fN.

This piece of band is held in equilibrium by the four forces F, $(F + dF)$, N, and fN. The summation of the horizontal and vertical components, respectively, gives the following equations:

$$-dF \cos(\tfrac{1}{2}d\theta) + fN = 0 \qquad \ldots (18.29)$$

and

$$-(2F + dF)\sin(\tfrac{1}{2}d\theta) + N = 0 \qquad \ldots (18.30)$$

Eliminating N, and substituting $\frac{1}{2}d\theta$ for $\sin(\frac{1}{2}d\theta)$ and 1 for $\cos(\frac{1}{2}d\theta)$, results in

$$fF\,d\theta - dF = 0 \qquad \ldots (18.31)$$

Separating the variables gives

$$\int_{F_2}^{F_1} \frac{dF}{F} = f \int_0^\theta d\theta \qquad \ldots (18.32)$$

Integrating gives the ratio of the tight tension to the loose tension. Thus,

$$\frac{F_1}{F_2} = e^{f\theta} \qquad \ldots (18.33)$$

The net tension is evidently equal to the tangential braking force, or

$$F_1 - F_2 = F_t \qquad \ldots (18.34)$$

Eliminating F_2 from equations 18.33 and 18.35 gives

$$F_1 = \frac{F_t e^{f\theta}}{e^{f\theta} - 1} \qquad \ldots (18.35)$$

From equations 18.33 and 18.35,

$$F_2 = \frac{F_t}{e^{f\theta} - 1} \qquad \ldots (18.36)$$

This equation shows that for a required F_t the magnitudes of F_1 and F_2 decrease with an increase of the friction coefficient f, and particularly with an increase of the angle of contact θ. Thus, for an average value of $f = 0.3$, the following results are obtained: For $\theta = \pi$, $F_1/F_t = 1.64$ and $F_2/F_t = 0.64$; for $\theta = 1.5\pi$, $F_1/F_t = 1.32$ and $F_2/F_t = 0.32$; and for $\theta = 3.5\pi$, $F_1/F_t = 1.04$ and $F_2/F_t = 0.04$.

Considering the operating lever as a free body and taking moments about the fulcrum 1, and assuming clockwise rotation in Fig. 18.9, results in,

$$Fa + F_1 b_1 = F_2 b_2 \qquad \ldots (18.37)$$

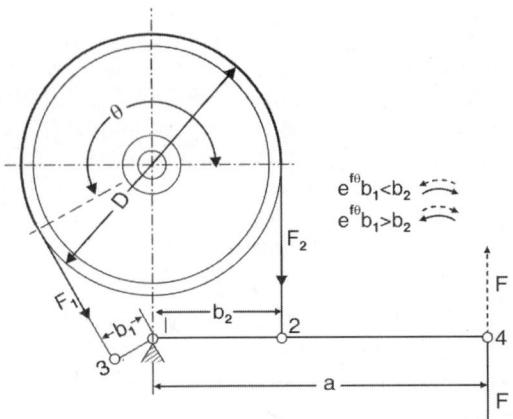

Fig. 18.9a: Differential brake

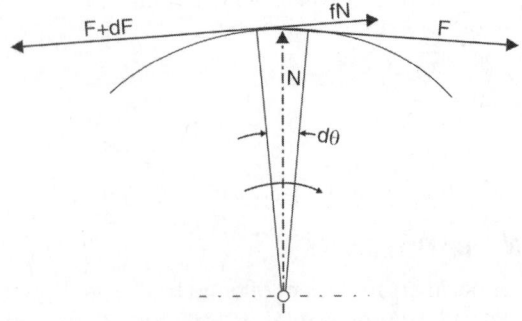

Fig. 18.9b: Band tension

Substituting the values for F_1 and F_2 from equations 18.35 and 18.36, and solving for F, gives

$$F = \frac{F_t(b_2 - e^{f\theta}b_1)}{(e^{f\theta} - 1)a} \quad \ldots (18.38)$$

The conditions represented in Fig. 18.9 require that $b_2 > e^{f\theta}b_1$, or

$$\frac{b_2}{b_1} > e^{f\theta} \quad \ldots (18.39)$$

The brake is self energizing as frictional force is helping or reducing the external force F to be applied.

If $b_2/b_1 = e^{f\theta}$, then F is zero and the brake becomes self-locking. This is undesirable and even dangerous because of the possible fluctuation of the friction coefficient f.

If $b_2/b_1 < e^{f\theta}$, the pull F becomes negative, the brake is applied automatically, and a pull must be applied in the opposite direction, as shown by dotted lines in Fig. 18.9, in order to allow the sheave to turn and thus to lower the load.

If the direction of rotation is reversed, or is counterclockwise, the greater tension F_1 will act at the right end of the band, and the smaller tension F_2 will act at the left end.

A similar analysis gives

$$F = \frac{F_t(e^{f\theta}b_2 - b_1)}{(e^{f\theta} - 1)a} \quad \ldots (18.40)$$

If b_2 does not differ much from b_1, the influence of $e^{f\theta}$, according to equations 18.38 and 18.40, is small. The main factor determining the magnitude of F for a given F_t is the average ratio of the lever arms, or the ratio $(b_1 + b_2)/2a$.

Simple band brake: A simple band brake is a special case of the differential brake, in which one of the band ends is fastened to the fixed fulcrum. One of these arrangements is shown in Fig. 18.10. The expression for F is obtained from equation 18.38 or equation 18.40, the proper one depending on the direction of rotation of the sheave. In this case, $b_1 = 0$ and

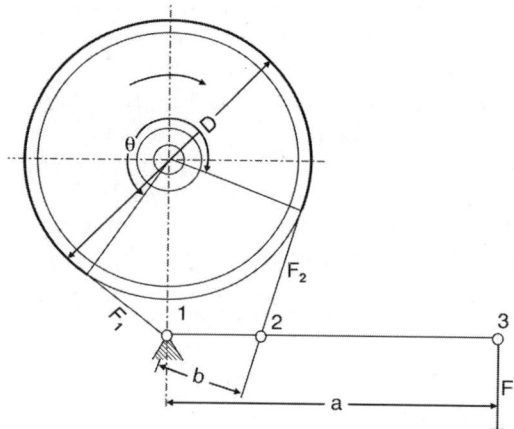

Fig. 18.10: Simple band brake

$b_2 = b$. Thus, for clockwise rotation, equation 18.38 gives

$$F = \frac{F_t b}{(e^{f\theta} - 1)a} \quad \ldots (18.41)$$

Equation 18.41 shows a certain advantage of a simple band brake over a differential brake. By making the angle of contact θ large, the necessary pull F can be made small without increasing unduly the lever arm a. *With this consideration in view, brakes are built in which the band makes several complete turns around the sheave (18.9).*

For counterclockwise rotation, equation 18.40 gives

$$F = \frac{F_t e^{f\theta} b}{(e^{f\theta} - 1)a} \quad \ldots (18.42)$$

In order not to increase F unnecessarily, the rotation should be clockwise. This means that the band end with the higher tension must be fastened to the fixed fulcrum.

Band brake for rotation in both direction: For hoist brakes in which the rotation is reversed, as in cranes, elevators, and mine hoists, it is desirable to have the same pull F, regardless of the direction of rotation. A study of Fig. 18.9 shows that this can be accomplished if both the moment of the tension F_1 and the moment of the tension F_2 act in the same direction and in the opposite direction to the moment of the pull F. To obtain this result the overhanging lever end 3–1 must be turned to another position and the lever arms b_1 and b_2 must be equal. A brake with this arrangement is shown in Fig. 18.11. By applying the same methods of analysis, or by using either equations 18.38 or 18.40, with proper substitutions, the required magnitude of the pull is determined to be

$$F = \frac{F_t b(e^{f\theta} + 1)}{(e^{f\theta} - 1)a} \qquad \dots (18.43)$$

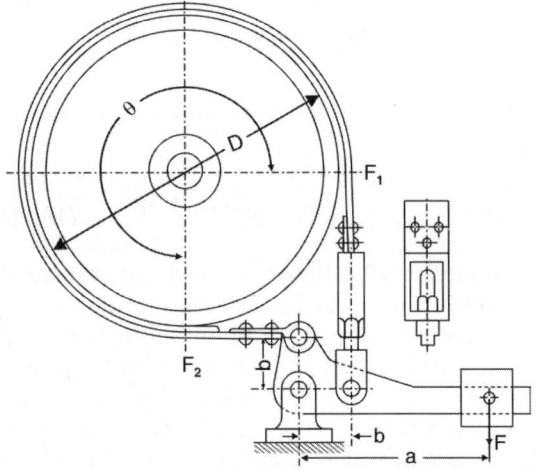

Fig. 18.11: Brake for rotation in both directions

With this arrangement the required pull for rotation in one direction is heavier than required by other brakes, but a considerably lighter pull is sufficient when rotation occurs in the opposite direction.

Pressure on band: The magnitude of the pressure p between the band and the brake sheave may be found by considering Fig. 18.12. The sum of the horizontal components of the pressures is equal to the product pD. Thus, the sum of the forces applied to the band ends, or $2F$, is equal to pDw, where w is the bandwidth. For rotating sheave, $2F = F_1 + F_2$ and the average pressure is

$$p = \frac{F_1 + F_2}{Dw} \qquad \dots (18.44)$$

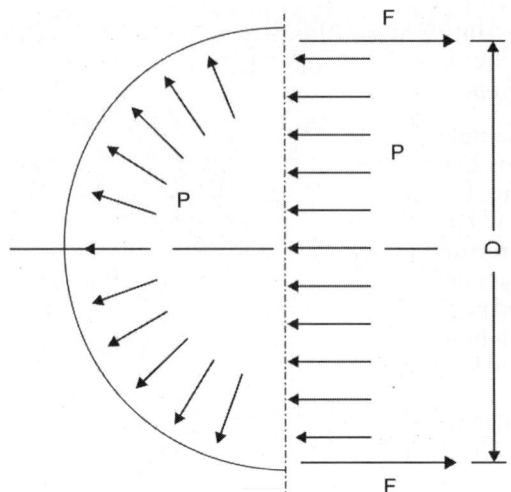

Fig. 18.12: Forces acting on a brake band

Because of stretching of the band, the pressure around the sheave is not constant. It is highest at the tight end of the band, where it is

$$p_1 = \frac{2F_1}{Dw} \qquad \dots (18.45)$$

It decreases gradually toward the other end to the minimum value of

$$p_2 = \frac{2F_2}{Dw} \qquad \dots (18.46)$$

Design remarks: The best material for brake drums and sheaves is cast iron. Particularly suitable is hard-wearing martensitic cast iron having a nickel content of 5 to 6 per cent and a Brinell hardness of 400 to 450. The proper material for brake bands is mild steel.

A practical rule is to determine the *band* thickness h by the relation

$$h = 0.005D \qquad \dots (18.47)$$

After h has been selected, the width w is determined from considerations of strength by applying the equation

$$wh = \frac{F_1}{S_d} \qquad \dots (18.48)$$

In selecting S_d, a safety factor n of $2 \times 2 = 4$ should be used because of the possible sudden application of a brake load. The width w must be checked by equation 18.45.

The brake-band ends are fastened to the lugs either by welding or by riveting with 10 mm rivets.

Example 18.4: A differential band brake needs a force F of 250 N to be applied at the end of the lever 3 (Fig. 18.9). The coefficient of friction $f = 0.35$. Angle of wrap = 200°. Find the maximum and minimum force in the band if clockwise torque of 500 Nm is applied to the drum. What is the maximum torque and power that the brake may apply for anticlockwise rotation. Find the width of band if the allowable load is 180 N/mm.

Take $a = 220$ mm, $b_1 = 40$ mm, $b_2 = 100$ mm, $D = 200$ mm

For clockwise rotation, check for the brake if it is self locking

$$e^{f\theta} = e^{0.35 \times \frac{\pi \times 200}{180}} = 2.71^{1.2217} = 3.38$$

Ratio $\dfrac{b_2}{b_1} = \dfrac{100}{40} = 2.5$ which is less than 3.38 so brake is self locking. Therefore $\dfrac{F_1}{F_2} = e^{\mu\theta}$ does not hold.

Taking moments about the fulcrum
$F_1 \times 40 - F_2 \times 100 = -250 \times 220$
$100(F_1 - F_2) = 500 \times 1000$
$F_2 = 4250$ N
$F_1 = 9250$ N

For anticlockwise rotation, the brake is not self locking so $\dfrac{F_1}{F_2} = e^{f\theta}$ will hold.

$-F'_2 \times 40 + F'_1 \times 100 = -220 \times 250$
or $2.5 F'_1 - F'_2 = 1375$

$\dfrac{F'_1}{F'_2} = 3.38$

$F'_2 = 184.56$ N
$F'_1 = 623.81$ N
Maximum force = 9250 N

Minimum force = 623.81 N

The torque capacity $= \dfrac{100(623.81 - 184.56)}{1000}$

$= 43.92$ Nm

Power required $= \dfrac{2\pi \times 750 \times 43.92}{60 \times 1000}$

$= 3.45$ kW

Taking the thickness of band = 0.005 D = 1 mm

The width of band $= \dfrac{9250}{180 \times 1} = 51.38$ say 52 mm

18.4 CONE BRAKES

A semidiagrammatic drawing of a cone brake is shown in Fig. 18.13a. The outer cone o may form a part of the hoist drum or be attached to it, while the inner cone i is splined to a shaft which can rotate in only one direction, being prevented from running in the opposite direction by a ratchet and pawl.

Force analysis: The magnitude of the force F at the end of the operating lever may be computed as follows: The axial force F_a applied at the cone surface can be resolved, as shown in Fig. 18.13b, into a normal force N and a radial force R. The normal force is

$$N = \frac{F_a}{\sin \alpha} \qquad \ldots (18.49)$$

The radial force is

$$R = \frac{F_a}{\tan \alpha} \qquad \ldots (18.50)$$

In a conical surface the radial forces balance each other. The tangential force, or braking force, F_t, is equal to the normal force multiplied by the friction coefficient, or

$$F_t = fN = \frac{fF_a}{\sin \alpha} \qquad \ldots (18.51)$$

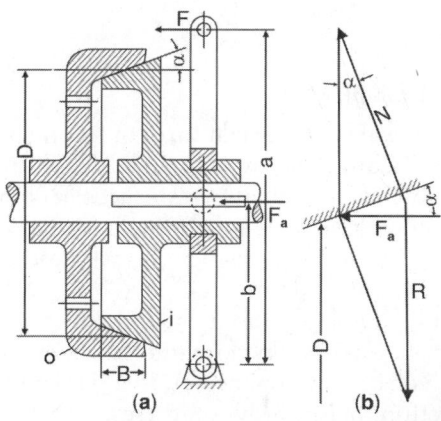

Fig. 18.13: Cone brake

The braking torque is then

$$T = \frac{fF_a D}{2\sin\alpha} \qquad \ldots (18.52)$$

where, D is the mean diameter of the cone.

Owing to the leverage,

$$F_a = \frac{Fa}{b} \qquad \ldots (18.53)$$

The relation between the operating force F and the braking force F_t may be obtained by combining equations 18.51 and 18.53

$$F = \frac{F_t b \sin\alpha}{fa} \qquad \ldots (18.54)$$

The area A of the contact surfaces can be determined, with the designations of Fig. 18.13, by the relation

$$A = \frac{\pi DB}{\cos\alpha} \qquad \ldots (18.55)$$

The average pressure between the contact surface is

$$p = \frac{N}{A} = \frac{F_a}{\pi DB \tan\alpha} \qquad \ldots (18.56)$$

Design remarks: The female cone is usually made of cast iron. The inner cone is also of cast iron, but it is often lined with wood or asbestos blocks in order to increase f. The angle α is made from 10 to 18 deg.

The axial width B, Fig. 18.13, is made from $0.12D$ to $0.22D$, and both D and B are so selected that the pressure p found by equation 18.56 does not exceed the value given in Table 18.1.

18.5 DISK BRAKES

A disk brake is a special form of a cone brake with the angle α equal to 90°. A disk brake usually has more than two surfaces in contact. A multidisk brake is shown diagrammatically in Fig. 18.14. The housing c is fastened to the hoist drum and runs freely with it on the shaft g. The flange f rotates with the shaft, but it can slide axially on the feather key e. The shaft is driven by a motor and can rotate in only one direction, being prevented from rotating in other direction by a ratchet mechanism.

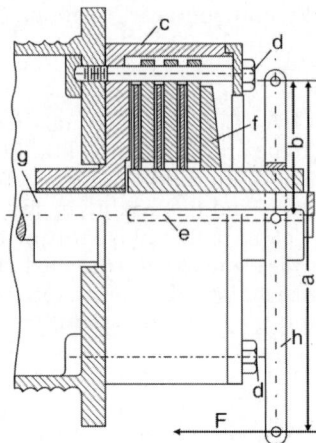

Fig. 18.14: Multidisk brake

Between the inner faces of the housing c and the flange f are assembled a number of friction disks. The larger disks are of cast iron and rotate with the housing c, having holes through which bolts d pass. The smaller disks are of steel, and each has in the center a square hole which fits over the square hub of the flange f. When the load is hoisted, this flange is locked to the housing and to the hoist drum by the pull F on the lever, and it acts as a clutch. When the load must be lowered, the motor is stopped. The shaft cannot run backward, and the operator, by decreasing the pull F, allows the drum to run backward.

Pressure distribution: A reasonable assumption is that the pressure p between the disks is distributed uniformly, as indicated in Fig. 18.15a. This is true with new plates. However, after a certain length of time in operation, the plates become thinner toward the outside diameter, as shown exaggerated in Fig. 18.15b. Such a condition can be explained if it is assumed that the wear is proportional to the work of friction. This work at an element of area is proportional to the product of the pressure and the velocity of rubbing. Therefore, the initial wear of a plate increases from the inside radius R_1 toward the outside radius R_2. After the plate is worn in, further wear is uniform, as indicated in Fig. 18.15c. This indicates that the product pR is constant.

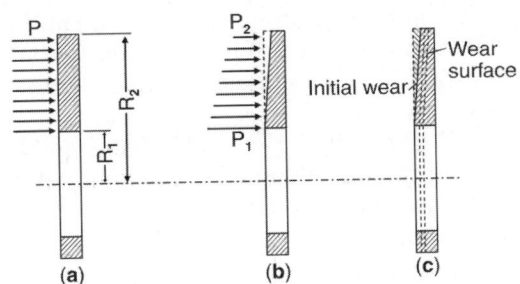

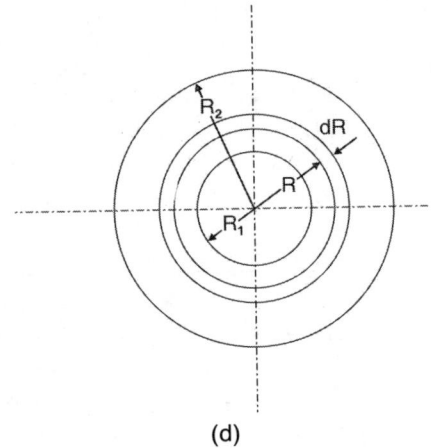

(d)

Fig. 18.15: Disk pressure and wear

For new surfaces, uniform pressure distribution is valid. After initial wear, the uniform wear takes place.

Considering uniform wear

Hence, in general,

$$pR = p_1 R_1 = p_2 R_2 \qquad ... (18.57)$$

Force analysis: The torsional resistance of one pair of disks may be found by integrating the elemental torque which is equal to the product of the tangential force dF_t and the radius of friction R. In this case,

$$dF_t = fp\, dA \qquad ... (18.58)$$

where, from equation 18.57,

$$p = \frac{p_1 R_1}{R} \qquad ... (18.59)$$

and the elemental ring area Fig. 18.15d is

$$dA = 2\pi R\, dR \qquad ... (18.60)$$

Thus the torque expression becomes

$$T = \int_{R_1}^{R_2} fp\, dA\ R = \int_{R_1}^{R_2} \frac{fp_1 R_1}{R} \times 2\pi R\, dR \times R$$

$$= 2\pi f p_1 R_1 \int_{R_1}^{R_2} R\, dR = \pi f p_1 R_1 (R_2^2 - R_1^2)$$

If $\frac{1}{2}D_1$ and $\frac{1}{2}D_2$ are substituted for R_1 and R_2, the torque for i pairs of surfaces is

$$T = \frac{\pi i f p_1 D_1 (D_2^2 - D_1^2)}{8} \qquad ... (18.61)$$

The axial force F_a transmitted to the disks from the flange f, Fig. 18.14, may be obtained from the relation

$$F_a = \int_{R_1}^{R_2} p\, dA \qquad ... (18.62)$$

Substituting the values from equations 18.59 and 18.60 in equation 18.62 and integrating, the result is

$$F_a = \frac{1}{2}\pi p_1 D_1 (D_2 - D_1) \qquad ... (18.63)$$

Now if $T = f\, F_a \dfrac{D_m}{2}$

From equations 18.61 and 18.63

$$\frac{T}{F_a} = \frac{D_2 + D_1}{4} = \text{Mean radius } R_m$$

i.e. $\qquad T = if\, F_a D_m \qquad ... (18.64)$

Because of leverage, the pull (Fig. 18.14) is

$$F = \frac{F_a}{b}$$

Considering constant pressure, i.e. pressure is uniform over the whole surface. It is a similar expression to equation 18.64 for new plates by using a number of springs spread over the contacting surface.

$$F_a = \int_{R_1}^{R_2} p\, 2\pi R\, d R = \pi p (R_2^2 - R_1^2)$$

i.e. pressure multiplied by area of the disk.

Torque is given by

$$T = \int_{R_1}^{R_2} f\, p\, 2\pi R\, dR\ R = f\pi p \frac{2}{3}(R_2^3 - R_1^3) \ ... (18.65)$$

Again $\dfrac{T}{f\,F_a}$

$$= \frac{2}{3}\left[\frac{R_2^3 - R_1^3}{R_2^2 - R_1^2}\right] = \text{Mean radius } R_m$$

Substituting for diameter

$$R_m = \frac{1}{3}\left[\frac{D_2^3 - D_1^3}{D_2^2 - D_1^2}\right]$$

or $\quad T = \dfrac{2}{3} f\,F_a \left[\dfrac{D_2^3 - D_1^3}{D_2^2 - D_1^2}\right]$ or $= f\,F_a\,R_m$ ... (18.66)

In multidisk brake if i is the number of pairs of surfaces

$$T = if\,F_a\,R_m \qquad\qquad \text{... (18.67)}$$

This expression is similar to the expression of 18.65, but the value of R_m is different.

Design procedure: First, the outside diameter D_2 is selected in accordance with load and speed conditions, and the corresponding inside diameter D_1 is estimated $\dfrac{D_1}{D_2} = 0.4$ to 0.7.

Next, the mean tangential effort F_t is determined by equation 18.5, and the torque requirement T is found by equation 18.6. The value of the mean radius R_m may be taken as

$$R_m = 0.25(D_2 + D_1) \qquad \text{... (18.68)}$$

However, a more accurate value is obtained as above, i.e.

$$R_m = \frac{\left(D_2^3 - D_1^3\right)}{3\left(D_2^2 - D_1^2\right)} \qquad \text{... (18.69)}$$

The remaining steps in the design are to decide what friction materials will be used, to select the corresponding values of f and p_1, and to determine from equation 18.61 the number i of pairs of friction surface. If the number i becomes too small, say under 4, the diameter D_2 was selected too large; on the other hand, if the number i becomes too large, the diameter D_2 was selected too small. In either case, a new value of D_2 should be selected and the calculations repeated.

In order to obtain smooth operation and to have a little wear, it is advisable to take for f and p_1 the lower limits for the respective materials that are given in Table 18.1.

The lever pull F also can be decreased by increasing the number of plates. However, more than seven pairs of surfaces are seldom used because of poor heat dissipation.

Example 18.5: Determine the axial force required for operation and power transmitting capacity of a multidisk brake consisting of 4 steel plate and 3 bronze plates. The outer and inner diameter of the plates are 200 mm and 120 mm respectively. The coefficient of friction between plates is 0.11 and pressure intensity of 0.35 N/mm^2.

Assume uniform wear and uniform pressure. The maximum rpm = 4000

(a) Uniform wear

Axial force required

$$F_a = \frac{1}{2}\pi p_1 D_1 (D_2 - D_1)$$

$$= \frac{1}{2}\pi \times 0.35 \times 120 \times (200 - 120)$$

$$= 5277.8 \text{ say } 5280 \text{ N}$$

No. of pairs $n = 6$

Torque capacity

$$T = \frac{\pi\,i\,f\,p_1 D_1(D_2^2 - D_1^2)}{8} = \frac{F_a\,f(D_2 + D_1)i}{4}$$

$$= \frac{5280 \times 0.11 \times (200 + 120) \times 6}{4}$$

$$= 278784 \text{ Nmm}$$

Power transmitted

$$= \frac{2\pi \times 4000 \times 278784}{1000 \times 60 \times 1000} = 116.7 \text{ kW}$$

(b) Uniform pressure

Axial force

$$F_a = \pi p(R_2^2 - R_1^2) = \pi \times 0.35(100^2 - 60^2) = 7037.16 \text{ N}$$

Torque transmitted $T = if\,F_a\,R_m$

$$R_m = \frac{2}{3}\left[\frac{100^3 - 60^3}{100^2 - 60^2}\right] = 81.66 \text{ mm}$$

$$T = 6 \times 0.11 \times 7037.16 \times 81.66$$

$$= 379271.96 \text{ or } 379.27 \text{ Nm}$$

Power to be transmitted $= \dfrac{2\pi NT}{1000 \times 60}$

$$= \frac{2\pi \times 4000 \times 379.27}{60 \times 1000}$$

$$= 158.86 \text{ kW}$$

Caliper disc brake or dual piston disc brake: In most of the heavy cars or automobiles, a dual piston disc brake is quite common. It consists of two pads either circular Fig. 18.16b or a circular strip one on each side of a metallic plate Fig. 18.16c. The pads are forced on the plate against the rotating disc by means of hydraulic pressure increased by operating the brake pedal. The two normal forces cancel each other but the tangential forces are additive and control the motion of disc and vehicle. As the area of the brake lining is restricted by a small width and depth, the pressure intensity may be taken as uniform. However, the variation in pressure is also considered by considering constant wear.

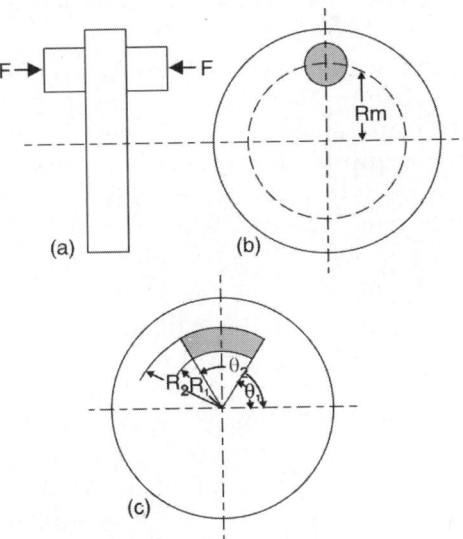

Fig. 18.16: Caliper brakes

(i) Uniform pressure with circular pad of radius r

Area of pad $= \pi r^2$

Normal force $F = \pi r^2 . p$

Frictional force $F_t = f\pi r^2 p$

Resisting torque $T = 2f\pi r^2 p . R_m$

(ii) Uniform pressure with a circular strip.

Area of element $dA = r\, d\theta dr$

Area $A = \displaystyle\int_{R_1}^{R_2}\int_{\theta_1}^{\theta_2} r\, dr\, d\theta$

$$= \frac{(\theta_2 - \theta_1)(R_2^2 - R_1^2)}{2}$$

$$= \frac{\theta}{2}(R_2^2 - R_1^2) \text{ where } \theta = \theta_2 - \theta_1$$

Axial force $F = p\dfrac{\theta}{2}\left(R_2^2 - R_1^2\right)$

Resisting torque

$$T = 2fp\frac{\theta}{2}(R_2^2 - R_1^2)R_m = 2fFR_m$$

Considering pads to be on both sides Alternatively

$$T = 2\int_{R_1}^{R_2}\int_{\theta_1}^{\theta_2} fpr^2\, dr\, d\theta$$

$$= 2\,fp(\theta_2 - \theta_1)\left(\frac{R_2^3 - R_1^3}{3}\right)$$

and normal force $F = \displaystyle\int_{R_1}^{R_2}\int_{\theta_1}^{\theta_2} pr\, dr\, d\theta$

$$= p(\theta_2 - \theta_1)\left(\frac{R_2^2 - R_1^2}{2}\right)$$

Substituting for F in torque equation

$$T = 2fF\frac{2}{3}\left(\frac{R_2^3 - R_1^3}{R_2^2 - R_1^2}\right) = 2fFR_m$$

$$R_m = \frac{2}{3}\left(\frac{R_2^3 - R_1^3}{R_2^2 - R_1^2}\right)$$

(iii) Uniform wear. It requires that $p_1R_1 = pr$ = constant

Normal force $F = \int\limits_{R_1}^{R_2} \int\limits_{\theta_1}^{\theta_2} pr\, dr\, d\theta$

$$= p_1R_1 (\theta_2 - \theta_1)(R_2 - R_1)$$

Resisting torque $2\int\limits_{R_1}^{R_2} \int\limits_{\theta_1}^{\theta_2} f\, pr^2\, dr\, d\theta$

$$= fp_1R_1(\theta_2 - \theta_1)(R_2^2 - R_1^2)$$

or $\qquad\qquad T = fF R_m$

These expressions are similar to ones for disc brakes.

Example 18.6: Find the piston force required for actuating a dual piston brake having a shoe strip of inner radius 75 and outer radius of 135 mm $f = 0.35$. Maximum pressure of 2.8 N/mm² making an angle of 40° at the centre and symmetrically placed with respect to the vertical axis of the disc of 300 mm diameter for the front wheels of a car. What is torque and power capacity of the front wheels if the car cruises with a maximum velocity of 110 km/hr. The wheel diameter is 520 mm. Consider uniform wear.

Axial force $F = p_1R_1 (\theta_2 - \theta_1)(R_2 - R_1)$

$$= 2.8 \times 75 \times \frac{\pi \times 40}{180} \times 60 = 8796.45 \text{ N say } 8800 \text{ N}$$

Force required at the pistons is $1.25 \times 8800 = 11$ kN
Torque capacity of one wheel $= 2 \times f \times F \times R_m$
$= 2 \times 0.35 \times 8800 \times 105 = 646800$ Nmm
$= 647$ Nm

Velocity of wheels $= \dfrac{110 \times 1000}{60 \times 60} = 30.55$ m/s

Velocity of disc $= \dfrac{30.55 \times 300}{520} = 17.625$ m/s

Power capacity $= \dfrac{647 \times 17.625}{1000} = 11.4$ kW

Total power capacity of front wheels
$$= 2 \times 11.4 = 22.8 \text{ kW}$$
Considering uniform pressure
Axial force

$$F = 2.8 \times \frac{40\pi}{180}\left(135^2 - 75^2\right) = 12315 \text{ N}$$

Torque capacity of one wheel

$$= 2 \times 0.35 \times 2.8 \times \frac{40\pi}{180}\left[\frac{135^3 - 75^3}{3}\right] = 929.8 \text{ N}$$

$$\text{Power} = \frac{929.8 \times 17.625}{100} = 16.38 \text{ kW}$$

Total power $= 2 \times 16.38 = 32.77$ kW.

18.6 COMPARISON OF BRAKES

Three factors can be used as a basis for comparison of brakes:
  (a) The ratio of the effort F to the tangential braking force F_t
  (b) The motion of the operating lever necessary to apply a brake or to release it, and
  (c) The heat dissipation.

Of the great variety of types, only the six more-typical brake arrangements, listed in Table 18.3, will be discussed.

Force ratio: For the sake of convenience, the expressions for the ratio F/F_t are shown in Table 18.3. In computing the numerical values for F/F_t, the coefficient of friction f was assumed to be 0.3. In compliance with actual conditions, the leverage a/b was assumed as 5:1, with the exception of the band brakes, in which it was taken as 10:1, as actually made. For the multidisk brake, seven pairs of surfaces were assumed.

The line of relative values shows that the simple band brake requires the smallest effort, and the multidisk brake comes next. The block brake requires the largest effort. Since this brake has certain advantages for heavy loads, it is used quite extensively; but the ratio F/F_t is made more favorable by additional leverage as indicated in Fig. 18.5.

Lever-end travel: In Table 18.3 the expressions for the travel of the lever end are given as functions of the normal distance h between the sheave and the stationary braking surface to prevent dragging. In computing the numerical values a minimum magnitude of 1.5 mm was assumed for h. For the disk brake the total spreading ih' was taken as 1.5 mm. Naturally, the small travel of the ordinary block brake will

Table 18.3: Comparison of hoist brakes

Brake characteristics	Block brakes		Band brakes		Axial brakes	
	Double block	V-grooved sheave	Simple	Both directions or rotation	Cone	Multidisk
Force ratio $\dfrac{F}{F_t}$	$\dfrac{b}{fa}$	$\dfrac{b\sin\alpha}{fa}$	$\dfrac{b}{a(e^{f\theta}-1)}$	$\dfrac{b(e^{f\theta}+1)}{a(e^{f\theta}-1)}$	$\dfrac{b\sin\alpha}{fa}$	$\dfrac{b}{nfa}$
Average numerical value	0.667	0.282	0.0323	0.165	0.161	0.097
Relative value	20.6	8.7	1	5.1	5.0	3.0
Travel at lever end	$\dfrac{ha}{b}$	$\dfrac{ha}{b\sin\alpha}$	$\dfrac{ha\theta}{2\pi b}$	$\dfrac{ha\theta}{4\pi b}$	$\dfrac{ha}{b\sin\alpha}$	$\dfrac{ih'a}{b}$
Average travel, mm	8	18	75	38	33	6
Maximum capacity, kW	1500	20	225	75	37.5	90

be increased in the same proportion as the force ratio is decreased if an additional lever system is used. Usually they are well lubricated for better dissipation of heat.

Heat dissipation: The heat is dissipated best with axial conical brakes, and block brakes with cast-iron shoes come next. Band brakes, especially those with asbestos lining, have a smaller radiating area and therefore, other conditions being equal, run hotter. Nevertheless, because of their powerful action they are used on hoists and are also used very extensively on automotive vehicles. Multidisk brakes are the poorest and therefore are not used for heavy loads. *Usually they are well lubricated for better dissipation of heat.*

PROBLEMS

18.1 What is a self-locking and self-energizing brake? Which is preferable?

18.2 What is a leading shoe and a lagging shoe?

18.3 Which type of brake is useful for heavy loads?

18.4 Why is the angle subtended by friction lining at the centre of the brake drum not made greater than 120°?

18.5 List the comparative advantages and disadvantages of external shoe brake, internal shoe brake, band brake and disc brake.

18.6 Name the different friction materials used for lining the brakes.

18.7 What purpose is served by providing holes in a brake drum?

18.8 Why are disc brakes lubricated though the coefficient of friction of metallic surfaces is less?

18.9 Name the methods for decreasing the operating effort in band brakes.

18.10 Sketch an internal shoe brake in which both the shoes are leading.

18.11 Does the frictional torque depend on the number of collars in a multidisc brake?

18.12 Do the number of collars effect the intensity of pressure?

18.13 Does the braking torque depend on number of collars?

18.14 (a) Determine the total energy which must be absorbed to slow down to 2.25 m/s a mine-hoist cage descending at a rate of 7.5 m/s. The weight of the cage with the load is 20 kN, the hoist-drum diameter is 1.8 m and the brake-sheave diameter is 1.4 m. The rotative speed of the brake sheave is one-third that of the driving engine, and the speed of the hoist drum is one-tenth that of the engine. The weight of the hoist drum is 75 kN and the weight of other rotating parts may be neglected, (b) Determine the normal and tangential forces which must be applied to the brake sheave to slow down the cage from 7.5 m/s to 2.25 m/s in 15 sec.

18.15 Assume that operation is intermittent; that case-iron shoes run dry on a steel brake sheave which has a diameter of 1m and runs at 87 rpm; and that the area of the heat-dissipating surface is 50 per cent larger than the braking surface. Determine the braking area for absorbing 96000 Nm in 30 sec.

18.16 A block brake similar to that in Fig. 18.3 is used for lowering a load of 6 kN. The main dimensions are: D = 90 mm, a = 1.6 m and b = 0.18 m. Assume that the rotation is clockwise and that n = 90 rpm, and that the coefficient of friction is f = 0.25. The hoist drum and brake drum are on the same shaft and diameter of the hoist drum is 50 mm. Find the effort F for each of the following positions of the fulcrum: (a) c = 200 mm, (b) c = 25 mm and (c) c' = 200 mm.

18.17 A double-block brake built according to the scheme given in Fig. 18.4 is used to lower a load of 10 kN at a speed of 4 m/s. Assuming for asbestos on cast iron a friction coefficient of 0.33, determine the necessary effort F and the tension F' of the spring s, if the bell-crank leverage is 50 mm to 450 mm, the diameter of the brake is D = 750 mm, and the diameter of the hoist drum is D_2 = 200 mm.

18.18 A double-block brake, Fig. 18.5, is used for lowering loads. The brake diameter is 800 mm, and other dimensions may be assumed: (a) Determine the power which this brake can absorb at 300 rpm of the sheave with a pull F of 400 N, (b) Determine the necessary width of the sheave drum from consideration of heat dissipation, (c) Check whether the specific pressure on the asbestos blocks with a

angle of contact of 90° will not require an increase of the drum width.

18.19 (a) Determine the capacity in kW, at 175 rpm of the brake sheave, of a differential band brake, Fig. 18.9. The principal dimensions are a = 1.05 m, b_1 = 50 mm, b_2 = 125 mm, D = 450 mm, the distance from the fulcrum 1 to the sheave centre is 300 mm, and the line from the fulcrum to the sheave centre forms a right angle with the centre line 1–4 of the operating lever. The band is asbestos-lined and can stand a tensile load of 18 kN, (b) State the direction of the force F, upward or downward, for a clockwise rotation of the sheave, (c) Find the magnitude of the force F.

18.20 (a) For a simple band brake as shown in Fig. 18.10, determine the magnitude of the effort F necessary to hold the load on the drum for clockwise rotation. The principal data for the brake are: a = 1.4 m, b = 140 mm, the sheave diameter is 0.7 m, and the angle of contact is = 225°. The coefficient of fraction is 0.25, the diameter of the hoist drum is 500 mm; and the load is 7 kN and is suspended directly from a 10 mm steel-wire rope, (b) determine the thickness and width of the steel brake band, considering stiffness and strength and assuming that the end lugs are welded to the band, (c) Determine the width of the band and the size and number of rivets if the end lugs are riveted to the band, (d) Show a sketch of the rivet spacing.

18.21 A band brake, Fig. 18.10, is geared to the hoisting motor of a traveling crane. The gear ratio between the motor and the brake shaft is 6 to 1, and the motor can develop 6 kW at 900 rpm. The principal dimensions are a = 530 mm, b = 76 mm, D = 300 mm, and θ = 270°. Assume that the friction coefficient is f = 0.27. Determine (a) the magnitude of the effort F to hold the maximum load which the motor can lift, (b) the width of the brake sheave, (c) the width and thickness of the brake band made of 7C4 or SAE 1010 steel, and (d) the size, number, and spacing of rivets to fasten the band to the end lugs (*see* Fig. 18.11).

18.22 A cone brake, Fig. 18.13, is mounted on a shaft which transmits 4.5 kW at 225 rpm. The small diameter of the cone is 225 mm, and the cone face is 50 mm wide; $\alpha = 15°$; the friction coefficient is 0.33; and the lever dimensions are $a = 0.6$ m and $b = 125$ mm. Find (a) the effort F necessary to stop the shaft and (b) the specific normal pressures on the cone surfaces.

18.23 Find the torque which can be absorbed by the brake of problem 18.21, using cork lining and such an effort F that the

normal unit pressure on the cone surfaces will reach the permissible high limit.

18.24 A hoist drum is bolted to a multiple-disk brake, Fig. 18.14. The drum diameter is 450 mm; the tangential load is 5 kN; a 12 mm wire rope is used; the outside diameter of the small steel friction disks is 300 mm, the inside diameter of the larger cast-iron disk is 125 mm; and the lever dimensions are $a = 750$ mm and $b = 100$ mm. Determine (a) the force F required on the lever when applying the brake, (b) the number of friction disks required, assuming that the permissible unit pressure is 0.5 N/mm^2 and that the disks run in oil, and (c) the number of friction disks required if F must not exceed 220 N.

19

Screws for Power Transmission

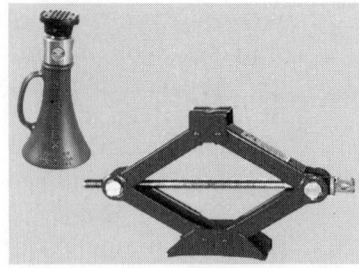

19.1 POWER THREADS

Screws are used to produce uniform, slow, and powerful motion, such as is required in presses, jacks, lathes, valves, and other machinery. *The efficiency of V threads is very low, partly because of the large angle of the profile (19.2). By making this angle small or zero, threads with a higher efficiency are obtained (19.1). Those employed for power screws are the square, buttress, acme, and round-groove threads.*

Square threads: The square thread, Fig. 19.1, has the highest efficiency and therefore is extensively used in spite of its higher manufacturing cost. There is no national standard for square threads. However, Sellers' standard is used quite widely. Its threads are shallower than those of a real square thread. The proportions are shown in Fig. 19.1.

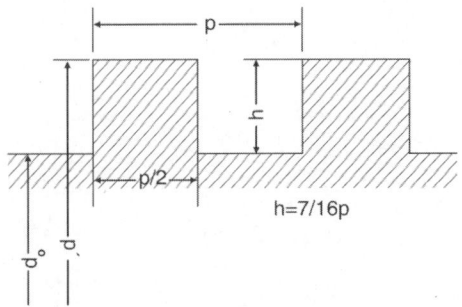

Fig. 19.1: Square thread

Buttress threads: The buttress thread, or trapezoidal thread, Fig. 19.2, is used for the transmission of power in one direction only.

The efficiency of this thread is the same as that of the square thread, while its cost of manufacturing is slightly smaller (19.5). There are no standard proportions for these threads, but those commonly used are given in Fig. 19.2. The nominal diameter varies by 2 mm from 10 to 50 mm sizes and a pitch of 2 mm steps is used for 10 to 20 mm and 3 mm steps from 20 to 62 mm.

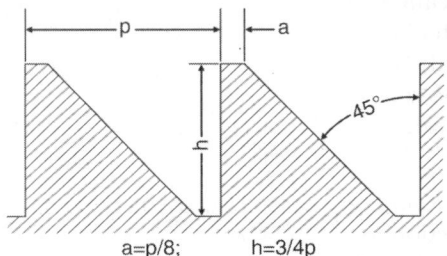

$$a=p/8; \qquad h=3/4p$$

Fig. 19.2: Buttress thread

Acme thread: The acme thread, Fig. 19.3, is used chiefly for lead screws and similar service where lost motion is objectionable. *Such lost motion, caused by wear, can be eliminated by means of a nut split lengthwise. The arrangement makes the screw and nut self adjusting to wear (19.4).* The efficiency of the Acme thread is slightly lower than that of a square thread, but its cost of production is smaller since it can be cut by dies also. The number of threads or pitch is commonly made the same as that of Sellers' thread, and all proportions can be computed from the data of Fig. 19.3.

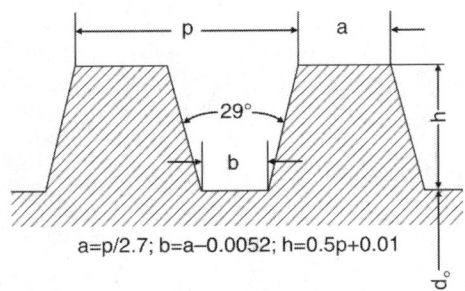

a=p/2.7; b=a−0.0052; h=0.5p+0.01

Fig. 19.3: Acme thread

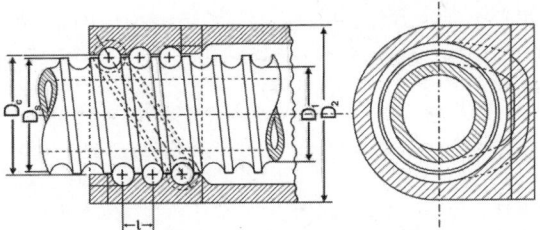

Fig. 19.4: Screw and nut with balls

Multiple threads: Power screws with two or three parallel threads are employed to increase the travel of the nut per revolution. The mechanical advantage of a multiple thread is correspondingly smaller, but the efficiency is higher because of the increase of the helix angle. The pitch of a screw is the distance p between consecutive threads, whether it is a single-thread or multiple-thread screw. The distance l which the nut of a multiple-thread screw advances for one revolution of the screw is called the lead. In a single-thread screw the pitch and the lead are equal. The helix angle at the mean diameter is called the lead angle. The relation between l, p, and the number of parallel threads i evidently is

$$l = pi \qquad \ldots (19.1)$$

Round thread with balls: In order to reduce friction in the threads and thus to increase the efficiency of power screws, a special round thread with steel balls inserted between the screw and the nut has been developed. The thread grooves have approximately semi-circular cross section. The nut contains $1\frac{1}{2}$ or $2\frac{1}{2}$ rows of bearing balls and a return groove, as shown in Fig. 19.4. These are also called ball bearing screws.

The efficiency of such a screw is of the order of 90 per cent and higher. So far these screws are used chiefly for automobile steering gears and power actuators and CNC machines and earlier in robotic arm. However, there are many more places where they may be used to advantage.[1]

19.2 FORCE ANALYSIS

If the notations and the reasoning in section 11.7 are used, the relation for the tangential force H_1, Fig. 19.5a, which must be applied to a square thread to exert an axial force F, is

$$H_1 = F \tan (\lambda + \phi) \qquad \ldots (19.2)$$

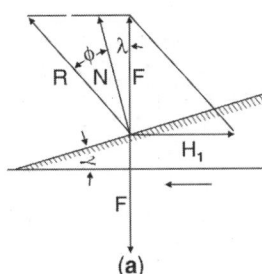

(a)

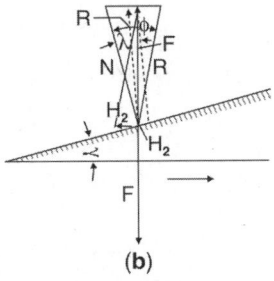

(b)

Fig. 19.5: Forces acting on a square thread

---

[1] Design data and procedure for screws with ball nuts are given in V. L. Maleev, *Machine Design*, rev. ed. (Scranton: International Textbook Company, 1946), pp. 248 ff.

Expressing tan $(\lambda + \phi)$ by functions of the component angles λ and ϕ, and using the designation $f_1 = \tan \phi$, gives

$$H_1 = \frac{F(\tan \lambda + f_1)}{1 - f_1 \tan \lambda} \qquad \ldots (19.3)$$

If the motion is reversed, as indicated in Fig. 19.5b, the screw is used for lowering the load F. The friction changes the normal force N to the resultant force R, which is inclined in the direction of the screw motion. In this case

$$H_2 = F \tan (\phi - \lambda) \qquad \ldots (19.4)$$

or

$$H_2 = \frac{F(f_1 - \tan \lambda)}{(1 + f_1 \tan \lambda)} \qquad \ldots (19.5)$$

If $\phi < \lambda$, H_2 becomes negative, as shown by the dotted lines in Fig. 19.5b, no force is required to lower the load F, and the load will begin to turn the screw. The screw is not self-locking and is reversible.

If d_m designates the mean diameter of the screw $= \dfrac{d + d_0}{2}$, the torque necessary to overcome the load F is

$$T = \frac{Fd_m}{2} \tan(\lambda + \phi) \qquad \ldots (19.6)$$

In the absence of friction, the torque would be

$$T_o = \frac{Fd_m}{2} \tan \lambda \qquad \ldots (19.7)$$

Hence the efficiency is

$$e = \frac{T_o}{T} = \frac{\tan \lambda}{\tan(\lambda + \phi)} \qquad \ldots (19.8)$$

Another equation for efficiency may be obtained by using equations 11.14 and 11.18. Thus

$$e = \frac{\tan \lambda(1 - f_1 \tan \lambda)}{\tan \lambda + f_1} \qquad \ldots (19.9)$$

The torque and the efficiency can also be expressed in terms of the screw dimensions by using the relation

$$\tan \lambda = \frac{l}{\pi d_m} \qquad \ldots (19.10)$$

As in a power screw, there will be torque wasted as friction collar torque, it will also be added for calculation of efficiency,

$$\text{Collar torque} = T_e = f_2 W \frac{dm}{2} \qquad \ldots (19.11)$$

In the preceding equations only the friction in the threads is taken into account. The pivot friction, such as that between the cap and the screw in a screw jack or in a thrust collar of a lead screw, can be determined by an analysis similar to that used in deriving equation 11.22 for V-shaped screws. The influence of the helix angle λ and friction on the efficiency of a screw is shown in Fig. 19.6.[2] The efficiency increases very rapidly as λ increases up to 15° to 20°; it increases more slowly until it reaches a maximum for values of λ between 40° and 50°, it then begins to decrease. In single-thread screws λ varies from 4° to 7° and therefore their efficiency is below 50 per cent. *In a screw with coefficient of friction 0.15, the efficiency is about 30% with $\lambda = 7°$. As the screw moves in the nut, the coefficient of friction may decrease to 0.1 and the efficiency will become >50%, which is not desirable in lifting machines so the maximum value of $\lambda \leq 7°$ (19.6, 19.7). In multiple-thread screws, a higher efficiency can be obtained but the screws lose the self-locking characteristic.*

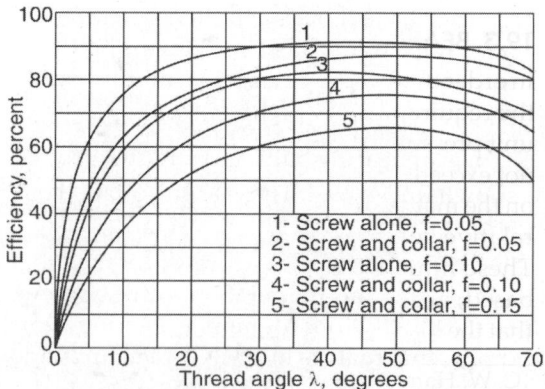

Fig. 19.6: Efficiencies of square screws

---

[2] D. S. Kimball and J. H. Barr, *Elements of Machine Design*, 3rd ed. (New York: John Wiley & Sons, Inc., 1935), p. 290

Since an Acme thread is a V thread with a small angle, $\beta = 14\frac{1}{2}°$, the tangential force may be determined by equation

$$H_1 = F \frac{\tan\lambda + f\sec\beta}{1 - f_1 \tan\lambda \sec\beta} \qquad \ldots (19.12)$$

and torque $T = \dfrac{Fd_m}{2} \dfrac{\tan\lambda + f\sec\beta}{1 - f_1 \tan\lambda\sec\beta} \quad \ldots (19.13)$

efficiency $\quad e = \dfrac{\tan\lambda(1 - f_1 \tan\lambda\sec\beta)}{1 - f_1 \tan\lambda\sec\beta}$

$$= \frac{\cos\beta - f_1 \tan\lambda}{\cos\beta + f_1 \cot\lambda} \qquad \ldots (19.14)$$

The *coefficient of friction* may be taken from Table 11.2. *For the sake of lowering both the friction and the wear, it is advisable to use different materials for the screw and the nut (19.10).* Since the nut is usually smaller and less expensive than the screw, the nut is made of softer materials, such as brass or bronze so screw will wear less. More recent investigations show that the materials of the screw and nut have less influence on the coefficient of friction than does workmanship.[3] For an average-quality material and workmanship, and heavy-oil lubrication, the coefficient of friction in motion may be taken as 0.125. For inferior workmanship it is safer to take the coefficient as 0.15. The starting friction is about 35 per cent higher.

19.3 BEARING PRESSURE

In order that the threads of a screw may transmit the required work without excessive wear, the unit pressure upon the surfaces in contact must not exceed certain values. These values depend on the materials of the screw and nut and on the relative speed between the rubbing surfaces. These limit values, upon which the design is based, are determined under the assumption that the load is distributed uniformly among all threads. This assumption is not correct, as a simple analysis will show.[4]

In Fig. 19.7 two threads of a screw working in tension are replaced by collars, and the corresponding part of a nut is replaced by a sleeve with two inner collars. The figure shows that if each pair of collars carries one-half the load, part a of the screw is in tension and part b of the nut is in compression. Since a becomes longer and b becomes shorter, the upper collar, of the nut, will move away from the corresponding screw collar. This being impossible under our assumption of load distribution, it follows that the lower pair of collars carries all the load, and that the upper pair only stays in contact, without carrying any load.

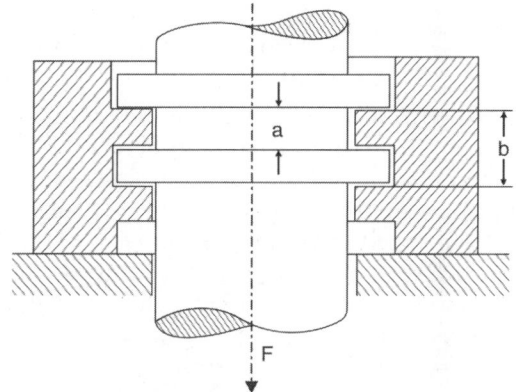

Fig. 19.7: Load distribution in a screw and nut

The explanation of the necessity of using more than one thread to carrying a certain load is in the elastic and plastic deformation of the threads which transmit part of the load from the lower thread to the next threads. Nevertheless the load distribution is far from uniform.

In Fig. 19.7 the screw is in tension and the nut is in compression. The results will be the same if the situation is reversed, the screw being stressed in compression and the nut in tension. If both the screw and the nut work in tension, or if both work in compression, the

[3] C. W. Ham and D. G. Ryan, *An Experimental Investigation of the Friction in Screw Threads*, Bulletin No. 247, University of Illinois Engineering Experiment Station (June, 1932).

[4] W. Trinks, "Things That Are Commonly Wrong in Textbooks on Machine Design," *The Journal of Engineering Education*, Vol. 23 (March, 1933), p. 523.

results will be different. Such an arrangement is shown in Fig. 19.8. Let it be assumed that each pair of collars carries one-third the total load F. Then the force distribution in the screw and in the nut is indicated by the values 0, $\frac{1}{3}F$, $\frac{2}{3}F$, and F. It is evident that the upper part of the nut between the upper and middle collars compresses more than does the corresponding part of the screw, while the lower part of the nut contracts less than does the corresponding part of the screw. Consequently the upper and lower pairs of collars remain in contact while the middle pair becomes disengaged, because the upper part of the nut contracts more than the upper part of the screw. In an actual screw all threads will be in contact, but the inner threads will carry a smaller load than the top or bottom threads depending on the load, i.e. compressive or tensile.

This analysis shows that the bearing pressure is not uniform on all threads which are in contact, and that the pressure is more uniformly distributed if the screw and nut are both subjected to the same stress, either both being in tension or both being in compression. These theoretical conclusions are supported by tests with undercut nuts, Fig. 11.34d. Ordinary nuts which transmit motion under heavy pressure are subject to great wear because of unequal pressure. By using a variable-section of nut, Fig. 19.9, the equation for which is a

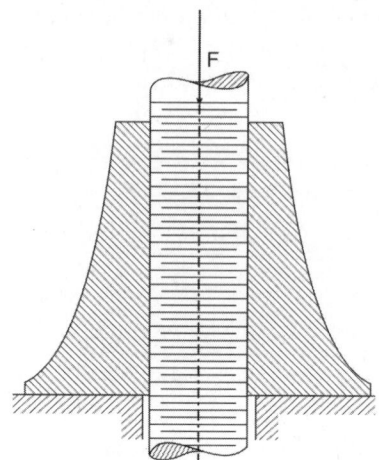

Fig. 19.9: Nut with uniform pressure on threads

parabola, a uniform pressure and reduced wear can be obtained.[5]

19.4 DIFFERENTIAL AND COMPOUND SCREWS

In some cases a very slow advance of the screw is desired, whereas in other cases a very fast movement is required. With a single-thread screw, slow motion can be obtained by using a small pitch. This, however, results in a weak thread. A fast motion is obtained by using a multiple-thread screw, but this means expensive machining and loss of the self-locking feature.

Differential screws: To a certain extent the difficulties just mentioned can be overcome by using the arrangement as shown in Fig. 19.10a. The screw has two threads of the same hand but of different pitches. As a result each revolution of the screw causes the nuts m and n to move toward or away from each other a distance equal to the difference of the pitches. If one nut and two screws are used, and the screws are prevented from turning, the arrangement in Fig. 19.10b results. In a third modification, shown in Fig. 19.10c, the hub of the handwheel, a has a coarse-pitch outside thread and a finer-pitch inside thread; and each

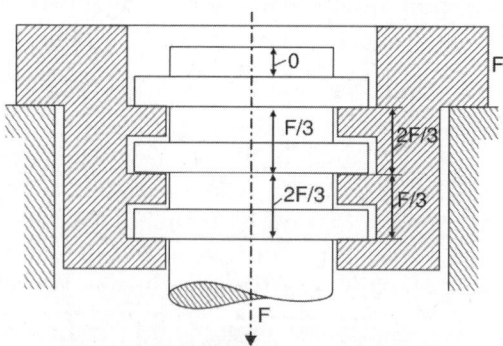

Fig. 19.8: Screw and nut loaded in tension

[5] S. Timoshenko and J. M. Lessels, *Applied Elasticity* (East Pittsburgh: Westinghouse Technical Night School Press, 1925)

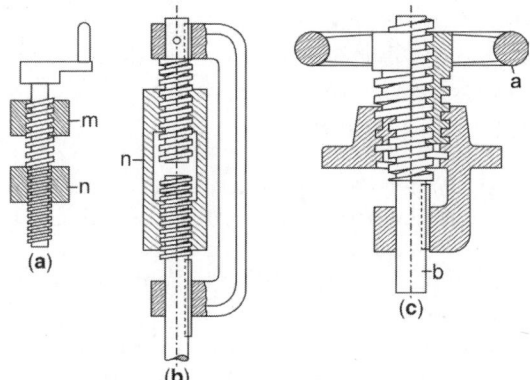

Fig. 19.10: Differential screws

revolution of the handwheel a moves the screw b a distance equal to the difference of the pitches.

If the coarser pitch is p_1 and the finer pitch is p_2, the motion h produced by a differential screw is

$$h = p_1 - p_2 \qquad \ldots (19.15)$$

Differential screws and nuts are cut in a lathe. There are no standards for pitches, but Table 19.1 may be used as a guide. The numbers of threads can be determined by using the desired motion h as a basis and assuming the difference $(p_1 - p_2)$ for a very coarse pitch and gradually increasing this difference up to 1 for a fine pitch. In general the increments between consecutive pitches should be used.

Example 19.1: Determine the size of differential screw (Fig. 19.10b) needed to exert a force of 50 kN and to produce a motion of about 0.6 mm for each revolution of the nut.

The minimum minor diameter d_o may be found by assuming a low design stress S_d of 55 N/mm^2 in order to take care of any possible column action. The necessary cross-sectional area is

$$A = \frac{50,000}{55} = 909 \text{ mm}^2$$

and

$$d_o = \sqrt{\frac{909}{\pi/4}} = 34 \text{ mm}$$

The approximate pitches may be found by using equation 19.15 and assuming, that $p_1 - p_2 = 0.6$. Thus

$0.6 = p_1 - p_2$, taking $p_1 = 8.6$ mm, $p_2 = 8$ mm

The major screw diameter may be determined now by referring to Fig. 19.1. Assuming $d_o = 34$ mm.

$$d = 34 + 2 \times \frac{7}{16} \times 8.6 = 41.525 \text{ mm}.$$

However, using the nearest stock size gives $d = 42$ and

$$d_o = 42 - 2 \times \frac{7}{16} \times 8.6 = 34.475 \text{ mm}$$

The minimum number of threads i in engagement may be found by assuming a bronze nut and a permissible bearing pressure $p = 17$ N/mm^2 from Table 19.1. Then

$$i = \frac{50,000 \times 4}{\pi(42^2 - 34.475^2) \times 17} = 6.5$$

Compound screws: If the threads in Figs 19.10a, b, or c are of different hands, the motion of the nut or screw, as the case may be, is equal to the sum of the pitches for every revolution of the actuating part. Such screws are called compound screws. With compound screws the pitches usually are made equal.

Efficiency: The expression for the efficiency of a differential screw may be found by noticing that when the coarse threads is raising the load, the finer thread lowers it. Each torque is computed for the tangential force found by taking the sum of H_1 from equation 19.3, for the coarser thread, and H_2 from equation 19.5, for the finer thread. The numerator in the expression for the efficiency is the sum of these torques with $f = 0$. If the respective helix angles are designated by λ_1 and λ_2, and the mean thread diameters by d_1 and d_2, the efficiency is

$$e = \frac{d_1 \tan \lambda_1 - d_2 \tan \lambda_2}{\dfrac{(\tan \lambda_1 + f)d_1}{1 + f \tan \lambda_1} + \dfrac{(f - \tan \lambda_2)d_2}{1 + f \tan \lambda_2}} \qquad \ldots (19.16)$$

When equation 19.16 is applied with some typical values of $\lambda_1, \lambda_2, d_1, d_2$, and f, it will be found that the efficiency of differential screws is very low, ranging from 2.5 to 8.8 per cent at the most. Tests confirm this conclusions.[6]

The efficiency of a compound screw is found in a similar way, but only the tangential force

---

[6] R. T. Kent, *Mechanical Engineers'* Handbook.

H_1 found by equation 19.3 is used in determining the torque for all terms. Thus

$$e = \frac{d_1 \tan\lambda_1 + d_2 \tan\lambda_2}{\dfrac{(\tan\lambda_1 + f)d_1}{1 - f\tan\lambda_1} + \dfrac{(\tan\lambda_2 + f)d_2}{1 - f\tan\lambda_2}} \quad \ldots (19.17)$$

Numerically the efficiency of a compound screw is of the same order as that of a single screw. In fact, if both screws have the same diameter and pitch, equation 19.17 can be simplified to the form of equation 19.9.

19.5 DESIGN CONSIDERATION

Screws used for the transmission of power are subjected to the following stresses: tension or compression in the central section; shear in the threads due to the axial load; shear due to the external torque; and bearing pressure. There is likelihood of bending stress in the screw due to eccentricity of load in tension or compression in addition to direct stresses.

Tension or compression: The service of the screw and the method of mounting it determine the kind of stress resulting from the direct load. If the stress is tension, its nominal magnitude is equal to the load divided by the area at the root of the thread. The absence of fillets in the threads invokes a localized stress, with a form stress factor K of 4.5 to 5. For ductile materials and static load the factor of sensitivity q may be assumed to be about 0.15, giving a stress-concentration factor K' of 1.5 to 1.6. The safety factor n may be taken as 1.5 to 2 on S_e or 4 on S_u.

If the load produces a compressive stress, the stress concentration, although present, is not so dangerous and may be neglected. If the length of the screw exceeds six to eight times the root diameter, the screw must be treated as a short column and checked for buckling load.

Shear: The shear stress in the threads, which is due to the axial load, is usually not dangerous, and the number of threads resisting the shear action is determined from consideration of wear rather than strength. Nevertheless a check of the average stress should be made. Stress concentration may be neglected. The threads of the screw may shear at the root diameter; the threads of the nut may shear at the major diameter, which is equal to the outside diameter of the screw. Since the screw and the nut are of different materials, the stresses and safety factors for both should be computed.

Torsional shear is induced in the screw by the external turning moment T, which can be found by equation 19.6. The stress is found by equation 2.17, in which d_0 is the minor diameter of the screw.

Bearing pressure: As shown in section 19.3, the pressure between the threads of a screw and nut is not uniform. However, an assumption of uniform pressure simplifies the design and is always made. The inaccuracy of this assumption is taken into account by using low values for the bearing pressure p.

The necessary number of threads i is then determined from the equation

$$F = \frac{\pi}{4}(d - d_0^2)\, p_b \quad \ldots (19.18)$$

where, d is the outside diameter of the screw and d_0 is the inside diameter of the nut. The allowable values of p_b determined from actual service are given in Table 19.1.

Computer programme and design procedure for design of a screw jack. As it may not be always possible to align the load F along the axis of the screw, an eccentricity e may be assumed approximately 1 mm for every 10 kN load. d_0 is the root diameter.

Stresses in the screws

Direct compression stress

$$S_1 = \frac{4F}{\pi d_0^3} \quad \ldots (19.19)$$

Bending stress due to eccentricity

$$S_2 = \frac{32Fe}{\pi d_0^3} \quad \ldots (19.20)$$

There is also bending stress due to the effort applied to raise or lower the load depending on the height of tommy bar from the top surface of the nut and is equal to $\dfrac{32Fh}{\pi d_0^3}$ which is very small and is neglected as h is quite small.

Table 19.1: Safe bearing pressures in power screws

Service	Material		Safe bearing pressure p N/mm²	Remarks
	Screw	Nut		
Hand press	Steel	Bronze	17–25	Low speed, well-lubricated
	Steel	Cast iron	12–17	Low speed, not over 2.4 m/m
Screw jack		Bronze	10–17	Low speed, not over 3 m/m
		Cast iron	4–7	Medium speed, 20 to 12 m/m
Hoisting screw	Steel	Bronze	5–9	Medium speed, 6 to 12 m/m
		Bronze	5–9	Medium speed, 6 to 12 m/m
Lead screw	Steel	Bronze	1–1.5	High speed, 15 m/m and over

If p is the pitch, assumed i is the number of threads in the nut calculated as

$$F = \frac{\pi}{4}\left(d_2 - d_o^2\right)ip_b \qquad \ldots (19.21)$$

Direct shear stress in the threads

$$S_{s1} = \frac{F}{\pi\, d_o\, \frac{p}{2}\, i} \qquad \ldots (19.22)$$

This stress is also occurring in the screw in the plane of top and bottom surface of the nut. Shear stress due to torque T_1

$$S_{s2} = \frac{16T}{\pi\, d_o^3} \qquad \ldots (19.23)$$

Total compressive stress $S = S_1 + S_2$
Total shear stress $S_s = S_{s1} + S_{s2}$
Maximum principle stress

$$= \frac{1}{2}\left[S + \sqrt{S^2 + 4S_s^2}\right] \qquad \ldots (19.24)$$

Maximum shear stress

$$= \frac{1}{2}\sqrt{S^2 + 4S_s^2} \qquad \ldots (19.25)$$

If these stresses are equated to the allowable stresses to find d_o, it is not convenient to calculate. d_o is first found from direct compression stress only and then check is made for safety factor as described below:

Design Procedure (Refer to Fig. 19.11)

1. Assume a safety factor n to find allowable stress

$$S_d = \frac{\text{UTS or yield stress}}{n}$$

2. Find $d_o = \left[\dfrac{4F}{\pi S_d}\right]^{1/2}$

Assume a suitable value of pitch p to find the nominal diameter $d = d_o + p$ rounded off to mm size and then $d_o = d - p$

Mean diameter $d_m = \dfrac{d + d_o}{2}$

3. Check for proper lead angle $\alpha = \tan^{-1}\dfrac{p}{\pi d_m}$

and it should be between $4°$ to $7°$. If not, change p and find d and d_o again.

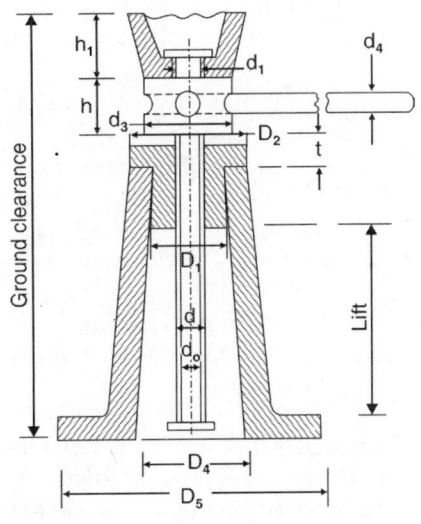

Fig. 19.11

4. Find lifting torque $T_1 = \dfrac{1}{2} F d_m \tan(\lambda + \Phi)$

5. Find number of threads i from bearing consideration equation (19.14), i.e. taking

$$d_2 \simeq d_o$$

$$F = \frac{\pi}{4}(d^2 - d_o^2)p_b \text{ or } i = \frac{4F}{\pi\left(d^2 - d_0^2\right)p_b}$$

Find length of nut $l = ip$. This should be preferably between d and $2d$ or revise i.

Checking for proper factor of safety.

6. Find direct compression stress $= \dfrac{4\,F}{\pi\,d_o^2}$

7. Bending stress $S_2 = \dfrac{32\,F_e}{\pi\,d_o^3}$

8. Shear stress in the threads of screw

$$S_{s1} = \frac{F}{\pi\, d_o\, \dfrac{p}{2}\, i}$$

9. Shear stress due to torque T_1.

$$S_{s2} = \frac{16\,T_1}{\pi\,d_o^2}$$

10. Find maximum principal stress, equation 19.24 or maximum shear stress equation 19.25 to check for final safety factor of 4 on S_u. If not suitable, go to step 1 and repeat. This procedure can also be used to write the computer programme.

11. Assume the diameter of head of screw d_3 at the top $= 1.75d$ to $2d$.

12. To find the dimensions of the tommy bar, find the frictional collar torque T_c by assuming cap dimensions with inner dia $d_1 \geq 20$ mm (or $d/4$) and outer diameter $= d_3$.

$$T_c = \frac{1}{3} F\, f_2 \left(\frac{d_3^3 - d_1^3}{d_3^2 - d_1^2}\right)$$

or taking $d_m = 0.75d$, $T_c = f_2 F d_m / 2$

The height of the cap may be kept minimum say 30 mm and is tapered to larger upper side. The top surface is serrated or provided with 4 semicircular grooves.

13. Total Torque, $T = T_1 + T_c$ which is to be produced by applying an effort P at the end of a bar of length L, $T = PL$

Total torque $T = T_1 + T_2$

14. Effort P up to 350 N can be conveniently applied and a bar up to a maximum of 0.75 m may be considered as a suitable length. If not suitable, consider the operation by two or four persons simultaneously applying effort for turning.

15. Then the diameter of a bar d_4 is found by considering maximum bending moment at the outside of the head of screw, i.e.

Maximum bending moment $= P\left(L - \dfrac{d_3}{2}\right)$

or safer value of PL.

By considering a low safety factor of 1.5 on yield stress or 2.5 on UTS for finding allowable stress S_{d1}, the diameter of bar d_4 is found.

$$d_4 = \left[\frac{32P(L - d_3/2)}{\pi\, S_{d1}}\right]^{\frac{1}{3}} \text{ or } \left[\frac{32PL}{\pi\, S_{d1}}\right]^{\frac{1}{3}}$$

16. The thickness of head of screw $h = 2\,d_4$.

17. The cup height $h_1 - 30$ mm

18. Size of nut. Outside diameter of nut $D_1 \geq 2d$, can be checked for tensile stress

$$= \frac{4\,F}{\pi(D_1^2 - d^2)}$$

The collar diameter $D_2 = 1.5\,D_1$ and can be checked from crushing stress

$$S_c = \frac{4\,F}{\pi\left(D_2^2 - D_1^2\right)}$$

The thickness of the collar t may be fixed 10 mm and is checked for shearing stress

$$S_s' = \frac{F}{\pi D_1 t}$$

19. Body dimensions. The body is made of cast-iron and its top diameter

$$D_2 = 1.5 D_1 \text{ to } 2D_1$$

The inside diameter at bottom of the body

$$D_4 = 1.5 D_1 \text{ to } 2D_1.$$

The outside diameter of flange

$D_5 = 3D_1$

The thickness of the body and base $= d/4$ or 10–15 mm depending on length.

The various diameters may be suitably changed depending on the height for proportional design.

20. Height of the body = Lift + height of the nut The thickness of collar of the nut will provided sufficient clearance below the screw for providing lift.

The ground clearance = height of the body + thickness of nut collar + $\dfrac{p}{2}$ + height of head + height of cap.

The screw below its head is provided with reduced diameter d_o for a length $\dfrac{p}{2}$ for ease of cutting threads (under cut).

Depending on the maximum length of screw above the nut, it may be checked for column action and buckling load P_b is calculated which should be much higher than F.

Example 19.2: Determine the main dimensions of the screw jack in Fig. 19.12 for a load of 155 kN and a lift of 360 mm.

The screw can be made of SAE 1030 (30C8) steel. The following values are taken from Table 4.2. In tensions S_e = 294 N/mm², in compression S_{ec} = 294 N/mm² and in shear S_{es} = 175 N/mm². The nut can be made of phosphor bronze for which S_e = 138 N/mm², S_{ec} = 122 N/mm², S_{es} = 115 N/mm².

For preliminary calculations a comparatively high safety factor n of 2.25 may be assumed to take care of size influence.

The minimum root diameter of the screw for simple compression is

$$d_o = \sqrt{\dfrac{155 \times 1000 \times 2.25}{294 \times \pi/4}} = 38.86 \text{ mm}$$

For square threads a pitch of 12 mm may be assumed the proper size seems to be $D = 38.86 + 12 = 51$ mm (say)

The root diameter $d_o = 51 - 12 = 39$ mm

Mean diameter $d_m = \dfrac{51 + 39}{2} = 45$ mm

The lead angle $\alpha = \tan^{-1}\dfrac{12}{\pi \times 45} = 4.85°$ is ok

The length of the protruding screw should be taken at least one diameter more than the lift to allow for the head part, say $2d$. Thus $l = 360 +$ $2 \times 51 = 462$ mm, since $\dfrac{l}{d_o} = \dfrac{462}{39} = 11.8$, short column equation may be used to check for diameter. If Ritter's formula is used (equation 2.41).

$$A = 0.7854 \times 39^2 = 1194.6, k = 0.25d_o = 9.75$$

$$\dfrac{l}{k} = \dfrac{l}{0.25\,d_o} = \dfrac{462}{0.25 \times 39} = 45$$

$n = 0.25$, $E = 207 \times 1000$ N/mm². The size influence can be taken into account by use of equation 5.8 and Table 5.1. In this case

$$e_{se} = 1 - \dfrac{(1 - 0.82)(51 - 12)}{63} = 0.89$$

Then $S_e' = 294 \times 0.89 = 262$ N/mm² and design stress with a safety factor $n = 2$ is

$$S_d = \dfrac{262}{2} = 131 \text{ N/mm}^2$$

From equation 2.41

$$S = \dfrac{155000}{1149.6}\left(1 + \dfrac{45^2 \times 262}{\pi^2 \times 0.25 \times 207 \times 1000}\right)$$

$$= 274 \text{ N/mm}^2$$

is not satisfactory. Increasing the root diameter to 48 mm and outside diameter to 60 mm, the size factor

$$e_{sc} = \dfrac{(1 - 0.82)(60 - 12)}{63} = 0.8629$$

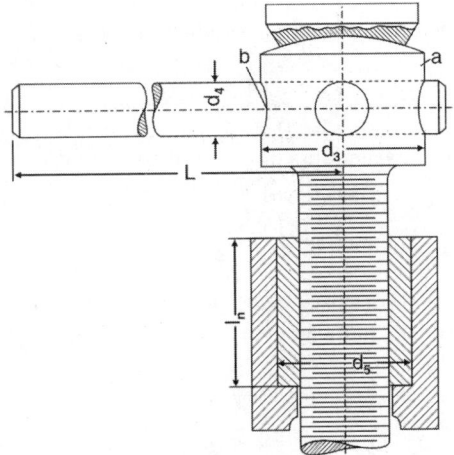

Fig. 19.12: Screw jack

The design stress

$$= \frac{294 \times 0.8629}{2} = \frac{254}{2} = 127 \text{ N/mm}^2$$

and $S_e' = 254 \text{ N/mm}^2$.

Now $A = 0.7854 \times 48^2 = 1809.5 \text{ mm}^2$.

$$\frac{l}{k} = \frac{462}{0.25 \times 60} = 30.8$$

Therefore,

$$S = \frac{155000}{1809.5}\left(1 + \frac{30.8^2 \times 254}{\pi^2 \times 0.25 \times 207 \times 1000}\right)$$

$$= 126 \text{ N/mm}^2$$

So the diameter of 60 mm is satisfactory.

The number of threads i necessary to obtain a proper bearing pressure is found from equation 19.18 with $P_b = 17 \text{ N/mm}^2$

$$i = \frac{F}{\pi/4(d^2 - d_o^2)p_b} = \frac{155000 \times 4}{\pi \times (60^2 - 48^2) \times 17}$$

$$= 8.9 \text{ say } 9$$

Length of nut = 12 × 9 = 108 mm o.k.

Now the lead angle may also be checked again

$$\alpha = \tan^{-1}\frac{12}{\pi \times 54} = \tan^{-1} 0.07 = 4.05° \text{ also o.k.}$$

If the pivot friction between top of the screw and cap can be taken into account by using a relation similar to equation 11.20. Thus

$$H_2 = \frac{Ff_2d_m'}{d_m}$$

The corresponding torque becomes

$$T_c = \frac{1}{2}d_m H_2 = \frac{1}{2}F f_2 d'm$$

and the equation for overall efficiency, becomes

$$e_2 = \frac{\tan\alpha}{\tan(\alpha + \Phi) + \dfrac{f_2 d_m'}{d_m}}$$

Where friction coefficient may be taken as 0.10 and mean diameter of cap dm' may be taken as 0.75d, with

$$\Phi = 5.71, \alpha + \Phi = 9.76, \tan(\alpha + \Phi) = 0.172$$

$$f_2 \frac{d_m'}{d_m} = \frac{0.10 \times 0.75 \times 60}{54} = 0.083$$

$$e_2 = \frac{0.07}{0.172 + 0.083} = 0.274$$

In the threads the shear stress in the threads can be found as

$$S_s = \frac{F}{\pi d_o\, i\, p/2} = \frac{2 \times 155000}{\pi \times 48 \times 9 \times 12} = 19.03 \text{ N/mm}^2$$

In the threads of the nut, the stress is still smaller.

The torsional moment transmitted by the screw is given by equation 19.6

$$T_1 = \frac{1}{2}F\, d_m \tan(\alpha + \Phi)$$

$$= \frac{1}{2} \times 155000 \times 54 \times 0.172 = 719820 \text{ Nmm}$$

The torsional shear stress is

$$S_s' = \frac{719820}{\pi/16 \times 48^3} = 33.14 \text{ N/mm}^2$$

Total shear stress = 19.03 + 33.14 = 52.17 N/mm²
Direct compression stress

$$= \frac{155000}{\pi/4 \times 48^2} = 85.65 \text{ N/mm}^2$$

The maximum principal stress

$$= \frac{1}{2}\left[86.65 + \sqrt{85.65^2 + 4 \times 52.17^2}\right]$$

$$= 110.32 \text{ N/mm}^2$$

which gives a factor of safety of $\dfrac{254}{110.32} = 2.3$

which is satisfactory for hand operated machinery.

The maximum shear stress

$$= \frac{1}{2}\sqrt{85.65^2 + 4 \times 57.17^2} = 67.49 \text{ N/mm}^2$$

The corresponding safety factor is

$$n = \frac{175 \times 0.8629}{67.49} = 2.29$$

which is also satisfactory.

The diameter d_3 at the top of the screw should be made equal to 1.75d to 2d. Thus

$$d_3 = 2 \times 60 = 120 \text{ mm}$$

The length L of the turning bar can be found by equating the torque T_1 plus torsional resistance T_c at the cap to the turning moment LF_t'. By equation (19.15).

$$T_c = \frac{1}{2} \times 155 \times 1000 \times 0.1 \times 0.75 \times 60 = 348750 \text{ Nmm}$$

If it is assumed that four men will act and F' per person is 350 N as the load is very large.

Total torque $= 719820 + 348750 = 1068570$ Nmm

$$L = \frac{1068570}{4 \times 350} = 763.26 \text{ say } 765 \text{ mm}$$

To give a grip for hands, the length will be made 0.8 m.

Finally the diameter d_4 of the turning bar should be determined. When the external bending moment is equated to the moment of the resistance in the bar at the point b; the result is

$F_t(L - 0.5\,d_s) = Z S_d$. Taking SAE 1020 (20C8) and a safety factor of $n = 1.5$,

$$S_d = \frac{275}{1.5} = 183 \text{ N/mm}^2$$

$$d_4 = \left(\frac{350(800 - 60) \times 32}{\pi \times 183} \right)^{\frac{1}{3}}$$

$$= 24.33 \text{ say } 24 \text{ or } 27 \text{ mm}$$

Example 19.3: Design completely a screw jack for lifting a load of 30 kN. The ultimate tensile or compressive stress of the material of screw is 650 MN/m². The allowable bearing pressure is 16 N/mm². The coefficient of thread friction is 0.14 and coefficient of collar friction is 0.2. The ground clearance is 250 mm. Take eccentricity of load 1.5 mm.

Taking a safety factor n of 6 on UTS, the design stress

$$S_d = \frac{650}{6} = 108 \text{ N/mm}^2$$

The core diameter

$$d_o = \left[\frac{30 \times 1000 \times 4}{\pi \times 108} \right] = 18.8 \text{ mm}$$

Let pitch $p = 5$ mm, $d = 18.8 + 5 = 23.8$ say 24 mm

Then $d_o = 19$, $d_m = \dfrac{19 + 24}{2} = 21.5$ mm

$$\tan \alpha = \frac{p}{\pi d_m} = \frac{5}{\pi \times 21.5} = 0.074,$$

$\alpha = 4.2°$ rather low

By taking $p = 6$, $d_o = 19$, $d = 25$

$$\alpha = \tan^{-1} \frac{6}{\pi \times 22} = 4.96° \text{ o.k.}$$

$\tan\varphi = 0.14$, $\varphi = 7.97°$, $\tan(\alpha + \varphi)$

$$= \tan 12.93° = 0.2296$$

Torque $T = 1/2\, F d_m \tan(\alpha + \varphi)$

$$= 1/2 \times 30{,}000 \times 22 \times 0.226$$

$$= 75762 \text{ Nmm}$$

Direct comp. stress

$$S_1 = \frac{4 \times 30 \times 1000}{\pi \times 19^2} = 105.8 \text{ N/mm}^2$$

Bending stress

$$S_2 = \frac{30 \times 1000 \times 1.5 \times 32}{\pi \times 19^2} = 66.82 \text{ NM/mm}^2$$

Number of threads

$$i = \frac{4 \times 30 \times 1000}{\pi [25^2 - 19^2] \times 16} = 9$$

Direct shear stress

$$S_{s1} = \frac{30{,}000 \times 2}{\pi \times 19 \times 6 \times 9} = 18.61 \text{ N/mm}^2$$

Shear stress due to torque

$$S_{s2} = \frac{16 \times 75762}{\pi \times 19^3} = 56.25 \text{ N/mm}^2$$

Total compression stress $S = 105.8 + 66.82 = 172.62$ N/mm²

Total shear stress $S_s = 18.61 + 56.25 = 74.86$ N/mm²

Principal stress

$$S_p = \frac{1}{2}\left[172.62 + \sqrt{172.62^2 + 4 \times 74.86} \right]$$

$$= 200.56 \text{ N/mm}^2$$

Safety factor $= \dfrac{650}{200.56} = 3.24$ which is low

Taking an initial safety factor of 9

$$S_d = \frac{650}{9} = 72 \text{ N/mm}^2$$

$$d_o = \left[\frac{4 \times 30{,}000}{\pi \times 72} \right]^{\frac{1}{2}} = 23.03 \text{ mm}$$

Taking $p = 8$, $d = 32$ mm, $d_o = 24$ mm, $d_m = 28$ mm

$$\alpha = \tan^{-1} \frac{8}{\pi \times 28} = \tan^{-1} 0.09 = 5.2° \text{ o.k.}$$

$\tan(\alpha + \phi) = 0.234$

$$T_1 = \frac{1}{2} \times 30{,}000 \times 28 \times 0.234 = 98280 \text{ Nmm}$$

Number of threads

$$i = \frac{4 \times 30{,}000}{\pi [32^2 - 24^2] \times 16} = 5.3 \text{ say } 6$$

length of nut $l = 6 \times 8 = 48$ o.k.

Direct compression stress

$$S_1 = \frac{4 \times 30,000}{\pi \times 24^2} = 66.31 \text{ N/mm}^2$$

Bending stress,

$$S_2 = \frac{32 \times 30,000 \times 1.5}{\pi \times 24^3} = 33.16 \text{ N/mm}^2$$

Direct shear stress

$$S_{s1} = \frac{30,000 \times 2}{\pi \times 24 \times 8 \times 6} = 16.58 \text{ N/mm}^2$$

Shear stress due to torque

$$S_{s2} = \frac{16 \times 98280}{\pi \times 24^3} = 36.20 \text{ N/mm}^2$$

Total compression stress = 66.31 + 33.16 = 99.47 N/mm².

Total shear stress = 16.58 + 36.20 = 52.78 N/mm²

Principal stress

$$= \frac{1}{2}\left(99.47 + \sqrt{99.47^2 + 4 \times 52.78^2}\right) = 122.24$$

Safety factor $= \dfrac{650}{122.54} = 5.3$

which is high

Now taking a safety factor of 7.5

$$S_d = \frac{650}{7.5} = 87 \text{ N/mm}^2$$

$$d_o = \left[\frac{4 \times 30,000}{\pi \times 87}\right]^{\frac{1}{2}} = 20.95$$

Taking $p = 7, d = 28$ mm, $d_o = 21$ mm, $d_m = 24.5$ mm

$$\alpha = \tan^{-1}\frac{7}{\pi \times 24.5} = 5.2° \text{ o.k. } \tan(\alpha + \Phi) = 0.234$$

$$T = \frac{1}{2} \times 30,000 \times 24.5 \times 0.234 = 85995 \text{ Nmm}$$

Number of threads

$$i = \frac{4 \times 30,000}{\pi[28^2 - 21^2] \times 16} = 6.96 \text{ say } 7$$

Length of nut $P = 7 \times 7 = 49$ mm or 50 mm o.k.

Direct compression stress

$$= \frac{4 \times 30,000}{\pi \times 21^3} = 86.61 \text{ N/mm}^2$$

Bending stress

$$= \frac{32 \times 30,000 \times 1.5}{\pi \times 21^3} = 49.49 \text{ N/mm}^2$$

Direct shear stress

$$= \frac{30,000 \times 2}{\pi \times 21 \times 7 \times 7} = 18.56 \text{ N/mm}^2$$

Shear stress due to torque

$$= \frac{16 \times 85995}{\pi \times 21^3} = 47.29 \text{ N/mm}^2$$

Total compression stress

$$= 86.61 + 49.49 = 136.10 \text{ N/mm}^2$$

Total shear stress = 18.56 + 47.29 = 65.85 N/mm²

Maximum principal stress

$$= \frac{1}{2}\left[136.1 + \sqrt{136.1^2 + 4 \times 65.85^2}\right]$$

$$= 162.74 \text{ N/mm}^2$$

safety factor $= \dfrac{650}{162.74} = 3.99$ which is satisfactory.

Dimensions of nut

Outside diameter of nut $D_1 = 2d = 56$ mm

Maximum tensile stress

$$= \frac{4 \times 30,000}{\pi[56^2 - 28^2]} = 16.24 \text{ N/mm}^2$$

For soft phospher bronze, UTS = 180 N/mm², it is quite safe.

The collar diameter $D_2 = 1.5 D_1 = 1.5 \times 56 = 84$ mm

Crushing stress

$$= \frac{4 \times 30,000}{\pi[84^2 - 56^2]} = 9.7 \text{ N/mm}^2 \text{ o.k.}$$

The thickness of the collar may be taken = 10 mm

Shearing stress in the collar

$$= \frac{30,000}{\pi \times 56 \times 10} = 17 \text{ N/mm}^2$$

Tommy bar dimensions.

Cap may have an inside diameter of 20 mm and outside diameter

$$= d_3 = 2d = 56 \text{ mm}$$

Collar frictional torque

$$T_c = 0.2 \times \frac{1}{3} \times 30,000\left[\frac{56^3 - 20^3}{56^2 - 20^2}\right]$$

$$= 122526 \text{ Nmm}$$

Total torque = 85995 + 122526 = 208521 Nmm

Assuming effort to be applied by one person is 350 N, length of the bar

$$= \frac{208521}{350} = 595.7 \text{ mm or 596 mm}$$

Consider half of gripping length, about 50 mm the length of bar be taken as 650 mm.

Maximum bending moment on the bar

$$= 350 \times \left(596 - \frac{56}{2}\right) = 198800 \text{ Nmm}$$

Considering material of bar 20C8 with UTS = 440 N/mm² with a safety factor of 2.5,

$$S_d = \frac{440}{2.5} = 176 \text{ N/mm}^2$$

The diameter of the bar

$$d_1 = \left[\frac{32 \times 198800}{\pi \times 176}\right]^{1/3} = 22.09 \text{ say 24 mm}$$

The height of the top of screw $h = 2 \times 24 = 48$ mm or say 50 mm

The cap can have a thickness of 12 mm and it tapers from 50 mm to a diameter of 70 mm on the top.

The top diameter of the body $D_2 = 1.5D_1 = 84$ mm

Inside diameter at bottom $D_4 = 1.75\ D_2 = 147$ or 150 mm

Outside diameter at bottom $D_5 = 3D_2 = 252$ mm

Thickness of the body = 28/4 or say 10 mm

If the ground clearance is 250 mm

Lift = $GC - h_1 - h - 10$

so lift = 250 – 30 – 50 – 10 = 110 mm.

Including the length of head and height of cap.

Maximum length of screw above the nut = 110 + 4 + 50 + 30 = 194 mm

$\dfrac{l}{d} = \dfrac{194}{21} = 9.52$, so the screw needs to be checked for buckling.

$$k = \sqrt{\frac{I}{A}} = \sqrt{\frac{\pi \times d^4 \times 4}{64 \times \pi \times d^2}} = \frac{d}{4} = \frac{21}{4} = 5.25$$

$$\frac{l}{k} = \frac{194}{5.25} = 36.95$$

Consider one end fixed and other free, $a = \dfrac{1.95}{25000}$

$S_c = 430$ N/mm²

$$\text{Buckling load} = \frac{S_c A}{1 + a\left(\dfrac{l}{k}\right)^2}$$

$$= \frac{430 \times \dfrac{\pi}{4} \times 21^2}{1 + \dfrac{1.95}{25000}(36.95^2)}$$

= 134600 N which is very safe with a F. O. S of > 4.

Example 19.4: A C-clamp has the following data:

Outside diameter of the screw	= 12 mm
Core diameter	= 10 mm
Pitch	= 2 mm
Coefficient of thread friction	= 0.12
Coefficient of friction of collar	= 0.25
Load to be handled	= 4 kN
Length of nut	= 25 mm
Effort applied at the handle	= 80 N
Mean collar radius	= 6 mm

(a) Find the stresses at section AA, BB, and CC (b) length and diameter of the handle, (c) bearing pressure, (d) stress at section CC of the body.

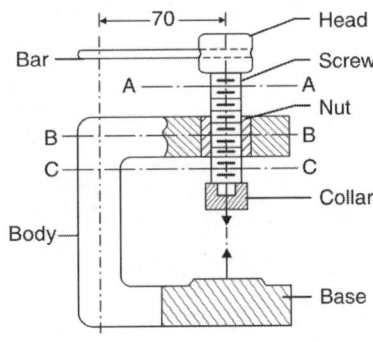

Fig. 19.13

1. Direct compression stress in the screw

$$= \frac{4W}{\pi d_o^2} = \frac{4000 \times 4}{\pi \times 10^2} = 50.92 \text{ N/mm}^2$$

Mean diameter $= \dfrac{12 + 10}{2} = 11$ mm

Lead angle $\tan \alpha = \dfrac{p}{\pi d_m} = \dfrac{2}{\pi \times 11} = 0.0578, \alpha = 3.3°$

$\phi = \tan^{-1} 0.12 = 6.84°$

Torque required to lift the load

$$T_1 = F\frac{d_m}{2}\tan(\alpha + \phi) = 4000 \times \frac{11}{2} \tan 10.14$$

$$= 3935.7 \text{ Nmm}$$

Collar torque,

$$T_c = fWr_m$$

$$= 0.25 \times 4000 \times 6 \doteq 6000 \text{ Nmm}$$

Total torque = $T = T_1 + T_c = 3935.7 + 6000 = 9935.7$

Number of thread in the nut $i = \dfrac{l}{p} = \dfrac{25}{2} = 12.5$

2. Direct shear stress in threads

$$= \dfrac{W}{\pi d_o \times \dfrac{p}{2} \times n} = \dfrac{4000}{\pi \times 10 \times 1 \times 12.5}$$

$$= 10.18 \text{ N/mm}^2$$

3. Shear stress due to torque T_c, $S_{s1} = \dfrac{6000 \times 16}{\pi \times 10^3}$

$$= 30.55 \text{ N/mm}^2$$

4. Shear stress due to torque $T = \dfrac{9935.7 \times 16}{\pi \times 10^3}$

$$= 50.60 \text{ N/mm}^2$$

5. Shear stress due to torque $T_1 = \dfrac{3935.7 \times 16}{\pi \times 10^3}$

$$= 20.04 \text{ N/mm}^2$$

Section AA

Max. shear stress = Shear stress due to torque T + direct shear stress = 50.60 + 10.18 = 60.78 N/mm²

Section BB

Shear stress = Shear stress due to torque T_1 + direct shear stress

$$= 20.04 + 10.18 = 30.22 \text{ N/mm}^2$$

Section CC

Shear stress = 30.55 + 10.18 = 40.23

Compress stress = 50.92

Maximum principal stress

$$= \dfrac{1}{2}\left[50.92 + \sqrt{50.92^2 + 4 \times 40.73^2} \right]$$

$$= 73.62 \text{ N/mm}^2$$

Length of handle $= \dfrac{T}{F} = \dfrac{9935.7}{80} = 124.19$ say 125

+ 50 = 175 mm

Diameter of handle

$$= \left[\dfrac{\text{Max B } M \times 32}{\pi \, S} \right]^{\frac{1}{3}} = \left[\dfrac{9935.7 \times 32}{\pi \times 80} \right]^{\frac{1}{3}}$$

$$= 10.8 \text{ mm say 12 mm (standard size)}$$

Bearing pressure $f_b = \dfrac{F}{\dfrac{\pi}{4}\left[d^2 - d_o^2 \right]n}$

$$= \dfrac{4000 \times 4}{\pi\left(12^2 - 10^2\right) \times 12.5}$$

$$= 9.26 \text{ N/mm}^2$$

Stress at section CC of the body

Direct tensile stress $= \dfrac{F}{\text{Area}}$

$$= \dfrac{4000}{25 \times 25} = 6.4 \text{ N/mm}^2$$

Maximum bending moment $= Fe = 4000 \times 70$

Bending stress $= \pm\dfrac{M}{Z} = \pm\dfrac{4000 \times 70 \times 6}{25 \times 25^2}$

$$= 107.52 \text{ N/mm}^2$$

Max tensile stress on outer the side

$$= 107.52 - 6.4 = 101.12 \text{ N/mm}^2$$

Max tensile stress on the inner side = 107.52 + 6.4 = 114.92 N/mm².

Example 19.5: Check stresses in all parts of the scissor screw jack (Toggle screw jack) shown in Figure 19.14. It has been designed for 7.5 kN to be used for cars. At maximum height distance AB = 75 mm. At minimum height, AB = 292 mm. Links are channel sections 40 × 20 and 2.5 mm thick. All the pins are 12 mm diameter. Material of links, pins and screw is 40C8 having $\sigma_u = 620$ N/mm, $\sigma_s = 380$ N/mm². The screw is having a core diameter of 15 mm and outside diameter 18 mm of square threads. Coefficient of thread friction = 0.1 and coefficient of collar friction = 0.12, length of nut = 22 mm. Design a convenient crank handle.

At any height of link (Fig. 19.14b).

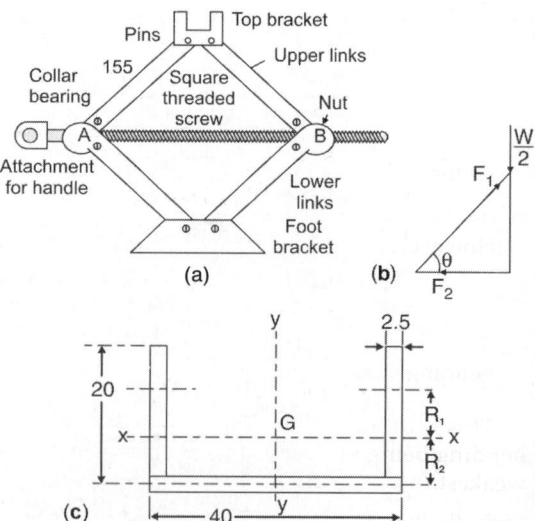

(a)

(b)

(c)

Fig. 19.14

Force in link $F_1 = \dfrac{W/2}{\sin\theta}$

Force in screw $F_2 = \dfrac{W/2}{\tan\theta}$

Links are subjected to buckling and screw is subjected to tension.

Both F_1 and F_2 will be maximum when θ is minimum, i.e. when the screw is having minimum height.

At the minimum, height $\cos\theta = \dfrac{292/2}{155}$, $\theta = 19.62$,

$\sin\theta = 0.3357$, $\tan\theta = 0.3564$

$$F_1 = \frac{7500/2}{0.3357} = 11170 \text{ N}$$

$$F_2 = \frac{7500/2}{0.3564} = 10522 \text{ N}$$

Pins: All pins are subjected to bearing pressure and double shear. The possibility of bending does not exist. As $\dfrac{l}{d}$ is quite high, it will be checked for bending also.

The pins have a length of 40 mm = width of the channel

Bearing pressure $f_b = \dfrac{11170}{40 \times 12} = 23.27 \text{ N}$ o.k.

As the usage is limited, the allowable pressure for such use is 25 N/mm²

Shear stress in pins

$$= \frac{11170}{2 \times \dfrac{\pi}{4} \times 12^2} = 49.38 \text{ N/mm}^2 \text{ o.k.}$$

As the $\dfrac{l}{d} = \dfrac{40}{12} = 3.33$, the possibility of bending is less

However, bending moment

$$= \frac{11170 \times 40}{8} = 55850 \text{ Nmm}$$

Bending stress $= \dfrac{55850 \times 32}{\pi \times 12^3} = 329.24 \text{ N/mm}^2$

This is quite high but in case of overload and bending being not possible, still it makes the pin weakest.

Links-Position of C.G.: Figure 19.14c

$\bar{x} = 20$ mm from the left or right edge

$$\bar{y} = \frac{2 \times 2.5 \times 20 \times 12.5 + 2.5 \times 40 \times 1.25}{2 \times 2.5 \times 20 + 2.5 \times 40}$$

$$= \frac{1375}{200} = 6.875$$

$R_1 = 5.625$, $R_2 = 5.625$

$$I_{xx} = \frac{2 \times 2.5 \times 20^3}{12} + 2 \times 2.5 \times 20 \times 5.625^2$$

$$+ \frac{40 \times 2.5^3}{12} + 40 \times 2.5 \times 5.625^2$$

$$= 3333.33 + 3164.0625 + 52.0833 + 3164.0625$$

$$= 9713.54 \text{ mm}^4$$

$$I_{yy} = \frac{2 \times 20 \times 2.5^3}{12} + 2 \times 50 \times 18.75^2 + \frac{2.5 \times 40^3}{12}$$

$$= 52.083 + 35156.25 + 13333.33$$

$$= 48541.66 \text{ mm}^4$$

As the channel has hinged ends about xx axis and fixed ends about yy axis, ratio of

$$\frac{I_{yy}}{I_{xx}} = \frac{48541.66}{9713.54} = 4.99$$

Section xx is weaker and buckling about section xx is checked.

$$k = \sqrt{\frac{I}{A}} = \sqrt{\frac{9713.54}{200}} = 6.969$$

Using Rankine's equation for hinged ends.

$$\sigma_c = \frac{P_c}{A}\left[1 + a\left(\frac{l}{k}\right)^2\right] = \frac{11170}{200}\left[1 + \frac{1}{6250}\left[\frac{155}{6.969}\right]^2\right]$$

$$= 60.27 \text{ N/mm quite safe as FOS on yield stress}$$

$$= \frac{324}{60.27} = 5.37 \text{ o.k.}$$

Screw

$$d_m = \frac{15 + 18}{2} = 16.5 \text{ mm}$$

Lifting torque $T_1 = \dfrac{d_m}{2} W \tan(\alpha + \phi)$

$\tan\alpha = \dfrac{3}{\pi \times 16.5}$, $\alpha = 3.31°$, $\theta = \tan^{-1} 0.1 = 5.71°$

$\tan(\alpha + \phi) = \tan 9.02 = 0.1585$

F_2 is force in screw by considering left side of link. Considering the reaction on the base $= W$, the left side link will also be subjected to F_2 force in the screw. Total force in screw is $F_2 + F_2 = 2 F_2$. The right side links will also produce a force $2 F_2$ is the opposite direction so screw is subjected to a force $2F_2 = 2 \times 10522 = 21044$ N.

$$T_1 = \frac{16.5}{2} \times 21044 \times 0.1585$$

$$= 27517.66 \text{ Nmm}$$

Shear stress $S_{s1} = \dfrac{27517.66 \times 16}{\pi \times 15^3} = 41.52 \text{ N/mm}^2$

Number of threads in the nut

$$n = \frac{22}{3} = 7.1 \text{ say } 8$$

Bearing pressure

$$f_b = \frac{21044}{\dfrac{\pi}{4}(18^2 - 15^2) \times 8} = 33.83 \text{ N/mm}^2$$

This is quite high, so taking $n = 12, f_b = 22.5$ N/mm^2 (o.k.)

Direct shear stress

$$= \frac{21044}{\pi \times 15 \times \dfrac{3}{2} \times 12} = 24.89 \text{ N/mm}^2$$

Tensile stress $= \dfrac{21044 \times 4}{\pi \times 15^2} = 119.08 \text{ N/mm}^2$

Total shear stress $= 41.52 + 24.81 = 66.33$ N/mm^2

Maximum principal stress

$$= \frac{1}{2}\left[119.08 + \sqrt{119.08^2 + 4 \times 66.33^2}\right]$$

$$= 148.67 \text{ N/mm}^2$$

Factor of safety in tension $= \dfrac{620}{148.67} = 4.17$ o.k.

Maximum shear stress $= \dfrac{1}{2}\sqrt{119.08^2 + 4 \times 66.33^2}$

$$= 89.13 \text{ N/mm}^2$$

Factor of safety in shear $= \dfrac{380}{89.13} = 4.26$ o.k.

Crank handle

Assume mean collar radius $= 0.5 \times 18 = 9$ mm.

Friction collar torque $= 0.12 \times 22044 \times 9$

$$= 22727.52 \text{ Nmm}$$

Total torque required $= 22727.52 + 27517.66$

$$= 50245.18 \text{ Nmm}$$

Consider a crank radius of 150 mm.

Effort required $= \dfrac{50245.18}{150} = 334.9$ say 335 N

Diameter of rod of crank $= \left[\dfrac{50245.18 \times 32}{\pi \times 260}\right]^{\frac{1}{3}}$

$$= 12.53 \text{ say } 12.5 \text{ mm}$$

Example 19.6: Determine the dimensions of screw, head of screw, bar, length of nut for the screw clamp shown in Fig. 19.13. Check for stresses in sections above and below the nut surface. The clamp can be used for gripping loads up to 10 kN. Coefficient of thread friction $f_1 = 0.1$ and collar friction $f_2 = 0.12$. The force at the lever end is limited to 200 N. The bearing pressure in the nut is limited to 16 N/mm^2. The maximum stress in the screw is not to exceed 150 N/mm^2.

Taking the compression stress in the screw
$$= 90 \text{ N/mm}^2$$

The core diameter of screw $d_o = \left[\dfrac{10,000 \times 4}{\pi \times 90}\right]$

$$= 11.89 \text{ mm say } 12 \text{ mm}.$$

Assuming pitch of 4 mm, outside diameter
$$= 12 + 4 = 16 \text{ mm}.$$

The mean diameter $= \dfrac{12 + 16}{2} = 14$ mm.

Checking for lead angle $\alpha = \tan^{-1}\dfrac{4}{\pi \times 14}$

$$= 5.19° \text{ o.k.}$$

$\phi = \tan^{-1} 0.1 = 5.71°$, $\tan(\alpha + \phi) = \tan 10.9° = 0.192$

Lifting torque $T_1 = W \dfrac{dm}{2} \tan(a + \phi)$

$$= 10,000 \times \frac{14}{2} \times 0.192 = 13440 \text{ Nmm}$$

Number of threads in the nut $= \dfrac{10,000}{\dfrac{\pi}{4}(16^2 - 12^2)16}$

$$= 7.1 \text{ say } 8$$

Length of nut $= 8 \times 4 = 32$ o.k $(= 2d)$

Direct shear stress $= \dfrac{10,000}{\pi \times 12 \times \dfrac{14}{2} \times 8}$

$$= 16.57 \text{ N/mm}^2$$

Actual compression stress

$$= \frac{10,000}{\frac{\pi}{4} \times 12^2} = 88.42 \text{ N/mm}^2$$

Shear stress due to torque $T_1 = \dfrac{13440}{\dfrac{\pi}{16} \times 12^3}$

$$= 39.61 \text{ N/mm}^2$$

Mean collar diameter $= 0.75 \times 14 = 10.5$ mm

Collar friction torque $T_c = 0.12 \times 10,000 \times 10.5$

$$= 12600 \text{ Nmm}$$

Total torque $= T_1 + T_c = 13440 + 12800$

$$= 26040 \text{ Nmm}$$

Length of bar $L = \dfrac{26040}{200} = 130.2$

Adding 50 mm for gripping length, $L = 180$ mm

Assuming 60 mm of screw above the nut

Shear stress due to torque

$$T = \frac{26040 \times 16}{\pi \times 12^3} = 76.74 \text{ N/mm}^2$$

Bending stress due to effort $= \dfrac{200 \times 60 \times 32}{\pi \times 12^3}$

$$= 70.73 \text{ N/mm}^2$$

Stresses above the nut, at the upper surface of nut

Total shear stress $= 76.74 + 16.57 = 93.31 \text{ N/mm}^2$

Maximum shear stress $= \dfrac{1}{2}\sqrt{70.73^2 + 4 \times 93.31^2}$

$$= 99.78 \text{ N/mm}^2$$

Maximum principal stress

$$= \frac{1}{2}\left[70.73 + \sqrt{70.73^2 + 4 \times 93.31^2} \right]$$

$$= 143.99 \text{ N/mm}^2$$

Stresses below the nut, at the lower surface of nut

Maximum shear stress $= \dfrac{1}{2}\sqrt{88.42^2 + 4 \times 16.57^2}$

$$= 47.21 \text{ N/mm}^2$$

Maximum principal stress

$$= \frac{1}{2}\left[88.42 + \sqrt{88.42^2 + 4 \times 16.57^2} \right]$$

$$= 91.42 \text{ N/mm}^2$$

Maximum bending moment on the bar

$$= 180 \times 200 = 36000 \text{ Nmm}$$

Assuming an allowable stress $= \dfrac{440}{2.5}$

$$= 176 \text{ N/mm}^2$$

Diameter of bar $= d_3 = \left[\dfrac{26040 \times 32}{\pi \times 176} \right]^{1/3}$

$$= 12.7 \text{ say } 14 \text{ mm}$$

Diameter of head of screw $= 2d = 2 \times 16 = 32$ mm

Height of screw head $= 14 \times 3.5 = 49$ say 50 mm

So maximum shear stress and maximum principal stress acts on the top part of screw and are safe.

PROBLEMS

19.1 How does a power screw differ from an ordinary screw?

19.2 Why is v-thread not suitable for power screw?

19.3 How does the stress concentration effect the design of power screw, when (a) it is used in tension (b) it is used in compression?

19.4 What are the advantages and disadvantages of the square threads? In what way Acme threads are superior to square threads?

19.5 What is the advantage of buttress thread over Acme and square threads when the load is in one direction only?

19.6 How does the efficiency of square thread vary with lead angle? What is the maximum lead angle which should be used when the screw has to be irreversible?

19.7 Why is it not desirable to use lead angle more than 7°?

19.8 How does the efficiency of ball bearing screw compare with the square threads? What is the harm of such a high efficiency? Why then it is used in power transmission?

19.9 Why is the length of the nut calculated from bearing pressure consideration rather than shear stress consideration?

19.10 Name the applications of power screws.

19.11 Why is the nut made of softer material than that of screw?

19.12 Why is it preferable to have a separate nut instead of making it integral with the body?

19.13 Why should the tommy bar be made the weakest part in the screwjack?

19.14 What is the value of thread friction and lead angle for self-locking property?

19.15 What purpose is served by a turn buckle?

19.16 A sluice gate weighting 600 kN is raised and lowered by means of two 70 mm square thread screws. The screws are operated by an electric motor at 600 rpm. A ball thrust bearing is used, reducing the apparent friction coefficient to 0.003 on 50 mm radius. Bronze nuts and fair lubrication are used. If the gate must be raised at the rate of 0.01 m/s, determine (a) the number of revolutions per minute of the screw. (b) the power of the motor required to raise the gate assuming a mechanical efficiency of 0.85 for the speed reduction mechanism and also for motor and (c) the power required at the gate.

19.17 70 mm screw with Sellers' square threads is used in a press. It has a maximum unsupported length of 450 mm. Using 35C8/SAE 1035 steel for the screw and phosphor bronze for the nut, determine (a) the safe capacity of the press, (b) the proper length of the nut, (c) the necessary torque, assuming 0.13 as the friction coefficient in the threads and 0.15 at the thrust collar, which has an outside diameter of 90 mm and an inside diameter of 25 mm. S_e for steel 300 N/mm^2.

19.18 (a) Determine the dimensions d_1, d_2, d_3, d_4, and l_1, Fig. P19.1, for a steel screw used in a hand punch to make 11 mm holes in a 1.5 mm 0.10 per cent carbon steel plate. Use a trapezoidal thread. The punch body is a steel casting. (b) Determine the length L of the lever to be used if

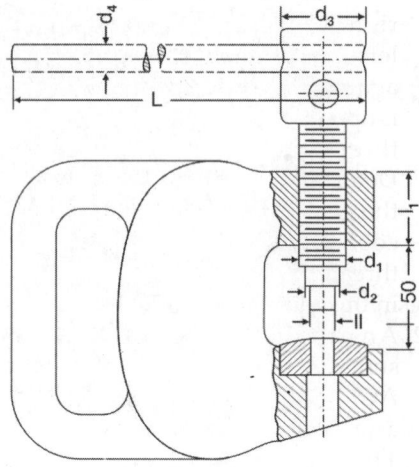

P19.1

the force at its end must not exceed 270 N. The bearing pressure is limited to 15 N/mm^2. The coefficient of thread friction is 0.15. Maximum compression stress is limited to 175 N/mm^2.

19.19 (a) Find the capacity of a screw jack, Fig. 19.11, made with a 38 mm. ACME thread and having a lift of 350 mm (b) Assuming that the friction coefficient f_1 in the threads is 0.15, the coefficient f_2 in the upper thrust support is 0.2, and the outside diameter of the thrust collar is 34 mm, determine the efficiency of the screw jack (c) Determine the length of the lever L required to raise the maximum load that can be lifted by applying a force of 2 × 270 N.

19.20 Determine the main dimensions of a screw jack for a load of 50 kN and a lift of 250 mm. Make the screw of 30C8/SAE 1030 steel, and use a cast iron nut.

19.21 The screw of a toggle press is driven by a gear g, Fig. P19.2, and turns at 75 rpm. The crossheads c move along the screw in opposite directions, since one has a

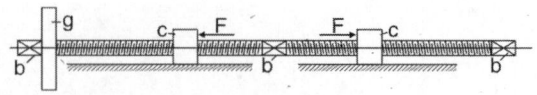

P19.2

right-hand thread and the other has a left-hand thread. The crossheads move against the axial forces *F* when the press is operated. The screw has Sellers' square threads with 56 mm major diameters. Determine the power required to drive the gear *g* if *F* = 20 kN. Assume that the coefficient of friction in the lubricated threads is *f* = 0.11, and neglect the friction in the bearings.

19.22 An automatic machine has a power screw with right-hand and left-hand ACME threads and nonrotating nuts, an arrangement similar to that in Fig. P19.2. The nuts must be moved against forces *F* of 8 kN. The screw has an outside diameter of 50 mm. Assume a friction coefficient f_1 of 0.12, and neglect the friction in the ball bearing *b*. Determine (a) the efficiency of the screw, (b) the power required to turn the driving gear g to move the nuts with a speed of 0.05 m/s and (c) the necessary length of the bronze nuts.

19.23 The 75 mm screw of 30 kN shop press, Fig. P19.3, has a Sellers' type thread. There are two parallel threads, to obtain a greater mechanical advantage. The

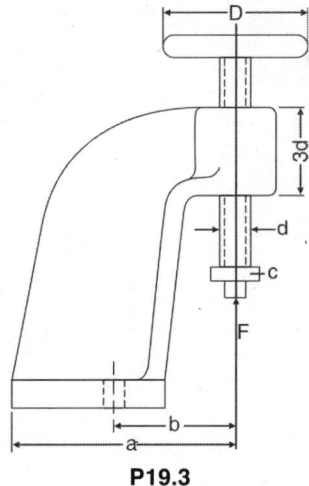

P19.3

handwheel has a diameter *D* of 1.65 m, and the mean diameter of the thrust collar *c* is 65 mm. Determine (a) the force that must be applied to the handwheel, assuming that the coefficient of friction in the threads is f_1 = 0.12 and that at the thrust collar is f_2 = 0.125, (b) the efficiency of the press, (c) the maximum compressive stress in the screw, and (d) the maximum shear stress and the maximum bearing pressure in the threads.

Shafts

20.1 INTRODUCTORY CONSIDERATIONS

A *shaft* is a rotating member which transmits power. An *axle* is a machine part which is loaded chiefly in bending and carries such rotating parts as wheels and gears. An axle may be stationary, or it may rotate. Short shafts and axles in machines are called *spindles*. However, the term shaft is often applied to all machine parts here mentioned, irrespective of the type of loading or size.

A *headshaft*, or *stubshaft*, is directly connected to an engine or motor. A *line shaft*, often called a transmission shaft, is a comparatively long shaft which receives its motion and power from a motor and transmits it to various machines. A *countershaft*, or *jackshaft*, is a short shaft placed between a prime mover and a line shaft or a driven machine.

Materials: Shafts are made of Bessemer steel, open-hearth steel, and alloy steels—and some copper alloys, when resistance to corrosion is desired.

Ordinary shafts are made of Bessemer or open-hearth steel with a carbon content of 0.15 to 0.40 per cent. Such steel is commonly called medium carbon, or machinery, steel. Steel shafting is sometimes finished by cold rolling or cold drawing. This produces a somewhat stronger bar than hot rolling. However, cold-finished shafts have several disadvantages. The diameter tolerances are not very close; the shaft does not come precisely straight; and the fibers at and near the outer surface are under stress which, when partially released through keyseating, causes a distortion of the shaft. The straightening of a distorted or twisted shaft is a difficult and expensive operation.

Most commercial shafting, after being hot-rolled, is turned and polished. The best is known as T and G, or *turned-and-ground*, shafting or CRTG-cold rolled turned and ground. It is made of open-hearth mild steel that is similar to 15C8/SAE 1015 but has a slightly higher content of manganese, phosphorus, and sulfur. On special order, turned-and-ground bars are furnished with a higher carbon content and of stainless steel.

When greater strength is required, as in high-speed machinery, shafts are made of alloy steels, the most common being nickel, nickel-chromium, and chrome-vanadium steels. Alloy-steel shafts are always heat-treated. For spline shafts subjected to severe shocks, 45 Ni 2Cr1/SAE 3245 steel that is oil-quenched from 815°C and tempered to 425°C is particularly suitable.

Shafts for special purposes, especially those having integral connecting flanges or gears, are forged, as are most hollow shafts.

The standard shaft diameters are given in Table 20.1.

Table 20.1: Standard shaft diameters

6	8	10	12	14	16	18	20	22
25	28	32	36	40	45	50	56	63
71	80	90	100	110	125	140	160	180
200	220	240	260	280	300	320	340	360
380	400	420	440	450	480	500	530	560
600								

Both turned-and-ground (T and G) and cold rolled shafting are also available in 0.5 mm increments up to 25 mm diameter, in 1 mm increment from 25 to 50 mm; 2 mm increments from 50 to 100 mm and 5 mm increments from 100 to 200 mm.

20.2 SHAFTS WITH STEADY LOADING

The loads to which shafts may be subjected cause simple torsion, simple bending, combined torsion and bending, or combined torsion and compression with or without bending.

Strength in simple torsion: For a given torsional moment T, in Nm and for an allowable shear stress S_d, the shaft diameter is found from equation 2.18. Thus,

$$D = \sqrt[3]{\frac{16T}{\pi S_d}} \qquad \dots (20.1)$$

Torque can be calculated from power as:

$$P = \frac{2\pi NT}{60 \times 1000} \qquad \dots (20.2)$$

Rigidity: In case where the torsional deflection must be taken into consideration, the angle θ of torsion can be determined from equation 2.19. Multiplying by $180/\pi$ to change from radians to degrees, the following equation is used:

$$\theta = \frac{32lT}{\pi D^4 G} \quad \text{or} \quad \alpha = \frac{584lT}{D^4 G} \qquad \dots (20.3)$$

In drive shafts of machine tools the angle α should be very small, not over 0.25°/m of shaft length. In camshafts of internal-combustion engines the total angle should not exceed 0.5 deg, irrespective of the length. In most shafts, it is a good practice to limit the torsional angle to 1 deg in a length of $L = 20\,D$.

The shaft diameter corresponding to a prescribed deflection α can be computed by solving equation 20.3 for D.

Strength in simple bending: For a given bending moment M and an allowable stress S_d, the shaft diameter may be determined from equation:

$$D = \left[\frac{32M}{\pi S_d}\right]^{1/3} \qquad \dots (20.4)$$

Safety factor: For stationary shafts the safety factor n may be from 1.5 to 2. In a rotating shaft the bending stress varies from a maximum tension to a maximum compression, and n should be taken 50 per cent higher, if referred to the elastic limit S_e, i.e. 2 to 3.

Stiffness: In many cases the stiffness of a shaft is not less important than its strength, and therefore the transverse deflections must be limited. Excessive transverse deflections may create uneven wear of the bearings if the bearings are not self-adjusting. These deflections depend on the distribution of loads acting upon the shaft as well as on the method of supporting it. For slow or moderate speeds they can be calculated in a given case from the general equation 2.27 or, for shafts of a constant diameter, by formulas used for beams and listed in Table 2.6.

Bending moments: In calculating the bending moment coming on a shaft it is customary to measure each moment arm to the middle of the bearing. It is assumed that the clearance between the shaft and the bearing permits the shaft to deflect up to the middle of the bearing. This is a reasonable assumption and, besides, gives results which are on the safe side.

When a machine part with a hub is forced or shrunk upon a shaft, the stress-concentration effect of the hub may cause the shaft to fail near the edge of the hub. This possibility must be taken into account in determining the maximum stress. It is better to reduce stress concentration by using a tapered hub, as shown in Fig. 5.28 or Fig. 13.2.

Combined torsion and bending: A revolving shaft transmitting power and carrying pulleys, gears, sprockets, or sheaves is subjected to simultaneous torsion and bending, usually no shaft transmits pure torque. Its design must be based on a significant stress which is a resultant of the stresses in torsion and bending.

The design of steel shafting usually is based on the maximum-shear theory of failure.[1] From equation 2.13, with $Z_o = \frac{1}{16}\pi D^3$, the relation between the equivalent torsional moment and the combined stress is

$$T_e = \frac{1}{16} S_s' \pi D^3 \qquad \text{...(20.5)}$$

The combined shear stress in this case is

$$S_s' = \sqrt{\left(\frac{1}{2}S\right)^2 + S_s^2}$$

where, S is the normal stress, S_s is the simple shear. When proper substitutions are made in equation 20.5, the result is

$$T_e = \sqrt{M^2 + T^2} \qquad \text{...(20.6)}$$

The necessary shaft diameter is found from equation 20.5 by using for S_s' the design stress $S_d = S_e'/n$, where S_e' is the elastic limit in shear, and using for T_e the value found by applying equation 20.6.

A shaft is always subjected to fatigue loads as it is rotating and there will always be a shock load transmitted when it is transmitting power so a preferable value of

$$T_e = \sqrt{(K_m M)^2 + (K_T T)^2} \qquad \text{...(20.6a)}$$

The values of K_m and K_T are given in Table 20.2.

Effect of keyways: A keyway lowers both the strength and the rigidity of a shaft. For a shaft made of mild steel and having a keyway of approximately standard proportions the lowering of the strength based on static tests[2] in torsion can be taken into account by introducing a factor similar to a stress-concentration factor, namely

$$K'' = 1 + \frac{0.2b + 1.1h}{D} \qquad \text{...(20.7)}$$

where, b and h are then width and depth of the keyway, respectively, and D is the shaft diameter. If there is a keyway, the design stress S_d must be divided by K''. For a standard keyway, with $b = \frac{1}{4}D$ and $h = \frac{1}{6}D$, $K'' = 1.23$.

In general, when a shaft is provided with a key, the allowable stress is reduced by a factor of 0.75, i.e. $k'' = 1.33$.

According to these tests the length of the keyway does not affect the strength of the shaft. Also, bending applied simultaneously with torsion does not affect the strength additionally, so far as the key is concerned.

Rigidity: The same tests[3] indicate that the increase of the angle of twist α, equation 20.3, may be expressed by a coefficient. Its value is

$$K_1 = 1 + \frac{0.4b + 0.7h}{D} \qquad \text{...(20.8)}$$

Naturally, this increase applies only to the keyseated length of the shaft.

20.3 DETERMINATION OF MOMENTS

A shaft is very seldom of uniform strength. A small shaft is usually made with a constant diameter. In the case of a large shaft, the diameter is varied in steps, and the shaft approaches uniform strength only at certain points where the diameter is changed. In either case the determination of the bending moments, and particularly the maximum values, is of great importance. For the sake of simplicity, it is advisable to find the moments separately in the horizontal and vertical planes and then to find the resultant moments.

Generally speaking, the torsional moment also varies at different shaft sections, and its maximum value may not occur at the section at which the bending moment is greatest. In this case, it may be advisable to determine the equivalent moment for two sections—the section with the maximum bending moment and the section with the maximum torque. If the shaft diameter is to be constant, the larger

---

[1] *Code for the Design of Transmission Shafting*, Engineering and Industrial Standards (New York: American Society of Mechanical Engineers).

[2] H. F. Moore, *The Effect of Keyways on the Strength of Shafts*, Bulletin No. 42, University of Illinois Engineering Experiment Station (1909).

[3] *Ibid.*

value is used for the design. Two methods of determination of moments may be applied—the analytical and the graphical.

No attempt will be made here to discuss the principles of mechanics involved. However, it seems advisable to refresh the memory of the designer in regard to the proper procedure by means of an illustrative example.

Example 20.1: A 1220 mm belt pulley B, Fig. 20.1, receives 75 kW from a horizontal drive, and it runs at 225 rpm. Of this power, a spur gear G with a pitch diameter of 400 mm delivers 45 kW to a gear located to its right, and lower, the line connecting their centres forming an angle of 55° with the horizontal line; and the balance of the power is delivered by a 210 mm bevel gear H, the normal tooth pressure being vertical and upward. The pulley and the gears are keyed to the shaft, which is supported by bearings C and D. The belt pulley weights 1800 N. Determine all bending moments and torsional moments necessary for designing the shaft.

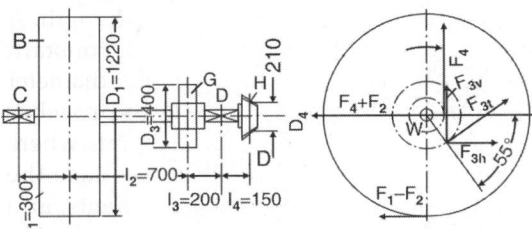

Fig. 20.1: Shaft loaded in bending and torsion

The procedures for determining the forces acting upon the shaft are explained in chapters 27, 30 and 31. Only the bare calculations will be given here.

The net belt tension is

$$F_1 - F_2 = \frac{10^6 \times 60\, P}{\pi\, Dn} = \frac{10^6 \times 60 \times 75}{\pi \times 1220 \times 225} = 5218 \text{ N}$$

The pull of the belt acting upon the bearings, in this case horizontally is $F_1 + F_2$. Approximately, it is equal to $3(F_1 - F_2)$ or

$$F_1 + F_2 = 5218 \times 3 = 15654 \text{ N}$$

Similarly, the tangential component of the tooth pressure in the spur gear is,

$$F_{3t} = \frac{10^6 \times 60 \times 45}{\pi \times 400 \times 225} = 9549 \text{ N}$$

The normal tooth load with a tooth-pressure angle β of approximately 15° ($\beta = 14\frac{1°}{2}$) is,

$$F_3 = \frac{9549}{\cos\beta} = \frac{9549}{0.9660} = 9885 \text{ N}$$

This load must be resolved into a vertical component and a horizontal component.

The horizontal component acting upon the shaft is

$$F_{3h} = F_3 \sin(\alpha - \beta) = 9885 \sin 40°$$
$$= 9885 \times 0.643 = 6356 \text{ N}$$

and the vertical component acting upon the shaft is

$$F_{3v} = F_3 \cos(\alpha - \beta) = 9885 \times 0.766 = 7572 \text{ N}$$

The tar.gential component of the tooth load in the bevel gear

$$F_{4t} = \frac{10^6 \times 60 \times 30}{\pi \times 210 \times 225} = 12126 \text{ N}$$

The normal tooth pressure which is acting upwards is

$$F_u = \frac{F_4}{\cos\beta} = \frac{12126}{0.966} = 12553$$

In this case there is no horizontal component and the vertical component acting upon the shaft is $F = F_t = 12553 \text{ N}$.

The horizontal reaction in the left bearing C is

$$R_{1h} = \frac{15654 \times (70 + 20)}{30 + 70 + 20} - \frac{6356 \times 20}{120} = 10681 \text{ N}$$

The horizontal reaction in the right bearing D is

$$R_{2h} = \frac{15654 \times 30}{120} - \frac{6356 \times 100}{120} = -1383 \text{ N}$$

These values may be checked as follows:

$$R_{1h} + R_{2h} = 10681 - 1383 = 9298 \text{ N}$$

and $(F_1 + F_2) + F_{2h} = 15654 - 6356 = 9298$

The vertical reactions are:

$$R_{1v} = \frac{1800 \times 90}{120} - \frac{7572 \times 20}{120} + \frac{12553 \times 15}{120}$$
$$= 1657 \text{ N}$$

$$R_{2v} = \frac{1800 \times 300}{120} - \frac{7572 \times 20}{120} + \frac{12553 \times 135}{120}$$
$$= -19982 \text{ N}$$

As a check, $1657 - 19982 = -18325$ N and

$1800 - 7572 - 12553 = -18325$ N

The bending moment in the horizontal plane at the middle of the pulley hub is

$$M_{bh} = 10681 \times 30 = 320430 \text{ Ncm}$$

and that at the middle of the spur gear hub is

$$M_{gh} = -1383 \times 20 = -27660 \text{ Ncm}$$

The bending moments in the vertical plane are:

$$M_{bv} = 1657 \times 30 = 49710 \text{ Ncm}$$

$$M_{gv} = -19982 \times 20 + 12553 \times 35 = 39715 \text{ Ncm}$$

When several forces act in the same plane as in the case of reaction on bearing D and a force outside the support, it is well to remember that the resulting bending moment is obtained by superimposing the simple moments. The force on the overhung bevel gear produces a bending moment at the bearing D, which is

$$M_{dv} = 12553 \times 15 = 188295 \text{ Ncm}$$

The axial thrust on the bevel gear causes a bending moment in the horizontal plane. This thrust, from other data of bevel gear is say $F_a = 6400$ N and the moment is

$$M_{dh} = 6400 \times 0.5 D_4 = 6400 \times 0.5 \times 0.21 = 672 \text{ Nm}$$

The resultant moments at those dangerous points are as follows:

$$M_b = \sqrt{320430^2 + 49710^2} = 324231 \text{ Ncm}$$

$$M_g = \sqrt{27660^2 + 39715^2} = 48393 \text{ Ncm}$$

$$M_d = \sqrt{67200^2 + 188295^2} = 199927 \text{ Ncm}$$

Thus the maximum bending moment is at the pulley and is equal to 324231 Ncm. The torsional moment in the left end, up to pulley hub is $T = 0$.

Between the pulley and spur gear, the torque is

$$T = \frac{1}{2}(F_1 - F_2)D_1 = 0.5 \times 5218 \times 122$$

$$= 318298 \text{ Ncm}$$

Between the spur gear and the bevel gear, the torque is,

$$T' = T - \frac{1}{2}F_3 D_3 = 318298 - \frac{1}{2} \times 9885 \times 37.5$$

$$= 132954 \text{ Ncm}$$

The dangerous section is at the pulley hub. To compute the equivalent torsional moment, apply equation 20.5 and take into account the stress concentration at the keyway by introducing the factor $K'' = 1.2$. Then

$$T_c = \sqrt{324231^2 + (1.2 \times 318298)^2}$$

$$= 501016 \text{ Ncm}$$

Shaft diameter

$$d = \left[\frac{16 \times 501016}{\pi \times 6000}\right]^{1/3} = 7.52 \text{ cm}$$

and the standard size of shaft may be taken as 75 mm.

20.4 SHEAR DIAGRAM

The procedure for finding a maximum bending moment is considerably simplified if the point at which it acts is known. This point can be easily found by means of a shear diagram, which also gives a graphical check of the values of the bearing reactions. The shear diagram is constructed by taking for the abscissas, distances along the shaft; and for the ordinates, all forces acting in the plane investigated including, the bearing reactions with their respective direction.

Figure 20.2a shows the shear diagram for the forces in the horizontal plane in Example 20.1. Point a, where the shear diagram intersects the axis for the first time, is at the shaft section where the bending moment reaches the maximum values; point b indicates the position of the minimum bending moment. The shear diagram for the forces in the vertical plane is shown in Fig. 20.2b. In this case, points a and c indicate two sections of maximum bending moment, and point b indicates the position of a minimum bending moment.

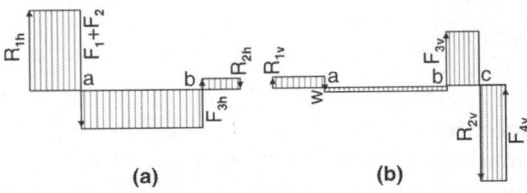

(a)   (b)

Fig. 20.2: Shear diagrams for shaft

If the reactions are computed correctly, the shear diagram, with the axis, must form a continuous contour. Also, if the construction of the shear diagram is correct, the sum of the diagram areas above the axis must be equal to that of the areas below it. However, to find the point of the maximum bending moment, the shear diagram does not have to be drawn accurately.

20.5 SHAFTS WITH IMPACT LOADING

Shafts are subjected to bending impact as well as to torsional impact, and often a shaft is subjected to both simultaneously. Generally the impact has no initial velocity and thus is equivalent to a sudden load, which, as has been shown, creates a stress twice as high as that due to a steady load. For design purposes it is convenient to consider a sudden application of the load and to multiply by 2 the steady bending moment M, or the torque T, in the design equations.

Heavy shocks, particularly those caused when the direction of rotation of the shaft is reversed under load, may increase the stress up to three times that due to a steady load; and allowance for such shock can be made by multiplying by 3 the moments M and torque T in the design equations.

It is a general rule that the resistance to impact is measured by the resilience of a part and that the resilience reaches its maximum when all sections of the part are stressed equally and as high as possible, consistent with strength. This rule applies in full to shafts. If a shaft is made with a constant diameter, as many shafts are, the diameter must be determined by the maximum stress due to impact. Heavy shafts, especially short ones, should be made to approach uniform strength and should not be more rigid than is necessary for proper operation.

The designer must keep in mind the size coefficient e_{sz} (equation 5.8), the stress-concentration factor K' (equation 5.9), and the remarks in section 5.6 in regard to the index of sensitivity q.

20.6 REPETITIVE LOADING

Any rotating shaft carrying a transverse load is subjected to repeated stresses.

Bending: The bending stress in any outside fiber of such a shaft changes at every revolution from a maximum tension to a maximum compression. In order to avoid failure through progressive fracture, the stress must be kept below the endurance limit.

The shaft diameter necessary to resist bending is computed from equation 20.4. If the shaft has no discontinuity at the design section, the design stress to be used should be determined in the following manner. Take the endurance limit S_{en} in bending from the endurance diagram for the shaft material, and multiply this limit by the size factor e_{sz} (equation 5.8), and then divide the result by the safety factor n. If the shaft has a discontinuity, the stress-concentration factor K_r (equation 5.41) and the index of sensitivity q (Table 5.2) must be used instead of the surface factor e_{sr}.

Torsion: The shaft diameter necessary to resist torsion must be computed from equation 20.5. To determine the design stress in this case, use the endurance limit S_{en} in torsion instead of the elastic limit S_e; and introduce all other factors mentioned in the preceding paragraph, using the surface factor e_{sr} from Fig. 5.8 i.e.

$$e'_{sr} = 0.425 + 0.57\, e_{sr}$$

Torsion seldom varies from a maximum value in one direction to the same value in the opposite direction. Usually, the torque fluctuates back and forth from a maximum value T_1 to a smaller and T_2. In this case the actual stress amplitude S_a may be found from the corresponding endurance diagram by using equation 5.38. The design amplitude is found by dividing S_a by the safety factor n, which in this case may be taken as 1.25 to 1.5.

Simultaneous torsion and bending: In this case the amplitudes of the bending and torsional moments should be combined by applying equation 20.6. The shaft diameter is computed by equation 20.5, in which the equivalent moment is replaced by the moment amplitude and in which the stress is replaced by the design stress amplitude. The design stress amplitude is computed by using an equation similar to equation 5.26 and by introducing the stress-concentration factor K_r. Thus,

$$S_{da} = \frac{e_{sz} e'_{sr} S_a}{n K_r} \qquad \ldots (20.9)$$

where, e'_{sr} is computed by equation 5.23 and S_a is found from the endurance diagram for torsion.

Discontinuities which must be taken into account in the design—or, if possible, avoided—are rough surfaces, abrupt changes of the diameter with an insufficient fillet, press and shrink fits with high pressure, and keyways. Numerical data for the evaluation of the corresponding stress-concentration factors are given in sections 3.7, 3.8 and 5.8 to 5.10.

Example 20.2: Determine the diameter of the shaft for the data of example 20.1. Assume that the torque constantly fluctuates between a maximum value and zero. The work delivered is as indicated in example 20.1.

If the average, or mean, torque is designated by T_m, with $T_2 = 0$, the maximum torque is $T_1 = 2T_m$. From example 20.1, $T_m = 3183$ Nm. The torque amplitude, according to its definition, is $T_a = \frac{1}{2}(2T_m)$ = 3183 Nm.

Since the bending moment is produced by the same forces which produce the torque, it is evident that the bending moment will fluctuate in the same proportion as the torque. However, in a revolving shaft the amplitude will be measured from the maximum value in one direction to the same value in the opposite direction. Thus it can be assumed that $M_1 = 2M$ and that $M_2 = -2M$. The maximum bending moment was found in example 20.1 to be $M_{bh} = 3204$ Nm. Hence $M_1 = 6408$ Nm. Also,

$$M_a = \frac{1}{2}[6408 - (-6408)] = 6408 \text{ Nm; and } M_m = 0.$$

The amplitude of the equivalent torsional moment, by equation 20.5, is

$$T_{ae} = \sqrt{6408^2 + 3183^2} = 7155 \text{ Nm}$$

From Fig. 4.2, which is the nearest to SAE 1020, $S_a = 110$ MN/m². The size coefficient may be taken from Table 5.1 as $e_{sz} = 0.84$. The surface coefficient, from Fig. 5.8, is $e_{sr} = 0.92$; and by equation 5.23,

$$e'_{sr} = 0.425 + 0.575 \times 0.92 = 0.954$$

The keyway factor $K_r = 1.14$, and the safety factor n can be taken as low as 1.5 because all other adverse influences are taken into account separately. Thus, by equation 20.9, the design stress amplitude is

$$S_{da} = \frac{0.84 \times 0.954 \times 110}{1.5 \times 1.14} = 51.54 \text{ MN/m}^2$$

The minimum shaft diameter, from equation 20.1, is

$$D = \sqrt[3]{\frac{16 \times 7155}{\pi \times 51.54 \times 10^6}} = 0.089 \text{ m}$$

The standard size may be taken as 90 mm.

Conclusion: Many machines, including various prime movers, cannot perform their work without shafts. Whether a shaft is large or small in diameter and length, it is important and indispensable for the functioning of the whole mechanism. Since shafts on the average are the most stressed machine parts, no effort should be spared in designing them properly. For a revolving shaft, the corresponding endurance diagram should be used. Transmission shafts are probably the only shafts which are sufficiently standardized to permit them to be designed directly by formulas.

To fix bearings, gears, pulleys, it is desirable to make the shaft stepped. Gears and Bearings need steps for fixing to avoid axial movement in a machine shaft (20.1). In line shafts, the position of various machines can change and all machines may not be working all the time, so line shafts are made of constant diameter (20.2). The distance between hangers should be limited to avoid large deflection specially when heavy pulleys are mounted (20.3).

20.7 DESIGN OF TRANSMISSION SHAFTING

Transmission shafts are used so often that the American Society of Mechanical Engineers has worked out a special procedure for their design.[4] The general formula for the outside diameter D, in mm, given in the ASME Code, is

$$D = \left[\frac{16\sqrt{\left\{K_m M + \frac{1}{8}aFD\left(1 + K^2\right)\right\}^2 + (K_T T)^2}}{\pi S_d\left(1 - K^4\right)}\right]^{1/3}$$

$$\dots (20.10)$$

The notations not used before, or needing explanation, are as follows:

S_d is the design stress, taken as the allowable shear stress N/mm²

---

[4] Code for the Design of Transmission Shafting (Westing house code)

K_m is the numerical combined shock and endurance factor to be applied to the computed bending moment;

K_t is the numerical combined shock and endurance factor to be applied to the computed torsional moment;

$K = D_1/D$ is the ratio of the inside diameter to the outside diameter of a hollow shaft;

a is the ratio of the maximum intensity of stress to the average intensity, resulting from the axial loading only;

F is the axial tensile or compressive load, in N

Values of K_m and K_t are given in Table 20.2. In most cases equation 20.10 will be simplified by the elimination of factors that do not apply. Thus, in the absence of an axial force, $F = 0$ and the corresponding term disappears; and for pure torsion, $M = 0$.

Table 20.2: Shock and endurance factors

Nature of loading	K_m	K_t
Stationary shafts		
Gradually applied load	1.0	1.0
Suddenly applied load	1.5–2.0	1.5–2.0
Rotating shafts		
Steady or gradually applied loads	1.5	1.0
Suddenly applied loads, minor shocks only	1.5–2.0	1.0–1.5
Suddenly applied loads, heavy shocks	2.0–3.0	1.5–3.0

The presence of the unknown diameter D in the second term in brackets under the radical sign in equation 20.10 compels the use of the cut-and-try method. First D must be assumed or calculated by equation (20.5)*. The approximate value of D is increased by one step is then inserted under the radical sign, and checked for correctness.

Allowable stresses for simple torsion: According to the ASME Code, for commercial-steel shafting, the allowable shear stress for simple torsion is $S_s = S_d = 55 \text{ MN/m}^2$. For steel shafting purchased under definite physical specifications $S_d = 0.3 \, S_e$, where S_e is the elastic limit in tension; but S_d

should not be higher than $0.18 \, S_u$, where S_u is the ultimate tensile strength.

Combined torsion and bending: The allowable stresses for combined torsion and bending are the same as those for simple torsion.

Simple bending: For commercial shafting the allowable stress in simple bending is twice that for simple torsion, or $S_d = 110 \text{ N/mm}^2$. For shafting with definite specifications $S_d = 0.6 \, S_e$, but it should not be higher than $0.36 \, S_u$.

Keyways: In shafts with keyways the allowable stresses are 75 per cent of the value just given.

Axial loading: The maximum stress s resulting from an axial force F is

$$s = a\frac{F}{A} \qquad \ldots (20.11)$$

where, A is the cross-sectional area of the shaft, in square mm, and a is the column-action factor. For short columns, or when $l/k < 115$, a is found by the relation

$$a = \frac{1}{1 - 0.0044\dfrac{l}{k}} \qquad \ldots (20.12)$$

For long columns, or when $l/k \geq 115$, Euler's formula must be used. From equation 2.42,

$$a = \frac{S_e}{\pi^2 nE}\left(\frac{l}{k}\right)^2 \qquad \ldots (20.13)$$

The ASME Code gives $n = 1$ for hinged ends, and $n = 2.25$ for fixed ends. For both ends pinned, guided, and partly restrained, as they are in bearings, $n = 1.6$.

For tensile load $a = 1$

Hollow shafts: The terms $(1 + K^2)$ and $(1 - K^4)$ in equation 20.9 take into consideration the weakening of the shaft by making it hollow with an inside diameter $D_1 = KD$. For solid shafts, $K = 0$.

Graphical solutions: The ASME Code gives two charts which may save time in computing the shaft diameters. However, these charts are of real value only when a great number of shafts must be designed.

---

* Then by increasing the diameter D to the next higher step. Check is made by using equation 20.10.

Transverse shear: In a shaft subjected to bending on a short span by a heavy transverse load F, the maximum shear stress occurs in the fibers in the neutral plane, and its magnitude may be computed for a solid circular section from the expression in Table 2.5 for case a. Thus,

$$S_s = \frac{16F}{3\pi D^2} = \frac{1.69F}{D^2} \qquad \dots (20.14)$$

For a hollow circular shaft, equation 2.18 gives

$$S_s = \frac{1.69F(D^3 - D_1^3)}{3\pi(D^4 - D_1^4)(D - D_1)}$$

$$= \frac{1.69F(1 - K^3)}{D^2(1 - K^4)(1 - K)} \qquad \dots (20.15)$$

Example 20.3: Using the recommendations of the ASME Code, find the shaft diameter for the conditions of example 20.1.

For a solid shaft without an axial load, equation 20.9 becomes

$$D = \sqrt[3]{\frac{16}{\pi S_d} \sqrt{(K_m M)^2 + (K_t \tau)^2}}$$

From example 20.1, the maximum bending moment is $M = 3242$ Nm, and the maximum torque is $T = 3183$ Nm.

For commercial steel shafting the nominal stress $S_d = 55$ Mn/mm². Because of the presence of a keyway, the design stress should be lowered to $S_d = 0.75 \times 55 = 41.25$ MN/m². From Table 21.2, for a revolving shaft with a steady load, $K_m = 1.5$ and $K_t = 1$. Thus

$$D = \sqrt[3]{\frac{16}{41.25 \times \pi \times 10^6} \sqrt{(1.5 \times 3242)^2 + (1.0 \times 3183)^2}}$$

$$= \sqrt[3]{\frac{16 \times 5812}{41.25 \times 10^6 \times \pi}} = 0.0895 \text{ m say 90 mm}$$

Example 20.4: Determine the diameter of a shaft, using data of example 20.1 but assuming that the torque fluctuates from a maximum value to zero. The work delivered is as indicated in example 20.1.

The procedure is similar to that in example 20.3. However, the load should be considered as applied suddenly but without heavy shocks, and the higher values from Table 20.2, $K_m = 2, K_7 = 1.5$.

$$D = \sqrt[3]{\frac{16}{415 \times 10^6 \times \pi} \sqrt{(2 \times 3242)^2 + (1.5 \times 3183)^2}}$$

$$= 0.0997 \text{ m say 100 mm}$$

[5] Hütte, *op. cit.*

However, with the more accurate method, the low safety factor n of 1.5 could be used. With a safety factor n of 2, such as used in the Code, the size would become $d = 85.7 \sqrt[3]{2/1.5} = 93.5$ mm which is practically the same as the value found in example 20.4.

20.8 PRACTICAL DESIGN CONSIDERATION

In addition to the information given at the end of section 20.1, it may be stated that standard bearings, set collars, clutches, and pulleys, and other parts carried in stock, are bored 1.5, undersize. Therefore, transmission shafts should be made from T and G shafting turned with diameters in 5 mm increments and 1.5 under the even sizes, beginning with 25 mm and running up to 150 mm.

Simplified design procedure:

1. In practice, the bending moment diagram is made to know maxima values.
2. Maximum torque is calculated from power by using suitable load factors.
3. Equivalent torque is calculated by taking values of K_m and K_t.
4. By taking allowable shear stress S_d as explained above, shaft diameter is calculated.
5. If there is an axial load, then ASME code is used as a check.

Rigidity in torsion: Usual practice is to limit the angle of torsion to 1 deg in a length L of 20D. For shafts over 6 m long it is recommended that the angle should not exceed 0.33° per m, or even 0.5° for any shaft length.[5]

Stiffness: The transverse deflection of line shafts and countershafts is usually limited to 1 mm per m of length. Sometimes a greater stiffness may be desired. Instead of computing the transverse deflection, the maximum distance between bearings in m may be computed by an empirical formula to restrict transverse deflection to 0.8 mm per m of length by equation,

$$L = \frac{1500 CD^{2/3}}{N + 1500} \qquad \dots(20.16)$$

Values of C are given in Table 20.3.

Tabel 20.3: Data for shaft calculations, constant C

Type of shaft loading	Coefficient C in Eq. 20.16	Allowable stress S_d (MN/m²)
Shaft heavily loaded, subjected to shock or reversed under full load	0.82	20.0
Line shaft and countershafts loaded in bending but not reversed	1.1	28.0
Line shafts bare or with pulleys close to the bearings	1.56	45.0

Load factor: The rotation of most machines is not absolutely uniform. Only steam turbines and gas turbines have a real uniform rotation, owing to the large flywheel effect of their rotors and their high speeds. The speed of electric motors fluctuates a little because of the intermittent impulses given to their rotors. Prime movers with reciprocating pistons have still less uniformity of rotation. At the same time, the torque required to operate machinery also varies. In addition the rotating parts of the driven machines have masses and flywheel effects. Therefore a machine part such as a coupling, clutch, or shaft that transmits power from a prime mover to a driven machine, is subjected to a fluctuating torque the maximum value of which may exceed the average torque by as much as 100 per cent or more.

The ratio of the maximum torque to the average, or nominal torque is called the *load factor* and is designated as k_l. Some typical values of the load factor are given in Table 20.4.

If, because of lack of more accurate data, a shaft is designed by the simplified procedure, the transmitted torque T or power P should be multiplied by the load factor k_l, data for which are shown in Table 20.4.

Example 20.5: Find the diameter of a transmission shaft connected to a six-cylinder, 75 kW oil engine through a belt drive. The shaft is driving woodworking and metalworking machinery and runs at 225 rpm. Assume that the maximum load is equal to the rated engine power.

The required diameter may be found by applying equation 20.2 and using $S_d = 28\,\text{MN/m}^2$. The maximum power is found by using Table 20.4. For a load consisting of metalworking and woodworking machinery, the average value of k_1 is 1.37, would be used for an electric-motor drive. A six-cylinder oil engine has a high coefficient of steadiness and a factor of 1.3 would be sufficient. Since the shaft is not directly connected to the engine but is driven through a belt, which absorbs a considerable amount of speed variation, the factor may be lowered to about 1.2. The final load factor is therefore $k_i = 1.375 \times 1.2 = 1.65$. Thus, by equation 20.2.

$$D = 36.51\sqrt[3]{\frac{75 \times 1.65}{225 \times 28}} = 0.09854 \text{ m or } 9.854 \text{ cm}$$

Table 20.4: Load factors for various machines

Driver	Driven machinery	Factor k_l
Steam turbine	Electric generator, steady load; turbine blower	1.00
	Electric generator, uneven load; centrifugal pump	1.25
	Induced-draft fan; line shaft., gear drive	1.50
	Rolling mill, gear drive	2.00
Electric motor	Turbine blower; metalworking machinery	1.25
	Centrifugal pump; woodworking machinery	1.50
	Line shaft; ship propeller; double-acting pump	1.75
	Triplex single-acting pump; elevator; crane	1.75
	Compressor, air or ammonia	1.75
	Rolling mill; rubber mill	2.50
Steam engine	Values for electric-motor drive multiplied by 1.2 to 1.5	
Gas and oil engines	Values for electric-motor drive multiplied by 1.3 to 1.6, the factor depending on the coefficient of steadiness of the flywheel.	

Example 20.6: A horizontal shafting is supported by two bearing 85 cm apart. A keyed gear, 20° involute, 250 mm diameter is driven by a gear on a parallel horizontal shaft and 600 mm diameter pulley weighting 1.2 kN is fixed to the shaft as shown in Fig. 20.3. The slack side is on the top and the tight side tension is three times the slack side tension. The drive transmits 30 kW at 280 rpm. The permissible stress in the shaft is limited to 60 N/mm². The pulley drive is horizontal.

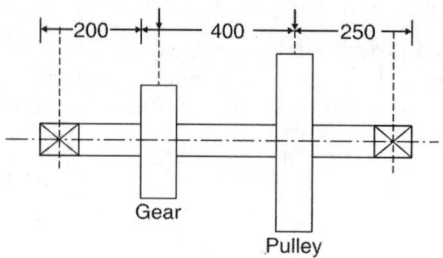

Gear

Pulley

Fig. 20.3

Torque transmitted

$$T_1 = \frac{30 \times 1000 \times 60}{2\pi \times 280} = 1023.14 \text{ Nm}$$

As $T_1 = 3T_2$ and $r = 300$ mm

$(T_1 - T_2) r = 1023.14$

$$T_2 = \frac{1023.14 \times 1000}{300 \times 2} = 1705.2 \text{ N}$$

$T_1 = 3 \times 1705.2 = 5115.6 \text{ N}$

$T_1 + T_2 = 6820.8 \text{ N say } 6821 \text{ N}$

Tangential force transmitted by gear

$$F_t = \frac{1023.14 \times 1000}{250/2}$$

$= 8185.12 \text{ N say } 8186 \text{ N}$

Normal component F_r of this force

$= 8185.12 \tan 20° = 2979.14 \text{ N say } 2979 \text{ N}$

Considering forces in the horizontal and vertical planes, the force diagram is shown in Fig. 20.4

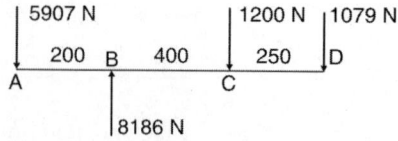

5907 N   1200 N   1079 N

   200   B   400   250   D

A   C

8186 N

Fig. 20.4

B.M. at B = 5907 × 500 = 1181,400 Nmm

B.M. at C = 1079 × 250 = 269750 Nmm (269800)

Considering force in the horizontal plane, the force diagram is shown in Fig. 20.5. The pull of belt is away from shaft while the radial load is towards the shaft.

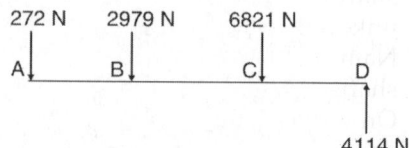

272 N   2979 N   6821 N

A↓   B↓   C↓   D

4114 N

Fig. 20.5

B.M. at B = 272 × 200 = 54400 Nmm

B.M. at C = 4114 × 250 = 1025500 Nmm

Resulting B.M. at B

$= \sqrt{1181400^2 + 54400^2} = 1182652$ Nmm

Resulting BM at C

$= \sqrt{269750^2 + 1025500^2} = 1060384$ Nmm

Considering the maximum B.M. at B, the shaft diameter is

$$d^3 = \frac{16}{\pi S_s} \sqrt{(K_b M)^2 + (K_T T)^2}$$

$$= \frac{16}{\pi \times 600 \times 75} \sqrt{(2 \times 1182652)^2 + (1.5 \times 1023140)^2}$$

$$d = \left(\frac{16 \times 2820 \times 1000}{\pi \times 60 \times 0.75}\right)^{1/3} = 68.3 \text{ mm}$$

The standard diameter available is 71 mm.

PROBLEMS

20.1 Explain why are machine shafts seldom made of constant diameter from end to end.

20.2 Why are line shafts delivering power at various intervals throughout their length seldom made to vary in diameter according to power transmitted?

20.3 Why is the distance between hangers for a long line shaft limited? Why is the distance between hangers shortened when the shafts carry pulleys?

20.4 Why are shafts subjected to principally torsion action of rare occurrence? Why a great many shafts that are not simple torsion members are so treated?

20.5 In what way is a hollow shaft superior to a solid shaft? Why are the hollow shafts not so common?

20.6 Why is the working stress for a keyed shaft taken as 0.75 of the value for an unkeyed shaft? Give reasons.

20.7 Name the materials commonly used for shafts.

20.8 On what theory does the ASME Code depend?

20.9 Find the diameter of a ship-propeller shaft to transmit 1200 kW at 95 rpm. The material of the shaft is equivalent to SAE 1025/25C4. The thrust from the propeller is 160 kN and the torque is uniform.

20.10 (a) Find the outside diameter of a hollow shaft for the data of problem 20.9, assuming that the inner diameter is about 0.4 of the outside diameter, (b) Find the saving in weight as compared with the solid shaft.

20.11 Find the diameter of the shaft in problem 20.9, using SAE 2340/40Ni3 nickel steel.

20.12 A revolving shaft is subjected to a maximum torque of 1700 Nm, and to a maximum bending moment of 3600 Nm. Determine the theoretical and commercial-size diameter for the shaft for the following cases, using analytical formulas and assuming 20C8/SAE 1020 steel: (a) The load is steady. (b) The load is applied gradually, (c) The load is applied suddenly, but with a small shock only, (d) The load is applied suddenly and with heavy shocks.

20.13 An electric motor transmits 11 kW to a centrifugal pump through a train of gears with a speed ratio of 2.5 : 1. The motor speed is 1,800 rpm. Determine (a) the diameter of the motor shaft, made of commercial steel shafting, neglecting bending but taking into account the keyway, and (b) the diameter of the pump shaft, made of Tobin bronze, also neglecting bending but not keyseating.

20.14 A line shaft rotating at 150 rpm must transmit 45 kW with a torsional deflection not to exceed 1° in a length l of 20D. For the shafting material the elastic limit in shear is $S_{es} = 150$ N/mm^2 and $G = 11 \times 10^3$ N/mm^2. Determine (a) the commercial diameter of the shaft and (b) the actual safety factor of the suitable commercial size.

20.15 A long line shaft is driven in a machine shop at 210 rpm by a belt drive from a electric motor and must transmit 60 kW. Find the commercial size of shafting to be used.

21 Couplings and Positive Clutches

21.1 GENERAL CONSIDERATIONS

A *coupling* is used to connect two shafts permanently; it is disconnected only for repairs or to make a change in the installation. A *clutch* permits easy and quick connection and disconnection of two shafts. A *clutch* is also used instead of a key to connect the shaft with a revolving part, such as a pulley or a gear.

As shafts have limited lengths for ease of transport, it is necessary to join the shafts with couplings. Other uses of couplings are:

1. To connect two units which are made separately, viz. motor and pump or generator and also to have easier repair and maintenance.
2. To take care of misalignment in shafts or to introduce flexibility.
3. To reduce the transmission of shock loads from one side to the other, e.g. a reciprocating pump.
4. To provide for protection against over-loads.
5. To change the vibration characteristics of rotating units.

Couplings are of two main types, *rigid* and *flexible*.

i. When the axes of shafts are perfectly colinear, *rigid flange couplings are used* (21.7a).
ii. When the axes of shafts are slightly inclined or slightly offset, bushed pin type flexible couplings are used (21.7c).
iii. When the axes of shafts are sufficiently inclined, universal couplings are used (21.7d).
iv. When the axes of shafts are offset by large distance, Oldham's couplings are used (21.7b).

Clutch constructions are based on the *positive-action* and *friction* principles.

21.2 FLANGE COUPLING

The flange coupling, Fig. 21.1, consists of two cast-iron flanges, keyed to the shaft ends and bolted together. The best practice is to press or shrink the flanges on the shafts, in which case straight keys should be used. If the flanges are put on only with a push fit, taper keys must be used. After the connection has been assembled, it is advisable to true up the faces of the flanges in a lathe to make sure that they are

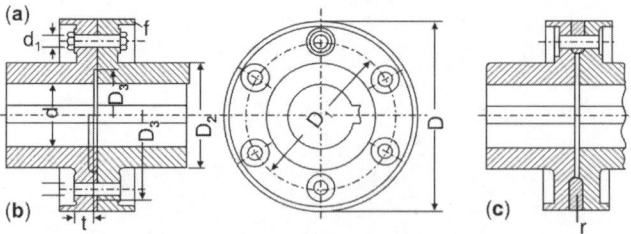

Fig. 21.1: Types of flange couplings

normal to the axis of rotation. The shafts are virtually collinear.

Register: A register is necessary to insure correct alignment of the two shafts (21.6). It is usually made by turning a projection on one flange and boring a slightly deeper female register in the other flange. The register is made either with a small diameter D_3, as in Fig. 21.1a, or with a large diameter, as in Fig. 21.1b, lower half of the Fig. 21.1 in which case the inner part of the face may be relieved or left rough. Another method is to make the faces of both flanges flat and to obtain a register by protruding one shaft end about 6 mm beyond the face and having the shaft end in the mating flange shorter, about 8 mm inside the face.

A third type of register uses a separate ring *r*, Fig. 21.1c. The ring is usually split into two halves. This construction permits an easy disconnection of the shafts when one of them must not rotate.

Steady torque that always acts in the same direction may be transmitted by friction between the flange faces; and holes slightly larger than the bolt shank may be drilled. *A better practice is to ream the holes to fit the bolts when assembling the coupling, and to let the bolts work in shear (21.8). This also ensures resistance to misalignment (21.2).* In either case the surfaces under the bolt heads and nuts must be normal to the bolt axis. *Either the surfaces may be spotfaced, as in the upper half of Fig. 21.1c, in which case slightly raised bosses, Figs 21.1a and 21.1b, are advisable to avoid bending of bolts; or the flanges are finished all over, as in the lower half of Fig. 21.1c (21.5).*

Couplings transmitting a large and not very steady torque, such as in boat propeller shafts, are made with tapered headless bolts fitted into reamed holes. A taper of 1:20 or 1:24, referred to the diameter, is used.

The projecting bolt heads and nuts must be covered by a safety flange *f*, Fig. 21.1, which also increases the rigidity of the flanges.

Dimensions: The torque acting on a flange coupling is transmitted through the bolts. The size of the bolts depends on the number of

bolts *n* and on the bolt-circle diameter D_1. There is no direct relation between these two factors and the torque. Both *n* and D_1 must be selected more or less arbitrarily; the bolt size and the other dimensions can then be computed.

The dimensions of flange couplings are standardized. However, the dimensions of couplings built by different manufacturers do not vary much. Since the torque capacity of a coupling is limited by the torque capacity of the shaft, it is logical, in establishing empirical expressions for the main dimensions, to present them as functions of the shaft diameter *d*.

The commonly used approximate number of bolts is

$$n = 0.02\,d + 3 \qquad \text{...(21.1)}$$

The preliminary value for the bolt diameter d_1 may be determined by the empirical formula

$$d_1 = \frac{0.5d}{\sqrt{n}} \qquad \text{...(21.2)}$$

The average value of the diameter of the bolt, circle, as found in stock couplings, is

$$D_2 = 2(d + 25) \text{ or } 3d \qquad \text{...(21.3)}$$

A suitable value for the hub diameter D_2 is the same as that used for belt pulleys, or

$$D_1 = 1.5\,d + 25 \text{ or } 2d \qquad \text{...(21.4)}$$

Since the bolts should be located halfway between the hub and the outside of the flange, the outside diameter *D* should be taken as

$$D = 2.5\,d + 75 \text{ or } 4d \qquad \text{...(21.5)}$$

This is in good agreement with data for stock couplings in catalogues.

The hub length *l* is usually established by the relation

$$l = 1.25\,d + 20 \text{ or } 1.25\,d \text{ to } 1.5\,d \quad \text{...(21.6)}$$

Design procedure: The first step in design is to determine the main dimensions *n*, D_1, D_2, *D*, and *l* from equations 21.1 to 21.6. In deciding on the number of bolts *n*, the nearest larger even number must be used, except for shafts under 40 mm which are made with three bolts.

The next step is to determine the tangential force $F_t \left(= \dfrac{2T}{D_1} \right)$ necessary to transmit the required torque T, referred to the selected bolt-circle diameter D_1. After this the diameter d_1 determined from equation 21.2 should be checked to make certain that a sufficient friction force F_t is produced. Here it is a good practice to assume that only one-half the total number of bolts is properly tightened. If the bolts are fitted, they should be checked for shear and crushing. Shear of bolts may occur between the flange faces. If the bolts are not fitted into reamed holes, it is advisable to consider again that only half the total number of bolts are doing the work.

Crushing of bolts may take place if the projected bearing surface td_1, Fig. 21.1, is not sufficient. According to equation 5.14 the allowable stress S_d for steel in bearing can be taken about twice as great as in ordinary compression.

The flange thickness t is found by considering that the flange may shear at the hub. As a general rule, t should be slightly greater than the bolt diameter d_1 or $d/4$ to d.

The shaft keys must be checked in shear and crushing. If necessary, two keys located 90 deg apart are used in each hub.

Finally, it is necessary to determine from practical consideration those dimensions that cannot be computed by stress analysis, such as the type, diameter, and depth of the register, and the thickness and height of the safety sleeve to cover the nuts and bolt heads.

Steps of design: Checks.

Torque transmitted by the shaft

$$T = \frac{P \times 60 \times 1000}{2\pi N}$$

Tangential force transmitted at the bolt circle diameter per bolt is

$$F_t = \frac{T}{n D_1/2}$$

Shear stress in the bolt $= \dfrac{F_t}{\frac{\pi}{4} d_1^2}$

Crushing stress in the bolt $= \dfrac{F_t}{td_1}$

Shear force at the hub surface $F_s = \dfrac{T}{D_2/2}$

Shear stress in the flange $S_s = \dfrac{F_s}{\pi D_2 t}$

Taking register diameter $D_3 = 1.2\, d$

Depth of register = 6 to 8 mm

Projection of register = 4 to 6 mm

Mean friction diameter $d_m = \dfrac{1}{3}\left[\dfrac{D^3 - D_3^3}{D^2 - D_3^2}\right]$

Friction torque transmitted $= f\dfrac{d_m}{2}Wn$

W is the initial tightening force/bolt
Key dimensions

Width of key $w = \dfrac{d}{4}$ to $\dfrac{d+13}{4}$

Depth of key $t = \dfrac{d}{6}$ to $\dfrac{d+13}{6}$

Length of key = length of hub, $l = 1.25\, d$ to $1.5\, d$

Tangential force at the shaft surface

$$F_{t1} = \frac{T}{d/2}$$

Shear stress in the key $S_{s1} = \dfrac{F_{t1}}{w \times l}$

Crushing stress in the key $S_{c1} = \dfrac{F_{t1}}{lt/2}$

Muff coupling: A muff coupling is the simplest device in the form of a hollow shaft connecting two shafts with a common key, Fig. 21.2. It is assumed that the shaft and coupling are equally strong with the muff

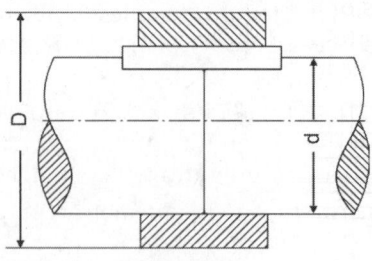

Fig. 21.2

made of cast iron having allowable shear stress equal to half of that of the shaft. By comparing the torque transmitted by the two with outer diameter of muff equal to D and $\dfrac{d}{D} = k$

$$T = \frac{\pi}{16} S_s d^3 = \frac{\pi}{16} \frac{(D^4 - d^4)}{D} \times \frac{S_s}{2}$$

$$= \frac{\pi D^3 (1 - k^4)}{16} \frac{S_s}{2}$$

$$2d^3 = D^3 (1 - k^4)$$

$$2k^3 = 1 - k^4 \text{ giving } \frac{1}{k} = 1.37 = \frac{D}{d}$$

So D is about $1.5d$ which forms the basis of hub diameter of various elements, e.g. coupling, gears, pulleys, etc.

21.3 CLAMP COUPLING

The coupling shown in Fig. 21.3a is made in two parts which are bolted together with through bolts. During the boring operation the two halves are separated by a thin shim. A one-piece square key assists in transmitting the torque. For diameters up to 100 mm, six bolts are used; for larger sizes, up to 180 mm, eight bolts are used. For a 28 mm shaft, the length L should be about $4.5d$, and the outside diameter D should be about $3.5d$; with an increase in size these dimensions are made relatively smaller, being reduced to $L = 3.2d$ and $D = 2.7d$ for a 180 mm shaft.

In the smaller sizes the coupling is made of cast iron; in larger sizes of cast steel. In sizes of 110 mm and larger, some manufacturers make the couplings of cast iron with reinforcement f, Fig. 21.3a, instead of the axial rib e.

Parts of a shaft subjected to an axial force tending to pull them apart must be equipped

with split rings r, Fig. 21.3b, ground on the sides to fit accurately into the grooves.

Compression couplings are based on the pressure which can be created by wedge action produced by a hollow cone. In Fig. 21.4 the cast-iron split cones c are pulled together by three bolts b. Square keys make the coupling still more positive. The square shape of the shanks of the bolts b facilitates the tightening of the nuts. The inspection opening o permits the mechanic to see whether the cones are pulled together.

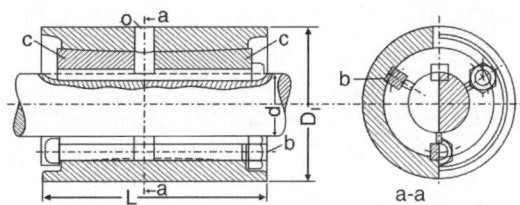

Fig. 21.4: Sellers cone-vise coupling

The sellers coupling is used for shafts with diameters from 36 mm up to 220 mm. The length L is made about $4d$; the outside diameter D_1 is made $2.6d$ for values of d up to 110 mm and is made relatively smaller, down to about $D_1 = 2.4d$, for the large sizes. The taper is made about 1 in 30, three times as much as in standard taper keys. Other dimensions may be found from usually strength consideration.

Figure 21.5 shows a coupling in which the split double cone is stationary and the two outside cones are pulled together by bolts b. In

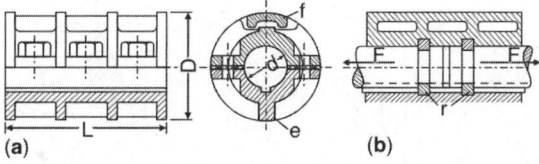

(a)   (b)

Fig. 21.3: Types of ribbed clamp couplings

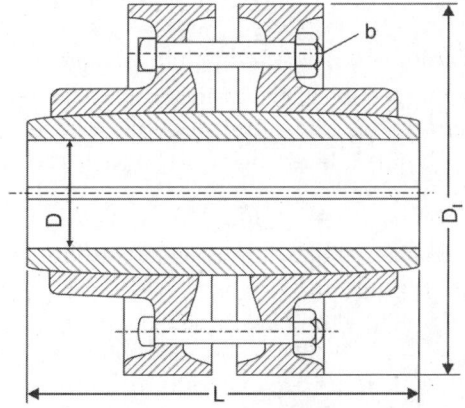

Fig. 21.5: Flange compression coupling

sizes up to $d = 56$ mm, four bolts are used; in larger sizes, up to 100 mm, it is better to use six bolts. The outside dimensions D_1 and L are approximately the same as those for the flange coupling, Fig. 21.1.

Finally, Fig. 21.6 shows a compression coupling in which the wedge action is obtained by driving on, usually by hammer blows, to two steel rings a. This type, in addition to being rather inexpensive, has the advantage that it can be used for outdoor service where the coupling is likely to become rusted. This coupling is made for shafts up to 160 mm.

In compression couplings, $L = 4d$. For the coupling in Fig. 21.6, $D_1 = 2.5d$, $D_2 = 1.9d$, and $b = 0.5d$. For the coupling in Fig. 21.5, the outside diameter and the bolt-circle diameter are made the same as those for a plain flange coupling.

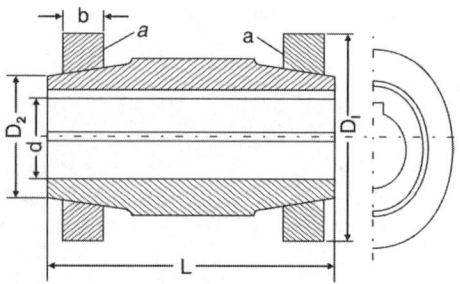

Fig. 21.6: Ring compression coupling

21.4 FLEXIBLE COUPLINGS

The purpose of a flexible coupling is to allow for imperfect alignment of two joining shafts, or to absorb impact from the fluctuation of torque or of angular speed (21.3). In practice, perfect alignment of axes is difficult to achieve and after it has been accomplished it is difficult to maintain because of the following factors:

1. Unequal setting of foundation of two machines.
2. Unequal deflection of supports.
3. Unequal deflection of shafts under load.
4. Temperature changes.
5. Unequal wear in bearings.
6. Affects of shock and vibrations.

If two shafts are even slightly misaligned and are connected by a rigid coupling, they are subjected to continuous reversal of bending stresses, which eventually leads to a failure through progressive fracture. In addition, there is always excessive friction and wear of the bearings. Therefore, in most cases of joining shafts which transmit power, it is necessary or at least desirable to use a flexible coupling of one type or another.

From the standpoint of design, all flexible couplings may be divided into two classes; couplings whose flexibility is obtained kinematically by the use of rigid members in which constraint is absent in certain directions; and couplings with incorporated flexible members. The first type counteracts misalignment only, while the second type anticipates both misalignment and impact. A few of the more typical constructions will be briefly discussed.

21.5 COUPLINGS WITH KINEMATIC FLEXIBILITY

Several types of couplings that provide flexibility kinematically will be described here.

Oldham coupling: Figure 21.7 shows the Oldham coupling. The tongues of the centerpiece b, located at right angle to each other, are fitted to grooves in each hub a. In the position shown, the left-hand tongues can slide up and down while the right-hand tongue can slide to and fro. The combined action produces a flexible connection which is especially adapted to shafts which are parallel but not colinear. This connection also takes care of slight angular misalignment. The Oldham coupling is intended for low speeds.

The design of the Oldham coupling must be based on the allowable pressure between the

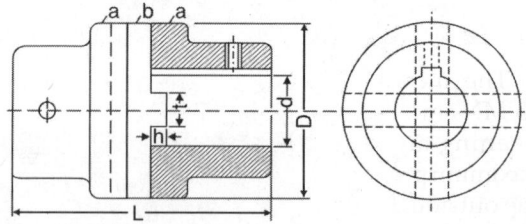

Fig. 21.7: Oldham flexible coupling

faces of the grooves and the tongues. Because of the elasticity of the material, the pressure distribution for perfectly flat surfaces can be assumed to change on each side, as shown in Fig. 21.8, from a maximum value p at the periphery of the coupling to 0 at the center line. The total pressure on each side is then $F = \frac{1}{4} pDh$, where h is the axial dimension of the contact area. The distance to the pressure-area centroid from the center line is $l = \frac{2}{3} \times \frac{1}{2} D = \frac{1}{3} D$. Thus the torque transmitted by both sides of the tongue is

$$T = 2 Fl = \frac{1}{6} pD^2h \qquad \ldots (21.7)$$

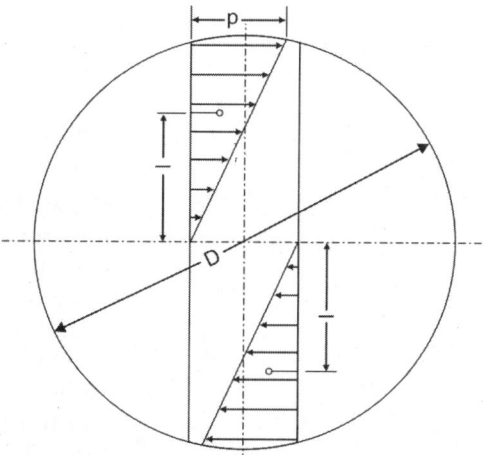

Fig. 21.8: Pressure distribution in an Oldham coupling

The power in kW which can be transmitted at n rpm is

$$P = \frac{2 \pi n T}{1000 \times 60} \qquad \ldots (21.8)$$

The allowable pressure p must not exceed $8.5 \ \text{MN/m}^2$. For a considerable misalignment, meaning an appreciable sliding travel, the maximum pressure should not exceed $7 \ \text{MN/m}^2$. The outside diameter D is made equal to about $3d$ to $4d$ and the dimension t of the tongue is made about $0.45d$. Other dimensions may be proportionately taken.

American flexible coupling: Figure 21.9 shows the American flexible coupling, which operates on the same principle as the Oldham coupling. It consists of two identical hubs f, Fig. 21.10a, turned at right angles to one another, and a floating centerpiece b between them. The floating member b is a square cast-iron block with screwed-on bearing strips c, made of Bakelite reinforced with canvas and impregnated with graphite. A hole in the center provides clearance for the shaft ends. The piece b is hollow, and the cavity is filled with lubricant, which reaches the surface of the bearing strips c through porous reeds in the block and felt pads in the strips.

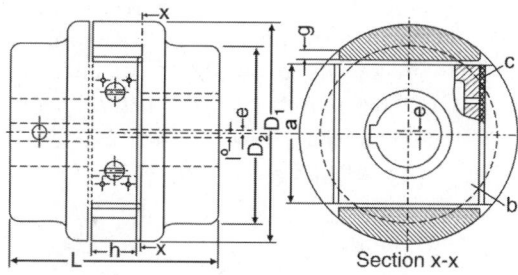

Fig. 21.9: Assembled American flexible coupling

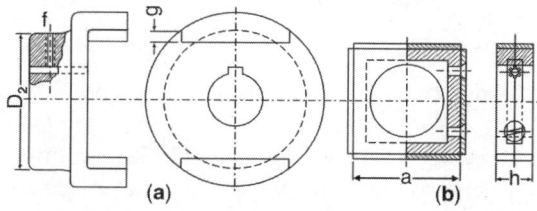

Fig. 21.10: Parts of American flexible coupling

The coupling may be used with an angular misalignment up to 1 deg and a parallel shaft eccentricity e, Fig. 21.9, not greater than $0.003d$. Its design can be based on equations 21.7 and 21.8 by substituting $a\sqrt{2}$ for D.

The allowable pressure is $p = 7 \ \text{MN/m}^2$ for light service, and it may be increased to $10 \ \text{MN/m}^2$ for heavy-duty service. The width h of the block is made equal to $0.3a$. With this value, equation 21.7 becomes

$$T = 0.1pa^3 \qquad \ldots (21.9)$$

The design procedure for a given torque T is as follows: The first step is to select the pressure p and then to determine the side a from equation 21.9. Next, the outside diameter D_1 may be determined graphically by allowing from 1.5 mm to 3 mm. for the Bakelite strips c and slightly more for thickness g of the edges. The hub diameter D_2 may be computed by using equation 21.4, and the length L is made about 3.5d.

Fast's self-aligning coupling: The coupling known as Fast's self-aligning coupling consists of two hubs a, Fig. 21.11, with gear teeth. These hubs are enclosed in a casing composed of two parts, c and d, bolted together by bolts e. Each part of the casing contains an integral gear, the teeth f of which mesh all around with the teeth on the hub. The clearance between the engaging teeth takes care of misalignment; g is a lubricating-oil hole closed with a pipe plug.

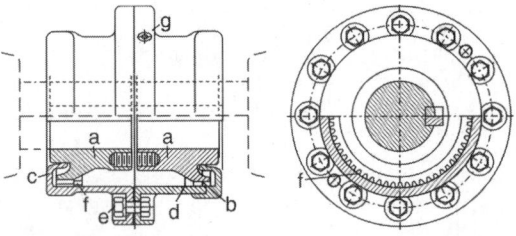

Fig. 21.11: Fast's self-aligning coupling

Clark chain coupling: The Clark chain coupling consists of two sprockets a, Fig. 21.12, connected by a roller chain. Some manufacturers use a double-row chain. Couplings employing silent chains are also used.

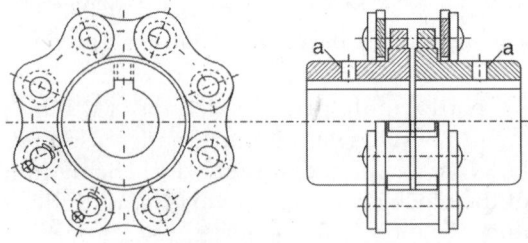

Fig. 21.12: Clark chain coupling

21.6 FLEXIBLE COUPLINGS WITH RESILIENT MEMBERS

The *Ajax flexible coupling* or *bushed pin type flexible coupling* consists of two cast-iron flanged hubs a, Fig. 21.13, with rubber bushings b and steel pins c. The rubber bushings are cemented into the holes and have thin bronze bushings cemented inside them. This process eliminates the wear of the rubber bushings due to sliding between them and the pins. The bronze bushings are impregnated with graphite and thus are self-lubricating. Standard couplings are made with 6 to 12 pins, the number depending on the load conditions.

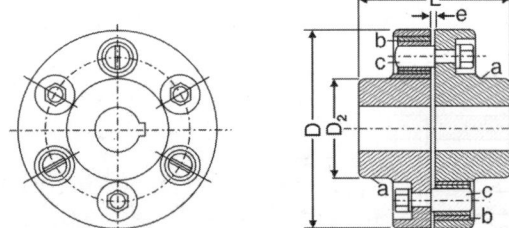

Fig. 21.13: Ajax flexible coupling

Heavy-duty couplings are built with 16 pins, eight in each flange, the rubber bushings and the pins alternating in the flanges.

The torque capacity of rubber-bushing couplings can be based on a specific pressure of $p = 0.2$ MN/m^2 on the projected area of the bronze bushing.

The *Francke coupling*, Fig. 21.14, uses laminated steel pins a, which are relatively flexible. One end of each pin is fastened to the flange by a spring-retaining ring r. The other end of each pin can slide in a bronze bushing peened in the flange b. Thus the coupling can care for some endwise motion, as well as for angular misalignment.

The *Westinghouse-Nuttall coupling*, Fig. 21.15, uses helical springs c as connecting members between the two cast-steel halves a and b. These springs are compressed between case hardened springs seats d. The latter rest against the twin arms of the spider b, and transmit the torque which comes from hub a and fingers e. A longitudinal clearance between the two

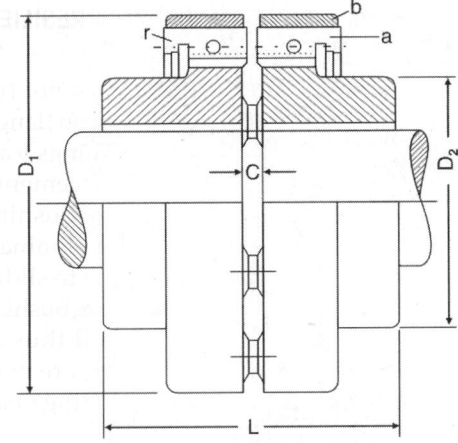

Fig. 21.14: Francke coupling

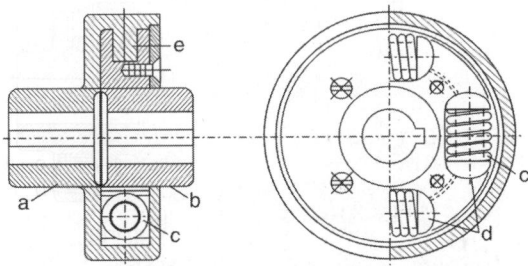

Fig. 21.15: Westinghouse-Nuttall flexible coupling

halves of the coupling permits a certain end play of the connected shafts. Because of the small clearances, the coupling is not suitable for taking care of misalignment. However, it is used to absorb impact, and particularly to remove the critical speed from the operating range of shafts driven by diesel engines.

The *Falk flexible coupling*, Fig. 21.16, consists of two flanged steel hubs *a* and *b*, and a ring *d* which is made of tempered spring-steel strip

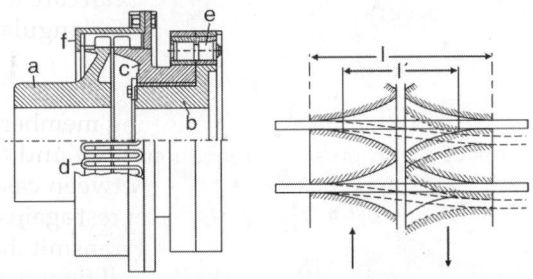

Fig. 21.16: Falk flexible coupling

and which forms a complete cylindrical grid. Figure 21.16 shows a coupling combined with a safety shear pin *e*, the right-hand hub being divided into the hub proper *b* and a ring-shaped flange *c*. The peripheries of the flanges have radial fingers formed by milling. The grooves between the fingers widen inwardly toward each other, and the spring grid *d*, through which the torque is transmitted, is inserted into them. When the torque is applied, the spring is bent along the arcs of the fingers. The curvature of the arcs is such that with an increase of the torque the active spring length gradually decreases from *l* to *l'*, Fig. 21.16b, making the spring stiffer. The steel cover *f* encloses the coupling tightly and holds the lubricant.

This coupling can withstand heavy shocks, as well as all kinds of misalignment.

Load factor: The torque capacities of flexible couplings given in catalogues, or computed by using the pressures or stresses indicated in the foregoing explanations, refer to uniform rotation. In most cases the rotation is not quite uniform and the couplings are subjected to impact. This condition must be taken into account by dividing the nominal coupling capacity by a load factor k_l, Table 20.3.

Speed: The torque capacity of a coupling can be considered to be practically independent of the rotative speed of the shafts.

21.7 UNIVERSAL JOINTS

The object of a universal joint is to transmit the torque between two shafts whose center lines form a rather large angle, as 5 to 15 deg, or even 30 deg.

Figure 21.17 shows a universal joint used for speeds not over 100 rpm and for shafts up to 125 mm. The hubs *a* and the inner yoke *b* are made of cast iron, and pins are made of steel.

The curves in Fig. 21.18 show the relation, found by actual tests, between the efficiency and the operating angle in a set of two universal joints. Both joints were at the same angle. Curve *a* was obtained with yokes on the intermediate shaft in the same plane; curve *b* was obtained with yokes at 90 deg.

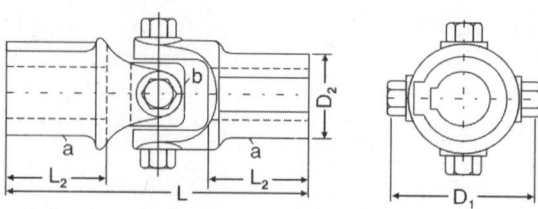

Fig. 21.17: Universal joint

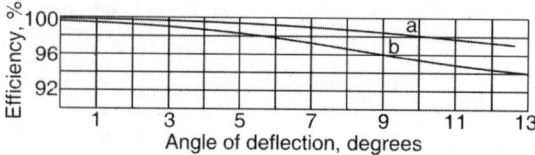

Fig. 21.18: Efficiencies of a double universal joint at different shaft angles

Figure 21.19 shows a universal joint which at the same time is a flexible coupling and consists of a rubberized canvas disk c to which are fastened, straddling each other, the fingers of the tubular shaft ends a and b. This joint allows an angle up to 4 or 5 deg. If a greater angle is necessary, the ends must have only two fingers instead of three, the coupling becoming a regular universal joint.

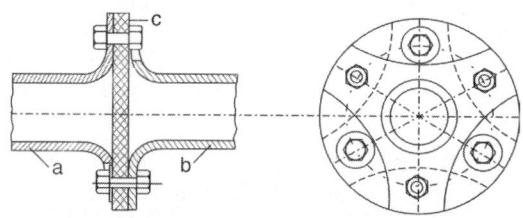

Fig. 21.19: Flexible coupling with a rubber disk

21.8 SLIP COUPLING

The object of a slip coupling is to permit relative rotation, or slip, between the driving shaft and the driven shaft. A slip coupling is a safety device that prevents damage to rotating parts because of overloading. The slip coupling is adjusted so that it will begin to slip if the transmitted torque exceeds a predetermined value. Usually the slip begins if the load exceeds by 10 to 20 per cent the maximum load for which the shafts and other parts are designed.

A typical slip coupling is shown in Fig. 21.20. The hub a is keyed to the driving shaft; the driven hub b has a flange faced on both sides with friction material; a number of small but heavy springs c are loaded by screwing down the nuts, and thus create the necessary pressure on the friction surfaces.

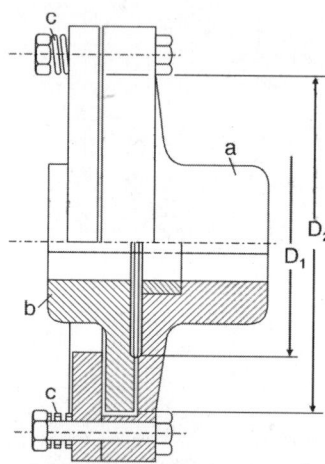

Fig. 21.20: Slip coupling

Force analysis: Slip in such a coupling occurs infrequently, and no measurable wear takes place. Therefore, it can be assumed that the unit pressure p between the disks is uniform and that the axial force exerted by the springs, with the designation of Fig. 21.20, is

$$F_a = \frac{\pi}{4} (D_2^2 - D_1^2)p \qquad \ldots(21.10)$$

With two pairs of friction surfaces, the tangential force is equal to

$$F_t = 2 F_a f \qquad \ldots(21.11)$$

Also, the radius of its application, with sufficient accuracy, is

$$0.5D_m = \frac{D_2 + D_1}{2 \times 2} \qquad \ldots(21.12)$$

Substituting these values in the general expression of torque, such as equation 18.12, gives

$$T = \frac{\pi}{8} (D_2^2 - D_1^2)(D_2 + D_1)pf \qquad \ldots(21.13)$$

Design: The values of p and f may be taken from Table 18.1. The diameters D_2 and

D_1 may be selected so that, approximately, $D_2/D_1 = 1.6$.

The number of springs i ranges from 6, for small couplings and for a shaft diameter $d = 56$ mm, up to 16 for $d = 160$ mm.

21.9 POSITIVE CLUTCHES

A jaw clutch is the commonest type of positive clutch. Jaw clutches are made either with square jaws, as shown in Fig. 21.21, for driving in both directions, or with spiral jaws, as in Fig. 21.22, for driving in one direction only. *The jaw clutch is used for quick and easy engagement and disengagement (21.1).*

Jaw clutches: When a clutch is used to connect a belt pulley, sprocket, or gear to a rotating shaft, the hub of the driven member has jaws to mesh with those on the sliding half of the clutch. The hub is held in the axial directions by two set collars and usually has a bronze bushing.

Jaw-clutch coupling: When a clutch is used for connecting two shafts, as in Figs. 21.21 or 21.22, it is called a clutch coupling (21.1). In a clutch coupling, one half of the clutch, usually the driver, is keyed fast to one shaft, while the other half, the follower, is attached to the other shaft by means of a feather key or spline which permits it to slide along the shaft when the shifting lever is operated. Approximately, the main proportions of a cast-iron jaw clutch, Fig. 21.20, are as follows:

$a = 2.2d + 25$ $g = d + 5$ $j = 0.2d + 4$
$c = 1.2d + 30$ $h = 0.3d + 13$ $k = 1.2d + 20$
$f = 1.4d + 8$ $i = 0.4d + 6$ $l = 1.7d + 60$

The counterbore b is made larger than the shaft diameter to clear the feather keys.

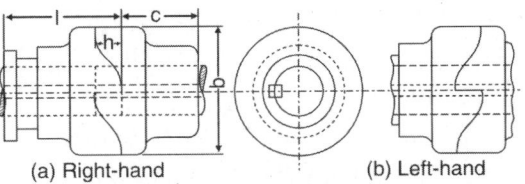

Fig. 21.22: Spiral-jaw clutch

Some of the dimensions of the coupling can and should be checked by considering stresses. Thus, it is advisable to check the jaws in shearing; the sides of the jaws in crushing; and the feather key in shear and crushing.

Tests show that cast-iron blocks loaded in compression fail by shearing along a plane making an angle α of about 35° with the direction of pressure. Figure 21.21c shows a development at the mean diameter of the jaws. The area in shear is $0.5(a - b)h/\sin \alpha$; and the component of the tangential force that will act in the plane of shear is evidently $F_t/\cos \alpha$. It should be assumed, for the sake of safety, that only one-half the total number of jaws i is in actual contact. For $\sin \alpha/\cos \alpha = \tan \alpha = 0.7$, the shear stress is

$$S_s = \frac{F_t \sin \alpha}{0.5 \cos \alpha (a-b)h \times 0.5i} = \frac{2.8 F_t}{(a-b)hi} \quad \dots (21.14)$$

A safety factor n of 2 should be used for the allowable shear stress.

The allowable bearing pressure between cast-iron jaws should be taken as $p = 20$ MN/m² for small clutches, but this value may be increased, as the size increases, up to $p = 40$ MN/m² for a 100 mm shaft.

The number of square jaws is usually two or three, but more are sometimes used for large shafts. If it is necessary to preserve a relative

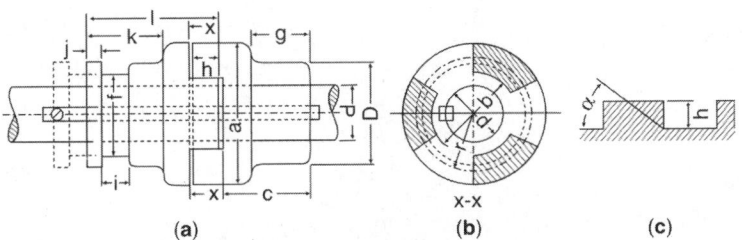

(a)   (b)   (c)

Fig. 21.21: Square-jaw clutch

angular position of the shafts, one of the jaws is made slightly smaller or larger than the others.

Speed limitations: Jaw clutches and couplings can be engaged only at low speeds, below *60 rpm. However, when engaged, the shaft can run at much higher speeds. The spiral jaws are engaged more easily and are used when engagement is frequent. They are also very convenient for presses, punches, and shears in which they are automatically thrown out of engagement when the direction of rotation of the shaft is reversed after completion of the working cycle.*

Clutches with cast jaws are used for shafts not over 100 mm in diameter. Clutches with finished square jaws are used for shafts up to 250 mm in diameter.

Shifting mechanism: Collars, also called bands, are made of brass. Their usual shape is shown in Fig. 21.23a. Figure 21.23b shows a horseshoe collar which can be used only on a horizontal shaft with a shifting lever in the vertical plane.

Shifting levers are made of flat iron with the fulcrum *f* at the end, as in Fig. 21.24a, or fork-shaped with the fulcrum *f* between the fork and handle, as in Fig. 21.24b.

Design procedure for Ajax or bushed pin flexible coupling or steps in computer programme. Determine number of pins as two more than the rigid coupling, pin diameter d_1 same as that for rigid coupling.

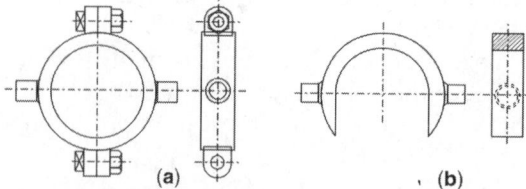

(a)   (b)

Fig. 21.23: Clutch-shifting collars

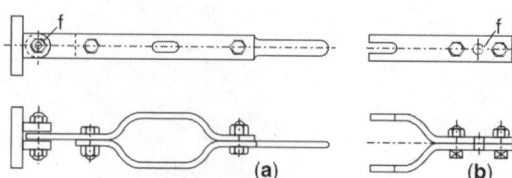

(a)   (b)

Fig. 21.24: Shifting levers

1. $n = 0.02d + 5$

2. $d_1 = \dfrac{0.5d}{\sqrt{n}}$, d_1 is to be a standard size.

   The pins are subjected to shear stress and bending stress (21.8).

3. Larger diameter of pin $d_2 = d_1 + 2 \times$ step size t_1 (Fig. 21.25).

   The step size t_1 is 3 or 4 mm to give rigidity to the pin.

   The bronze bush is having thickness $t_1 = 1.5$ to 2 mm, t_2, the rubber bush thickness is 6, 8, 10 mm.

4. Outer diameter of bronze bush $d_3 = d_2 + 2t_1$

5. Outer diameter of rubber bush or hole diameter $d_4 = d_3 + 2 \times t_2$

   The length of the bush is calculated from bearing pressure f_b consideration.

6. Bolt circle diameter $D_1 = 3d$

7. The force acting on each pin is

$$F = \frac{T}{\dfrac{D_1}{2} \times n}$$

8. Length of pin $L = \dfrac{F}{d_3 \times f_b}$

   By assuming a clearance $c = 4$ to 6 mm to allow for bending of the flange.

9. The bending moment on the pin

$$M = F\left[\frac{l}{2} + c\right]$$

If there is no clearance, there will be no bending of flange and no flexibility (21.10) in addition to adjustment of rubber bush inside, the hole when there is angular or axial misalignment, i.e. rubber bush and clearance make the coupling flexible (21.4).

10. The bending stress $S_b = M/Z$

    where, $\qquad Z = \pi d_1^3 / 32$

11. The shear stress in the pin $S_s = \dfrac{F}{\pi/4 \times d_1^2}$

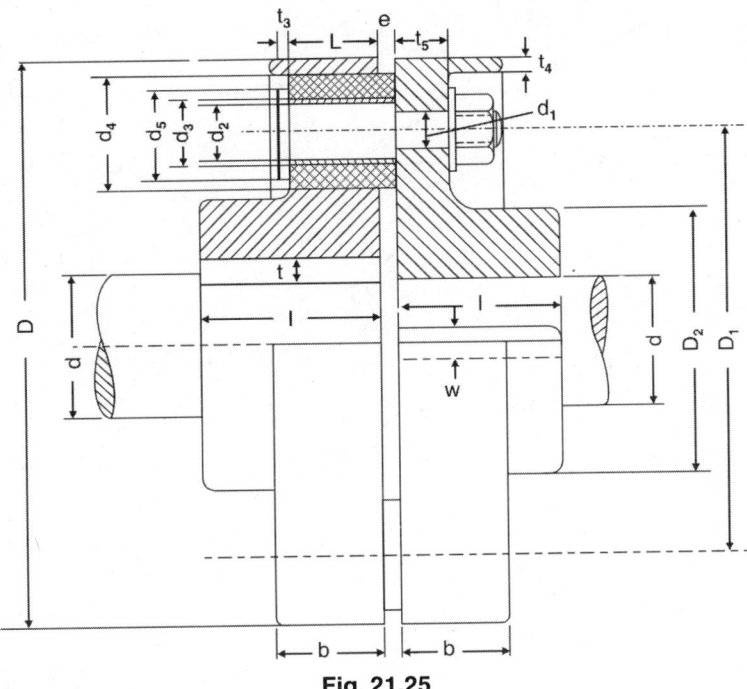

Fig. 21.25

12. The principal stress

$$S_p = \frac{1}{2}\left[S_b + \sqrt{S_b^2 + 4\,S_s^2} \right]$$

If S_p is more than the allowable stress in the pin, then the calculations are repeated with the next higher diameter of the pin from step 3 till $S_p \le$ allowable stress. These steps will also facilitate making the computer program.

13. Outside diameter of flange = $D = 4d$

14. Hub diameter $D_2 = 2d$

15. Hub length $l = 1.25d$ to $1.5d$

16. Head diameter of pin $d_5 = d_4 - 5$

17. Thickness of head $t_3 = 5$ mm

18. Width of key $w = \dfrac{d}{4}$ thickness $t = \dfrac{d}{6}$

19. Thickness of flange $t_5 = 0.5d$ to d

20. Thickness of Rim $t_4 = 0.1d$

21. $b = L + 10$

22. Force acting on key $F_1 = \dfrac{T}{d/2}$

23. Shear stress in key $= \dfrac{F_1}{wl}$

24. Crushing stress in key $= \dfrac{F_1}{lt/2}$

After increasing the diameter of pins, the pin circle diameter D_1 needs to be checked for available space.

Example 21.1: Design a rigid flanged coupling for shafts 50 mm diameter rotating at 300 rpm and transmitting a power of 55 kW. Allowable shear stress in bolts 50 N/mm² C.I. flange 10 N/mm².

Considering the register to be provided by the shaft the two surfaces of the flanges are providing frictional resistance obtained by prestressing of the bolts. Calculate the frictional torque which the flanges resist, assuming the initial tightening load per bolt = 15 kN and coefficient of friction between the flanges = 0.15.

Torque transmitted, $T = \dfrac{55 \times 1000 \times 60}{2\,\pi \times 300}$

$= 1750.7$ Nm

No. of bolts, $n = 0.02\,d + 3 = 0.02 \times 50 + 3 = 4$ bolts

Bolt circle diameter, $D_1 = 3d = 3 \times 50 = 150$ mm

Diameter of bolts, $d_1 = \dfrac{0.5d}{\sqrt{n}} = \dfrac{0.5 \times 50}{\sqrt{4}} = 12.5$ mm

Taking the next standard size M14 and checking for shear stress.

Force on the bolt $F_t = \dfrac{1750.7 \times 1000}{4 \times 150/2} = 5835.67$ N

Shear stress $= \dfrac{F_t}{\pi/4\, d_1^2} = \dfrac{4 \times 5835.67}{\pi \times 14^2}$

$$= 37.9 \text{ N/mm}^2$$

which is satisfactory

Outside diameter of flange $D = 4 \times 50 = 200$ mm

hub diameter, $D_2 = 2d = 2 \times 50 = 100$ mm

hub length, $l = 1.25d + 20 = 1.25 \times 50 + 20 = 83$ mm

flange thickness $d/4 = 12.5$ or 13 mm

Key width $= d/4 = 12.5$ mm say 13 mm

key depth $= d/6 = 8.5$ mm say 9 mm

Checking the flange for shear stress

$$= \dfrac{T}{D_2/2\, \pi\, D_2 \times t}$$

$$= \dfrac{1750.7 \times 1000 \times 2}{\pi \times 100^2 \times 13} = 8.57 \text{ N/mm}^2$$

which is satisfactory

For finding the frictional torque, inner diameter = 50 mm and outer diameter is 200 mm.

Frictional radius $= \dfrac{2}{3}\left[\dfrac{100^3 - 25^3}{100^2 - 25^2}\right] = 70$ mm

Frictional torque $= fWR \times 4$

$$= \dfrac{0.15 \times 15 \times 1000 \times 70 \times 4}{1000} = 630 \text{ Nm}$$

So the bolts will be more safe to transmit the remaining torque.

Force transmitted by key $= \dfrac{1750.7 \times 1000}{25}$

$$= 7008 \text{ N}$$

Shear stress in key $= \dfrac{70028}{13 \times 83} = 64.9 \text{ N/mm}^2$

As it is quite high take $w = 16$ mm, $t = 16$ mm and $l = 90$ mm.

Shear stress $= \dfrac{70028}{16 \times 90} = 48.6 \text{ N/mm}^2$ ok.

Crushing stress $= \dfrac{70028}{8 \times 90} = 93.2 \text{ N/mm}^2$ ok.

Example 21.2: Design a bushed pin type flexible coupling to transmit 20 kW at 1000 rpm. The shaft diameters are 50 mm each. The allowable stress in the pins is limited to 65 N/mm². The allowable bearing pressure is 0.3 N/mm².

Torque transmitted, $T = \dfrac{20 \times 1000 \times 60}{2\,\pi \times 1000}$

$$= 190.98 \text{ say } 191 \text{ Nm}$$

Number of pins $n = \dfrac{50}{50} + 3 = 4$

Taking 6 pins for the flexible coupling

Diameter of pins $d_1 = \dfrac{0.05 \times 50}{\sqrt{6}} = 10.2$ say M 12

Enlarged diameter of pins $d_2 = 12 + 6 = 18$ mm

Brass or bronze bushing of 2 mm thickness and rubber bush of 8 mm thickness is assumed to give a hole diameter.

$$d_3 = d_2 + 2t_1$$

$$= 18 + 2 \times 2 = 22$$

Rubber bush diameter d_4

$$d_4 = 22 + 2 \times 8 = 38$$

Pitch circle diameter $D_1 = 3d = 150$ mm

Force on each pin $F = \dfrac{T}{n \times D_1/2} = \dfrac{191 \times 1000}{6 \times 150/2}$

$$= 424.4 \text{ N}$$

The length L of pins $= \dfrac{F}{p \times d_4} = \dfrac{424.4}{0.3 \times 38}$

$$= 37.2 \text{ say } 38 \text{ mm}$$

The head of pin has a diameter d_5, $38 - 5 = 33$ mm and thickness t_3 of 5 mm.

Shear stress in the pins $S_s = \dfrac{424.4}{\dfrac{\pi}{4} \times 12^2}$

$$= 3.75 \text{ N/mm}^2$$

Bending moment on the pin

$$= F\left[\dfrac{L}{2} + e\right] = 424.4\left[\dfrac{38}{2} + 5\right]$$

$$= 10185.6 \text{ Nmm}$$

where e is the clearance say 5 mm

Bending stress $S = \dfrac{M}{Z} = \dfrac{10185.6 \times 32}{\pi \times 12^3}$

$$= 60 \text{ N/mm}^2$$

Maximum principal stress $= \dfrac{1}{2}\left[S + \sqrt{S^2 + 4S_S^2}\right]$

$$= \dfrac{1}{2}\left[60 + \sqrt{60^2 + 4\times 3.75^2}\right] = 60.2\,\text{N/mm}^2$$

Which is satisfactory, so a pin dia of 12 mm is a suitable size.

Length of hub l on each side $= 1.5d = 1.5 \times 50 = 75\,\text{mm}$

Diameter of hub $D_2 = 1.75d = 1.75 \times 50 = 87.5$ say 88 mm

Overall flange diameter $D = pcd + 2 \times$ hole dia $+ 10 \times 2$

$= 150 + 2 \times 38 + 20 = 246$ say 250 mm

Thickness of flange on the left (rubber) side $= 38\,\text{mm}$

Length of flange on the left side $= 38 + 5 + 5$
$= 48$ say 50 mm

Depth of flange at top $= 6$ mm

Thickness of flange on the right side $= d/4$ and is also taken $= 38$ mm.

key width $w = \dfrac{d}{4} = 12.5$ say 13 mm

key depth $= t = \dfrac{d}{6} = 8.5$ say 9 mm

length of key $l = 1.5 \times 50 = 75$ mm

The key is checked for shear stress

Shear stress $= \dfrac{2\times 191\times 1000}{50\times 13\times 75} = 7.83\,\text{N/mm}^2$

which is quite low.

Crushing stress $= \dfrac{2\times 191\times 1000\times 2}{50\times 9\times 75} = 22.63\,\text{N/mm}^2$

which is also satisfactory.

PROBLEMS

21.1 When for convenience it is desirable to engage and disengage two shafts easily and quickly, what kind of coupling is used?

21.2 Why should finished bolts be fitted into reamed holes of the flanges in a rigid coupling?

21.3 What is the necessity of a flexible coupling?

21.4 What makes a flexible coupling flexible?

21.5 Why is it necessary to spot face the areas where bolts or nuts rest in a rigid coupling?

21.6 What purpose is served by register in a rigid coupling?

21.7 Name the couplings used for the following:
   (a) When the axes of shafts are colinear
   (b) When the axes of shafts are parallel but off set.

   (c) When the axes of shafts are slightly inclined.
   (d) When the axes of shafts are sufficiently inclined.
   (e) When output shaft may rotate relatively with different speed as compared to input shaft.

21.8 What type of stress is set in the bolt of a rigid coupling?

21.9 What is the nature of stresses in the pin of bushed pin type flexible coupling?

21.10 Which is the weakest element of shaft coupling?

21.11 What purpose is served by clearance between two flanges of a pin bushed flexible coupling?

21.12 Determine the main dimensions of a flange coupling, Fig. 21.1, for a 110 mm shaft to transmit the full torque capacity of the shaft with an allowable stress of 55 N/mm².

21.13 (a) Determine the main dimensions of a clamp coupling, Fig. 21.3, for a 125 mm shaft to transmit the full torque capacity of the shaft, (b) Make a freehand sketch of the assembled coupling more or less to scale.

21.14 Determine the main dimensions of a compression coupling, Fig. 21.5, for a 100 mm shaft to transmit the full torque capacity of the shaft.

21.15 (a) Determine the main dimensions of a compression coupling, Fig. 21.4, for a 90 mm shaft to transmit the full torque capacity of the shaft, (b) Make freehand sketches, more or less to scale, of all details.

21.16 Determine the main dimensions and give freehand sketch to scale of an Ajax flexible coupling, Fig. 21.22, for a 100 mm shaft for heavy-duty service.

21.17 Determine the main dimensions and give freehand sketches of all details for a Westinghouse-Nuttall coupling, Fig. 21.15, for a 125 mm shaft of an electric generator with fluctuating load, driven by an oil engine.

22

Friction Clutches

22.1 GENERAL CONSIDERATIONS

The object of a friction clutch is to connect a stationary machine part to a rotating part, to bring it up to speed, and to transmit the required power with a minimum of slippage. In certain cases a friction clutch serves as a safety device by slipping when the torque transmitted through it exceeds a safe value, thus preventing the breakage of parts in the transmission train.

Friction clutches are based on the same principles as brakes, and their construction is also similar to that of a brake. In fact, every time a brake is locked, it becomes a friction clutch. Equations and experimental data given in Chapter 18 can be used for friction clutches either directly or with certain modifications explained at various points in the following discussion.

Fluid clutches: Fluid clutches, also called *hydraulic clutches*, or rather *hydraulic couplings*, are coming into extensive use for connecting internal combustion engines (or electric motors) to driven machinery. Their main advantage is that they absorb impact loads imposed by the driven machine, thus eliminating excessive stresses and increasing the service life of the whole equipment. Another advantage is that a hydraulic coupling cuts a long shaft into two shorter lengths, each of which has a higher natural frequency of vibration. This reduces the danger of torsional vibration.

However, a fluid coupling is essentially a piece of hydraulic machinery, and its design is therefore outside the scope of this book.

Design requirements: The following consideration must be observed in the design of a friction clutch:

(a) Selection of a type suitable for given operating conditions.

(b) Selection of suitable materials forming the contact surfaces.

(c) A sufficient torque capacity.

(d) Engagement and acceleration without shock.

(e) Quick disengagement without drag.

(f) Provision for holding the contact surfaces together by the clutch itself.

(g) Low weight in order to keep down the inertia, especially in high-speed service.

(h) Balancing of all moving parts, especially in high-speed service.

(i) Provision for taking up wear of the contact surfaces.

(j) Accessibility and provisions for facilitating repairs.

(k) In an industrial clutch, protection of projecting parts.

(l) In a clutch frequently operated, provision for carrying away the heat generated at the contact surfaces.

Classification: Friction clutches may be divided into two main groups, according to the direction of the acting force: (1) axial clutches and (2) radial clutches, or rim clutches. A friction clutch used to connect two shafts is often termed a clutch coupling, or friction coupling.

An axial clutch is one on which the contact pressure is applied in a direction parallel to the axis of rotation. Axial clutches, in turn, can be subdivided into (a) cone clutches, (b) disk clutches, and (c) combined cone and disk clutches. Magnetic clutches are usually of the disk types.

In a rim clutch the contact pressure is applied upon a rim in a radial direction. These clutches, like brakes, may be subdivided into band clutches and block clutches, or into external, internal, and combined internal-and-external clutches.

Only a few examples of the more representative clutch types will be discussed. These examples and the field of their application in accordance with their torque capacity and speed may serve as a guide for the analysis and design of other arrangements.

22.2 SELECTION OF TYPE

The factors which must be taken into consideration in deciding what type of clutch is to be used are torque, rotative speed, available space, and frequency of operation.

Torque: To transmit a heavy torque or a fluctuating torque a clutch must have sufficient gripping power. *This condition is fulfilled in general by multidisk clutches (22.1)* and for low-speed service by cone and rim clutches of large diameter. The use of special materials for the contact surfaces may increase the torque capacity of a clutch.

Speed: Light, compact, and internally balanced clutches, such as the double-cone and *multidisk types, are best suited for high rotative speeds* (22.2).

Space limitations: Multidisk, twin-cone, and double-cone clutches are more compact than other types (22.3).

Frequency of operation: Clutches which are in frequent or continuous operation should have a small travel, a simple engaging mechanism, and large cooling areas to dissipate the heat generated during the engaging of the clutch under load. *In this, single-disk clutches with metal*

contact surfaces, and cone clutches, are the most suitable (22.4).

22.3 MATERIALS FOR CONTACT SURFACES

A material suitable for use as a friction surface must meet the following conditions:

(a) It must have a high coefficient of friction
(b) It must not be affected by moisture and oil
(c) It must resist wear
(d) It must be capable of resisting high temperatures caused by slippage.

Materials in common use are wood, asbestos, cork, leather, and various metals.

Wood: Maple, elm, and pine, in lighter service, are used with cast iron in many clutches. All woods have a high coefficient of friction (*see* Table 18.1). Maple and elm resist wear fairly well, but neither complies with condition b or condition d.

Asbestos: Asbestos fabric and molded blocks have the same coefficient of friction and, according to Table 18.1, comply well with condition a. They also comply well with conditions b, c, and d. In addition, their friction coefficient is less affected by oil or grease than that of any other material.

Asbestos-metallic blocks are molded under a very high pressure into any desired shape.

Cork: Cork is used in the shape of round plugs or inserts in connection with some other material. The surface covered by cork inserts varies from 10 to 40 per cent of the total friction area. Because of its higher coefficient of friction, cork increases the torque capacity of the clutch. In general, the cork inserts project slightly over the harder surface and are operative chiefly in engaging the clutch. After full engagement, the harder material–metal or wood–forms the surface in contact. In a lubricated clutch, cork inserts also help to keep the friction surface lubricated.

Leather: Most small cone clutches are faced with leather. Oak-tanned leather and so-called chrome leather seem to be equally serviceable.

Before the leather facing is fastened to the cone, it should be soaked in caster oil or neat's-foot oil or boiled in tallow. The excess oil or grease is removed by passing the facing between rolls.

Metals: If one of the friction surfaces is faced with a fibrous material, the other surface must be of cast iron, cast steel, or rolled steel. In some clutches, both contact faces are made of metal. Cast iron bears against cast iron, cast steel, or bronze; or hard saw steel bears against steel or bronze. *In order to obtain smooth operation in a clutch with metal-to-metal surfaces, these surfaces must be lubricated to dissipate heat even though, the coefficient of friction decreases and also wear (22.6).*

Other requirements for good clutch design will be discussed simultaneously with the analysis of different clutch types.

The *load factors* must be used as given in Table 20.3 for shafts.

22.4 CONE CLUTCHES

Figure 22.1a shows schematically a cone cutout clutch in which the outer cone *a* is the driving member and the inner cone *c* can be moved axially.

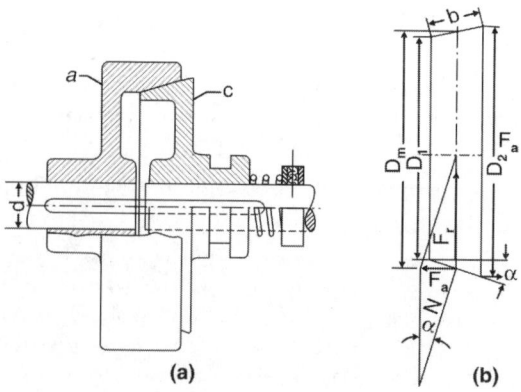

(a)   **(b)**

Fig. 22.1: Cone-clutch coupling

Figure 22.2 shows an industrial clutch with cast-iron contact surfaces. The cone *c* is keyed to the shaft, while the pulley *p* with the cone cup rotates the cone hub and carries with it the levers *l*. When the thimble *t* is forced under the rollers *r*, by means of the sliding ring *s*, the levers *l* bring the cone surfaces into contact. Heavy springs at *s*, not shown in the illustration, throw the surfaces apart when the thimble is shifted to the right. The wear is taken up by adjusting the collar *g* held by a locknut *h*. If the clutch is operated frequently, the cone should be lined with wood or asbestos blocks. *It is also useful for synchronising speeds in automobile gear box having constant mesh gears (22.7).*

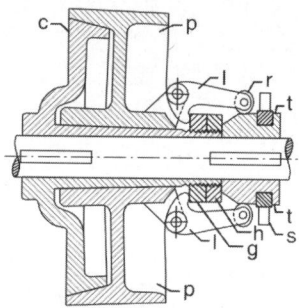

Fig. 22.2: Industrial cone clutch

A double-cone clutch having cones faced with asbestos lining is shown in Fig. 22.3. Because of the symmetrical arrangement of the cones, their angle can be large enough (35° to 45°) to eliminate sticking and drag. The cast-iron friction ring *f* is held to the housing *h* by a feather key *i*; the right cone *c* slides on the feather key *k*. When the cones are pulled together by the lever mechanism, the lever *l* is pushed slightly past the vertical or deadline position, and there is no need for an outside force for holding the contact surfaces together. The ring *r* with the fulcrum forks serves to take up the wear.

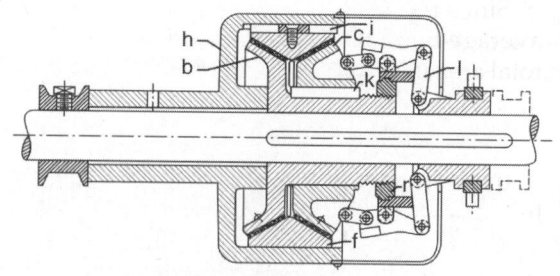

Fig. 22.3: Double-cone clutch

Force analysis: The forces acting upon the inner cone in Fig. 22.1a are shown in Fig. 22.1b. In finding the expression for the torque T which the clutch can transmit when a certain axial force F_a is applied, the following designations also will be used; p for the unit normal pressure at the contact surface, and f for the coefficient of friction.

The torque which the clutch will transmit is equal to the torsional moment of the frictional resistance between the inner and outer cones. For an analysis, the male cone c may be considered as a free body. It is acted upon by the axial force F_a, the normal force N created by F_a and distributed around the cone, and the tangential force F_t due to friction. From Fig. 22.1b,

$$F_a = N \sin \alpha \qquad \ldots (22.1)$$

The friction force F_t is equal to fN and

$$F_t = \frac{fF_a}{\sin \alpha} \qquad \ldots (22.2)$$

The torque transmitted through friction is

$$T = \frac{F_t D_m}{2} = \frac{fF_a D_m}{2 \sin \alpha} \qquad \ldots (22.3)$$

where the mean diameter is approximately

$$D_m = \frac{1}{2}(D_1 + D_2)$$

The power P that a cone clutch will transmit is found by substituting the value of T from equation 22.3 in the general equation 2.17 and taking into account the load factor k_l, Table 20.3. This gives

$$P(kW) = \frac{\pi f F_a D_m n}{60 \times 1000 \sin \alpha \, k_l} \qquad \ldots (22.4)$$

Since the normal force is also equal to the average normal pressure p multiplied by the total area in contact

$$N = \pi D_m bp = \frac{F_a}{\sin \alpha} \qquad \ldots (22.5)$$

Combining equations 22.4 and 22.5 results in

$$p = \frac{\pi^2 f p D_m^2 \, bn}{60 \times 1000 \, k_l} \qquad \ldots (22.6)$$

The relation between the axial force F_a and the clutch dimensions, as found from equation 22.5, is

$$F_a = \pi D_m bp \sin \alpha \qquad \ldots (22.7)$$

The axial force F_a exerted by the spring, as determined by equations 22.1 or 22.7, is sufficient to transmit the torque. For engaging the clutch the force must be greater in order to overcome friction when cone c is pressed into cone a. By reasoning similar to that used in deriving equation 18.19, it will be found that the force F_a' necessary to engaged the clutch is

$$F_a' = N(\sin \alpha + f \cos \alpha) \qquad \ldots (22.8)$$

or, with equation 22.5,

$$F_a' = \pi D_m bp (\sin \alpha + f \cos \alpha) \qquad \ldots (22.9)$$

The actual force required to engage a clutch is slightly greater than F_a' because of friction in the joints. If a spring is used to engage a clutch, it must be able to exert a still greater force in order to have a certain reserve strength.

Design data: The friction coefficient may be taken as 0.15 for cast iron on cast iron, 0.2 for a surface faced with leather or wood, 0.3 for a surface faced with asbestos, and 0.25 for cone surfaces with cork inserts.

Slip: In normal operation all friction clutches have a slip of about 2 per cent.

Cone-face angle: The SAE rules recommend, for cone clutches faced with leather or asbestos or having cork inserts, a standard angle α of 12.5°. For industrial clutches faced with wood, the angle α may range from 15° to 25°.

Mean diameter: If the ratio D_m/b is designated by q, and if this substitution is made in equation 22.6 and the equation is solved for D_m, the result is

$$D_m = \sqrt[3]{\frac{Pk_l q \times 60 \times 1000}{\pi^2 fpn}} \qquad \ldots (22.10)$$

The permissible pressure p in equation 22.10 must be selected by using Table 18.1 as a guide. To prevent excessive wear of the contact surfaces, it is recommended that p can be taken near the lower limit, or even slightly below that limit. The value of q varies in existing designs from 4.5 to 8. The greater q is made, the smaller will be the difference in the peripheral speeds at the outer and inner diameters of the cone, and hence the less the wear will be. On the other hand, the greater q is made, the larger D_m and the clutch in general will be larger.

After D_m is found from equation 22.10, it may be checked by comparison with the shaft diameter d. In commercial clutches D_m varies from $5d$ to $10d$. Another check is made by computing the peripheral velocity v. For high-speed, leather-faced clutches v should be from 10 to 25 m/s. For clutches not balanced inherently, this speed must be considerably lower; for clutches with metal-to-metal contact surfaces, the speed limits are 300 to 1,000 rpm.

Shaft sleeves: The hub of the clutch half in which the shaft turns when the clutch is not engaged and which does not move axially must have a sleeve, as shown in Fig. 22.3, which can be replaced when it is worn. The type of sleeve depends on the speed. Cast-iron, wick-oiled sleeves can be used for speeds up to 450 rpm, and bronze-bushed sleeves can be used up to 800 rpm. Sleeves for higher speeds should have ball bearings, one on each end of the sleeve.

Example 22.1: (a) Determine the main dimensions of a cone clutch similar to that in Fig. 22.1. It is to be faced with leather and is to transmit 30 kW at 750 rpm from an electric motor to an air compressor. (b) Also find the axial force that must be produced by the spring.

(a) Shaft size needed to transmit the given torque is by equation 20.2, with $S_d = 45$ N/mm^2 and $k = 1.75$ (Table 20.3).

$$T = \frac{30 \times 60 \times 1000}{2\pi \times 750} = 381.97 \text{ Nm}$$

$$d = \left[\frac{16 \times 1.75 \times 381.97 \times 1000}{\pi \times 45} \right]^{1/3} = 42.29$$

The next-larger standard size is 45 mm. However, for the sake of rigidity it is advisable to take a still larger size and to use $d = 50$ mm.

The mean cone diameter D_m may be determined by selecting the following average values: $q = 6$, $f = 0.2$, and $p = 0.1$ N/mm^2. Then, by equation 22.10.

$$D_m = 10 \sqrt[3]{\frac{30 \times 1.75 \times 6 \times 60 \times 1000}{0.2 \times 0.1 \times 750 \times \pi^2}} = 503 \text{ mm}$$

A check shows that $D_m/d = 503/50 = 10$, which is within the usual limits. The peripheral velocity is

$$v = \frac{\pi \times 503 \times 750}{60 \times 1000} = 19.75 \text{ m/s}$$

This is also satisfactory.

The trial value of the face width may be taken as

$$b = \frac{503}{6} = 83.8 \text{ say } 85 \text{ mm}$$

The smaller cone diameter, with $\alpha = 12.5°$, is then

$D_1 = D_m - b\sin\alpha$

$\quad = 503 - 85 \times 0.2164 = 484.6$ say 484 mm

The outside cone diameter is

$D_2 = 503 + 85 \times 0.2164 = 521.4$ say 522 mm

The corresponding face width is

$$b = \frac{D_2 - D_1}{2\sin\alpha} = \frac{522 - 484}{2 \times 0.2164} = 87.8 \text{ say } 88 \text{ mm}$$

(b) The minimum axial force F_a found from equation 22.7 is

$F_a = \pi \times 503 \times 88 \times 0.1 \times 0.2164 = 3009$ N

22.5 DISK CLUTCHES

Figure 22.4 shows a heavy-duty friction coupling that is engaged. In order to disengage the clutch, the collar c is moved to the right and the steel lever l turns about the fulcrum a and pushes the bolts b to the left. The wear of the asbestos lining is taken up by the nuts on the bots b.

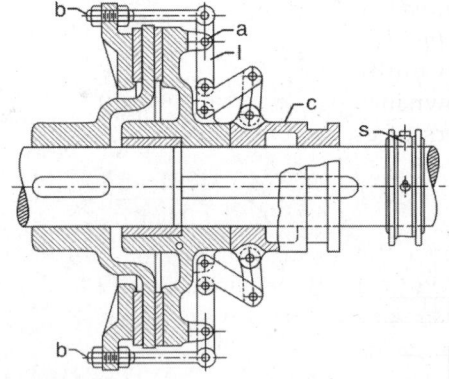

Fig. 22.4: Heavy-duty disk clutch coupling

A single-disk dry-plate automobile clutch coupling, commonly called a clutch, is shown in Fig. 22.5. The clutch disk d consists of a thin steel plate faced on both sides with asbestos lining.

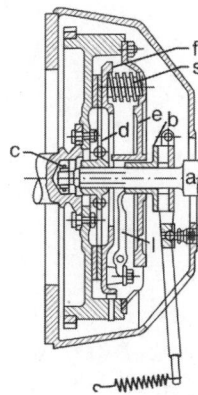

Fig. 22.5: Automobile clutch coupling

The clutch is kept engaged by springs *s* and is released by pushing toward the flywheel the bearing *b*, which is in contact with the inner ends of the release levers *l*. These levers are pivoted on pins fastened to the clutch cover *e*. The outer fork-shaped ends of levers *l* engage lugs on the pressure plate *f* and pull this plate away from the disk *d*, compressing the springs *s*.

The clutch comes in contact, heat is generated and it is dissipated by circulation of air through a number of radial slots or grooves in the surface of the plate. For transferring heat to outside the assembly, ventilation is provided through openings in the clutch housing. Torsional damper springs are provided circumferentially in the plate around the hub and below the lining for damping of vibrations produced in the engine.

A multidisk industrial clutch coupling is shown in Fig. 22.6. As a result of the use of four pairs of contact surfaces, this coupling is compact and is suitable for comparatively high speeds. The asbestos-lined rings *r* can slide on the feather key *l* and can rotate with the clutch housing *h*, which is keyed to the left shaft. The inner disk *a* and the right disk *b* slide on a feather key fastened to the hub of the left disk *c*, which itself is keyed to the end of the right shaft. When the clutch collar *d* is shifted to the left in order to engage the clutch, the link *g* must be pushed past a position parallel to the shaft axis, thus locking the mechanism. The threaded yoke collar *e* serves to take up wear.

A multidisk clutch coupling often used for driving boats by internal combustion engines is shown in Fig. 22.7. It is similar in construction to that in Fig. 22.6, but both the driving disks *a* and the driven disks *b* are comparatively thin cast-iron plates without any facing, and they run in an oil bath. The driving disks *a* have square holes fitted to the square-shaped hub *c*. The driven plates *b* are connected to the housing *h* by pins *l*, instead of by a key. For smooth operation, the clutch is partially filled with heavy lubricating oil.

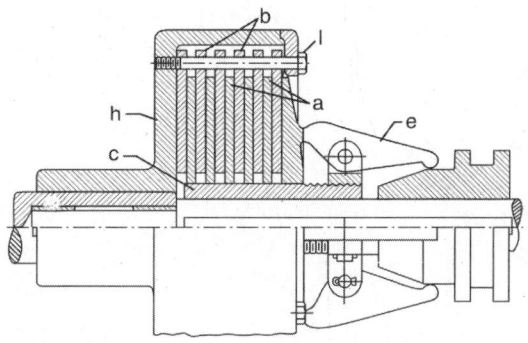

Fig. 22.7: Multidisk marine-type clutch coupling

A multidisk industrial clutch that is suitable for high speeds is shown in Fig. 22.8. Its construction is similar to the marine clutch in Fig. 22.7, but it is fully enclosed and is actuated by a very powerful double-toggle mechanism, shown schematically in Fig. 22.17. The smaller disks *a*, Fig. 22.8, are connected to the hub by a feather key. The necessary adjustment for taking up wear is made by turning nuts *n* and

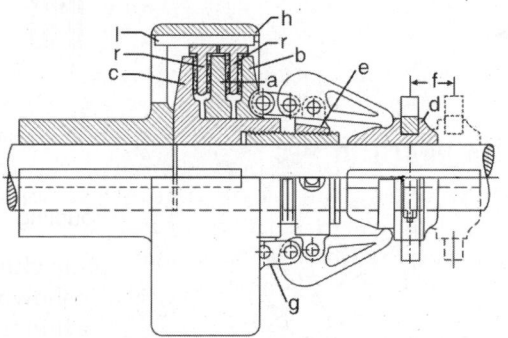

Fig. 22.6: Multidisk clutch coupling

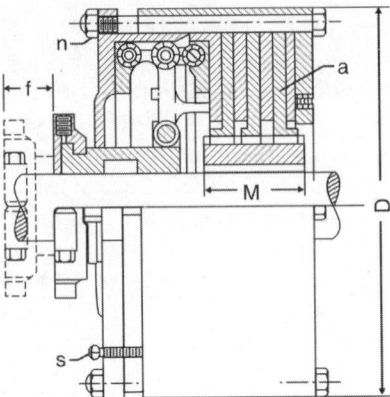

Fig. 22.8: Multidisk industrial clutch

locking them by screw *s*. The plates are lubricated to obtain smooth engagement and to reduce wear.

Finally, Fig. 22.9 shows an industrial clutch with hardwood or asbestos blocks. The surface of the block in contact with the flange *d* is flat, while the other end, in contact with the ring *e*, has the form of a double cone in order to increase friction. The clutch is engaged by a double-toggle mechanism. Counterweights *c* are put on to balance part of centrifugal force acting on the lever *g*. The springs between the parts *d* and *e* help to disengage the clutch and to prevent excessive wear of the blocks when the clutch is disengaged. Bolts *i* are provided to take up wear.

Basic data: Equation 18.53, which was derived for the torque in a brake with *i* pairs of friction surfaces, can be used without any changes

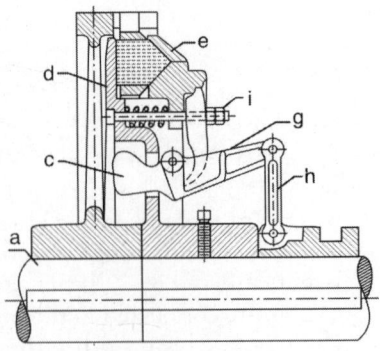

Fig. 22.9: Cone-disk clutch

for the torque on a disk clutch. The power that the clutch can transmit may be obtained by substituting this value for *T* in the general equation and introducing the load factor k_1.

$$T = \frac{\pi^2 i \, f \, p_1 D_1 (D_2^2 - D_1^2)}{8 k_l} \qquad \ldots (22.11)$$

Similarly, the axial force F_a necessary to create this torque may be found by equation 18.55. Thus,

$$F_a = \frac{1}{2} \pi \, p_1 D_1 (D_2 - D_1) \qquad \ldots (22.12)$$

Design: Both the procedure outlined and special remarks given in section 18.5 apply in full to disk clutches. The values for *f* and *p* may be taken from Table 18.1.

Speed: Industrial clutches are designed for a torque at 100 rpm. When the clutch is used at a higher speed, it can be considered that the torque is not changed if the clutch is so designed that the centrifugal force does not affect the engaging mechanism and the pressure on the friction surface. Otherwise, its rating should be divided by a speed factor k_s when the clutch is used at a higher speed of *n* rpm. Some designers determine the speed factor by the relation

$$k_s = 0.9 + 0.001 \, n \qquad \ldots (22.13)$$

In actual practice, however, it is better to determine k_s from actual tests.

Example 22.2: A certain clutch is designed to transmit 50 kW at 100 rpm. Determine its expected minimum capacity at 600 rpm.

By equation 22.13,

$k_s = 0.9 + 0.001 \times 600 = 1.5$

Therefore, the clutch rating at 600 rpm should be

$$P = \frac{50 \times 600}{100 \times 1.5} = 200 \text{ kW}$$

22.6 RIM CLUTCHES

There are a number of different types of rim clutches which vary in the method of gripping the rim and in the shape of the rim. Rim clutches can therefore be classified as block clutches, band clutches, slip-ring clutches, and roller clutches.

Centrifugal Clutch

A centrifugal clutch operates by utilization of centrifugal force for making contact between driving and driven members automatically and the drive is by frictional force. The driven member is fixed to a drum mounted on the shaft Fig. 22.10. The driving member shaft carries a spider having usually four shoes which have friction lining on the outer surface. The shoes are connected to the spider by means of springs. The spring in designed for a centrifugal force corresponding to a speed of 75 per cent of the running speed. As the speed increases beyond this speed, the shoes flyout and make contact with the drum. At the running speed, the frictional force generated is responsible for transmitting the torque.

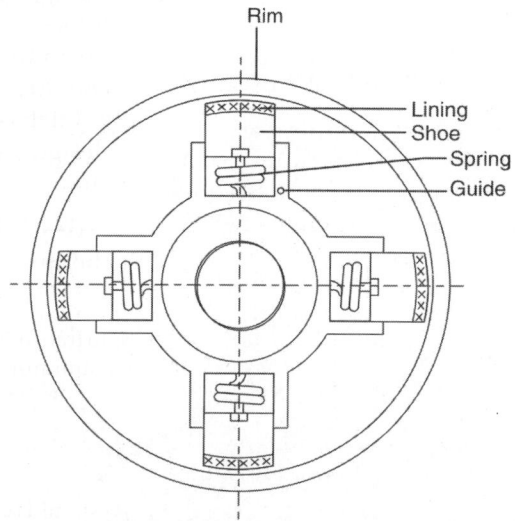

Fig. 22.10: Centrifugal clutch

w_1 speed at which spring is designed $= 0.75\, w_2$

w_2 running speed

F_1 centrifugal force at speed $w_1 =$ spring force

F_2 centrifugal force at running speed

R radius of the drum

R_g radius of centre of gravity of the shoes

i no. of shoes

$$F_1 = \frac{W}{g} w_1^2 R_g$$

$$F_2 = \frac{W}{g} w_2^2 R_g$$

Force acting on the drum per shoe $= F = F_2 - F_1$

$$= \frac{W}{g} R_g (w_2^2 - w_1^2)$$

Frictional force $F_t = fF$

Frictional torque $T = F_t R$

$$= f \frac{W}{g} R_g (w_2^2 - w_1^2) R \quad \ldots (22.14)$$

$$\text{Power transmitted} = \frac{2\pi N T\, i}{60 \times 1000} \quad \ldots (22.15)$$

Example 22.3: A centrifugal clutch has four blocks which slide radially in a spider keyed to the shaft of the driving engine and make contact with the inner surface of a drum of 300 mm diameter keyed to the main shaft. When at rest, the spring pulls each block against a stop. The clutch transmits 35 kW at 1000 rpm with engagement starting at 75 per cent of running speed. Determine the weight of each shoe.

Assume the radius of C.G. of shoe

$$R_g = 0.8 \times 150 = 120 \text{ mm}$$

Weight of each shoe $= W$

Spring force $F_1 = \dfrac{W}{g} w_1^2 R_g$

$$w_2 = \frac{2\pi \times 1000}{60} = 104.72 \text{ rad/s}$$

$$w_1 = 0.75 \times 104.72 = 78.54 \text{ rad/s}$$

$$F_1 = \frac{W}{g} \times (78.54)^2 \times 120 = 75.45\, W$$

$$F_2 = \frac{W}{g} \times (104.72)^2 \times 120 = 134.14\, W$$

Normal force $F = F_2 - F_1$

$$= 134.14\, W - 75.45\, W = 58.69\, W$$

Frictional force $= fF$

$$= 0.25 \times 58.69\, W = 14.67\, W$$

Frictional torque $= 14.67\, W \times 150 = 2200.5\, W$

Torque requirement $T = \dfrac{35 \times 60 \times 1000}{2\pi \times 1000}$

$$= 334.225 \text{ Nm}$$

$$\frac{2200.5\, W \times 4}{1000} = 334.225$$

$$W = 37.97 \text{ N}$$

If this weight is to be reduced, changes can be made as: (i) increase the radius of centre of gravity and (ii) also decrease the speed for start of engagement of $R_g = 130$, $w_1 = 0.7$ per cent of running speed

$$w_1 = 104.72 \times 0.7 = 73.3 \text{ rad/s}$$

$$F_1 = \frac{W}{g} (73.3)^2 \times 130 = 71.2 \, W$$

$$F_2 = \frac{W}{g} (140.72) \times 130 = 145.32 \, W$$

$$F = F_1 - F_2 = 145.32 \, W - 71.2 \, W = 74.12 \, W$$

$$\text{Torque} = \frac{0.25 \times 74.12 W \times 150 \times 4}{1000} = 334.225$$

OR $\quad 11.118 \, W = 334.225$

$$W = 30.06 \text{ N}$$

Figure 22.11 shows a typical block clutch with a grooved rim surface to increase the torque. The clutch has four, or in larger sizes six, cast-iron blocks b which are pressed against the rim either by adjustable push rods made up to parts c, d, and e, or by flat S-shaped springs s shown in dotted lines. Centrifugal force helps to keep the clutch engaged. Similar clutches are made with flat rims and blocks that are lined with leather or asbestos fabric as described earlier.

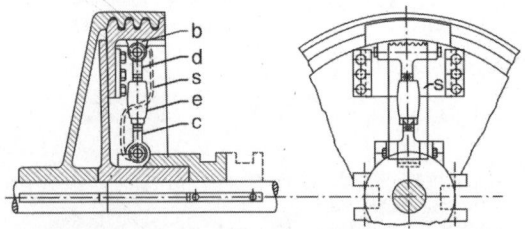

Fig. 22.11: Leblanc block clutch coupling

A rim clutch with a double grip, internal and external, is shown in Fig. 22.12. The double-lever engaging mechanism in conjunction with the double row of wooden blocks gives the clutch a high torque capacity.

Force analysis: There are no experimental data regarding the actual distribution of the contact pressure on a rim clutch. For each block, or shoe, of the clutch with a grooved rim, the angle β, Fig. 22.13, is relatively small.

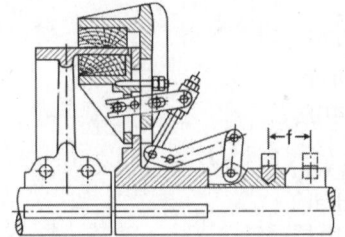

Fig. 22.12: Hill rim clutch coupling

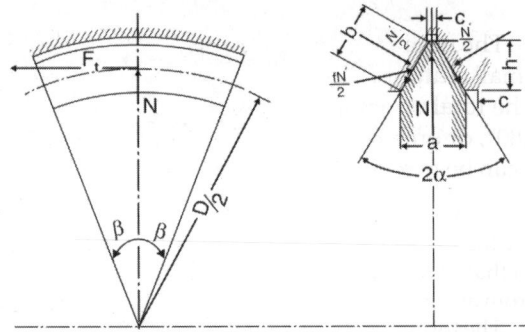

Fig. 22.13: Grooved rim clutch

Therefore the simplest assumption is justified and the normal pressure p may be considered uniform over the whole contact area of each block. When the clutch is being engaged, the equation for equilibrium of forces along the vertical axis is

$$N = N'' (\sin \alpha + f \cos \alpha) \qquad \ldots (22.16)$$

After the block is pressed on firmly,

$$N = N' \sin \alpha \qquad \ldots (22.17)$$

Since $F_t = fN'$ and $N' = 2\beta Dbp$, the torque which the clutch can transmit is

$$T = \frac{1}{2} i_1 i_2 F_t D = i_1 i_2 f \beta D^2 bp \qquad \ldots (22.18)$$

where, i_1 is the number of grooves in the rim, i_2 is the number of shoes, b is the inclined face, Fig. 22.12, and β is expressed in radians. If another assumption is used for the pressure distribution, such as uniform wear, the torque equation will be changed, but numerically the difference is smaller than the uncertainty in regard to the friction coefficient.

If the clutch has a flat rim, $i_1 = 1$ and the number of sides b is only one-half that of a grooved rim. Thus equation 22.16 becomes

$$T = \frac{1}{2} \, if\beta D^2 bp \qquad \ldots (22.19)$$

Design data: In order to prevent grabbing, the groove angle α should not be made smaller than 15° with cast-iron blocks, or smaller than 20° with wood or asbestos blocks. An angle α greater than 25° decreases the torque too much. The width b of the inclined face, in mm, may be determined by the equation

$$b = 0.01 \, D + 6 \qquad \ldots (22.20)$$

The number of grooves varies from three for small clutches up to six or seven for large ones. The total arc of contact, $2i_2\beta$, should be about 180°, or one-half the circumference. The space c can be made between $0.15h$ and $0.25h$.

Example 22.4: Determine the main dimensions and the force of the S-shaped springs for a clutch similar to that in Fig. 22.10 to transmit 500 kW at 220 rpm from an electric motor to a mine hoist.

The torque to be transmitted is, by equation 2.19,

$$T = \frac{600 \times 1000 \times 60}{2\,\pi \times 220} = 26043.5 \text{ Nm}$$

According to Table 20.3, a load factor of 1.75 should be used. The design torque is then $T = 45576$ Nm

If the peripheral velocity is assumed as 6.5 m/s at 100 rpm, or 14.3 m/s at 220 rpm, the mean diameter D_m may be selected as

$$D_m = \frac{14.3}{\pi \times 220} = 1.24 \text{ m say } 1.25 \text{ m}$$

For cast-iron shoes on a cast-iron rim the angle α may be taken as 15°. From Table 18.1, safe values would be $f = 0.15$ and $p = 1$ N/mm² the groove face, by equation 22.20, is

$$b = 0.01 \times 1250 + 6 = 18.50 \text{ mm say } 20 \text{ mm}$$

With these values the arc of contact of each shoe, by equation 22.18, is

$$2\beta = \frac{45576 \times 1000 \times 2}{0.15 \times 1.25^2 \times 10^6 \times 20 \times 1 \times i_1 \times i_2} = \frac{19.455}{i_1 \times i_2}$$

If seven grooves are made in the rim, $i_1 = 7$, and six shoes are used, or $i_2 = 6$, then the arc required is $2\beta = 0.4629$ radians, or 26.5°. This is satisfactory, as the total arc is $26.5 \times 6 = 159°$, which is somewhat less than 180°.

The depth of the grooves is

$$h = b \cos \alpha = 20 \times 0.966 = 19.3 \text{ or } 20 \text{ mm}$$

The inside rim diameter is then $D = 1250 - 20 = 1230$ mm. The distance between the grooves is

$$c = 0.25 \, h = 0.25 \times 20 = 5 \text{ mm}$$

The width of each groove is

$a = 2h \tan\alpha + c = 2 \times 20 \times 0.268 + 5 = 15.77$ or 16 mm

and the width of the rim is

$$B \geq 7 \, (16 + 5) + 5 = 152 \text{ mm}$$

To find N, or the pressure which each spring s must exert, it is necessary first to compute the magnitude of N', which is

$$N' = 2 \, i_1 \, \beta \frac{Dbp}{i_2} = \frac{7 \times 0.4629 \times 1.25 \times 20 \times 1 \times 1000}{6}$$

$$= 13501 \text{ N}$$

Now, by equation 22.17,

$$N = 13501 \times 0.259 = 3494 \text{ N}$$

Band clutches: The force analysis and design of band clutches can be conducted along the lines of band brakes. These clutches are rather bulky, and are not balanced; they can be used only for low speeds and at present are seldom used.

22.7 EXPANSION CLUTCHES

There exist a great variety of expansion-ring, or slip-ring, clutches. Some have cast-iron rings, and some have heavy steel bands lined with asbestos fabric. They differ chiefly in the mechanism for expanding the ring.

The cast-iron split ring a in Fig. 22.14 is fastened to the hub b that is keyed to the shaft c and is fitted into the outer shell s, which runs loose. When the nosepiece n is pushed toward the flange, the action of a pair of levers l with

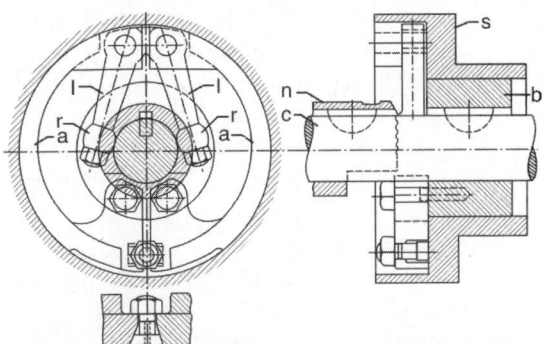

Fig. 22.14: Machine-tool expansion clutch

rollers r on one end expands the half rings and thus engages the shell s to which is fastened a gear or a pulley.

Moment of friction: There are no data available in regard to the actual distribution of the pressure exerted by the ring of an expansion clutch upon the clutch shell. If the average pressure is designated by p, the torsional moment of friction force acting upon the elementary area $\frac{1}{2}bDd\theta$ in Fig. 22.15 is

$$dT = \frac{1}{4}fpbD^2d\theta \quad ...(22.21)$$

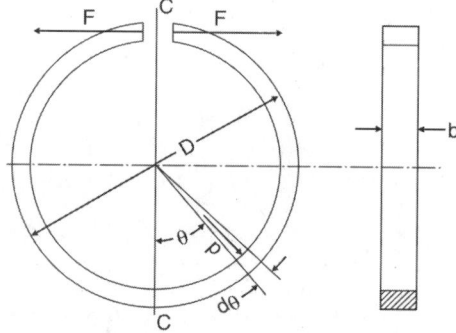

Fig. 22.15: Clutch expansion ring

The angle of contact to the ring may be assumed to be equal to 2π. Assuming the coefficient of friction f to be constant, and integrating equation 22.21, gives

$$T = \frac{1}{2}f\pi pbD^2 \quad ...(22.22)$$

Power capacity: The power P transmitted by the clutch at n revolutions per minute may be found by substituting the value of T from equation 22.22 in equation 2.19 and introducing the load factor k_l. Thus,

$$P = \frac{\pi^2 fpbnD^2}{60 \times 1000\, k_l} \quad ...(22.23)$$

Force to spread the ring: The outside diameter of the split ring is made very little smaller than the inside diameter of the shell. Even with a comparatively heavy ring, the force necessary to spread the ring so as to bring it

into contact with the shell is negligibly small as compared with the force F, Fig. 22.15, which must be exerted by the operating mechanism to produce the necessary pressure p against the shell.

The force acting upon an elementary length $\frac{1}{2}Dd\theta$ of the ring is $\frac{1}{2}pbDd\theta$, and the moment of this force about the section at c, Fig. 22.15, is

$$dM = \frac{1}{4}pbD^2\sin\theta\, d\theta \quad ...(22.24)$$

Integrating from 0 to π gives, for the total bending moment upon the ring,

$$M = \frac{1}{2}pbD^2 \quad ...(22.25)$$

Since this moment must be equal to that due to the force F,

$$FD = \frac{1}{2}pbD^2$$

or

$$F = 0.5pbD \quad ...(22.26)$$

When equations 22.22 and 22.26 are combined, the result is

$$F = \frac{T}{f\pi D} \quad ...(22.27)$$

Design data: The outside diameter of a split cast-iron ring is usually made 0.4 to 0.8 mm smaller than the inner diameter of the shell. If an asbestos-lined ring of flat steel is bent, the difference is made 0.8 mm for a small clutch and up to 6 mm for a large one.

The width b may be determined from equation 22.26 by using for p values given in Table 18.1. The coefficient of friction f may be assumed as 0.125 for cast iron on cast iron and as 0.25 for asbestos-lined rings on cast iron.

22.8 ROLLER CLUTCHES

In Fig. 22.16 is represented a roller clutch which is used as a free-wheeling clutch in power transmission. This clutch is in fact a friction ratchet. The cam a is keyed to the drive shaft and consists of several recesses which form inclined planes. The rollers c, impelled up these inclined planes by friction between them and

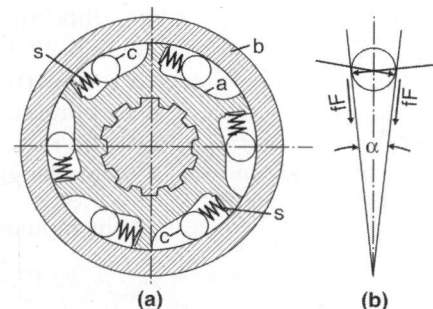

Fig. 22.16: Roller clutch

the shell b, wedge themselves between a and b. This action causes the shell b, which is keyed to the driven shaft, to rotate with the drive shaft. The rollers are held in place by a cage. In some industrial clutches the cage is used to engage and disengage the clutch while the drive shaft is running. *In automobile use, the clutch is engaged automatically when the rotary speed of the driving shaft is greater than that of the driven one, and it is disengaged when the speed relation is reversed (22.10).*

The condition for the operation of the clutch is that the angle α, Fig. 22.16b, between the tangents to the curves of the cam and the shell at points of roller contact, must be smaller than twice the angle of friction ϕ; that is, $\alpha < 2\phi$, where $\tan\phi = f$. In order that the roller will not stick and the clutch may be disengaged readily, the difference between 2ϕ and α must be small.

Design data: The friction coefficient f between hardened and polished steel surfaces may vary from 0.03 to 0.05, its value depending on the material and lubrication. For $\tan\phi = 0.03$, the angle $\phi = 1° 43'$; thus, α must be less than $3° 26'$.

The force F crushing the roller is evidently

$$F = \frac{F_t}{\tan\alpha} \qquad \dots (22.28)$$

where, F_t is tangential force necessary to transmit the torque at a pitch diameter D. This torque is $\frac{1}{2} F_t D$.

When three rollers are used, it is safe to assume that the torque is distributed evenly among all three rollers. In practice a greater

number of rollers are used, and extreme accuracy in workmanship is required in order that all rollers may carry the torque evenly.

The allowable load upon a roller depends on the roller diameter d. For i rollers it may be computed by the equation[1]

$$F \leqq iS_b k'ld \qquad \dots (22.29)$$

where k' is the coefficient of the flattening of the roller, l is the length of the roller, and S_b is the allowable crushing stress. For high-grade hardened chrome-steel, S_b may be taken as high as 1030 N/mm³. Numerically

$$k' = \frac{4.64S_b}{E} \qquad \dots (22.30)$$

The roller diameter d is made from $0.1D$ to $0.25D$.

The outer shell not only must be strong enough to withstand the tendency of the force F to split it but also must be rigid enough not to show any distortion.

Example 22.5: Determine the main dimensions of a roller clutch, Fig. 22.16, to transmit the full torque capacity of a 32 mm shaft.

For steel for which the elastic limit in shear is $S_e = 180$ N/mm² and for a sudden load for which the safety factor is $n = 4$, the torque found from equation 20.1, without considering the size factor, is

$$T = \frac{\pi \times 32^3 \times 45}{16 \times 1000} = 289.5 \text{ Nm}$$

If the cam diameter is taken as $D = 2D_o$, or 64 mm, the tangential force is

$$F_t = \frac{289.5 \times 1000 \times 2}{64} = 9046.8 \text{ N}$$

By equation 22.28, in which the cam angle α is assumed to be $3° 20'$, or $\tan\alpha = 0.0582$, the crushing force F is

$$F = \frac{9046.8}{0.0582} = 155445 \text{ N}$$

The coefficient k' is, by equation 22.30,

$$k' = \frac{4.64 \times 1030}{207 \times 1000} = 0.023$$

The diameter of the rollers may be $0.20D = 0.20 \times 64 = 12.8$ or 13 mm. If the clearance between each pair of rollers is approximately equal to the roller

[1] Hutte, *Des Ingenieurs Tashenbuch*, 26th ed., Vol. II (Berline: Wilhelm Ernst and Son, 1931), p. 131.

diameter, the number of rollers is found from the relation

$$i = \frac{\pi(64+13)}{2\times13} = 9.3 \text{ (use 9)}$$

The roller length, from equation 22.27, is

$$l = \frac{155445}{9\times1030\times0.023\times13} = 56 \text{ mm}$$

22.9 CLUTCH LINKAGES

Clutch-operating linkages are used to obtain a relatively large force F_p necessary to press the friction elements against one another, by applying a moderate external force F_e. The mechanical advantage F_p/F_e is obtained by different combinations of links and levers. However, all these arrangements are based on three simple principles: toggle mechanism, leverage, and inclined plane.

Toggle mechanism: Figure 22.17a shows a typical four-bar linkage with two cranks a and c, one connecting rod b, and a stationary link d. The relation between the produced force F_p and the acting force F_a is

$$F_p = \frac{F_a s \cos\alpha}{t} \qquad \dots (22.31)$$

A modification of the mechanism is obtained by substituting a slide a, Fig. 22.17b, which is the equivalent of an infinitely long crank. The relation between F_p and F_a is again expressed by equation 22.31. If $\beta = 0$, the

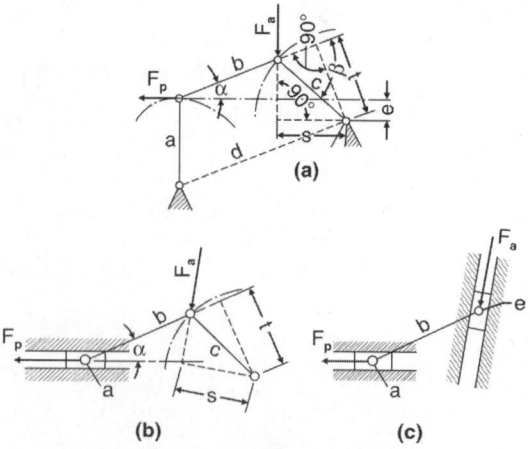

(a)

(b)   **(c)**

Fig. 22.17: Toggle mechanisms

distance t theoretically must be 0, and $F_p = \infty$. Actually, because of friction and elasticity of the links, F_p merely becomes very large. When the link b is pushed further down, the angle β is negative and the force F_p will become slightly smaller than its maximum value. However, the mechanism becomes locked, which is very desirable when a clutch is engaged. Usually $e = 0$, and $\beta = 0$ when $\alpha = 0$.

If both cranks are changed to slides, the arrangement in Fig. 22.17c is obtained. For this case

$$F_p = \frac{F_a}{\tan\alpha} \qquad \dots (22.32)$$

Figure 22.18a shows a combination of two toggle joints with a common slide a and crank b. The mechanical advantage gained by such an arrangement is illustrated in Fig. 22.18b, from which it is easy to obtain the relation

$$F_p = \frac{F_a}{2\tan\alpha \ \tan\beta} \qquad \dots (22.33)$$

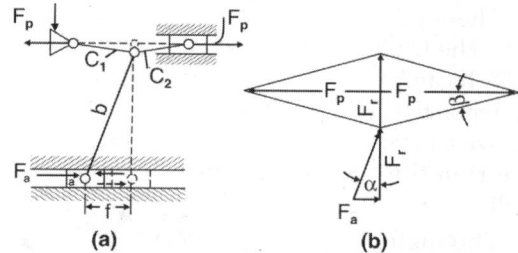

(a)   **(b)**

Fig. 22.18: Toggle mechanism

The toggle joint is a very powerful arrangement which gives a mechanical advantage up to 20, and even more.

Lever mechanisms: Two types of lever mechanisms used in clutch linkages are shown in Figs. 22.19a and b. For both cases the force relation is

$$F_p = F_a \frac{a}{b} \qquad \dots (22.34)$$

In a clutch linkage the ratio a/b varies from 3 to about 5. A bell-crank lever, as shown in Figs 22.2 or 22.4, is equivalent to a straight lever as far as the mechanical advantage is concerned.

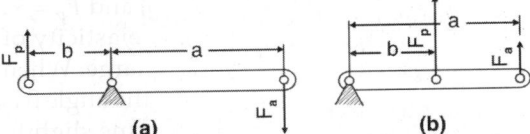

Fig. 22.19: Lever mechanisms

Inclined plane: Actually, a clutch uses a conical surface. The analysis is the same as that given in section 11.7 for a helical thread. With the designations of Fig. 22.20,

$$F_p = \frac{F_a}{\tan(\lambda + \phi)} \qquad \dots (22.35)$$

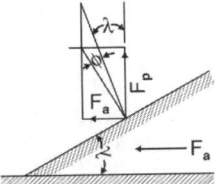

Fig. 22.20: Inclined-plane action

The angle λ usually is made between $15°$ and $20°$. The friction angle ϕ corresponds to about $8.5°$. It can be decreased to about $4°$ by using a roller on the lever, as shown in Fig. 22.2. In such a case λ may be increased up to $25°$ in order to shorten the axial travel of the engagement collar.

The inclined-plane arrangement gives a mechanical advantage of 1.5 to about 2.7.

Combination linkage: In most cases, in order to increase the mechanical advantage, several mechanisms are used in series. The mechanical advantage of such a combination linkage is equal to the product of the mechanical advantages of the individual mechanisms.

Thus the combination in Fig. 22.2 consists of an inclined plane and a lever; that in Fig. 22.3 consists of a toggle joint, a lever, and another toggle joint; and that in Fig. 22.6 consists of an inclined plane, a lever, and a toggle joint. The clutch of Fig. 22.8 uses a double toggle mechanism as illustrated by Fig. 22.18; and the clutch in Fig. 22.11 uses a single toggle joint, as illustrated by Fig. 22.17c.

Example 22.6: An automobile engine has an output of 114 kW at 3600 rpm. The disc of the clutch is to have a mean diameter of 220 mm and a permissible pressure of 0.2 N/mm². If $f = 0.2$ for the asbestos facing, what should be the inner diameter of the disc, considering both the sides of the plate with the friction lining. Also design the springs.

$$\text{Torque to be transmitted} = \frac{114 \times 1000 \times 60}{2\pi \times 3600}$$

$$= 302.4 \text{ Nm}$$

$$T = f\pi d_m b p \, \frac{d_m}{2} n$$

$$b = \frac{302.4 \times 1000 \times 2}{0.2 \times \pi \times 220^2 \times 0.2 \times 2} = 49.7 \text{ or } 50 \text{ mm}$$

So inner diameter of the disc $= 220 - \dfrac{50}{2}$

$$= 195 \text{ mm}$$

Axial force $F = \pi d b p$

$$= \pi \times 220 \times 50 \times 0.2 = 6911.5 \text{ N}$$

Allowing for an overload of 25 per cent for design of springs, the total force $= 1.25 \times 6911.5 = 8639.37$ say 8640 N

Assuming 8 springs, $\text{load/spring} = \dfrac{8640}{8} = 1080 \text{ N}$

Taking $S_s = 414$ N/mm² for the spring wire and mean coil diameter $= 42$ mm, $k = 1.22$ for $c = 6$, the wire diameter d_w

$$d_w = \left[\frac{8 \times 1080 \times 42 \times 1.22}{\pi \times 414}\right]^{1/3} = 6.99 \text{ say 7 mm}$$

Which gives a value of $c = \dfrac{42}{7} = 6$

Which is same as assumed.

Pitch of the spring $= \dfrac{\pi \times 220}{8} = 8.65$. There is sufficient distance between the springs to provide bosses, etc. and will also be sufficiently close to give uniform pressure distribution on the lining.

Assuming 5 coils, the deflection is given by

$$\delta = \frac{8FD^3 n}{Gd_w^4} = \frac{8 \times 1080 \times 42^3 \times 5}{8 \times 10^4 \times 7} = 16.6 \text{ mm}$$

Considering 2 end turns

Free length = solid length + deflection + clearance
$= 7 \times 7 + 16.6 + 0.1 \times 6 \times 7 = 69.8$ say 70 mm.

PROBLEMS

22.1 Which type of clutch transmits heavy torque?

22.2 Which type of clutch is suitable for high rotational speeds?

22.3 Which type is the most compact?

22.4 If the frequency of operation is high, which clutch is preferable?

22.5 Name the properties required for materials of friction surfaces.

22.6 Why are surfaces of metal clutches usually lubricated when the coefficient of friction is already low?

22.7 Which type of clutch is suitable for synchronising device for noiseless engagement of teeth in an automobile gear box?

22.8 Describe the features commonly found in an automobile clutch which would not be commonly found in clutches used with industrial machinery.

22.9 Show that a cone clutch will engage itself in engagement if the tangent of the cone angle is less than that of the coefficient of friction.

22.10 In which situation a roller clutch is used?

22.11 Which type of clutches are suitable for low starting torque?

22.12 Which of the theories (uniform wear and uniform pressure) give a higher value of torque? Is this theory commonly used?

22.13 Does the friction torque depend on the number of collars/discs in a multidisk clutch?

22.14 Determine the main dimensions and the necessary force of the spring for a cone clutch, Fig. 22.1a, to transmit 34 kW at 2,000 rpm, (a) with a leather facing, (b) with an asbestos-fabric facing, and (c) with cast-iron contact surfaces and cork inserts. Use $\alpha = 12\frac{1}{2}°$

22.15 (a) Determine the main dimensions, the necessary axial force at the cone surface, and the effort which must be applied to the shifting collar s, Fig. 22.2, for this clutch to transmit 56 kW at 300 rpm, using cast-iron contact surfaces. (b) Work the problem, using contact surfaces lined with asbestos blocks. (c) Make a sketch, to scale, of the clutch and the engaging mechanisms.

22.16 Using the data of problem 22.15, design a double-cone clutch, Fig. 22.3, applying asbestos-fabric lining.

22.17 Using the data of problem 22.15, design a disk clutch, Fig. 22.4, applying asbestos blocks on the friction surfaces.

22.18 Determine the main dimensions, the necessary axial force at the friction surfaces, the proper lever dimensions of the engaging mechanism, and the effort which must be applied to the shifting collar for a disk clutch similar to Fig. 22.4 to transmit 260 kW at 275 rpm from a diesel engine to a boat propeller.

22.19 Using the data of problem 22.14, design a disk clutch similar to that in Fig. 22.5 of the text.

22.20 Using the data of problem 22.15, design a disk clutch, Fig. 22.6, with asbestos-fabric lining.

22.21 Determine the main dimensions, the necessary axial force at the friction surfaces, the dimensions of the toggle mechanism and the effort that must be applied to shift the collar when engaging a multidisk clutch similar to Fig. 22.8 to transmit 90 kW from an electric motor running at 600 rpm to a line shaft driving a metal working machinery. The outside diameter of disks is 290 mm and inside diameter is 125 mm. Find the number of disks also.

23 Bearings with Sliding Contact

23.1 GENERAL CONSIDERATIONS

A bearing is a machine-part which supports a moving part and confines its motion. That part of a shaft which rotates in a bearing is called a journal and the bush enclosing the journal is called journal bearing. Bearings in which one rubbing surface slides over another are called plain bearings and may be divided into two classes: those with a continuous rotary motion and those with an intermittent motion. To the first class belong journal bearings, which carry a load acting at right angles to the shaft axis, and thrust bearings, which take a load acting in the direction of the shaft axis. To the second class belong bearing of parts having a rocking motion, as wrist pins, or a linear reciprocating motion, as crossheads. A crosshead may be considered a rocking part with an infinitely large radius of the bearing surface.

Bearings with a continuous rotary motion form the great majority of all bearings. They are also the only ones in which an oil film pressure sufficient to support the journal can be created by the journal itself. Bearings with an intermittent motion must depend for proper operation either on an outside source for obtaining the necessary oil pressure or on an abundant oil supply and a low specific bearing pressure.

The failure of a bearing with a sliding contact, or the need of replacement, may be due to excessive wear of the bearing surfaces, overheating, or cracking of the bearing metal (23.11).

Excessive wear: Wear is caused by metal-to-metal contact. Wear cannot be entirely eliminated, but it can be appreciably reduced by providing sufficient bearing area and adequate lubrication.

Overheating is primarily caused by metal-to-metal contact because of an excessive load or improper lubrication. Unless overheating is stopped in time, it may cause either seizing of the journal or melting of the bearing surface. Seizing will occur if the journal runs in a bearing of hard metal, such as a copper alloy. The bearing surface will melt if the bearing is lined with a metal, such as babbitt, which melts at a low temperature. Lubrication decreases the danger of overheating. However, overheating may occur even with proper lubrication if the heat dissipation of the bearing is not adequate.

Cracking: The bearing metal may crack if it is subjected to heavy shock loads, such as are taking place in the running gear of internal-combustion engines when the compressive stresses in the bearing metal exceed its endurance limit. The remedies are to lower the specific bearings pressure or to use a bearing metal with a proportionately higher endurance limit. However, even in this case the presence of an oil film is useful as a shock absorber.

Design basis: From the discussion of causes of bearing failure it is evident that reducing friction by proper lubrication is one of the main problems of bearing design. Proper lubrication means making provision to interpose a liquid film between rubbing surfaces and thus to substitute fluid friction for frictional resistance between the metal surfaces of the journal and its bearings. Factors which must be considered in the design of a bearing are specific pressure, rubbing speed, viscosity of the lubricant, and heat dissipation.

23.2 THEORY OF JOURNAL LUBRICATION

Most bearings supporting rotating machine parts are lubricated with oil. Some are lubricated with heavy greases; others are lubricated with water; and a few can run apparently dry if the surface is impregnated with oil or graphite.

According to the hydrodynamic theory of lubrication of rotating journals with oil, the following three phases of lubrication may be considered: (a) starting of the journal from rest; (b) operating with imperfectly lubricated surfaces; (c) running with perfectly lubricated surfaces.

Starting: If two lubricated surfaces are pressed together by a load, the pressure tends to expel the lubricant between the surfaces. When machinery stands at rest, a large amount of the lubricant will be squeezed out and the metal surfaces will come more or less into contact, as in Fig. 23.1a at point *a*. When the shaft starts to rotate, the friction of the journal against the bearing is high and a certain amount of abrasion will always occur.

Imperfect lubrication: As the journal begins to rotate, Fig. 23.1b, it tends to roll up toward the right side of the bearing, the contact being at some point *a*. Because of molecular attraction,

the wedge-shaped oil film is drawn in between the rubbing surfaces, at first in the shape of a thin film. In *thin-film lubrication*, or *imperfect lubrication*, there exists an unstable condition, and the metal surfaces may therefore touch each other from time to time. Under certain conditions, such as low rubbing speeds or high unit loads, thin-film lubrication may continue indefinitely.

Perfect lubrication: The rotating journal acts as a pump, and the oil pressure increases with an increase of the speed. If the surface velocity of the journal increases above that of imperfect lubrication and there is a sufficient oil supply, enough oil pressure will be developed to raise the shaft completely off the bearing and to make it float upon the lubricant. Oil carried by the pumping action pushes the shaft to the left, and the point of nearest approach to the bearing surface moves to *b* (Fig. 23.1c). The point *c* of maximum pressure is somewhere between point b and the bottom of the bearing.

Where there is no metal-to-metal contact, lubrication is termed *thick-film lubrication*, or *perfect lubrication. Since clearance must exist in order to obtain the eccentric position of the shaft necessary for the pumping action, or converging shape (23.12) part of the oil is discharged at the ends of the bearing, and the oil pressure decreases from the middle of the bearing toward each end.* Figure 23.2 shows the pressure distribution in the plane of maximum pressure for four different loads with the same oil and constant speed.[1] Theoretically each pressure curve should be a parabola. The deviations from parabolas are due to deflection of the shaft, Fig. 23.3a gives the approximate pressure distribution along the length (23.3).

Typical pressure distribution around the circumference at different planes of rotation is shown in Fig. 23.3. The curve marked 0.05L gives the pressure around the journal at a distance of 0.05L from the end of the bearing, where L is the length of the bearing; the curve marked 0.5L represents the pressure

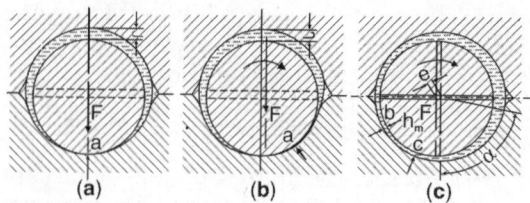

(a)   (b)   (c)

Figs 23.1a to c: Oil film at various phases of rotating of a journal

[1] L. J. Grunder and L. J. Bradford, Oil-Film Pressure in a Complete Bearing, Bulletin No. 39, Pennsylvania State College Engineering Experiment Station (Sept. 8, 1930)

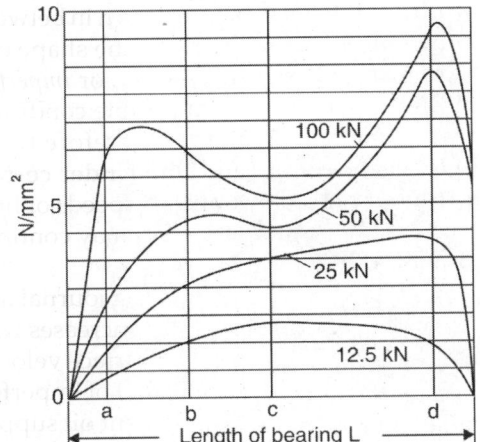

Fig. 23.2: Longitudinal distribution of oil pressure

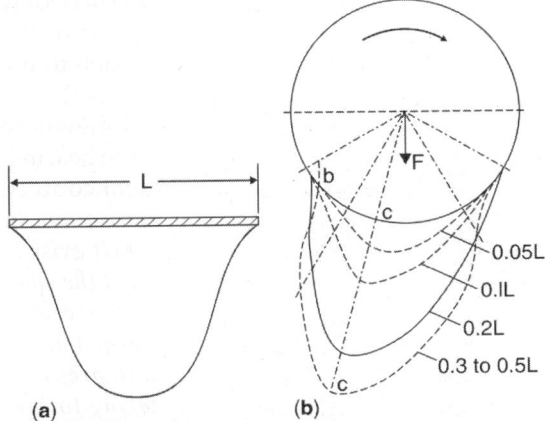

Figs 23.3a and b: Oil-pressure distribution around the circumference and parallel to axis of a bearing

distribution at the center of the bearing. The point of maximum pressure along the bearing lengths are not in the same plane with the maximum pressure at point *c* near the middle of the bearing. The pressure drops to atmospheric or below at some point in the diverging channel after passing point *b* of minimum approach.

Surface conditions: In order to insure perfect lubrication, the bearing surfaces must be true and smooth. Bearings are finished with reamers or broaches, and in large sizes

are hand-scraped. Journals are polished or accurately ground. However, even the most carefully finished bearings and journals are improved by running in.[2] The improvement is particularly noticeable in the region of thin-film lubrication.

Oiliness: Oiliness is a joint property of a lubricant and metal surfaces in contact. It is an important factor for boundary and thin-film conditions, when actual metal-to-metal contact is prevented only by the absorbed oil film. There is no absolute measure of oiliness—values are only comparative. Lard oil has better oiliness than mineral oils. Babbitts favor the establishment of an absorbed film.

Journal bearings are also classified to various types depending on the type of friction as given below:

1. *Hydrodynamically lubricated or perfect or thick film bearings*: In this there is always a thick film of lubricant which separates the two surfaces moving at some relative velocity with respect to each other. The relative velocity between the surfaces generates positive pressure by creating a converging space which supports the load. As there is not direct contact between the two surfaces, the coefficient of friction is due to the viscosity of the lubricant only.

2. *Boundary lubricated or thin film or imperfect bearing*: As it is not always possible to have a thick film bearing because of starting, stopping, variation of speed and sever operation, so boundary lubrication occurs, i.e. the intervening lubricant film does not completely separate the surfaces. Sometimes the lubricant is not sufficient. Such bearings are suitable for low speeds and low loads as there is a metal to metal contact causing wear of surfaces. The coefficient of friction is depending on the surfaces.

3. *Hydrostatic lubricated bearings*: In such bearings, no relative motion is required to

---

[2] S. A. McKee, "The Effect of Running-in on Journal Bearing Performance," *Mechanical Engineering*, Vol. 49 (1927), p. 1335.

cause a thick film of lubricant to separate the surfaces. The two surfaces are separated by supplying lubricant under pressure which supports the load. A hydrostatic bearing consists of a pad with or without a recess and is supplied with oil under pressure from a pump through hydraulic restrictor. Hydrostatic bearings are used where hydrodynamical conditions are not possible and give an extremely low coefficient of friction -0.75×10^{-6}. The lubricant can be liquid, gas or air, in the later two cases the bearings are known as gas bearings or air bearings.

4. *Dry bearings*: Cast iron sintered bearings do not need any lubricant. However, oil or copper is impregnated into the surface which contains graphite in free form. The coefficient of friction is between the two surfaces and is quite high. These are also known as self-lubricated bearings. Certain plastics like nylon also do not need any lubrication.

5. *Oil less bearings*: *When solid lubricants like graphite, molybdenum disulphide are used, the bearings are known as oilless bearings (23.7).*

6. *Magnetic bearings*: When a bearing is separated from contacting the shaft by means of magnetic forces, by suitably locating the magnets, the bearings are known as magnetic bearings. The surfaces do not come into contact.

These bearings are used in a variety of applications, e.g. turbo molecular pumps, compressors, turbo generators, semiconductor equipment, cryogenic conditions and space application. They levitate the shaft by inducing a controlled magnetic field so there is no contact between the shaft and the bearing. The system senses the shaft position and adjusts the force in real time keeping the shaft at the required position.

Some of the advantages are:
(i) No contamination from wear
(ii) Lubrication free
(iii) Operate in severe environments, e.g. extreme high and low temperatures, ultra high vacuum or submerged applications
(iv) Minimum vibration transferred to housing
(v) Built in conditions monitoring of rotor dynamics of vibration and forces
(vi) Precision control and elimination of shaft runout caused by unbalance.

23.3 VISCOSITY

The most important property of the lubricating oil is its internal friction, or viscosity (23.2). Viscosity is expressed numerically by a coefficient which represents the resistance offered by a layer of the liquid of unit area to motion parallel to the area of another layer of the liquid, at unit distance, moving with unit velocity with respect to the first layer. In CGS units this coefficient is known as the absolute viscosity, and it is equal to the force in dynes per sq. cm at a velocity of 1 cm per sec at a distance of 1 cm. The unit of absolute viscosity is termed a *poise*. For practical purposes the unit is too big, so one-hundredth of it, the *centipoise*, is commonly used. It so happens that the viscosity of water at 20°C is exactly one centipoise. In English units viscosity is measured in pounds seconds per sq. in. of film thickness, and the unit is called a *reyn*.

Newton's law of viscous flow states that the shear stress in the fluid is proportional to the rate of change of velocity with respect to y, i.e.

$$S_s = \frac{F}{A} = Z\frac{du}{dy} \qquad \text{... (23.1)}$$

where Z is the absolute or dynamic viscosity of oil. The units of viscosity from the above equation is Ns/m² or kg/ms or Pa.s. Considering rate of shear is constant, i.e.

$$S_s = Z\frac{u}{y} \qquad \text{... (23.2)}$$

Viscosity is measured by an instrument called the Saybolt Universal Viscometer. The method consists of measuring the time in second for 60 ml of lubricant at a specific temperature to run through a tube 17.6 mm in diameter and 12.25 mm long. The viscosity determined is called the kinematic viscosity Z_k, and its unit square centimeter per second which is defined as centistoke. The kinematic viscosity based upon seconds Saybolt, also called Saybolt Universal viscosity S (SUA) in seconds is

$$Z_k = \left(0.22S - \frac{180}{S}\right) \qquad \dots (23.3)$$

In S.I. units, the Kinematic Viscosity has the unit of square meter/second (m²/s). Thus

$$Z_k = \left(0.22S - \frac{180}{S}\right) \times 10^{-6}$$

To convert to absolute viscosity, v is multiplied by the specific gravity ρ

$$Z = \rho\left(0.22S - \frac{180}{S}\right) \times 10^{-6} \quad \dots (23.4)$$

where, Z is in Ns/m² or kg/ms or Pascal-seconds (Pas).

Values of specific gravity of several representative minerals are given in Table 23.1 at 15°C compared to that of water of 15°C. In order to find the specific gravity ρ_t at that temperature, the conversion may be made using the equation,

$$\rho_t = \rho_{15} - 0.000657\, t \text{ or}$$

Table 23.1: Specific gravity of oils at 15°C

No.	Oil characteristic	ρ
A	Turbine oil, ring-oiled bearing	0.8877
B	Turbine oil, ring-oiled bearing, SAE 10	0.8894
C	All-year automobile oil, SAE 20	0.9036
D	Ring-oiled bearing oil, high-speed machinery	0.9346
E	Automobile oil, SAE 20	0.9254
F	Automobile oil, SAE 30	0.9263
G	Automobile oil, SAE 40, medium-speed machinery	0.9275
H	Airplane oil 100, SAE 60	0.8927
I	Transmission oil, SAE 110, spur and bevel gears	0.9328
J	Gear oil, slow-speed work gears	0.9153
K	Transmission oil, SAE 160, slow-speed gears	0.9365

$$\rho_t = \rho_{15} - 0.000657\,(t - 15) \quad \dots (23.5)$$

ρ_{15} is given in Table 23.1.

23.4 BEARING CHARACTERISTIC NUMBER

As shown in section 23.2, the oil pressure between a bearing and rotating journal varies both axially and circumferentially, the variation following a parabolic curve more or less. The design of bearings is materially simplified if the variation in pressure is taken into account by introducing the concept of an average pressure p. This average pressure is found by dividing the bearing load F by the projected area of the bearing, which area is the product of its length and its diameter. Thus,

$$p = \frac{F}{ld} \qquad \dots (23.6)$$

Considering a vertical shaft supported in a bearing which supports a load F. The Newton's law of viscosity states that the shearing stress is proportional to the rate of change of velocity (Fig. 23.4).

$$S_s \alpha \frac{du}{dy}$$

As the clearance c is considered uniform.

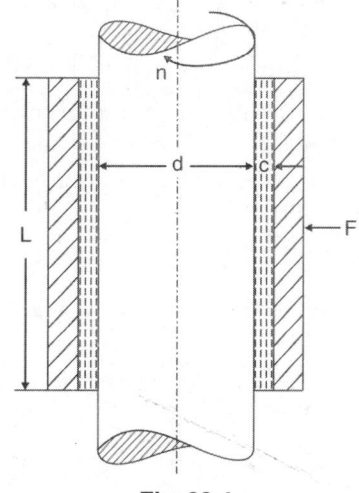

Fig. 23.4

$$S_s = Z\frac{u}{c} \qquad \ldots (23.7)$$

where, Z is the absolute oil viscosity, at the temperature of the oil in the bearing, in kg/ms or Ns/m^2.

u is the peripheral velocity of the journal, in m/s

c is the radial clearance between the journal and the bearing, in m

If the peripheral velocity u is replaced by its value in terms of the journal speed n revolutions per minute, i.e. $u = \dfrac{\pi dn}{60}$

Shearing force, $F_s = S_s \times \text{Area} = S_s \times \pi dl$

Shearing torque, $F_s \times \dfrac{d}{2}$

$$= \frac{Z\pi dn \times \pi dld}{60 \times c \times 2} \qquad \ldots (23.8)$$

This is equated to the frictional torque with equivalent coefficient of friction f, i.e. $fFr = fpld\,\dfrac{d}{2}$

$$f = \frac{\pi^2}{60}\left(\frac{Zn}{p}\right)\left(\frac{d}{c}\right) \qquad \ldots (23.9)$$

This is known as Petroff's equation.

$$s = \left(\frac{Zn}{p}\right)\left(\frac{d}{c}\right)^2 \text{ is called Sommerfeld number.}$$

The value of $\dfrac{Zn}{p}$ is very helpful in designing

bearings of the relative clearance $\dfrac{c}{d}$ or the clearance per meter of diameter, and of the bearing construction in general. The minimum

value of $\dfrac{Zn}{p}$ is called bearing modulus or bearing

characteristic.

23.5 FRICTION

A very important factor in the design of bearings is the coefficient of friction f. It has

been established that f is a function of the bearing characteristic number Zn/p and the relative clearance c/d. If a given bearing is operated with oils of different viscosities under various conditions of speed and load, and the values of the observed friction coefficients are plotted against the corresponding values of Zn/p, a curve such as $abcd$ in Fig. 23.5 is obtained. The portion ab corresponds to thick-film lubrication. The portion cd corresponds to thin film condition. The segment bc represents a transition conditions.[3] The actual value of f depends on many other factors, such as the materials of the bearing and journal, the conditions of the surfaces, and the oil viscosity, which in turn depends on the temperature of the surfaces. The curve in Fig. 23.5 indicates neither a thin-film conditions cd nor the transition states bc is stable. Therefore, it is advisable to have the value of Zn/p, for a bearing, far enough above the value corresponding to the minimum value of f. On the other hand, it is desirable to have the value of Zn/p as low as possible in the thick-film region in order to have a low friction coefficient and a small loss of power, usually the values of ZN/p should be 3 times the value corresponding to the minimum value. It is called the bearing characteristic number and is a function of relative clearance.

Neglecting oil leakage, Petroff's equation gives the coefficient of journal friction based on

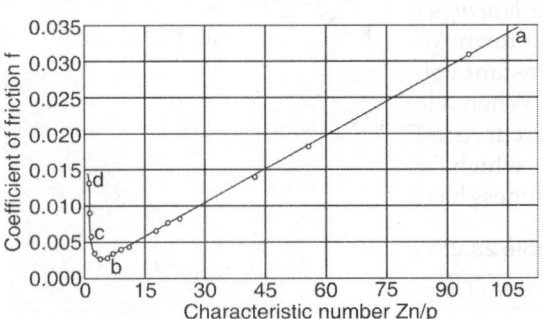

Fig. 23.5: Journal friction for a bearing with $l/d = 1.0$ and $c/d = 0.001$.

---

[3] S. A. McKee and T. R. McKee, "Friction of Journal Bearings as influenced by Clearance and Length," *Transactions of the American Society of Mechanical Engineers*, Vol. 51 (1929). APM–51–51, p. 164.

a journal concentric with the bearing as, with c as diametrical clearance

$$f = \frac{\pi^2}{60}\left(\frac{Zn}{p}\right)\frac{d}{c}$$

Experiments have shown that the coefficient of friction decreases with an increase of the relative clearance c/d. By using data obtained from these tests and plotting the friction coefficient f against the ratio of Zn/p and also against c/d, it was established that the relation is represented by an inclined straight line.

The equation of this line may be taken as (McKee-equation).

$$f = 0.326\left(\frac{Zn}{p}\right)\left(\frac{d}{c}\right) + \Delta f \quad \dots (23.10)$$

The first term of the second member of this equation is the expression obtained by the hydrodynamic theory. The second term is a correction which takes into account the effect of side leakage and eccentricity and the ratio l/d. Therefore, although the tests were conducted only with small-bore full-journal bearings having perfect lubrication, it seems justified for practical purposes to use this equation wherever a bearing operates with thick film the characteristic number Zn/p specified in the Table 23.4 and $f > 0.001$.

The term Δf in equation 23.10 takes into account the effect of end leakage of lubricant. *For bearings for which l/d ranges from 0.75 to 2.8,* the term Δf may be considered to have the constant value 0.002.[4]

When a journal and its bearing are run in, the curve in Fig. 23.5 is altered so that the point for which f is a minimum is moved to the left. Oiliness has the effect of making the hook in the curve flatter and of moving the portion cd further to the left. A practical value of $\dfrac{d}{c}$ is 1000.

Work of friction: The work of friction W_f is equal to the product of the frictional force and the distance through which the force moves. The frictional force may be taken as the load on the bearing times the coefficient of friction, or $fpld$. The distance through which force moves in a unit time is the velocity of a point on the surface of the journal, or $\dfrac{\pi dn}{1000 \times 60}$ m/s. Thus,

$$W_f = fFv = \frac{fpld \times \pi dn}{60 \times 1000} = \frac{fF\pi dn}{60,000} \quad \dots (23.11)$$

Considering d is in mm.

23.6 IMPERFECT LUBRICATION

When the characteristic number Zn/p is too low, or when the supply of lubricant to the bearing is insufficient to maintain fluid-film lubrication, thin-film lubrication and metal-to-metal contact will exist. This condition is represented by the part of the curve in Fig. 23.5 from b to c or to d. The coefficient of friction in this case may be estimated by the following formula:[5]

$$f = 0.004 C_1 C_2 \sqrt[4]{\frac{p}{v_m}} \quad \dots (23.12)$$

where, C_1 and C_2 are factors; p is the pressure on the projected area, in N/mm^2 and v_m is the rubbing velocity, in m/s. Values of C_1 are given in Table 23.2, and values of C_2 are given in Table 23.3.

If the load varies, the average pressure is used for p, but it must not be less than one-half the maximum pressure.

Table 23.2: Values of factor C_1 in equation 23.12

Method of lubrication	Workmanship	Attendance	Operating conditions	C_1
Bath; flooded; oil ring	Very good	Good	Clean and dust-free	1
Oil, by constant drop feed	Good	Satisfactory	Ordinary conditions	2
Oil, by intermittent feed; grease cup	Fair	Poor	Exposed to dirt	4

[4] Louis Illmer, "High-Pressure Bearing Research," *Trans.* ASME, Vol. 46 (1924), p. 833.
[5] *Ibid.*

Table 23.3: Values of factor C_2 in equation 23.12

Types of examples of bearings	C_2
Rotating journals in rigid bearings and crankpins	1
Oscillating journals, such as rigid wrist pins and block pintles	1
Rotating journals, not very rigid, such as eccentrics	2
Rotating surfaces lubricated from center, such as annular step and pivot bearings	2
Reciprocating crosshead shoes wiping over ends of long guides	2
Reciprocating crosshead shoes wiping over ends of short guides	3
Wiping surfaces, such as marine thrust bearings and worm gears	3–4
Sliding parts, such as long nuts for power screws	4–6

A journal in sintered porous bearings is an example of thin-film lubrication. At low and moderate speeds and low loading, these bearings are quite satisfactory. The bearing clearance c in such bearings should be slightly greater than that in the bearings with thick film lubrication. *Bearings with thin-film lubrication are often called oilless bearings, or self-lubricating bearings (23.7).*

23.7 FACTORS IN BEARING DESIGN

Three factors or ratios that must be considered carefully in designing a bearings are the following: (a) the bearing pressure, or the ratio of the load to the projected bearing area: (b) the ratio of the bearing length to its diameter; and (c) the relative clearance, or the ratio of the clearance to the bearing diameter.

Bearing pressure: Experience shows that the pressure p is important in bearing design not only because it affects the characteristic member Zn/p but also because there are limits for p which should not be exceeded if thick-film lubrication is to be maintained. These limits values are given in Table 23.4 for various types of machinery and bearings. In some instances, as for automobile engines, locomotives, and

punching and shearing machinery, there is not enough room for larger bearing surfaces, and the pressure limits used in practice are higher than those corresponding to thick-film lubrication. These high pressures result in thin-film lubrication. However, they are comparatively safe, as far as friction and wear are concerned.

In the design of motor and generator bearings, the General Electric Company has successfully used the equation[6]

$$p = 0.622\sqrt[3]{v_m} \qquad \dots (23.13)$$

This equation corresponds to the use of a safety factor of about 2.

Finally, Table 23.5 gives safe bearing pressures for crosshead and trunk pistons of various machines as found in actual operation.

Bearing length: Experiments have shown that for a given characteristic number Zn/p, and for values of the ratio of the bearing length l to its diameter d greater than 0.75, changes in l/d do not materially affect the coefficient of friction. However, the bearing length may affect friction indirectly. If the bearing is long, the deflection of the shaft may result in points of excessively high bearing pressures, which will cause a breakdown of the oil film and therefore cause metal-to-metal contact. This undesirable condition may be remedied by the use of self-aligning bearing with spherical seats. Although an end bearing a, Fig. 23.6, may adjust itself to the deflection, a spherical seat in an inner bearing b may be of little advantage. Therefore, increasing the ratio l/d in order to lower the pressure with a given diameter d is advisable only with a rigid shaft.

Commonly used values of the ratio l/d are given in the last column of Table 23.4. It should be noted that this ratio is governed chiefly by

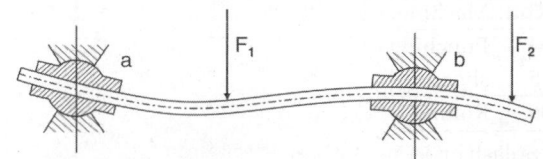

Fig. 23.6: Self-aligning bearings

[6] D. S. Kimball and J. H Barr, Elements of Machine Design, 3rd ed. (New York: John Wiley & Sons, Inc., 1935), p. 139.

Table 23.4: Design data for bearings

No.	Machinery	Bearing	Maximum p MN/mm^2	Suitable $Z \times 10^{-3}$ NS/m^2 kg/ms	Minimum $\dfrac{Zn}{p} \times 10^6$	$\dfrac{c}{d}$	$\dfrac{l}{d}$
1	Automobile	Main	5–12å	7–8	2.2	–	0.8–1.3
2	and aircraft	Crankpin	10–24å		1.5	–	0.7–1.4
3	engines	Wrist pin	16–35å		1.2	–	1.5–2.2
4	Gas and oil	Main	52–8å	20–65	3	0.001	0.6–2.0
5	engines,	Crankpin	9–12å		1.5	<0.001	0.6–1.5
6	four-stroke	Wrist pin	12–15å		0.7	<0.001	1.5–2.0
7	Gas and oil	Main	42–6å	20–65	4	0.001	0.6–2.0
8	engines	Crankpin	7–10å		1.8	<0.001	0.6–1.5
9	two-stroke	Wrist pin	8–12å		1.5	<0.001	1.5–2.0
10	Marine	Main	3.5	30	3	<0.001	0.7–1.5
11	steam	Crankpin	4	40	2	<0.001	0.7–1.2
12	engines	Wrist pin	10	30	1.5	<0.001	1.2–1.7
13	Stationary	Main	1.5–3	60	3	<0.001	1.0–2.0
14	slow-speed	Crankpin	4–1.0	80	2.2	<0.001	0.9–1.3
15	steam engines	Wrist pin	12	60	1.5	<0.001	
16	Stationary	Main	2	15	3.7	<0.001	1.5–3.0
17	high-speed	Crankpin	4	30	0.9	<0.001	0.9–1.5
18	steam engines	Wrist pin	12	25	0.8	<0.001	
19	Reciprocating	Main	2ç	30–80	4.4	0.001	1.0–2.2
20	Pumps and	Crankpin	4ç	30–80	0.3	<0.001	0.9–1.7
21	compressor	Wrist pin	7ç*	30–80	1.5	<0.001	1.5–2.0
22		Driving axle	4	100	4.4	0.001	1.6–1.8
23	Steam	Crankpin	14	40	0.8	<0.001	0.7–1.1
24	locomotives	Wrist pin	28	30	0.8	<0.001	0.8–1.3
25	Railway cars	Axle	3	10	7.3	0.001	1.0–2.0
26	Steam turbines	Main	0.5–1.2ç	2–16	15	0.001	1.0–2.0
27	Generators, motors, centrifugal pumps	Rotor	0.5–1.5ç	25	30	0.0013	1.0–2.0
28	Gyroscope	.Rotor	6	30	8	0.0013	–
29		Light, fixed	0.2ç	25–60	145	0.001	2.0–3.0
30	Transmission	Self-aligning	1.0ç	25–60	4.4	0.001	2.5–4.0
31	Shafting	Heavy	1.0ç	25–60	4.4	0.001	2.0–3.0
32	Cotton mill	Spindle	0.007	2	1450	0.005	–
33	Machine tools	Main	2	40	5.8	0.001	1.0–4.0
34	Punching and	Main	28ç	100	–	0.001	1.0–2.0
35	shearing machines	Crankpin	56ç	100	–	0.001	1.0–2.0
36	Rolling mills	Main	21	30	1.5	0.0015	1.1–1.5

* Splash or scraper lubrications.

å Force-feed lubrication.

ç Ring or drop oiler.

space limitations, rather than by some other factor.

Clearance: The number in the seventh column in Table 23.4 indicates the class of running clearance recommended for each bearing. The corresponding value of c may be computed by using data of Table 23.4.

There is no definite relation between the relative clearance c/d and the shaft diameter d. Some engineers believe that the relative clearance can be gradually decreased with an increase of the journal diameter, in order to decrease oil leakage. Curves 1 and 2 in Fig. 23.7 show clearances and corresponding tolerances as used for journals of different machines with babbitted bearings in one plant of the Westinghouse Electric Corporation. In another plant of the same concern the relation $c/d = 0.002$ is used for the maximum clearance for all shaft sizes. Another manufacturer also uses minimum clearances corresponding to curve 1 of Fig. 23.7, but follows closer tolerances that result in curve 3 for the maximum clearances. Still another manufacturer even recommends that the relative clearance be increased slightly with an increase in shaft size. Usual values of d/c are 750–1000.

Table 23.6 gives running clearances as used in various industrial applications by the majority of manufacturers, for five typical shaft sizes. This table shows the large tolerances used in practice. For other diameters the clearances may be obtained by interpolation.

23.8 HEAT DISSIPATION

The work of friction in a bearing is transformed into heat which raises the temperature of the oil and of the journal and bearing. This heat must

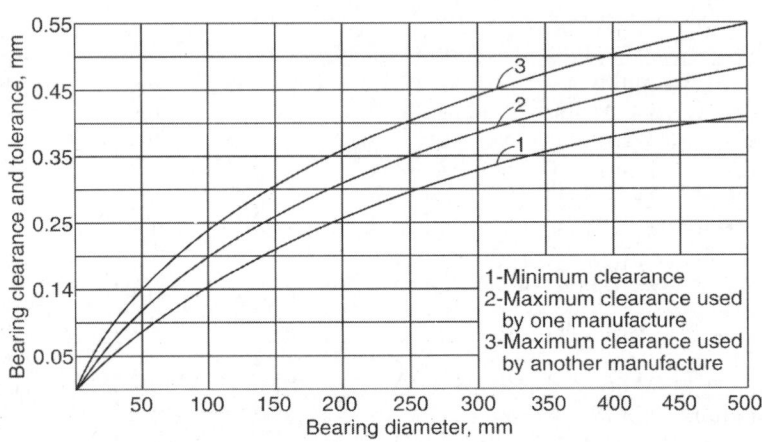

Fig. 23.7: Clearances used in babbitted bearings

Table 23.5: Allowable bearing, pressure, reciprocating motion

Type of bearing	Type of machinery	Pressure p N/mm²
Crosshead	Steam engine, stationary	2.5–4.2
	Steam engine, marine	4.0–7.0
	Steam engine, locomotive	5.0–6.0
Trunk piston	Gas and oil engines, stationary	3.0–5.0
	Compressor and pumps	4.5–6.0
	Gas and oil engines, stationary	1.5–2.0
	Automotive and aircraft engines	1.5–2.0

Table 23.6: Bearing clearances in industrial applications

Types of service, material, and finish of journal and bearing	Running clerance C, in mm × 10⁻³ for shaft diameter d mm				
	$d = 12$	$d = 25$	$d = 50$	$d = 90$	$d = 140$
Precision spindle, hardened and ground steel, lapped-in bronze bearing, rubbing speed under 2.5 m/s, pressure $p < 3.5$ N/mm	7–19	19–38	38–63	63–88	88–125
Precision spindle, hardened and ground steel, lapped-in bronze bearing, rubbing speed over 2.5 m/s, pressure $p > 3.5$ N/mm	13–25	25–50	50–75	75–113	113–163
Electric motors and generators, ground journals in broached or reamed bronze bearing or reamed babbit bearings	13–38	25–50	38–85	50–100	75–150
General machinery, continuous rotating or oscillating motion, turned or cold-rolled steel journals in bored and reamed bronze bearing or poured and reamed babbit bearing	50–100	63–113	75–125	100–175	125–200
Rough machinery, turned or cold-rolled steel journals in poured babbit bearing	75–150	125–200	200–300	300–400	400–500

be dissipated in order to prevent an excessive temperature rise. The amount of heat that a bearing will dissipate depends on the difference in temperature of the bearing and the surrounding air, the area of the exposed surface of the bearing and its pedestal, and the rate of movement of air or other gases over the bearing, or of water if cooling by water circulation is used. When a journal begins to rotate, the heat generated is greater than the heat dissipated, and the temperature of the bearing rises until a steady state is reached. The temperature of a bearing should be kept below 65°C if possible, and it should never exceed 82°C. If the equilibrium temperature exceeds the safe limit, artificial cooling must be resorted to. It is possible to pass a stream of air across the bearing or to provide coils circulating cooling water in its body.

The heat-dissipating capacity Q of a bearing, in watts, may be computed by the general expression

$$Q = hA\,(t_b - t_a) \qquad ...(23.14)$$

where, h is the film coefficient, in watts per square meter degree centigrade

A is the exposed area of bearing housing = 20 dl of bearing area,

t_b is the temperature of the exposed surface of bearing in °C;

t_a is the temperature of the surrounding air, in °C;

The film coefficient can be computed by the relation[7]

$$h = mv^{0.89} \qquad ...(23.15)$$

where the constant m may be taken for the selected units of h as 14.82 and v is the air velocity in m/s. The air velocity v may be taken as 0.75 m/s for normal conditions without ventilation, and about 2.5 m/s for a well-ventilated bearing. These values of h are in agreement with data obtained from special bearing tests.[8] However, the following values of h are recommended.

h = 11.4 W/m² °C for still air;

= 15.3 W/m² °C for average design conditions;

= 33.5 W/m² °C for air moving at 2.5 m/s

Equation 23.14 may be used only when approximate values are required.

[7] V. L. Maleev, *Internal Combustion Engines*, 2nd ed. (New York: McGraw-Hill Book Company, Inc., 1995), p. 376.

[8] G. B. Karelitz, "Performance of Oil Ring bearing," *Trans. ASME*, Vol. 52 (1930), APM–52–5, pp. 57–70.

Bearing temperature: The temperature of the bearing is lower than t_o the oil temperature, because of heat dissipation. The difference between the bearing-wall temperature t_b and the ambient temperature t_a may be taken from the chart of Fig. 23.8, which gives curves for the three main types of lubrication—by oil bath, by an oil ring, and by waste pack or drop feed. The chart gives two limit conditions for the surrounding air. Data for intermediate conditions may be estimated by interpolation.

Exposed area: In computing the exposed area A of a bearing, it may be assumed that this area includes all parts of the bearing which lie within three diameters of the shaft axis. In a ring-oil bearing the exposed area is the part within 100 mm of the bottom of the oil reservoir.

Use of projected journal area: Another method of computing the heat-dissipating capacity of a bearing is to apply the relation

$$Q = cld \qquad \dots (23.16)$$

where, ld is the projected journal area and c is a coefficient found by using the chart[9] in Fig. 23.9. To determine c, the expected temperature rise $(t_b - t_a)$ is selected from Fig. 23.8 and is taken as the ordinate in Fig. 23.9; and the corresponding abscissa for the proper curve is the desired value of c.

Instead of using Fig. 23.9, the coefficient c may be computed from the relation[10]

$$c = \frac{(t_b - t_a + 18)^2}{k}$$

i.e. $$Q = \frac{(t_b - t_a + 18)^2}{k} ld \qquad \dots (23.17)$$

k = 0.273 for bearing of heavy construction well ventilated °C m^2/W.

k = 0.484 for bearing of light or medium construction in still air °C m^2/W.

t_o = operating temperature.

$$t_b - t_a = \frac{1}{2}(t_o - t_a)$$

23.9 OIL-FILM THICKNESS

The smallest oil-film thickness h_m, Fig. 23.1c, when the journal is turning, can be expressed in terms of the clearance c, Fig. 23.1a, and the

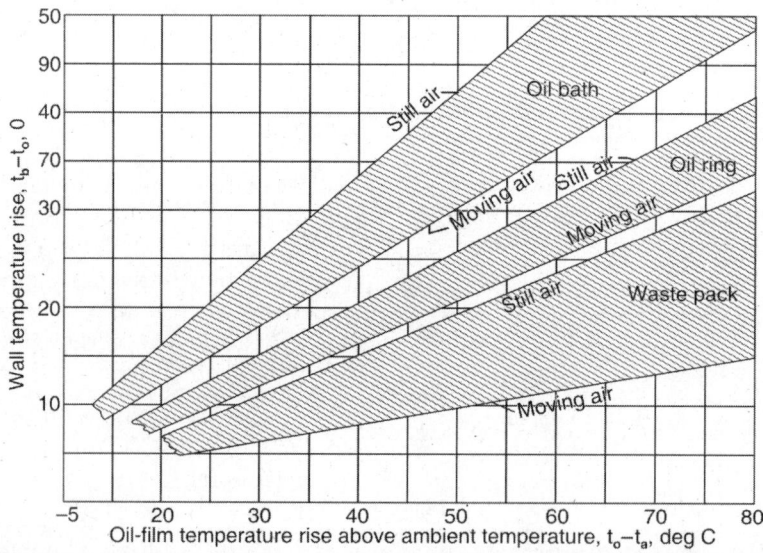

Fig. 23.8: Relation between oil-film temperature and bearing-wall temperature

[9] O. Lasche and W. Kieser, *Materials and Design in Turbo-Generation Plant* (London: Oliver and Body, 1927).
[10] Axel K. pederson, "Charts for Journal Bearing." *American machinist*, Vol. 37 (October 10, 1912), p. 599.

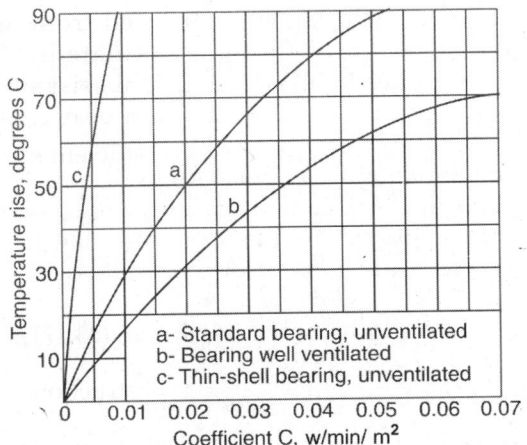

Fig. 23.9: Heat dissipation by bearing

eccentricity e, Fig. 2.1c, which is the distance between the center of the journal and the center of the bearing. From the diagram, $h_m = 0.5c - e$, or

$$h_m = 0.5c\left(1 - \frac{2e}{c}\right) \qquad ...(23.18)$$

The magnitude of *e* is function of the pressure *p* on the projected actual bearing area, the relative clearance *c/d*, the oil viscosity *Z*, the rotative speed *n* of the journal, and the angle α, Fig. 23.1c, between the edge of the bearing relief where the oil film starts and the direction of the load. The ratio *e/c*, which is termed the *eccentricity coefficient*, can be found by means of the chart in Fig. 23.10, in which the abscissas are values of a quantity designated as *K*. The latter, determined from theoretical considerations and corrected by experimental data, is computed by the relation[11]

$$K = \frac{28.7\left(1000\dfrac{c}{d}\right)^2}{Zn/p} \qquad ...(23.19)$$

Comparison of equations 23.19 and 23.9 shows that $K = 28.7 \times 10^6 / k'$. When the quantity *K* has been computed by equation 23.19 for a given bearing with a certain angle α, Fig. 23.1c, and a certain oil viscosity *Z*, the ordinate of the corresponding curve in Fig. 23.10 gives the eccentricity coefficient *e/c* directly. Equation 23.18 can then be applied to find the expected smallest oil-film thickness h_m for this bearing.

As speed increases the value of *K* decreases so *e* decreases causing oil film thickness to decrease.

The minimum thickness h_m' that must be maintained for perfect lubrication depends on the surface finishes of the journal and the bearing. For a finely bored small bronze bushing and a polished journal, this thickness is of the order of 0.0025 mm. In the case of a steel journal in an ordinary babbitted bearing, the thickness h_m' should be at least 0.02 mm if *n* is high and should not be less than one-half this value for lower speeds.[12] With large steel shafts of turbogenerators, fans, and the like, h_m should be from 0.007 to 0.012 mm. For journals of diesel engines running with peripheral speeds of 2.5 to 6 m/s, a thickness, h_m of 0.001 to 0.0015 mm is recommended for 120 mm to 290 mm bearing.[13]

In general the safe thickness for a bearing in good condition and $v_m \geq 1$ m/s is[14]

$$h_m' = 0.018\, v_m^{0.4} A^{0.2} \qquad ...(23.20)$$

where, v_m is the peripheral, or rubbing velocity of the journal, in m/s and *A* is the projected area *ld* of the bearing, in m².

Oil supply: The approximate amount of oil *G*, in litres per minute, that must be supplied to a bearing to make up for leakage and maintain thick-film lubrication may be determined by the formula[15]

$$G = 785\, cldn \qquad ...(23.21)$$

[11] Karelitz, *loc. cit.*, p. 59.

[12] *Ibid.*

[13] E. S. Dennison, "Film-Lubrication Theory and Engine-Bearing Design," *Trans. ASME*, Vol. 58 (1936), p. 25.

[14] Albert Kingsbury, "Optimum Conditions in Journal Bearing," *Trans. ASME*, Vol. 54 (1932), pp. 123 ff. These data were coordinated with other data available in the literature.

[15] S. J. Needs, "Effects of Side Leakage in 120° Centrally Supported Journal Bearing," *Trans. ASME*, Vol. 56 (1934), p. 721, and Vol. 57 (1935), p. 135.

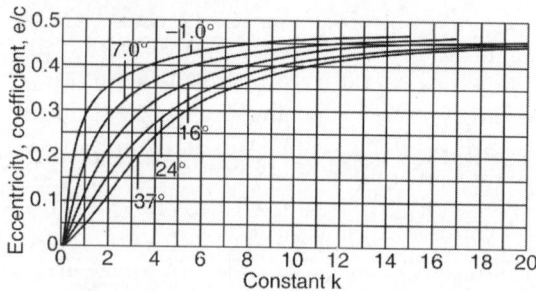

Fig. 23.10: Chart for eccentricity coefficient

where, n is the journal speed, in revolutions per minute, and $c, l,$ and d are bearing dimensions, in m.

23.10 METHODS OF LUBRICATING BEARINGS

Bearings may be lubricated intermittently, continuously with a limited supply of lubricant, or continuously with an abundant amount of lubricant.

(i) Intermittent lubrication: The lubricant for intermittent lubrication may be oil or grease. Oil may be applied by dropping it from an oil can into an oil hole in the cap of the bearing or by dropping it from an oil can into a felt plug or a wad of cotton waste covering the oil hole. Grease may be applied by forcing it from a compression grease cup screwed into the bearing cap or by forcing it into a hollow

space in the bearing by means of a pressure gun.

Any of these methods, generally speaking, provides only thin-film lubrication. The coefficient of friction is variable and uncertain. To be on the safe side, it may be assumed to be 0.12 to 0.15.

In designing bearings that must work under conditions of thin-film lubrication the allowable mean pressure, $p = F/ld$, may be taken from Table 23.7.

(ii) Limited continuous lubrication: Continuous lubrication with a limited supply of lubricant may be obtained by use of a grease cup with spring action; by use of an oil reservoir with a wick which carries the oil by capillarity, as in Fig. 23.11; or by use of a sight-feed drop oil cup, as in Fig. 23.12.

Both methods are suited to light duty only. Only the third method may approach thick-film lubrication, and the computed value of the coefficient of friction should be used very cautiously.

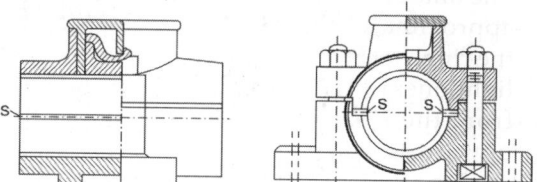

Fig. 23.11: Two-piece bearing with wick lubrication

Table 23.7: Allowable bearing pressures for semifluid lubrication

Journal material	Bearing material	Allowable pressure p N/mm²
Hardened tool steel	Lumen of phosphor bronze	17.5
Hardened alloy steel	Hardened steel	14.5
SAE 1050 steel 50 C4	Hard babbitt	10.3
Hardened alloy steel	Bronze	9.0
Cast iron	Cast iron	7.5
Alloy steel	Bronze	6.0
SAE 1040 steel 40 C8	Babbitt, soft	5.0
Mild steel, smooth finish 20 C8	Bronze	4.0
Mild steel, ordinary finish	Bronze	3.0
Cast iron	Bronze	3.0
Mild steel	Cast iron	2.5
Mild steel	Lignum vitae, water-lubricated	2.5

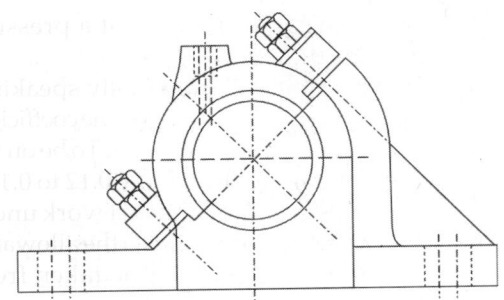

Fig. 23.12: Angle bearing

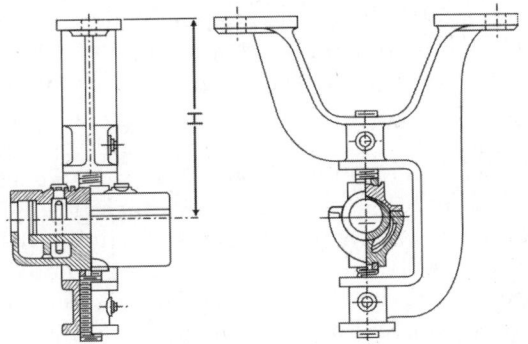

Fig. 23.14: Sellers oil-ring bearing in a hanger

(iii) Abundant lubrication: Only continuous lubrication with an abundant supply of lubricant insures thick-film lubrication. It may be obtained by means of ring, chain, or collar oiling; splash lubrication; bath lubrication; or flooded lubrication, under pressure or without it.

Ample quantities of oil at low and medium speeds up to a peripheral journal speed of 600 to 750 mpm are furnished by *ring oiling*. The amount of oil delivered to the journal is approximately proportional to the width of the ring, heavy rings delivering more oil than light ones. At high speeds the oil is thrown from the ring by centrifugal force on the upward journey, and special grooves must be used to collect the oil and return it to the journal. Typical bearings with ring oiling are shown in Figs 23.13 and 23.14.

A variation of the idea of the ring oiler, with a chain substituted for the ring, is *chain oiling*.

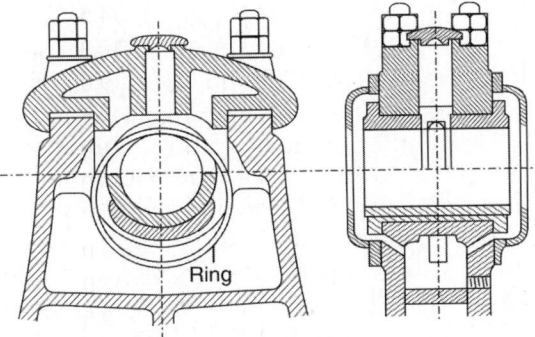

Fig. 23.13: Oil-ring bearing

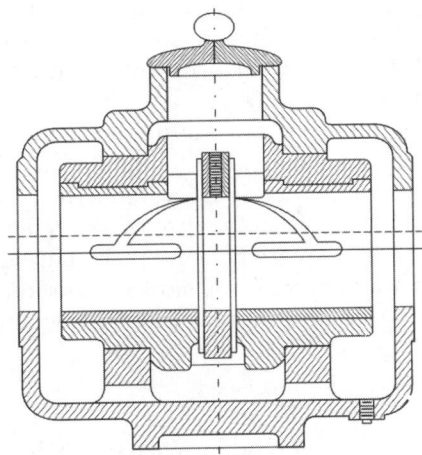

Fig. 23.15: Rigid collar-oiled bearing

With *collar oiling* (a bearing with such oiling is shown in Fig. 23.15), a split collar that is clamped to the shaft dips into the oil reservoir and carries oil to the top of the bearing, where the oil is wiped off and led to the journal. The disadvantage of the collar is that it divides the bearing into two parts, making the bearing equivalent to two bearings as far as end leakage and longitudinal pressure distribution are concerned.

The method known as *splash lubrication* may be used in a fully enclosed mechanism. A crank or a similar part dips into an oil reservoir at each revolution.

In *bath lubrication* the journal is partly submerged in an oil reservoir. It is particularly suitable for a bearing that carries the load on the top half, because it admits oil at the point of minimum pressure.

In *flooded lubrication* the oil flows by gravity from an overhead tank or is pumped continuously to the bearing under a moderate head. In pressure lubrication the oil is pumped from an oil sump to the bearing, from which it flows back to the sump. Since the high pressures in the oil film are produced by the pumping action of the rotating shaft, the pressure under which the oil is supplied does not need to be high; a pressure of 0.04 to 0.20 N/mm² is sufficient. Generally, the oil may be admitted at any point of the bearing. However, if the shaft must be floated before it starts to rotate, the oil must be introduced at the bottom into the high-pressure region.

When oil is circulated under pressure, an oil cooler, with water or air as the cooling medium, usually is included in the system. The cooling medium takes away the heat that is generated in the bearing by friction and picked up by the oil. The oil temperature is thus lowered to its initial value and the desired oil viscosity is maintained.

23.11 DESIGN PROCEDURE

Because the design of a journal bearing involves many variables, the proper procedure is first to make certain assumption and later to check the assumed values and, if necessary, to correct them. The following procedure is suggested, when the bearing load F, in Newton, the journal diameter d, in mm, and its speed n, in revolutions per minute, are known

(a) Select the mean bearing pressure p by using Table 23.4 and equation 23.13 as guide; determine the necessary length l by applying the equation $p = F/ld$; and check the obtained ratio l/d by Table 23.4.

(b) Select a value of the relative clearance c/d in accordance with the recommendations in Table 23.4, and find the clearance c.

(c) Select from Table 23.1 a suitable lubricating oil; assume its operating temperature t_o in the bearing as about 60°C to 65°C, or slightly higher; determine the viscosity Z of the oil, in kg/ms at this temperature by using Fig. 23.4 and equations 23.2 to 23.5; and compare this value of Z with the value given in Table 23.4.

(d) Compute the characteristic number Zn/p and check it with the minimum data of Table 23.4.

(e) Select the angle of relief γ which gives the angle of support $\alpha = 90 - \gamma$, Fig. 23.1c; compute the friction coefficient f by equation 23.10; and compute the work of friction W_f by equation 23.11. Alternatively find coefficient of fiction using McKee's equation

$$f = 0.326\left(\frac{2n}{p}\right)\left(\frac{d}{c}\right) + 0.002$$

and work of fiction

$$w_f = f\, Fv = \frac{f\, F\pi dn}{60 \times 1000}$$

(f) Determine the heat-dissipating capacity Q of the bearing at the assumed oil temperature t_o by using equations 23.14 and 23.15 or 23.17.

i.e. $\qquad Q = hA\, (t_b - t_a)$

or $\qquad Q = \dfrac{\left(t_b - t_a + 18\right)^2 ld}{k}$

(g) The heat-dissipating capacity Q must be equal to the work of friction W_f. If it is not equal, the desired equilibrium can be obtained by assuming a new oil temperature t_o and finding the new value of Z, Zn/p, and f.
A method of finding the exact equilibrium temperature is shown in Fig. 23.16. For the first assumption of $t_o = 60°C$; $W_f = a$; and $Q = b$; for the second assumption of $t_o = 62.5°C$; $W_f = c$ and $Q = d$. The intersection of a line through points a and c and line through b and d is point e. At this point the temperature t_o is 60.9°C, and $W_f = Q$.

Alternatively, the amount of additional heat generated $(W_f - Q)$ needs to be dissipated by artificial means, e.g. additional quantity of oil required for an expected 10°C rise of temperature.

$$W_f - Q = mst$$

where, m is mass of oil

s is the specific heat of oil

t is the rise in temperature $=10°C$

(h) Check the expected minimum oil-film thickness h_m, found by equation 23.18, against the safe thickness h_m', found by equation 23.20.

If the equilibrium temperature t_e is found to be too high, or the film thickness h_m too small, the designer must either use another oil with a higher viscosity or increase the bearing dimensions l, c, α, and even d.

The outlined procedure is simple and gives satisfactory results. There exist more elaborate procedures,[16] but their advantage is doubtful, mainly in view of unavoidable inaccuracies in machining and of deflection in actual bearings and journals.

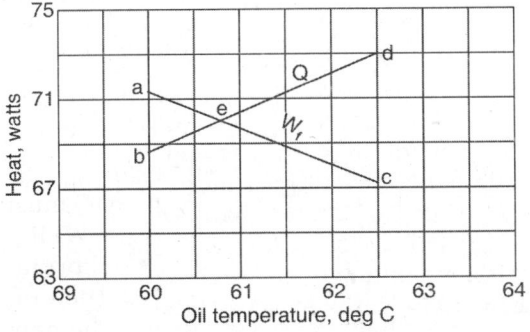

Fig. 23.16: Finding of equilibrium oil temperature

Example 23.1: Determine the length and clearance for an oil-ring bearing to support a 95 mm belt-driven jackshaft. The shaft transmits 90 kW at 230 rpm, and the belt pulley is close to the bearing and has a diameter of 1.32 m.

The net belt tension, found as shown in example 20.1, is

$$F = F_1 - F_2 = \frac{90 \times 1000 \times 60}{\pi \times 1.32 \times 230} = 5662 \text{ N}$$

and the pull of the belt on the bearing is approximately

$$F_1 + F_2 = 3F = 3 \times 5662 = 16986 \text{ N}$$

In accordance with the recommendations in Table 23.4, an average pressure of $p = 0.76$ N/mm^2 is selected for heavy shafting, maximum is 1.0 N/mm^2. The required length, from equation 23.6, is

$$l = \frac{16986}{0.76 \times 95} = 236 \text{ mm}$$

The ratio $236/95 = 2.48$ is within the limits for l/d in Table 23.4. The final pressure, with $A = 95 \times 236 = 22420$ mm^2

$$p = \frac{16986}{22420} = 0.7576 \text{ N/mm}^2$$

The diametral clearance, by Table 23.4, should be

$$c = 0.001 \times 95 = 0.095 \text{ mm}$$

Oil D, Table 23.1, should be used. For $t_a = 60°C$, equation 23.5 gives $\gamma_{60} = 0.000657 (60-15) = 0.905$.

From Fig. 23.4, the Saybolt viscosity at 60°C is SSU = 110 sec. by equations 23.2 and 23.4, the absolute viscosity is

$$Z = \left(122 \times 110 - \frac{180}{110}\right) \times 10^{-3} \times 0.905$$

$$= 0.0204 \text{ kg/ms}$$

Therefore, the bearing-characteristic number is

$$\frac{Zn}{p} = \frac{0.0204 \times 230}{10^6 \times 0.7576} = 6.19 \times 10^{-6}$$

This is higher than the minimum value 4.4×10^{-6} given in Table 23.4 and therefore seems to be satisfactory.

Example 23.2: For the bearing of example 23.1, determine (a) the equilibrium temperature and (b) the oil-film thickness.

(a) With a horizontal pull by the belt, an oil-ring bearing, Fig. 23.13, will not have any side relief, and angle $\alpha = 90°$. The coefficient of friction will be, by equation 23.10,

$$f = 0.326 \times 6.19 \times 10^{-6} \frac{95}{0.095} + 0.002 = 0.004017$$

[16] R. T. Kent, *Mechanical Engineers' Handbook*, 12th ed., Vol. 2, *Design and production*, ed. by Colin Carmichael (New York: John Wiley & Sons, Inc., 1950), pp. 2–14 to 12–31.

Heat generated

$$H_g = \frac{0.004017 \times \pi \times 95 \times 230 \times 16986}{1000 \times 60} = 78.06 \text{ W}$$

The heat-dissipating capacity is found as follows: From a sketch of the bearing, the exposed area, extending 100 mm down from the oil reservoir, is about 20 dl. Also, from Fig. 23.8, with $t_o - t_a = 60 - 15 = 45°C$, Therefore, by equation 23.14, the heat-dissipating capacity, without ventilation, is

$$\frac{11.4 \times 22.5 \times 20 \times 95 \times 236}{10^6} = 115 \text{ W}$$

By using eq. 23.17,

$$Q = \frac{(22.5 + 18)^2 \times 95 \times 236}{0.273 \times 10^6} = 134.7 \text{ W}$$

Both are in satisfactory agreement.

(b) In order to find the minimum oil-film thickness, it is first necessary to compute the constant K by equation 23.19. Thus,

$$K = \frac{28.7 \left(1000 \times \dfrac{0.095}{95} \right)}{6.74 \times 10^6} = 4.26$$

From Fig. 23.10, for $\alpha = 90°$, $e/c = 0.275$. By equation 23.18,

$$h_m = 0.5 \times 0.095 \, (1 - 2 \times 0.275) = 0.0214$$

The rubbing velocity is

$$v_m = \frac{\pi \times 95 \times 230}{1000 \times 60} = 1.144 \text{ m/s}$$

and the safe thickness, by equation 23.20 is

$$h_m = 0.018 \times (1.144)^{0.4} \times (95 \times 236)^{0.2} = 0.014 \text{ mm}$$

Hence the whole design is satisfactory.

23.12 JOURNAL BEARINGS CONSTRUCTION

There exist so many different bearing constructions that only a few of the more typical ones can be discussed here. *Bushings* for small bearings are of either brass or bronze and are often made in one piece. The bushing is pressed into place and reamed to size. After excessive wear has taken place, the bushing is replaced.

A two-piece babbitted bearing with wick lubrication is shown in Fig. 23.11. The shim s serve to take care of the wear. They are made of strips of wood, or of brass or steel shim

stock. The line of action of the load should be normal to the plane of separation. If this is not feasible, a satisfactory expedient is to place the bearing so that the line of action of the load is not more than 30 deg from the bottom of the bearing. If the line of action of the load forms a large angle with the vertical, an angle bearing is used. Figure 23.12 is shown such a bearing, with a hole in the cap to take a grease cup or sight-feed oiler.

Figure 23.13 shows a rigid bearing with ring oiling. It is used for heavy shafts and as an outboard bearing for an engine stub shaft.

A self-aligning bearing with ring oiling is shown in Fig. 23.14. It is used for transmission shaftings.

A rigid bearing with collar oiling is shown in Fig. 23.15. It has cast-iron babbitted shells.

Bearing supports: A sole plate which takes practically any type of bearing is shown in Fig. 23.17. The horizontal shaft alignment is made by means of the screws s. The vertical alignment is obtained by means of steel shims between the plate and the bearing.

A drop hanger for a two-point adjustable pivoted bearing is shown in Fig. 23.14. The hanger is made of cast iron and has different drops H for bearings of different sizes up to 125 mm bore.

A *Michell multipad journal bearing* is made up of several pivoted segments a, Fig. 23.18, of the same curvature as the journal. Five are shown in the illustration. Because of the unsymmetrical shape of the segments, the oil pressure built up by the rotating shaft tips the segments slightly in their seats and thus forms

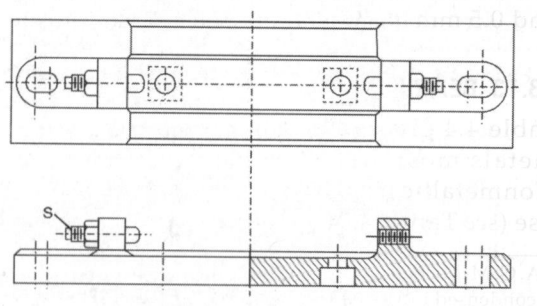

Fig. 23.17: Plain sole plate

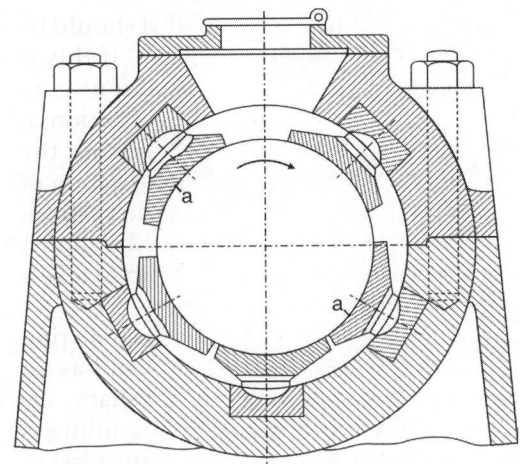

Fig. 23.18: Michell multipad bearing

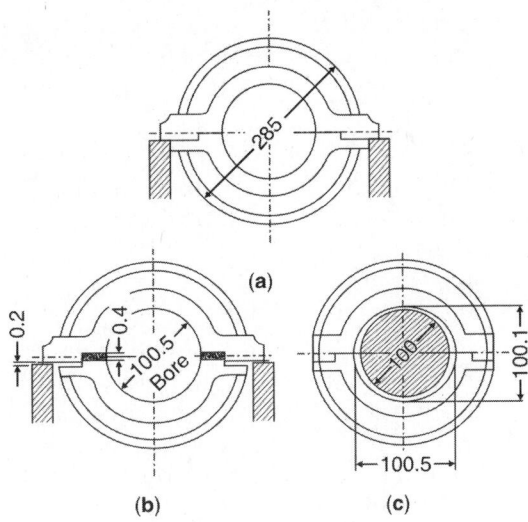

Fig. 23.19: Boring of a bearing shell for wedge lubrication

the converging oil film characteristic of perfect lubrication. A bearing of the type shown in Fig. 23.18 was built for a 525 mm shaft of a certain turbogenerator, with the result that the overall length of the machine was shortened from 9.6 m to 6.75 m.[17]

For smaller journals, similar results may be obtained much more cheaply by eccentric boring of the bearing.[18] Figure 23.19 illustrates the method for a 100 mm journal. The two halves of the bearing are bolted firmly together, as shown in Fig. 23.19a, and are machined to final dimensions, except for the inside surface. They are then separated, shims 0.4 mm thick are inserted, and the bearing shells are bored to a diameter of 100.5 mm, as indicated in Fig. 23.19b. When the shims are taken out and the shells are put in place, Fig. 23.19c, the clearances become 0.1 mm vertically and 0.5 mm horizontally.

23.13 BEARING MATERIALS

Table 4.4 gives information regarding the metals most generally used for bearing. Nonmetallic materials are now coming into use (*see* Table 4.8).

Requirement of bearing materials: To serve as a good bearing material, it should have the following requirements:

1. *Low coefficient of friction: Every bearing will not always be working as thick film lubricated bearing so it is preferable that the coefficient of friction be low (23.5).*
2. *High compression strength*: The material should be of high compressive strength to withstand maximum hydrodynamic pressure to avoid deformation.
3. *High fatigue resistance*: The material has to withstand repeated and varying loads without developing surface fatigue, so should have high fatigue strength.
4. *Low coefficient of thermal expansion*: If the bearing is to operate at high temperatures or over wide range of temperatures, then its coefficient of expansion should be low so that the clearance is not effected.
5. *High thermal conductivity*: To allow for rapid removal of heat, the material should have high thermal conductivity.
6. *Corrosion resistance*: In certain applications such as internal combustion engines, the material should be corrosion resistant.

---

[17] A. G. M. Michell, "Progress of Fluid-Film Lubrication," *Trans. ASME*, Vol. 51 (1929), MSP–51–21, p. 156. The same paper is condensed in *Mechanical Engineering*, Vol. 52 (1930), p. 115.
[18] E. A. Kraft, *The Modern Steam Turbine* (Berlin: VDI-Verlage, 1931), p. 63.

7. *Embedability*: As the lubricant gets contaminated with dust and metal particles, the bearing material should be soft enough so that these particles get embedded to avoid scoring and wear.

8. *Conformability*: The material should be soft with low modulus of elasticity so that the material is able to conform to slight misalignment and geometrical errors in the journal dimensions.

9. *Bondability*: As the material itself will not have sufficient strength to resist heavy loads, the bearings are made by bonding with high strength steel shell. Hence the material should be able to form a good bond with hard backing.

10. *Low cost*: The cost of a bearing should be low.

As no one material can have all the above requirements, it is composed of a bearing material and a high strength backing shell.

To satisfy some of the above requirements, the following materials are commonly used for construction of bearing:

Babbitts: Tin base babbitt is used for bearings which are subjected to heavy pressures. It is very fluid in molten state, and can be applied for thin linings of bronze-backed or steel-backed bearing shells like those used in automotive and aircraft engines.

Cadmium base bearing metal is an alloy about 50 per cent stronger than tin-base babbits, and it may be used for bearings subjected to shock action.

Lead-base babbitts may be used for larger bearings when the maximum pressure are below 3.5 N/mm^2 However, a lead-base alloy with the trade name magnolia metal, the analysis of which is about the same as that of lead base babbit, seems to give good service even in heavy-duty bearing if they are not subjected to pounding.

The advantage of babbitts is that if lubrication fails, they simply melt out without scoring the journal.

Brass and bronze: Brass is used where the pressure is too high for babbitt but where the service is not severe enough to call for a more expensive bearing metal. Bronzes are used where the pressures are so high that thin-film lubrication may occur.

Hard bronzes, known under the trade names Nida-brone[19] and Carobronze,[20] are melted in electrical induction furnaces. Their allowable working pressures are so high—up to 55 N/mm^2—that bearings made of them are interchangeable with ball and roller bearings. With special precautions in regard to lubrication these bearings seem to work with a perfect oil film and do not show any wear.

Copper-lead alloys: Copper alloys having a high lead content—20 to 50 per cent—are of special interest. These alloys, put on the market under different trade names, have a low coefficient of thin-film-lubrication friction, about 0.005; and like babbitts, they do not score a journal when lubrication fails. The allowable pressure is 7 to 10 N/mm^2. In order to obtain a good copper-lead bearing alloy, it must be cast centrifugally. Such bearing shells and bushings have a uniform structure with a fine lead distribution and form a good bond with preheated steel shells.[21] With such bearings, particularly good results have been obtained in Germany. The allowable bearing pressure is 35 N/mm^2 at a rubbing speed as high as 10 m/s.[22] The bearings are especially suitable where heavy pounding occurs, as in connecting-rod bushings of aircraft engines.

Aluminium alloys: Aluminium alloys are remarkable for their great resistance to scuffing, their low friction, and their high wear resistance under conditions of boundary and thin-film lubrication. The high coefficient of heat conductivity k helps to carry away the

[19]Vereingte Deutsche Metall-Werke A. G., *Nida Bronze* (Frankfurt am Main-Heddernheim; 1938)
[20]Carobronze G.m.b.H., *Carbonize* (Berlin: 1937)
[21]F. R. Hensel and L. M. Tichvinsky, "Copper-lead Bronze for Bearings," *trans ASME*, Vol. 54 (1932), IS–54–3, p. 11.
[22]Manufactured according to patents of Glyco-Metall-Werke, Wiesbaden-Schierstein, Germany.

heat of friction. However, aluminium begins to lose its strength at about 105°C and therefore the temperature of aluminium shells should not exceed 105°C or 120°C if they have been cold-worked. Otherwise, there is a danger of losing the necessary pressure between the contact surfaces of the two shells. Aluminium bearings can operate with pressures up to $20 \, N/mm^2$, or even $28 \, N/mm^2$, and with peripheral velocities up to $10 \, m/s$.

Cast iron: For hardened steel journals, cast iron is a very good bearing material in regard to friction and wear, even if lubrication is in the thin-film region. However, this combination is suited only for light service where the pressures do not exceed $3 \, N/mm^2$.

Oilless bearing: An oilless bearing depends on a lubricant incorporated in the bearing during its manufacture. This is done in several ways.

(a) In some types, flaked graphite is inserted into the metallic (usually bronze) surface in the shape of spirals or studs.

(b) Another development consists in suspending graphite in a babbitt alloy under a high pressure at the fusing temperature of the babbitt.

(c) In other types, oil is impregnated into wood or some other porous or fibrous carrier.

(d) In powder-metal bearings, pulverized graphite and bearing bronze or iron are mixed with a binder and are pressed into molds with the application of heat. This process is called *sintering*. A sintered bearings is porous. It can absorb an amount of lubricating oil up to 30 per cent of its own weight, and can give up oil very slowly in operation. Sintered bearings maintain a thin oil film for a long time and are correctly called *self-lubricating* bearings, not oilless bearings. Sintered bearings are also impregnated with graphite or oil to serve as lubricant.

Oilless and self-lubricating bearings may be used in places where a little or no attention can be given to the bearings and when the load and speed are low. Even under such conditions, however, the wear is relatively rapid. The clearance on bronze bearings with graphite-filled grooves and on sintered bearings should be 0.05 mm greater than for ordinary journal bearings.

Plastics: Celoron, formica, teflon and micarta bearings are made from a special woven duck impregnated with phenolic resin. The materials are fused together under a very high pressure and at a high temperature. These materials have good mechanical properties (see Table 4.8) and great resilience. Such bearings can be lubricated with oil, grease, or water. They are used on heavy rolling mills as a replacement for bearings of bronze or lignum vitae without any changes in the roll stand itself. With water lubrication they can stand pressures up to $35 \, N/mm^2$ and peripheral velocities of $10 \, m/s$, with a friction coefficient f less than 0.007.[23] In general, the coefficient of friction is of the same order as that of tin-base babbitt.[24] Their drawback is a low heat conductivity, which necessitates the use of water or oil to cool them.

Rubber: In hydraulic turbines, and in stern bearings of ships, and in other machines where water is available, rubber bearings lubricated with water, are used with increasing success instead of bearings with lignum vitae lining. The bearing is made of soft rubber and the inner surface is either fluted, as in Fig. 23.20, or has helical grooves. The softness of the rubber enables these bearings, to stand up in the presence of sand and grit. They have a low coefficient of running friction, 0.02 to 0.04[25] can carry loads up to $6 \, N/mm^2$, and run at speeds up to $22 \, m/s$ with very little wear, if any. Rubber bearings are particularly suitable for use on shafts running at high speeds. Because

[23]O. K. Graef, "Phenolic Plastics—Design's Latest Bearing Materials," *Machine Design*, Vol. 8 (September, 1936), p. 34
[24]L. M. Tichvinsky, " Properties and Performance of Plastic Bearing Materials," *Trans. ASME*, Vol. 62 (1940), p. 461.
[25]W. F. Busse and W. H. Denton, "Water Lubricated Soft Rubber Bearings," Trans ASME, Vol. 54 (1932), IS–54–2, p. 3.

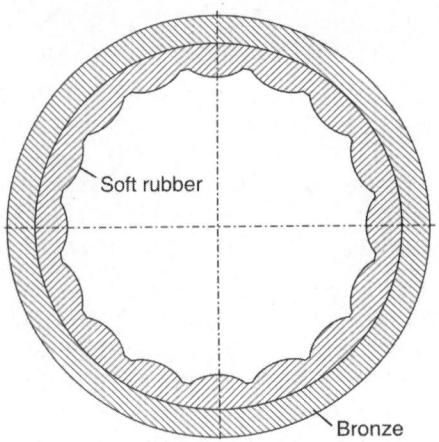

Fig. 23.20: Fluted rubber bearings

the soft rubber allows the shaft to turn about the axis going through its center of gravity, even though this axis differs slightly from the geometrical axis. As a result the dynamic load on the bearing is reduced and there is less vibration in the machine. The main requirements for insuring small friction and low wear are the circulation of enough water to keep its temperature below 98°C, and a very smooth surface of the shaft. Rubber bearings are obtainable in stock sizes for shafts from 25 mm to 560 mm in diameter.

23.14 MECHANICAL DESIGN

In addition to determining the diameter, length, and clearance of a bearing, proper design involves:

    (a) Selection of a suitable bearing material.

    (b) Provision for sufficient lubrication.

    (c) Prevention of excessive oil leakage.

    (d) Dissipation of heat generated by friction.

    (e) Sufficient strength and rigidity of shells, caps, and bolts.

    (f) Small wear and provision for taking it up in large bearings.

    (g) Preservation of proper shaft alignment.

Bearing material: The right-hand column in Table 4.4 and information given in section 23.13 may serve as a basis for selection of the bearing material for any specific case.

Lubrication: Oil admission must be in the region of low pressure for journal rotating in one direction and in the top position when it rotates in both directions (23.9). As shown in Fig. 23.21, a circumferential groove *g* on the outside of the bearing shell with holes *h* drilled into the reliefs *r* will deliver oil to the crankshaft-journal surface where there is no pressure in the oil film. The detail in Fig. 23.22 illustrates how oil delivery from the crankpin to the piston through a rifle bore *a* is increased by an outside groove with two cross-drilled oil holes in the big end of the connecting rod.

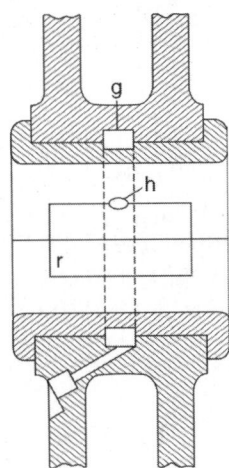

Fig. 23.21: Oil admission to main bearing of a crankshaft

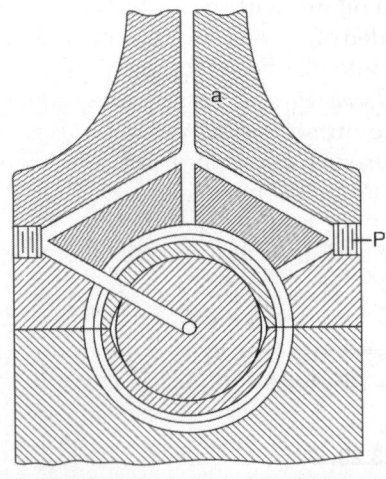

Fig. 23.22: Oil delivery to a piston pin

Oil grooves are useful in bearings that operate all or part of the time with thin-film lubrication as in bearings with oscillatory motion in which the relative speed between the journal and bearing is low. In such cases properly cut oil grooves in the load-carrying surface help to distribute oil to the entire bearing surface. With thick-film lubrication, oil grooves placed in the high-pressure region are only harmful, since they interrupt the continuity of the oil film and may thus cause thin-film lubrication. The full-line curve in Fig. 23.23a shows the lengthwise pressure distribution in a normal journal bearing; the dotted-line curves show the pressures when an oil groove is cut around the middle of the bearing. In order to support a given load, the pressure goes up. Theoretically this should decrease the oil-film thickness. However, tests indicate that actually the load capacity increases somewhat, probably because of an increase in oil flow which lowers the oil temperature and raises the oil viscosity.

A lengthwise oil groove at the point of maximum pressure is more harmful, as indicated in Fig. 23.23b, and may cause metal-to-metal contact. *However, a longitudinal groove cut in the low-pressure region helps to distribute the oil over the whole bearing (23.10).* Side reliefs a, Fig. 23.25, are useful, as they help to wedge in the oil film and act as oil reservoirs.

All oil grooves should have the edges well rounded off, in order to facilitate the admission of oil into the pressure region.

Oil seals: Provisions for preventing oil leakage to the outside, and penetration of foreign matter, such as dust or fumes, to the inside of a

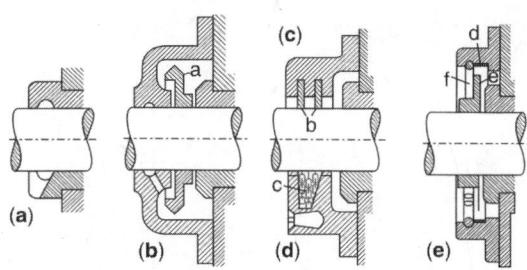

Fig. 23.24: Oil seals for bearings

bearing, are shown in Fig. 23.24. In Fig. 23.24a oil leakage about the shaft is stopped by a circular groove with sharp edges acting as a scraper; in Fig. 23.24b the main part of the oil is thrown off by centrifugal force by means of the throw ring a; in Fig. 23.24c oil is stopped by the sharp-edged brass rings b of the end cap; in Fig. 23.24d there is a split cap with a felt or leather ring c, which seals the bearing in both directions; in Fig. 23.24e the bearing is sealed by a stationary flanged tin ring d held in place by a retainer spring f and slightly pressed against the tin ring e acting as a throw ring.

Heat dissipation: The rate of dissipation of heat is materially increased by grinding and scraping the shells to a snug fit in the bearing body. Caps with ribs have a greater area and better heat dissipation than box-shaped caps. Water-cooling of the bearings is used only in special cases, as in roll-neck bearings of rolling mills or in main bearings of two-stroke-cycle oil engines. Bearings with pressure-feed lubrication often have an oil cooler inserted in the oil-circulation system.

Strength and rigidity of bearing shells: The thickness h of a small brass or bronze shell can be taken as $0.08d + 2.5$ mm. Shells of medium-size bearings, about 75 mm and up, are made as shown in Fig. 23.25. They have a cast-iron layer with a thickness h_1 equal to $0.20d$, and a babbitt layer having a thickness h_2 equal to $0.02d + 3$ mm. Engine main-bearing shells for 300 mm journals and up are made of a cast-steel layer for which $h_1 = 0.15d$, and a babbitt layer for which $h_2 = 0.01d + 3$ mm.

Lengthwise, swallow-tailed, babbitt anchor grooves are shown in Fig. 23.25, while

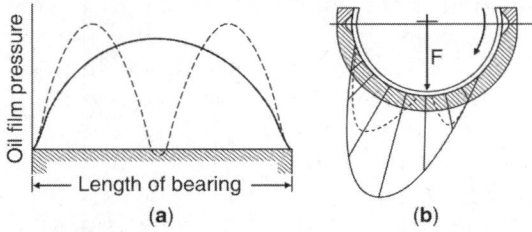

Fig. 23.23: Oil-pressure distribution without and with oil grooves

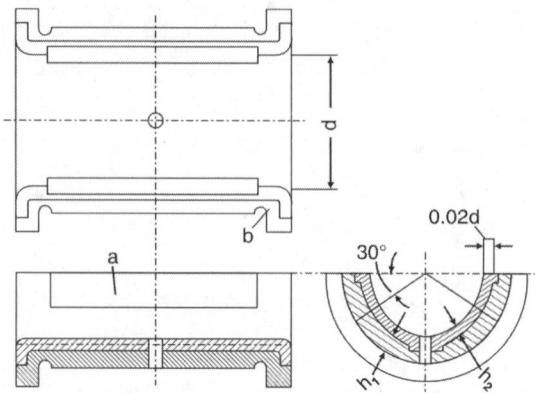

Fig. 23.25: Main-bearing shell

Fig. 23.15 shows button-type anchors on the cylindrical part and round swallow-tail anchors at the ends.

An aluminium shell, when made solid and subjected to dynamic loads, should have a thickness h equal to $0.044d + 0.5$ mm. When the shell must be prevented from turning or from moving longitudinally by means of a dowel pin, the minimum thickness h is 4 mm. A bearing with a concentric bore should have a relief of about 0.05 mm deep and extending about 6 mm from the parting line and blending into the bearing surface to compensate for a slight distortion which may be caused by the crush. The recommended relative clearance is $c/d = 0.002$ for small bores, up to 40 mm; $c/d = 0.00175$ for $d = 50$ mm; $c/d = 0.0015$ for $d = 130$ mm; and $c/d = 0.00125$ for $d = 200$ mm

In order to maintain the desired clearance c and uniform pressure distribution, it is essential that the shell be rigid and be supported rigidly.

Strength and stiffness of cap: The cap of a bearing is not usually subjected to a heavy load. However, sometimes the load acts upon the cap, as in the case of automobile main bearings. In such a case the cap may be regarded as a simple beam loaded at the center and supported by the holding-down bolts, Fig. 23.26.

The deflection of the cap may be computed by applying the expression for case c in Table 2.4 and substituting $\frac{1}{12}bh^3$ for I. Usual practice

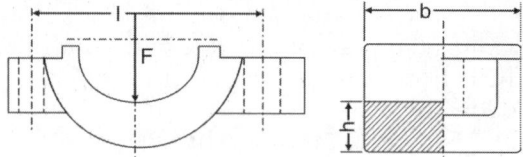

Fig. 23.26: Bearing cap

limits the permissible deflection y to 0.025 mm; in small sizes it should be only 0.012 mm.

Holding-down bolts: The bolts, studs, or screws which hold down the cap may be assumed to be subjected to simple tension. Each bolt must be designed for a load $1.33\,F/i$, where i denotes the number of bolts. The coefficient 1.33 takes into account the uneven load distribution on the bolts due to friction of the journal.

Take-up of wear: With thick-film, or perfect, lubrication, theoretically there should not be any wear at all. *Practically, most bearings eventually run with imperfect lubrication and metal-to-metal contact, which results in wear of the rubbing surfaces.* In order to reduce wear to a minimum, it is desirable to keep the bearing pressure low. It should never be higher, and if possible should be lower, than the value given in Table 23.4.

In a two-piece bearing the simplest means of taking up the unavoidable wear is by removing some of the shims inserted between the bearing halves. Shims are also used in an angle bearing, Fig. 23.12. If the force acts horizontally, a four-piece shell is used in which the wear may be taken up either by means of a jackscrew, as in Fig. 23.27a, or by means of wedges, as in Figs 23.27b and 23.27c.

Shaft alignment: Transmission bearings and pillow blocks are usually fastened to sole plates, the adjustment being by means of jackscrews in the horizontal direction, as in Fig. 23.17, and by shims between the plate and the bearing in the vertical direction. In a hanger bearing, both adjustments may be obtained by means of screws. In a large engine main bearing the horizontal adjustment may be obtained by means of two wedges. A vertical adjustment is shown in Fig. 23.27c.

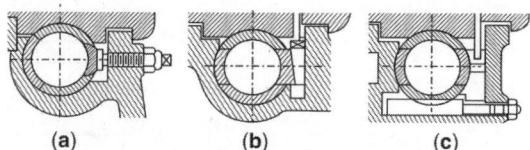

(a)   **(b)**   **(c)**

Fig. 23.27: Adjustment of four-piece bearing shells

23.15 STEP, VERTICAL, AND COLLAR BEARINGS

A horizontal shaft is often subjected to an axial force, and a vertical shaft always. This force is taken by thrust bearings of various types.

Step bearings: The simplest thrust bearing is the plain step bearing, Fig. 23.28, which is used for vertical shafts. The greatest wear in such a bearing occurs at the outer radius because the maximum velocity occurs there. The center portion is thus left higher. This wear eventually produces an excessive pressure at the center and causes overheating and failure of lubrication. In order to eliminate this trouble, the thrust disk *a* can be made with a hole in the center, or the shaft can be counterbored with a shallow hole at the end. Another method of reducing the danger of lubrication failure is to use a number of bronze and steel disks arranged alternately, in which case each disk rotates at a fraction of the shaft speed and reduces and distributes the wear. Should a pair of adjacent disks stick because of faulty lubrication, the

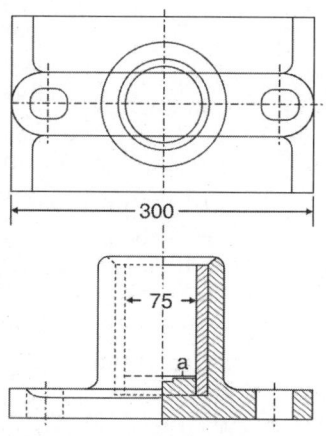

Fig. 23.28: Plain step bearing

bearing will not be damaged since the remaining disks are free to turn.

An adjustable step bearing which is used when side adjustment is required is shown in Fig. 23.29. A bronze bushing receives the lateral wear from the shaft, while the thrust is carried on a removable hardened-steel disk. The bearing can also adjust itself if the axis of the shaft is not quite normal to the plane of the bearing base.

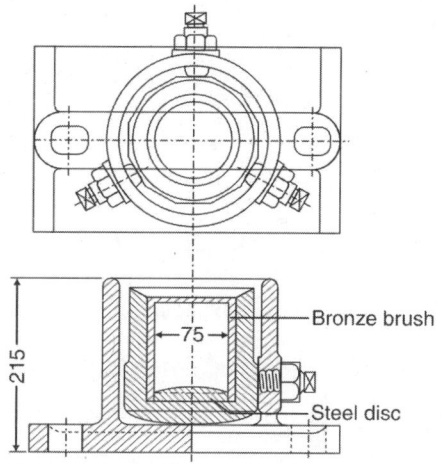

Fig. 23.29: Adjustable step bearing

Vertical bearings: For supporting the upper end of a vertical shaft any rigid bearing similar to a standard bearing may be used, but a lubrication system must supply the oil to the top part. This may be done by means of a wick-feed or sight-feed oiler. The loss of power is due only to the oil friction and may be computed in *kW* by the expression [26]

$$P = \frac{Zd^3 ln^2}{c} \qquad \ldots (23.22)$$

where the notations are the same as previously used.

If the journal and the bearing are eccentric and the distance between their axes is *e*, the power loss will be

---

[26] A. I. Ponomareff and E. D. Howe, "Some Problems in Lubrication of Vertical Bearings," *Trans. ASME*, Vol. 55 (1933), PME–55–4, pp. 27 f.

$$P = \frac{Zd^3 \, ln^2}{\sqrt[c]{1 - \left(2\frac{e}{c}\right)^2}} \qquad \dots (23.23)$$

Collar bearings: Collar bearings are used chiefly on a horizontal shaft which must carry a large axial load, such as is created by a ship's propeller. In Fig. 23.30 is shown a thrust bearing with wick lubrication for a small propeller shaft. The lower half of the bearing is water-cooled.

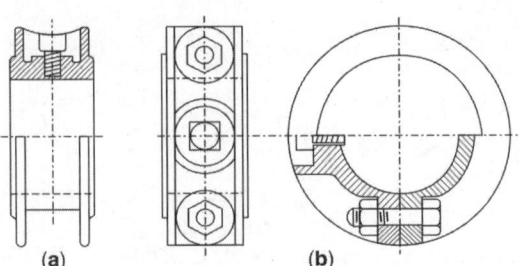

Fig. 23.31: Safety set collars

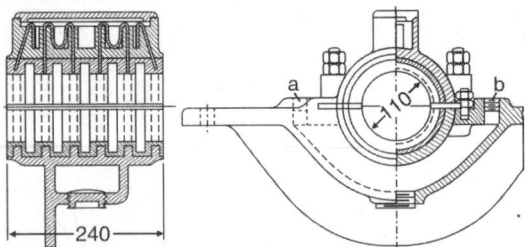

Fig. 23.30: Marine multicollar thrust bearing

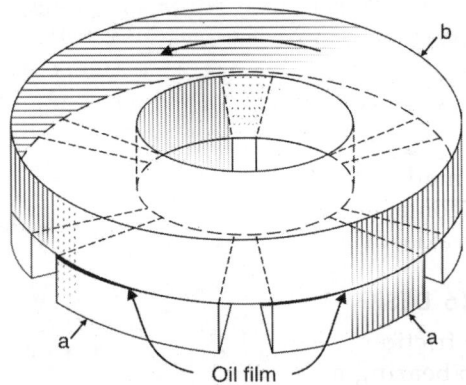

Fig. 23.32: Kingsbury thrust bearing

If feasible, the collars should be placed near the point of application of the axial load in order to avoid a column action upon the shaft. The allowable bearing pressure for the collars should be slightly smaller than for a step bearing, since the load is not likely to be evenly distributed among all collars. The diameter of the collars is made equal to 1.3d to 1.7d, where d is the shaft diameter.

Set collar: A small axial thrust can be taken by a set collar pressed against the end of a journal bearing. Set collars are used chiefly on transmission shafting. The head of the setscrew which fastens the collar to the shaft should be below the outside diameter of the flange, in which case the collar is termed a *safety collar*. Set collars are made of cast iron, malleable iron, or pressed steel. A solid collar is shown in Fig. 23.31a, and a split collar is shown in Fig. 23.31b.

Kingsbury thrust bearing: The principle of the Kingsbury thrust bearing is similar to that of the Michell bearing, Fig. 23.18. The Kingsbury bearing, Fig. 23.32, consists of several stationary pivoted segments, or shoes, *a*, against which is pressed the thrust collar *b* fastened to the rotating shaft. The shoes rest on spherical supports, or pivots, and are thus free to tilt in any direction. The pivots support the segments not at the center of gravity but slightly forward in the direction of rotation, as explained for the segments in Fig. 23.18. Therefore, the pressure on the back edge of each segment is lower than that on the forward edge. The collar runs in an oil bath, and its rotation draws an oil film into the spaces between it and the shoes, which tilt automatically so that a fluid wedge is formed at the back edge.

A Kingsbury thrust bearing for a vertical suspended shaft with a self-aligning spherical seat is shown in Fig. 23.33. The bearing is self-lubricating but is arranged for external oil circulation and cooling. A uniform distribution of load among the shoes is very important for proper operation of the bearing. It may be obtained by making the shoe supports adjustable, or by resting them upon a ring formed by equalizing levers. These bearings

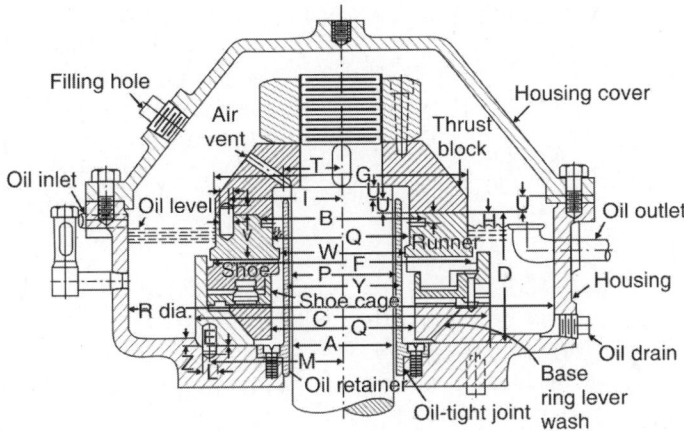

Fig. 23.33: Kingsbury thrust bearing

are made in types suitable for vertical or horizontal shafts, and for carrying thrusts in both axial directions, or in only one direction.

23.16 DESIGN OF THRUST BEARINGS

The friction coefficient of a well-lubricated step bearing may be taken as 0.015 to 0.02. The average friction coefficient of a collar bearing depends on the oil viscosity, speed, and load as expressed by the characteristic number Zn/p. A relation based on actual tests[27] is

$$f = 0.008v^{0.5} p^{-0.67} \qquad ...(23.24)$$

where, v is the average peripheral speed, in m/s, and p is the specific pressure, in N/mm².

The friction coefficient of a Kingsbury bearing depends on the characteristic number Zn/p. However, because of the presence of a perfect oil film and the allowable high specific pressure, it is about $\dfrac{1}{10}$ the friction coefficient in a collar thrust bearing, as found from tests. Under favorable conditions, Kingsbury thrust bearings have values of f as low as 0.0015, and 0.003 is a conservative figure.[28]

Friction loss in a pivot bearing: If a uniform pressure distribution is assumed in a pivot bearing, the friction torque, referred to the mean diameter d_m, is

$$T_f = \frac{fF_a d_m}{2} \qquad ...(23.25)$$

where, F_a is the axial force. Since $d_m = \dfrac{2}{3}d$, the power absorbed by friction, by equation 2.17, is

$$P_t = \frac{\pi f F_a d n}{1000 \times 60} \qquad ...(23.26)$$

Friction loss in a collar bearing: Equation 23.25 may be used for the friction loss in a collar bearing. The collars are either integral parts of the shaft or rigidly fastened to it, and the axial force F_a may be assumed to be evenly distributed over all collars. If $d_2 \leqq 1.5d_1$, where d_2 is the outside diameter of the collars and d_1 is the inside diameter, then it is sufficiently accurate to assume that $d_m = \dfrac{1}{2}(d_2 + d_1)$. For this condition

$$T_f = \frac{fF_a(d_2 + d_1)}{4} \qquad ...(23.27)$$

[27] J. E. Hamilton, "Ball versus Tapered Roller Bearings," *Journal of the American* Society of Naval Engineers, Vol. 44 (1932), pp. 407 – 29.

[28] Kraft, *op. cit.*, p. 59.

The friction power may be found by the equation

$$P_f = \frac{\pi f F_a (d_2 + d_1) n}{2 \times 1000 \times 60} \qquad \ldots (23.28)$$

The average pressure, with i collars, is

$$p = \frac{4 F_a}{\frac{\pi}{4}(d_2^2 - d_1^2) i} \qquad \ldots (23.29)$$

With proper lubrication and moderate pressures the wear of the bearing collars is small, and the assumption of uniform wear, used for brakes and friction clutches, is not justified.

Allowable pressures: For step bearings, single-collar bearings, and water-cooled multicollar thrust bearings, and for rubbing speed v ranging for 0.25 to 1.0 m/s, the allowable pressure p may be taken so that

$$pv \le 7 \qquad \ldots (23.30)$$

For very low speeds, the pressure may be as high as 14 N/mm², for intermittent service, pressures up to 10 N/mm² may be used; and for speeds over 1 m/s, the pressure should not exceed 7 N/mm².

For multicollar thrust bearing that are not water-cooled, the above values for p should be divided by 2.

Kingsbury pivoted thrust bearings operate satisfactorily with pressures from 2 to 7 N/mm².

Oil circulation: The usual practice is to allow 7°C for the temperature rise of oil circulating through low-speed and medium-speed thrust bearings. The amount of oil Q, in liters per second can be calculated by assuming that 1 liter oil weighs 0.88 kg, the specific heat of oil is 555 cal/s, and 1 kW = 239 cal/s. Thus

$$G = \frac{239}{0.88 \times 555 \times 7} = 0.0699$$

$$= 0.07 \text{ L/s}$$

The approximate amount of oil is 0.07 L/s per kW for friction loss.

In bearings with high rubbing speeds, where turbulence is an important factor, the fiction losses may be reduced by supplying less oil and allowing an oil temperature rise up to 14°C.

The passages in the bearings and in the oil pipes should be amply large, the velocity of oil going to the bearings should not be more than 1.2 m/s, and the velocity of the oil returning to the oil cooler should not be more than 0.75 m/s.

23.17 DESIGN PROCEDURE FOR A MULTICOLLAR BEARING

For a given shaft diameter d_1, select d_2 not greater than $1.5 d_1$. Take the mean diameter d_m as $\frac{1}{2}(d_1 + d_2)$, and find the mean rubbing velocity v_m in meter/minute.

Select the safe average pressure p and check it by equation 23.30. If several collars are used, it is advisable to use a value for p not more than one-half the limit given by equation 23.30.

Find the force F_a that each collar can take, and determine the total number of collars i required. Compute the friction coefficient f by equation 23.24. The loss of power is computed by equation 23.28.

If the bearings is not water-cooled, the heat-dissipating capacity should be estimated by equation 23.14 in conjunction with equation 23.15.

If water cooling is used, determine the amount of water that must be circulated to carry away the heat of friction. Allow a rise in water temperature of 5 to 11°C.

Example 23.3: Determine the main dimensions of a multicollar thrust bearing for a propeller shaft of a 445 kW power marine oil engine. The engine makes 220 rpm; the shaft diameter is 150 mm; the propeller has a pitch of 2.5 m.

If a slip of 25 per cent is assumed, the boat is driven 1.875 m, for each revolution of the engine, a distance of 412.5 mm. The total axial force multiplied by the speed represents the work done. Therefore

$$F_a = \frac{445 \times 1000 \times 60}{412.5} = 64727 \text{ N}$$

For $d_2 = 1.5\, d_1 = 1.5 \times 150 = 225$ mm, the mean diameter is

$$d_m = \frac{1}{2} \times (150 + 225) = 187.5 \text{ mm},$$

The mean rubbing speed is

$$v = \frac{\pi \times 187.5 \times 220}{1000 \times 60} = 2.16 \text{ m/s}$$

If the thrust bearing is water-cooled, a pressure p of 0.6 N/mm^2 may be assumed. The force that each collar can carry is

$$F_a = \frac{\pi}{4} \times (225^2 - 150^2) \times 0.6 = 13254 \text{ N}$$

and the number of collars should be

$$i = \frac{64727}{13254} = 4.88 \text{ mm say } 5$$

The actual pressure would be

$$p = \frac{64727}{\pi/4 \left[225^2 - 150^2 \right] \times 5} = 0.586 \text{ N/mm}^2$$

The coefficient of friction computed by equation 23.24 is

$$f = 0.008 \times 2.16^{0.5} \times 0.586^{-0.67} = 0.018$$

By equation 23.12, with $C_1 = 2$ and $C_2 = 4$ from Tables 23.2 and 23.3,

$$f = 0.0037 \times 2 \times 4 \sqrt[4]{\frac{0.586}{2.16}} = 0.021$$

This is a good agreement. To be safe, use $f = 0.021$. The loss of power, by equation 23.28, in which $F_a = 64727$ N, is

$$P_f = \frac{\pi \times 0.021 \times 64727 \times 375 \times 220}{2 \times 1000 \times 1000 \times 60}$$

$$= 2.935 \text{ kW}$$

Example 23.4: Check the suitability of the journal bearing having diameter of 75 mm and length of 125 mm. The load transmitted is 20 kN by the journal rotating at 1000 rpm. Assume a d/c ratio of 1000 and an oil having viscosity of 0.01 kg/ms at the operating temperature. The bearing surface temperature is 75°C and surrounding atmosphere temperature is 30°C.

Average pressure

$$p = \frac{F}{ld} = \frac{20,000}{0.075 \times 0.125} = 2.13 \times 10^6 \text{ N/m}^2$$

Frictional coefficient

$$f = 0.002 + 0.326 \left(\frac{0.01 \times 1000}{2.13 \times 10^6} \right) \times 1000 = 0.00353$$

heat generated or work of friction

$$W_f = \frac{fF\pi dn}{60} = \frac{0.00353 \times 20,000 \times \pi \times 0.075 \times 1000}{60}$$

$$= 277.25 \text{ W}$$

Heat dissipated

$$Q = \frac{(75 - 30 + 18)^2}{0.484} \times 0.075 \times 0.125 = 65.18 \text{ W}$$

Amount of heat to be dissipated = 277.25 – 65.18 = 212.07 W

This must be dissipated by artificial cooling or by calculation of quantity of oil needed for dissipation of heat by assuming a temperature rise of 5 to 10°C and by equating it to *mst*, where *m* is the mass of oil, *s* is the specific heat and *t* is the rise in temperature.

PROBLEMS

23.1 Describe the mechanics of a thick film lubrication.

23.2 Which is the most important property of oil as far as thick film lubrication is concerned?

23.3 Make sketches to show the distribution of pressure is a bearing with thick film lubrication along the circumstance and along the axis.

23.4 If the projected area of a bearing is found to be insufficient, is it desirable to increase the diameter or length?

23.5 Name the most important characteristics of a good bearing material.

23.6 Why is the material of a bearing of any concern when the lubrication is such that a film of oil separates the surfaces of shaft and bearing?

23.7 What are oil-less bearings?

23.8 Make a sketch for a self aligning bearing.

23.9 Explain the best location of oil holes in the bearing when the journal rotates in one direction and when it rotates in both the directions.

23.10 How are oil grooves located in a bearing for proper distribution of oil in the bearing?

23.11 What are the types of failure of journal bearing?

23.12 What is the shape of film for supporting a load in a journal bearing?

23.13 Using equation 23.10, calculate the safe load on a bearing with a 100 mm diameter and 150 mm length, if the viscosity Z at the operating temperature is 0.03 kg/ms and the journal rotates at 750 rpm.

23.14 The radial load on a 80 mm oil-ring bearing 100 mm long is 5 kN. The bearing clearance is 0.1 mm, and the journal turns at 300 rpm. Oil D, Table 23.1, is used, and the operating temperature is 55°C. Determine the load the bearing can carry when running at 480 rpm, taking into account (a) heat dissipated, and (b) the heat generated by friction at an oil temperature of 68°C.

23.15 The main bearings of a steam engine are 175 mm in diameter, the load coming upon each bearing in 66 kN and the engine speed is 135 rpm Determine (a) the length of the bearing, (b) the necessary running clearance, using SAE 60 oil, and (c) the power lost in friction at each bearing.

23.16 (a) Determine the necessary length of the bearing, running clearance, side relives, and loss of power due to friction for a journal which has a 63 mm diameter and rotates at 175 rpm. The total load on the bearing due to belt pull and weight of pulleys is 6 kN, and the load is steady. Use a lubricating oil which has a Saybolt viscosity of 255 at 54°C and for which $p = 0.905$ at 15°C. (b) Determine the temperature of the oil film and the necessary area to dissipate the heat of friction.

23.17 A locomotive weight 600 kN and 15 per cent of this weight is carried by four truck wheels 0.9 m in diameter having journals 110 mm in diameter and 150 mm long. The rest of the weight is distributed uniformly over eight 1.8 m driving wheels having journals 200 mm in diameter and 240 mm long. (a) Find the loss of power due to friction in the journals of the truck wheels with SAE 110 lubricating oil at a speed of 9 m/s. (b) Check the minimum oil-film thickness. The heat-dissipating area of each bearing is 0.278 m², and the air temperature is 27°C.

23.18 A journal bearing 63 mm in diameter and 95 mm long carries a load of 8 kN at 1,200 rpm. Assume a cooling area of 0.28 m² a cooling rate of 11 W/m²/°C, and a room temperature of 22°C. Use oil B, Table 23.1. Compute the oil-film temperatures and the minimum film thicknesses for relative clearances c/d of 0.001, 0.00133, 0.00167, and 0.002. Tabulate the calculations and plot the results.

23.19 The maximum load which is taken by the bearings of a punch press as the punch enters a steel plate is 760 kN. Assuming that this load is distributed equally between the two bearing, one on each side of the eccentric, determine the minimum diameter and length of the bearings.

23.20 A walking beam is 3.6 m long, and the bearing fulcrum is 1.2 m from one end and 2.4 m from the other. The long end pulls a vertical load of 70 kN. The short end is connected by a 1.8 m rod to a crankpin with a 0.3 m radius of rotation. The diameter of the fulcrum in is 150 mm. Assuming that oil H, Table 23.1, is used, and that thin-film lubrication exists, determine (a) the length of the fulcrum pin necessary to support the load and (b) the work lost in friction, per min, when the crankpin turns at 30 rpm.

23.21 A multicollar bearing similar to that shows in Fig. 23.30 has six collars which have an inside diameter of 65 mm and an outside diameter of 90 mm. Determine the thrust load which the bearing can carry and the power lost through friction if the shaft speed is (a) 200 rpm, (b) 600 rpm, (c) 1,200 rpm, and (d) 20 rpm (for slow and intermittent service).

24 Bearings with Rolling Contact

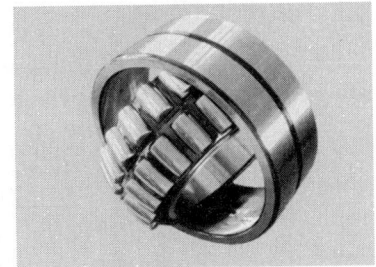

24.1 GENERAL CONSIDERATIONS

In a bearing with rolling contact the shaft is supported on rollers or balls. *These bearings were also called antifriction bearings which is a misnomer as the bearings have friction (24.1). A bearing of this type has the theoretical advantage of reduced friction (23.4)*, but to be practical it must fulfill the following conditions:

(a) Unavoidable sliding should be reduced to a minimum.
(b) The rolling elements must be properly guided in their motion.
(c) All rolling elements should be of exactly the same size.
(d) The rolling elements and their guides, or raceways, must be extremely hard and very smoothly polished.
(e) The pressure should be approximately normal to the surface of contact.
(f) The rolling elements must not be over-loaded.

Advantages: Well-manufactured bearings with rolling contact in properly designed applications have the following advantages over bearings with sliding contact:

(a) They maintain accurate shaft alignment over long period of time.
(b) They can carry heavy momentary over-loads without failure.
(c) Power loss caused by friction is small except at high speeds.
(d) They are particularly suitable for very low speeds.

(e) Starting friction is very low.
(f) Lubrication is simple and requires a little attention.
(g) Replacement in case of failure is easy.
(h) *Highly suitable for varying speeds as in a gear box (24.7).*
(i) Also suitable where lubrication is not convenient.
(j) Some varieties are suitable against electric currents.
(k) Equally suitable for any type of thrust loads.

Disadvantages: Against the advantages just mentioned, rolling-contact bearings have the following disadvantages:

(a) The design of the shaft and housing is more complicated.
(b) The first cost is higher.
(c) The housing diameter is larger, except with some needle bearings.
(d) The resistance to shock loads is lower.
(e) There is more noise, especially at higher speeds.
(f) They are sensitive to dirt and grit.

Classification: Bearings with rolling contact may be divided into two main classes: *Ball bearings*, and roller bearings with cylindrical, conical, spherical, or concave rollers.

Each of these classes may be subdivided into the following types: (a) radial bearing; (b) thrust bearings, and (c) radial-thrust bearings, or angular contact bearings, which can take both radial and axial forces.

24.2 BALL BEARINGS

Each radial ball bearing consists of four elements: an inner ring, or *inner race*, grooved on its outer surface; an *outer race*, grooved on its inner surface; steel *balls*; and a *ball retainer*, or cage, for spacing the balls so that they do not touch each other, in order to reduce wear and noise.

The assembly is done by keeping the two races eccentric to each other and the balls are inserted at the maximum clearance space. After inserting the balls, the balls are slowly shifted so that the races are made concentric and the balls are fitted with cages to retain their position.

Types: The various types of ball bearings are shown in Fig. 24.1. The single-row radial bearing, Fig. 24.1a, is usually made with a deep groove. The angular-type bearing, Fig. 24.1b, can take an axial load in addition to the radial load. In Fig. 24.1c is shown a double-row bearing for increased load capacity. Figure 24.1d shows a self-aligning double-row bearing, the inner surface of the outer race is part of a sphere.

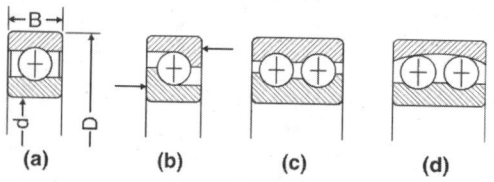

| (a) | (b) | (c) | (d) |

Fig. 24.1: Types of ball bearings

Standardization: Through the efforts of the Society of Automotive Engineers, the manufacturers of radial ball bearings have adopted the international standard dimensions, according to which the bearings are divided into three series called the *light, medium,* and *heavy* series. The light series is designated by the number 200; the medium, by 300; and the heavy, by 400. The adapter type is designated by 500 or 600.

In addition there is extra light series designated by number 100.

The inside diameter d, Fig. 24.1a, the outside diameter D, and the width B are in millimeters.

Some manufacturers make special, non-standard bearings with the dimensions d, D, and B in inches. The last two digits of a standard bearing number designate the bore d, which comes in multiples of 5 mm, beginning with 10 mm, except for two intermediate size, 12 and 17 nm. Beginning with $d = 20$ mm, the last two digits in the bearing number multiplied by 5 give the bore d. The number before the serial number is given by manufacturers in accordance with their systems. Thus, an SKF bearing No. 6309-A has a single row of balls, its bore is $9 \times 5 = 45$ mm, and it is of the medium series with $B = 25$ mm. This bearing is interchangeable with the bearing No. 1309, which is a self-aligning, double-row bearing; that is, both bearings have the same value of d, D, and B. The bearing mentioned is also interchangeable with any bearing of another make with has the last three digits 309. The international standards do not specify either the size or the number of the steel balls.

Load capacity: Theoretically, the contact between a ball and a race is a point. Actually, however, because of the elastic deformations of the material, a small area in the form of an ellipse supports the load. Races of radial ball bearings are always grooved, and the groove increases the area of contact of a ball under load. The groove has a radius about 4 per cent larger than the ball radius. The increased contact area permits the bearing to sustain larger loads with the same stress. *Under normal conditions the failure of a ball bearing is caused by a flaking of the surfaces under repeated high stresses (24.2).* Therefore, the larger the load, the shorter will be the bearing life, because it takes fewer repetitions of larger stresses than of lower stresses to produce an endurance failure.

In the beginning of motion a clean lubricant is used. *The failure of bearing can start in the form of pit either in rolling element or in the races. The pit will be responsible for producing noise and vibration and also introduce hard particles in the lubricant causing further deterioration of the surfaces (24.2).* More noise and vibration will be felt. It may result in grinding and finally stopping motion and causing damage to other parts.

There does not exist any standard method of rating the load capacity of ball bearings. According to research conducted by Stribeck,[1] a rational expression for static load capacity F is

$$F = KCid_o{}^2 \qquad \ldots (24.1)$$

where K is a factor which depends on the material and is determined from compression tests; C is the conformity factor and depends on the curvature of the ball relative to both the transverse curvature of the race and its curvature in the plane normal to the axis of the bearing; i is the number of balls; and d_o is the ball diameter. The determination of the factors K and C is rather involved; only a specialist is in a position to use equation 24.1 correctly. The manufacture of ball bearings is a highly specialized field requiring both technical knowledge and special precision production methods. The problem confronting the general machine designer is to select the most suitable bearing from the catalogue of a reliable manufacturer in accordance with the instructions given in the catalogue.

As already mentioned, neither the diameter d_o of the balls nor their number i is standardized. Moreover, there are some differences in the materials used and in the factor of safety assumed. These differences, when considered in conjunction with equation 24.1, explain the variations in the rating of ball bearings of different manufacturers. Thus, the load ratings of some manufacturers, for interchangeable bearings with the same diameters d and D, are sometimes 100 per cent higher than those of other manufacturers.

Basic static load rating: The magnitude of the basic static load rating is

$C_o = f_o\, i\, d^2 \cos \alpha$

i = number of balls in bearing

α = nominal angle of contact

= nominal angle between the line of action of the ball load and plane perpendicular to the bearing axis.

z = Number of balls per raw

d = ball diameter

Values of the factor f_o for different kinds of bearings are given in Table 24.1.

Static equivalent load: The magnitude of the static equivalent load F for radial bearings under combined radial and thrust load is the greater of the loads given below.

$F = (X_o F_r + Y_o F_a)$

$F = F_r$

where, X_o = a radial factor

F_r = the radial load

Y_o = a thrust factor

F_a = the thrust load

Values of X_o and Y_o are given in catalogues.

Table 24.1: Factor f_o for various types of bearings

Bearing type	$f_o\, Nm$
Self-aligning bearing	3.34×10^6
Radial and angular contact deep groove ball bearing	12×10^6

Life of bearings: When a group of identical ball bearings is run under identical conditions of load and speed until they all fail by fatigue of the surfaces, it will be found that a few bearings will fail early, others will last longer, and some will last about four times as long as the average life L_m of the whole tested group. This dispersion in their lives is caused by unavoidable variations in the steel properties, heat treatment, surface characteristics, and dimensions of the balls.

Ball bearings fail due to repetition of compressive stresses created in the surfaces of the balls and races. Therefore the term *life* may be defined as the number of million revolutions or hours of operation at a given speed.

The rating life of a bearing is the life which 90% of a group of bearing will reach or exceed before

---

[1]S. Stribeck, "Kugellager für beliebege belastung," zeitschrift Verein Deutscher Ingenieure, Vol. 45 (1901), p. 121.

the first evidence of fatigue develops. The life which 50 per cent of the group of ball bearings will complete or exceed is approximately five times this rating life.

If the number of bearings tested is 30 or more, it is found that 90 per cent of the bearings have a life longer than one-fifth of the average life L_m. Thus the average life is a good criterion of the quality of the bearings and may be used for estimating the cost of replacement of bearings in a plant.

Requisite bearing life for different types of machines: When the size of the bearing necessary for a particular machine is to be ascertained, difficulty is often experienced in deciding the bearing life to be aimed at. In such cases experience must be used as guide. Values that are regarded as normal are given in the Table 24.2. (Given in SKF catalogue).

Influence of load: Tests have shown that the life L of bearings is a function of the load. Tests also show that with sufficient accuracy, L may be considered to be inversely proportional to the third power of the load. Thus, if one group of test bearings of a certain type and size are run under a constant load F_1 and have a life L_1, and another group are run under a load F_2, but otherwise under identical operating conditions, and have a life L_2, then

$$\frac{L_1}{L_2} = \left(\frac{F_2}{F_1}\right)^3 \qquad \ldots (24.2)$$

If L_2 equal to one million revolutions under a constant load C, called the *dynamic specific capacity* of a bearing, then the life L_n of the bearing, also in millions of revolutions, under a constant load F can be presented by the more general equation

$$L_n = \left(\frac{C}{F}\right)^3 \text{ for ball bearings and}$$

$$\left(\frac{C}{F}\right)^{\frac{10}{3}} \text{ for roller bearings} \qquad \ldots (24.3)$$

The numerical values of C, in kgf or N, are given for all bearings in the SKF catalogue.

Table 24.2: Bearing life

Class of machine	Life in working hours L_h
1. Instruments and apparatus that are used only seldom. Demonstration apparatus, mechanisms for operating sliding doors.	500
2. Household machines, instruments, technical equipment for medical use.	300–3000
3. Aircraft engines.	1000–2000
4. Machines used for short periods or intermittently and whose breakdown would not have serious consequences: Hand tools, lifting tackle in workshops, hand operated machines, generally agricultural machines, cranes in errecting shops, domestic machines.	4000–8000
5. Machines working intermittently and where breakdown would have serious consequences. Auxiliary machines in power stations, conveyor plant for flow production, lifts, cranes for piece goods; machines tools used infrequently.	8000–12000
6. Machines for use 8 hours per day and not always fully utilized: Stationary electric motors, general purpose gear units.	12,000–25,000
7. Machines for use 8 hours per day and fully utilized: Machines for engineering industry generally; cranes for bulk goods, ventilating fans, countershafts.	20,000–30,000
8. Machines for continuous use 24 hours per day: Separators, compressors, pumps, mine hoists, stationary electric machines, machines in continuous operation on board naval ships, rolling mills.	40,000–60,000
9. Wind energy machinery: Main shaft, yaw, gear box, generator bearing.	30,000–100,000
10. Machines, required to work with high degree of realiability 24 hours per day: pulp and paper machinery; public power plants; mine pumps, pumps in water works, machinery in continuous operation on board merchant ships.	100,000–200,000

Basic dynamic capacity or dynamic specific capacity is a constant load under the influence of which the bearing will attain a calculated life of one million revolutions. In the case of radial bearings, the basic capacity refers to pure radial load and in the case of thrust bearings to a centrally acting pure thrust load.

Basic static capacity or rating is defined as that static load which corresponds to a total permanent deformation of ball and race at the most heavily stressed contact of 0.0001 of ball diameter. Static equivalent load is defined as that static radial load or static axial load which if applied would cause the same total permanent deformation at the most heavily stressed ball and race contact as that which occurs under actual load conditions, i.e. combination of radial and axial loads.

With a speed of 2000 rph, 1,000,000 revolutions are obtained in 500 hr, and equation 24.3 may be presented graphically, as shown in Fig. 24.2. *Curve SKF shows that if the load is decreased to one-half, the life is increased from 500 to 4000 hr (8 times).* The manufacturers on New Departure ball bearings used in equation 24.2 is an exponent of 4. Also, they take as average life, 3,800 hr. As a result the curve *ND* has a different slope and runs higher.

The ratio of the actual load F to the basic load F_c given in a catalogue may be called the load factor K_l.

Thus

$$K_l = F/F_c \qquad \ldots (24.4)$$

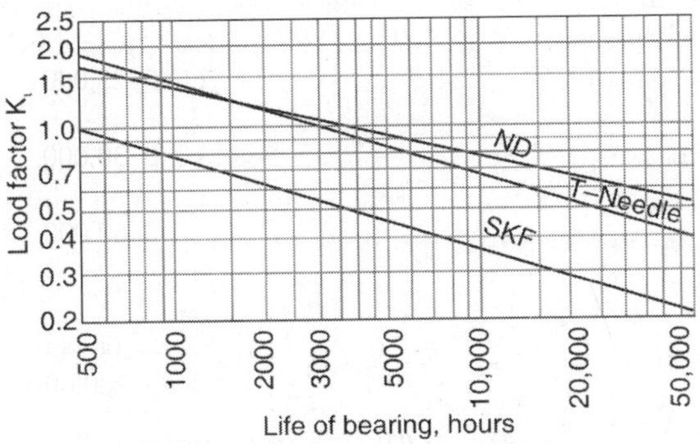

Fig. 24.2: Life curves of ball bearings

Influence of speed: The number of load repetitions increases with an increase of the rotative speed, thus resulting in a shortened bearing life expressed in revolutions or the time units at a certain speed. The relation between the allowable load and the speed for a certain bearing life may be represented by the same curve, Fig. 24.2, but with a change in the scale of the abscissas. However, most bearing catalogues give the allowable load ratings directly for different speeds, from a minimum to the highest speed advisable for a given bearing.

The SKF engineers use a different procedure, introducing a speed factor f_n and a life factor f_h. With these factors given in tables of bearing catalogue the permissible load F is determined by the relation

$$F = f_n C / f_h \qquad \ldots (24.5)$$

Example 24.1: Determine the load capacity of a single row deep groove SKF bearing—C = 4700 N medium series that must run for 2000 hrs. at 700 rpm.

From catalogue, f_n =0.362, f_h = 1.585. Hence by equation 24.5,

$$F = \frac{0.362 \times 4700}{1.585} = 1073 \text{ N}$$

Also by equation 24.3,

$$\frac{2000 \times 700 \times 60}{1,000,000} = \left[\frac{4700}{F} \right]^3$$

$$F = 1073 \text{ N}$$

Materials: Under normal load conditions the magnitude of the compressive stresses occurring at the contact areas of balls and races varies from 1300 to 2000 N/mm². It is, therefore, evident that only special alloy steels, with nickel and either chromium or molybdenum, vanadium, that have been given extraordinary and uniform hardness and toughness by heat treatment can render satisfactory service.

In special cases: Ceramics, silicon nitride is used for rolling elements. The cages are made from brass, low carbon steel, stainless steel, polyamide and glass reinforced polyether therketone (PEEK).

Accuracy: The balls must be true spheres, and to secure an even load distribution the actual size should not vary from the nominal size by more than

0.00125 mm. All other dimensions, such as the diameters d and D, and the radii of fillets, are machined to tolerances expressed in microns. Balls and races are ground and highly polished.

24.3 SELECTION OF RADIAL BALL BEARINGS

In selecting a ball bearing from a trade catalogue for a specific installation, three main points must be considered:

(a) The bearing must be of the series best suited to the installation, in regard to both capacity and dimensions.

(b) The type of bearing selected—radial, thrust, or combined—must be suitable for the type of imposed load.

(c) The size of the bearing must be such as to give the required length of service with sufficient assurance.

Light series: Light-series bearings should be used where loads are moderate and shaft sizes are comparatively large, as for hollow shafts or for long shafts whose diameter is influenced not by strength but by requirements of rigidity. They should also be used where the housing space requires the narrowest bearing width and the smallest outside diameter available for a given bore size.

Medium series: Medium-series bearings provide a capacity increase of approximately 30 to 40 per cent. They should therefore be employed where loads are heavy and considerable bearing capacity in proportion to the shaft size is desirable.

Heavy series: Heavy-series bearings have capacities approximately 20 to 30 per cent greater than those of the medium series. They are applied only to specially proportioned shafts and are recommended only for special installations.

Bearing types: Single-row bearings are used when the main load acts radially. However, they can also carry a certain axial load, usually not more than 50 per cent of the applied radial load and suitable for light varying load.

Deep-groove single-row bearings have a slightly smaller radial capacity because of the use of fewer balls, but their axial capacity is about 75 per cent of the rated radial capacity.

Angular-contact bearings, Fig. 24.1b, have an axial capacity of 100 to 200 per cent of the rated radial capacity, but they resist axial thrust in only one direction. If an axial thrust occurs in both directions, then two angular-contact bearings must be used, as shown in Fig. 24.3a. If great radial rigidity only is desired, as in machine-tool spindles, the mounting shown in Fig. 24.3b is used. A tandem mounting, Fig. 24.3c, is used for an extremely large thrust load in one direction.

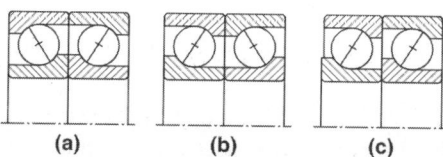

Fig. 24.3: Methods of mounting two angular-contact bearings

These bearings are not recommended for a light axial load.

Double-row bearings are used either to obtain a larger radial and axial capacity, Fig. 24.1c, or to have the self-aligning feature, Fig. 24.1d, if the shaft is not very rigid.

Insocoat bearings: Bearings in electric motors, generators and similar applications are subjected to electric current which can damage the surfaces of rolling elements and cages and degrade grease rapidly. This type of damage is called electric erosion. It gets increased if these machines are controlled by frequency converters. There is also a risk for high frequency bearing currents occurring in applications due to stray capacitances in electric machinery. Insocoat bearings are economically designed by SKF. Electric insulation is integrated with the bearing and it gives increased reliability. This is obtained by SKF plasma technique to provide 100 μm thick layer of aluminium oxide on the external surfaces of outer and inner races. This can withstand 1000 V, DC.

Hybrid bearings: Hybrid bearings have rings of bearing steel and rolling elements of bearing grade silicon nitride, Si_3N_4. In addition to being excellent electric insulators, these bearings have a higher speed capability and provide larger service life than all steel bearings. Excellent electrical insulation property is one of the essential features of the silicon nitride. It protects the rings from electric current damage and so gives higher service life. An additional advantage is that it has less weight as the density is 40% of that of bearing steel. This results in lower inertia and so less stresses at sudden starts and stops. It has also less friction at high speeds so run at lower temperature and larger lubricant service life. Silicon nitride has a higher hardness and higher modulus of elasticity than that of steel resulting in higher bearing stiffness and longer life. It has a lower thermal expansion than steel so less effect of temperature changes.

No wear bearings: To withstand severe operating conditions such as risk of smearing, boundary lubrication, sudden speed and load variation, low load or high operational temperature. No wear bearings are the solution offered by SKF.

No wear bearings are equipped with low friction ceramic coating on contact surfaces.

No wear coating material. A physical vapour deposition process applies the low friction ceramic coating which provides the resilience of the underlying material, low friction and wear resistance but with hardness and can withstand temperatures up to 350°C. No wear bearings are not intended for vacuum or other completely dry running applications.

To furnish a general idea of the loads which ball bearings can carry, SKF bearings are taken as an example. Table 24.3 gives the values of C/F for ball and roller bearing.

Size of bearing: In the majority of applications, ball bearings have to resist some combinations of radial and thrust loads. The bearing rating is always referred to a radial load. Therefore, the combined load must be reduced to an equivalent radial load. Various catalogues give different methods of reduction. In general, the equivalent load F_e may be computed by the equation

$$F_e = (VX F_r + Y F_a)C \qquad \dots (24.6)$$

where, F_r is the actual radial load, F_a is the actual axial load, and X and Y are coefficients which depend on the ratio $\dfrac{F_a}{F_r}$. Values of X and Y are given in catalogues. For most bearings, it may be assumed that $X = 1$ and $Y = 1.5$. For angular-contact bearings, $X = 0.5$ and $Y = 2.5$.

V = is a rotational factor

= 1 for inner ring rotating in relation to load.

= 1.2 for inner ring stationary in relation to load.

n_s = Shock factor / safety factor given in Table 24.4.

C = Service or safety factor (Table 24.4)

Equivalent load under condition of varying load is given by constant cubic mean load or mean effective load F_m which gives the same life as the variable loads.

(a) If load F_1 acts for N_1 revolutions or $= \mu_1 L$ where μ_1 is the fraction of the life L. F_2 acts for N_2 revolutions or $= \mu_2 L$ and so on

$$F_1^3 N_1 = \mu_1 C^3 \left[\text{as } \mu_1 L = N_1 = \mu_1 \left(\frac{C}{F_1} \right)^3 \right]$$

$$F_2^3 N_2 = \mu_2 C^3$$

$$F_n N_n = \mu_n C^3$$

Adding these equations

$$\Sigma F^3 N = \Sigma (\mu) C^3 = C^3$$

Because $\quad \Sigma \mu = 1$

$$L_n = C^3 \bigg/ \left[\frac{\Sigma F^3 N}{L_n} \right]$$

Where

$$L_n = N_1 + N_2 + \dots + N_n$$

$$F_m = \sqrt[3]{\frac{\Sigma F^3 N}{L_n}} \qquad \dots (24.7)$$

Table 24.3: Loading ratio $\frac{C}{F}$ for different lives

Life in millions of revolutions L	$\frac{C}{F}$		Life in millions of revolutions L	$\frac{C}{F}$	
	Ball bearings	Roller bearings		Ball bearings	Roller bearings
0.5	0.793	0.812	600	8.43	6.81
0.75	0.909	0.917	650	8.66	6.98
1	1	1	700	8.66	7.14
1.5	1.14	1.13	750	9.09	7.29
2	1.26	1.24	800	9.28	7.43
3	1.44	1.39	850	9.47	7.56
4	1.59	1.52	900	6.65	7.70
5	1.71	1.62	950	9.83	7.82
6	1.82	1.71	1000	10	7.94
8	2	1.87	1100	10.3	8.17
10	2.15	2	1200	10.6	8.39
12	2.29	2.11	1300	10.9	8.59
14	2.41	2.21	1400	11.2	8.79
16	5.52	2.30	1500	11.4	8.97
18	2.62	2.38	1600	11.7	9.15
20	2.71	2.46	1700	11.9	9.31
25	2.92	2.63	1800	12.2	9.48
30	3.11	2.77	1900	12.4	9.63
35	3.27	2.91	2000	12.6	9.78
40	3.42	3.02	2200	13	10.1
45	3.56	3.13	2400	13.4	10.3
50	3.68	3.23	2600	13.8	10.6
60	3.91	3.42	2800	14.1	10.8
70	4.12	3.58	3000	14.4	11
80	4.31	3.72	3500	15.2	11.6
90	4.48	3.86	4000	15.9	12
100	4.64	3.98	4500	16.5	12.5
120	4.93	4.20	5000	17.1	12.9
140	5.19	4.40	5500	17.7	13.2
160	5.43	4.58	6000	18.2	13.6
180	5.65	4.75	7000	19.1	14.2
200	5.85	4.90	8000	20	14.8
250	6.30	5.24	9000	20.8	15.4
300	6.69	5.54	10000	21.5	15.8
350	7.05	5.80	12500	23.2	16.9
400	7.37	6.03	15000	24.7	17.9
450	7.66	6.25	17500	26	18.7
500	7.94	6.45	20000	27.1	19.5
550	8.19	6.64	25000	29.2	20.9

or
$$L_n = \left(\frac{C}{F_m}\right)^3 \qquad \text{... (24.8)}$$

where, F_m is the constant cubic mean load or mean effective load which gives the same life as the actual variable load.

(b) If the load diagram has a complicated form, it is necessary to divide the load diagram into infinite number of small parts, i.e.

replace $\Sigma F^3 N$ by $\int_0^{L_n} F^3 dN$, then

$$F_m = \sqrt[3]{\frac{\int_0^{L_n} F^3 dN}{L_n}} \qquad \text{... (24.9)}$$

This can be easily solved by graphical method.

If the load varies periodically, the cubic mean load for all periods is obviously same as the cubic mean load for one period.

It is then possible in equation 24.9 to let L_n represent the number of revolutions in one period.

(c) If F varies from a minimum F_{min} to maximum F_{max} reasonably continuously and gradually, the following approximate formula can often be used

$$F_m = \frac{1}{3}F_{min} + \frac{2}{3}F_{max} \qquad \text{... (24.10)}$$

or $= 0.68 F_{max}$

(d) If the speed of rotation is constant, the load varying with time

$$F_m = \sqrt[3]{\frac{\Sigma F^3 t}{T}} = \sqrt[3]{\frac{\int_0^T F^3 dt}{T}} \qquad \text{... (24.11)}$$

where,

F = the force at any instant of time t

T = time for one cycle of load variation.

(e) If the load is constant, and the speed varies, the coverage speed may be used since fatigue occurs in bearings after a certain number of stress repetitions.

Shock: The load capacities given in trade catalogues are based on a steady load. In the event of shock action, the equivalent load F_e must be multiplied by a safety factor or service factor C_1. Values of C_1 recommended by the Marvin-Rockwell Corporation are given in Table 24.4. It should be noted that these values also include life-load factors. If the latter are taken into account separately, the column for intermittent service should be used.

Speed influence: The wear of a ball bearing increases, and its life decreases, with an increase of the speed of one race relative to that of the other race. If n_1 is the speed of the inner race and n_2 is the speed of the outer race, the *effective speed* n_e which determines the life of the bearing may be found from the relation[2]

$$n_e = n_1 \pm n_2 \qquad \text{... (24.12)}$$

In which the plus sign must be used when the races rotate in opposite directions and the minus sign must be used when they rotate in the same direction.

Temperature: A ball bearing should not be exposed to a temperature over 95 to 105°C. A

Table 24.4: Safety factors for ball bearings C_1

Load conditions	Safety factor, life of 5 to 10 years			Safety factors life of 10 to 20 years Continuous
	Intermittent	10 hr per day	Continuous	
Steady load	0.5–1	1.5	2	3
Light shock	1–2	2.5	3	4
Moderate shock	2–3	3.5	4	5
Severe shock	3–4	4.5	5	6

[2] General Motors Corporation, New Departure Division, New Departure Handbook, 18th ed. (Bristol, Conn.: 1946), p. 9.

higher temperature may not affect the load-carrying capacity of the bearing, but at a higher temperature the lubricating oil will begin to evaporate and there will be danger that the bearing may run dry. With a special lubricant a ball bearing may operate, without any loss of load capacity, at temperatures up to 150°C or 160°C.

Bearing life: The life-load factor K_l, equation 24.4, may be taken from the corresponding curve, Fig. 24.2, New Departure bearings, or it may be taken from catalogue tables for bearings of other manufacturers.

Selection of size: After the equivalent load F_e, the safety factor C_1 for shock, and the life-load factor K_l are established, the proper size of a bearing will be that whose capacity rating F_c, at the proper speed, as given in the manufacturer's catalogue, satisfied the condition

$$F_c \geq \frac{C_1 F_e}{K_l} \qquad ...(24.13)$$

Corresponding to bearing life, loading ratio $\dfrac{C}{F}$ is selected from SKF catalogue for ball bearing or roller bearings. Knowing the value of $\dfrac{C}{F}$, the basic dynamic capacity C is calculated as $C = \dfrac{C}{F} \times F_e$. For this value of C and d the shaft diameter, an appropriate bearing is selected.

Procedure for selections of bearing

(i) For a given machine, life in hours is selected from Table 24.2.

(ii) The life is calculated in millions of

$$\text{revolutions } = \frac{L \times n \times 60}{10^6} \text{ mR}$$

(iii) For this life find $\dfrac{C}{F}$ from Table 24.3.

(iv) Knowing the reaction on the shaft where bearing is to be fitted, calculate equivalent bearing load $F_e = (V \times F_r + Y F_a)C_1$ and find the proper capacity rating

$$F_e' = \frac{F_e}{k_l}$$

(v) Find basic dynamic capacity

$$C = \frac{C}{F} \times F_e'$$

(vi) Select appropriate bearing for this value of C and diameter of shaft, with bearing designation.

Example 24.2: An SKF bearing must be selected for the end of a 90 mm shaft which can be reduced to 70 mm, if necessary. The shaft speed is 230 rpm, and the shaft carries a radial load of 17 kN with light shocks, and also an axial load of 5.4 kN. A bearing life of 1 years with 10 hr of service a day, 5 days per week, is desired. Take $X = 1$, $Y = 1.5$

The equivalent radial load, by equation 24.6, is

$$F_e = 17000 + 1.5 \times 5400 = 25100 \text{ N}$$

The desired minimum bearing life is $1 \times 52 \times 5 \times 10 = 2600$ hr, for light shock C_1 may be taken as $= 1.5$.

$$LmR = \frac{2600 \times 230 \times 60}{10^6} = 35.88$$

For this life $\dfrac{C}{F} = 3.3$

$$F_e = 1.5 \times 25100 = 37650$$

and $\qquad C = 3.3 \times 37650 = 124245 \text{ N}$

According to SKF catalogue, a single-row deep-groove bearing No. 7314, with a bore $d = 70$ mm and $C = 127$ kN, may be used, requiring the shaft to be of 70 mm diameter at the end.

Bearings with reliability of more than 90%: The catalogues give the data for bearings of life having reliability of 90%. The life is defined as the life that 90% of bearings of a sufficiently large group of apparently identical bearings can be expected to attain or exceed before fatigue failure.

The reliability is defined as:

$$R = \frac{\text{No. of bearings which successfully completed LmR}}{\text{No. of bearings tested}}$$

The reliability of bearings in the catalogue is 90% or 0.90. In many applications as in space, there is need for much higher reliability. The relation of life and reliability is given by Wiebull as

$$R = e^{-\left(\frac{L}{a}\right)^b}$$

where, a and b are constants. These are evaluated as

$$\frac{1}{R} = e^{\left(\frac{L}{a}\right)^b}$$

$$\log_e \frac{1}{R} = \left(\frac{L}{a}\right)^b$$

If L_{90} represents a reliability of 90%, then

$$\log_e\left(\frac{1}{R_{90}}\right) = \left(\frac{L_{90}}{a}\right)^b$$

Dividing these two

$$\frac{\log_e \dfrac{1}{R}}{\log_e \dfrac{1}{R_{90}}} = \frac{\left(\dfrac{L}{a}\right)^b}{\left(\dfrac{L_{90}}{a}\right)^b} = \left(\frac{L}{L_{90}}\right)^b$$

$$\frac{L}{L_{90}} = \left[\frac{\log_e \dfrac{1}{R}}{\log_e \dfrac{1}{R_{90}}}\right]^{\frac{1}{b}}$$

If L_{50} is the life which 50% of bearings will complete or exceed, then

$$L_{50} = 5\, L_{90}$$

For this condition, value of $a = 6.54$ and $b = 1.17$. This equation can be used for calculating the life for any reliability.

Example 24.3: It is desired to install a bearing in a component of aircraft with a reliability of 99%. The bearing is subjected to a radial load of 7.5 N and is expected to have a life of 3000 hrs at 1000 rpm. Find the dynamic load capacity of the bearing to be selected from S.K.F. Catalogue. If the reliability required in 99.2%, what is the dynamic load capacity as compared to 99% reliability.

$$\frac{L_{99}}{L_{90}} = \left[\frac{\log_e \dfrac{1}{99}}{\log_e \dfrac{1}{R_{90}}}\right]^{\frac{1}{1.17}} = 0.1342$$

$$L_{99} = \frac{60 \times 3000 \times 1000}{10^6} = 180 \text{ mR}$$

$$L_{90} = \frac{180}{0.1342} = 1341.3 \text{ mR}$$

$$C = 7500[1341.3]^{\frac{1}{3}} = 82712.3 \text{ N}$$

$$\frac{L_{99.2}}{L_{90}} = \left[\frac{\log_e \dfrac{1}{99.2}}{\log_e \dfrac{1}{90}}\right]^{\frac{1}{1.17}} = 0.111$$

$$L_{90} = \frac{180}{0.111} = 1621.62 \text{ mR}$$

$$C = 7500[1621.62]^{\frac{1}{3}} = 88113.8 \text{ N}$$

24.4 INSTALLATION OF BALL BEARINGS

In the installation of ball bearings, the following points must be considered separately:

(a) The shaft must be designed to take the inner race.

(b) A suitable mounting must be designed for the outer race.

(c) Lubrication must be provided for the bearing.

(d) Methods of sealing the lubricant and preventing penetration of foreign matter must be provided.

Shaft design: It is very important that the inner race be a light press fit on the shaft, about k6. If the fit is too loose, the inner race will slip; if it is too tight, the shaft may stretch the race, and the life of the bearing may be shortened.

The inner race must be clamped between a shoulder and a nut, as in Fig. 24.4a. The nut may be secured against unscrewing by the washer a with tongues, the outer one being bent

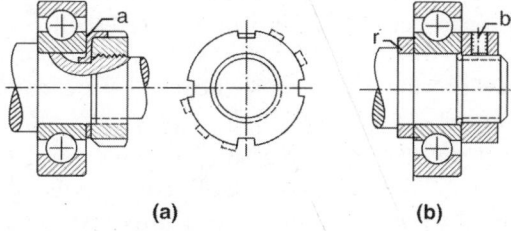

(a)   **(b)**

Fig. 24.4: Methods of fastening ball bearings to a shaft

into a slot of the nut; or a setscrew may be used, as in Fig. 24.5b at *b*. It is essential to provide a sufficiently high shoulder on the shaft to locate the bearing positively. If the shoulder is so low that it would enter the corner radius of the bearing, a special shoulder ring must be made, as in Fig. 24.5b. This ring should be a tight fit, k7, and should have a corner radius *r* exactly equal to the fillet radius on the shaft.

Corner fillet: For maximum strength the shaft requires a definite fillet at the junction of the bearing seat and the locating shoulder. However, it is very important that the shaft-fillet radius *r*, Fig. 24.5a, be smaller than the corner radius *R* of the bearing. If the bearing is located at a point of the shaft where the latter is under stress, it is imperative to have in the shaft a large fillet radius *r* to reduce stress concentration. In such a case a special shoulder ring *e*, Fig. 24.5b, should be used.

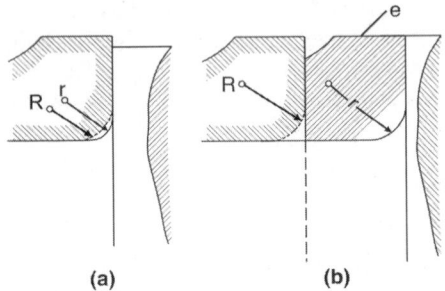

(a)   **(b)**

Fig. 24.5: Fitting of inner race to a shaft

Shaft rigidity: Ball bearings give the best service where the shaft and its supports are rigid. When a shaft is comparatively long, its diameter between bearings must be large enough to properly resist bending. In general, the use of more than two bearings on any single shaft should be avoided, because of the difficulty of securing accurate alignment. With bearings mounted close to each other, inaccurate alignment may produce dangerously heavy bearing loads.

Bearing adapters: For shafts without shoulders, such as transmission shafts, bearing adapters are used. In Fig. 24.6a, a split conical sleeve *a* is tightened by a nut *b* and secured by a steel-wire snap ring *c*. In Fig. 24.6b, the inner race is

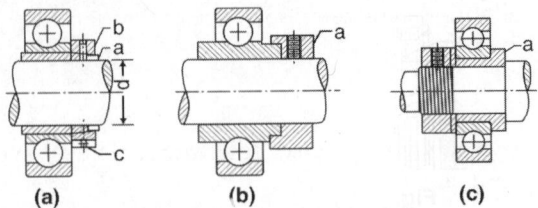

(a)   **(b)**   **(c)**

Fig. 24.6: Types of bearing adapters

extended and the end is turned eccentrically; the ring *a* has an eccentric counterbore fitted accurately over the end of the race; and a slight turn of this ring creates enough pressure between the bearing race and the shaft to hold the race firmly by friction. Figure 24.6c shows an adapter to be used if the bearing bore is larger than the shaft diameter.

Housing design: Where two radial bearings are to be used, it is usually desirable to locate the shaft axially by clamping one of the bearings both on the shaft and in the housing. When this is done, the other bearing should have an unrestricted axial clearance, Fig. 24.7, of 0.25 to 0.4 mm. In this manner, shaft expansion and variations in housing and shaft machining cannot so combine as to place the bearings under a useless, and possibly an injurious, thrust. This method of fixing in shaft and housing is known as axial retainment of bearing (24.7).

In many cases where there is no thrust load and the axial location of the shaft need not be closely maintained, it is entirely practicable to bore both housings straight through without shoulders, and so machine the closure caps that the total axial movement of the bearings in the housing is from 0.4 mm to 0.5 mm. In such a mounting the bearings may be simply pressed on the shaft.

Clamping to the housing: If there is an axial thrust or a fixed shaft location, the outer bearing race must be clamped to the housing. The simplest method of clamping is by means of a machined locating shoulder *s*, Fig. 24.7. The clamping cover must have a narrow register fitting into the housing bore. This register is very important when the clamping piece must also act as a closure about the shaft to insure a leakproof joint.

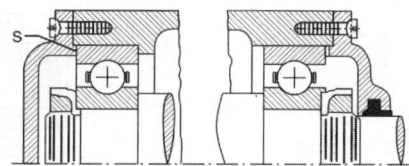

Fig. 24.7: Ball-bearing mounting

Lubrication: The main functions of ball bearing lubricants are: (a) to assist in the dissipation of heat caused by the deformation of load-carrying members; (b) to provide a lubricating film between the balls and the cage pockets; (c) to support any contact between the load-carrying members that is not pure rolling; (d) to protect the polished surfaces of the balls and races from rust and corrosion; (e) to assist the housing closure in excluding dust, dirt, and other foreign matter from the bearing.

In general, either mineral oil or grease may be used. Vegetable oils should never be used, since they turn rancid and tend to become acid. Where an economical means of supplying oil can be used and where a slight leakage past the bearing closure is not objectionable, oil is the most suitable lubricant. However, where a minimum of attention is essential, the use of grease is indicated, provided the speed is not too high.

Oil lubrication: For moderate speeds, the simplest method of lubricating a bearing of a horizontal shaft is to provide an *oil bath*, with the level reaching slightly above the center of the lowest ball. Another method is *wick-feed lubrication*. For higher speeds, to prevent churning of oil and the ensuing frictional heat generation, the oil should be supplied in small quantities, as by a *sight-feed drop oiler*, which can be adjusted to supply a drop every hour. The drop is broken up into mist by the rapidly rotating parts and thus reaches all points.

In enclosed assemblies where rotating parts require oil for their own lubrication, as in gearboxes, the ball bearings may be conveniently lubricated by the splash or mist of oil coming from these parts.

Grease lubrication: With grease lubrication, the use of closures and bearing housings of

proper design reduces the loss of lubricant to a minimum. Renewal should not be required in less than six months, and the interval may be even greater.

Bearings with solid oil: Ordinary greases and lubricating oils provide satisfactory lubrication in most applications. However where there is lack of accessibility (i.e. re-lubrication is impossible) or where good contaminant exclusion is required, solid oil—third choice of lubrication is an answer. It provides lubrication for life and good sealing, e.g. outdoor lifting arrangement and in vertical shafts. Solid oil consists of a polymer matrix which is saturated with lubricating oil. The polymer material has a structure with millions of micro-pores which hold the oil. The pores are so small that oil is retained in the material by surface tension. Oil represents an average of 70% by weight of the material. The oil filled polymer is molded into the bearing. A very narrow gap will form around the rolling elements and races during the molding process. The oil which seeps into the gap, provides good lubrication and is better than grease. Solid oil uses the cage as a re-inforcement element and rotates with it. A consistent oil film is provided between metallic surface and solid oil. Though solid oil keeps contaminants out of bearing but seals are needed for proper protection.

Sealing devices: In general the devices employed for sliding bearings, Fig. 23.24, may also be used for ball bearings. Because of absence of bearing wear, they all give much better results. Some manufacturers of ball bearings make bearings with closures as part of the bearing.

24.5 THRUST BALL BEARINGS

Figure 24.8 shows the various types of thrust ball bearings. That Fig. 24.8a shows a one-direction thrust bearing with flat parallel seats; that Fig. 24.8b shows a self-aligning one-direction thrust bearing; and that Fig. 24.8c shows a two-direction thrust bearing with an inside shoulder. Single direction thrust bearing can accommodate axial loads in one direction and locate a shaft axially in one direction.

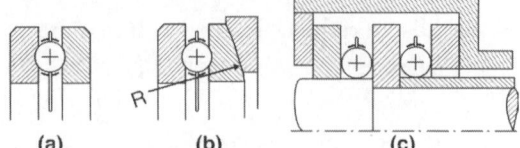

Fig. 24.8: Types of thrust ball bearings

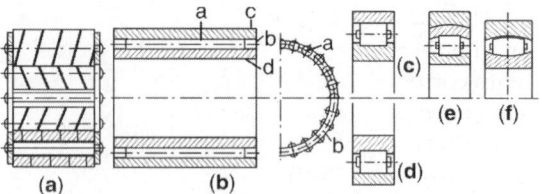

Fig. 24.9: Types of radial roller bearings

The outside dimensions of one-direction thrust bearings have been standardized by the Society of Automotive Engineers and SKF for both flat-seat and self-aligning types, for light and medium series, with the dimensions given in mm. Two-direction thrust-bearing sizes are standardized by SKF. However, some manufacturers do not make any disk-race thrust bearings at all, since angular-contact radial bearings have greater axial-thrust capacity than do thrust bearings of the types shown in Fig. 24.8.

All data necessary for the selection and mounting of thrust bearings is similar to ball bearings.

24.6 ROLLER BEARINGS

According to the load application, roller bearings may be classified as radial, angular, and thrust. In this section, only radial roller bearings will be discussed.

The *Hyatt bearings*, Fig. 24.9a, have cylindrical hollow rollers that are wound helically from flat-strip chrome-nickel steel and are ground off on the ends to the proper length. The windings are alternately right- and left-hand. The rollers are flexible and adjust themselves to slight irregularities on the surfaces of the races. The hollow spaces within the rollers are used as containers for the lubricant, which the helical channels distribute uniformly over the bearing surfaces. In the *Standard type* of Hyatt bearing, the rollers are in direct contact with the shafts, the outer race being split. In the *heavy-duty type* the specific pressures are higher, and the shaft must be hardened by heat treatment at the place of contact with the rollers, or else a hardened-steel sleeve must form the bearing surface. Dimensions of these bearings and load

capacities may be found in the manufacturer's catalogues or in handbooks.

Hyatt bearings with helical rollers are not affected by temperatures up to 175°C. At higher temperatures their load rating should be multiplied by a factor designated as k_t. It is 0.87 for 200°C, 0.73 for 220°C, and 0.60 for 260°C.

A roller bearing with long rollers is shown in Fig. 24.9b. The long rollers *a* are held in alignment by a bronze or steel cage *b*, and they roll between a solid inner and outer race *c* and *d*. The rollers and races are made of alloy steel that is hardened and ground. The inner race is sometimes omitted, and the rollers bear directly on the shaft. In this instance, however, the load capacity is lowered by the comparative softness of the shaft material.

The Norma-Hoffman bearings, SKF bearings and FAG bearings usually have short solid cylindrical rollers with lips on the inner race, as shown in Fig. 24.9c. There are very large number of types of roller bearings. Some are: taper roller single row bearings, taper roller double row bearings, four row roller bearings, spherical roller bearings, self aligning spherical roller bearings. A special one: CARBTROIDAL roller bearings described below.

To carry a small one-direction thrust, there may also be lips on one side of the outer race, as in Fig. 24.9d; and for a small two-direction thrust, there may also be lips on both sides of the outer race. These bearings can be used up to 1,500 rpm for all sizes, and up to 5,000 rpm for the smaller sizes. The overall dimensions are the same as those of the corresponding series of ball bearings, whereas the load capacities are about 100 per cent higher. Figure 24.9e shows a self-aligning Hoffman bearing.

The *SKF roller bearings* are built similar to those in Figs 24.9d and e. They are interchangeable with corresponding ball bearings and have load capacities from 50 to 100 per cent greater than the latter.

CARB TOROIDAL roller bearings: These were introduced by SKF in 1995. It is used where axial displacement is large and also shaft may be misaligned. It combines the self-aligning capability of spherical roller bearing with unconstrained axial displacement of cylindrical roller bearings. It is also compact, light, with reduced vibration characteristic, low friction and high load carrying capacity. The angular misalignment can be up to 0.5° between inner and outer rings and axial displacement of up to 2 mm with a bore of 25 mm and up to 46 mm with a bore of 1 m.

The CARB bearing Fig. 24.9f is a single row bearing with long, slightly crowned symmetrical rollers. The raceways of both inner and outer rings are concave and symmetrically placed with respect to bearing center, Fig. 24.9f.

Selection: Information necessary for the selection and application of roller bearings may be obtained from catalogues of manufacturers.

Comparison of roller and ball bearings: Compared with ball bearings, roller bearings that are interchangeable with them have the advantage of greater load capacity (24.14). Against this advantage there are the following disadvantages: considerably smaller axial thrust capacity, higher cost, and greater sensitivity to misalignment, dirt, and grit.

Needle rollers: Because of their comparatively small radial dimensions and their exceptionally high load capacity, particularly at low peripheral speeds, the use of needle bearings is rapidly increasing. Needle rollers are comparatively long rollers of hardened alloy steel of small diameter—2 to 4 mm—which are used mostly without a cage. They may be used without a race, as in Figs. 24.10a and b; with inner and outer races, as in Figs. 24.10c and d; or with only an outer race.

Table 24.5 gives the stock sizes of needle rollers that can be obtained from FAG and several other manufacturers.

The load capacity F, in N is calculated on a projected-area basis. Thus,

$$F = K_h K_l p l D \qquad \dots (24.14)$$

where, K_h is the hardness factor, given in Table 24.6.

K_l is the life-load factor, taken from the curve in Fig. 24.2 marked T-needle;

p is the allowable pressure, in N/mm^2;

Table 24.5: FAG needle-roller sizes in mm

Diameter	Length	Diameter	Length	Diameter	Length	Diameter	Length
2	7.8	2.5	7.8	3	9.8	3.5	29.8
2	9.8	2.5	9.8	3	11.8	3.5	34.8
2	11.8	2.5	11.8	3	13.8	4	39.8
2	13.8	2.5	13.8	3	15.8	5	49.8
2	15.8	2.5	15.8	3	17.8		
2	19.8	2.5	19.8	3	23.8		
		2.5	23.8	3	27.8		

Table 24.6: Hardness factors for needle-roller bearings

Rockwell C hardness of raceway	Approximate Brinell hardness (bhn)	Hardness factor K_h	Rockwell C hardness of raceway	Approximate Brinell hardness (bhn)	Hardness factor K_h
63	660	1.00	54	545	0.83
60	620	0.98	52	515	070
58	595	0.96	50	490	0.50
56	570	0.92			

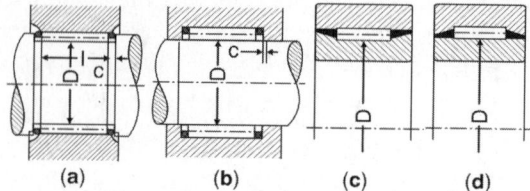

Fig. 24.10: Methods of applying needle rollers

l is the roller length, in mm;

D is the shaft or inner race diameter, in mm.

For wrist pins, rocker arms, and similar oscillating mechanisms, p may be as high as 35 N/mm². For rotary motion, p may be computed from the relation.

$$p = \frac{111}{\sqrt[3]{D_1 n}} \qquad \ldots (24.15)$$

where, D_1 is the diameter of the revolving race, in mm, and n is its speed, in revolutions per minute.

The total circumferential clearance can vary from 0.5 mm. up to the diameter d_r of one roller, and it may be checked by the formula

$$c = \pi(D + d_r) - id_r \qquad \ldots (24.16)$$

where, i is the number of needles.

Table 24.7 gives the recommended needle diameters and the total radial clearances for various diameters of the shaft or inner race.

Table 24.7: Design data for needle-roller bearings

Journal race diameter mm	Recommended	
	Total radial clearance mm	Needle diameter mm
9.50–19.00	0.0125–0.040	2.0
19.00–32.00	0.0180–0.050	2.5
32.00–51.00	0.020–0.055	3.0
51.00–76.00	0.0255–0.065	3.0
76.00–127.00	0.0305–0.075	3.5
127.00–177.00	0.0355–0.085	4.0–5.0

Needle rollers are also used assembled with ground steel races as in other roller bearings. The main advantage of needle-roller bearings is their small outside diameter. In addition, needle-roller bearings are used rather extensively assembled in an outer pressed-steel race, either with both end open, as in Fig. 24.11a, or with one end closed, Fig. 24.11b. Such bearings are available for shafts with diameters from 4 mm to 70 mm, and lengths from about one-half to one shaft diameter. Dimensions and load capacities should be taken from manufacturers' catalogues.[3] The advantages of these bearings are small weight, compactness, and high radial-load capacity. However, the coefficient of friction of needle bearings is considerably higher than that of short roller bearings. It may be six times as high.

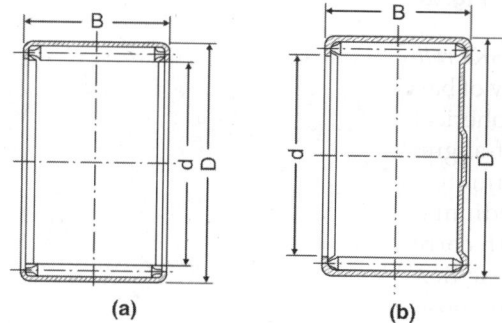

Fig. 24.11: Needle bearings

Large rollers: For rollers of large diameters, such as those used on bridge turntables and in other places where the speeds are very low, the safe load is

$$F = Cild_r \qquad \ldots (24.17)$$

where, C is an empirical coefficient. The value of C is 2.48 for hard cast-iron roller running on hard cast-iron tracks, and 5.86 for hard-steel rollers on hard-steel tracks.[4] For very large hardened steel rollers with $d_r > 250$ mm, on hardened-steel tracks, $C = 903840/d_r + 220$. If $l > 5d_c$, smaller values of C should be used.[5]

---

[3] The Torrington Company, Torrington, Conn.; Roller Bearing Company of America, Trenton, N. J.; Orange Roller Bearing Company, Orange N. J.; and others.

[4] Lionel S. Marks, ed., Mechanical Engineers' Handbook, 3d ed. (New York: McGraw-Hill Book Company, Inc., 1930), P. 1004.

[5] W. M. Wilson, the Bearing Value of Rollers, Bulletin No. 263, University of Illinois, Engineering Experiment Station (1934), p. 4.

24.7 THRUST ROLLER BEARINGS

Bower roller bearings, in Fig. 24.12a, are designed to take both radial loads and thrust loads. The cylindrical rollers take the radial load, and the flanged roller heads take the thrust load, the total axial load being divided among all the rollers. The maximum thrust load is equal to one-third the radial rating capacity. Bower roller bearings are made interchangeable with single-row and double-row ball bearings.

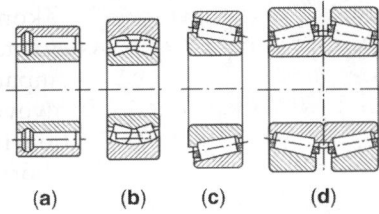

| (a) | (b) | (c) | (d) |

Fig. 24.12: Types of thrust roller bearings

SKF roller bearings, Fig. 24.12b, have a double row of barrel-shaped rollers. The rollers run in a spherical outer race, which makes the bearing self-aligning. These bearings are built for shaft diameters from 80 to 300 mm in light and medium series, and they are interchangeable with corresponding ball bearings. Their load capacities are from 50 to 100 per cent greater than those of ball bearings.

Timken, SKF, FAG roller bearings, Fig. 24.12c, have taper rollers whose races are frustums of cones with their apexes coinciding on the axis of the bearing. Taper roller bearing should be used in pairs, each bearing supporting the end thrust in one direction. Axial adjustment is usually provided for in the cone, or inner race. The axial-trust capacity depends on the thickness of the cup, or outer race, and is approximately equal to the radial capacity.

Timken bearings are also made with two rows of rollers, as shown in Fig. 24.12d. The bearing may have a double cup and standard cones, or a double cone and standard cups. In either case the cones may be turned either toward or away from each other. This latter arrangement provides ease of assembling and is used in roller pillow blocks.

Timken bearings are made either with a straight bore, both in millimeter sizes and in inch sizes, or with a conical bore to be used with a split adapter, such as that shown in Fig. 24.6a, for use with ball bearings.

Slewing bearing: These are ball or cylindrical roller bearings that can sustain axial, radial and moment loads either singly or in combination and in any direction. Slewing bearings can perform both oscillating movements as well as rotating movements. The outside diameter vary from 400 mm to 7200 mm even up to 14000 mm and are standardised. These bearings are made either in single row or double row or triple row.

Magnetic bearings: Magnetic bearings are used in turbomolecular pumps, compressors, turbogenerators and semiconductor equipment. They levitate the shaft by inducing controlled magnetic field. These have the benefit of no contact, and no lubricant and minimum vibrations.

24.8 ROLLER BEARING APPLICATION

An operating temperature up to 175°C does not affect the load capacity of roller bearings. The higher temperatures affect mainly the problem of lubrication. Up to 85°C either oil or special greases may be used. Above 85°C greases are not recommended; only mineral oils with a viscosity of at least 100 SSU at 98°C should be used.

Standardization: Roller bearings are made in standardized series. The SAE standards include light and medium series corresponding to outside dimensions to the same series in the wide-type radial bearings. Load ratings vary considerably for different makes, as do the sizes and numbers of rollers, and therefore data must be taken from catalogues of manufacturers.

In the Indian standards, the designations are given as follows. The first two digits give the series code corresponding to capacity as 02, 03 and 04. The next two places are type of bearings as deep groove ball bearing is indicated by BC, self-aligning by B5, angular contact by BA; cylindrical roller bearing by

RN, single thrust bearing by TA. The last two or three places indicate directly the bore diameter as 80BC02 indicates 80 mm bore, deep groove ball bearing and medium load capacity.

Design of mountings: The general rules for mounting of ball bearings also apply to roller bearings. The revolving race must be a tight fit on the shaft and securely fastened in place; the stationary race should be a snug fit on its seat but need not have lateral play. The rollers must run in a bath of lubricant contained within the housing and protected from leakage and contamination.

Bearing failures: Ball and roller bearings can be damaged and become unusable for various reasons. Because of fatigue load at the point of contact or line of contact. Though in the beginning a clean lubricant is used. It does get contaminated. Following are the various types of failures which can develop

1. Flaking.
2. Cracks and fractures.
3. Cavities and indentations.
4. Smearing.
5. Wear and creep.
6. Crater and wash board formation.
7. Corrosion.
8. Cage failure.

1. *Flaking*: Flaking starts with development of small fatigue crack in the rolling element or cage. The cracks cause material fragments to break loose from the races. It starts with small flakes and may spread to the entire surface. Its appearance is like a small bruise and when it spreads to entire surface, it gives a craggy appearance. This type of failure can be due to misalignment of cages or incorrect shape of shaft or bearing housing or faulty mounting, etc. Flaking can also develop as a result of continuing use of the bearing after failure of another kind has occurred.

2. *Cracks and fractures*: Cracks are formed because of heavy overloads or if outer cage is supported at edges or due to rubbing or sliding of outer race against stationary part or it can be due to abnormal heating of the surface during grinding or errors in hardening.

3. *Cavities and indentations*: When foreign particles get in the bearing with lubricant and are pressed between rolling elements and cages or subjected to excessively heavy or impact loads or small movements or vibrations in the stationary condition can develop indentations.

4. *Smearing*: Smearing as a kind of seizing caused by two surfaces sliding on each other. It can develop between rollers and races due to the bearing being completely filled with a hardened lubricant which retards the roller rotation in the unloaded zone. When they are suddenly accelerated to normal speed, smearing develops in the contact between rollers and ring flanges. It is also possible with gyratory moment on balls sliding in the direction diagonal to the direction of rolling. This causes spiral-shaped smearing streaks.

5. *Wear and creep*: Normally wear does not occur in ball and roller bearing but can be caused by defects in application. Abrasive particles can enter the bearing and cause wear of races, flanges and cages. Lack of lubrication and poor lubrication can also cause wear. A special kind of wear may develop due to creep between inner race and shaft. The race may creep when it is mounted with too tight a fit. Creep can cause wear of shaft and bearing housing.

6. *Crater and wash board formation*: If an electric current passes through a bearing, the current sparks through the thin lubricant film when there is drop of more than 0.4 v to 0.5 v of potential across the bearing. The spark causes annealing and due to rapid cooling, a rehardening of material points where spark occurs. These spots show either small crater or etching. After a large increase of these points, a wash board formation develops on the surface. Heavy sparks cause larger craters

indicating the melting of material. If the load is vibrating and the lubricant is contaminated, it can also cause washboard formations between.

7. *Corrosion*: Humidity and entry of water in bearings can cause corrosion. Rusting is possible due to unsuitable lubricants. Contact erosion can lead to special form of rust. Contact erosion is possible due to deflection of shaft or bearing housing caused by heavy load.

8. *Cage failure*: Cages are usually sensitive to poor lubrication. Wear develops initially where cages touch the rolling elements and finally resulting in the breaking of cages and then blocking the bearing. In exceptional case, cages are subjected to fatigue cracks.

Comparison of coefficient of friction: The coefficient of friction or the resistance to motion of the rolling element is influenced by the following characteristics.

(i) The elastic properties of material.

(ii) The condition of the surfaces.

(iii) The shape and positions of the surfaces.

(iv) The magnitude of the perpendicular force.

(v) The speed of rolling.

(vi) The temperature.

(vii) The condition of the lubricant and the atmosphere.

The frictional torque in a rolling bearing is given by:

$$T = T_r + T_s + T_{sc} + T_d$$

T_r = rolling frictional torque = $f F d / 2$

T_s = sliding frictional torque

T_{se} = frictional torque of the seals

T_d = frictional torque of drag losses, churning, splashing, etc.

Approximate values of coefficient of friction are given in Table 24.8.

General comparison of variation of coefficient of friction of ball bearings, roller bearings and journal bearings with speed is given in Fig. 24.13.

Table 24.8: Coefficient of friction f

Bearing type	f
Deep groove ball bearings	0.0015
Angular contact ball bearings	
– Single row	0.0020
– Double row	0.0024
– Four point contact	0.0024
Self-aligning ball bearing	0.0010
Cylindrical roller bearing	0.0011
Taper roller bearing	0.0018
Spherical roller bearings	0.0018
CARB Toroidal bearings	0.0013
Cylindrical roller thrust bearings	0.0050
Thrust ball bearings	0.0013
Spherical roller thrust bearings	0.0018

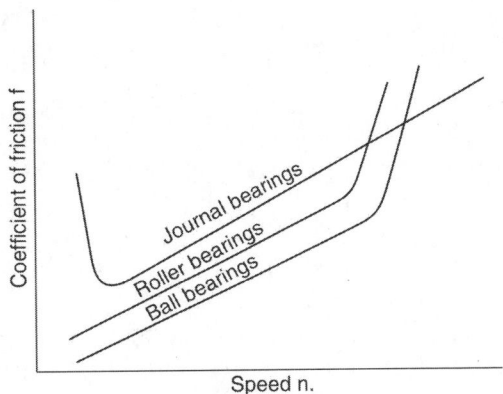

Fig. 24.13

PROBLEMS

24.1 Give the relative advantages and disadvantages of ball and roller bearings as compared to journal bearings.

24.2 Describe the mechanics of failure of a ball and roller bearing and discuss the effect of speed and desired life on the load carrying capacity of the bearing.

24.3 Describe the phenomenon of 'stress corrosion' and how it affects balls or roller bearing subjected to vibratory loads with no rotation.

24.4 Why are ball and roller bearings called anti-friction bearings?

24.5 Of what material are balls, races and cages made?

24.6 Make a sketch for fixing a ball bearing to the shaft and in the housing (support and axial retainment).

24.7 Why is it preferable to use ball and roller bearings in a gear box of an automobile?

24.8 Why is it preferable to use ball bearings in ceiling fan and journal bearings in table fans?

24.9 Why is it preferable to use geese lubricated bearings in ceiling fan?

24.10 What are the main functions of lubricant for ball bearings?

24.11 If load on a ball bearing is doubled, what is the effect on the life of bearing?

24.12 What type of load can be carried by the bearings? (1) ball bearing (2) taper roller bearing (3) spherical roller bearing (4) needle bearing.

24.13 State the conditions for which a ball bearing should be used in preference to a sliding bearing (a) for a radial load and (b) for an axial thrust load.

24.14 State the conditions for which a roller bearing should be used in preference to a ball bearing.

24.15 Determine the type and size of a ball bearing for a 75 mm shaft. The shaft speed is 325 rpm, the radial load is 9 kN, with very light shocks, and the axial load is 3.5 kN. The installation is a temporary one, to serve not over 1 year with 8-hr service per day. The bearing is to be placed 0.9 m from one end of the shaft.

24.16 Sketch the method of mounting the bearing of problem 24.14, showing the methods of lubricating it and preventing dust from entering the housing.

24.17 Determine the radial capacity of an SKF 6211-Z bearing running at 500 rpm and carrying an axial load of 2.2 kN for 2 years of continuous service.

24.18 A ball bearing is to be used on a drill press operating at 3,000 rpm with a 1.1 kN maximum thrust and a 2.2 kN radial load. The press will be operated 8-hr, five days a week but will be idle 20 per cent of the time. Determine the type and size of the bearing (a) if it should last 1 year and (b) if it should last 2 years.

24.19 Select suitable ball bearings for the spindle of a woodworking machine revolving at 1,200 rpm. One bearing is subjected to a radial load of 2.6 kN and a thrust load of 2 kN; the other carries only a radial load of 2.9 kN. The machine is to be used 8 hr. per day and 5 days a week, and a service life of 10 years is desired. The diameter of the spindle is 50 mm, and it can be turned down slightly.

24.20 A shaft is supported by two bearing 400 mm apart and carries a bevel gear of 195 mm pitch diameter, 150 mm from one end. The gear produces a radial load of 9.5 kN and a thrust load of 2.8 kN when rotating at 525 rpm. Determine (a) the shaft diameter if the shaft is made of 45C8/SAE 1045 steel and (b) the proper type and size of ball bearings to be used on each end of the shaft. The desired life is 2 years, at 50 hr. per week.

24.21 How are ball bearings assembled?

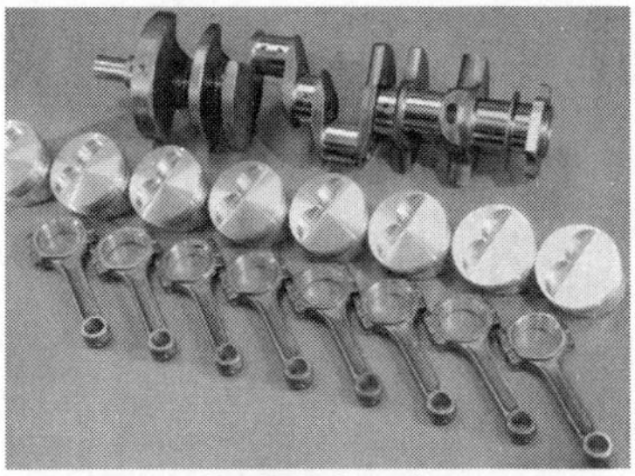

25 Crankshafts Connecting Rod and Piston

25.1 GENERAL CONSIDERATIONS

A shaft with a crank is used to transform a reciprocating motion into a rotary one, or vice versa. Regardless of its type and shape, a crankshaft may be considered as a beam with two or more supports. In addition to bending stresses, there are also shear stresses due to the torsional moment on the shaft. Both the bending moment and the torsional moment are caused by the main forces acting upon the crosshead connected to the crank, and by intertia and centrifugal forces of the moving parts. In addition, there may exist bending moments caused by the weight of the flywheel, the pull of the belt, or the weight of the rotor of an electrical generator.

Type: There are two main crankshaft types. The shaft may have a side crank, or overhung crank, as in Fig. 25.1a, or may have a center crank, as in Fig. 25.1b. A shaft is often made with two side cranks, one on each end, or with two or more center cranks.

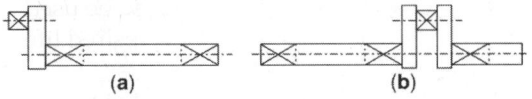

| (a) | (b) |

Fig. 25.1: Types of crankshafts

Materials: One-piece crankshafts are usually forged of open-hearth steel, similar to SAE 1030/30C8 with an elastic limit in tension of 260 N/mm² and an elongation of 25 per cent in 50 mm length. Crankshafts of marine engines are forged of steel with a lower carbon content, such as SAE 1020/20C8 to SAE 1025/25C8. Automobile and airplane-engine crankshafts are made of chrome-nickel steel, such as 5Ni 1cr60/SAE 3140 or 40Ni2Cr1, SAE 3240 heat-treated to obtain an elastic limit of 655 to 750 N/mm². Automobile crankshafts have of late been cast of a special iron alloy, No. 4 in Table 4.1 and iron copper alloy.

Built-up crankshafts have crankpins made of steel with a carbon content of 0.45 to 0.55 per cent, and the cranks are made either of cast steel or of a nickel cast iron.

Stresses: Since the failure of a crankshaft is likely to cause a serious engine wreck and neither all acting forces nor all stresses can be determined accurately, and high safety factor, from 3 to 4, should be used. Stress concentration due to various discontinuities must be taken into account. Where possible, their effect should be reduced to a minimum. Radial holes and keyseats with sharp corners should be avoided; and if two different cross sections must be joined, they should be blended with a large fillet r, as in Figs 25.2a or b. If possible, the radius r should not be less than 0.2d. *A gradual change of section, as shown by dotted lines in Fig. 25.2a, is still better. If there is no room for a regular fillet, it is advisable to undercut the cheek in order to obtain the fillet r, Fig. 25.2c (25.1). A weakening of the cheek is avoided by increasing its width b. Such undercut fillets in a built-up crankshaft of a large gas engine are shown in Fig. 25.3.[1]*

[1]H. Dubbel, *Oel und Gas Maschienen* (Berlin: Julius Springer, 1926), p. 371.

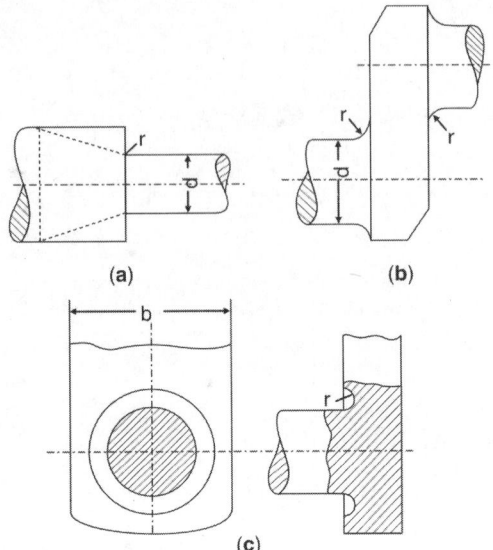

Fig. 25.2: Fillets at junctures of different sections

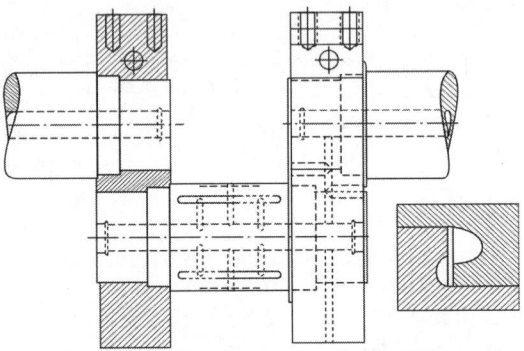

Fig. 25.3: Built-up crankshaft

Repeated stresses: Most crankshaft failures are caused by a progressive fracture due to repeated bending or reversed torsional stresses. Reversed bending stresses exceeding the endurance limit may be produced in a crank-shaft by one or more bearings being lower or higher than the rest. Dangerous torsional stresses may be produced in multithrow crank-shafts because of torsional vibration at or near critical speeds. In single-throw or double-throw engine crankshafts these stresses may occur from counterweights set into the rim of the flywheel instead of being fastened to the crank cheeks.

In a revolving shaft the bending stress is the most-stressed fibers changes from compression to tension, and vice versa. The maximum compressive stress in general differs from the maximum tensile stress. However, to be on the safe side, it is advisable to use the endurance limit S_{en}, which corresponds to a complete reversal of the stresses and a mean stress $S_m = 0$. The same rule applies to the torsional stress.

Example 25.1: Select the material and determine the allowable stresses in bending and torsion for a crankshaft of a medium-size gas compressor.

SAE1025/25C8 steel is commonly used for similar applications, because it is relatively inexpensive and is easily forged and machined.

Since in a gas compressor the change of the load on the crankshaft is rather gradual, a safety factor n of 3 is sufficient. The endurance stress of 25C8 is $S_{en} = 200$ N/mm^2, and the allowable nominal stress is

$$S_d = \frac{S_{en}}{n} = \frac{200}{3} = 67 \text{ N/mm}^2$$

In torsion

$$S_{en} = \frac{110}{3} = 37 \text{ N/mm}^2$$

The actual allowable stresses must be smaller because of the size influence, as given by equations 5.7 and 5.8.

25.2 DESIGN PROCEDURE

The first step in the design of a crankshaft is to determine the magnitudes of the various loads on the crankshaft. The next step is to determine the distances between the supports and their position with respect to the loads. For the sake of simplicity and also for safety, the shaft is considered supported at the centers of the bearings, and all forces and reactions are assumed to be acting at these points. The distances between the supports depend on the lengths of the bearings, which in turn depend on the diameter of the shaft because of the allowable bearing pressures. Therefore certain maximum bearing pressures p and the length-to-diameter ratios l/d are selected from Table 23.4, and these dimensions l and d are determined by considering the acting loads. Values thus obtained are checked by computing the bearing characteristic number Zn/p, as explained in section 23.4. If necessary, dimensions are changed to insure better operation. After this the thickness h of the cheeks, or webs, is selected, being made about $0.4d$ to $0.6d$. The width of web b is taken as $1.1d$ to $1.2d$. With these values determined, the distances between the supports are found; the main dimensions of the crankshaft are fixed.

When computing the bending moments at various cross sections of a crankshaft, two methods may be used. One method is to find, at the supports, the net reactions due to all loads, and then to determine the moments at the various points. Another method is to compute the reactions and the bending moments due to each load separately, and then to combine all moments either graphically or analytically. After all shaft dimensions have been determined, the shaft should be checked for rigidity by adopting a procedure similar to that outlined in section 20–2, i.e. $\theta \leq 0.25°/m$.

25.3 SHAFT WITH SIDE CRANK

This shaft type is used for medium-size and large horizontal engines. Its main advantage is that it requires only two bearings, in either the single-crank or two-crank construction. Two bearings present less danger of misalignment, which causes most shaft failures, than do three or more supports.

Construction: Cranks for pumps, compressors, and other machines are either steel forgings, as in Fig. 25.4a, steel castings, as in Fig. 25.4b or Fig. 25.5. Only in small, cheap machinery is the crank made of cast iron. The crankpin has a cone seat and is held by a cotter, as in Fig. 25.4a, or by a nut, as in Fig. 25.4b. If the crankpin is not expected to wear much, it may be pressed in or shrunk in, and the end may be riveted over, as in Fig. 25.5. A pin with a collar, Fig. 25.4a, requires a connecting rod with an open end and a cap. A straight pin with a collar plate p, Fig. 25.5, takes a connecting rod with a closed end. Sometimes the crank and pin are

a one-piece steel casting. The drawback of this construction is that it requires an expensive steel for better wear of the crankpin.

The crank-to-shaft connection is a press or shrink fit with a rectangular or round key, Figs 25.4a or b.

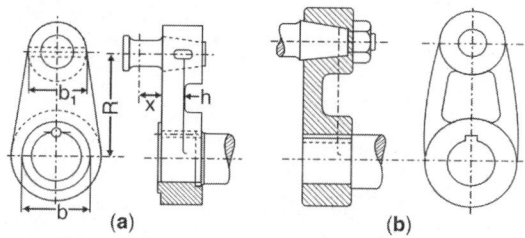

Fig. 25.4: Side cranks for built-up crankshafts

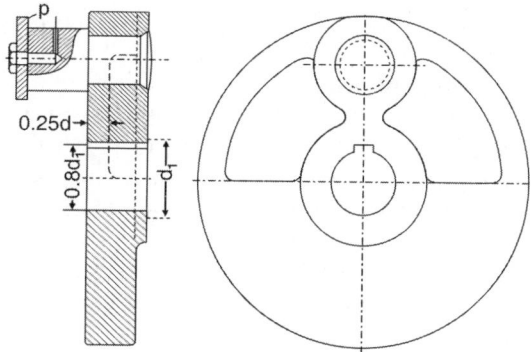

Fig. 25.5: Disk crank

Smaller steam and internal-combustion engines usually have a one-piece forged crankshaft with a cast-iron counterweight bolted to the crank, as shown in Fig. 25.6. In larger engines the shaft, crank, and pin are made separate, as in Fig. 25.7, and are forced together with a hydraulic press. The crank and counterweight are usually a single steel casting. The key K is merely a safety device, as the press fit must be sufficient to transmit the maximum torque. The usual main proportions of a separate crank are indicated in Fig. 25.7.

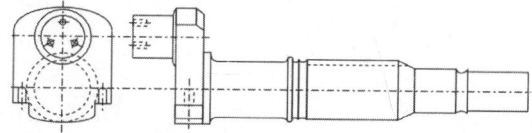

Fig. 25.6: One-piece crankshaft with side crank

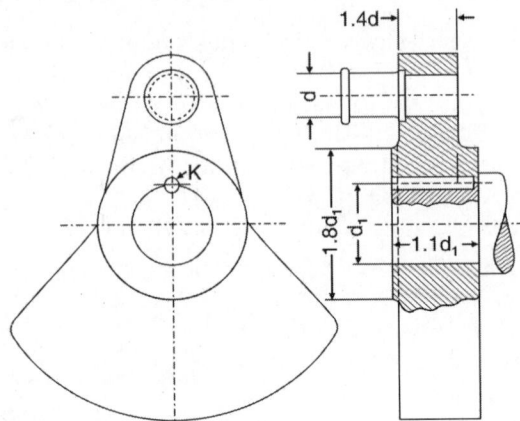

Fig. 25.7: Three-piece crankshaft with side crank

Design calculations: Every crankshaft must be designed or checked for at least two crank position—one when the bending moment is a maximum, and the other when the torsional moment is a maximum. In computing the bending moments, the crankshaft can be considered to be a straight shaft, even for close, long cranks and short pins.[2] In both positions the additional moments due to the flywheel weight, belt tension, and other forces must be considered. If a new crankshaft is designed, the first step is to determine the approximate values of the diameters and lengths of the crankpin and the journals and then checked.

Force Analysis of Overhanging Crank and Crankshaft

Crank pin: Figure 25.8 the crank pin is subjected to bearing pressure and bending moment due to gas forces P. As the dimensions are decided empirically, i.e. $d = 0.67\,D$ to $0.73\,D$ and $\dfrac{l}{d} = 0.5$ to 1.5. It is usual to have $\dfrac{l}{d} = 0.7$ to 1.1 to reduce the overall dimensions of the engine.

$P = ldf_b$ and $f_b = \dfrac{P}{ld}$ and f_b should be within limits of bearing pressure depending on the type of engine.

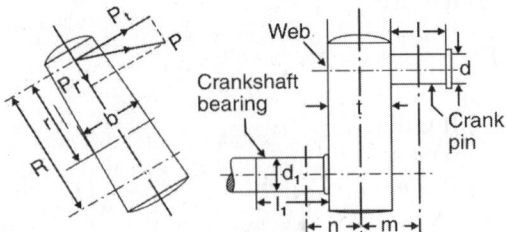

Fig. 25.8

The maximum bending moment $M = \dfrac{Pl}{2}$ due to gas force P.

In the overhanging crank pin, there is likely to be some deflection of pin. This will cause the centre of force likely to shift outwards, it is desirable to take the bending moment $M = 0.75\,Pl$.

The maximum bending stress

$$S_b = \frac{M}{\pi/32\,d^3}$$

Crank web: The width of web $b = 1.1$ to $1.2\,d$.

The thickness $t = 0.4d$ to $0.6d$

The gas force P acts eccentrically to the crank web. P is resolved into components:

(a) Radial component P_r and (b) Tangential components P_t.

(a) Effect of P_r.

    (i) It will induce compressive stress or tensile stress

$$S_c = \frac{P_r}{bt}$$

The web is considered as a cantilever acted upon by force P_r at a distance m.

    (ii) It will induce bending stress,

$$S_b = \frac{P_r \times m}{b \times t^2/6}$$

(b) Effect of P_t.

    (iii) It will cause shear stress as web is like a cantilever

$$S'_s = \frac{P_t}{b \times t}$$

---

[2]F. Rötscher, ''Die Ermittlung der Spannungsverteilung in Konstructionsteilen durch Dehnungsmessungen,'' *Zeitschrift Verein Deutscher Ingenieure*, Vol. 77 (1933), p. 375.

(iv) It will cause bending stress with arm r.

$$S_b' = \frac{P_t \times r}{b^2 \times t/6}$$

(v) It will also have a tendency to rotate the web so cause shear stress due to torque $P_t \times m$

$$S_s' = \frac{P_t \times m}{b \times t^2/4.5}$$

As the stresses are varying and act in different planes, it is convenient to consider two critical positions of the crankshaft.

(a) When the crank is in dead centre position.
$P_t = 0$, $P_r = P$ and the stresses are:

$$S_c = \frac{P}{b \times h}$$

$$S_b = \frac{P \times m}{b \times t^2/6}$$

Maximum compressive stress

$$S = S_c + S_b = \frac{P}{b \times h} + \frac{P \times m}{b \times t^2/6}$$

and maximum tensile stress

$$S_b = S_c - S_b = \frac{P}{b \times h} - \frac{P \times m}{b \times t^2/6}$$

(b) When the crank is at right angles to the connecting rod $P_r = 0$ and P_t can be taken equal to P. The magnitudes of P has decreased but due to obliquity,

$$P_t = \frac{P}{\cos \alpha} \simeq P$$

The stresses are

$$S_s' = \frac{P}{b \times t}$$

$$S_b' = \frac{P \times r}{b^2 \times t/6}$$

and

$$S_s' = \frac{P \times m}{b \times t^2/4.5}$$

Total shear stress

$$= S_s = \frac{P}{b \times h} + \frac{P \times m}{b \times h^2/4.5}$$

Maximum principal stress

$$S_p = \frac{1}{2}\left[S_b' + \sqrt{S_b'^2 + 4S_s^2} \right]$$

and maximum shear stress

$$= \frac{1}{2}\sqrt{S_b'^2 + 4S_s^2}$$

Crankshaft journal: The journal will be critically stressed when the crank is right angle to the correcting rod, i.e. $P_r = 0$ and $P_t = P$

If L_1 is the distance of centre of pin to centre of journal, then $L_1 = m + n$

Maximum bending moment

$$= P_t \times L_1 = P \times L_1$$

Maximum torque

$$= P_t \times R = P \times R$$

Equivalent torque

$$T_e = \sqrt{(P\,L_1)^2 + (P\,R)^2}$$

The diameter of the journal is given by:

$$T_e = \frac{\pi}{16}S_s d_1^3$$

$$d_1 = \left[\frac{16T_e}{\pi S_s} \right]^{1/3}$$

Length of the journal L to diameter D_1 is

$$\frac{l_1}{d_1} = 1.5 \text{ to } 1.75$$

As the stresses are complex so the dimensions are first assumed empirically and stresses are checked.

25.4 SHAFT WITH CENTER CRANKS

The general procedure for the design of a shaft with center cranks is similar to that for a shaft with side cranks. To make the computations simpler without seriously impairing accuracy, it is customary to assume that the influence of any bending force does not extend beyond the two bearings between which that force is applied.

The shaft of a stationary engine with a center crank is usually a one-piece forging. With progress in the metallurgy of alloy cast irons and steels, cast crankshafts have begun to be used, even in larger machines. Built-up crankshafts with cast cranks similar to those in Figs 25.4b and Fig. 25.5, with shrunk-in steel shaft ends and steel crankpins, are used for small pumps and compressors.

25.5 MULTITHROW CRANKSHAFTS

To determine the stresses in a multithrow crankshaft, it is necessary to make certain assumptions in regard to the shaft axis, and the analysis is rather laborious if it is theoretically accurate. In these computations it is assumed that the crankshaft is correctly aligned and its axis is a straight line. When the engine begins to operate, one or several of the bearings will wear faster than the others; the crankshaft center line will sag at these bearings; and the bending stresses will be increased greatly, but indefinably, because the deflections are not known. Similar conditions arise if the bedplate or crankcase is not rigid. Thus the theoretically correct procedure does not insure the safety of the shaft against progressive fracture.

For this reason actual shaft dimensions are often based on experience, and the only calculations involve simple proportion. Another common method is to determine the dimensions of the crankpin, the crank webs, and the journals, as for a single-throw shaft, and to disregard the influence of the other throws except for the torsional moments. The use of a comparatively high factor of safety takes care of the simplifying assumption.

Empirical formulas: Large insurance companies have set up certain empirical formulas pertaining to the main dimensions of crankshafts of marine oil engines. Shafts complying with these rules give very satisfactory service, possibly because these rules give a stronger shaft than do usual computations. These formulas differ one from another, but all give practically the same dimensions. Therefore only one set, that of the American Bureau of Shipping, will be given here.

The diameter d of the crankpins and journals is found from the relation

$$d = a\sqrt[3]{\frac{D^2 pc}{S_d}} \qquad \dots (25.1)$$

where a is a coefficient the values of which are given in Table 25.1;

D is the cylinder bore, in mm;

p is the maximum gas pressure, in N/mm^2;

c is the distance over the crank web plus 25 mm, shown in Fig. 25.9;

S_d is the allowable fiber stress, in N/mm^2.

The value of S_d may be taken as 48 N/mm^2 for cast steel, 50 N/mm^2 for open-hearth steel, and 55 N/mm^2 for the best grade of forged steel. In Table 25.1, L is the engine stroke, in mm.

The thickness t and the width b of the crank cheeks, also called *webs*, can be selected in the same way as indicated for a single-throw shaft, but they must satisfy the following conditions:

$$bt^2 \geq 0.4d^3 \qquad \dots (25.2)$$

Table 25.1: Coefficient *a* in the American Bureau of Shipping Formula

Type	Number of cylinders		Ratio of stroke to distance over crank webs = L/c							
	Four-stroke	Two-stroke	0.7	0.8	0.9	1.0	1.1	1.2	1.3	1.4
Explosion engines	1, 2, 4	1, 2	1.17	1.17	1.17	1.17	1.17	1.17	1.17	1.17
	3, 5, 6,	3	1.17	1.17	1.17	1.17	1.19	1.20	1.22	1.24
	8	4	1.17	1.19	1.21	1.23	1.25	1.28	1.30	1.32
	10, 11, 12	5, 6	1.18	1.20	1.23	1.25	1.28	1.31	1.33	1.35
Air-injection diesel engines	1, 2, 4	1, 2	1.17	1.19	1.22	1.25	1.28	1.31	1.34	1.36
	3, 5, 6,	3	1.19	1.22	1.25	1.28	1.32	1.35	1.38	1.41
	8	4	1.20	1.24	1.27	1.30	1.33	1.37	1.40	1.43
	12	5, 6	1.22	1.25	1.29	1.32	1.36	1.39	1.42	1.45
	16	8	1.25	1.29	1.33	1.36	1.40	1.44	1.47	1.50

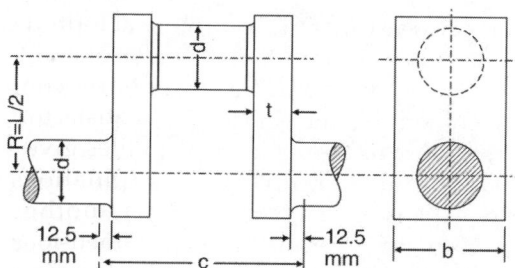

Fig. 25.9: American Bureau of Shipping method

and

$$b^2 t \geq d^3 \qquad \dots (25.3)$$

The relations given here were set up for marine internal-combustion engines. However, they can be used satisfactorily for preliminary calculations of crankshaft dimensions for other reciprocating machinery, such as compressors or pumps.

Example 25.2: Design an overhanging crankshaft for a diesel engines having cylinder bore 80 mm and stroke 90 mm. Maximum allowable shear stress in the shaft is 90 N/mm², maximum explosion pressure = 2.5 N/mm².

For a bore $D = 80$ mm.

Crank pin diameter $d = 0.67 \times 80 = 51.6$ say 50 mm
Crank pin length $l = 1 \times 50 = 50$ m
Web thickness $t = 0.6 \times 50 = 30$ mm
Maximum web width $b = 1.1 \times 50 = 55$ mm
Journal length $l_1 = 1.5 \times 50 = 75$ say 76 mm
Maximum gas force

$$P = \frac{\pi}{4} D^2 \times p = \frac{\pi}{4} \times 80^2 \times 2.5 = 12566.37 \text{ N}$$

1. *Crank pin*

Bearing pressure in the crank pin

$$f_b = \frac{12566.37}{50 \times 50}$$

$$= 5.03 \text{ N/mm}^2 \text{ O.K.}$$

Maximum bending moment on the pin

$$M = 0.75 \times 12566.37 \times 50 = 471238.87$$

Bending stress

$$= \frac{471238.87 \times 32}{\pi \times 50^3} = 38.39 \text{ N/mm}^2$$

Maximum shear stress

$$= \frac{12566.37 \times 4}{\pi \times 50^2} = 6.39 \text{ N/mm}^2$$

Maximum principal stress

$$= \frac{1}{2} \left[38.39 + \sqrt{38.39^2 + 4 \times 6.39^2} \right]$$

$$= 40.46 \text{ N/mm}^2$$

Maximum shear stress

$$= \frac{1}{2} \sqrt{38.36^2 + 4 \times 6.39^2} = 20.23 \text{ N/mm}^2$$

2. *Journal of crankshaft*

Distance of centre of crank pin to center of journal

$$L_1 = \frac{l}{2} + t + \frac{l_1}{2} = 25 + 30 + 38 = 93 \text{ mm}$$

Maximum bending moment on the journal

$$= PL$$

$$= 12566.37 \times 93$$

$$= 1168672.41 \text{ N mm}$$

Maximum torque on the journal, $R = 45$ mm.

$$= 12566.37 \times 45$$

$$= 565486.65 \text{ N mm}$$

Equivalent torque

$$T_e = \sqrt{(1168672.41 \times 2)^2 + (565486.65 \times 1.5)^2}$$

$$= 2486.5 \times 1000 \text{ mm}$$

Shaft diameter

$$D_1 = \left[\frac{2486.5 \times 1000 \times 16}{\pi \times 90} \right]^{1/3} = 52.01$$

say 55 mm OK.

3. *Web*:

$$m = \frac{50}{2} + \frac{30}{2} = 40 \text{ mm}, r = 45 - 28 = 17 \text{ mm}$$

When the connecting rod is in dead center position
Compression stress

$$S_c = \frac{12566.37}{55 \times 30} = 7.61 \text{ N/mm}^2$$

Bending stress

$$S_b = \frac{12566.37 \times 40 \times 6}{55 \times 30^2} = 60.92 \text{ N/mm}^2$$

Maximum tensile stress

$$= 60.92 - 7.61 = 53.31 \text{ N/mm}^2 \text{ OK}$$

Maximum compression stress

$$= 60.92 + 7.61 = 68.53 \text{ N/mm}^2 \text{ OK.}$$

When connecting rod is at right angle to crankshaft
Shear stress

$$S_s' = \frac{12566.37}{55 \times 30} = 7.61 \text{ N/mm}^2$$

Shear stress due to torque

$$S_s' = \frac{12566.37 \times 40 \times 4.5}{55 \times 30^2} = 45.69 \text{ N/mm}^2$$

Bending stress

$$S_b' = \frac{12566.37 \times 17 \times 6}{55^2 \times 30} = 14.12 \text{ N/mm}^2$$

Total shear stress

$$= 7.61 + 45.69 = 53.3 \text{ N/mm}^2$$

Maximum principal stress

$$= \frac{1}{2}\left[14.12 + \sqrt{14.12^2 + 4 \times 53.3^2}\right]$$

$$= 60.82 \text{ N/mm}^2 \text{ OK.}$$

Maximum shear stress

$$= \frac{1}{2}\sqrt{14.12^2 + 4 \times 53.3^2}$$

$$= 53.76 \text{ N/mm}^2 \text{ O.K.}$$

Construction: Figure 25.10 shows a drawing of a six-throw crankshaft for an oil engine of 222 × 267 mm. The shaft is drilled for delivering pressure lubrication to the crankpins and wrist pins. All junctions of different section have fillets of 1.5 mm radius, which according to Figs 3.17 and 3.31 still leave high stress

concentration. One of the journals, that between cylinders No. 3 and 4, locates the shaft lengthwise by means of an accurately machined distance between the cheeks. The other journals are made slightly longer in order that, with all bearing shells of the same length, free expansion of the shaft may be permitted.

Multithrow crankshafts of large engines are built up of several pieces, either shrunk together, as in Fig. 25.3, or bolted together by means of flanges.

25.6 TORSIONAL VIBRATION

In general a crankshaft is made up of parts having different cross sections. The computations are simplified if the shaft is replaced by a shaft with a constant cross section but the same frequency of vibration. Such a shaft is called an *equivalent shaft*.

Equivalent shaft: From equations 2.14 and 2.16, the angle of torsion of a solid round shaft is directly proportional to the length l and inversely proportional to the fourth power of the diameter d, or d^4. Thus a portion of shaft of length l and diameter d can be replaced by a portion of length l_o and diameter d_o, where

$$l_o = l\left(\frac{d_o}{d}\right)^4 \qquad \ldots (25.4)$$

The length h_o equivalent to each crank web may be found by the expression

$$h_o = \frac{RC}{B} \qquad \ldots (25.5)$$

where, R is the crank radius, as shown in Fig. 25.9;

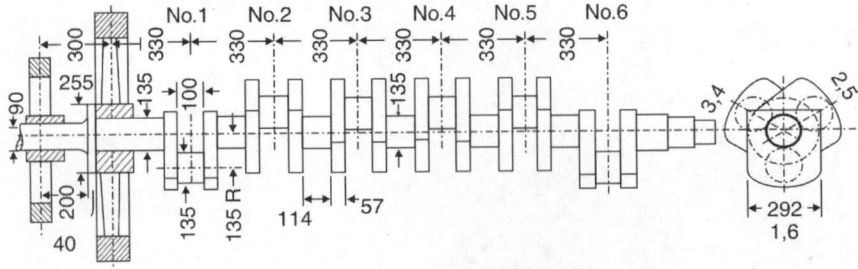

Fig. 25.10: Crankshaft for six-cylinder oil engine

$C = \dfrac{1}{32}\pi d_o^4 G$ is the torsional rigidity of the crankpin;

$B = \dfrac{1}{12}tb^3 E$ is the flexural rigidity of the web.

Equations 25.4 and 25.5 give a correct equivalent length if the clearances in the bearings are such that free angular and axial displacements of the journals during twist are possible. Small bearing clearances and axial constraints increase the rigidity of the shaft and give a shorter equivalent shaft. However, the computations for a constrained crankshaft are rather involved.[3] since the bearing clearances are variable because of wear, and their numerical influence upon the total length of the equivalent shaft is small, this influence may be neglected.

After the equivalent length of the whole shaft is found, equations 5.45 to 5.49 can be applied. However, it should be remembered that they give only approximate values.

Example 25.3: Analyze the crankshaft of a six-cylinder mechanical-injection oil engine, Fig. 25.9, in regard to danger from torsional vibration. Additional data are:

Engine speed	1,200 rpm
Weight of revolving parts for each crank	660 N
Weight of reciprocating parts for each crank	355 N
Weight of flywheel	1850 N
Weight of generator rotor	800 N
Radius of gyration of cranks	135 mm
Radius of gyration of flywheel	460 mm
Radius of gyration of rotor	300 mm

The journals and the crankpins have the same diameter. The equivalent length of a crank web reduced to a straight shaft 135 mm diameter, as found by equation 25.5, is

$$h_0 = \frac{135\pi \times 135^4 \times 82000 \times 12}{57 \times 292^3 \times 20700 \times 32} = 14.74 \text{ mm}$$

The equivalent length of the part of the extension shaft 200 mm long and 90 mm in diameter is, by equation 25.4,

$$l_0 = 200 \times \left(\frac{135}{90}\right)^4 = 1012.5 \text{ mm}$$

By equation 25.4, the equivalent length of the flywheel flange, for which $D = 255$ mm and $l = 40$ mm, is

$$l'_0 = 40 \times \left(\frac{135}{255}\right)^4 = 3.148 \text{ mm}$$

If it is considered that one-half the reciprocating mass of a crank is added to the revolving mass, the polar moments of inertia, or flywheel effects, Wk^2 are as follows:

Cranks: $(660 + 0.5 \times 355)135^2 = 1526$ MNmm

Flywheel: $1850 \times 460^2 = 391.46 = 392$ MNmm

Rotor: $80 \times 300^2 = 72$ MNmm

First crank length from flywheel = $330 - 57 + 3.14$ = 276.14 mm

The equivalent lengths of portions of a straight 135 mm shaft with flywheel effects, are shown in Fig. 25.11. If the distance from the end to the node is designated by c, its value can be found by proceeding as explained for equation 5.1. Taking the sum of the products of the flywheel effects and the corresponding distances from the left end over the sum of the flywheel effects, the result is

$$c = \frac{72 \times 0 + 392 \times 1075.64}{72 + 392 + 6 \times 15.26}$$

$$+ \frac{15.26\big[(1075.64 - 276.14)\times 6 + 222.28 \times 15\big]}{72 + 392 + 6 \times 15.26}$$

$$= 1073.33 \text{ mm}$$

The node is between the rotor and the flywheel.

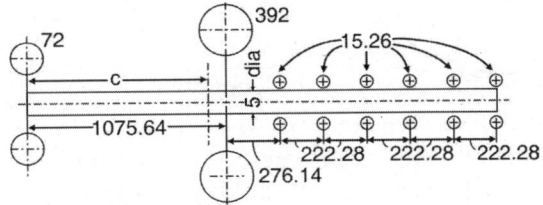

Fig. 25.11: Diagram of equivalent shaft

[3]S. Timoshenko, Vibration problems in Engineering (New York: D. Van Nostrand Company, Inc., 1928), p. 157; also Lionel S. Marks, ed., Mechanical Engineers' Handbook, 5th ed. (New York: McGraw-Hill Book Company, Inc., 1951), P. 494.

The natural frequency can be found from equation 5.49. For the sake of simplicity, the part of the shaft to the left of the node will be considered.

For this part,

$$J = \frac{\pi}{32} \times 135^4 = 32583976 \text{ mm}^4$$

and

$$\Sigma(Wk_0^2l) = 72 \times 10^6 \times 1073.33 = 77279 \times 10^6 \text{ N/mm}^3$$

Remaining crank distances = $330 - 2 \times 57 + 2 \times 3.14 = 222.28$ mm

Hence,

$$f = \frac{1}{2\pi}\sqrt{\frac{gGJ}{\Sigma\left(\omega k_0^2 l\right)}}$$

$$f = \frac{1}{2\pi}\sqrt{\frac{82000 \times 32583976 \times 9810}{77279 \times 10^6}} = 92.69 \text{ vibr/s}$$

A six-cylinder four-stroke engine has three firing events per revolution. Therefore, the main critical speed will be $\frac{1}{3} \times 92.69 \times 60 = 1854$ rpm which is 54.5 per cent above the normal speed of 1,200 rpm. This is usually considered to be a safe margin. However, a reliable evaluation of the safety of such a system can be made only if the harmonic components of the engine torque are known.[4]

Elimination of resonance: To obtain a greater difference between the critical speed of a shaft and the engine speed, the natural frequency of the revolving shaft must be increased. As indicated by equation 5.49, the desired increase can be attained by one or more of the following methods:

(a) Increasing the polar moment of inertia of the shaft.

(b) Decreasing the flywheel effect, Wk_o^2, of the rotating masses.

(c) Decreasing the length of the shaft.

The first method means an increase of the shaft diameter. This method can be combined with that of making the shaft hollow. For a hollow shaft with an outside diameter d_1 and an inside diameter d_2, the polar moment of inertia is

$$J = \frac{\pi}{32}\left(d_1^4 - d_2^4\right) \qquad \dots (25.6)$$

It is evident that removing even a considerable part at the center has a little effect on J. Thus a hollow shaft with $d_2 = 0.31d_1$ has moment of inertia only 1 per cent smaller, but a weight 10 per cent smaller, than a solid shaft of the same outside diameter (25.2).

The second method means that with a long shaft the flywheel should be made only heavy enough to obtain the required uniformity of rotation, and all other rotating masses should be made as light as possible.

The third method can be applied to a crankshaft connected to a large extension shaft. By inserting a flexible coupling between the crankshaft and the driven shaft, the length of each is decreased and the natural frequency is raised.

25.7 CONNECTING ROD

The connecting rod converts the reciprocating motion of the piston or crosshead to rotary motion at the crankshaft. As described in chapter 3, the motion of the connecting rod is a combination of the rotation of the crank and reciprocating motion of the wrist-pin end or small end. The forces acting on the connecting rod are:

1. Gas force or steam force.
2. *Inertia force due to rotation (25.8).*
3. *Inertia force due to the reciprocating motion (25.8).*
4. Frictional force due to piston rings, piston and crosshead in the case of steam engine.
5. Friction force due to the two end bearings.

The friction force due to the piston rings is given by,

$$F_f = \pi D \, bi \, p_r f \qquad \dots (25.7)$$

D = cylinder bore or piston diameter

b = width of rings

i = number of rings

---

[4] Applying the accurate Holzer method, it will be found that $c = 1019$ mm, and the corresponding value of f is 95 vibr per sec. Thus the approximate method is in error by only 2.4 per cent.

p_r = pressure between rings and cylinder and is of the order of 0.025–0.045 N/mm² and is due to the normal component of gas force and the obliquity of the connecting rod.

f = coefficient of friction = 0.06 to 0.1

This force is less than 0.5 per cent of the piston force and therefore is neglected. The friction force due to two end bearings is not over 8 per cent of gas force and is also neglected and the resulting design becomes safe, the variation of gas force and inertia force due to reciprocating masses is shown in Fig. 25.12a four strokes.

The combined effect is shown as (3), of gas force (1) and inertia force (2) are shown in Fig. 25.12b with the crank angle base.

Inertia force due to rotation: The magnitude of inertia force due to rotation equation (3.7) and the bending moment is given by equation 3.9, i.e.

$$M = \frac{2F_i' l}{9\sqrt{3}} = \frac{Wv^2 l \sin\alpha}{9\sqrt{3}\,gr} \qquad \text{... (25.8)}$$

The bending stress: $S = M/z$ (from 25.8) is drawn to crank angle base in Fig. 25.12c. If the net force P-F_i is used to calculate stress (4) in buckling, the bending stress (5) is also plotted to the same crank angle base, the resulting stress (6) is given in Fig. 25.11c which shows that the maximum value is approximately at an angle of 60°–70°.

The cross-section of the connecting rod is assumed as I-section *as it is lighter and economical (Same as hollow rectangular sections) as compared to a rectangular section (25.6)*. Figure 25.13 with width = $4t$ and depth $5t$, t being the thickness of the section. The connecting rod is treated as a strut having hinged ends in the plane of motion and fixed ends in the perpendicular plane Y. Euler's equation gives $I_x = 4\,I_y$ for equal strength in the two planes. However, the l/k is lower so Rankin's equation requires $I_x = 3.5\,I_y$ for equal strength in the two perpendicular planes. The dimension chosen, gives a value of $I_x = 3.2\,I_y$ and also will be lighter in comparison with a rectangular section. The net force acting on the piston P-F_i is used to calculate the section corresponding to x-plane as x-plane of the assumed I-section is weaker than y-plane with a safety factor of 6 on yield stress. The maximum stress including bending stress due to F_i will be checked for the final factor of safety. The dotted line in Fig. 25.13 indicates the actual section for proper forging of the connecting rod with proper draft of 7°. *If for a long connecting rod for a vertical engine, a central hole for lubrication is needed (25.10)*, then a further modification of the section at the center is also provided with dotted lines. This location will have a lower stress as compared to other locations.

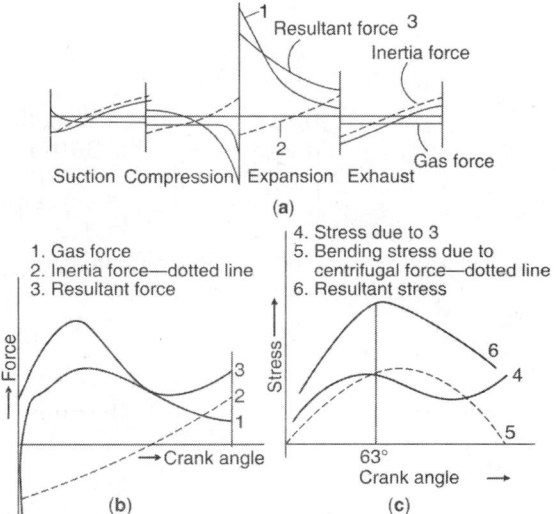

Resultant force 3
Inertia force
2
Gas force

Suction Compression Expansion Exhaust

(a)

1. Gas force
2. Inertia force—dotted line
3. Resultant force

→ Crank angle

(b)

4. Stress due to 3
5. Bending stress due to centrifugal force—dotted line
6. Resultant stress

63°
Crank angle →

(c)

Fig. 25.12

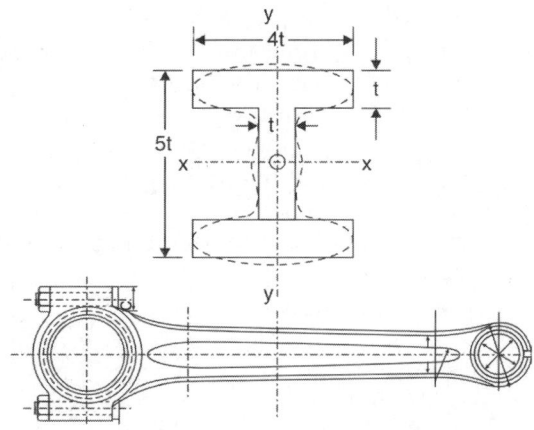

Fig. 25.13: Connecting rod

Small end and big end: There is a small relative motion between the small end bearing and piston pin as compared to the big end which rotates over the crank pin. The wear in the big end is more than that in the small end. *To equalise wear, the bearing pressure in the big end is reduced which results in the larger dimension of big end (15.7).*

The small end is to be accommodated in the piston and the two piston hubs also serve as a bearing. The length l of the small end $\leq 0.45\,D$, its diameter $d = 0.35\,D$, and $l/d = 1.5$ to 2 so that the small end and hubs both are two equal bearings to be accommodated in a length D.

The bearing pressure will depend on the type of engine and is usually 8.5–35 N/mm³.

Big end: The big end diameter d_1 at crank pin is 0.67D to 0.73D and length $l_1 = 0.5\,d_1$ to 1.5 d_1. The bearing pressure is 5 to 25 N/mm².

The bolts of the big end are subjected to inertia force due to reciprocating masses F_i. *To prevent unlocking of nuts, castle nut is recommended to be tightened to a predetermined initial tightening load by a torque wrench (25.13). The uniform strength bolt along with castle nut have less weight in comparison with a bolt and nut and lock nut. It also has more resistance to (25.11, 25.12) shock loads. To reduce weight, studs are forged solid in the palm end of connecting rod (25.14).*

Three types of big ends are found in practice.
1. Separate top-half bolted to palm end, the bearing metal being run direct into bearing halves.
2. One piece forged solid with rod and the bearing metal is carried in loose renewable shells of steel or bronze. The steel shell carries bearing metal lined inside.
3. Same as (1) but the bearing metal being run directly into the rod end and cap. This is rarely used as replacement in case of wear is very costly.

The friction of the two end bearings produces bending moment which tends to bend the connecting rod into S-shape. For the big end, the friction at the crank pin creates a reaction R at the piston pin normal to the connecting rod centre line,

$$R = \frac{Pfr_c}{l} \qquad \qquad \dots (25.9)$$

r_c = crank pin radius
l = length of the connecting rod

The bending moment due to this reaction, $M = R(l - a)$, a is the distance of section from the end, and the stress will be M/z. Similarly the stress created by friction of the piston pin at a section b from the piston pin is M_1/z, where

$M_1 = R_1(l - b)$ and $R_1 = \dfrac{Pfr_c}{l}$, r_c is the radius of

piston pin. f is 0.02 to 0.03 and = 0.05 in case of lubrication failure. As these values are very small and happen to be near the ends, so these are not taken into consideration.

Type (1) is convenient to use for assembly with centre cranks, while types (2) and (3) can be used with overhanging type of crankshaft.

The thickness of steel shells is about 0.05 cylinder bore while the thickness of bearing metal 0.025 mm to 1.5 mm and in some cases up to 3 mm.

Lubrication of small end is by splash for horizontal engines and for small vertical engines. For large size of connecting rod, lubrication is through a small hole drilled through the connecting rod. The oil under pressure travels from big end to the small end. The big end lubrications is with a sump of oil in the crank case (25.9).

The inertia force due to reciprocating masses: The reciprocating parts are—piston, piston pin and about one third of the weight of con-necting rod. The magnitude of this force, equation 3.10, is

$$F_i = \frac{W\,v^2}{g}\left(\cos\alpha + \frac{r}{l}\cos 2\alpha\right) \quad \dots (25.10)$$

The bolts are placed close to the bearing to reduce weight.

The bolts are subjected to inertia force due to reciprocating masses F_i.

Force per bolt $F = \dfrac{F_i}{n}$ where n is the number of bolts

$$F = \frac{\pi}{4}d_c^2 S_d$$

where d_c is the core diameter of bolt

Nominal diameter $d = 1.2\,d_c$ of standard size.

The bearing cap thickness is calculated from bending moment due to the inertia force F_i. The bending moment is approximated as $\dfrac{F_i\, l_3}{6}$ and the section modulus $\dfrac{l_1\, t_1^2}{6}$, where t_1 is the thickness of the cap.

l_3 is the distance between centres of bolts.

Design steps for connecting rod: These steps can be used for making computer programme also.

1. Calculate the maximum gas force

$$P = \frac{\pi}{4} D^2 p$$

   where, p is the maximum explosion pressure.

2. Calculate the inertia force due to reciprocating masses F_i equation 25.10 to get the net force $P - F_i$.

3. Assuming the I-section as in Fig. 25.12 find thickness and establish the section using Rankin's equation using safety factor 6 on yield stress in compression.

$$P - F_i = \frac{S_y\, A}{1 + a\left(\dfrac{l}{k}\right)^2}$$

4. Find the maximum bending stress S_b due to centrifugal force F'_i. Equation 25.8a and check for the safety factor with total stress $S_y/6 + S_b$.

5. Assume the small end dimensions $l \le$ 0.45D, $d_1 = \dfrac{D}{3}$, $l_1/d_1 = 1.5$ to 2 and check for bearing pressure.

6. Assume big end dimension $d_1 = 0.67$ to 0.73D and $l_2/d_2 = 0.5$ to 1.5 and check for bearing pressure.

7. Find the bolt diameter d_b from force/bolt from F_i.

$$\frac{F_i}{2} = \frac{\pi}{4} d_c^2 S_d \text{ and } d_b \approx 1.2\, d_c$$

8. Decide the distance from centre to centre of bolts l_3 = bearing diameter $d_2 + 2 \times$ (thickness of bearing shell + thickness of

bearing) + bolt diameter d_b + allowance (1 to 5 mm).

9. Find the maximum bending moment

$$M = \frac{F_i l_3}{6} \text{ and find the thickness } t_1 \text{ of cap}$$

by equating $M = \dfrac{l_2 t_1^2}{6} S_d$.

Example 25.4: Design a connecting rod for a petrol engine with the following data:

Diameter of piston	= 110 mm
Stroke length	= 150 mm
Weight of reciprocating parts	= 20 N
R.P.M	= 1500
Compression ratio	= 8 : 1
Length of connecting rod (centre to centre)	= 325
Maximum explosion pressure	= 2.5 N/mm^2
Compression yield stress	= 320 N/mm^2

$$w = \frac{2\pi n}{60} = \frac{2\pi \times 1500}{60} = 50\pi \text{ rad/s}$$

$$\frac{l}{r} = \frac{325}{75} = 4.34$$

Force on the piston

$$= \frac{\pi}{4} D^2 p = \frac{\pi}{4} \times 110^2 \times 2.5 = 23758 \text{ N}$$

Inertia force due to reciprocating masses

$$= \frac{W}{g} w^2 r \left(1 + \frac{1}{n}\right)$$

$$= \frac{20}{9810} \times (50\pi)^2 \times 75 \times \left(1 + \frac{1}{4.34}\right) = 4642 \text{ N}$$

Net force on the piston = 23758 – 4642 = 19116 N
Consider $\Theta = 0$ force along the connecting rod = 19116 N

Considering I section of Fig. 25.12

$$I_x = \frac{419}{12} t^4, \quad A = 11t^2, \quad I_y = \frac{131}{12} t^4$$

$$\frac{I_x}{I_y} = 3.2 \quad \text{O.K.}$$

As $\dfrac{I_x}{I_y} < 3.5$, X plane is weaker so design for X plane

$$K_x^2 = \frac{I_x}{A} = \frac{419 t^4}{12 \times 11 t^2} = 3.18 t^2$$

Taking a factor of safety of 6 and applying Rankin's equation 2.41a

$$6 \times 19116 = \frac{320 \times 11t^2}{1 + \frac{1}{7500}\left(\frac{325^2}{3.18t^2}\right)}$$

Solving for t^2, $t^4 - 0.31t^2 - 0.0136 = 0$

$t^2 = 35$ and $t = 5.91$ say 6 mm

Width $= 4 \times 6 = 24$ mm, depth $= 5 \times 6 = 30$ mm

Checking for bending stress due to centrifugal force

$$F_i' = \frac{W\omega^2 r}{2g} = \frac{wAl\omega^2 r}{2g}$$

$$= \frac{78000 \times 11 \times (0.006)^2 \times 0.325 \times (50\pi)^2 \times 0.075}{2 \times 9.81}$$

$$= 946.8 \text{ N}$$

Maximum bending moment

$$= \frac{2F_i l}{9\sqrt{3}} = \frac{2 \times 946.8 \times 0.325}{9\sqrt{3}} = 39.4 \text{ Nm}$$

Modulus of section

$$= \frac{I_x}{y} = \frac{419t^4 \times 2}{5t \times 12} = \frac{419t^3}{30}$$

$$= \frac{419 \times 6^3}{30} = 3017 \text{ mm}^3$$

Bending stress

$$S_b = \frac{M}{Z} = \frac{39.4 \times 1000}{3017} = 13.05 \text{ N/mm}^3$$

Maximum compression stress = buckling stress + bending stress

$$\frac{320}{6} + 13.05 = 66.38 \text{ N/mm}^2$$

which gives a factor of safety of $\frac{320}{66.38} = 4.8$ on yield stress, so O.K.

Big End Bolts

Consider two bolts to hold the cap,

Force per bolt $= \frac{4642}{2} = 2321$ N

For bolt with core diameter d_c

$$d_c = \left[\frac{4 \times 2321}{\pi \times 80}\right]^{1/2} = 6.07$$

Nominal bolt diameter $1.2 \times 6.07 = 7.28$ mm

M 10 may be adopted

Big End

Big end bearing diameter $d_2 = 0.67 \times 110 = 73.7$ say 74 mm

Taking $\frac{l_1}{d_1} = 0.5$, $l_1 = 37$ mm

Bearing pressure

$$S_b = \frac{19116}{74 \times 37} = 6.98 \text{ N/mm}^2$$

which is satisfactory

Considering gas force,

$$S_b = \frac{23758}{74 \times 37} = 8.6 \text{ N/mm}^2 \text{ also ok.}$$

Small End

Length of small end $l = 0.45 \times 110 = 48.5$ or 50 mm

$$d = \frac{1}{3} \times 110 = 36.6 \text{ or } 37 \text{ mm}$$

Bearing pressure

$$f_b = \frac{23758}{37 \times 50} = 12.8 \text{ N/mm}^2$$

which is satisfactory.

Cap

Bolt centre distance = bearing diameter + 2 (Thickness of bearing) + bolt diameter

$$= 74 + 2 \times 6 + 10 = 96 \text{ mm}$$

Maximum bending moment

$$= \frac{Wl}{6} = \frac{4642 \times 96}{6}$$

$$= 74272 \text{ Nmm}$$

Cap Thickness

$$t_1 = \left[\frac{74272 \times 6}{37 \times 80}\right]^{1/2} = 12.2 \text{ or say 13 mm}$$

25.8 PISTON

General consideration: One of the most important element in the I. C. Engine is the piston. After ignition of gases the piston is pushed because of large force due to explosion and also carries the heat for dissipation.

The following need consideration:

1. The piston receives the large impulse and transmits it through the connecting rod to the crank.
2. It transfers large amount of heat of combustion chamber to cylinder walls and other parts.
3. It acts as a movable wall for variable volume of gas chamber.
4. It has to have sufficient strength to receive the gas force or fluid power.
5. It reciprocates without noise and resistance.
6. It resists the thermal and mechanical distortion.

It is desirable to absorb as little heat as possible. In order to obtain good heat flow to the cylinder walls, the piston crown should be of uniform thickness and the barrel where it joins the crown should have the same thickness. The rate at which heat is transferred from the piston head to cylinder walls is directly proportional to the temperature difference, cross-section of heat path, coefficient of heat transfer and inversely proportional to the length of heat path.

The flat top gives least exposed surface to gases and absorbs less heat. *If the crown is not flat but concave or convex, it is obvious that it being less rigid, can stand higher temperatures by changing its shape and thus to a certain extent equalizing the stresses.* Figure 25.14 illustrates this point.

It will also improve turbulence. *At the same time this explains why many designers advise against connecting crown with the cylindrical part by ribs which make the construction rigid. However, the ribs carry a considerable part of heat*

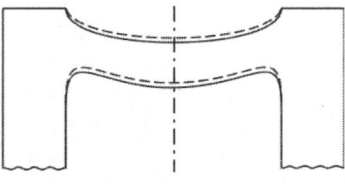

Fig. 25.14.

from the centre direct to piston rings and to the liner, thus quicker heat transfer. To reduce heat flow to piston skirt a partial circumferential groove between head and top ring is also desirable.

Piston materials: Cast iron, cast steel, forged steel, cast aluminium alloy, forged aluminium alloy with silicon. Cast iron and cast steel have high strength, good wear resisting properties at high temperature, low thermal expansion but relatively low thermal conductivity. Aluminium alloys containing silicon, copper, nickel, magnesium have high thermal conductivity, five times that of cast iron but coefficient of thermal expansion is about twice of cast iron. It also has loss of strength and wear resistance at high temperature. However, the maximum and minimum temperature of aluminium alloy piston is 260° and 140°C respectively.

Construction: The crown or head of piston carries high temperature and load. The skirt acts as bearing for connecting rod side thrust. The piston pin acts as a crosshead and delivers the thrust to connecting rod at various inclinations. The piston rings act to seal the cylinder gases and transfer heat to cylinder walls. The scraps ring scraps the lubricating oil so that it does not mix with gases. *The ribs inside the piston make for large surface area for convection and faster transfer of heat (25.17).* The lower crown temperature therefore prevents pre-ignition. The piston rings must be narrow for good heat transfer.

The variation of temperature (25.15) on the crown and along the length of piston are indicated for cast iron and aluminium alloys, cast iron retains elasticity at high temperature.

The temperature of the gases is quite high. As the rate of heat dissipation is very effective, the material does not melt (25.15).

The piston pin is responsible for transmitting the thrust of piston to the connecting rod and also allowing the obliquity of connecting rod for turning the crankshaft. To obtain uniform distribution of pressure between piston and cylinder during slap, The piston pin should he half way between edge of last ring

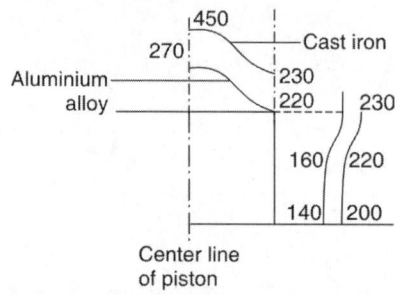

Fig. 25.15: Temperature variation on the piston

groove and end of piston. For the shorter height of engines, it is usually moved nearer to top. In vertical engines, the location of piston pin is nearer to the top and makes the piston run quieter.

Material of piston pin: It is case-hardened or nitrided carbon steel or alloy steel.

The piston diameter has to be less than cylinder bore so that it does not cease as the temperature increases. The liners are jacketed with coolant for faster heat dissipation. *The head of piston will expand more as campared to the open end as temperature of head is much higher than the open end so the diameter of head of piston must be less as compared to the open end (25.16)* as the piston has to have a free movement in the cylinder so it cannot seal the high pressure of gases. To reduce heat flow to piston skirt, circumferential groove is provided between top ring and head. Also or alternately, a slot is made after the lost ring with circular hole at the ends.

Piston rings: To prevent leakage of gases through piston.

Material of piston rings: To have spring proper-ties, high strength, high wear resistance and desired friction coefficient, alloy cast iron of the composition as silicon 2.5 to 2.8%, phosphorus 0.5–0.7%; manganese 0.6 to 0.8% sulphur not

over 0.1%, combined carbon 0.6 – 0.8% and total carbon 3.5–3.8% is used.

The open ends having shapes as shown in Fig. 25.16, with adjacent rings placed at 180° to prevent leakage (25.18, 25.19). However, it is not possible to prevent some leakage.

For special liners, rings are made of nickel-chromium iron or nickel, chromium molyb-denum.

Design consideration: Figure 25.17. Piston head or crown is considered as a flat circular plate fixed around the edge and uniformly loaded with gas pressure. Its thickness is given by:

$$t_1 = 0.43\, D \sqrt{\frac{p}{s_t}}$$

D is cylinder diameter

s_t is allowable tensile stress

p is the maximum gas pressure

Values of s_t are 38 N/mm² for CI having maximum temperature of 444°C

60 N/mm² for nickel cast iron-semisteel having a maximum temperature of 600°C.

84 N/mm² for forged steel and having a temperature of 875°C.

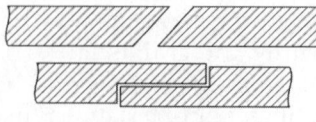

Fig. 25.16

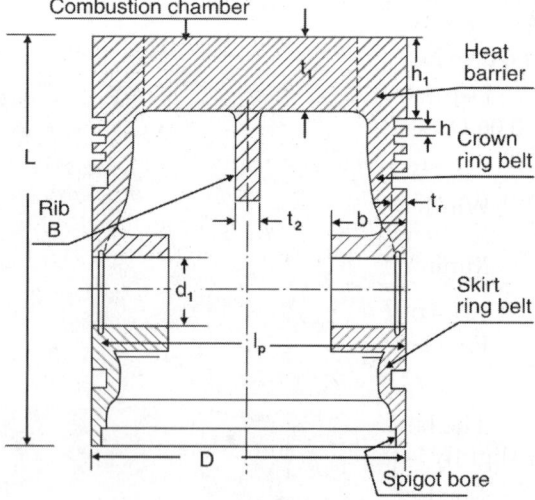

Fig. 25.17

50–70 N/mm² for aluminium alloy having maximum temperature of 275–300°C.

The empirical value of $t_1 = 0.032\,D + 1.5$

Considering heat flow

$$t_1 = \frac{D^2 q}{1600 K\left(T_c - T_e\right)}$$

q = heat flow from gases depending upon piston material, m.e.p and stroke/bore ratio J/s m².

     = 32000 to 12800 for CI in 4 stroke engine

     = 64000 to 256000 for aluminium piston

D = cylinder diameter

K = heat conductivity

     = 460 J/sm²°C/mm for cast iron

T_c = Temperature at the center of the head, 444° for cast iron; 275°C for aluminium alloy.

T_e = Temperature at the edge of the head 270°C for cast iron and 160°C for aluminium alloy.

$t_c – t_e$ = 222°C for cast iron

     = 111°C for aluminium alloy.

Radial width of ring groove

$$t_r = D\sqrt{\frac{3p_r}{s_1}} \quad \text{or} \quad \frac{D}{20}$$

s_1 = allowable stress = 82 N/mm² for cast iron

p_r = radial pressure on piston rings = 0.025 to 0.06 N/mm² for various types of engines.

Depth of ring groove $h = t_r + 0.4$ or $0.03\,D$ to $0.06\,D$.

i = number of rings [2 to 7]

Width of piston rings $h = 0.7\,t_r$ to t_r

Minimum thickness $h = \dfrac{D}{10i}$

i = number of rings

First land or distance of first ring

$h_1 = t_1$ to $1.2\,t_1$

The lands between the ring grooves = h or slightly less than h.

Maximum thickness of piston barrel

$t_3 = 0.03D + h + 4.5$

Wall thickness towards open end of piston

$t_4 = 0.25\,t_3$ to $0.35t_3$

Length of piston

$L = 0.75\,D$ for aero-engines

     = D to $1.5\,D$ for other engines and upto $2.5\,D$ for marine engines

Rib thickness $t_2 = \dfrac{t_1}{4}$ there will be 4 ribs.

Length of stroke = $1.3\,D$ to $1.4\,D$

A large piston runs more quietly and has less slap, i.e. lateral movement in the clearance between piston and cylinder.

Piston pin: To obtain uniform distribution of pressure between piston and cylinder, the piston pin should be half way between to edge of last ring groove and end of piston. To reduce length of piston, the pin is moved nearer to top. In vertical engines, piston pin nearer to top is quieter. *Usual materials are case hardened or nitrided carbon steel or alloy steel as it is subjected to varying loads (25.26).*

Length of piston pin $l_p = 0.88\,D$ to $0.933\,D$ for automobile engine and $0.8\,D$ to $0.9\,D$ for floating pin. A floating pin is one which has relative motion in both small end and piston bosses. Both the bearings will have bronze bushes. *To prevent the pin against axial movement, bronze or aluminium plugs or spring retainers or circlips are used (25.22).* These will prevent axial movement and scoring of cylinder walls of liners. The liners are made of mild steel or nickel venadium steel hardened and ground.

Length of small end bearing

$l_1 = 0.43$ to $0.45\,D$ or maximum $0.62\,D$

Diameter of piston pin

$d_1 = 0.22\,D$ to $0.28\,D$ for petrol engines

     = $0.3\,D$ to $0.38\,D$ for diesel engine

Maximum gas pressure

$$P = \frac{\pi}{4}D^2\, p_{max}$$

$$= l_1\, d_1 f_b$$

For piston bosses of length b

$$P = 2 \, b \, d_2 f_b$$

$$\frac{l_1}{d_1} = 1.5 \text{ for petrol engines}$$

$$= 2 \text{ for other engines}$$

f_b, bearing pressure

$$= 12.5 \text{ N/mm}^2 \text{ for gas engines}$$

$$= 15 \text{ N/mm}^2 \text{ for oil engine}$$

$$= 16 \text{ N/mm}^2 \text{ for automotive engines}$$

Piston pin is made hollow for reducing weight (25.21).

Inside diameter of piston pin $= d_i = 0.6 \, d_1$.

Consider the bearing pressure as concentrated load at the center of pin

Maximum bending moment

$$= \frac{Pl_1}{4} \square \frac{PD}{8} = \frac{\pi}{32} \left[\frac{d_1^4 - d_i^4}{d_1} \right] s_1$$

$s_1 = 85 \text{ N/mm}^2$ for case hardened carbon steel

$= 135 \text{ N/mm}^2$ to 250 N/mm^2 for case hardened nitrided alloy steel

The skirt is subjected to slap in which case the side pressure should not exceed 0.25 N/mm² but can be up to 1 N/mm². Considering coefficient of friction $m = 0.03$ to 0.1, the frictional force $= \mu P$

This will cause an average pressure

$$f_b = \frac{\mu P}{l_s D}$$

where, l_s is the skirt length.

Number of holes in the oil groove is 6 to 12

Oil hole diameter $= (0.3 - 0.5)h$

The clearance of the piston head $= 0.006 \, D$ to $0.008 D$

The clearance of the open end $= 0.001 \, D$ to $0.002 \, D$

The piston has a least area of cross-section at the ring groove. The suitable thickens at this cross-section can be checked by finding maximum compressive stress due to gas pressure and also tensile stress due to inertia

force of piston mass above the piston scrapring groove.

Example: 25.5: Design a trunk piston for a 4-stroke diesel engine having the following data:

Power	110 kW
Maximum speed	3600 rpm
Maximum torque	380 N at 1800 rpm
Bore	86 mm
Stroke	86 rpm
Compression ratio	16.3 : 1
Maximum explosion pressure	9.5 N/mm²

There are three compression rings and one oil ring.

Considering head as circular disc fixed along the circumference subjected to uniform pressure, crown thickness is

$$t_1 = 0.43D \sqrt{\frac{p}{s_t}} = 0.43 \times 86 \sqrt{\frac{9.5}{38}}$$

$$= 18.49 \text{ say } 19 \text{ mm}$$

Considering heat flow

$$t_1 = \frac{D^2 q}{1600 K (t_c - t_e)} = \frac{(86)^2 \times 128000}{1600 \times 4600 \times 220}$$

$$= 5.8 \text{ mm}$$

Empirical value of $t_1 = 0.032 \, D + 1.5$ up to 0.2 D, i.e. 17.2 so 19 mm is O.K.

Distance of first land $= t_1 = 19$ mm.

Radial depth of ring groove

$$t_r = D \sqrt{\frac{3 p_r}{s_1}}$$

$p_r = 0.04$ and $s_1 = 82 \text{ N/mm}^2$.

$$t_r = 86 \sqrt{\frac{3 \times 0.04}{82}} = 3.28 \text{ say } 3.5 \text{ mm}$$

Empirical value is $t_r = 0.04 \, D = 0.04 \times 86 = 3.44$ or 3.5 is O.K.

Width of piston rings

$$h = 0.7 \, t_r \text{ to } t_r \text{ say } 3.5 \text{ mm}.$$

Land between each ring

$$l_1 = h = 3.5 \text{ mm}$$

Depth of ring groove

$$h_1 = t_r + 0.4 = 3.9 \text{ say } 4 \text{ mm}$$

Maximum thickness of piston barrel

$$t_3 = 0.33\,D + h + 4.5 = 10.5 \text{ mm.}$$

Wall thickness of barrel towards open end

$$t_4 = 0.25\,t_3 = 2.5 \text{ mm}$$

Rib thickness

$$t_2 = \frac{t_1}{4} = 4.5 \text{ mm}$$

Piston skirt length

$$0.6\,D = 0.6 \times 86 = 51.6 \text{ say } 52 \text{ mm}$$

Piston height above piston pin

$$= 0.50 \text{ to } 0.65\,D = 43 \text{ to } 55$$

also $= 18 + 4 \times 3.5 + 4 \times 3.5$

$$= 46 \text{ mm OK } (>.5\,D)$$

Total length of piston

$$L = 46 + 52 = 98 \text{ mm OK } (0.75\,D \text{ to } 1.5\,D)$$

Piston pin:

Length of piston pin

$$= 0.88\,D \text{ to } 0.93\,D$$

$$= 75.68 \text{ say } 76 \text{ mm}$$

Length of small end of connecting rod

$$= \leqslant 0.45 \leqslant 38.7 \text{ mm}$$

say 34 mm.

Length of end bosses

$$= \frac{86 - (34 + 2)}{2} = 25 \text{ mm}$$

2 mm is the total clearance on both sides.

Diameter of piston pin

$$d_b = 0.22\,D \text{ to } 0.28\,D = 25 \text{ mm}$$

Inner diameter or piston pin

$$d_i = 0.6 \times 25 = 15 \text{ mm}$$

Maximum gas force

$$F = \frac{\pi}{4} \times 86^2 \times 9.5 = 55184 \text{ N}$$

Angular velocity of crank

$$= \frac{2\pi \times 6000}{60} = 628 \text{ red/s}$$

Assuming total weight of reciprocating parts = 20 N

Inertia force due to reciprocating masses

$$= \frac{20 \times 628^2 \times 0.043(1.32)}{9.81}$$

$$= 45638 \text{ N}$$

Net force acting on the piston pin

$$= 55184 - 45638 = 9546 \text{ N}$$

Section modulus of pin

$$= \frac{\pi}{32}\left[\frac{d_p^4 - d_i^4}{d_p}\right] = \frac{\pi}{32}\left[\frac{25^4 - 15^4}{25}\right]$$

$$= 1335 \text{ Nmm}^3$$

Maximum bending moment

$$= \frac{Pl}{8} = \frac{9546 \times 76}{8} = 90687 \text{ Nmm}$$

Bending stress

$$= \frac{90687}{1335} = 68 \text{ N/mm}^2$$

Shear stress in the pin

$$= \frac{9546}{2 \times \dfrac{\pi}{4}(25^2 - 19^2)} = 23 \text{ N/mm}^2$$

Maximum principal stress

$$= \frac{1}{2}\left[68 + \sqrt{68^2 + 4 \times 23^2}\right]$$

$$= 75 \text{ N/mm}^2 \text{ ok as it is } <85 \text{ N/mm}^2$$

Bearing pressure in the small end

$$f_b = \frac{9546}{25 \times 34} = 11.2 \text{ N/mm}^2$$

This pressure is quite low so ok.

Bearing pressure due to slap on the skirt

$$f_{b_1} = \frac{\mu P}{l_s D}$$

Taking

$$\mu = 0.03$$

$$f_{b_1} = \frac{0.03 \times 9546}{52 \times 86}$$

$$= 0.064 \text{ N/mm}^2 \text{ O.K.}$$

Oil groove:

Assume 8 holes on the circumference [6–12]

Diameter of holes

$$= 0.3h = 0.3 \times 3.5 = 1.0 \text{ mm}$$

Section passing through center of holes is having the minimum area, so it is a critical section. This area is to be checked for compression and tensile stresses.

Thickness of wall = 10.5 mm.

Inner diameter of barrel = 86 – 2 × 10.5 = 65 mm.

Outer diameter, i.e. outer diameter at the root of grooves = 86 – 2 × 4 = 78 mm.

Net area of cross-section

$$A = \frac{\pi}{4}\left[78^2 - 65^2\right] - 8 \times 6.5 \times 1$$

$$= 1460 - 52 = 1408 \text{ mm}^2$$

Compression stress

$$S_c = \frac{9546}{1408} \pm 6.78 \text{ N/mm}^2$$

Inertia force due to weight of piston beyond the ring groove. Mass of piston = 10N

Inertia force

$$= \frac{W\omega^2 r}{g}[1 + 0.32]$$

$$= \frac{10 \times 628^2 \times 0.043}{9.81} \times 1.32 = 22819 \text{ N}$$

Maximum tensile stress

$$= \frac{22819}{1408} = 16.20 \text{ N/mm}^2$$

Diameter of piston at the open end

$$= 86 - 0.001 \, D$$

$$= 85.914 \text{ mm}$$

Diameter of the piston crown

$$= 86 - 0.006 \, D$$

$$= 85.484 \text{ mm}$$

PROBLEMS

25.1 Explain with sketches methods of reducing stress concentration in crankshafts.

25.2 If there is resonance in the crankshaft, which parameters should be changed?

25.3 Beyond what ratio of length to diameter, it is unsafe to treat a bar as a simple compression member?

25.4 For a given area of cross-section, which shape of a column is preferable?

25.5 Which values of slenderness ratio, Euler's formula and Rankine-Gordon formula are used?

25.6 Why are connecting rods usually of I-section.

25.7 Why are the big ends of connecting rod made bigger than the small end when the small end is satisfactory in strength?

25.8 How many types of inertia forces act on the connecting rod?

25.9 How is the lubrication of small end and big end effectively done?

25.10 Why is it necessary to locate the oil hole in the centre of the section of connecting rod of large lengths? Why is the oil hole not located in the flanges?

25.11 Why is it preferable to use a uniform strength bolt in the connecting rods?

25.12 Why is it preferable to use a castle nut instead of a nut and locknut for holding the cap?

25.13 Why is it desirable to tighten the nuts with a torque wrench instead of an ordinary wrench?

25.14 Why are studs usually forged in the palm end of a connecting rods used in preference to through bolts?

25.15 When the gases burn, the temperature of gases is very high, but the piston does not melt. Explain.

25.16 Why is diameter of the top of piston slightly smaller than the diameter of open end?

25.17 Why are ribs provided inside the piston though it makes the head more stiff?

25.18 The piston needs sufficient clearance to move inside the cylinder. How is leakage of gases prevented past the piston?

25.19 Why is it necessary to use at least two piston compression rings?

25.20 Name the materials used for I.C.E. piston.

25.21 Why is the piston pin usually made of hollow section?

25.22 How is the axial movement of the piston pin prevented?

25.23 Why are helical grooves made in the piston?

25.24 What purpose is served by scrap ring?

25.25 Why is the gudgeon pin usually surface hardened?

25.26 What is the material of piston pins?

25.27 The temperature of piston is highest on the head and lowest at the open end. How is variation of temperature taken care of in the design of the piston?

25.28 Describe the process of heat dissipation from the piston.

25.29 Determine the main dimensions of a crankshaft with center cranks for a single acting 178 × 178 mm air compressor running at 360 rpm. Maximum air pressure: 0.7 N/mm².

25.30 Determine the main dimensions of a crankshaft for a 88 × 95 × 150 mm automobile engine. The maximum pressure is 2.5 N/mm² at speeds of 2,000 to 3,600 rpm. The center lines of the cylinders are 120 mm apart.

25.31 Determine the main dimensions of a crankshaft for a 200 × 240 × 100 mm two-stroke gas engine running at 480 rpm. The maximum pressure is 2.6 N/mm². The distance between cylinder center lines is 290 mm.

25.32 Find the equivalent length of a shaft system consisting of the following: an oil-engine crankshaft 200 mm in diameter and 3.3 m long; and intermediate shaft 140 mm in diameter and 10 mm long; and a propeller shaft 160 mm in diameter and 6 m long.

25.33 Find the natural frequency of torsional vibration of the combined shaft of problems 25–29 and 25–30, assuming that there is between the crankshaft and the intermediate shaft a flywheel that weighs 12 kN and has a radius of gyration of 0.64 m. The propeller weighs 3 kN and has a radius of gyration of 480 mm. The weight assumed is concentrated at each crankpin is 2.9 kN.

25.34 Design the piston for a single acting four stroke engine for the following data:

Cylinder bore	125 mm
Stroke	150 mm
Maximum gas pressure	5N/mm²
Indicated mean effective pressure	0.75 N/mm²
Mechanical efficiency	82%
Fuel consumption	15 kg kW of brake power/hour
H.C.V. of fuel	42.5×10³ kJ/kg
Speed	1800 rpm.

25.35 Calculate suitable thickness of the piston crown for a 4-stroke diesel engine which runs at 350 rpm. The brake mean effective pressure is 0.7 MPa while maximum pressure is 5.0 MPa. Following data may be used:

Specific fuel consumption	3.0 kg/kw/hr
Calorific value of fuel	46300 kJ/kg
Temperature at the centre of piston	425°C
Temperature at piston edge	200°C
Heat conductivity factor	1775 J/m²°C. hr. mm
Heat conducted through top	5% of heat produced
Permissible tensile stress for the material of piston	27 MPa
Cylinder diameter	300 mm
Length of stroke	450 mm

Flywheels

26.1 FLYWHEEL ACTION

The purpose of a flywheel is to keep the speed of a machine between given limits while the machine is doing work or receiving energy at a variable rate (26.1). A flywheel stores up energy when energy is supplied more rapidly than it is used, and it gives out energy when the reverse is the case. While a flywheel is storing energy, its speed is increasing; while it is giving out energy, its speed is decreasing. The allowable amount of variation of speed in any particular problem depends on the conditions of the problem.

In some machines, such as shears, presses, or punches, the work is done during a small part of the cycle. If such a machine is driven directly by an electric motor or a belt, the latter must be powerful enough to supply all the energy consumed by the machine during the working portion of the cycle, and will run idle during the remaining part of the cycle. By inserting a flywheel between the driving device and the driven machine, a much smaller motor or belt may be used to supply energy at a practically constant rate throughout the cycle.

Flywheel effect: The kinetic energy in the rim of a rotating flywheel is

$$K = \frac{Wv^2}{2g} \qquad \ldots (26.1)$$

where, W is the rim weight, in Newtons, v is the mean velocity of the rim, in m per second, and $g = 9.81 \, m/s^2$. The amount of energy in the arms

and the hub is so small that it may be neglected. If the velocity changes from v_1 to v_2, the energy released or absorbed during this change, which is called *excess energy*, is

$$E = K_1 - K_2 = W \frac{\left(v_1^2 - v_2^2\right)}{2g} \qquad \ldots (26.2)$$

From equation 26.2, the rim weight is

$$W = \frac{2gE}{v_1^2 - v_2^2} \qquad \ldots (26.3)$$

The velocity may be expressed by the relation

$$v = \frac{2\pi k n}{60} \qquad \ldots (26.4)$$

where, k is the polar radius of gyration of the rim, in m, and n is the speed, in revolutions per minute. Then

$$Wk^2 = \frac{30 \times 60 gE}{\pi^2 \left(n_1^2 - n_2^2\right)} = \frac{182.4 gE}{n_1^2 - n_2^2} \qquad \ldots (26.5)$$

The product Wk^2 is the polar moment of inertia. It is also known as the *flywheel effect*.

Speed fluctuation: The relative speed variation is determined from the relation

$$m = \frac{n}{n_1 - n_2} \qquad \ldots (26.6)$$

where, n is the mean speed of the flywheel, and n_1 and n_2 are the maximum and minimum speeds, respectively. The number m is called the coefficient of steadiness, and its reciprocal $u = 1/m$ is called the *coefficient of fluctuation of speed or rotation*. The value of m, or u, depends on the nature of the service for which the machine is built; that is, on the permissible variation between the highest and lowest speeds during each operating cycle of the driven machine and on the method of connecting it to the driving motor. With a flexible connection, such as a belt drive, the coefficient m may be smaller than with a less flexible connection. An electric generator requires a more uniform drive, and a greater m, than a pump. Two or more generators operating in parallel require a still greater value of m. Table 26.1 gives minimum values of m for a number of typical drives. The maximum limits, as used in practice, is about 25 per cent higher.

The mean speed may be determined by the relation

$$n = \frac{n_1 + n_2}{2} \qquad \ldots (26.7)$$

Equation 26.5, in which $n_1^2 - n_2^2$ is replaced by $(n_1 - n_2)(n_1 + n_2)$, and in which the values from equations 26.6 and 26.7 are substituted, becomes

$$Wk^2 = \frac{1800\, Em}{2\pi^2 n^2} = \frac{91.2 gEm}{n^2} \qquad \ldots (26.8)$$

Excess energy: The excess energy E, defined by equation 26.2, may be found either analytically or graphically. Often the graphical method is simpler, particularly when the work is done at a variable rate and sufficient information is available to draw a work diagram.

Example 26.1: Determine the necessary flywheel effect for a horizontal-press drive for which the plan view is shown in Fig. 26.1. The belt pulley a drives the press and makes 250 rpm; its diameter is 0.8 m; the speed reduction between the spur gears c and d is 6.5 to 1; the actual work done during each revolution of the crankshaft e is equal to 9220 Nm and it is done during four-tenths of a revolution. Assume that the mechanical efficiency of the machine is 82 per cent.

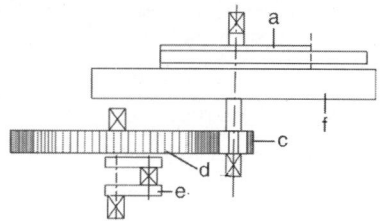

Fig. 26.1: Drive for a horizontal press

The crankshaft speed is

$$n = \frac{250}{6.5} = 38.45 \text{ rpm}$$

The power needed to drive the press is

$$P = \frac{9220 \times 38.45}{1000 \times 60} = 5.9 \text{ kW}$$

The belt speed is

$$v = \frac{\pi \times 0.8 \times 250}{60}$$
$$= 10.47 \text{ m/s}$$

Table 26.1: Required coefficient of steadiness m

Driven machinery	Type of drive	m
Hammers, crushers, punch presses	Belt	5
Compressors, concrete mixers, excavators	Belt	7–10
Pumps, shears	Belt or flexible coupling	20–25
Metalworking and woodworking machinery	Belt	30
Flour, paper, and textile mills	Belt	40–50
Compressors, pumps, and similar machines	Gears	50
Spinning machinery, coarse to fine	Belt	50–65
D-C generators, single or parallel	Belt	35
D-C generators, single or parallel	Direct-coupled	70
A-C generators, single or parallel	Belt	60
A-C generators, single or parallel	Direct-coupled	100

The net belt pull is

$$F_1 = \frac{P \times 1000}{v} = \frac{5.9 \times 1000}{10.47} = 563.5 \text{ N}$$

The total work to be done during one cycle is

$$W_a = \frac{9220}{0.82} = 11244 \text{ Nm}$$

The distance which the belt travels during the working portion of the cycle is

$$\frac{10.47 \times 0.4 \times 60}{38.45} = 6.535 \text{ m}$$

and the work done by the belt during this period is

$$W_b = 563.5 \times 6.535 = 3682.6 \text{ Nm}$$

The energy to be stored by the flywheel is

$$W_a - W_b = 11244 - 3682.6 = 7561.4 \text{ Nm}$$

From Table 26.1 the required coefficient of steadiness is $m = 5$. By equation 26.8, the required flywheel effect is

$$Wk^2 = \frac{91.2 \times 9.81 \times 7561.4 \times 5}{(250)^2} = 541.2 \text{ Nm}^2$$

26.2 STRESSES IN FLYWHEELS

The stresses which are created by the centrifugal force in the rim and arms of a flywheel are rather complicated, and no entirely satisfactory method of computing them can be offered. In a wheel cast in one piece, unknown shrinkage stresses of great magnitude may exist, making futile, any refined calculations. In general, the stresses in both the rim and the arms are tensile. However, since the rim and the arms are rigidly fastened together and stretch to different degree, bending stresses are produced in the rim, as indicated in Fig. 26.2. Speed variations and consequent inertia forces act tangentially and produce bending stresses in the arms.

Stresses in the rim: Tensile and bending stresses are caused by centrifugal force in the rim (26.3). The tensile stress created in each cross section of the rim by the centrifugal force is found by the equation 3.5, which is

$$s_1 = \frac{\pi^2 W r^2 n^2}{900g}$$

where, r is the mean radius of the wheel, in metres.

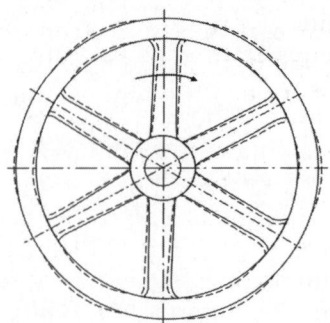

Fig. 26.2: Strains in a flywheel

The stress due to bending may be found by considering a strip of the rim with a face 1 unit wide (*see* Fig. 26.3). Since the area of the cross section for a strip of rim 1 unit wide is h, the centrifugal force per unit of rim width is, applying equation 3.2 with equation 3.3,

$$C' = \frac{2\pi^2 W r^2 n^2 h}{900g} \qquad \ldots (26.9)$$

The rim is loaded by the force C' uniformly over the whole circumference, as indicated in Fig. 26.3.

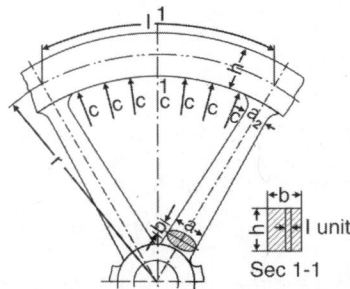

Fig. 26.3: Bending of flywheel rim

The maximum bending moment in the rim occurs at the arms. If the rim is considered as a straight beam, this bending moment may be found by the expression in Table 2.6 for case *i*. It is

$$M = \frac{C' l^2}{12} \qquad \ldots (26.10)$$

where, l is the length of the rim between arms, in mm. The stress due to the moment M is

$$s_2 = \frac{M}{Z} \qquad \ldots (26.11)$$

Substituting for M its value found by applying equations 26.10 and 26.9, replacing Z by $h^2/6$, and taking l as $2\pi r/i$, where i is the number of arms, results in

$$s_2 = \frac{0.216\, Wr^3n^2}{i^2hg} \qquad \dots (26.12)$$

The stretch of the arms may be taken as three-fourths that necessary for free expansion of the rim.[1] The combined tensile stress is then

$$s = 0.75s_1 + 0.25s_2 \qquad \dots (26.13)$$

Experience has shown that for cast-iron wheels the total stress s may be as high as 35 N/mm², or even 40 N/mm², but the tensile stress s should not exceed 7 N/mm². For cast-steel wheels, s_1 should not exceed 28 N/mm².

Stresses in arms are: Tensile and bending stresses due to centrifugal force and bending stress due to torque and acceleration (26.3). When a flywheel is accelerated from rest, or when the energy supply is suddenly cut off, the arms may have to carry the full torque load. Each arm of a wheel with a rigid rim may be considered to act as a cantilever beam that is fixed at the hub end and carries a concentrated load at the free end at the rim. The bending moment in this case is a function of the transmitted torque T. Thus, $M = T(D-d)/iD$, where d is the hub diameter. The stress is

$$s_1 = \frac{T(D-d)}{iZD} \qquad \dots (26.14)$$

where, Z is the section modulus of the arm cross section at the hub. For cast iron this stress should not exceed 14 N/mm² because of the uncertainties of the material and possible shrinkage stresses. In the case of a sudden load a still lower stress should be used, it should be reduced to 7 N/mm² for very severe load conditions. However, the constraint of the rim reduces the stress s_1. A very heavy rim may reduce it almost to 0.5 s_1, where s_1 is found by equation 26.14.

If the flywheel is used as a belt pulley, the arms are bent not only by the variation in speed but also by the net belt tension ($F_1 - F_2$). The moment due to this belt action is

$$M = \frac{(F_1 - F_2)(D - d)}{2i} \qquad \dots (26.15)$$

where, D is the flywheel diameter and d is that of the hub. The stress at the hub is

$$s_2 = \frac{(F_1 - F_2)(D - d)}{2iZ} \qquad \dots (26.16)$$

In a thin-rim wheel, because of the absence of rigidity, the load is not distributed equally among the arms.[2] In such a case it is safer to assume that the maximum stress is twice as great as its average magnitude, i.e. the load is shared by half the number of arms. Thus

$$s_2' = \frac{(F_1 - F_2)(D - d)}{iZ} \qquad \dots (26.17)$$

Finally, the arms are subjected to a tensile stress s_3 due to the centrifugal force acting upon the rim when the wheel is running at its maximum speed. This stress is evidently equal to $s_3 = \dfrac{Wv^2}{g}$, determined by equation 3.5. The maximum tensile stress in an arm is at the hub end, and is

$$s_{max} = s_1 + s_2 + s_3 \qquad \dots (26.18)$$

For cast iron this stress should not exceed 20 N/mm².

Stress due to acceleration: When a large load slows down a machine, the flywheel tends to maintain its speed, and this action throws a considerable bending stress into the arm. A quick stopping of the machine may be considered as a limit case. The stress depends on the number of seconds t in which the wheel is topped. The force F necessary to stop the wheel—whose rim weight W—in that time is

$$F = \frac{Wa}{g} \qquad \dots (26.19)$$

---

[1] D. S. Kimball and J. H. Barr, *Elements of Machine Design*, 3rd ed. (New York: John Wiley and sons, Inc., 1935), p. 440.
[2] Lionel S. Marks, ed., *Mechanical Engineers' Handbook*, 5th ed. (New York: McGraw-Hill Book company, Inc., 1951), p. 913.

where the negative acceleration a, or deceleration, is

$$a = \frac{v}{t} \qquad \ldots (26.20)$$

The force F is applied at the center of gravity of the rim, and at the hub the bending moment due to this force is $M = \frac{F(D-d)}{2}$. The stress produced is found by proceeding as indicated for the stress s_1 in equation 26.14.

Example 26.2: Find the stress produced in the arms of a flywheel when it is stopped in two revolutions from a speed of 250 rpm. The rim weighs 1700 N, its radius of gyration is $k = 0.5$ m, and the arms have an elliptical section 56 × 28.

The normal velocity of the center of gravity is

$$2\pi \times 0.5 \times \frac{250}{60} = 13 \text{ m/s}$$

The time of deceleration, which is the distance travelled divided by the mean velocity, is

$$t = \frac{2\pi \times 0.5 \times 2}{(13+0)/2} = 0.967 \text{ s}$$

By equation 26.20, the deceleration is

$$a = \frac{13}{0.967} = 13.44 \text{ m/s}^2$$

Since the weight of the rim is $W = 1700$ N, the force F is, by equation 26.19,

$$F = \frac{1700 \times 13.44}{9.81} = 2329 \text{ N}$$

If the hub diameter is $d = 150$ mm, the moment is

$$M = 2329 \times \left(0.5 - \frac{1}{2} \times 0.150\right) = 989.2 \text{ Nm}$$

The section modulus of the arm section is, from case i in Table 2.5,

$$Z = \frac{\pi \times 28 \times 56^2}{32} = 8620.5 \text{ mm}^3$$

With six arms, the stress is

$$s = \frac{M}{Zi} = \frac{989.8 \times 1000}{8620.5 \times 6} = 19.14 \text{ N/mm}^2$$

For cast iron this is a fairly high stress, but it does not reach the danger point.

26.3 FLYWHEEL DESIGN

In finding the weight of a flywheel, the mass of the rim only is considered; the additional small effect of the arm and the hub is neglected. First the radius of gyration k in the expression Wk^2 is determined by assuming a proper rim speed $v = 2\pi kn$.

Rim speed: The rim speed should not exceed 25 m/s for cast-iron wheels of machines *as at higher velocity the centrifugal force may cause failure (26.2)* under 75 kW, and 30 m/s for larger machines. With a special rim design which prevents blowholes in the rim, v may be increased to 35 m/s; and with special arm designs rim speeds as high as 50 m/s are used.[3] Cast-steel wheels can operate with speeds up to 50 m/s. Large flywheels for steel mills, having a rim speed of 75 m/s at 375 rpm, are assembled by using a cast-steel spider with arms and a laminated rim made of rolled steel plates. In automotive engines, rim speeds of cast-iron flywheels reach 50 m/s; those of semisteel wheels, 75 m/s; and those of cast-steel wheels, 100 m/s.

Rim dimension: The relation between k, in m and the outside diameter D of the rim, in m, is

$$k^2 = 0.125\left[D^2 + \left(D\frac{2h}{1000}\right)^2\right] \qquad \ldots (26.21)$$

If the rim thickness h, in mm, is small compared to D, it may be neglected. Then $k = 0.5\,D$.

Considering stresses in rim

$$S = \frac{v^2}{135} \text{ for cast iron}$$

$$S = \frac{v^2}{125} \text{ for steel}$$

After the weight W is found from equation 26.8, the cross-sectional area A of the rim, in square mm, can be computed from the equation

$$W = 2\pi k \times 4w \qquad \ldots (26.22)$$

where, for cast-iron, $w = 70600$ N/m^3, and for steel, $w = 76500$ N/m^3. The most common cross

[3] F. A. Halsey, Handbook for Machine Designers, 2nd ed. (New York: McGraw-Hill Book Company, Inc., 1916), pp. 69–71.

section is a rectangle with a width b and a height h, for which $A = bh$. The ratio b/h is selected between 0.65 and 2. If the flywheel is to be used also as a belt pulley, the rim face b must be made at least 25 to 50 mm wider than the belt. The outside diameter, in m, can be found from equation 26.21; or its approximate value, in m, may be taken as

$$D_0 = 2k + h \qquad \dots (26.23)$$

The *hub diameter* may be made equal to two shaft diameters, and the hub length may be 2 to 2.5 shaft diameters.

Arms: The standard number of arms on a flywheel is six. Very wide flywheels serving as belt pulleys are made with two rows, or 12 arms; and flywheels with very large diameters are sometimes made with 8, 10, or even 12 arms. Some oil engines have flywheels with webs, as indicated in Fig. 26.6, instead of arms. The holes are made to facilitate handling.

The arms usually have an elliptic section, as shown in Fig. 26.3, with the major axis twice the minor, or $a = 2b$. *With the elliptical arms, air resistance is reduced as compared to rectangular or I-section and has more moment of inertia as compared to circular section for the same weight or it has less weight for the same strength (26.4)*. From the hub down to the rim the arms taper from 10 to 25 per cent. Arms with H and I sections are used very rarely, but in very large flywheels hollow elliptic sections are sometimes used to prevent porous castings. For an elliptic section, $Z = \pi ba^2/32$. After Z has been found by equation 26.14, $a/2$ can be substituted for b, and the major axis can be computed from the relation

$$a = \sqrt[3]{\frac{64Z}{\pi}} \qquad \dots (26.24)$$

These values of a and b are at the hub.

These values are reduced near the rim to make them approach uniform strength.

Example 26.3: Find the weight and the main dimensions of the flywheel of example 26.1. The distance from the center of the shaft to the wall is 0.6 m.

The maximum outside diameter of the wheel is $D < 1.2$ m. The radius of gyration must be assumed

slightly smaller than $D/2$, and k will be taken as 0.5 m. Since $Wk^2 = 541.2$ Nm^2.

$$W = \frac{541.2}{0.5^2} = 2164.8 \text{ or } 2165 \text{ N}$$

By equation 26.22, the area of a cast-iron rim should be

$$A = \frac{W}{2\pi k \times w} = \frac{2165}{2\pi \times 0.5 \times 70600}$$

$$= 0.00976 \text{ m}^2 = 9760 \text{ mm}^2$$

A suitable rim section would be a rectangle 140×70 mm. The outside diameter may be found by applying equation 26.21, in which $k = 0.5$ m and $h = 70$ mm, and solving for D. This results in $D = 1.072$ m. The approximate value, by equation 26.23, is

$$D = 2 \times 0.5 + 0.07 = 1.07 \text{ m}$$

The difference is negligible and taking the value of $D = 1.07$ m.

The arm section may be found by using equation 26.14. The moment is equal to

$$F_t(D - d)/2 = \frac{563.5 \times (1.07 - 0.15)}{2} = 259.2 \text{ Nm}.$$

If s_1 is taken as 7 N/mm^2, since shock action is present, and if six arms are used.

$$Z = \frac{M}{is_1} = \frac{259.2 \times 1000}{6 \times 7} = 6171 \text{ mm}^3$$

with $b = a/2$, the major axis of the ellipse, by equation 26.24, should be

$$a = \sqrt[3]{\frac{64 \times 6171}{\pi}} = 50 \text{ mm}$$

Then $b = a/2 = 25$. At the rim, if $a' = 0.9a$, the dimensions are $a' = 45$, and $b' = 22.5$.

Construction: Flywheels up to 2.5 m diameter are cast in one piece; above this size they are usually made in halves. A split hub, Fig. 26.4., bored 0.025 to 0.05 mm smaller than the shaft, clamps the shaft and prevents the key from working loose. A steel wedge driven into the hub slot is used to put the wheel on the shaft. The diameter of the clamping bolts is made about one-sixth the shaft diameter. Sometimes the hub is split clear through, as shown by the dotted lines below the horizontal center line in Fig. 26.4. This procedure has the additional

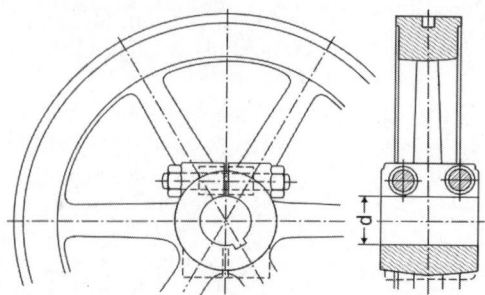

Fig. 26.4: Flywheel with a split hub

advantage of relieving the shrinkage stresses in the arms to certain extent.

A flywheel made in two halves should be parted on an arm rather than between arms, the latter method giving a joint only half as strong as the former, as shown in Table 26.2. The halves are connected by bolts through the hub and near the rim, and also by shrink links, as shown in Fig. 26.5, or by shrink anchors, as in Fig. 26.6. The anchor connection has the advantage of easier and more accurate machining, which assures that the desired force will be created when the anchor is shrunk into place. If the rim section is made I-shaped, as in Fig. 26.7, the anchors can be so proportioned that the joint will be as strong as the rim proper.[4]

An anchor connection with wedge-shaped cotters, as in Fig. 26.8, eliminates the troubles encountered with shrink fits, especially if the flywheel must be taken off eventually.

While the relative strengths of a rim joint and the solid rim depend on the exact proportions

used, average values confirmed by tests are given in Table 26.2.[5]

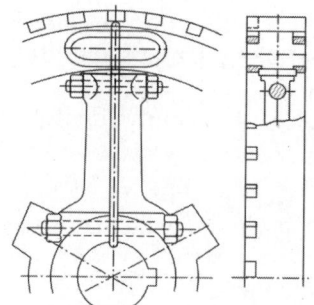

Fig. 26.5: Split flywheel

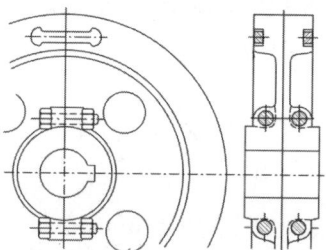

Fig. 26.6: Split disk flywheel

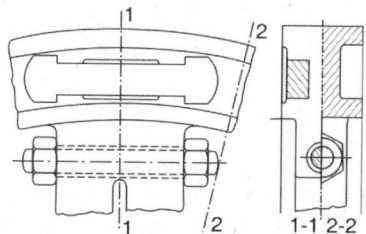

Fig. 26.7: The haight rim joint

Table 26.2: Relative strengths of flywheel rim

Type of construction	Relative strength
Solid rims	1.00
Flanged joint, bolted, rim parted between arms	0.25
Flanged joint, bolted, rim parted on an arm	0.50
Shrink-link joint (Fig. 26.5)	0.60
Anchor joints (Figs 26.6 and 26.8)	0.70
Haight joint (Fig. 26.7)	1.00

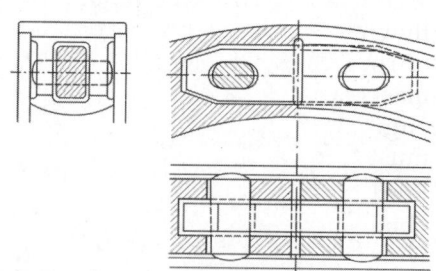

Fig. 26.8: Anchor with cotters for a split-flywheel rim

[4] H. V. Haight, "A High Efficiency Flywheel Joint," *American Machinist*, Vol. 51 (February, 1907), p. 267; also Halsey, *op. cit.*, p. 73.
[5] C. H. Benjamin, 'The Bursting of Small Cast-iron Flywheels,' *Transactions of The American Society of Mechanical Engineers*, Vol. 20 (1899), p. 209, and Vol. 23 (1902), p. 168.

The force F which acts upon a rim connection is

$$F = 2sA \qquad \text{...(26.25)}$$

where, s is the stress due to the centrifugal force, determined by equation 3.5, and A is the cross-sectional area of the solid rim.

Welded wheels: Large flywheels and those having high peripheral speeds are fabricated by welding. The hub is made of a steel forging; the rim is obtained by bending a rectangular bar with a suitable cross section into a ring and welding the ends together; and the arms are built up along the lines of Figs 30.14 and 30.15.

PROBLEMS

26.1 What purpose does a flywheel serve?

26.2 Why should the diameter of a flywheel be determined from the constraint of velocity?

26.3 Name the stress in the rim and arms of a flywheel.

26.4 Why is it preferable to use elliptical cross-section of the arms?

26.5 Why are arms usually tapered?

26.6 How many flywheels may be used to minimize the speed of an engine?

26.7 In a certain engine the mean flywheel diameter is 1.8 m. The weight of the rim is 6.6 kN. The maximum and minimum instantaneous wheel speeds are 195 and 185 rpm, respectively. Determine (a) the excess energy stored and given up by the flywheel, (b) the coefficient of uniformity of rotation and (c) the kind of driven machinery that can be connected to this engine, and the type of drive which must be used in each case.

26.8 Find the flywheel effect necessary to obtain a coefficient of steadiness m of 50 with a four-stroke, four cylinder 300 × 380 mm gas engine running at 300 rpm. The 150 mm length of the torque diagram covers 180 deg of the crank travel, its

scale of ordinates is 0.18 N/mm²/cm, and the excess area is $e = 1750$ mm².

26.9 Find the main dimensions of the flywheel in problem 26.8. Assume that the maximum cylinder pressure is $p_{max} = 2$ N/mm², and that, at the crank angle of maximum torque, $p = 0.4\,p_{max}$.

26.10 Find the stresses in the rim of the flywheel of problem 26.9.

26.11 Compute the stress in the arms of the flywheel of problem 26.9 when the engine load is increased so suddenly that the engine slows down to 240 rpm after three revolutions.

26.12 Determine the weight of the flywheel required for a punch press for the following conditions: The press requires 11 kW power; the complete cycle consists of 7 revolutions of the flywheel, only two and one-half of which take place during the working portion of the cycle; there are 30 complete cycles per minute; and the desirable mean diameter is 1.3 m.

26.13 Design a flywheel for a punch press which must be brought to rest by one punching operation if the power has been shut off. The maximum work required consists of punching a 30 mm hole in 20 mm mild-steel plate. The shear strength of the plate is 300 N/mm². The punch capacity is 24 holes per minute, and the speed ratio of the driving shaft to the eccentric shaft operating the punch is 9:1. In order to clear the floor, the wheel diameter cannot be larger than 1.06 m. The wheel is keyed to the driving shaft, and the mechanical efficiency of the press and drive is 72 per cent. Determine (a) the weight of the flywheel rim, (b) the cross section of the rim, (c) the maximum stress in the rim, assuming six arms, and (d) the coefficient of steadiness of the driving shaft, if the duration of the working stroke is one-half that of the idle cycle.

Belt Drives

27.1 BELT MATERIALS

Belt drives are used to transmit power from one shaft to another when the shafts are some distance apart and it is not required that their velocity ratio be absolutely constant. The shafts may have any speed ratio, within reasonable limits; and while they are generally parallel, other arrangements are also used. With an open belt, Fig. 27.1a, the shafts turn in the same direction; with a crossed belt, Fig. 27.1b, they turn in opposite directions. Crossed belts are subjected to greater wear and tear and should be used only where the speed is low and small power is transmitted.

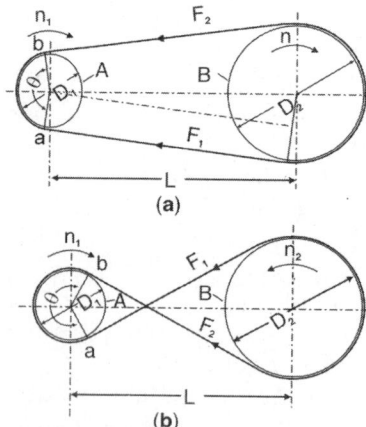

Fig. 27.1: Open and crossed belts

Leather belts: Most belts, especially those for high speeds, are made of leather strips cemented together, the best quality being obtained from the parts of steerhides near the backbone. Leather for belting may be oaktanned or chrome-tanned. Both classes are used for general service, but chrome leather is less affected by dampness, acid, fumes, or oil. The thickness at different places of a hide varies from about 3 to 6 mm. The various belt thicknesses are obtained by using hides of different thicknesses and also by gluing several layers together, giving what are known as single, double, and triple belts. The thicknesses and widths in which belts can be obtained are shown in Table 27.1. The hair side of leather is smoother and harder than the flesh side, but the flesh side is considerably stronger. The hair side of a single belt should run in contact with the pulley face, since the flesh side is better adapted to stand the greater tensile stress when going around the pulley. The modulus of elasticity varies from 1400 to 3000 N/mm². The deformation of leather does not conform to Hooke's law, and the modulus of elasticity increases with an increase of the stress. The weight of leather for belting is 970 kg/m³ or 9500 N/m³.

Rubber belting: Rubber belting consists of cotton duck and is normally from three to ten plies in thickness, although heavier belting, up to 15 plies, can be obtained on special order. The duck is impregnated with rubber and is provided with a wearing surface of rubber on both faces. Table 27.2 gives data for thicknesses and widths of commercial rubber belting. Compared with leather belts of the same

Table 27.1: Leather belt data

Grade of belt	Approximate thickness (mm)				Width increments mm	Ultimate strength N/mm^2
	Single	Double	Triple	Quadruple		
Light	3	6	–	–	12 to 24 by 3 24 to 102 by 6 102 to 198 by 12	Oak tanned 20 to 30 Chrome-tanned
Medium	4	8	12.5	17.5	200 to 800 by 25 800 to 1400 by 50	28 to 40
Heavy	5	10	15	20	1500 to 2100 by 100	

Table 27.2: Rubber and balata belt data

Thickness	Number of plies	Width mm	Ultimate strength N/mm of width	
			Rubber	Balata
3.9	3	25–100	8	5
5.1	4	25–200	11	7
6.4	5	76–250	13	9
7.7	6	100–250	16	11
9.1	7	100–250	19	12
10.4	8	200–400	23	14

power-transmitting capacity, rubber belts are cheaper but have shorter life. Because they stand up better than leather belts under adverse atmospheric conditions, they are the most preferable for outdoor service, but they are quickly ruined by oil or grease. The modulus of elasticity of rubber belting varies from 900 to 1450 N/mm^2 and the belting weights 1250 kg/m^3 or 12300 N/m^3.

Balata belting: Balata is a gum similar to rubber and is used in the manufacture of fabric belts to render them acidproof and waterproof. However, balata belts cannot be used at temperatures above 38°C, at which temperature balata begins to soften.

Textile belts: Textile belts are made of cotton, being built up of three to ten layers of duck stitched together or solidly woven into a strip of the desired width and thickness. The fabric is treated with linseed oil to make it waterproof. Textile belts are used for temporary, rough service. They are more durable than either leather or rubber belts if they have to run in contact with dirt and grit or in dusty atmosphere. The tensile strength of cotton belting varies from 35 to 50 N/mm^2, and the

modulus of elasticity may vary from 480 to 1400 N/mm^2. the material weighs 6870 to 13700 N/m^3 (700 to 1400 kg/m^3).

27.2 BELT FASTENINGS

The *efficiency* of a joint is the ratio of its strength to the full strength of the belt.

Cemented joint: The best form of belt fastening is the scarfed splice glued under pressure, which makes a very strong lap joint with an average efficiency of 98 per cent. Its great disadvantage is that in most cases it has to be made with the belt in place. If it is not necessary to make the splice in place, a very accurate measurement of the required endless belt must be taken in order to obtain just the right tension when the belt is stretched over the pulleys.

Metal hinges: A strong joint, used rather extensively for narrow belts, is made by securing a row of wire loops in the belt ends and locking them together with a special steel-wire pin or a rawhide pin. If the loops are applied by special machine, this joint has an efficiency of about 60 to 90 per cent. It is easily put together and taken apart. Also, since it is very flexible, it is suitable for small pulleys.

Rawhide lacing: A popular form of joint is made with rawhide lacing threaded through holes punched in the ends of the belt. It is easily made and is durable, but it is rather stiff and has an efficiency of only 40 to 60 per cent.

Metal clamps: Heavy rubber or canvas belts subjected to shock loads, such as are encountered in oil-field work, are often joined by clamps formed of two steel plates with holes and through-bolts. The efficiency of such a joint is from 50 to 70 per cent.

27.3 STRESSES IN A BELT

All stresses produced in a belt are tensile stresses.

Tensions in a running belt: Flat belts are put on with a certain initial tension to transmit power, otherwise the belt will slip over the pulleys (27.11). When the driving pulley A, Fig. 27.1a, turns without transmitting any power, the tension on both the running-on and running-off sides of the pulley remains unchanged, if the small frictional resistance in the bearing is neglected. As soon as power is transmitted, the tension F_0 in the pulling side will increase to F_1, and that in the running-off side will decrease to F_2. The force causing the driven pulley to rotate is the difference between these tensions, or $F_1 - F_2 = F_t$. This value is sometimes termed the *net pull*, or *effective pull*.

Slip: Any belt transmitting a load will slip on the surface of the pulleys. This means that the belt will move somewhat slower than the face of the driving pulley and somewhat faster than the face of the driven pulley.

Creep: Creep action adds to the effect of slip, but it is entirely different. *Because of the difference in tension in the various sections of the belt, a unit length of the belt in passing from the point a, Fig. 27.1a, to the point b, decreases in length owing to its elasticity. Therefore the driver A delivers a shorter unit length of belt at b than it receives at a, and the average velocity of the belt is slightly lower than that of the pulley surface. A similar action occurs on the driven pulley B, but here the average belt velocity is slightly higher than that of the pulley face (27.5). This action,* known as creep, reduces the speed of the driven pulley and also causes some loss of power.

In experiments it is easier to measure the combined action of slip and creep, which is called simply slip. Most figures referring to percentage slip include creep as well. The normal range of slip is considered to be from 1.5 to 2 per cent. When a belt is overloaded, its slip increases. If excessive slip occurs, the heat generated by friction may damage the belt.

Belt section: If the allowable belt stress is designated by S_d, and the thickness and width of the belt, in mm, are represented by h and b, respectively, then for lower belt speeds up to 10 m/s the cross-sectional area of the belt may be found from the equation

$$hb = \frac{F_1}{S_d} \qquad \ldots (27.1)$$

Centrifugal force: For belt speeds above 10 m/s, the centrifugal force must be taken into account. This force increases the tension in the belt without increasing its driving power. The centrifugal force of a piece of belt 1 m long may be determined from the general relation of equation 3.1. Thus,

$$c = \frac{h\,b\,w\,v^2}{gr} \qquad \ldots (27.2)$$

where, w is the specific weight of the belt material, in N/m^3 or $\left(\dfrac{kgf}{m^3}\right)$

v is the belt velocity, in m/s;

g is the acceleration of the force of gravity, 9.81 m/s^2;

r is the radius of the smaller pulley, in m. The projected length of the belt on the pulley is $2r$, and the component C of the centrifugal force parallel to the line connecting the centers of both pulleys is

$$C = 2rc = \frac{2hbwv^2}{g} \qquad \ldots (27.3)$$

This force is taken up evenly by both sides of the belt, creating on each side an additional stress given by the relation

$$s = \frac{2hbwv^2}{g \times 2hb} = \frac{wv^2}{g} \qquad \ldots (27.4)$$

This stress must be deducted from the nominal allowable stress S_d, and equation 27.1 becomes

$$hb = \frac{F_1}{Sd - \dfrac{wv^2}{g}} \qquad \ldots (27.5)$$

Allowable stresses: For leather belts the allowable stress S_d is taken equal to one-tenth to one-eighth of the ultimate strength as indicated in Table 27.1, or 2 to 4 N/mm². Rubber and balata belting have a more uniform structure, and the safe stress may be taken equal to one-eighth to one-sixth of the ultimate strength in Table 27.2, or 1 to 1.75 N/mm². These values must be multiplied by the efficiency of the joint used.

Belt fatigue: Different sections of belt are subjected to various stresses. The stresses are; tensile stresses due to slack side tension F_2, tight side tension F_1 which includes the effect of initial tension F_0. There is additional tension due to centrifugal force C when the belt passes over the pulleys. In addition when the belt passes over the pulleys, there is bending stress inversely proportional to the diameter of pulleys, i.e. $\dfrac{Et/2}{D_1/2} = \dfrac{Et}{D_1}$ or $\dfrac{Et}{D_2}$. The complete effect is illustrated in Fig. 27.2.

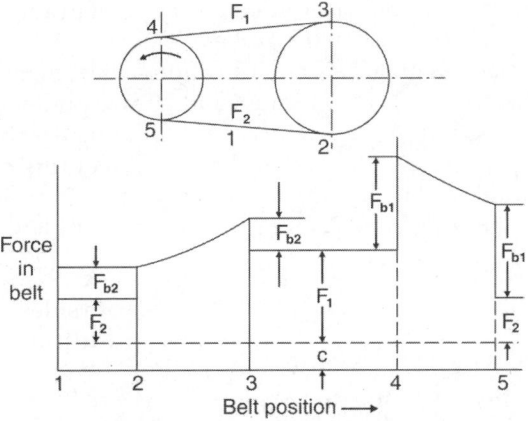

Fig. 27.2: Variation in tensions in belt

Stress due to $F_2 = S_2 = \dfrac{F_2}{A}$

Stress due to $F_1 = S_1 = \dfrac{F_1}{A}$

Stress due to centrifugal force = C

Stress due to bending

$$Sb_1 = \frac{Et}{D_1} \text{ and } F_{b_1} = S_{b_1} \times A$$

Stress due to bending

$$Sb_2 = \frac{Et}{D_2}, \ F_{b_2} = S_{b_2} \times A$$

Therefore the variation of tensile stress is

$\dfrac{F_0}{A}$ at 1 to $\dfrac{F_0}{A} + \dfrac{F_2}{A} + \dfrac{Et}{D_2}$ at 2.

At 3, it changes to $\dfrac{F_0}{A} + \dfrac{F_1}{A} + \dfrac{Et}{D_2}$

At 4, it is $\dfrac{F_0}{A} + \dfrac{F_1}{A} + \dfrac{Et}{D_1}$

At 5, it is $\dfrac{F_0}{A} + \dfrac{F_2}{A} + \dfrac{Et}{D_1}$

These changes are responsible for fatigue.

27.4 BELT CAPACITY

A belt transmits power through its friction upon the faces of the pulleys. The transmitting capacity depends on the allowable maximum tension in the belt; the belt speed; the coefficient of friction between the belt and the pulleys; the angle of wrap, or contact, on the smaller pulley; and the service conditions as expressed by the load factor.

Tension: Belt tension is discussed in section 27.3.

Belt speeds: Based on equation 27.5, the limit speed for a leather belt with $S_d = 2.8$ N/mm² and $w = 970$ kg/m³ (9500 N/m³) is about 54 m/s. At this speed the centrifugal force is so great that there is no traction left between the pulley face and the belt. However, experiments have shown that a large amount of power can be transmitted at speeds up to 60 m/s.

For maximum power and economy, speeds of 25 to 30 m/s[1] may be used. It so happens that this is also about the limit of safety for ordinary cast-iron pulleys. For a longer belt life combined with efficiency, a speed between 15 and 20 m/s is advisable. Speeds of shafts, and pulley sizes, often limit the belt velocity to considerably lower values.

Coefficient of friction: The coefficient of friction *f* is a function of the material of the pulley, the material of the belt, the rate of slip of the belt on the pulley, and the belt speed.

The influence of the material of the pulley on a leather belt may be seen from Fig. 27.3. The coefficient of friction of rubber belts is about 20 per cent lower than that of leather belts. The coefficient of friction increases with an increase of the rate of slip, but the latter should be limited to 1.5 or 2 per cent.

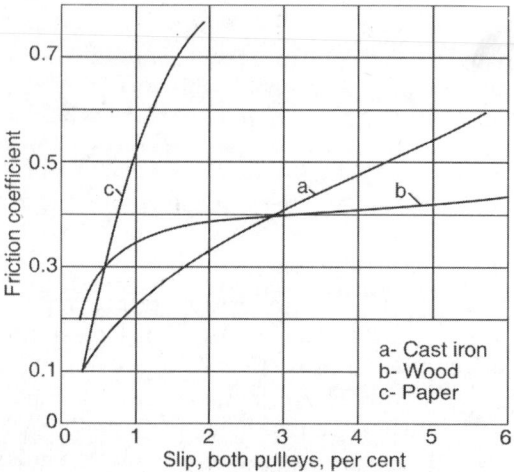

Fig. 27.3

For leather belts on cast-iron pulleys the influence of speed is given by Barth's empirical formula, which is

$$f = 0.54 - \frac{0.712}{2.542 + v_m} \qquad \ldots (27.6)$$

where, v_m is the belt speed in m/s. This formula does not take into account the slip and should be considered only as giving the limit values

which may be attained in operation under favorable conditions.

Values of the friction coefficient, often used in practice, are given in Table 27.3.

Table 27.3: Average coefficients of friction for leather and rubber belts

Belt material	Pulley material		
	Cast iron	Wood	Paper
Leather	0.35	0.38	0.50
Rubber	0.30	0.33	0.42

Angle of contact with tight and slack tensions: If the influence of centrifugal force is neglected, the ratio of the tight and slack tensions may be determined by equation 18.25, which is

$$\frac{F_1}{F_2} = e^{f\theta}$$

Also, the relation between the maximum pull F_1 and the net pull F_t may be found by equation 18.27. Thus

$$F_1 = \frac{F_t e^{f\theta}}{e^{f\theta} - 1}$$

Combining this equation and equation 27.5, and solving for F_t, gives

$$F_t = hb\left(S_d - \frac{wv^2}{g}\right)\left(\frac{e^{f\theta} - 1}{e^{f\theta}}\right) \qquad \ldots (27.7)$$

The angle of contact θ which must be used in equation 27.7, if the pulleys are of the same material, is the smaller of the two as the smaller angle restricts the power to be transmited. In installations without an idler pulley, θ is the angle on the smaller pulley. If the pulleys are made of materials having different coefficients of friction, or if the angle of belt wrapping on the small pulley is increased by the use of an idler pulley, the design should be based on the pulley with the smaller value of $f\theta$.

For an *open belt*, Fig. 27.1a, the angles of contact, in radians, are approximately

$$\theta = \pi \pm \frac{D_2 - D_1}{L} \qquad \ldots (27.8)$$

where the plus sign applies for the larger pulley and the minus sign applies for the smaller one.

[1]C. H. Norman, High-Speed Belt Drives, Bulletin No. 83, Ohio State University Engineering Experiment Station (May, 1934)

For a *crossed belt*, Fig. 27.1b, the angles of contact are the same in both pulleys. Each is

$$\theta = \pi + \frac{D_2 - D_1}{L} \qquad \ldots (27.9)$$

Equation 27.8 applies for a horizontal belt with the slack side on top. *The angle of contact increases with the slack side on the top (27.2).* For belts with the slack side on the bottom, and for vertical or inclined belts, equation 27.8 gives values somewhat too large. For an inclined belt with an angle of 50° or more, the belt width b found by means of equation 27.7, in which θ is determined by equation 27.8, must be divided by a capacity coefficient given in Table 27.4.

Table 27.4: Capacity coefficient of an inclined belt drive

Angle with horizontal (deg)	Capacity coefficient
0	1.00
50	0.90
60	0.80
70	0.70
80	0.60
90	0.50

With idler pulleys, whether on a horizontal or vertical drive, the angle θ, Fig. 27.4, may be found graphically from a layout of the drive.

Sag and stiffness: Equation 27.7 is derived by taking into account only what happens when the belt passes over the pulley, and it must be considered a rough approximation, although on the safe side. Actually a centrifugal force acts upon the belt between the pulleys and sets up a tension equal to and oppositely directed from that set up on the pulley.[2] It is a known fact that fan belts can operate with an astonishing degree of slackness, the net pull F_t often being much greater than the standstill tension. Under such conditions the looseside tension F_2, without regard to centrifugal force, might sink to zero, in which case F_1 should also become zero according to equation 18.27. This

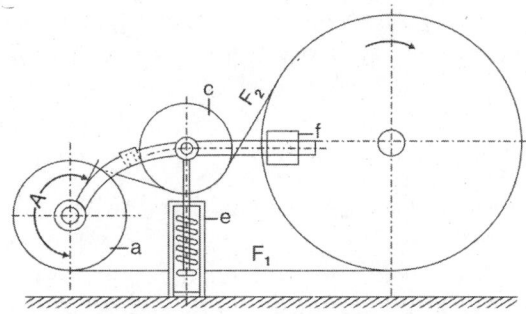

Fig. 27.4: Short-center drive with an idler pulley

contradition can be overcome by introducing in equation 27.7 a correction ΔS_1 to take into account the influence of belt sag, and a correction ΔS_k to allow for the stiffness of the belt.[3] However, for usual operating conditions these corrections are negligibly small.

Initial tension: In order to transmit power, the belts must be given an initial tension, otherwise the belt will slip without transmitting any power. It has been found that there is a definite relation between the tensions F_1 and F_2 and the initial tension F_0. This relation is expressed by the equation[4]

$$\sqrt{F_1} + \sqrt{F_2} = 2\sqrt{F_0} \text{ or } \frac{F_1 + F_2}{2} = F_0 \quad \ldots (27.10)$$

Later experiments have brought out the value of this approximate equation, which seemingly can be applied for case where F_2 becomes very small and nevertheless, contrary to equation 18.27, the pull F_1 is comparatively great. For leather belts a suitable initial tension is 1.4 to 1.5 N/mm², and for rubber belts it is about 0.5 to 0.7 N/mm².

When there is centrifugal tension c, then the value of initial tension F_0 may be = $\frac{F_1 + F_2 + 2C}{2}$. *As the initial tension increases, the power transmitted also increases up to a certain value after which there will be a decrease in the power transmitted (27.13).*

---

[2]G. Schulze-Pillot, *Neue Riementhevrie* (Berlin: Julius Springer, 1926).
[3]Normal, *op. cit.*
[4]C. Barth, "Transmission of Power by Leather belting," *Transactions of the American society of Mechanical Engineers*, Vol. 31 (1909), pp. 29–203.

Load factor: A belt absorbs a large amount of shock, and the load factor K' for given conditions may be computed by the relation

$$K' = 1 + \frac{K_l - 1}{5} \qquad \ldots (27.11)$$

where K_l is a factor the value of which may be taken from Table 20.4.

Power: The power that can be transmitted with a belt running at v_m m/s can be determined by taking into account the load factor K' of equation 27.11 and the force F_t of equation 27.7. Thus,

$$P = \frac{F_t \, v_m}{1000 \, K'} \qquad \ldots (27.12)$$

27.5 DESIGN OF BELT DRIVES

The design of a belt drive is based to some extent on empirical rules.

Belt thickness: The selected belt thickness depends on the pulley size. For *leather belts* a good practical rule is to determine the thickness h by the relation

$$h \leqq 0.02D \qquad \ldots (27.13)$$

where, D is the diameter of the small pulley. *When the pulley diameter is too small for a given belt thickness, the bending of the belt will generate excessive heat and cause its rapid deterioration. It is desirable to use large pulley diameters as large as possible as that the smaller pulley is not very small to avoid this type of damage.* If for some reason a smaller pulley diameter must be used, the limit is given by the equation

$$D = \frac{h}{0.03} \qquad \ldots (27.14)$$

Also, in order to extend the belt life, a lower working stress, and hence a greater belt width, should be used in such a case.

A single belt should not be over 200 or 250 mm wide. Another rule is not to use a single belt where its width is more than four-thirds the diameter of the smallest pulley.[5]

For *rubber belts* a good practical rule is to have 75 mm of pulley diameter per ply of belt at lower speeds, and at higher speeds to have 100 mm of pulley diameter per ply. If the foregoing proportion cannot be maintained, a shorter belt life must be accepted, unless a special very flexible belt is used.

Belt width: The width of a belt can be determined from equation 27.1 or 27.5 or 27.7 after the belt thickness h has been selected as just explained and the other quantities have been computed.

Center distance: A good distance L between shaft centers is 6 to 7.5 m. For single leather belts and 3-ply and 4-ply rubber belts, the distance may be cut down to 3.6 m. A further decrease of L with pulleys of different diameters decreases the angle θ and requires an increased initial tension F_0. The minimum value is sometimes given as $L \geq 3.5 \, D$ where D is the diameter of the larger pulley. When the belt has to stand shock action, an increase of belt length will give a longer belt life.

Example 27.1: Determine the diameters of cast-iron pulleys and the thickness and width of a leather belt to transmit 130 kW from a shaft that is direct-connected to a steam engine turning at 300 rpm to a centrifugal pump running with a speed ratio of 1:3.5.

Assume a belt velocity of 25 m/s. Then the diameter of the driving pulley is

$$D_1 = \frac{v_m}{\pi n} = \frac{25 \times 60}{\pi \times 300} = 1.59 \text{ or } 1.6 \text{ m}$$

With an assumed total slip of 1.5 per cent, the diameter of the driven pulley must be

$$D_2 = \frac{1.6}{3.5 \times 1.015} = 0.450 \text{ m}$$

The general load factor K_l, from Table 20.4, is $1.5 \times 1.25 = 1.875$, and the load factor K' for the belt drive is, by equation 27.11,

$$K' = 1 + \frac{1.875 - 1}{5} = 1.175$$

[5]O. A. Leutwiler, *Elements of Machine Design* (New York: McGraw-Hill Book Company, Inc., 1917), p. 66.

The maximum net pull, by equation 27.12, is

$$F_t = \frac{130 \times 1000 \times 1.175}{25} = 6110 \text{ N}$$

By equation 27.13, the thickness of the belt should be $h \le 0.02 \times 0.45 = 0.009$ m or 9 mm. From Table 27.1, a medium double belt with a thickness of 8 mm may be selected. The allowable stress may be taken as $30/9 = 3.4 \text{ N/mm}^2$ (say). For metal-hinge joint with an efficiency of 78 per cent, the design stress is $S_d = 3.4 \times 0.78 = 2.65 \text{ N/mm}^2$.

The term wv^2/g is

$$\frac{9500(25)^2}{9.81 \times 10^6} = 0.605 \text{ N/mm}^2$$

The pulley center distance may be taken as 7 m. With an open belt, the angle of belt contact θ, by equation 27.8, is

$$\theta = \pi - \frac{1.6 - 0.45}{7} = 2.98 \text{ radians}$$

The coefficient of friction, by Barth's formula (equation 27.6), is

$$f = 0.54 - \frac{0.712}{2.542 + v_m} = 0.45 - \frac{0.712}{2.542 + 25} = 0.51$$

If this is considered as a high limit, and the value of 0.35 given in Table 27.3 for a cast-iron pulley as a rather low limit, a conservative design figure will be the mean value.

Thus,

$$f = \frac{0.51 + 0.35}{2} = 0.43$$

The term $e^{f\theta} = e^{0.43 \times 2.98} = 3.63$, and the width b, from equation 27.7, is now

$$b = \frac{6110 \times 3.63}{8 \times (2.65 - 0.605)(3.63 - 1)} = 515.47 \text{ mm}$$

say 525 mm

Efficiency: The losses of power in a belt drive are due to slip; creep; bending over the pulleys; windage, or air resistance to the motion of the belt and the pulleys; and bearing friction. The loss due to slip and creep combined is, under normal conditions, about 2 per cent, and not over 3 per cent of the total power transmitted. The losses due to bending and windage are usually negligible. The bearing loss is about 1 per cent, and not over 2 per cent. Thus the losses are from 3 to 5 per cent, and the overall efficiency is from 97 to 95 per cent.

27.6 SHORT-CENTER DRIVES

A shortening of the center distance of an open-belt drive with pulleys of different sizes decreases the angle of contact θ and increases the belt slip (27.3). By using an idler pulley c, Fig. 27.3, near the smaller pulley a, the angle of contact θ can be made greater than $180°$ (27.4). The idler pulley may maintain the necessary small tension F_2 either by its own weight or by additional springs e or by a weight f. In order to produce a smooth operation of the belt, idler pulleys must be machined all over and carefully balanced. The disadvantage of idler pulleys is the bending of the belt in two directions, which shortens its life. On the other hand, a smaller slack tension F_2 tends to lengthen the life.

The *Rockwood drive*, Fig. 27.5a, is a short-center drive which is superior to an idler drive because it eliminates the bending of the belt in opposite direction and automatically reduces the bearing pressure produced by the belt tension when the net belt pull decreases. It was originally designed as a mounting for electric motors, but it can be used for other drives as well. The platform f is so pivoted on the axis e that the weight of the motor produces the necessary belt tension. This tension is adjusted by moving the motor with respect to e and thus changing the moment arm a. The axis e, in turn, can be moved with respect to the axis of the driven shaft. The drive is made in sizes up to 75 kW. It can be used also for an inclined or vertical drive. To make the installation more compact, the position of the platform can be made vertical, as in Fig. 27.5b.

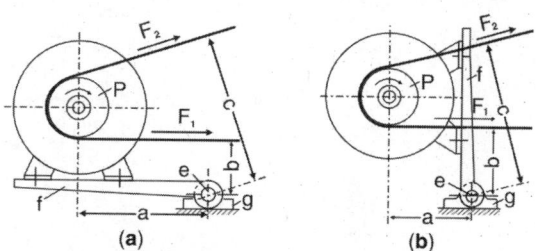

Fig. 27.5: Rockwood drives

In designing a Rockwood drive, it is first necessary to assume a suitable value of the ratio F_1/F_2. For a given power to be transmitted, the higher the value of this ratio, the lower will be the belt tensions and the bearing loads. For oaktanned belts, recommended values of F_1/F_2 vary from 3 for $\theta = 120°$ up to 5 for $\theta = 180°$. For high-strength belts these values may be increased about 25 per cent.[6]

If F_t is the required net pull, and W is the weight of the motor, and a, b, and c are distances indicated in Fig. 27.4, and if moments are taken with respect to the pivot e, the following equations may be obtained for the tensions F_1 and F_2:

$$F_1 = \frac{aW + cF_t}{b+c} \qquad \ldots (27.15)$$

and

$$F_2 = \frac{aW - cF_t}{b+c} \qquad \ldots (27.16)$$

On the other hand, the required pivot-arm length a may be computed by dividing equation 27.15 by equation 27.16 and solving the resulting equation. Thus,

$$a = \frac{F_t\left(bF_1/F_2 + c\right)}{W\left(F_1/F_2 - 1\right)} \qquad \ldots (27.17)$$

The distances b and c must be taken from a general layout; the approximate value for a is then computed from equation 27.17; and a final layout of the drive is made. This layout will give an accurate distance c. It is then advisable to check the arm length a by equation 27.17, and the tensions F_1 and F_2 by equations 27.15 and 27.16. In any case, the arm length a and the position of the pivot e must be made adjustable to take care of a possible variation in the weight W and a stretching of the belt.

Example 27.2: Design a Rockwood drive for a 20 kW motor having a speed of 1,750 rpm and weighing 1700 N. The motor must drive a jacksaft at 600 rpm.

A belt speed v_m of 22.5 m/s may be selected. The motor-pulley diameter is then

$$D_1 = \frac{22.5 \times 60}{\pi \times 1750} = 0.245 \text{ say } 0.25 \text{ m}$$

The net pull is found from the usual relation and is

$$F_t = \frac{P}{v_m} = \frac{20 \times 1000}{22.5} = 888.8 \text{ say } 889 \text{ N}$$

From a preliminary sketch, with $b = 30$ mm and $D_1 = 250$ mm, it is found that $c = 65$ mm. Next, select the ratio F_1/F_2 as 5. Then, by equation 27.17

$$a = \frac{889 \times (30 \times 5 + 65)}{(5-1) \times 1700} = 28.1 \text{ mm}$$

A final layout can be drawn by assuming the centre distance of pulleys and if there is change in a and c, it can be done.

The tensions found by equations 27.15 and 27.16 are

$$F_1 = \frac{28.1 \times 1700 + 65 \times 889}{30 + 65} = 1111 \text{ N}$$

and

$$F_2 = \frac{28.1 \times 1700 + 30 \times 889}{30 + 65} = 222 \text{ N}$$

A check gives $F_t = F_1 - F_2 = 1111 - 222 = 889$ N and

$$\frac{F_1}{F_2} = \frac{1111}{222} = 5.0045$$

These results are in satisfactory agreement with the assumed design data.

27.7 QUARTER-TURN DRIVES

A quarter-turn belt is used to connect two shafts whose axes are in different planes. Usually one shaft is horizontal and the other is vertical.

Law of belting: In order to stay on the pulleys, a belt must approach each pulley in a plane normal to its axis of rotation. If the approaching side is deflected, the belt will run off. In Fig. 27.6a this law is illustrated on a quarter-turn belt used for connecting two shafts with axes in planes normal to each other. A greater freedom in the location of pulleys is obtained by using guide pulleys, as c and d in Fig. 27.6b, and the drive may be made reversible.

In a horizontal quarter-turn drive, such as encountered in centrifugal deep-well pumps,

---

[6] R. R. Tatnall, "The pivoted Motor Drive," *Mechanical Engineering*, Vol. 57 (1935)

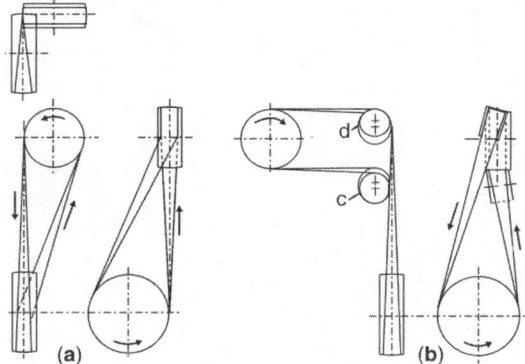

Fig. 27.6: Quarter-turn drives

Fig. 27.7, the height h at which the belt rides on the vertical pulley depends on the tension and stretching of the belt and also on the least deviation of the axis of the vertical pulley from a strictly vertical position. Thus if the axis assumes position a, the height h decreases and the belt goes up; if the axis moves to position b, the height h increases and the belt drops. A rule of thumb, good for normal conditions only, is to make h equal to 1 in 20 of center distance L. Thus,

$$h = \frac{L}{20} \qquad \ldots (27.18)$$

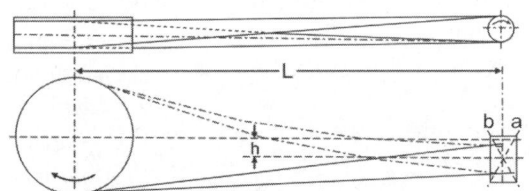

Fig. 27.7: Horizontal quarter-turn belt drive

Because of this uncertainty in regard to h, the vertical pulley must have a very large face. The installation is much simplified by the use of a double-pulley idler, Fig. 27.8. The pulley c, which is adjustable vertically, helps to bring the belt to the desired location on the driven pulley b; the idler pulley d is a regular idler pulley which, in this case, only maintains the necessary tension in the slack side of the belt.

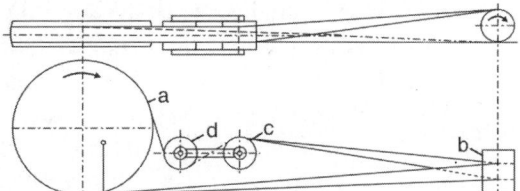

Fig. 27.8: Horizontal quarter-turn drive with idler pulley

27.8 MATERIAL FOR PULLEYS

Cast iron is almost an ideal material for belt pulleys since it can be cast in any desired shape, is readily machined, and gives pulleys which do not change their form. Solid cast-iron pulleys can be used with a rim speed up to 25 m/s; when they are made of a better grade of cast iron, the speed may be 27.5 m/s. Split pulleys with the split through the arms as in Fig. 27.10c, can be used up to 22.5 m/s; and split types of ordinary construction, as in Fig. 27.10b, can be used up to 20 m/s.

Pressed-steel pulleys are lighter and less expensive than cast-iron pulleys but may not run quite as truly.

Wooden pulleys have a rim built up of maple segments. They are usually of the split type and can be obtained in sizes up to 1.2 m in diameter, and up to a 0.3 m face. They are apt to warp from change of atmospheric conditions and are therefore not suitable for high speeds.

Paper pulleys are used rather extensively where a higher transmitting capacity is essential. Such a pulley consists of a web built up of thin sheets of straw fiber cemented together under a high pressure and bolted to a cast-iron hub. They are obtainable in sizes up to 450 mm in diameter.

Cork-insert pulleys, made by pressing cork inserts into countersunk holes in cast-iron pulleys, are used to take advantage of the high coefficient of friction between a belt and cork. The inserts should not protrude over 1 mm. These pulleys are generally made in small sizes, up to 0.35 m in diameter. Cork inserts do not help wood pulleys and paper pulleys.

27.9 DESIGN OF CAST-IRON PULLEYS

Only the features of cast iron pulley design will be considered here.

Face: The pulley diameter should be as large as possible to reduce bending stress in the belt (27.6). The diameter is decided on the basis of allowable centrifugal stress and velocity (27.7). After the diameter of a pulley has been determined by considering the selected belt velocity and taking the slip into account, the face B of the pulley, in mm, may be computed by Barth's formula. This formula is

$$B = 1.094b + 0.185 \qquad \ldots (27.19)$$

where, b is the belt width corrected to a commercial size. For simplicity, B may be taken as $1.1b$.

Crown: If a belt is led to a revolving conical pulley, it will tend, because of its lateral stiffness, to climb higher and higher upon the cone. This tendency is utilized in the principle of crowning, to keep belts in position (27.8). A crown may be formed either in a very flat inverted V-form, as in Fig. 27.9a, or with a convex curve, as in Fig. 27.9b. The V-form is easier to machine, while the curved face has the advantage of stretching the belt more uniformly. The usual crown height is $c = 1$ in 100 of face with, but pulleys for very wide belts are given a relatively smaller crown height. Barth's empirical formula is

$$c = \frac{1}{32} B^{2/3} \qquad \ldots (27.20)$$

This gives practically the same results as the first, simpler rules.

For rubber belts on well-aligned shafts,

$$c = \frac{B}{200} \qquad \ldots (27.21)$$

For poorly aligned shafts,

$$c = \frac{B}{120} \qquad \ldots (27.22)$$

Or $\qquad c = 0.003\,D$

An idler pulley should never be crowned, since its crown would bend the belt in the

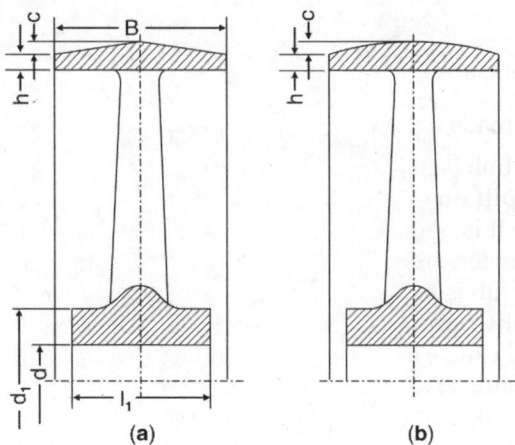

Fig. 27.9: Crowned pulleys

opposite direction. The pulley next to an idler should also have a straight face or only a small crown. When a pair of pulleys are connected by a belt, it is sufficient to crown only one pulley, usually the driven one. If both pulleys are crowned, the pulleys must be lined up very carefully so as to have both crowns in one plane and thus avoid sidewise distortion of the belt. Pulleys carrying shifting belts are not crowned.

The crown bends the belt and causes additional stress, so life is reduceed (27.8).

Rim: For a light pulley, the thickness h of the rim at the edge, in mm, may be determined by the relation

$$h = 0.25\sqrt{D} + 1.5 \qquad \ldots (27.23)$$

For a heavy-duty pulley for a triple belt, the thickness may be as great as

$$h = 0.375\sqrt{D} + 3 \qquad \ldots (27.24)$$

Or $\qquad h = \dfrac{D}{200} + 3$ for single belt

$$= \frac{D}{200} + 6 \text{ for double belt}$$

The stress in the rim should be checked by equation 26.13 in conjunction with centrifugal stresses. If it is high, the thickness h, and perhaps also the number of arms, should be increased.

Hub: The diameter of the hub, in mm is given by the relation

$$d_1 = 1.5d + 25 \qquad \dots (27.25)$$

where, d is the shaft diameter. The hub length l_1 is made equal to $\dfrac{2B}{3}$, but not less than $1.5d$. The hub is made either solid, with a taper key, or split on one side, as in Fig. 27.10a. In this case it is bored slightly smaller than the shaft diameter, and a wedge is used to open it when the hub is being put on the shaft. Sometimes a split-clamp hub, Fig 27.10b, is used with a solid rim.

Table 27.5 gives the type of key to be used with belt pulleys, according to German industrial standards (DIN).

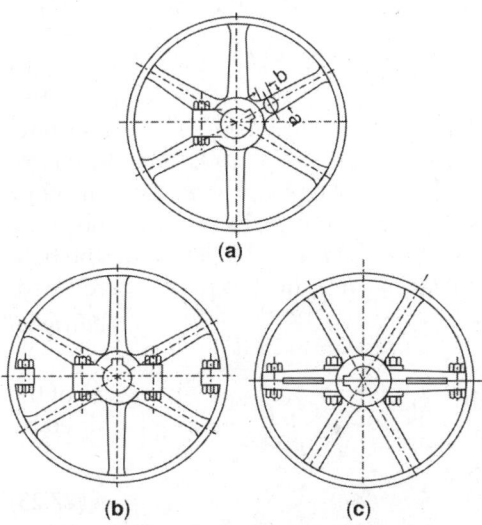

(a)

(b) **(c)**

Fig. 27.10: Types of cast-iron pulleys

Arms: Small pulleys are made with a web instead of arms. Pulleys from 150 to 450 mm in diameter have four arms; those from 450 to 1500 mm in diameter have six arms; and those larger than 1.5 m have eight arms. Pulleys with a face width up to 550 to 600 mm have one set of arms. Pulleys having a face of 600 mm or wider are made with two sets of arms.

The necessary arm section at the hub may be calculated by assuming that the arms act as cantilevers subjected to a load equal to the net belt pull F_t. If it is assumed that only one-half the total number i of the arms carry the load F_t, the bending moment on each arm is approximately

$$M = \frac{F_t \times 0.5D}{0.5i} = \frac{F_t D}{i} \qquad \dots (27.26)$$

By equation 2.22, $S_d = M/Z$, where Z is the section modulus of the arm at the hub. Therefore

$$Z = \frac{F_t D}{iS_d} \qquad \dots (27.27)$$

In the usual elliptic section of the arm, Fig. 27.9a, the major axis a is twice the minor axis b, or $a = 2b$, and the section modulus is

$$Z = \frac{1}{32}\pi a^2 b = \frac{1}{64}\pi a^3 \qquad \dots (27.28)$$

Substituting this value in equation 27.27, and solving for a, results in

$$a = \sqrt[3]{\frac{64 F_t D}{\pi i S_d}} \qquad \dots (27.29)$$

Table 27.5: Recommended types of keys for belt pulleys

Diameter or pulley (m)	Split pulleys, hubs bored for clamping						Solid pulleys			
	Width of face (m)						Width of face (m)			
	0.025–0.1	0.1–0.2	0.2–0.3	0.3–0.4	0.4–0.5	0.5–0.7	0.025–0.1	0.1–0.2	0.2–0.3	Over 0.3
Under 0.5	No key, only setscrews						Saddle key			
0.5–0.6										
0.6–0.8								Flat key		
0.8–1.0										
1.0–1.25	Flat key				Square key					
1.25–1.6										
1.6–2.0								Square key		
Over 2.0										

This determines the dimensions at the hub. The arms are tapered, with a decreasing at the rate of 1 in 50 to 1 in 30 toward the rim. The design stress S_d may be taken from 15 to 20 N/mm².

Example 27.3: Determine the dimensions of a cast-iron pulley which is 1.2 m in diameter and transmits 75 kW at 200 rpm. It is to be used with a heavy double belt of width 350 mm.

The rim face, by equation 27.19, is

$$B = 1.094 \times 350 + 5 = 388 \text{ mm}$$

The thickness of the rim must be, by equation 27.24,

$$h = 0.375\sqrt{1200} + 3 = 16 \text{ mm}$$

If a 90 mm shaft is used, the hub diameter should be by equation 27.25,

$$d_1 = 1.5 \times 90 + 25 = 160 \text{ mm}$$

The hub length must be

$$l' = \frac{2}{3}B = \frac{2}{3} \times 388 = 258.6 \text{ say } 260 \text{ mm}$$

The pulley must have six arms. The belt velocity is

$$v_m = \frac{\pi \times 1.2 \times 200}{60} = 12.56 \text{ m/s}$$

and the net pull is

$$F_t = \frac{P}{v_m} = \frac{75 \times 200}{12.56} = 5971 \text{ N}$$

If the safe stress is taken as $S_d = 16$ N/mm², the major axis of the elliptic section of each arm at the hub is, by equation 27.29,

$$a = \sqrt[3]{\frac{64 \times 5971 \times 1200}{\pi \times 6 \times 16}} = 114.9 \text{ mm say } 120 \text{ mm}$$

Then

$$b = 0.5 a = 60 \text{ mm}$$

At the rim,

$$a' = 120 - \frac{600}{50} = 108 \text{ mm, and } b' = \frac{108}{2} = 54 \text{ mm}$$

Example 27.4: Check the stress in the rim in example 27.3.

By equation 3.5, the tensile stress is with

$$r = 600 - \frac{16}{2}$$

$$s_1 = \frac{\pi^2 W r^2 n^2}{900g} = \frac{\pi^2 \times 70600 \, (600-8)^2 \times 200^2}{900 \times 1000^2 \times 9.81}$$

$$= 1.1 \times 10^6 \text{ N/m}^2 = 1.1 \text{ N/mm}^2$$

By equation 26.12, the bending stress is

$$s_2 = \frac{0.216 \times 70600 \times 0.592^3 \times 200^2}{6^2 \times 9.81 \times 0.016}$$

$$= 22.39 \times 10^6 \text{ N/m}^2 = 13.3 \text{ N/mm}^2$$

In order to take into account the stiffening influence of the arms, the combined tensile stress is found by equation 26.13:

$$s = 0.75 \times 1.1 + 0.25 \times 22.39 = 0.825 + 5.597$$

$$= 6.422 \text{ N/mm}^2$$

The stiffening influence of the arms is greater on a pulley rim than on a heavy flywheel rim, and experience confirms that a pulley with the rim thickness just determined is entirely satisfactory.

Loose pulleys: By shifting a belt from a pulley a, Fig. 27.11a, fastened to the shaft, to a pulley b which is loose on the shaft, or back the same result may be obtained as with a friction clutch. The hub of the loose pulley must have a bushing c. To decrease the wear the loose pulley is often mounted on ball bearings or roller bearings. In order to decrease the tensions in the belt while it is standing still on the loose pulley, the latter's diameter may be made somewhat smaller than that of the tight pulley, as in Fig. 27.11b, and a taper flange may be provided for shifting the belt back to the tight pulley.

27.10 V BELTS

The gripping action between the belt and the groove enables ropes and wedge-shaped belts,

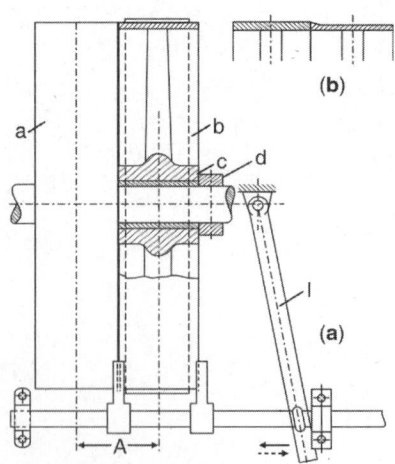

Fig. 27.11: Tight and loose pulleys

running in V-grooved pulleys, to transmit large amounts of power with a relatively small initial tension. However, only after the development of the so-called Texrope Drive, which consists of special fabric-and-rubber belts of trapezoidal cross section running on sheaves with grooves, did this method come into wide use. At present a number of rubber-belt manufacturers make V belts. Although the belts of different makes differ in details, the principle is the same. A cord core a, Fig. 27.12, transmits the power, c is a cushion and compression member, and e is the outer rubber and fabric wrapping. The belts are made in five standard section, as shown in Table 27.6. They are made endless at the factory in many standard lengths, the lower and upper limits also being given in Table 27.6, together with the maximum number of strands used.

The wedging effect between a V belt and the sheave increases the friction considerably, so that the drive can work with small arcs of contact and a low initial tension. This makes the drive particularly suitable for a shortcenter

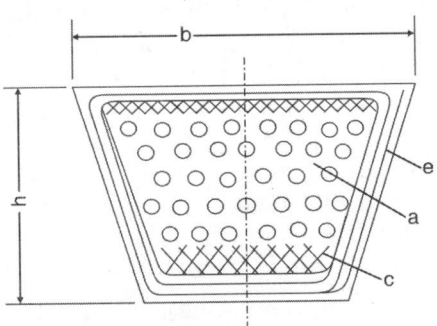

Fig. 27.12: V belt

arrangement of pulleys without an idler and with speed ratios as high as 10 to 1. The speed ratios are computed on the basis of the pitch diameters of the pulleys, each of which is equal to the outside diameter minus the belt thickness. Furthermore, the position of the driving pulley with respect to the driven one makes no difference; the drive operates equally well with horizontal or vertical belts and with the slack side on either the top or the bottom. Other advantages are absence of vibration and noise; ability to absorb shock; high efficiency (about 98 per cent), because of absence of slip; and great dependability, since the breakage of one of the strands merely shifts the load to the remaining ones until replacement can be made.

27.11 V-BELT DRIVES

In designing V-belt drives, the catalogues of manufactures of belts and pulleys should be consulted.

Belt selection: Catalogues of the manufacturers of V belts give the necessary instructions for selecting the belt size, pulley size, center distance, and number of belts to be used. For a preliminary design the permissible loads per belt can be taken from Fig. 27.13. The curves given there take into consideration the belt speeds, but they are plotted for an arc of contact of 180°. The capacities thus obtained must be multiplied by a correction factor found by the relation

$$C = \frac{\left(e^{f\theta} - 1\right)e^{f\pi}}{e^{f\theta}\left(e^{f\pi} - 1\right)} \qquad \ldots (27.30)$$

where, θ is the arc of contact of the smaller pulley and the friction coefficient f may be

Table 27.6: Standard sizes of V belts

Section number	Width b (mm)	Thickness h (mm)	Stock pitch length (mm) Minimum	Stock pitch length (mm) Maximum	Recommended Power range (kW)	Recommended maximum number of strands	Minimum diameter of smaller pulley mm
A	13	8	645	3693	0.4–4	6	75
B	17	11	932	5377	1.5–15	9	125
C	22	14	1351	9143	10–70	14	200
D	32	19	3127	16792	35–150	14	355
E	38	23	5426	16805	70–260	20	500

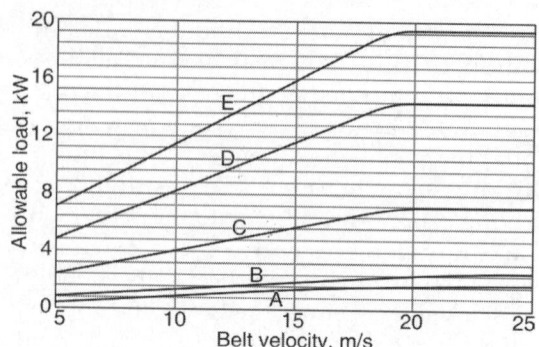

Fig. 27.13: Allowable loads per strand of V belts

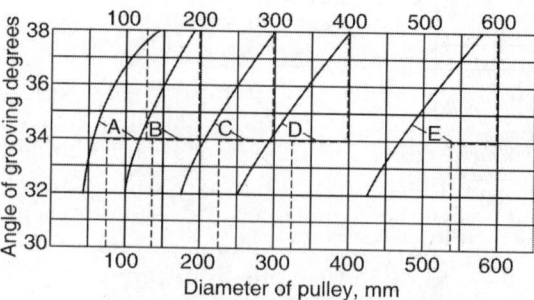

Fig. 27.15: Chart for determination of groove angle

taken conservatively as 0.1. The presence of a great starting torque, of shocks, and of possible overloads is taken into account by dividing the capacity by a load factor K'. The value of K' may be computed by equation 27.11.

Pulleys: Pressed and welded steel pulleys can be obtained from several manufactures in a great variety of diameters and number of grooves for the belt section No. A and No. B. Cast-iron pulleys for all belt sections are also carried in stock and can be made by any machine shop in accordance with data in the catalogues of belt manufacturers. The angle of grooving a, Fig. 27.14, is made from 30° to 38°. In Fig. 27.15 the solid curves give the recommendations of one belt manufacture, and the dotted lines give those of another one. In either case the angle depends on the pulley diameter. The width of the groove b, Fig. 27.14, is made equal to the width of the belt; the lands c are made 2.5 mm for the No. A section and gradually increase to about 6 mm for the No. E section; the depth h_1 is made 4.5 to 6 mm,

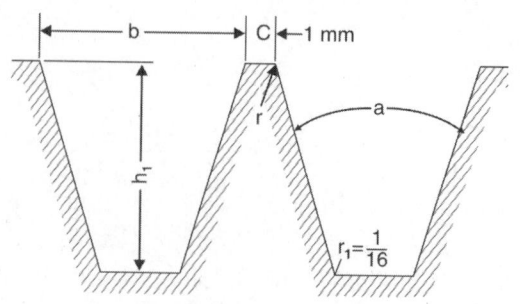

Fig. 27.14: Groove profile

greater than the thickness h of the belt. The sides of the grooves must be finished and smoothly polished to prevent excessive wear of the belt.

The recommended minimum pulley diameters, referred to the pitch line, are indicated in Fig. 27.15 for the five belt sizes. However, the use of larger pulleys will increase the belt life.

The center distance should be slightly larger than the diameter D_1 of the larger pulley and slightly smaller than the sum of the diameters of both pulleys. However, both shorter and longer center distances may be used if required. On high-speed drives short center distances give smoother running.

V flat drives: On V-belt drives with a speed ratio of 3 to 1 or more, and with short center distances, the grooving of the larger pulley can be omitted without sacrifice of power or efficiency. A belt has its maximum pulling power when the arc of contact on the large pulley is 240° to 250°. This angle is obtained when the following equation is satisfied:

$$\frac{D_2 - D_1}{L} \geqq 0.5 \qquad \dots (27.31)$$

where, D_2, D_1, and L have the meanings shown in Fig. 27.1a.

In calculating the speed ratios, it should be remembered that the pitch diameter of a flat pulley is equal to its outside diameter plus the belt thickness. The use of V flat belts is increasing very rapidly, particularly for low and medium torques.

Double-V belts: A double-V belt, Fig. 27.16a, is used when power must be transmitted to

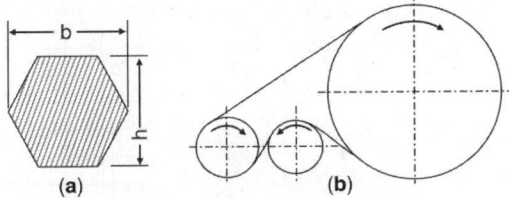

Fig. 27.16: Double-V belt drive

grooved pulleys from both the top and bottom of the belt, as in Fig. 27.16b. Such belts are made to fit standard V grooves.

Quarter-turn drives: Quarter-turn drives with V belts are being used both with and without idlers.[7] An idler may be omitted if the pulleys are relatively small and a take-up can be provided on one of the shafts. The minimum center distance must be equal to $6(D + B)$, where D is the diameter of the larger pulley and B is its width. With a larger speed ratio the shortest center distance becomes too long, and an idler should be used. The latter can also serve to obtain the necessary take-up for the belts. The center distance between the idler and the small pulley should be not less than $8B$. Flat-faced idler pulleys may be used when the belts move from the idler pulley toward the quarter turn. If the belts move from a quarter turn toward a flat-faced pulley, the belt strands will squeeze together and pile up. Therefore the idler must be grooved.

The power rating of a V belt used on a quarter-turn drive should be taken as 75 per cent of that of a straight drive.

27.12 ANALYSIS

The ratio of belt tension in V-belt is

$$\frac{F_1}{F_2} = e^{\frac{f\theta}{\sin\alpha}}$$

α is the semi angle of the belt section.

Tables of manufacturers manual usually suggest the diameter of the smaller pulley for various types of belts. The diameters of larger pulley can be calculated corresponding to the velocity ratio or torque requirements and efficiency as:

The diameters of larger pulley

$$D_1 = \frac{D_2 n_1 \eta}{n_2}$$

Nominal pitch length of belt

$$L = 2C + \frac{\pi}{2}(D_1 + D_2) + \frac{(D_1 + D_2)^2}{4C}$$

C being the centre distance.

The nominal inside length, the nominal pitch length and permissible length variations are given by manufacturers of V belts. The center distance for a given belt length and diameter of pulley is

$$C = \frac{L}{4} - \frac{\pi(D_1 + D_2)^2}{8} +$$

$$\sqrt{\left[\frac{L}{4} - \frac{\pi(D_1 + D_2)}{8}\right]^2 - \frac{\pi(D_1 + D_2)^2}{8}}$$

The usual initial tension is provided by stretching = 0.5% to 1% of L

Minimum centre distance = $0.55(D_1 + D_2)$

Maximum centre distance = $2(D_1 + D_2)$

Arc of contact $\theta = \pi \pm 2\alpha$.

where $\alpha = \dfrac{D_1 - D_2}{2C}$, + *ve* sign being for large

pulley and – *ve* sign for smaller pulley.

Number of belts

$$= \frac{\text{Power to be transmitted}}{\text{Power transmitting by one belt}}$$

Maximum power that can be transmitted by a single belt at 180° arc of contact is given for all types of belt:

For A section, power kW

$$= \left[0.5v^{0.9} - \frac{20v}{d_e} - \frac{0.8v^3}{10^4}\right],$$

maximum value of d_e is 125

---

[7] The Gates Rubber *Company, Complete Guide* for *Selecting or Designing V-Belt Drives* (Denver: 1950), p. 44.

For B section power kW

$$= \left[0.8v^{0.9} - \frac{51v}{d_e} - \frac{1.3v^3}{10^4} \right],$$

maximum value of d_e is 175

For C section power kW

$$= \left[1.5v^{0.9} - \frac{143v}{d_e} - \frac{2.4v^3}{10^4} \right],$$

maximum value of d_e is 300

For D section power kW

$$= \left[3v^{0.9} - \frac{510v}{d_e} - \frac{5v^3}{10^4} \right],$$

maximum value of d_e is 425

For E section power kW

$$= \left[4.5v^{0.9} - \frac{950v}{d_e} - \frac{7v^3}{10^4} \right],$$

maximum value of d_e is 700

d_e = equivalent pitch diameter $D_1 \times k_d$ mm.

D_2 = pitch diameter of smaller pulley.

k_d = Small diameter factor depending on velocity ratio (V.R):

V.R	k_d
1–1.5	1 to 1.1
2	1.13
3 and over	1.14

Proportionate values may be taken for various V.R within the range.

Example 27.4: A 15 kW, 720 rpm motor has a 20 cm diameter pulley with 15 cm facewidth. Design a leather belt and a cast iron pulley for the driven shaft which is to run at 180 rpm. The pulley is mounted on its shaft such that the center line of the pulley is at distance of 20 cm from the center line of the bearing. Assume maximum allowable tension of 200 N per cm width of the belt and belt slip of 4%. Take $f = 0.45$. Allowable shear stress in the shaft 55 N/mm². Allowable tensile stress of C.I. pulley 20 N/mm². Neglect centrifugal tension.

Belt design

As distance between the shafts is not given, the lap angle can be taken as 180°.

Peripheral velocity,

$$V = \pi D n = \frac{\pi \times 20 \times 720}{100 \times 60} = 7.52 \text{ m/sec}$$

Diameter of the driven shaft pulley

$$= 20 \times 0.96 \times \frac{720}{180}$$

$$= 76.8 \text{ cm say } 77 \text{ cm}$$

$$\frac{T_1}{T_2} = e^{0.45\pi} = 4.385, \quad T_1 = 4.385 \, T_2$$

$$T_1 - T_2 = \frac{15 \times 1000}{7.52} = 1994.6 \text{ say } 2000 \text{ N}$$

$$T_2 = \frac{2000}{3.385} = 591 \text{ N}$$

and $T_1 = 4.385 \times 591 = 2591$ N

Width of the belt $= \dfrac{2591}{200} = 12.95$ say 13 cm

Width of the pulley face $= 1.2 \times 13 = 15.6$ cm which is O.K

Because $\dfrac{Diameter}{Facewidth} < 7$

Thickness of rim $= \dfrac{D}{80} + 0.3 = 14$ mm (say) and at the crown it may be taken as 16 mm. The same value are obtained by using eq. 27.24.

Arms Design

Torque on the pulley shaft

$$= (T_1 - T_2)\frac{D}{2} = 2000 \times \frac{77}{2} = 77000 \text{ Ncm}$$

Assuming the pulley with 6 arms,

B. M. on each arm $= \dfrac{77000}{6} = 12833$ Ncm

considering elliptical section ($a = 2b$),

$$12833 = \frac{\pi}{32} a^2 b \times 20$$

$$b = \left[\frac{12833 \times 32 \times 10}{\pi \times 4 \times 20} \right]^{1/3} = 25.3 \text{ say } 26 \text{ mm}$$

and $a = 52$ mm

i.e. major axis = 52 mm and minor axis = 26 mm

These dimensions can be tapered to $\dfrac{2}{3}$ rd of these near the rim, i.e. 35 mm and 18 mm.

Shaft design

Max. B.M. on the shaft = $(2591 + 591) 200 = 636400$ Nmm

Neglecting weight of the pulley, the equivalent torque is given by:

$$T_e = \sqrt{6364^2 + 77000^2} = 77263 \text{ Ncm}$$

$$998950 = \frac{\pi}{16} S d^3$$

Taking $S = 55 \times 0.75$ due to keyway

$$d = \left[\frac{772630 \times 16}{\pi \times 55 \times 0.75}\right]^{1/3} = 45.69 \text{ say } 50 \text{ mm}$$

Diameter of the boss = $1.75 \times 50 = 87.5$ say 90 mm

Length of the boss = $1.25 \times 50 = 62.5$ say 65 mm

Example 27.5: A V-belt drive is used to transmit 20 kW at 750 rpm from a motor pulley having pitch diameters of 250 mm to a machine pulley of diameter 900 mm. The belt has a cross-section area of 695 mm, top width 38 mm, thickness 25 mm and bottom width 22 mm and angle of groove of pulley 40°. Coefficient of friction is 0.25. If the maximum allowable tension is limited to 1000 N, find the number of belts required. The center distance of pulleys is 1 m and weight of belt is 7.5 N/m.

Velocity of belt

$$v = \frac{\pi \times 250 \times 750}{1000 \times 60} = 9.817 \text{ m/s}$$

Distance of centroidal axis of belt from bottom

$$\bar{y} = \frac{\frac{2 \times 8 \times 25}{2} \times \frac{2 \times 25}{3} + 22 \times 25 \times \frac{25}{2}}{\frac{2 \times 8 \times 25}{2} + 22 \times 25}$$

$$= 13.6 \text{ mm}$$

Pitch radius of larger pulley

$$r_2 = 450 + 13.6 = 463.6 \text{ mm}$$

$$r_1 = 125 \text{ mm}$$

$$\sin\beta = \frac{r_2 - r_1}{C} = \frac{463.6 - 125}{1000}$$

$$= 0.3386$$

$$\beta = 19.79° \text{ say } 19.8°$$

$$\theta_1 = 180° - 2 \times 19.8° = 140.4°$$

$$\theta_2 = 180° + 2 \times 19.8° = 219.6°$$

Taking the smaller pulley

$$e^{\frac{\mu\vartheta_1}{\sin\alpha}} = e^{\frac{0.25 \times 140.4° \times \pi}{180 \times \sin 20}} = e^{1.79}$$

$$= 5.989$$

$$T_c = \frac{wv^2}{g} = \frac{7.5 \times 9.817^2}{9.81} = 73.68 \text{ N}$$

$$\frac{T_1}{T_2} = 5.989$$

$$T_1 + T_c = T_{max} = 1000$$

$$T_1 = 1000 - 73.68 = 926.32 \text{ N}$$

$$T_2 = \frac{926.32}{5.989} = 154.67 \text{ N}$$

Power transmitted by belt

$$= \frac{(T_1 - T_2)^2}{1000}$$

$$= \frac{(926.32 - 154.67) \times 9.817}{1000}$$

$$= 7.57 \text{ kW}$$

Number of belts

$$= \frac{20}{7.57} = 2.64 \text{ say } 3.$$

Example 27.6: Select a V-belt for transmitting 2.5 kW at 1500 rpm from an electric motor to an industrial fan to run at 500 rpm. The center distance is to be restricted to approximately diameter of largest pulley. The maximum tension is 500 N.

For 2.5 kW power type A belt is selected. It has a width of 13 mm and thickness of 8 mm and pulley diameter of 75 mm.

Larger pulley diameter = $3 \times 75 = 225$ mm

Centre distance $C = 225$ mm.

$$C_{min} = 0.55 [75 + 225] + 8 = 158 \text{ mm}$$

$$C_{max} = 3 (75 + 225) = 900 \text{ mm}$$

Therefore, $C = 225$ is OK.

Pitch length of belt

$$L = 2C + \frac{\pi(D_1 + D_2)}{2} + \frac{(D_2 + D_1)^2}{4C}$$

$$= 2 \times 225 + \frac{\pi(75 + 225)}{2} + \frac{(225 + 75)^2}{4 \times 225}$$

$$= 450 + 471.23 + 25$$

$$= 946.23 \text{ mm}$$

Nearest pitch length available is 950 mm

For this pitch length, the centre distance is

$$C = \frac{L}{4} - \frac{\pi(D_1 + D_2)}{8} +$$

$$\sqrt{\frac{L}{4} - \frac{\pi(D_1 + D_2)^2}{8} - \frac{\pi(D_2 + D_1)^2}{8}}$$

$$C = \frac{950}{4} - \frac{\pi(75+225)}{8} +$$

$$\left[\left\{ \frac{950}{4} - \frac{\pi(75+225)^2}{8} \right\} - \frac{\pi(225+75)^2}{8} \right]^{\frac{1}{2}}$$

$$= 237.5 - 117.80 + [107.31] = 227 \text{ mm.}$$

$$\sin\beta = \frac{r_2 - r_1}{C} = \frac{112.5 - 37.5}{225} = 0.33$$

$$\beta = 19.26°$$

$$\theta_1 = 180 - 2 \times 19.26 = 141.48°$$

$$e^{\frac{\mu\theta_1}{\sin\beta}} = e^{\frac{0.25 \times 141.48 \times \pi}{180 \times \sin 19.26}}$$

$$= 6.499$$

Neglecting centrifugal tension

$$T_1 = 500 \text{ N}$$

$$T_2 = \frac{500}{6.499} = 76.93 \text{ N}$$

$$\text{Power transmitted} = \frac{(T_1 - T_2) \times 25}{1000}$$

$$v = \frac{\pi D_1 N_1}{1000} = \frac{\pi \times 75 \times 1500}{1000} = 5.89 \text{ m/s}$$

$$P = \frac{(500 - 67.93) \times 5.89}{1000} = 2.49 \text{ kW}$$

This is safe.

PROBLEMS

27.1 Discuss the relative advantages and disadvantages of belt drives, chain drives and gears as means of transmitting power.

27.2 Why is it preferable to have the slack side up in a horizontal belt drive?

27.3 What is the effect of short center distance of the pulleys on the drive?

27.4 What arrangements are made if the centre distance is short?

27.5 How is creep responsible for slip in the belt drive?

27.6 Why pulley diameters should be as large as can be accommodated?

27.7 Why is the maximum linear speed of pulleys limited? On what factors does it depend?

27.8 Why is crowing of pulleys required? Has it any harmful effects?

27.9 Make a sketch for transmitting power between two shafts at right angles to each other in a belt drive.

27.10 Explain the phenomenon of fatigue in belts.

27.11 Why is initial tension required in the belt drive?

27.12 What is the value of initial tension required for belt drive?

27.13 What is the effect of increase of initial tension on the power transmitted?

27.14 A 10 mm chrome-tanned leather belt 350 mm wide is running over a 1.8 m cast-iron driving pulley and a 470 mm driven pulley. The larger pulley is keyed to the shaft of a gas engine and turns at 235 rpm. Assuming an endless belt and a center distance of 6.6 m, determine the power which this belt will transmit and the probable speed of the driven compressor pulley.

27.15 Using the data of problem 27.14, find the power and speed of the driven pulley if the center distance is reduced to 1.5 m and an idler pulley is used to increase the arc of contact.

27.16. A line shaft must be connected to the driving motor by a medium double leather belt. The cast-iron motor pulley is 430 mm in diameter and runs at 900 rpm. The actual speed of the line shaft should be reduced in the ratio of 4.75:1. If all the machines driven from the line shaft are operating simultaneously, the total maximum load is 84 kW. However, the actual load is never higher than 72 percent of the maximum load. Determine the necessary width of the belt, and the diameter and width of the line-shaft pulley. The center distance is 5.4 m and makes angle of 40° with the horizontal line.

27.17 A seven-ply 0.5 m balata belt is to be replaced by an oaktanned leather belt. The diameter of the driving pulley is 1.7 m and its speed is 215 rpm; the diameter of the driven pulley is 1 m; and the center distance is 6.6 m. Assume all other necessary data, determine the width and thickness of the leather belt.

27.18 A heavy triple leather belt that is 0.9 m wide and travels at the rate of 27.5 m/s over pulleys 1.06 m and 1.98 m in diameter has to be be replaced by a rubber belt. Determine the number of plies and the width of the rubber belt. The center distance is 7.5 m.

27.19 (a) Using the data of problem 27.18, and assuming that the smaller pulley is the driver, determine the speed of the pulleys and the power transmitted by the belt. (b) Find the initial tension which must be put in the belt to transmit the load.

27.20 A 5 kW motor running at 1,175 rpm is installed with a Rockwood mounting, Fig. 27.5a in the text. The diameter of the cast-iron motor pulley is 230 mm, the motor weights 1.2 kN the tight belt side is horizontal, and the angle of belt wrap is $\theta = 165°$, $b = 0.3$ m; and the starting torque is 200 per cent of the motor rating. Determine (a) the belt tensions F_1 and F_2 at rated load of the motor, (b) the distance a at full load, (c) the tensions F_1 and F_2 at one-half of motor load with unchanged distance a, and (d) the tensions F_1 and F_2 when the motor is running idle.

27.21 A 4 kW electric motor running at 1,180 rpm is installed with a Rockwood mounting, Fig. 27.4a. The diameter of the motor pulley is 230 mm, the motor weighs 890 N, the tight side of the belt is

horizontal, the driven pulley must run at 480 rpm, and the distance between the pulley centers is 500 mm. (a) Determine the distance a for full-load operation of the drive if $b = 250$ mm, (b) Compare the belt tensions F_1 and F_2 at full, three-quarter, and half-load operation if distance a is adjusted for full-load condition and kept unchanged.

27.22 (a) Design a leather-belt drive from a 120 kW gas engine running at 360 rpm to a vertical deep-well centrifugal pump to run at 1,150 rpm. (b) After all necessary data are obtained, make a layout for a horizontal quarter-turn drive with a long center distance. (c) Make a layout for a drive using a center distance of 5.4 m and a double-pulley idler, as in Fig. 27.8.

27.23 Design and make a sketch of a pulley 1 m in diameter to transmit 35 kW at 175 rpm by means of a leather belt. The pulley shaft is 90 mm in diameter.

27.24 Design a V-belt drive for a 90 kW centrifugal water pump running at 1,200 rpm and driven by an oil engine running at 350 rpm. Make the center distance as short as possible.

27.25 A pump to run at 450 rpm is to be driven by a motor running at 750 rpm transmitting 75 kW. The center distance is 0.75 m and the motor pulley diameter is 295 mm. Area of belt section is 230 mm² of section C, top width 22 mm, bottom width 13.5 mm and height 21 mm. Weight of belt is 8 N/m. Angle of groove is 40°. Coefficient of friction = 0.3. Allowable tensile stress is 2 N/mm².

28

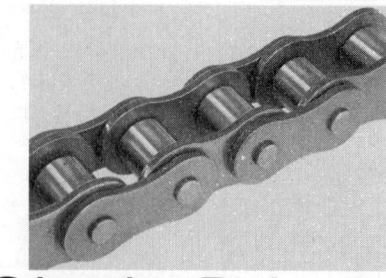

Chain Drives

28.1 GENERAL CONSIDERATION

The types of chains used for power transmission are the *block chain*, the *roller chain*, and the *inverted-tooth chain*, usually called the *silent chain*.

Block chains: Block chains, Fig. 28.1, are used only for transmitting power at a low rate and at moderate speeds, not exceeding 4.5 m/s. With a steady load they can be used up to 9 kW. With a fluctuating load, their limit is not over 2 to 3 kW.

Roller- and silent-chain drives: Advantages of roller-chain and silent-chain drives are as follows:

  (a) They are compact because they make it possible to use very short center distances.
  (b) They permit greater flexibility in locating sprocket shafts, as the shafts need not be aligned perfectly.
  (c) They have a positive speed ratio, with no slippage.
  (d) They permit the use of a large speed ratio—8 to 1 or even 10 to 1—in one step.

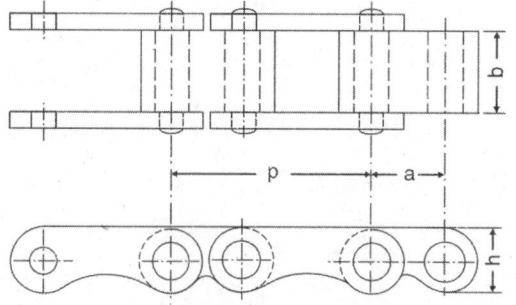

Fig. 28.1: Block chain

  (e) They cause small pressures on the bearings, because the tension on the slack side is produced only by the weight of the chain.
  (f) Their mechanical efficiency is high, seldom less than .98 per cent.
  (g) They are immune to the effects of temperature changes, moisture, or fumes.
  (h) Their maintenance cost is low, adjustments are rarely necessary, and lubrication is simple

Because of high speeds used in chain drives, all parts of these chains are accurately machined, and those parts which are subject to sliding action are hardened.

Chain drive is particularly used in bicycles, motor cycles and automotive industry and material handling equipment, e.g. chain and pulley blocks (28.1).

28.2 ROLLER CHAINS

The construction of a roller chain is shown in Fig. 28.2. In order to obtain interchangeability of roller chains and sprockets, regardless of make, ISO has developed standard dimensions for roller chains. However, this standard list includes more sizes than are actually necessary. Table 28.1 contains a list of standard sizes which are recommended by ISO. These will meet practically all requirements for the roller chain as a power-transmission medium.

Multiple-strand chains: In order to increase the load capacity of a chain with a certain pitch,

Table 28.1: Standard chains (all dimensions in mm)

ISO chain number	Pitch p	Roller diameter d₁ max	Width between inner plates b₁ min	Bearing pin body diameter d₂ max	Chain path depth h₁ min	Inner plates depth h₂ max	Measuring load, N			Breaking load, kN			Average weight N/m simple*
							Simple	Duplex	Triplex	Simple min	Duplex min	Triplex min	
05B	8.00	5.00	3.00	2.31	7.37	7.11	50	100	150	4.60	8.00	11.40	1.8
06B	9.525	6.35	5.72	3.28	8.52	8.26	80	150	220	9.10	17.30	25.40	4.1
08B	12.70	8.51	7.75	4.45	12.07	10.92	130	260	390	18.20	31.80	45.40	7.1
10B	15.875	10.16	9.65	5.08	14.99	13.72	200	400	600	22.70	45.40	68.10	9.1
12B	19.05	12.07	11.63	5.72	16.39	16.13	290	570	860	29.50	59.00	88.50	11.7
16B	25.40	15.88	17.02	8.28	21.34	21.08	510	1020	1520	43.10	86.20	129.30	27.0
20B	31.75	19.05	19.56	10.19	26.68	26.42	790	1590	2380	65.80	131.60	197.40	36.5
24B	38.10	25.40	25.40	14.63	33.73	33.40	1130	2270	3400	99.80	199.60	299.40	68.5
28B	44.45	27.94	30.99	15.90	37.46	37.08	1540	3080	4630	131.60	263.20	394.80	83.0
32B	50.80	29.21	30.99	17.81	42.72	42.29	2040	4080	6120	172.40	344.80	517.20	105.0
40B	63.20	39.37	38.10	22.89	53.49	52.96	3180	6350	9530	267.70	535.40	803.10	160.0
48B	76.20	48.26	45.72	29.24	64.52	63.88	4540	9070	13610	408.30	816.50	122.470	250.0
56B	88.90	53.98	53.34	34.32	78.64	77.85	6210	12430	–	553.40	110.680	–	–
64B	101.60	63.50	60.96	39.40	91.08	90.17	8120	16240	–	725.80	145.150	–	–
72B	114.30	72.39	68.58	44.48	104.67	103.63	10300	20590	–	916.30	183.260	–	–

* Average weight of duplex and triplex chain is twice and thrice of simple chain.

several chain strands are assembled side by side by using long through-pins. The capacity of a multiple-strand chain is equal to the capacity of a single strand multiplied by the number of strands. The usual limits for the number of strands are given in the last columns of Table 28.1, although some chain-makers go as high as eight strands. If the capacity of a multiple-strand chain is not sufficient, two chains may be run side by side.

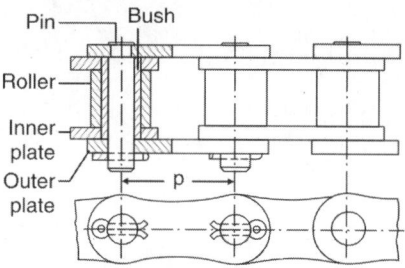

Fig. 28.2: Roller chain

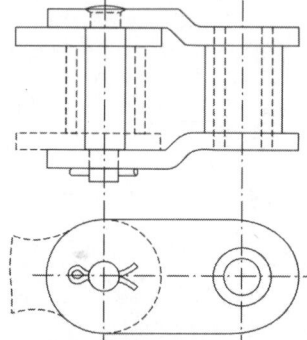

Fig. 28.3: Offset link

Chain construction: Most chains are assembled by riveting the pins from both sides in all links except the last one. In the last connecting link one end of each pin has a hole and a cotter pin. Sometimes the whole chain is assembled with cotter pins on one side, in order to make it easily detachable.

Offset link: Figure 28.3 shows a link used in a roller chain when an odd number of links is necessary.

Chain selection: Noisy operation and rapid chain wear are caused chiefly by the impact between the sprocket and the rollers as the rollers seat themselves. The following empirical formula gives good results:

$$p \leq \left[\frac{1920}{n}\right]^{2/3} \qquad \ldots(28.1)$$

where, p is the pitch, in mm, and n is the speed of the small sprocket, in revolutions per second. According to more recent investigations, the relation between n and p can be based on the allowable amount of impact between a roller and a sprocket,[1] as shown by the equation

$$n \leq \frac{1,920}{p}\sqrt{\frac{A}{w_f p}} \qquad \ldots(28.2)$$

where, A is the projected area of the roller, in square mm, which is equal to its diameter d times its width l; and w_f is the weight of the chain, in N/mm. All these values given in Table 28.1. A comparison of equations 28.1 and 28.2 shows that the former, in spite of its simplicity, gives values within 5 or 7 per cent of those obtained by equation 28.2. If v_{max}/zp is substituted for n, where v_{max} is the maximum allowable chain velocity, in m/s, and z is the number of teeth of the small sprocket, equation 28.2 gives

$$v_{max} \leq 160z\sqrt{\frac{A}{w_f p}} \qquad \ldots(28.3)$$

If the maximum speed is based on the energy of impact per tooth per minute, the equation is

$$n \leq \frac{55}{p}\sqrt[3]{\frac{A}{w_f}} \qquad \ldots(28.4)$$

Substituting equation 28.4 for n in the equation $v_{max} = npz$ gives

$$v_{max} \leq 55z\sqrt[3]{\frac{A}{w_f}} \qquad \ldots(28.5)$$

[1] G. M. Bartlett, "New Basis for Rating Roller-Chain Drives," Transactions of the American Society of Mechanical Engineers, Vol. 57 (April, 1935), MSP–57–1, p. 98.

Finally, if the maximum sprocket speed is based on the effect of centrifugal force, the equation is

$$n \leq \frac{66.4}{p} \sqrt{\frac{A}{zw_f}} \qquad \ldots (28.6)$$

and

$$v_{max} \leq 66.4 \sqrt{\frac{Az}{w_f}} \qquad \ldots (28.7)$$

In an important design the allowable speeds n and v_{max} must be calculated by equations 28.2 to 28.7, and the minimum values should be used. It may be noted that values of n and v_{max} are the same whether the chain consists of one strand or more than one. *As a general rule it is advisable to use the smallest pitch possible, in order to reduce surging and impact (28.5)*; if necessary, a multiple-strand chain should be used in preference to a heavier single-strand chain.

Chain pull: The allowable pull F_d should be taken from the catalogue of the chain manufacturer, as it depends on many variables, such as material, workmanship, speed, bearing pressure, and number of teeth of the smaller sprocket. For preliminary computations F_d may be determined by the relation

$$F_d = \frac{F_u}{n_o} \qquad \ldots (28.8)$$

Where, F_u is the ultimate strength or breaking load given in Table 28.1, and n_o is a working factor. For large sprockets having over 40 teeth, and for low velocities of about 0.5 m/s, n_o may be taken as 5. It should be increased up to 18 for small sprockets having 10 or 11 teeth, and for high velocities of about 6 m/s. However, the best procedure is not to base the load capacity of a roller chain on its strength but to consider the allowable bearing pressure to which roller pins can be subjected without undue wear. For low speed this specific pressure should be preferable less than 45 N/mm^2 but it may be as high as 50 N/mm^2; referred to the projected area A of the pins. More conservative limits are 35 to 40 N/mm^2.

The practice recommended by the American Gear Manufactures Association is based on a bearing pressure of 30 N/mm^2, velocity factor $c = 3/(3 + v)$. The allowable pull is then found by the equation

$$F_d = \frac{98.07 \, ld_2}{3+v} - \frac{v^2 w_f}{g} \qquad \ldots (28.9)$$

where, l is the length of the roller pin, d_2 is its diameter, and v is the actual chain velocity, in m/s

$$v = \frac{zpn}{1000} \qquad \ldots (28.10)$$

The second term in equation 28.9 is the influence of the centrifugal force.

28.3 SPROCKETS FOR ROLLER CHAINS
Variation in Velocity-polygonal Effect

A chain is consisting of links and is not continuous like a belt. *The movement of chain is not smooth as it moves over the sprocket teeth, its perpendicular distance from the centre of sprocket changes causing change of velocity (28.8)*. When the end of link is on the tooth, its distance from the centre is radius r, Fig. 28.4a, when it is

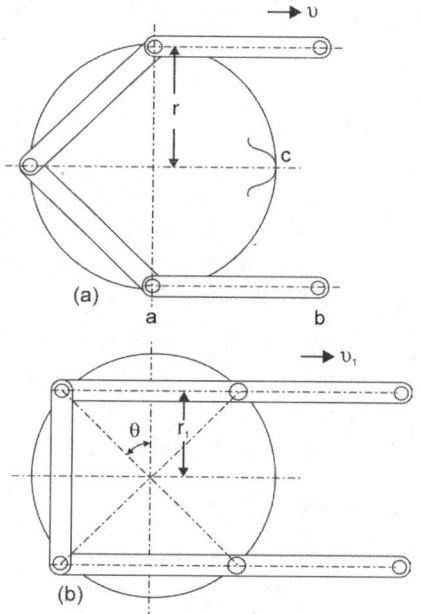

Fig. 28.4: Polygonal action

forming a chord, the distance is reduced to r_1, Fig. 28.4b. For simplicity a four teeth sprocket is shown.

Figure 28.4(a) the velocity of chain

$$v = rw = \frac{\pi Dn}{60}$$

Figure 28.4(b) the velocity of chain

$$v_1 = r_1 w = \frac{\pi Dn \cos \theta}{60}$$

$$\theta = \frac{360}{2z} = \frac{180}{z}$$

i.e.
$$v_1 = \frac{\pi Dn \cos \dfrac{180}{z}}{60}$$

The velocity varies from maximum value of

$\dfrac{\pi Dn}{60}$ to a minimum value of $\dfrac{\pi Dn \cos \dfrac{180}{z}}{60}$

This variation can be reduced by increasing the number of teeth z.

It is preferable to use sprockets with at least 17 teeth.

When number of teeth on specked is 17, the variation is 1.7% and reduced to 0.5% when the number of teeth is 30.

Impact of chain: When the sprocket turns with uniform velocity through 2θ, *the link of the roller chain ab strikes the next sprocket tooth c with impact velocity pw on the sprocket tooth as it moves along the arc bc. This causes damage to the chain and the tooth (28.7).*

A sprocket for a roller chain is shown in Fig. 28.5, with the dimensions used in design.

Number of teeth: The recommended number of teeth in the smaller sprocket is 15, 17, 19, 21,

or 23 for moderately high speeds. Even number of teeth will give less-uniform wear. *An odd number of teeth in each sprocket and even number of links in chain is desirable to reduce wear (28.4).* A smaller number of teeth, down to 9, or even to 7, may be used, if necessary, for low-speed drives.

Main dimension: For the notation in Fig. 28.4A, $2\alpha \times z = 360°$ and

$$\alpha = \frac{180°}{z} \qquad \ldots (28.11)$$

The pitch diameter is

$$D = \frac{p}{\sin \alpha} \qquad \ldots (28.12)$$

and the root diameter is $D_o = D - d$.

The width t_o is

$$t_o = l - 0.045p \qquad \ldots (28.13)$$

where, l is the roller width given in Table 28.1.

The radius r is

$$r = 0.545 \, p \qquad \ldots (28.14)$$

and the corner relief e is

$$e = \frac{1}{8} p \qquad \ldots (28.15)$$

According to the Diamond Chain Company, in order to obtain teeth giving maximum efficiency throughout the life of the drive, the angles and distances shown in Fig. 28.4 should have the following values:

The diameter d' of the seating curve, in mm, must be

$$d' = 1.005 \, d + 0.076 \qquad \ldots (28.16)$$

The required value of the angle β is

$$\beta = 35° + \frac{60°}{z} \qquad \ldots (28.17)$$

The distance from a to c is made equal to $0.8d$, and the radius r_1 from c to b is $0.8d + 0.5d'$. Also, the angle γ is found from the relation

$$\gamma = 18° - \frac{56°}{z} \qquad \ldots (28.18)$$

The height h is made equal to $0.3p$. With this value, the outside diameter D' is

$$D' = p \left[0.6 + \csc \left(\frac{180°}{z} \right) \right] \qquad \ldots (28.19)$$

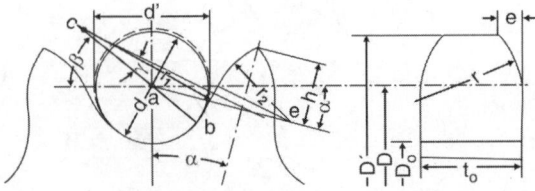

Fig. 28.5: Roller-chain sprocket teeth

Sprockets cut according to Fig. 28.4 allow the chain, as it stretches, to ride higher on the teeth without catching on them.

In order to obtain a sufficient hub diameter, the pitch diameter D of the smaller sprocket should be determined by the relation

$$D \geq 2d_o + p \qquad \qquad \therefore (28.20)$$

where, d_o is the shaft diameter. At the same time the diameter D should meet the requirement that

$$D \leq \frac{2424}{n} \qquad \qquad \dots (28.21)$$

28.4 CHAIN-DRIVE DESIGN

The most economical chain and sprocket size may be selected most quickly if the following procedure is used. First, the maximum allowable pitch is determined from equation 28.1, and the next smaller pitch is taken tentatively from Table 28.1. By using equation 28.10 the number of teeth z_1 and the chain speed v are selected. Simultaneously the number of teeth z_2 in the larger sprocket determined. After this the selected chain pitch and number of teeth are checked by one or all of equations 28.2 to 28.7.

The next step is to find the allowable chain pull F_d by equation 28.9. Also, the required chain pull F_t is found from the equation

$$P = \frac{F_t v}{1000\, K_1 K_2} \qquad \dots (28.22)$$

where, K_1 is a load factor and K_2 is a service factor. If the drive is used for a fluctuating load, the factor K_1 must be from 1.1 to 1.5, the exact value depending on the severity of the shock and being selected by using Table 20.3 as a comparative basis. The service factor K_2 depends on the length of daily operation. If the drive operates not more than 10 per day, $K_2 = 1$; for 24 hr operation, $K_2 = 1.2$.

If F_t from equation 28.22 is smaller than F_d found by equation 28.9, the drive may be improved by using a chain with the next smaller pitch, in accordance with the general rule given in section 28.2 that it is always

desirable to use the smallest pitch possible. Naturally, all computations must be repeated.

If F_t is greater than F_d, a chain with several strands must be used. Then the number of strands j evidently is found by the relation

$$j = \frac{F_t}{F_d} \qquad \dots (28.23)$$

Power: The power which a chain can safely transmit under normal load conditions is given by the chain manufacturers in their catalogues. Unfortunately, the data of various manufacturers differ considerably, being based on different assumptions. However, a safe value will generally be obtained by using equation 28.22, in which F_d found from equation 28.9 is used instead of F_t.

Center distance: The proper center distance usually is considered to be equal to the sum of the sprocket diameter D_1 and D_2. The minimum value is $30p$, but if necessary it can be made smaller, down to $20p$, provided proper shaft and sprocket alignment is secured.

For speed ratios greater than 3.5 to 1, the center distance should not be less than $D_1 - D_2$, in order to have a sufficient arc of wrap of the chain on the small sprocket. *This arc should not be less than 120°. This will give uniform distribution of load on teeth and smooth engagement and without causing much wear (28.2).*

Short center distances should be avoided, as a long chain has greater elasticity, to absorb irregularities of motion, and has a longer life.

Chain length: The length of a chain may be computed from formulas given in catalogues and by using various constants given in special tables. However, satisfactory results are obtained from the simple approximate formula

$$L_p = 2C_p + 0.5(z_1 + z_2) + \frac{0.026(z_1 + z_2)^2}{C_p}$$

$$\dots (28.24)$$

where, L_p is the chain length, in pitches, and C_p is the center distance, also in pitches. Naturally, the next whole number above the calculated value of L_p must be used. The chain length in mm is $L = pL_p$.

Steps of design procedure

1. Select a pitch $p = \left(\dfrac{1920}{n}\right)^{2/3}$

   Also pitch $p = \dfrac{C}{30 \text{ to } 60}$

   Take a standard value of pitch corresponding to the lesser value, C is centre distance.

2. Select number of teeth on the smaller sprocket depending on the velocity ratio as:

   Minimum number of teeth is 15

Velocity ratio	Number of teeth
1–2	30–27
2–3	27–25
3–4	25–23
4–5	23–21
5–6	21–17

   Find the number of teeth of a larger sprocket = $z_2 = z_1 \times V.R.$

3. Find the pitch diameter $D_p = \dfrac{p z_1}{\pi}$

   $$D_g = \dfrac{p z_2}{\pi}$$

4. Check the velocity $v = \dfrac{p z_1 n_1}{60 \times 1000}\ m/s$

   This should be 2.5 m/s and up to 7.5 m/s. If speed is higher than this, reduce pitch p.

5. Find allowable working load/strand

   $$F_a = \dfrac{\text{Breaking load}}{F.O.S \times k_s}$$

   Breaking load is found from table corresponding to pitch selected k_s service factor 1.2 to 2.8 depending on the type of service, i.e. from a few hours to continuous operation, Factor of safety is 5 to 15.

6. Find the total load F_t

   $$F_t = F_d + F_c + F_s$$

F_d = driving force = $\dfrac{1000\,F}{v}$

F_c = tension in chain due to centrifugal

force = $\dfrac{w v^2}{g}$

w = weight of chain N/m (Tables).

F_s = tension due to sagging = KwC

K = coefficient of sag = 1 for vertical

= 2–4 for inclined

= 6 for horizontal

$$F.O.S = \dfrac{F_u}{F_t}$$

F_u is the breaking load (Tables).

7. Diameter of roller from table or $d = \dfrac{5}{8}p$

   Pin diameter $d_p = \dfrac{5}{10}p$

   Width of chain $b_i = \dfrac{5}{8}p$

   Thickness of link plates $t_p = \dfrac{p}{8}$

   Length of roller $l = 0.9\,b_i - 0.15$

   Tooth width = $0.93\,w - 0.95$

   w-width between plates = $b_i + 2t_b$

   Maximum height of outer plates = $0.82\,p$

   Maximum height of inner plates = $0.95\,p$

   Minimum center distance $C = kC_1 < 80\,p$

   $$C_1 = \dfrac{do_1 + do_2}{2}$$

   Pitch diameter of sprocket

   $$D = \dfrac{p}{\sin\dfrac{180}{z}}$$

   Roller seating radius = $0.505d$.

Chain case and lubrication: Good lubrication should be provided for all chains. The oil drops should be led to the inside of the chain, because otherwise the centrifugal force will remove the oil before it has performed any service. If maximum life of the drive is essential, or if protection against dust or corrosive agents is important, the drive should be enclosed in a case.

A case must have provision of lubrication and for easy inspection of the drive. A chain case is also desirable as a safety guard.

28.5 SILENT CHAINS

Silent chains have several advantages over roller chains:

(a) They allow the use of higher linear velocities. Values up to 12.5 m/s are used even with large pitches, and values up to 25 m/s are used with some makes and under favorable conditions. To transmit power economically, however, the velocities should be from 7.5 to 10 m/s.

(b) They can be built in great widths, up to 750 mm, and one chain is thus able to transmit about 1250 kW power.

(c) The load is distributed equally among all sprocket teeth in contact with the chain, their wear thus being decreased.

Their main disadvantage, as compared with roller chains, is the absence of standardization and of interchangeability of different makes. They differ in the shape of the links and the profile of the sprocket teeth; in pitches; in the types of guides holding the chain on the sprocket; and finally in the types of pins which hold the links, and consequently in the capacity of chains of the same pitch and width.

Efficiency: Properly designed silent-chain drives have a very high efficiency, occasionally over 98 per cent. *The loss of power is due to chain stiffness, friction in bearings and resistance to motion (28.6).*

General requirements: For satisfactory operation of a silent-chain drive, several conditions are essential. The prime requisite is *true and rigid installation*. The shafts on which the sprockets are mounted must be parallel and properly supported so as not to tremble or pull together. The sprockets must be carefully aligned. *An odd number of teeth in each sprocket and an even number of links in the chains is desirable. An odd number of teeth distributes the working contact over the whole face of each tooth, the wear thus being reduced to a minimum (28.4).* If,

because of the exact speed ratio required, one or both sprockets have an even number of teeth, a hunting link as shown in Fig. 28.6, should be inserted to secure uniform wear.

The larger pitches should be used for large powers as well as low speeds. For moderate powers the smaller pitches are better suited. In most cases the largest pitch which can be used for the desired rotative speed will give a drive with the lowest first cost. *However, at a given linear speed a smaller pitch will provide a quieter drive due to reduced impact (28.5).* Naturally, it will require a wider chain.

A silent-chain drive should not be used between shafts on which the torque is not uniform. Silent chains can absorb a certain amount of shock, but more flexible drive, such as V-belt drive, is preferable in such a case, as far as both quietness and life of the drive are concerned. However, if a silent chain is used for a fluctuating load, it should be figured with a load factor K_1 from 1.2 to 1.6, the exact value depending on the severity of the shock. Table 20.3 may serve as a guide for the amount of shock action. The values of K_1 just mentioned are for electric-motor drives. For oil-engine and gas-engine drives, these values should be increased by 25 per cent. As for roller chains, a service factor K_2 must be introduced if a drive must operate more than 10 hr per day. For 24-hr operation, $K_2 = 1.2$.

Silent chains must run with a small slack. Whenever possible it is desirable to have an adjustable center distance to take care of the lengthening of the chain by wear. However, slack in a chain drive is not objectionable in itself, provided there is no uneven running or rapid starting, which might cause a very loose chain to jump.

Center distance: The center distance for a silent-chain drive may be selected by following the rules given in section 28.4 for roller-chain drives.

The chain length may be computed satisfactorily by equation 28.24.

Drive selection: For a given power and given speeds of the driving and driven shafts, it is possible to design several drives with different combinations of pitch, width, and numbers of sprocket teeth. Usually, however, only one of these drives will give the maximum service for the smallest cost. This best drive can be selected after several combinations are tried.

Instead of furnishing engineering data similar to the values in Table 28.1, most silent–chain manufacturers give in their catalogues the power that their chains of certain pitches can transmit, per mm or m of width, at various speed of sprockets and with certain numbers of teeth. These ratings differ considerably, one from another, because of a great difference in the construction and material used, chiefly in the pins and bushings.

28.6 MORSE CHAIN

The Morse silent chain has unsymmetrical links, as shown in Fig. 28.6, because of the use of a split pin. The half *a* of the pin has a T section, and forms the seat for the rocker pin *b* held in the other link. This construction introduces rolling friction instead of the sliding friction of common pins, but requires that the chain be run in one direction only, as indicated by an arrow on each outside link.

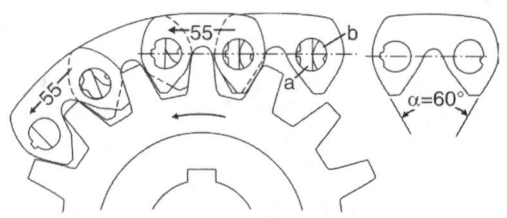

Fig. 28.6: Morse silent chain

Selection of pitch: Table 28.2 shows the normal and maximum rotative speeds recommended for each pitch. The use of a speed higher than the recommended normal speed shortens the life of the chain drive.

Number of teeth: The number of teeth on the driver should be as recommended in Table 28.2, so that a chain speed between 7.5 and 10 m/s is obtained. Under favorable conditions, such as uniform torque and good lubrication, a chain speed up to 15 m/s may be used.

Chain width: After selecting the pitch and number of the teeth in each sprocket, and the speeds of the sprockets, the total pull *F* necessary to transmit the power is found. The width *b* is found by dividing *F* by the allowable tension F_d from Table 28.2, corrected for the actual sprocket speed in accordance with Table 28.3. The actual width will be the next-greater one, as given in the catalogue.

The range of width for each chain is approximately from *p* to 10 *p*.

Hunting link: If the length L_p comes out in an odd number of pitches, or if a chain has stretched and one pitch must be taken out, an offset hunting link, Fig. 28.7, is used.

Example 28.1: Design a drive from an electric motor to a compressor to transmit 60 kW at 690 rpm of the motor and 135 rpm of the driven shaft. A short center distance is desired, and 1 m is the limit for the diameter of the driven sprocket.

Table 28.2 shows that a 32 pitch is as large as should be used normally at 690 rpm if a recommended

Table 28.2: Design data for Morse silent chain drives

Design item	Pitch of chain (mm)									
	10	12	15	19	23	25	30	32	38	50
Minimum teeth, driver	13	13	13	13	15	15	15	17	17	17
Desirable teeth, driver	17–25	17–25	19–25	19–25	19–25	19–25	21–25	21–25	21–27	21–27
Range of teeth, driver	21–120	21–130	23–150	23–150	23–150	23–150	25–150	25–150	27–150	27–50
Speed, normal (rpm)	3,000	2,400	1,800	1,200	1,100	1,000	800	800	600	450
Speed, maximum (rpm)	4,000	3,000	2,500	2,000	1,800	1,500	1,200	1,200	900	700
Normal tension (N/mm)	13	18	22	27	33	36	47	47	60	105
Center distance (mm), minimum	150	200	350	450	525	600	750	750	900	1200

Table 28.3: Influence of rotative speed of tension factor

Per cent normal speed	Tension factor K_t	Per cent of normal speed	Tension factor K_t	Per cent of normal speed	Tension factor K_t
10	3	60	1.40	110	0.84
20	3	70	1.25	120	0.68
30	2.35	80	1.15	130	0.54
40	2.0	90	1.05	140	0.40
50	1.65	100	1.00	150	0.25

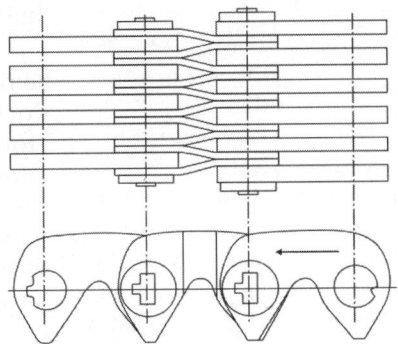

Fig. 28.7: Morse chain hunting link

21-tooth sprocket is used, the number of teeth on the driven sprocket becomes

$$z_2 = \frac{21 \times 690}{135} = 107$$

The corresponding sprocket diameter is

$$D_2 = \frac{z_2 p}{\pi} = \frac{107 \times 32}{\pi} = 1090 \text{ mm}$$

This is larger than can be used.

For the next-smaller odd number of driver teeth, or $z_1 = 19$, it is found that $z_2 = 97$, and thus

$$D_2 = \frac{97 \times 32}{\pi} = 988 \text{ mm}$$

The chain speed is

$$v_m = \frac{19 \times 32 \times 690}{1000 \times 60} = 6.992 \text{ m/s}$$

which is satisfactory. The total pull is

$$F = \frac{1000 \times P}{v_m} = \frac{1000 \times 60}{6.992} = 8581 \text{ N}$$

The allowable tension is 47 N/mm at normal speed of 800 rpm. However, the actual speed is

$(690/800) \times 100 = 86.2$ per cent of the normal speed. Therefore, according to Table 28.3, the tension may be increased 1.09 times. On the other hand, a fluctuating torque due to the compressor load calls for a load factor of about 1.4. Hence, the necessary width is

$$b = \frac{1.4 \times 8581}{47 \times 1.09} = 234.5 \text{ or } 235 \text{ mm}$$

From Table 28.2, the minimum center distance is 750 mm. Probably the shafts should not be brought so close together and a safer value would be

$$C = D_1 + D_2 = \frac{(z_1 + z_2)p}{\pi} = \frac{(19 + 97) \times 32}{\pi}$$

$$= 1182 \text{ mm}$$

By equation 28.24, in which $C_p = \frac{1182}{32} = 36.94$ pitches, the chain length is

$$L_p = 2 \times 36.94 + 0.5 (19 + 97) + \frac{0.026(97 - 19)^2}{36.94}$$

$$= 136.2 \text{ pitches}$$

If a hunting link is inserted, 137 pitches may be used. Ordinarily, however, 138 pitches, or 69 double links, would be ordered.

Example 28.2: Design a roller chain drive to transmit 35 kW from a gas engine running at 1175 rpm to a pump which is to run at 240 rpm. The possible centre distance is 0.6 m.

$$\text{The velocity ratio} = \frac{1175}{240} = 4.89$$

For this velocity ratio, the number of teeth on the smaller sprocket is 21.

Number of teeth on the larger sprocket is

$$= \frac{1175 \times 21}{240} = 102.8 \text{ say } 103 \text{ ok} < 110$$

The chain pitch

$$p = \frac{600}{30} = 20 \text{ mm to } \frac{600}{50} = 12 \text{ mm}$$

Also $\quad p = \left[\frac{1920 \times 60}{1175}\right]^{2/3} = 21 \text{ mm}$

Taking the smaller value, the nearest standard chain is 08 B and pitch is 12.7 mm.

Chain velocity $v = \dfrac{pn_1 Z_1}{60 \times 1000} = \dfrac{12.7 \times 1175 \times 21}{60 \times 1000}$

$$= 5.2 \text{ m/s ok.} < 7.5 \text{ m/s}$$

Total load $\quad F_t = F_d + F_c + F_s$

$$F_d = \frac{35 \times 1000}{5.2} = 6731 \text{ N}$$

So chain 08 B triplex will be suitable with a breaking load of 45400 N.

$$F_c = \frac{wv^2}{g} = \frac{20.2 \times 5.2^2}{9.81} = 55.61 \text{ say } 56 \text{ N}$$

$$F_s = 6 \times 20.2 \times 0.6 = 72.7 \text{ say } 73 \text{ N}$$

Total load $F_t = 6731 + 56 + 73 = 6860$

Factor of safety $= \dfrac{45400}{6860} = 6.61 \text{ ok. } 75$

Check for wear

Pin diameter $\quad = 4.45 \text{ mm}$

Length of pin $\quad = 11.81 \text{ mm}$

$$A = 3 \times 4.45 \times 11.81 = 157.66 \text{ mm}^2$$

$$p = \frac{6731}{157.66} = 42.69 \text{ N/mm}^2 \text{ ok.} < 45$$

28.7 LINK-BELT CHAIN

Link-belt silent chains are made in nine pitch sizes and have hardened-steel round pins working in hardened-steel split bushing, as shown in Fig. 28.8. The use of thin-leaf links produces a more uniform chain.

The recommended chain speeds lie between 7.5 and 15 m/s but encased drives with continuous lubrication can be designed to

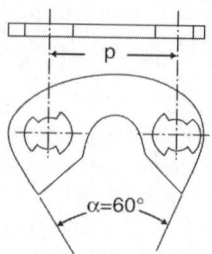

Fig. 28.8: Link-belt silent chain

operate at chain speeds up to 25 m/s. Under favorable conditions speed ratios as high as 25 to 1 may be used.

The necessary chain width must be found from catalogue tables.[2]

Chain widths from $2p$ to $6p$ should be preferred from a mechanical standpoint, although the chains are made in widths up to $12p$. Load and service factors should be used to take into account torque fluctuation and long-hour operation.

28.8 SILENT-CHAIN SPROCKETS

The outside contours of the individual links of the various makes of silent chains are all similar. The angle α included between the working faces of a link is made 60° by all manufacturers, as shown in Figs 28.6 and 28.8. Therefore the angle α, Fig. 28.9, included

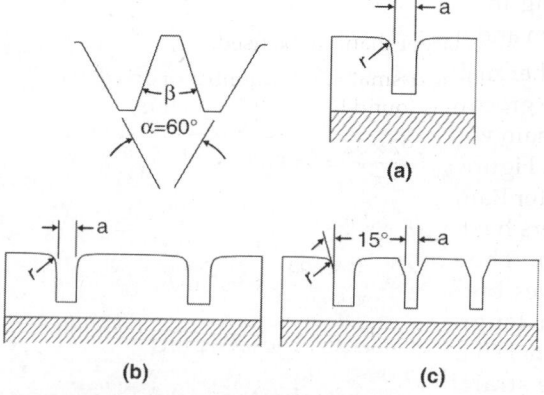

Fig. 28.9: Chain-guide grooves in sprockets

---

[2] A useful condensation of Link-bet's Catalogue 900, 1950 ed., can be found in R. T. Kent, *Mechanical Engineers' Handbook*, 12th ed., Vol. II, *Design and Production*, ed. by Colin Carmichael (New York; John Wiley & Sons, Inc., 1950), p. 15–78.

between the flanks of alternate teeth is also 60°, irrespective of the number of teeth in a sprocket. However, the angle β formed by the flanks of the same tooth varies with the number of teeth z. From the geometry of the triangle, it follows that

$$\beta = 60° - \frac{720°}{z} \qquad \dots (28.25)$$

Equation 28.25 shows that β decreases with a decrease of the number of teeth, and vice versa. In order to have sufficiently strong teeth, the smallest number of teeth in the small sprockets is usually limited to 15. Again, since the angle β increases with the number of teeth, it is necessary to limit the greatest number of teeth to about 150 because of the danger of the chain sliding over the teeth if β is greater than 55.2°

Chain guides: The difference in the shape of sprockets which renders them uninterchangeable for chains of different makes comes chiefly from the method used for guiding the chains. Most chains have one, two, or three lines of straight links, without a cutout, and the links run in grooves cut across the sprocket teeth. Figure 28.8a shows a single-groove guide used in Morse chain sprockets having up to 32 teeth; and Fig. 28.8b shows a double-groove guide for Morse chain sprockets having more than 32 teeth and a width of 75 mm and up.

Other manufacturers use one, two, or three guide grooves, the number depending only on the chain width, irrespective of the number of teeth. Figure 28.8c shows the shape of grooves used for Ramsey silent chains. Still other chain-makers have guide plates on each side of the chain. Then the sprocket teeth have no grooves, but only rounded end corners, and the tooth length is slightly smaller than the nominal chain width. Finally, some chains are made straight and without any guides, but guides are put on the sides of the sprockets, whose teeth are made slightly longer than the nominal chain width. Figure 28.10a shows a sprocket with screwed-on cast-iron ring

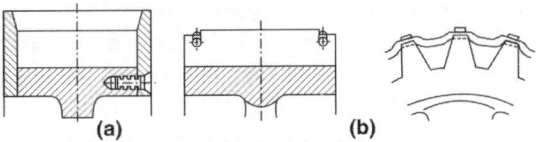

Fig. 28.10: Rim guides of link-belt sprockets

flanges; and Fig. 28.10b shows guides formed by cramped wire driven between the teeth.

Material: Small sprockets are cut of steel and sometimes they are made integral with shaft. Occasionally the rims are steel rings bolted to cast iron disks and hubs. In larger sizes, the sprockets are of cast iron, which is strong enough because the pull is taken by all the teeth in contact.

Fastening to shaft: The common method of fastening a sprocket to a shaft is to use a key between the hub and the shaft. A large-diameter sprocket with a tapered key must be handled with caution in order that the sprocket will not be distorted and will not wobble when running. Sprockets that are split and then clamped on the shaft are convenient, but they should be secured by one or two setscrews.

When heavy shocks are expected, the sprocket may be connected with the hub by means of a shear pin, as in Fig. 28.11a, or a breaking pin, as in Fig. 28.11b. Naturally, either a shear pin or a breaking pin *p* may be used with either of the hub designs shown. In case there is an excessive overload, the pin shears off or breaks and thus protects the chain and other machinery. A spare pin is readily inserted.

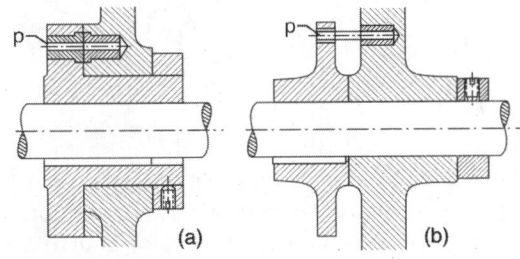

Fig. 28.11: Shear and breaking safety-pin connections

Idlers: In general, idlers which bend the chain in the opposite direction are not desirable. However, if the installation requires an idler, as in the case of a vertical drive which cannot be run slack, one made of hard vulcanized fiber with cast-iron flanges, as in Fig. 28.12, may be used. The idler should bear against the back of the chain. Such idlers are made for chains from 50 to 450 mm wide and with diameters of 65, 100 and 150 mm. The diameter should be selected so as to avoid an excessive speed of the idler. It should not be over 1,000 rpm for the largest diameter, and not over 1,500 rpm for the smallest.

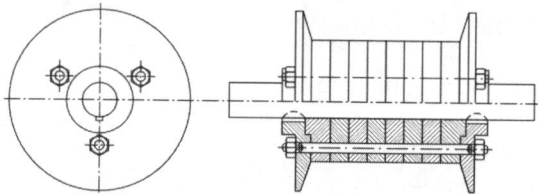

Fig. 28.12: Link-belt silent-chain idler

Casing: The best way to extend the life of a silent-chain drive is to have it properly lubricated. Running the chain in an oil bath is not feasible with normal chain speed, because of excessive resistance. The use of a continuous small oil stream on the inside of the chain is about the best method; the next-best scheme is to use a sight-feed oiler. In any event, the whole drive must be enclosed. In designing a casing, particular attention should be given to oiltightness, ease of inspection of the drive, and ease of removal of the casing for chain adjustments or repairs.

A special method may be used for lubricating chains running at high speeds, above 19 m/s. The chain drive is enclosed in a perfectly airtight casing in which an oil level is maintained, about 6 mm below the lowest point of the chain. When the chain is running at such high speeds, it imparts motion to the air over the oil surface, and the air whips the oil into a spray. The spray is carried around and strikes a baffle in the top of the case. From the baffle, oil flows into a trough that distributes it over the lower strand of the chain.

PROBLEMS

28.1 Name the applications where chain drives are most suitable.

28.2 Why should the arc of wrap on a smaller sprocket be preferably not less than 120°?

28.3 Give the advantages and disadvantages of a silent chain.

28.4 Why is it desirable to have odd number of teeth in each sprocket and even number of links in the chain?

28.5 Does a small pitch/large pitch give a quieter drive?

28.6 What are the sources of loss of power in chain drive?

28.7 What causes impact load on chain?

28.8 What causes variable velocity in the chain drive?

28.9 Design a roller-chain drive for a small fan. The motor speed is 850 rpm, the desired fan speed is 550 rpm, and the power requirement is 4 kW.

28.10 Design a roller-chain drive to transmit 35 kW from a gas engine running at 240 rpm to a pump with a speed of 75 rpm. Determine the length of the chain, using a center distance equal to the sum of the two sprocket diameters.

28.11 Determine the power which can be transmitted by a four-strand, 16 mm pitch roller chain driven by a motor at 1,750 rpm. The motor sprocket has 17 teeth, and the driven sprocket has 42 teeth.

28.12 A roller chain operates under a steady load and transmits 6 kW from a motor shaft rotating at 720 rpm to a shaft running at 950 rpm. Determine (a) the chain and sprockets required, (b) the pitch diameters of the sprockets, (c) the shortest suitable center distance, and (d) the length of the chain in number of links and in meters.

28.13 A 7.5 kW motor running at 1,170 rpm drives a line shaft at 240 rpm through a roller chain. The motor-shaft diameter is 40 mm. The starting torque is two times the running torque, and the load produces moderate shocks. Determine (a) the type and size of a suitable roller chain, (b) the pitch diameters of the sprockets, (c) the closest advisable center distance and the corresponding length of a chain, and (d) the longest permissible center distance.

28.14 A roller chain must be used to drive the camshaft of a four-stroke gasoline engine running at 720 rpm. The center distance is approximately 0.6 m. The crankshaft diameter is 140 mm. The camshaft drive requires 2.25 kW. Determine all necessary dimensions for the chain and sprockets, and check the chain for impact and centrifugal force.

28.15 An oil engine developing 185 kW at 1,200 rpm drives a wire-rope reel on an oil-well rig through a roller chain. The speed of the reel varies from 10 to 50 rpm. The engine can be slowed down to 240 rpm, and the torque remains approximately constant. The load is applied with heavy shocks. (a) Select a suitable roller chain. (b) Determine the numbers of teeth, the pitch diameters, and the outside diameters of the sprockets. (c) Determine a suitable center distance and length of chain.

28.16 A 45 kW truck engine uses a roller chain as the final drive to the rear axle. The driving sprocket runs at 225 rpm, the driven sprocket runs at 100 rpm, and the center distance is approximately 0.9 m. The efficiency of the transmission between the engine and the driving sprocket is 85 per cent, (a) Select a suitable chain, using a low velocity, not over 3.5 m/s, (b) Determine the number of teeth in the sprockets, their pitch diameters, and their outside diameters, and (c) Find the length of the chain.

28.17 Design a drive, using a Morse silent chain, to transmit 120 kW with a motor speed of 450 rpm and a speed of 88 rpm of the driven compressor. The drive must operate 15 hr per day.

29

Friction Gearing

29.1 GENERAL CONSIDERATIONS

Friction gears depend for their driving action upon the friction of the driving wheel, or *driver*, against its mate, or *follower*. The friction surface of a driver should be of a comparatively soft material, such as wood, fiber, leather, or rubber, while that of the follower is usually made of cast iron. This arrangement insures the maintenance of the correct shape of the friction surfaces. If the follower is of the softer material, its surface might be injured when the drive is started under load or when an excessive load brings it to a standstill.

Friction gears are used for light and medium powers in machinery which is frequently started and stopped, and also where provision must be made for a change of speed of the driven shaft, or for its reversed motion.

The disadvantages of friction gears are the thrust on the bearings, and slippage, resulting in a comparatively low efficiency (29.1). However, by using metal-to-metal-contact surfaces and ball or roller bearings, slippage and thrust drawbacks can be considerably reduced.

Experimental data: Tests conducted with strawboard driving wheels and a cast-iron follower gave the following results:[1]

(a) Slippage increases gradually as the load increases, up to 3 per cent; after

that it is likely to increase suddenly so much that the follower may stop.

(b) the coefficient of friction seems to be constant for all pressures up to a safe limit of slightly over 2.5 N per mm of face.

(c) *The coefficient of friction increases with slip, up to 2 per cent slip (29.2).*

(d) The coefficient of friction is not affected by the tangential velocity between the limits of 5.5 and 14 m/s.

Table 29.1 contains data for the design of friction gearing complied from various sources.[2] *In operation, leather fiber becomes glazed and hard, as a result of slippage, and its coefficient of friction drops (29.2).* Tarred fiber and straw fiber are better if the drive is started and stopped frequently.

Rigidity: In order to obtain an even contact pressure across the whole face of each wheel of friction gears, the wheels must be rigid and must be rigidly supported by closely located bearings.

Classification: Friction gears are of the spur, bevel, and disk types (29.3).

29.2 SPUR FRICTION GEARS

The wheels of spur friction gears may be plain or grooved.

---

[1]W. F. M. Goss, "Power Transmission by Friction Driving," *Transactions of the American Society of Mechanical Engineers*, Vol. 29 (1907), pp. 1093 ff.

[2]R. T. Kent, *Mechanical Engineers' Handbook*, 12th ed., Vol. II, Design and Production, ed. by Colin Carmichael (New York: John. Wiley & Sons, Inc., 1950), p. 15–83.

Table 29.1: Design data for friction gearing

Material of driver	Allowable pressure p kN/m	Coefficient f with cast iron	Material of driver	Allowable pressure p kN/m	Coefficient f with cast iron	Coefficient f with aluminium
Cast iron	50	0.15	Leather	2.50	0.09	0.13
Cork composition	0.8	0.21	Leather fiber	4.15	0.18	0.18
Paper	2.5	0.15	Straw fiber	2.50	0.15	0.16
Rubber	1.75	0.20	Sulfite fiber	2.41	0.20	0.19
Wood	2.5	0.15	Tarred fiber	4.35	0.28	0.28

Plain spur wheels: The simplest type of friction gearing consists of two plain cylindrical wheels, Fig. 29.1, held in contact with each other by two forces F. In the following equation,

P is the power transmitted in kW;

v_m is the mean circumferential velocity, in m/s;

b is the width of face of the gears, in m;

p is the permissible pressure, in N/m of face;

f is the coefficient of friction;

F_t is the tangential force due to the pressure $F = bp$, namely,

$$F_t = fbp \qquad \ldots (29.1)$$

and the power transmitted is

$$P = \frac{F_t v_m}{1000} \qquad \ldots (29.2)$$

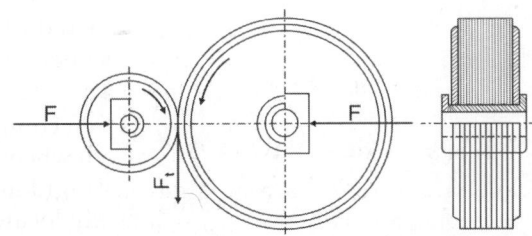

Fig. 29.1: Spur friction gears

By substituting for F_t its value from equation 29.1, and substituting for v_m its equivalent $\pi Dn / 60$, where D is in m, and solving for b, these result

$$b = \frac{60,000P}{\pi fpDn} \qquad \ldots (29.3)$$

Applications: Plain spur friction drives are used for driving light power hoists, coal

screens, gravel washers, driers, etc. both at an increased speed and at a decreased speed. The commercial limits for the sizes of fiber-face pulleys are as follows: D ranges from 0.15 m to 0.9 m , and b ranges from 0.1 m to 0.45 m.

Grooved spur wheels: Grooved spur wheels are capable of transmitting more power with the same radial force applied than the cylindrical type. They are therefore used where a greater power is required, as in hoisting machines. Such wheels are formed as shown in Fig. 29.2a. In Fig. 29.2b the radial force F is held in equilibrium by the normal pressures R, and friction forces fR. Summing up along the vertical axis gives, for each groove,

$$F = 2R \sin \alpha + 2fR \cos \alpha \qquad \ldots (29.4)$$

The tangential resistance F_t of each groove is equivalent to the normal pressure 2R multiplied by f, or

$$F_t = 2Rf \qquad \ldots (29.5)$$

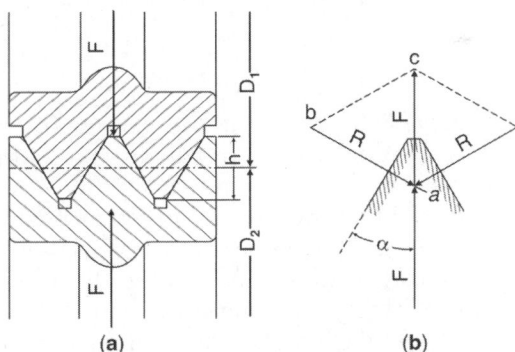

(a)   **(b)**

Fig. 29.2: Grooved spur friction gears

Also, from the geometry of the illustration, the normal pressure is

$$R = ph \sec \alpha \qquad \ldots (29.6)$$

where, p is the pressure per meter at the groove side, and h is the depth of the groove.

When the value of R from equation 29.6 is substituted in equations 29.4 and 29.5 and the results are simplified, the result is

$$F = 2ph(f + \tan \alpha) \qquad \ldots (29.7)$$

and

$$F_t = 2phf \sec \alpha \qquad \ldots (29.8)$$

By using the general equation 29.2, substituting iF_t for the total tangential force, where i is the number of grooves, and replacing F_t by its value from equation 29.8, we obtain the relation

$$ih = \frac{30,000P \cos \alpha}{fpDn} \qquad \ldots (29.9)$$

Since along the lines of contact the so-called *pitch point* is the only one at which the two gears have the same peripheral speed, at all other points there must be slippage. In order to avoid excessive friction and wear, the depth h should be made comparatively small, usually not over 6 mm, a practical value, in mm is

$$h = 0.006D + 3.8 \qquad \ldots (29.10)$$

where, D is the diameter of the smaller gear, in mm. Generally, α ranges from 12° to 18°, and it should not be over 20°. Both gears are made of cast iron. The drive works satisfactorily only with a sufficiently high peripheral speed.[3] A recommended value is

$$v_m \geq 6 + 0.02D \qquad \ldots (29.11)$$

Bearings of friction gears: A change of the center distance necessary in engaging and disengaging a pair of cylindrical friction gears is obtained by means of suitable bearings, as shown in Fig. 29.3. The bearing sleeve a is bored eccentrically and may be rotated through a small angel γ by means of the attached lever. The center c of the shaft is thus shifted through

a distance e. If axial displacement is needed, it can be obtained by means of a square thread on the outside of the bearing sleeve, as shown in Fig. 29.4.

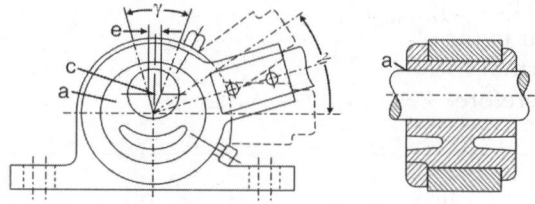

Fig. 29.3: Side-motion bearing

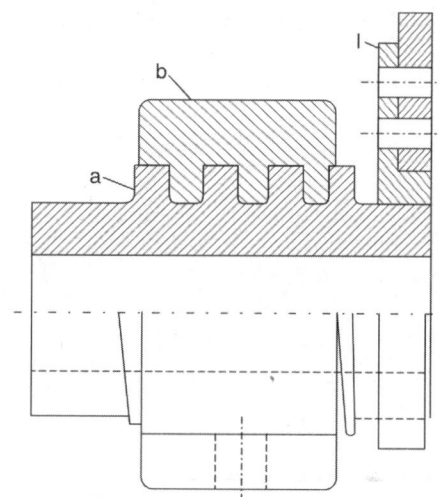

Fig. 29.4: Axial-motion bearing

29.3 BEVEL FRICTION GEARS

Bevel friction gears are suitable only for light power transmission, as it is difficult to maintain a uniform contact pressure. The larger gear is usually keyed rigidly to the shaft, while the smaller one is either splined to the shaft or can be moved axially together with the shaft. It should be noted that the starting conditions differ from the running conditions.

Starting: It will be assumed that the drive is started under full load, as is often the case. As for grooved spur gears, the reaction R', Fig. 29.5, is inclined from the normal by the angle of friction ϕ for these conditions, and therefore

$$R' = \frac{F_1'}{\sin(\alpha + \phi)} = \frac{F_2'}{\cos(\alpha + \phi)} \qquad \ldots (29.12)$$

The tangential force transmitted from one gear to the other is equal to the product of the normal force and the coefficient of friction. Therefore

$$F_t' = fR' \cos\phi = \frac{1000P}{v_m} \qquad \ldots (29.13)$$

Combining equations 29.12 and 29.13 gives, for the least axial thrust required,

$$F_1' = \frac{1000P(\sin\alpha + f\cos\alpha)}{fv_m} \qquad \ldots (29.14)$$

and

$$F_2' = \frac{1000P(\cos\alpha - f\sin\alpha)}{fv_m} \qquad \ldots (29.15)$$

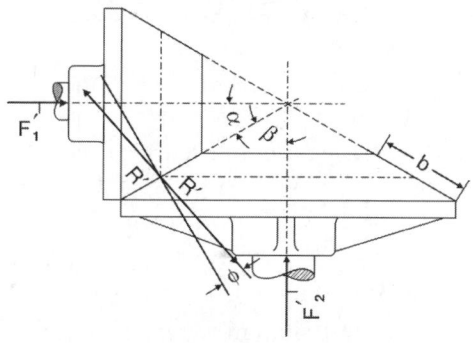

Fig. 29.5: Bevel friction gears

Running: After the drive gets up to speed, the relative motion between the gears along the line of contact ceases, the reaction between the two surfaces in contact is normal, and the angle ϕ is 0. If the reaction in this case is designated by R, the following equations are obtained:

$$R = \frac{F_1}{\sin\alpha} = \frac{F_2}{\cos\alpha} \qquad \ldots (29.16)$$

and

$$F_t = fR = \frac{1000P}{v_m} \qquad \ldots (29.17)$$

Combining these equations gives, for the least axial thrust,

$$F_1 = \frac{1000P\sin\alpha}{fv_m} \qquad \ldots (29.18)$$

and

$$F_2 = \frac{1000P\cos\alpha}{fv_m} \qquad \ldots (29.19)$$

Thus the thrust F_1 on the driver is less than F_1', and the thrust F_2 on the driven wheel is greater than F_2'. In equation 29.16, $R \leq bp$, where, p is the permissible unit pressure.

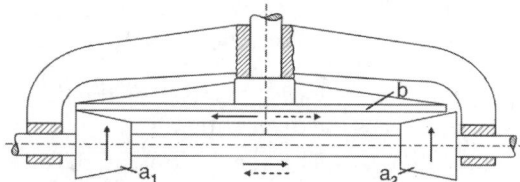

Fig. 29.6: Reversing-motion friction gearing

Reversing-motion drive: A drive for reversing motion by means of bevel gears is illustrated in Fig. 29.6. When the driver a_1 is pressed against the follower b, the latter rotates in a certain direction. If the driving shaft is shifted to the left, a_1 is disengaged and a_2 is pressed against b, and the follower begins to rotate in the opposite direction.

Bearings: To obtain the necessary pressure between the faces of two bevel gears, one of them must have a stationary thrust bearing, and the thrust bearing of the other gear must have provisions for an axial motion. In Fig. 29.4, for example, the bearing a has an Acme thread on the outside that meshes with a thread in the stationary support b, and the lever l serves to turn the bearing a and thus to move it axially.

29.4 DISK FRICTION GEARS

The disk type of friction gear, Fig. 29.7, is a special kind of bevel-gear drive in which the angle α is 0. It is used mostly as a variable-speed drive (29.5). In a disk friction gear, unlike the spur friction gear, the usual practice is to make the driver a of cast iron or aluminum, and to face the driven wheel c with a composite material. The speed of the wheel c is changed by moving it across the face of the disk a, and the direction of its rotation may be reversed by moving it on the other side of the center of the disk a.

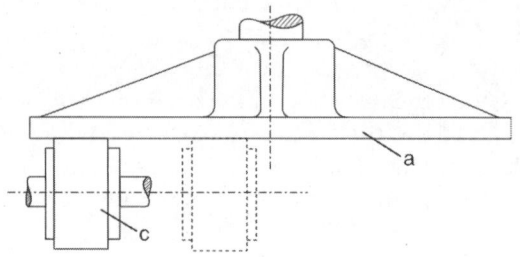

Fig. 29.7: Disk friction gearing

Sometimes the wheel c is used as the driver and the disk a is the driven member.

Force analysis: The torque on the driving shaft is $500P/n\pi$. Hence the tangential forces acting upon the driven wheel for the two limiting speeds corresponding to a contact at the minimum and maximum diameters D_1 and D_2 are as follows:

At the minimum speed,

$$F_{t1} = \frac{1000P}{\pi n D_1} \qquad \ldots (29.20)$$

At the maximum speed,

$$F_{t2} = \frac{1000P}{\pi n D_2} \qquad \ldots (29.21)$$

The thrust that may be applied to the disk for the two limiting speeds are found by dividing the values of F_{t1} and F_{t2} by the friction coefficient f. Therefore

$$F_1 = \frac{1000P}{\pi f n D_1} \qquad \ldots (29.22)$$

and

$$F_2 = \frac{1000P}{\pi f n D_2} \qquad \ldots (29.23)$$

Again $F_1 \leq bp$, where b is the face of the driven cylindrical wheel.

Since there is relative motion between the faces of parts a and c, the actual tangential forces available at the driven wheel are found by multiplying the values just stated by the efficiency e. Approximately,

$$e = \frac{D}{D+b} \qquad \ldots (29.24)$$

The efficiency varies from about 0.60 at low speeds, when $D = D_1$, to about 0.85 at high speeds, when $D = D_2$.

The axial force F_a required to shift the driven wheel c under load is found by multiplying the thrust F_1 or F_2 by the friction coefficient f and adding to the product the frictional resistance to axial motion of the shaft. Thus,

$$F_a = F_1(f + f_1) \qquad \ldots (29.25)$$

where, f_1 is the coefficient of friction between the shaft of wheel c and its bearings.

PROBLEMS

29.1 What are the disadvantages of a friction gearing?

29.2 Does the coefficient of friction effected by slip?

29.3 Name the types of friction gearing.

29.4 On what factors does the slippage depend?

29.5 What type of friction drive is used when the requirement is of variable speed?

29.6 Two shafts 0.6 m between centers are connected by a pair of plain spur friction wheels. The driver makes 380 rpm, and the follower makes 140 rpm. Select the materials for both wheels, and find the thrust and width of face necessary to transmit 9 kW.

29.7 Using data of problem 29.6, design wheels with grooved faces (a) if both wheels are of cast iron and (b) if the driver face is made of maple. (c) Also find the bearing pressures for both cases.

29.8 A small printing press is driven by a 2.5 kW motor running at 1,175 rpm. On the motor shaft is a rubber spur friction wheel 140 mm in diameter. Find (a) the necessary width of the wheel and (b) the required bearing pressure.

29.9 A pair of grooved spur friction wheels is used to connect two shafts. The driver makes 260 rpm, and the follower makes 200 rpm. The power to be transmitted is 35 kW. Using cast iron for both wheels, find the diameters of the wheels. Select the angle of grooves and their depth, and find their number and the least pressure on the bearings.

29.10 If the driver in problem 29.9 has three grooves 6 mm deep, with $\alpha = 16°$, find (a) the least pressure per meter of contact line and (b) the bearing load.

29.11 Find the power that can be transmitted by a grooved-face spur friction driver with a 0.5 m diameter running at 340 rpm. Both wheels are of cast iron, the groove angle is 14.5°, the wheels are pressed together with a force of 4.2 kN and the driver has two grooves with an effective depth of 5 mm.

29.12 A hoist driven by spur friction gears has two driving wheels 200 mm in diameter with tarred fiber faces, the two followers are 1.2 m in diameter and are keyed to the drum shaft, and the faces are 150 mm wide. Determine (a) the load which can be hoisted if the drum diameter is 0.6 m and a 20 mm cable of mild plow steel is used and (b) the power required if the drivers make 215 rpm.

29.13 A small fan requiring 1 kW is to be driven from a 1,200 rpm motor by means of bevel friction gears. (a) Using a speed ratio of 1.9 : 1, select the materials and determine the main dimensions of the wheels. (b) Determine the bearing pressures when starting and when running.

29.14 A grinding machine requires 0.9 kW and must operate at speeds varying from 300 to 2,400 rpm. The motor runs at 1,175 rpm. Select all materials for a suitable disk friction drive, and determine the main dimensions and all forces involved in the operation of the drive.

30 Toothed Spur Gearing and Helical Gearing

30.1 GENERAL CONSIDERATIONS

Toothed gears may be considered as friction gears in which the contact surfaces are provided with grooves and projections which prevent slip and insure positive means of transmitting rotation. Toothed gears are used when a constant speed ratio is desired and the distance between the shafts is relatively small. Toothed spur gears are used to connect parallel shafts and are suitable for transmission of any amount of power.

Methods used to manufacture gears depend on design requirements, machine tools available, quantity required, cost of materials, usage and tradition. The gear designer—studying gears as a whole—can get a good perspective of gear work by reviewing the methods of manufacture in each field. The following classifications is based on the methods of manufacture and the power to be transmitted.

1. *Small low-cost gears for toys, gadgets and mechanisms*: In all these cases a slow speed and very low power is to be transmitted. Pinions may be die cast or extruded and cut to lengths required. Gears are punched. For quietness, injection moulded gears and pinions may be used. In many instances, e.g. watches, cameras, etc. cycloidal profile punched gears or cold drawn gears are used for long life as wear is less, lubricant stays within convex and concave surfaces. The cost of production is less in mass production. Zinc alloy, brass or aluminium, plastics and tin sheet are often used.

2. *Appliance gears*: Home appliances like washing machines, food processors, fans, etc. use large numbers of small gears. Medium carbon steel gears cut by conventional cutting and sintered iron gears completely finished in sintering process and may be impregnated with oil or copper to improve its lubrication. These are very inexpensive, and wear less. Laminated gears using phenolic resins nylon, teflon wear little as they have lubricating property and run quietly.

3. *Machine tool gears*: Accuracy and power transmitting capacity are quite important in machine tools which are literally full of gears of all types. Mild alloy steels, cast iron and also alloy steels with high hardness are used where accuracy is desired.

4. *Control gears*: The instruments, guns on ships, airplanes, tanks are controlled by gear trains with backlash held to lowest possible limits. The gear may become inserviceable if it wears out more than 0.001 mm. Control gears are usually spur, bevel or worm and helical. Usually finer modules in the range of 1 and lower are common. Medium alloy and medium carbon steels are favoured. Both flame hardening and induction hardening have been widely used. Lapping, grinding and shaving are the finishing processes. In

many cases they are hardened to a medium hardness before final machining.

5. *Automotive gears*: The automobile uses spur, helical and bevel gears. Automobile gears are usually cut from low alloy steel to high alloy steel forgings. The gears are case carburized and quenched. Multistation shapers, hobbers and shaving machines are used. Newer automatic machines take a minute to cut a complete gear. Spur and helical gears are shaved after cutting. Automotive gears are heavily loaded for their size. However, the heaviest loads are of short duration, so the gears last many years.

6. *Transportation gears*: Buses, subways, mine cars, all use large quantities of gears. Plain carbon, low alloy steel to high alloy steels are usually used. In some applications severe but infrequent shock loads are encountered. Much of the gearing is through hardened and ground. Shallow hardening medium carbon steels seem to resist shock better than fully hardened carburized case. Both furnace and induction hardening techniques are used. Gear cutting is done mostly by conventional hobbing or shaping machines. Some gears are shaved and then heat treated while others are heat treated after cutting and then ground. Both generating grinders and form grinders are used to finish gears after hardening.

7. *Marine gears*: Merchant, navy ships and submarines use very large size of gears up to 4 m diameter and the power ratings are also very high 40,000 kW or more. Extreme accuracy in gears is required. Spur, bevel, helical-double helical are often used. These are made of both plain carbon and alloy steels. Gear noise on ship gears needs to be reduced to minimum. Extreme accuracy is desired. Most of the marine gears are fabricated by welding webs and tyre, the materials should have good welding properties.

8. *Aircraft gears*: There is a large application of gears on aircraft. Many accessories, viz generators, pumps, hydraulic regulators, techometers, landing wheels, control guns, etc. utilize gears. To get maximum load capacity with light weight, use of epicyclic gear trains and high pressure angle viz $22.5°, 25°, 27.5°$ are quite common. A gear failure can result in disaster. So gears are made of high alloy steel and fully hardened. Speeds and tooth loads are very high, so high strength and high wear resistance is required.

Selection of Gears

First functional requirement is depending on the geometric arrangement of the system. If parallel shafts are to be the drive, spur gears maybe used for slow speeds. The helical gears can be used for high speeds and high power (4500 kW). Bevel and worm gears can be used for shafts at right angles. If the axis are non-intersecting and non parallel, crossed helical gears and hypoid gears are used.

Bevel gears are ordinary used when high efficiency is needed, about 98%. Worm gears may not have an efficiency higher than 90% but high reductions even up to 500 may be obtained in a single step. Hypoid gears have less efficiency than bevel gears but are capable of transmitting high power.

Classification: Spur gearing may be classified as external, Fig. 30.1a; internal, Fig. 30.1b; and rack-and-pinion, Fig. 30.2. If the tooth elements are parallel to the shaft axis, the gears are termed *straight-tooth spur gears*; if the elements are helices, the gears are called *helical gears*. If the words spur gears are used without a

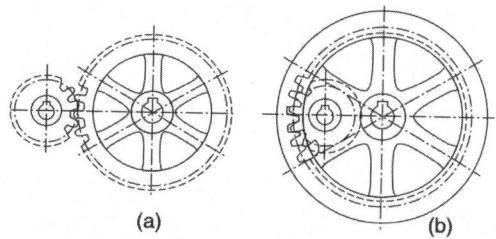

(a)   (b)

Fig. 30.1: External and internal spur gearing

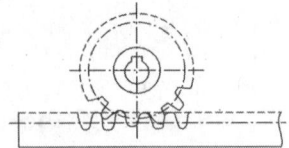

Fig. 30.2: Rack-and-pinion spur gearing

modifying adjective, the more commonly used straight-tooth gears are signified.

Requirements: Toothed spur gears must meet the following requirements:

(a) The teeth must have a profile which insures a constant velocity ratio.

(b) The relative motion of one tooth upon the other should be more of a rolling nature than of a sliding nature.

(c) The arc of engagement should be so long that at all times more than one pair of teeth is in mesh. In practice this is not always fulfilled.

(d) The tooth profile should approach a cantilever beam of uniform strength.

Definitions: The information in Fig. 30.3 will help to explain the meaning of different terms used in connection with toothed gears.

When two gears are in mesh, the larger one is called the *gear* and the smaller one is called the *pinion*, regardless of which one is the driver.

The rubbing surfaces of the friction wheels have become the *pitch surfaces* of the toothed gear and pinion. The intersection of the pitch surface with a plane perpendicular to the axis of a gear or pinion is called the *pitch circle*. The contact point *O*, Fig. 30.3, of two pitch circles is called the pitch point. However, in involute gears the nominal pitch circles may not be tangent. The pitch circle is the basis of measurement of

gears. The *size* of a gear is the diameter of this nominal pitch circle, in m or mm.

The *addendum circle* is the circle which bounds the outer ends of the teeth, the addendum *a* being the radial distance between the pitch circle and the addendum circle.

The *dedendum circle*, or *root circle* or *clearance circle*, bounds the bottom of the teeth, the dedendum *d* being the radial distance between the pitch circle and the dedendum circle.

Thus the *height* of the tooth is $h = a + d$.

The *clearance c* is the difference between the dedendum and the addendum; and for equal-addendum gears, $c = d - a$.

The *thickness* of the tooth is its thickness measured on the pitch circle, as shown in Fig. 30.3.

The *tooth space* is the width of the empty space on the pitch circle.

The *backlash* is the difference between the tooth space and the thickness of the tooth. Backlash is necessary to care for inaccuracies in the form and in the spacing of the teeth, and in the mounting of the gears. In gears with teeth cut very accurately, backlash may be practically zero.

The *face of the tooth* is the surface of the tooth between the pitch cylinder and the addendum cylinder; the *flank* is the surface between pitch cylinder and the root cylinder.

The *facewidth of the gear* is its width measured parallel to its axis.

The *line of centers* is the line connecting the centers of a pair of mating gears.

The *pressure angle*, or *angle of obliquity*, is the inclination of the line of action of the pressure between a pair of meshing teeth with respect to a line drawn tangent to the pitch circle at the pitch point, as angle β in Fig. 30.3.

The *base circle* is an auxiliary circle used in involute gearing to generate the tooth profile. It is tangent to the line representing the tooth profile.

The *describing circle* is an auxiliary circle used in cycloidal gearing to generate the tooth profile.

The *arc of approach* is the arc measured on the pitch circle from the position of the tooth at the

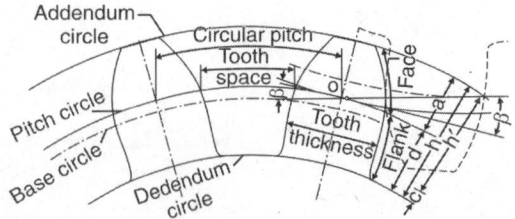

Fig. 30.3: Definitions and dimensions relating to spur gears

beginning of contact to the pitch point, as arc aO in Fig. 30.4. The *arc of recess* is the arc measured on the pitch circle from the pitch point to the position of the tooth where contact ends, as arc Or. The *arc of action* is the sum of the arc of approach and the arc of recess. Naturally it does not make any difference whether these arcs are measured on the pitch circle of the gear or that of the pinion. However, the corresponding angles of motion will be different.

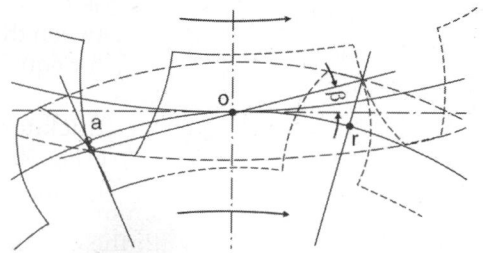

Fig. 30.4: Duration of engagement of a pair of gear teeth

By *velocity ratio*, or *speed ratio*, r_c is always meant the ratio of the number of revolutions of the driver to the number of revolutions of the driven gear. In a speed reducer, r_c is greater than 1; in a speed increaser, it is less than 1.

Pitch: The pitch is a measure of the size of a gear tooth. The pitches in common use are the circular pitch and the diametral pitch.

The circular pitch, designated p_c, is the distance, in mm, measured along the pitch circle from a point on one tooth to the corresponding point on an adjacent tooth, as shown in Fig. 30.3.

If D is the diameter of the pitch circle and z is the number of teeth, then evidently

$$p_c = \frac{\pi D}{z} \qquad \ldots (30.1)$$

Circular pitch is used for cast-tooth gears because of its convenience when laying out the pattern. It is also used to a certain extent for cut-tooth gears and especially for large gears, when $p_c > 3$ in or 7.5 mm.

The *diametral pitch p_d* of a gear represents the number of teeth per inch of pitch diameter. Thus

$$p_d = \frac{z}{D} \qquad \ldots (30.2)$$

Module m is reverse of diametral pitch and is equal to $m = \dfrac{D}{z}$ and is in mm.

Diametral pitches for which cutters may be obtained from stock are: from 1 to 4, by increments of $1/4$; from 4 to 6, by $1/2$; from 6 to 16, by 1; from 16 to 32, by 2. The sizes of modules: 1, 1.25, 1.5, 2, 2.5, 3, 4, 5, 6, 8, 10, 12, 16, 20, 25, 32, 40, 50 as first preference. The second choice is 1.125, 1.375, 1.75, 2.25, 2.75, 3.5, 4.5, 5.5, 7, 9, 11, 14, 18, 22, 28, 36, 45.

Pitch relation: Multiplying equation 30.1 by equation 30.2 gives the relation for converting one pitch into the other one. This relation is

$$p_c p_d = \pi \text{ or } p_c = \pi m \qquad \ldots (30.3)$$

30.2 GEAR TEETH

Two types of curves, the cycloidal and the involute, are now in general use for gear teeth. In regard to efficiency and strength, both forms are practically equal. *An advantage of the cycloidal tooth over the involute one is that a convex surface is always in contact with a concave one; consequently, the wear is not as fast as with involute teeth, whose surfaces are either convex or straight. For this reason, gears transmitting very large amounts of power are sometimes cut with cycloidal teeth. Also the wear is uniform and the life of gear teeth is large because the lubricant stays in. So these are also preferred for watches, gadgets, cameras, etc. where the teeth are obtained by punching operations (30.4). On the other hand, the involute teeth have a decided advantage over the cycloidal one in that the actual distance between the centers may deviate slightly from the theoretical distance without affecting the velocity ratio or general performance (30.3).* Because of this distinct advantage, gears with involute cut teeth are used much more than those with cycloidal teeth. Involute profile also gives constant pressure angle. *However, with the increase of centre distance, there is an increase in pressure angle (30.11).* In cycloidal profile, the center distance has to be exact.

Involute-tooth action: It should be noted that an involute can exist only outside of the base circle.

Therefore only that part of a tooth profile which lies outside the base circle is an involute and can mesh with another involute of the same pitch. The part of the flank of tooth inside the base circle does not come in driving contact with the mating tooth. It may have any form, provided it does not interfere with the face of the mating tooth. Formerly this part was made radial. Modern methods of generating gear teeth automatically undercut this part of the flank.

For the same pressure angle β, the radial distance between the pitch circle and the base circle increases with an increase of the pitch diameter, as shown in Fig. 30.5. For a given pitch circle this radial distance increases as the pressure angle increases, as shown in Fig. 30.6. Pressure angles in common use are $14\frac{1°}{2}$, 15° and 20°. Also $22\frac{1°}{2}$, 25° and $27\frac{1}{2}°$ are also used.

Since the pressure line coincides with the generating line of an involute and is normal to the involute, the point of contact of two mating teeth has the pressure line as its common normal. Therefore two gears in mesh must have the same pressure angle, or must be of the same degree of involute.

Law of gearing: To get constant velocity ratio, the shape of tooth must be such that common normal at the point of contact of two mating tooth (surfaces) must always pass through a fixed point—the pitch point lying on the line of centres (30.16), i.e. the

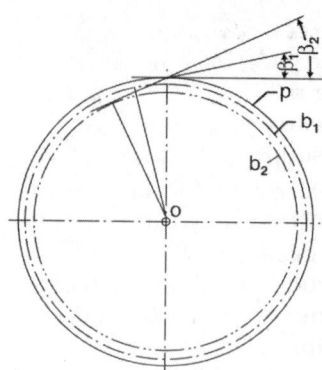

Fig. 30.6: Base circles and degrees of involute

profiles produce constant angular velocity ratio (30.1).

The gears then are said to produce conjugate action: For involute profile the common normal has a constant inclination with the common tangent to the two pitch circles and is the pressure angle. In the case of cycloidal profile the pressure angle varies from maximum value at the point of engagement to zero as the contact shifts to pitch point beyond which it again increases.

Interference: If, with a certain angle of obliquity, the number of teeth of a pinion is decreased below a certain minimum, the tooth tips of the gear or rack will begin to interfere with the part of the tooth flank of the pinion below the base line (30.10).

Thus, with a $14\frac{1°}{2}$ involute, the smallest pinion of interchangeable sets that will mesh correctly with a rack of the same pitch contains 32 teeth. This difficulty is eliminated by slightly correcting the points of all the teeth in a set of cutters, so that a pinion of 12 teeth may still mesh with any of the gears of the same pitch.

For a 20° full-depth involute the smallest pinion that can correctly mesh contains 17.1, i.e. 18 teeth; and for the 20° stub tooth, the pinion must have at least 14 teeth. For 25° involute, the minimum number of teeth is 12. For 20° full depth if pinion has 17 tooth, the number of gear tooth will be 1309 for no interference.

Methods of manufacture: Gear teeth are formed either by molding and casting or by

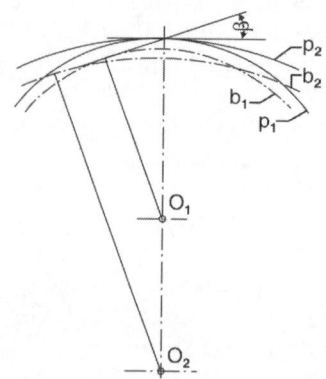

Fig. 30.5: Base and pitch circles

machine cutting. Cast teeth are molded from complete gear patterns, or by means of gear-molding machines. The latter method gives better results. Nevertheless, cast teeth are always somewhat rough and warped out of shape, so the gears are noisy and are not suited for higher speeds.

Most gears of ordinary size are cut with a milling cutter, or by special gear planers, or else by means of hobs. This hobbing method of generating spur and helical gear teeth gives very accurate teeth. Other advantages of hobbing are that only one hob is required for all numbers of teeth of a given pitch, and the production cost is lower.

Other methods of manufacture are forging, extrusion, punching, and rolling which is also used as a finishing process for conventional cut gears. They can also be milled, drawn, sintered. They can be finished by grinding, shaving, lapping or burnishing and rolling.

Interchangeability: In order to manufacture gears economically, it is necessary that any gear of a given pitch should work correctly with any other gear of the same pitch, thus making the gears interchangeable. *Other conditions for interchangeability are that the gears must have the same pressure angle and that the addendum must equal to the dedendum minus the clearance. These conditions are fulfilled by making the teeth according to certain standard proportions (30.6), which differ with the method of manufacture and the special requirements of strength.* Table 30.1 gives proportions for the more commonly used systems.

30.3 CAST TEETH

Cast-tooth gears are used for low-speed rough service, mostly in outdoor installations where noise is not objectionable. Because of the inaccuracy of forming and spacing the teeth, even with molding machines, it is safer to assume that the entire load is transmitted by one tooth.

It may be assumed that the load F_t acts at the upper corner of the tooth but is uniformly distributed along the length b of the tooth, as indicated in Fig. 30.7. If the bending moment at the root of the tooth is equated to the resisting moment and the stress is designated by the product S_0c, the ressult is

$$F_b h = \frac{1}{6} S_0 cbt^2 \qquad \ldots (30.4)$$

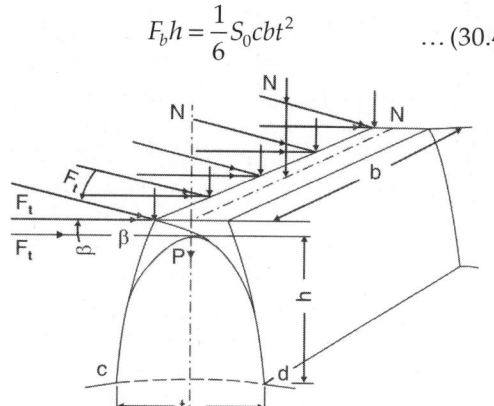

Fig. 30.7: Load action on a gear tooth

Table 30.1: Proportions of involute teeth

Tooth characteristics	Cast teeth	ISO composite 14.5° system	Full-depth 20° system	ISO stub teeth	BIS
Pressure angle (deg)	15	14.5	20	20	all
Addendum (mm)	0.943 m	m	m	0.8 m	m
Minimum dedendum (mm)	1.257 m	1.157 m	1.157 m	m	1.25 m
Minimum total depth (mm)	2.2 m	2.157 m	2.157 m	1.8 m	2 m
Minimum clearance (mm)	0.314 m	0.157	0.157 m	0.2 m	0.25 m
Thickness of the teeth (mm)	1.493 m	1.571 m	1.571 m	1.571 m	1.571 m
Backlash (mm)	0.157 m	0	0	0	0
Outside diameter (mm)	(z +2) m	(z +2) m	(z +2) m	(z +1.6) m	(z +2) m
Approximate fillet radius (mm)	0.209 m	0.209 m	0.209 m	0.209 m	0.4 m

in which S_o denotes the allowable stress in the material and c is the velocity factor.

According to Table 30.1, $h = 2.2 \text{ m} = \dfrac{2.2p_c}{\pi}$

and $t = \dfrac{1.493\,p_c}{\pi}$. Substituting these values in equation 30.4, and solving For F_t, gives

$$F_t = 0.054 S_o cbp_c \qquad \dots (30.5)$$

Instead of this equation, general Lewis equation (30.12) derived later can also be used.

The velocity factor c is introduced in order to take care of the impact when the teeth come in contact. It is computed by the following Barth's formula:

$$c = \frac{3.05}{3.05 + v_m} \qquad \dots (30.6)$$

where, v_m is the pitch-line velocity. For cast teeth it should not exceed 2.5 m/s. The stress may be taken equal to the elastic limit in tension divided by a safety factor $n = 1.5$. The average values used are as follows: For ordinary cast iron, $S_o = 56 \text{ MN/m}^2$; for high-grade or nickel cast iron, 80 MN/m^2; and for cast steel, 138 MN/m^2. The gear face b is made from $2p_c$ to $2.5p_c$. After p_c is found from equation 30.5, the next larger standard circular pitch is used. The circular pitch of cast teeth varies in larger sizes in increments of 6 mm and from 25 mm down in increments of 3 mm.

With p_c found, the number of teeth z of a gear which has a peripheral speed of v_m m/s rotating at n rpm is found from the relation

$$v_m = \frac{z\, p_c\, n}{60 \times 1000} \text{ m/s or } = \frac{\pi D n}{60 \times 1000} \quad \dots(30.7)$$

The pitch diameter D is found from equation 30.1. Then the outside diameter D', with the addendum $a = 0.943 \text{ m} = \dfrac{0.943\,p_c}{\pi}$ from Table 30.1, is

$$D' = D + 2\,m \qquad \dots (30.8)$$

30.4 CUT TEETH

Three main profile systems are used for cut teeth. They are known as the standard $14\dfrac{1}{2}°$, involute system, the full-depth 20° involute system, and the stub 20° involute system.

Standard $14\dfrac{1}{2}°$ involute system: The standard angle of obliquity adopted by the manufactures of gear cutters is 14° 28' 40", the sine of which is 0.25. The standard proportions of teeth for this system are given in Table 30.1.[1]

Full-depth 20° involute system: The tooth proportions for the full-depth 20° system are the same as those of the standard $14\dfrac{1}{2}°$ system, the only difference being in the pressure angle. The 20° angle reduces the interference, permitting the use of a 17-tooth pinion without any changes with a rack. Because of the increased pressure angle, the teeth also become slightly broader at the root, as shown in Fig. 30.8, and stronger.

Stub 20° involute system: The 20° stub tooth is subject to a still-smaller interference because of a shorter addendum (30.12), as shown in Table 30.1. The interfering portion of the tooth is thus removed. The minimum number of teeth can be 14. Stub teeth are also considerably stronger than full-height teeth because of a smaller moment arm of the bending force, Fig. 30.7.

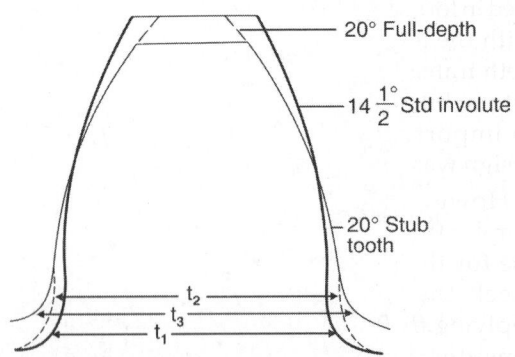

Fig. 30.8: Comparison of tooth shapes

---

[1]*Spur Gear Tooth Forms*, ASA B6.1–1932 (New York; American Standards Association 1932).

A third advantage of stub teeth is their lower production cost, as less metal must be cut away. The advantages of stub teeth apply chiefly to gears with a small number of teeth. With a large number of teeth, full-depth tooth gears having pressure angles of either $14\frac{1°}{2}$ or $20°$ preform better than stub-tooth gears.[2]

Several gear manufacturers have established their own standards for the proportions of stub teeth, particularly Nuttal (Westinghouse) and Fellows Gear Shaper Company. The American Standards Association has established proportions, given in Table 30.1, which make the gears interchangeable with other $20°$ stub-tooth gears.

Fellows stub-tooth system uses a 20-deg pressure angle and a fractional designation for the pitch. The numerator of this designation is the actual diametral pitch of the gear, whereas the denominator indicates the pitch of the cutter used in cutting the teeth and thus determines their depth.

The numerator is used in most equations, but the denominator must be used in the expression for the height of the tooth, as shown in Table 30.1, and in determining the outside diameter.

Design procedure: For many years the proportions of gear teeth were determined by the Lewis formula, which aims to obtain teeth sufficiently strong in regard to bending. With intermittent service, as gears were generally used in former days, the results were satisfactory. With the advent of machinery in which gear teeth transmit power continuously, such as in automobiles, the wear of the gear teeth became an important factor, and a new method of design was developed.

However, the older method is so much simpler that it seems advisable to continue its use for the preliminary design and later to check the dimensions thus obtained, by applying the newer formulas which take into consideration the service conditions and requirements.

30.5 DESIGN FOR STRENGTH

In the derivation of the Lewis equation it is assumed that at the beginning of contact the full load F is applied at the end of the tooth, with its line of action normal to the tooth profile, as shown in Fig. 30.7. In actual practice, the contact ratio is more than 1 so only a part of load is transmitted at the first point of contact. When maximum load is transmitted, moment arm reduces.

The normal force F is resolved at the end of the tooth into two components, N and F_t at P. The component N acts radially and produces pure compression, while F_t acts tangentially.

It is assumed the force F acts at P and tangential component acts at P and also the radial component N acts at P.

The dangerous section is at cd, where the stress may be determined by the equation (*The parabola shown is the outline of beam of uniform strength*), comparing the parabola with actual profile, cd is the weakest section (30.7), h is measured from P.

By omitting the compression stress produced due to N, the design is safer as the failure is assumed to be in tensile.

$$s = \frac{M}{Z} = \frac{6F_t h}{bt^2} = \frac{6h p_c F_t}{bt^2 p_c}$$

$$F_t = \frac{s b p_c t^2}{6 h p_c} \qquad \dots (30.9)$$

The factor $t^2/6hp_c$ is a geometrical property of the size and shape of the tooth. It may therefore be expressed as a function of the circular pitch p_c by the relation

$$F_t = sbp_c \cdot y \qquad \dots (30.10)$$

where, y is an abstract number known as the *Lewis form factor*.

From equations 30.9 and 30.10

$$y = \frac{t^2}{6hp_c} \qquad \dots (30.11)$$

---

[2]V. M. Faires, *Design of Machine Elements*, rev. ed. (New York: The Macmillan Company, 1942), p.199.

By substituting the product $S_o c$ for s, as was done in deriving equation 30.5, and solving for F_t, we obtain the original Lewis equation, which is

$$F_t = S_o cbyp_c \qquad \dots (30.12)$$

$$y = 0.124 - \frac{0.684}{z} \text{ for } 14\tfrac{1}{2}° \text{ pressure angle}$$

$$= 0.154 - \frac{0.912}{z} \text{ for } 20° \text{ pressure angle}$$

$$= 0.175 - \frac{0.95}{z} \text{ for } 20° \text{ stub teeth system}$$

Substituting πm for p_c, from equations 30.3 and 30.12 give the modified Lewis equation, which is

$$F_t = S_o c\, bp_c y = S_o cbmY \qquad \dots (30.13)$$

The factor Y is also called the *form factor*. The value of Y depends on the shape of the tooth profile, including the influence of the number of teeth, and may be taken from Table 30.2 for anyone of the various gear systems. Since the Lewis formula is used in designing gear teeth that are cut with standard modules, equation 30.13 is equally convenient as equation 30.12.

The product $S_o c$ bpy in the load which a gear can transmit, p, c and b will be same for pinion and gear. The product of $S_o y$ gives the capacity of the gears. To design the product $S_o y$ called the strength factor is calculated to find which is weaker of the two. The design is made for weaker gear.

The value of y given for 20° stub teeth in Table 30.2 can be used also for Nuttal stub-tooth gears.

In the derivation of Lewis equation, it is assumed that the maximum or full load is transmitted when the gears are stationary. However, the gears rotate and a factor c is

Table 30.2: Values of tooth form factor (Lewis) y

z	14.5-deg form	14.5-deg variable centre distance	20-deg full depth form	20-deg stub-tooth form	Internal gears Spur pinion	Internal gear
12	0.067	0.125	0.078	0.099	0.104	
13	0.071	0.123	0.083	0.103	0.104	
14	0.075	0.121	0.088	0.108	0.105	
15	0.078	0.120	0.092	0.111	0.105	
16	0.081	0.120	0.094	0.115	0.106	
17	0.084	0.120	0.096	0.117	0.109	
18	0.086	0.120	0.098	0.120	0.111	
19	0.088	0.119	0.100	0.123	0.114	
20	0.090	0.119	0.102	0.125	0.116	
21	0.092	0.119	0.104	0.127	0.118	
22	0.093	0.119	0.105	0.129	0.119	
24	0.095	0.118	0.107	0.132	0.122	
26	0.098	0.117	0.110	0.135	0.125	
28	0.100	0.115	0.112	0.137	0.127	0.220
30	0.101	0.114	0.114	0.139	0.129	0.216
34	0.104	0.112	0.118	0.142	0.132	0.210
38	0.106	0.110	0.122	0.145	0.135	0.205
43	0.108	0.108	0.126	0.147	0.137	0.200
50	0.110	0.110	0.130	0.151	0.139	0.195
60	0.113	0.113	0.134	0.154	0.142	0.190
75	0.115	0.115	0.138	0.158	0.144	0.185
100	0.117	0.117	0.142	0.161	0.147	0.180
150	0.119	0.119	0.146	0.165	0.149	0.175
300	0.122	0.122	0.150	0.170	0.152	0.170
Rack	0.124	0.124	0.154	0.175		

introduced, called the Barth velocity factors to take care of motion and impact.

If the diameters of gears are not known, the above equation is modified in terms of torque T as

$$cS_o m^3 = \frac{2T}{k\,\pi^2 zy} \qquad \text{...(30.13a)}$$

where, $k = \dfrac{b}{p_c} = \dfrac{b}{\pi\,m}$ has value from 2 to 5 and

usually 3 to 4 is used and $p_c = \pi m$.

$$F_t = \frac{2T}{D} = \frac{2T}{zm}$$

Gear face: There is no strict relation between the gear face b and the pitch. However, if the face is very long compared with the pitch, or the thickness of the tooth, it is likely that the pressure will not be uniformly distributed along the face. Since this is contrary to the assumption made in deriving the Lewis formula, the maximum stress probably will be greater than is assumed. Conversely, a very narrow face will require a coarse pitch and will give a less smooth action and poor wearing qualities. Practice indicates that the proper width of face is between $3p_c$ and $4\,p_c$, or in terms of the module pitch,

$$b > 9.6\,m \qquad b < 12.5\,m \qquad \text{...(30.14)}$$

However, under certain conditions, as where there are space limitations the width b may be made as narrow as 6.3 m or as wide as 19 m.

Design stress: The stress S_o in the Lewis formula should be taken, as for cast teeth, equal to the elastic limit of the material in bending divided by a factor of safety $n = 1.5$ and can be called static stress or $n = 3$ with ultimate stress.

The velocity factor for ordinary cut gears running with a pitch-line velocity up to 10 m/s may be computed by Barth's formula (equation 30.6).

For carefully cut gears with a pitch velocity up to 13 m/s. Barth's formula may be modified to the relation

$$c = \frac{4.5}{4.5 + v_m} \qquad \text{...(30.15)}$$

For very accurately cut and ground metallic gears having a pitch velocity v_m from 10 to 20 m/s. Barth's formula may be modified by increasing the constant. Thus

$$c = \frac{6}{6 + v_m} \qquad \text{...(30.16)}$$

For hardened-steel ground and lapped-in precision gears made for speeds over 20 m/s the American Gear Manufacturers Association recommends the use of velocity factor given by the relation

$$c = \frac{5.6}{5.6 + \sqrt{v_m}} \qquad \text{...(30.17)}$$

For convenience, values of safe static stress $S_o = S_e/n$ for materials commonly used for gears are given in Table 30.3. Alternately $n = 2.5$

Table 30.3: Allowable static stresses for use in Lewis formula

Material	$S_o\,N/mm^2$	Material	$So\,N/mm^2$
Ordinary cast iron	55	Forged steel, about SAE 1030, 30C8	170
Cast iron, about class No. 35	84	Steel, SAE 1030, 30C8 heat-treated	220
High-grade cast iron, about		Steel SAE 1040, 40C8 untreated	210
Class No. 50	104	Alloy steel, case hardened	345
Cast steel, 0.20% C, untreated	140	Cr-Ni steel, about SAE 3245,[3] 90 Ni$_2$ Cr 60	
Cast steel, 0.20% C, heat-treated	180	heat-treated	460
Bronze, SAE 62	70	Cr-Va steel, about SAE 6145 heat-treated	520
Phosphor gear bronze, SAE 65	84		
Manganese bronze SAE 43	140	Rawhide, Fabroil, etc.	40
Aluminium bronze, SAE 68	150	Laminated phenolic materials (Bakelite, Micarta, Celoron)	

[3]J. W. Sands and F J. Walls, "Nickel-Alloy Gear Materials and Their Heat-Treatments," *Product Engineering*, Vol. 6 (1935), p. 370; *Gear Materials and Blanks*, ASA B6.2–1933 (New York: American Standards Association, 1933).

to 3 corresponding to the ultimatic tensile stress of materials.

Noiseless, or silent, gears: In order to reduce the noise of the meshing teeth, especially at high pitch-line velocities, spur gears are made of nonmetallic materials such as rawhide, Fabroil, Bakelite, Textolite, or Celoron. Rawhide and Fabroil, which is cotton that is treated with oil under a high hydraulic pressure, are not self-supporting, and gears of these materials must be made with metal flanges, or shrouds, at both ends for teeth support. Bakelite, Celoron, Formica, Micarta, Textolite, and other materials which are made of laminated fibrous material impregnated with synthetic resin of the phenolic type, do not require a support for the teeth. The AGMA recommends for non-metallic teeth, the use of a velocity factor c found by the relation

$$c = \frac{0.7625}{1.0167 + v_m} \quad \ldots (30.18)$$

In cutting the teeth in phenolic materials, it should be remembered that their coefficient of expansion from heat is slightly more than twice that of cast iron, and they therefore require a greater backlash. With all non-metallic teeth, pitch-line speeds up to 15 m/s can be used; and with Celoron teeth, speeds up to 20 m/s are permissible.

The tooth load F_t is determined from the general equation

$$F_t = \frac{1000 PCs}{v_m} \quad \ldots (30.19)$$

where, C_s is a service factor, the value of which may be taken from Table 30.4. P is power transmitted.

30.6 DYNAMIC LOAD

Even with the best machine tools and most careful workmanship, gears will have inaccuracies in the teeth. These inaccuracies cause short period of acceleration and deceleration, which are increased by the elastic deflections of the teeth. Also, even with exactly-spaced teeth there is always a sudden load application when two teeth come into contact. Thus, while the power transmitted, may be constant, the load on the teeth varies.

A newer method of gear design considers that the maximum dynamic load F_d on the gear tooth consists of the transmitted, or useful, load F_t and an increment load F_i caused *by inaccuracies of the teeth, errors in spacing, tooth deflection, unbalance, or power-flow fluctuation*[4] *(30.15)*. Thus

$$F_d = F_t + F_i \quad \ldots (30.20)$$

The increment load F_i depends on the masses of the moving parts. For average conditions it may be found by the equation:

$$F_i = \frac{21v(bC + F_t)}{21v + \sqrt{bC + F_t}} \quad \ldots (30.21)$$

where values of the coefficient C are given in Table 30.5 and the probable error e in the tooth profile may be taken from Fig. 30.9.

For materials not shown in Table 30.5 the value C may be calculated from the relation

$$C = \frac{ae}{\dfrac{1}{E_p} + \dfrac{1}{E_g}} \quad \ldots (30.22)$$

Table 30.4: Service factor C_s for gears in equation 30.19

Type of load	Type of service		
	Intermittent, or 3 hr per day	8 to 10 hr per day	Continuous, 24 hr per day
Steady	0.80	1.00	1.25
Light shocks	1.00	1.25	1.50
Medium shocks	1.25	1.50	1.80
Heavy shocks	1.50	1.80	2.00

[4]*Dynamic Loads on Gear Teeth*, Report of the ASME Special Research Committee on the Strength of Gear Teeth (New York: American Society of Mechanical Engineers, 1931); E. Buckingham, *Manual of Gear Design* (New York: The Industrial Press, 1953), Section 2, pp. 141 ff.

where E_p and E_g are the moduli of elasticity of the pinion material and the gear material, respectively, and a may be taken as 0.107 for the $14\dfrac{1°}{2}$ tooth form, 0.111 for the 20° full-depth form, and 0.115 for 20° stub teeth.

Nonaverage conditions, such as gears used in aeronautical works where the rotating masses are less than average, or gears connected with heavy flywheels where the masses are greater than average, require special rather involved calculations.[5]

Silent gears, being made of materials whose resilience and flexibility are very high and whose specific weight is low, suffer from shock action much less than metal gears. Therefore their maximum load may be considered to be equal to the useful load divided by the velocity factor c, from equation 30.18 or

$$F_d = \frac{F_t}{c} \qquad \ldots (30.23)$$

Endurance strength: In order to determine the degree of safety, the maximum load may be compared with the strength of the tooth found by applying the Lewis formula and using the endurance limit S_{en} instead of the permissible stress $S_o c$. This strength is

$$F_{en} = S_{en}\, b\, \pi\, y\, m \qquad \ldots (30.24)$$

The permissible load computed by equation 30.24 is termed the *endurance strength* of the gear teeth. The use of the endurance limit S_{en} takes care of stress concentration at the base of the teeth. This method gives good result in practice. Values of S_{en} to be used in equation 30.24 are given in Table 30.6.

Safety margin: The endurance load F_{en} found by equation 30.24 must be greater than the maximum load F_d determined by equation 30.20 or equation 30.23, as the case may be. The difference $(F_{en} - F_d)$ may be called the *safety margin*. This safety margin should be about

Table 30.5: Values of deformation factor C in kN/m

| Material | | Involute | Tooth error—mm | | | | |
Pinion	Gear	tooth form	0.01	0.02	0.04	0.06	0.08
Cast iron	Cast iron	14.5°	55	110	220	330	440
Steel	Cast iron	14.5°	76	152	304	456	608
Steel	Steel	14.5°	110	220	440	660	880
Cast iron	Cast iron	20° full depth	57	114	228	342	456
Steel	Cast iron	20° full	79	158	316	474	632
Steel	Steel	depth	114	229	457	687	916
Cast iron	Cast iron	20° stub	59	118	237	356	475
Steel	Cast iron	20° stub	81	162	325	488	648
Steel	Steel	20° stub	119	238	476	714	952

Table 30.6: Endurance limits for checking gear teeth

Material	Core (bhn)	S_{en} N/mm^2	Material	Core (bhn)	S_{en} N/mm^2
Gray cast iron	160	84	Steel	200	345
Semisteel	200	124	Steel, normalized	240	415
Manganese bronze, SAE 43	100	118	Steel, SAE 3140, heat-treated	280	480
Gear bronze, SAE 65	100	165	Steel, SAE 3240, heat-treated	320	550
Nonmetallic		40	Steel, oil-tempered	360	620
Steel	150	260	Steel, nitralloy	400	690

SAE 3140 – 35 Ni Cr 60; SAE 3240 = 40 Ni 2 C$_r$1.

---

[5]*Ibid.*

25 per cent of the actual load F_d for a steady transmitted load, about 35 per cent for a pulsating load, and 50 to 60 per cent for shock loads.[6]

i.e. $F_{en} \geq 1.25\, F_d$ for steady load

$F_{en} \geq 1.35\, F_d$ for pulsating load

$F_{en} \geq 1.5\, F_d$ for shock load

Permissible errors: A very useful curve is given in Fig. 30.10. It shows the maximum error in the gear-tooth profile which a gear can carry for satisfactory operation at a given pitch-line speed. In connection with Fig. 30.9 the curve shows at a glance the class of workmanship required and the minimum value of m. Thus for a pitch-line speed of 15 m/s the permissible error is $e = 0.025$ mm, and it is necessary to use either carefully cut gears with $m = 5$ or lower, or precision gears with $m = 12$ or lower. These values are also given in Table 30.5.

30.7 CHECK FOR WEAR

Gears in continuous service lose their usefulness because of excessive wear, not because of a sudden failure. Wear occurs in five ways:

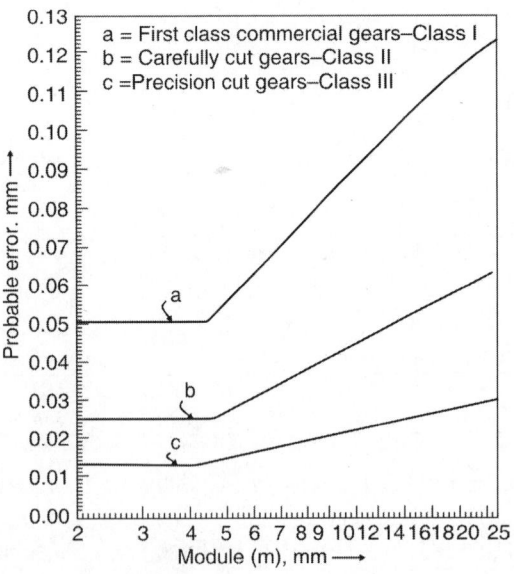

Fig. 30.9: Probable errors in tooth profiles and spacing

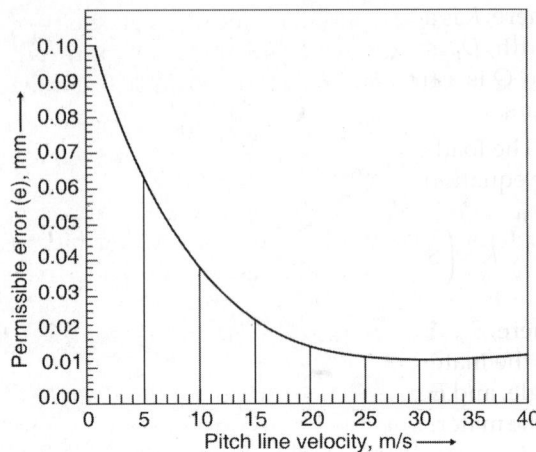

Fig. 30.10: Permissible errors in tooth profiles for quiet operation of gears

(a) by pitting of the tooth surface by repeated compressive stressing; (b) by abrasion caused by foreign matter; (c) by scoring caused by sharp and projecting edges and rough surfaces: (d) by scuffing, which results from the use of an improper lubricant; and (e) by seizing, due to a complete failure of the lubrication, accompanied by a locally generated heat sufficient to weld the surfaces one to another.

Tests have shown that in the order of increasing resistance to wear, the $14\frac{1}{2}°$ full-depth involute tooth comes first, the 20° involute stub tooth comes next, while the best is the 20° full-depth involute tooth.[7]

Pitting: The most serious cause of wear is pitting. It can be prevented only by proper design, while the other causes of wear can be overcome during operation of the gears. In order to prevent pitting, the compression endurance limit of the material must not be exceeded.

Pitting usually increases rather slowly. The *limiting load for wear*, F_w, is the load beyond which wear is likely to be rapid. It is found by the equation

$$F_w = Kb D_p Q \qquad \qquad ...(30.25)$$

---

[6]*Ibid.*

[7] G. H. Marx, L. E. Cutter, and B. M. Green, "Some Comparative Water Experiments on Cast-Iron Gear Teeth," *Mechanical Engineering,* Vol. 48 (1926), p. 35.

where, K is a load-stress factor, b is the gear face width, D_p is the pitch diameter of the pinion, and Q is a function of the relative size of the gears.

The load-stress factor may be calculated by the equation

$$K = \left(S^2_{en,c} \times \frac{\sin\beta}{1.4} \right)\left(\frac{1}{E_p} + \frac{1}{E_g} \right) \quad \dots (30.26)$$

where, $S_{en,c}$ is the compressive endurance limit for the material of the pinion, β is the pressure angle, and E_p and E_g are the moduli of elasticity of the materials of pinion and gear respectively.

The value of K makes allowance for the cold working received by the more plastic material for the harder mating material. For a number of more commonly used materials the values of K computed by equation 30.26 are given in Table 30.7. Table 30.7 shows that cast iron is a very good material as far as wear resistance is concerned.

In addition it must be stated that Table 30.7 confirms the general rule that less wear is incurred if the two surfaces in contact are of different materials. Exceptions to this rule are cast iron, which wears well on cast iron, and heat-treated steel, tempered to a very hard

Table 30.7: Values of K for computing the limiting wear load

Gear		Pinion			Value of K MN/m^2	
Material	Hardness (Bhn)	Material	Hardness (Bhn)	Endurance limit S_{es} MN/m^2	$14\frac{1}{2}°$ involute	20° involute
Steel	150	Steel	150	340	0.205	2.284
Steel	150	Steel	200	410	0.294	0.402
Steel	150	Steel	250	480	0.402	0.539
Steel	200	Steel	200	480	0.402	0.539
Steel	200	Steel	250	550	0.520	0.706
Steel	200	Steel	300	620	0.657	0.902
Steel	250	Steel	250	620	0.657	0.902
Steel	250	Steel	300	690	0.814	1.108
Steel	250	Steel	350	755	0.990	1.344
Steel	300	Steel	300	755	0.990	1.344
Steel	300	Steel	350	825	1.177	1.599
Steel	300	Steel	400	860	1.275	1.746
Steel	350	Steel	350	895	1.386	1.893
Steel	350	Steel	400	960	1.598	2.189
Steel	350	Steel	450	995	1.716	2.344
Steel	450	Steel	450	1167	2.363	3.226
Steel	450	Steel	500	1200	2.500	3.413
Steel	450	Steel	600	1236	2.648	3.609
Steel	500	Steel	500	1305	2.952	4.040
Steel	600	Steel	600	1570	4.325	5.913
Cast iron	180	Steel	150	345	0.304	0.412
Cast iron	180	Steel	200	393	0.598	0.814
Cast iron	180	Steel	250	620	0.990	1.344
Gear bronze	100	Steel	150	345	0.314	0.422
Gear bronze	100	Steel	200	345	0.628	0.853
Gear bronze	100	Steel	250	585	0.932	1.265
Cast iron	180	Cast iron	180	620	1.324	1.814
Metal	–	Nonmetallic	(34)	220	1.296	1.772

surface, against equally heat-treated hard steel.

The value of Q is found by the equation

$$Q = \frac{2z_g}{z_p + z_g} = \frac{2D_g}{z_p + D_g} \qquad \ldots (30.27)$$

where, z_p and z_g are the numbers of teeth and D_p and D_g are the pitch diameters of the pinion and gear, respectively.

If the gears are intended for continuous operation, the limit load for wear F_w must be greater than the maximum load F_d, i.e. $F_w > F_d$.

Scoring: Wear due to scoring progresses very rapidly if the operating conditions which cause it are not changed. Tests have shown that scoring starts when the tooth load reaches a certain value.[8] This limit value can be increased up to 10 to 15 per cent by using an oil with a higher viscosity.

30.8 DESIGN PROCEDURE

In the preliminary design by the Lewis equation, the following steps should be taken:

1. The peripheral velocity v_m should be selected. The value of v_m should be low—3 to 5 m/s—for a low rotary speed of the pinion and a low-to-moderate power transmitted. The velocity v_m should be increased for a higher pinion speed and greater power.

2. The materials of the pinion and gear and their static strengths S_o are selected for Table 30.3. Low-strength steel and cast iron are preferred for a low velocity v_m, and better-grade materials should be used for higher velocities and greater power.

3. The relative width of the face b is tentatively selected in terms of k or m in accordance with equation 30.14.

4. The tooth load and the velocity factor c are computed, the type of tooth profile is decided upon, and a preliminary value for the form factor y of the pinion is assumed. It may be taken as 0.08 for

a $14\frac{1}{2}°$ tooth form, 0.094 for a 20° full-depth form, and 0.115 of a 20° stub tooth. Strength factor $y \times S_o$ is calculated for both pinion and gear and the one which has lower strength factor is weaker of the two and is designed.

5. Equation 30.13 is solved for m, and the nearest standard value is taken.

6. The approximate pinion diamter D_p is determined and check for induced stress to be lower than allowable stress is made.

7. The width b of the face is computed for both the pinion and the gear from equation 30.13, in which the corrected values for y and c are used and the greater value is taken.

8. The dimensions of the pinion and gear are checked for dynamic load, endurance strength and wear.

When the diameters are not known, the following stepwise procedure may be followed. This also gives procedure for computer programming. Given stresses S_o.

1. Assume number of teeth for pinion and gear considering the velocity ratio.

2. Find the strength factor $S_o y$ for pinion and gear for a given or assumed pressure angle say 20°.

3. The minimum value of the product $S_o y$ gives the indication of the weaker element: pinion or gear. For the same material of pinion and gear, the pinion will be weak. Design the weaker one.

4. Evaluate $m^3 S$ using equation 30.13a, i.e.

$$m^3 S_o = \frac{2T}{k\pi^2 y z} = \text{say A.}$$

Take $c = 0.5$ first trial value

Then $S_a = 0.5\, S_o$ and find

$$m = \left(\frac{A}{S_a}\right)^{\frac{1}{3}}$$

---

[8] V.N. Borsoff, J. B. Accinell, and A. G. Cattaneo, "The Effect of Oil Viscosity on the Power Transmitting Capacity of Spur Gears," *Transactions of the American Society of Mechanical Engineers*, Vol. 73 (1951), p. 687.

Take the nearest standard value of m and start iterating as in step 6.

5. For computer programing, start iterating with module 1 onwards.

6. For the first value of module, calculate:

   (a) Diameter $D = mz$.

   (b) Velocity $v_m = \dfrac{\pi D n}{1000 \times 60}$ m/s

   (c) $c = \dfrac{3}{3 + v_m}$

   (d) Allowable stress $S_a = cS_o$

   (e) Induced stress $S_i = \dfrac{A}{m^3}$

   (f) Compare $S_i \le S_a$

   If not repeat with a new higher value of module.

7. The final value of module is established and facewidth b calculated.

8. Check for dynamic load and wear load.

9. Calculate all relevent dimensions for pinion and gear.

If the pitch diameter is specified, the following form of the Lewis equation may be used or (13.13a).

In step 4 for the weaker gear

$$m^2 = \frac{F_t}{k\pi^2 y S_a}$$

Where, $F_t = 2T/D$. As diameter is known, v is calculated and also c. Then $S_a = cS_o$ and module found. Find induced stress S_i and $S_i \le S_a$ if not, increases m.

In the above two equations, the smallest possible module will provide for the most economical design. In general, the design should be for the largest number of teeth possible when the diameters are known and for the smallest pitch diameters possible when the diameters are not known.

Example 30.1: Determine the proper pitch, face-width number of teeth and outside diameter of a pair of 14.5° spur gear to transmit 120 kW from a pinion running at 750 rpm to a gear running at 140 rpm. The service is intermittent with light shocks.

Since no limitations are given for the largest diameter of maximum centre distance, the way to proceed is to select the materials of the pinion and gear, make some assumptions, calculate the drive and if the results are not satisfactory, change the assumptions.

A suitable material for the pinion is forged steel 30C8 (SAE 1030). From Table 30.3, $S_o = 170$ N/mm^2.

It is next necessary to select a suitable pitch-line velocity. For gears operating at moderate speeds, a pitch velocity of 7.5 m/s seems to be suitable. The approximate pitch diameter is then

$$D_p = \frac{v_m}{\pi n_p} = \frac{7.5 \times 1000 \times 60}{\pi \times 750}$$

$$= 190.98 \text{ mm or } 191 \text{ mm}$$

If C_s is taken as 1 from Table 30.4, the tangential tooth load is

$$F_t = \frac{1000 \times 120}{7.5} = 16000 \text{ N}$$

The velocity factor for the allowable stress, by equation 30.6 is

$$c = \frac{3}{3 + 7.5} = 0.286$$

Then the allowable stress is

$$S = 170 \times 0.286 = 49.0 \text{ N/mm}^2$$

The relative gear face may be assumed as $b = 9.6$, m and y may be taken as 0.08 for 14½°pressure angle, using equation 30.13, module is

$$m = \sqrt{\frac{16,000}{49.0 \times 9.6 \times 0.08 \times \pi}} = 11.6$$

The nearest module is 12. With a 191 mm pitch diameter there should be 16 teeth or diameter will be modified to 12 × 16 = 192 mm. From Table 30.2, $y = 0.081$.

The facewidth found from equation 30.13 by using pitch line velocity, the load F_t, the factor c and the exact value for y.

This procedure gives

$$b = \frac{16000}{0.285 \times 170 \times 0.081 \times \pi \times 12} = 108 \text{ mm}$$

In accordance with equation 30.14, $b = 9.6 m = 115$ mm should be used.

The number of teeth required in the gear will be

$$Z_g = \frac{16 \times 750}{140} = 86$$

It is next necessary to find the gear facewidth. Here taking high grade *CI* with $S_o = 104$ N/mm^2 and for 86 teeth $y = 0.116$. Then, from equation 30.13,

$$b = \frac{16000}{0.285 \times 104 \times 0.116 \times \pi \times 12} = 123.4 \text{ mm}$$

A facewidth $b = 124$ mm on both the gear and pinion is permissible according to equation 30.14.

The outside diameters, can be determined by applying a relation similar to equation 30.8 and using data from Table 30.1.

Thus,

$$D'_p = (Z_p + 2)m = (16 + 2)12 = 216 \text{ mm}$$

$$D'_p = (Z_g + 2)m = (86 + 2)12 = 1056 \text{ mm}$$

Instead of finding facewidth for both gears, the weaker is designed as explained in Examples 30.3, 30.4 and 30.5.

Example 30.2: Find the safety margin and suitability for continuous operation of the pair of gears computed in example 30.1.

In order to determine the dynamic tooth load by equation 30.20, the increment load F_i must be found first.

For the probable error e from Fig. 30.9, is 0.5 the value of C is computed from Table 30.5, is 363

$$Cb + F_t = 363 \times 124 + 16000 = 61012 \text{ N}$$

Then by equation 30.21, the increment load is,

$$F_i = \frac{21 \times 7.5 \times 61012}{21 \times 7.5 + \sqrt{61012}} = 23756.2 \text{ N}$$

The total dynamic load by equation 30.20 is

$$F_d = 16000 + 23756.2 = 39756.2 \text{ N}$$

For steel with a Bhn of 150, S_{en} from Table 30.6 is 260 N

By equation 30.24 in which $y = 0.081$, the endurance strength.

$$F_{en} = 260 \times 124 \times 0.081 \times \pi \times 12 = 98449 \text{ N}$$

The safety margin $= 98449 - 39756.2 = 58693$ N or 147 per cent more, o.k.

In order to check the limit load for wear by equation 30.25, the value of K from Table 30.7 is found to be 0.402, the function Q is

$$Q = \frac{2 \times 86}{16 + 86} = 1.686$$

Therefore

$$F_w = 0.402 \times 124 \times 192 \times 1.686 = 16136.4 \text{ N}$$

Since F_w is less than F_d, the pinion is not suitable for continuous work. If the pinion is made of heat-treated steel for which Bhn is 300, Table 30.7 shows that $k = 1.108$. The $F_w = 44475.4$ and is greater than $F_d = 39756.2$.

Therefore this design is satisfactory.

Example 30.3: A spur gear reducer is to transmit 4 kW at 1500 rpm of the pinion of cast steel ($S_o = 138$ MN/m^2). The gear is made from phosphor bronze ($S_o = 83$ MM/m^2). The velocity ratio is 3.5 to 1.0. Assume a preseure angle of 20° Design the gears for strength only.

It is necessary to select number of teeth for pinion say $z_p = 16$. Then for a velocity ratio of 3.5, the number of teeth for gear $= 16 \times 3.5 = 56$.

Strength factor for gear $S_o y = 83 \times 0.1324 = 10.98$

For pinion $S_o y = 138 \times 0.094 = 12.97$.

The gear is weaker and gear is designed:

Torque

$$T = \frac{4000 \times 60 \times 3.5}{2\pi \times 1500} = 89.13 \text{ Nm}$$

As diameters are not known the equation (30.13a) is to be used, with $k = 3.5$

$$Sm^3 = \frac{2 \times 89.13 \times 1000}{3.5 \times \pi^2 \times 0.1324 \times 56} = 695.99$$

Assuming as a first trial, the allowable stress ($c = 0.5$), $S_a = 0.5 \times 83 = 41.5$ N/mm^2 to find an approximate module

$$m = \left[\frac{695.99}{41.5}\right]^{1/3} = 2.559$$

Taking standard module

$m_t = 2.5$, $D_1 = 2.5 \times 56 = 140$ mm,

$$v_1 = \frac{\pi \times 140 \times 1500}{1000 \times 60 \times 3.5} = 3.142 \text{ m/s}$$

$$c = \frac{3}{3 + 3.142} = 0.488, \text{ Allowable Stress}$$

$$S_a = 0.488 \times 83 = 40.5 \text{ N/mm}^2$$

and induced stress

$$S_i = \frac{695.99}{(2.5)^3} = 45.59 \text{ N/mm}^2$$

As induced stress is more than the allowable stress, another higher module is tried.

$m_2 = 3$, $D_2 = 56 \times 3 = 168$ mm,

$$v_2 = \frac{\pi \times 168 \times 1500}{1000 \times 3.5 \times 60} = 3.77 \text{ m/s}$$

$$c_2 = \frac{3}{3 + 3.77} = 0.443,$$

$$S_a = 0.443 \times 83 = 36.77 \text{ N/mm}^2$$

$$S_i = \frac{695.99}{3^3} = 25.77 \text{ N/mm}^2$$

Here the allowable stress is greater than induced stress but the difference is very large. So module is taken as 2.5 and then the facewidth will be modified as follows. With values of S_a and S_i with $m = 2.5$

$$b = k \pi m = 3.5 \times \pi \times 2.5 = 27.49$$

Actual width

$$= 23.56 \times \frac{45.59}{40.5} = 30.49 \text{ say } 31 \text{ mm}$$

$$D_p = 2.5 \times 16 = 40 \text{ mm};$$

$$D_g = 56 \times 2.5 = 140 \text{ mm}$$

Note: Change of facewidth has increased the value of $k = \frac{31}{\pi \times 2.5} = 3.96$ which is within the acceptable limits while increase of module to 3 will have made the design more costly as the cost is proportional to D^3 which is proportional to m^3.

Example 30.4: Check the design of example 30.2 for continuous operation of gears. The gears are to be checked for dynamic and wear loads with endurance load.

S_{en} for Table 30.6 is 165 N/mm^2.

Endurance load $F_{en} = S_e\, b\, y\, \pi\, m = 165.0 \times 31 \times 0.1324 \times \pi \times 2.5 = 5319$ N

Value of C from Fig. 30.9 for an error of 0.071, is 550.51 N/m.

$$F_t = \frac{89.13 \times 1000}{70} = 1273.3 \text{ N}$$

$$F_i = \frac{21 \times 3.142 \times (550.5 \times 31 + 1273.3)}{21 \times 3.142 + \sqrt{550.5 \times 31 + 1273.3}} = 6008 \text{ N}$$

Total dynamic load = 6008 + 1273.3 = 7281 N

This is higher than F_{en}, so gear material is not suitable.

The gear material should have a

$$S_e = \frac{7281}{31 \times 0.132 \times \pi \times 2.5} = 226.6$$

This corresponds to steel with Bhn of 250.

Now checking for wear load, value of K from Table 30.7 is 0.402.

$$Q = \frac{2 \times 56}{56 \times 16} = 1.56$$

$$F_w = 0.402 \times 31 \times 1.56 \times 140 = 2722 \text{ N}$$

which is less than the dynamic load so another material with

$$K = \frac{7281}{31 \times 1.56 \times 140} = 1.076$$

From Table 30.7, the materials for gear and pinion have to be steels with Bhn of 250 and 300 respectively.

Example 30.5: A 20° stub, 100 mm diameter spur pinion transmits 15 kw at 960 rpm to a gear which is to run at 400 rpm. The static stress of pinion and gear are 160N/mm^2 and 95 N/mm^2 respectively. Design for maximum number of teeth and find module and face with of gears.

Tentatively assuming number of teeth $Z_p = 20$ and $Z_g = 48$

For pinion $y = 0.175 - \dfrac{0.95}{20} = 0.1275$ and strength factor $= S_o y = 160 \times 0.1275 = 20.4$

For gear $y = 0.175 - \dfrac{0.95}{48} = 0.1552$ and strength factor $= S_o y = 95 \times 0.1552 = 14.74$ Nm

Gear is weaker and gear is designed. $D_g = 100 \times 2.4 = 240$ mm, $k = 3.5$

$$T = \frac{15 \times 1000 \times 60}{2\pi \times 400} = 358.09 \text{ Nm}$$

$$Sm^3 = \frac{2 \times 358.09 \times 1000}{3.5 \times \pi^2 \times 48 \times 0.1552} = 2783.04$$

$$v = \frac{\pi \times 240 \times 400}{60 \times 1000} = 5.02 \text{ m/s}$$

$$c = \frac{3}{3 + 5.02} = 0.374$$

$$S_a = 0.374 \times 95 = 35.53 \text{ N/mm}^2$$

$$m = \left[\frac{2784.04}{35.53} \right]^{1/3} = 4.27. \text{ Take } m = 5$$

$$S_i = \frac{2783.04}{5^3} = 26.68 \text{ N/mm}$$

It is quite less so a smaller module 4 is tried

$$S_i = \frac{2783.04}{4^3} = 43.48$$

As $m = 4$ will be more economical, the value of k is modified as $k = \dfrac{3.5 \times 43.48}{35.53} = 4.28$

This is within limit so $b = 4.28 \times \pi \times 4 = 53.78$ say 55 mm. $z_p = \dfrac{100}{4} = 25$, $z_g = 60$. With increased teeth stress will reduce.

30.9 AGMA FORMULAS

The American Gear Manufacturers Association has issued tentative standards for rating various toothed gearings. The rating for the beam strength of spur-gear teeth is a modified Lewis formula with several refinements. Factors for these refinements must be taken from numerical tables and graphs. The rating for wear is based on a formula similar to equation 30.25 but with numerous refinements.

For a more accurate final design, the AGMA procedure should be used.[9] However, the much simpler procedure presented in sections 30.5 to 30.8 gives results sufficiently safe and has the advantage of bringing out more clearly the interrelation of the various factors.

30.10 GEAR CONSTRUCTION

The dimension of the parts of a gear are determined largely by empirical rules.

Rim: The thickness h of the rim of a gear, Fig. 30.11, is made according to empirical rules from $0.5\,p_c$ to $0.65\,p_c$. The rim should taper at the rate of about 1 in 12 toward the center, to secure draft in molding or forging.

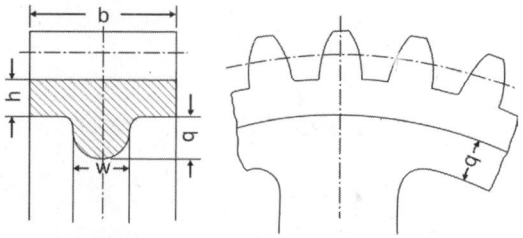

Fig. 30.11: Gear-rim dimensions

The Westinghouse-Nuttal formula is

$$h = m\sqrt{\frac{z}{2j}} \qquad \ldots (30.28)$$

where, z is the number of teeth and j is the number of arms. This formula gives good results for larger gears; if it gives $h < 0.5\,p_c$, make $h = 0.5\,p_c$ to $0.65\,p_c$

When arms are used, the stiffness of the rim should be increased by a rib having a depth $q = h$ and a width w equal to the thickness of the arms at this end.

Hub: Recommendations for the diameter and length of the hub of a gear are given in Table 30.8. However, the hub length should never be less than the gear face b. A small gear may be fastened to the shaft by a square key and a setscrew over it. A large gear requires a taper key which must be fitted very accurately to avoid excessive hammering that may distort the gear and result in noisy running and rapid wear.

Table 30.8: Dimensions of gear hubs

Type of service	Diameter		Length
	Cast iron	Steel casting	
Light load, no shock	1.75d	1.6d	$l \geq 1.5d$
Medium load and shock	1.85d	1.7d	$l \geq 1.75d$
Heavy load with shock	2d	1.8d	$l \geq 2d$

The minimum thickness t of metal permissible above the keyway of a pinion may be determined by the empirical relation

$$t = \sqrt{5\,m z_p} \qquad \ldots (30.29)$$

Solid web: In a pinion the pitch diameter is often so small that there is no space left between the rim and the hub. In such a case the pinion is made solid and of uniform thickness equal to b.

The diameter, for a solid pinion is given by the empirical equation

---

[9] A comprehensive section on gear design, including AGMA recommended procedures and a good list of references, will be found in R. T. Kent, *Mechanical Engineers' Handbook*, 12th ed., Vol. II, *Design and Production*, ed. by Colin Carmichael (New York: John Wiley & Sons, Inc., 1950)

$$D \leq 1.6 \text{ bore } (d) \qquad \dots (30.30)$$

Arms: The number of arms is usually four for gears with diameters up to 0.4 m; six for diameters from 0.4 to 3 m, in both the solid and split constructions; and eight for all larger diameters.

The usual cross section of the arms is an ellipse, as in Fig. 30.12a, with the major axis twice the minor one. Cross-shaped sections, as in Fig. 30.12b, and I-shaped sections, as in Fig. 30.12c, are also frequently used; and the H-shaped section, as in Fig. 30.12d, is used in very large gears.

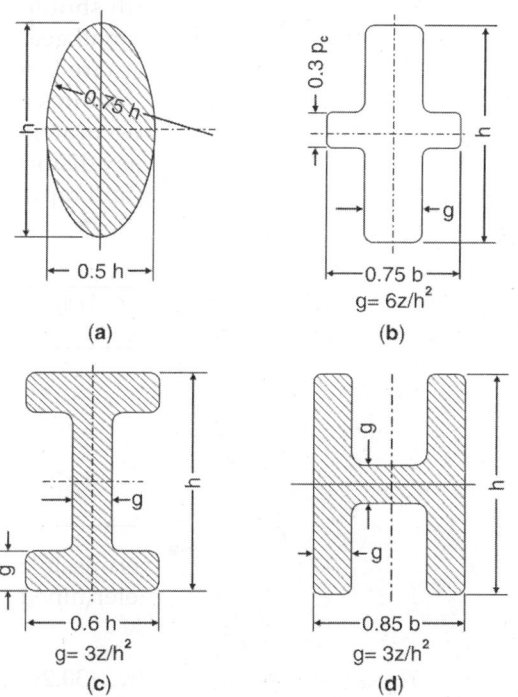

Fig. 30.12: Proportions of gear arms

In deducing a formula for calculating the dimensions of a gear arm, it is customary to assume that the rim has sufficient rigidity to distribute the load on the teeth equally among the arms, which are acting as cantilevers. In this case the required section modulus of the arm at the center of the hub may be determined by the relation

$$Z = \frac{F'D}{2jS_d} \qquad \dots (30.31)$$

where, F' is the stalling load on the teeth at zero velocity, j is the number of arms, and S_d is the allowable stress. If the arm has the cross section shown in Fig. 30.12a, the arm width h at the hub may then be found from the formula for the section modulus of an elliptic section with the minor axis one-half the major axis. Thus

$$h = \left[\frac{64Z}{\pi}\right]^{1/3} \qquad \dots (30.32)$$

The dimension h of any of the other cross sections of Fig. 30.12 is determined from the section modulus by using the proportions and the equation shown for the corresponding section. Care must be taken to obtain the necessary value of Z. The older practice of assuming that only one-half or one-third of all arms makes the load is unnecessarily conservative. The arms are tapered 1 in 15, the smaller end being toward the rim.

Nonmetallic-tooth pinions: Materials that are not self-supporting, such as rawhide or Fabroil, are fastened to steel flanges by rivets, as in Fig. 30.13a, or by threaded rods, as in Fig. 30.13b. The construction of Fig. 30.13 gives better support for the soft-materials teeth but requires a wider face, to make sure that the teeth of the gear will not come in contact with the steel plates. Laminated phenolic materials are used without metal reinforcement, but a bushing is provided in the center to take the key. In larger diameters the phenolic material is molded over a cast-iron or steel center, as in Fig. 30.13c, to reduce the cost.

Welded gears: Large gear blanks are made by welding. Figure 30.14 shows a spur gear whose rim is a steel plate bent into a ring and

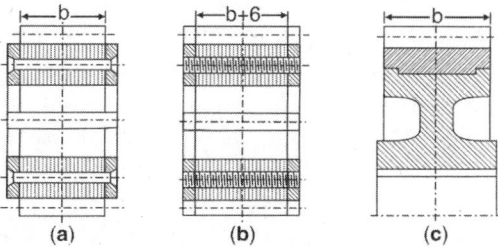

Fig. 30.13: Nonmetallic-tooth pinions and gears

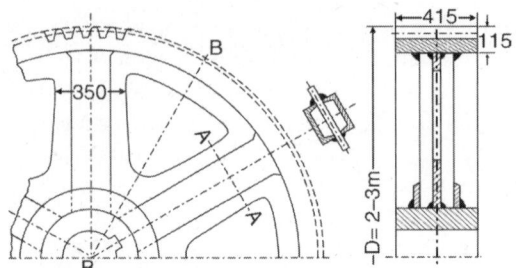

Fig. 30.14: Welded gear construction

welded, the weld coming at a tooth space. The arms are U-shaped channels welded to a flat plate and reinforced at the hub by welded-on flat rings, and the hub is a forging. A wide-face herringbone gear with box-shaped arms is shown in Fig. 30.15. The construction may be varied to suit conditions. For a given strength, a welded construction is much lighter than a cast-iron or cast-steel blank.

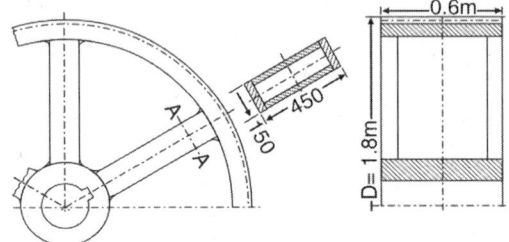

Fig. 30.15: Welded gear construction with wide face

Split gears: Large gears were formerly cast in several pieces—two to six and more—which were bolted together or held by shrink anchors. With the introduction of welding, this type of construction, being much heavier, more expensive, and difficult to manufacture, is now hardly ever employed.

30.11 OTHER FACTORS

There are a few other factors that should be considered in the design of gears.

Gear trains: When the velocity ratio is very high, it becomes desirable to reduce the speed in two or more steps. As an illustration, a triple reduction is shown in Fig. 30.16. Ordinarily a speed ratio of about 6 for straight tooth

spur gears and a ratio of 12 for helical and herringbone gears are considered practical limits for a single reduction. The velocity ratio for a train of gears is equal to the product of the ratios of all pairs. Thus, for the train in Fig. 30.16 the ratio is

$$r = \left(\frac{n_1}{n_2}\right)\left(\frac{n_3}{n_4}\right)\left(\frac{n_5}{n_6}\right)$$

$$= \left(\frac{D_2}{D_1}\right)\left(\frac{D_4}{D_3}\right)\left(\frac{D_6}{D_5}\right)$$

...(30.33)

As the speed is reduced, the useful load on the tooth is increased. Therefore, coarser teeth (i.e. a higher module) must be used. Because of the cumulative effect of lost motion, the shock action in the consecutive pairs of gears is increased, and a gradual decrease of the allowable stress, by about 10 to 15 per cent in each lower-speed step, is advisable.

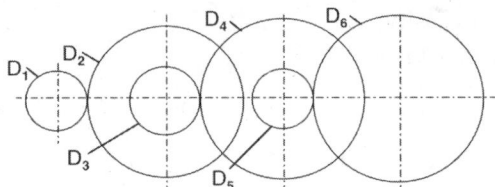

Fig. 30.16: Gear trains

Hunting tooth: If an exact velocity ratio, *r* is an integer and a slight variation is not undesirable, then an extra tooth is added to the number of gear teeth for equalizing wear of all the teeth. This extra tooth is called a hunting tooth (30.9).

Efficiency: There exist several theoretical formulas for expressing the efficiency of a toothed spur or bevel gear as a function of the coefficient of friction between the teeth, the pressure angle, and the number of teeth. However, they all are based on involved assumptions which are questioned by competent authorities. Experimental data form the most reliable guide. Considering tooth friction only, the efficiency of good unhardened spur gears is about 99 per cent. It is independent of the pitch-line velocity (at least within the range from 0.3 to 7.5 m/s, the pressure angle, the load transmitted, and the quantity of lubricating oil used, provided the quantity is sufficient to prevent cutting and heating.[10] The condition of the tooth surface is the most important factor. Tooth friction in accurately cut, hardened, and ground spur gears may be considered negligible. Table 30.9 shows the small losses that can be expected in various gear trains under the best construction and care.[11] The starting losses are rendered higher by the greater viscosity of the cold oil.

Gear bearings: By decreasing the pitch-line velocity v_m for a given torque, the tangential tooth pressure F_t is increased in inverse proportion to v_m. Hence the bearing load, which is equal to $F_t \sec \beta$, is increased, and the bending moment on the gear shaft is likewise increased.

With electric-motor drives, and speeds higher than 6.5 m/s, overhung gears give trouble, and experience shows that an outboard bearing is a necessity and a good investment in diminishing repairs and shutdowns.

In general, the more rigid the whole gear installation is, the more satisfactory will be its service.

30.12 STRENGTHENING OF GEAR TEETH

Cut gear teeth may be strengthened by one of the following methods: (a) increasing the pressure angle; (b) using short teeth; (c) using stub teeth; (d) using teeth with an unequal addendum and dedendum; (e) crowning the teeth; (f) using helical teeth.

Increasing pressure angle: The gain in strength obtained by increasing the pressure angle has already been mentioned (see Fig. 30.8).

In aircraft gears pressure angles of $22\frac{1}{2}^{\circ}$, 25°, $27\frac{1}{2}^{\circ}$ are used to reduce weight.

Short teeth: The Hunt and Logue standards use an addendum a of $0.25p_c$ and a dedendum d of $0.30p_c$, but Hunt has β equal to $14\frac{1}{2}^{\circ}$, whereas Logue has it equal to 20°. The main objection to this method is that the gears are not interchangeable with other more widely used standards.

Stub teeth: The stub-tooth method is actually a combination of the first two methods and is used rather extensively. In general there are two main systems of stub teeth: the Nuttall, which practically coincides with the ASA standard; and the Fellows system, with the fractional designation for the pitch. Each system has a pressure angle of 20°. A still greater pressure angle, as $\beta = 22\frac{1}{2}^{\circ}$, is not practical because of poorer wearing quality.

Unequal addendum and dedendum: A large part of the flank of the standard-tooth gears with small tooth numbers lies inside the base circle and is practically useless. It is therefore logical to use the same tooth height, but to change the shape of the pinion so as to reduce the dedendum and to increase the addendum. The mating gear must have its dedendum

Table 30.9: Friction loss in spur and helical gears in oiltight cases

Number of reductions for straight spur, helical, or herringbone gears	Roller or ball bearings		Journal bearings	
	Starting (per cent)	Running (per cent)	Starting (per cent)	Running (per cent)
Single	1–2	0.5–1	10–20	0.5–1.5
Double	2–4	1.0–2	15–25	1.0–3.0
Triple	3–6	1.5–3	20–35	1.5–4.0

[10] C. W. Ham and I. W. Huckert, *An Investigation of the Efficiency and Durability of Spur Gears*, Bulletin No. 149, University of Illinois Engineering Experiment Station (1925), p. 21.

[11] W. H. Himes, "Modern Gear Efficiency Exceeds Limits Used in Most Designs," *Machine Design*, Vol. 4 (February, 1932), p. 29.

increased and its addendum decreased, as shown in Fig. 30.17.

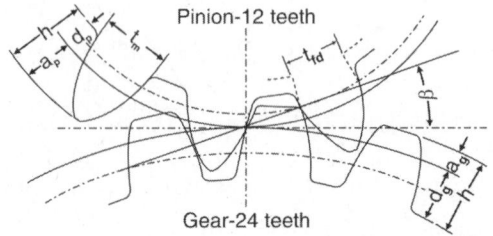

Pinion-12 teeth

Gear-24 teeth

Fig. 30.17: Teeth with unequal addenda and dedenda

The *Maag system* makes use of this method. However, it varies not only the addendum and dedendum but also the pressure angle, in order to obtain the best combination of all tooth elements, h, a, d, and b, in Fig. 30.17, in regard to running conditions. Therefore a 12-tooth gear which runs with another 12-tooth pinion is not the same as a 12-tooth pinion mating with a 24-tooth gear. No rules for proportions of Maag gears exist. Therefore gears based on this system are not interchangeable. However, this disadvantage is not serious when quantity production is maintained and replacement stocks are available. A comparison of the teeth of the standard $14\frac{1}{2}°$ full-depth system, the Fellows stub-tooth system, and the Maag system for a 12–24 tooth drive is given in Fig. 30.18. The chief advantages of the Maag system are:

  (a) The teeth are much stronger than ordinary teeth.

  (b) The teeth of the pinion and gear can be made equally strong.

  (c) High reduction ratios, up to 20 to 1, are possible.

  (d) The profiles of the teeth eliminate any interference.

  (e) The gears run more quietly and last longer.

Crowning: Inaccuracies in the teeth caused by machining and heat-treatment distortion, faulty assembly, and deflections at the shaft supports result in uneven pressures along the tooth face and throw toward the end of a tooth

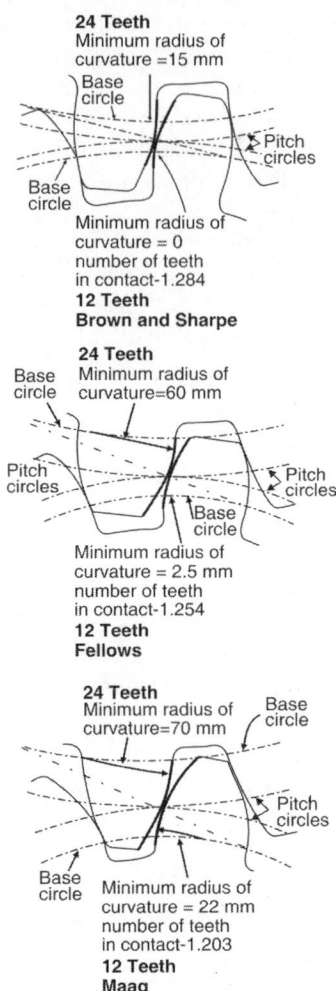

24 Teeth
Minimum radius of curvature =15 mm

Base circle

Base circle

Pitch circles

Minimum radius of curvature = 0
number of teeth in contact-1.284
12 Teeth
Brown and Sharpe

24 Teeth
Minimum radius of curvature=60 mm

Base circle

Pitch circles

Pitch circles

Base circle

Minimum radius of curvature = 2.5 mm
number of teeth in contact-1.254
12 Teeth
Fellows

24 Teeth
Minimum radius of curvature=70 mm

Base circle

Pitch circles

Base circle

Minimum radius of curvature = 22 mm
number of teeth in contact-1.203
12 Teeth
Maag

Fig. 30.18: Comparison of gear-teeth action. (Courtesy Niles Tool Works Company)

the maximum pressure on the tooth. The load can be moved toward the middle of each tooth by crowning or elliptoiding the teeth; that is, by making the ends slightly thinner, as shown to an exaggerated degree in Fig. 30.19. The teeth become 20 to 100 per cent stronger, and their service life is increased up to 40 times.[12]

Crowning is done by shaving cutters. The advisable amount of crowning c, Fig. 30.19, depends on the deflection of the teeth in operation. However, it is very small, being only 0.0003 to 0.0005 mm per mm of facewidth.

[12]National Broach & Machine Co., *Modern Methods of Gear Manufacture*. 3rd ed. (Detroit: 1950), p. 50

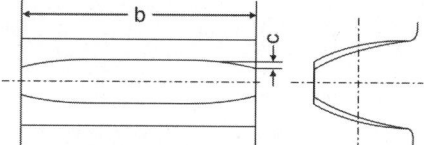

Fig. 30.19: Crowning of gear teeth

The bending strength of spur gears is increased by roll pressing on the fillets and by gear rolling by another gear.

30.13 HELICAL GEARS

A helical gear may be considered to be composed of an infinite number of infinitesimally narrow staggered spur gears. The result is that each tooth is slanting across the face, as shown in Fig. 30.20, so as to form a cylindrical helix.

With one end of the tooth advanced over the other, engagement takes place progressively. The line of contact is a diagonal line extending from some point on the face of the advanced end to a point on the flank of the trailing end. As a result, the engagement of the teeth on a helical gear (i) is much smoother than that on a straight-tooth spur gear, (ii) and the gears run more quietly and (iii) operate satisfactorily at much higher pitch-line velocities, up to 30 m/s and more. The smooth engagement results in (iv) very slight wear and (v) high efficiency, as may be seen from Table 30.9.

Strength: In straight-tooth spur gears there is a time in each period of contact when the load is concentrated at the upper edge of the tooth, thus acting with a leverage equal to the height of the tooth. With helical teeth the points of contact are at all times distributed over the entire working surface of the tooth. Therefore the mean lever arm of the bending action of the load is about one-half the height of the tooth. This makes (vi) helical-tooth gears much stronger than straight-tooth spur gears and (vii) also decreases the tooth deflection considerably, i.e. they have higher torque capacity.

Pitch: The *pitch of the gear* is the one in the diametral plane and is designated as p_d or p_c, as the case may be. The distance between the teeth measured along a normal to the helix is called the normal pitch and is designated as p_{cn} or p_{dn}. Evidently

$$p_{en} = p_c \cos \psi \qquad p_{dn} = \frac{p_d}{\cos \psi} \quad \dots (30.34)$$

Normal module

$$m_n = m \cos \psi \text{ or } \frac{p_{en}}{m_n} = \pi = \frac{p_c}{m}$$

If helical gears are cut with standard hobs, the normal pitch p_{dn} or module must be specified, and the gear pitch p_d will contain a decimal fraction, as will also the pitch diameter.

Axial thrust: A drawback of helical gears is the axial thrust F_a equal to the axial component of the tooth pressure F_t. As may be seen in Fig. 30.20, the amount of this thrust may be expressed as

$$F_a = F_t \tan \psi \qquad \dots (30.35)$$

This thrust must be taken by the bearings and affects slightly the efficiency of the gear action.

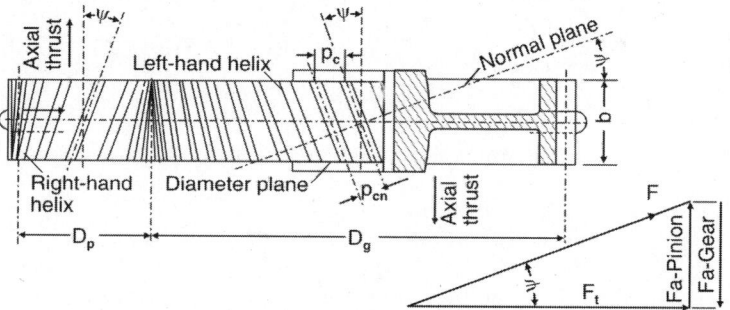

Fig. 30.20a: Helical gears

Helix angle and face: According to the definition of the AGMA, the helix angle ψ, Fig. 30.20a, is the angle between a tangent to a helix and an element of the cylinder. Unless otherwise specified, the helix is referred to the pitch circle. Since helical gears by their nature are not interchangeable, there were no standard helix and pressure angle. Current practice in helix-angle selection, for best overall result, is to make this angle 20° to 35°.

Since the axial thrust F_a increases with an increase of the helix angle ψ, it is desirable that ψ be not greater than is necessary to obtain the advantages of the helical tooth; its value depends on the width b. For smooth operation one end of the tooth should be advanced over the other end, a distance slightly greater than the circular pitch, $1.1p_c$ according to Fellows practice. Evidently, the narrower the face b, the greater the angle ψ must be to fulfill this requirement. The minimum value of b is given by the relation

$$b = \frac{1.15\pi m}{\tan \psi} = \frac{1.15\pi m_n}{\sin \psi} \qquad \ldots (30.36)$$

There is no definite limit for the maximum value of b. Sometimes it is suggested that b should not exceed $\dfrac{20m}{\tan \psi}$ but greater values are used. For helical gears, the face width may be taken 12.5 m_n to 20 m_n while for herring-bone gears, it is 20 m_n to 30 m_n, i.e. $k = 2$ to 5.

The number of teeth for which the cutter must be selected is equal to the number of teeth with a normal pitch in a circumference corresponding to the helix. This number is called the *formative number of teeth* and is computed by the relation

$$z_f = \frac{\text{Circumference}}{\text{Normal pitch}}$$

$$= \frac{2\pi d}{2\cos^2 \psi \, m\pi \cos \psi}$$

$$= \frac{d}{m} \times \frac{1}{\cos^3 \psi}$$

$$Z_f = \frac{Z}{\cos^3 \psi} \qquad \ldots (30.37)$$

Formative number of teeth is defined as the number of teeth that could be generated on the surface of a cylinder having radius equal to the radius of curvature of the tip of minor axis of an ellipse obtained by taking a section of the gear in the normal plane (Fig. 30.20b).

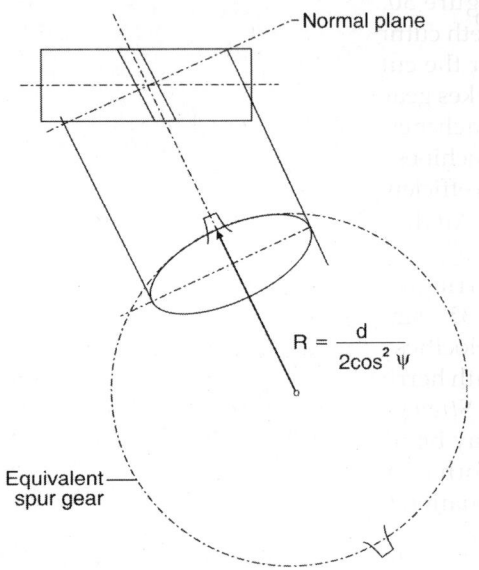

Fig. 30.20b

Some gear manufacturers make their cutters on the basis of a standard pitch in the diametral plane. Thus, helical gears can be obtained with either a standard normal pitch or a standard gear pitch or standard module.

30.14 HERRINGBONE GEARS

As shown in Fig. 30.21, a herringbone gear is a double helical gear. The combination of right-hand and left-hand helices absorbs the axial thrust within the gear itself and eliminates the

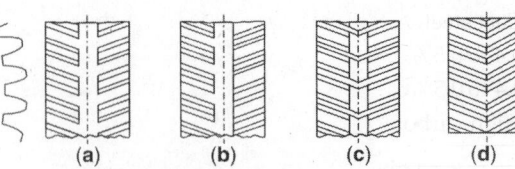

   (a)   (b)   (c)   (d)

Fig. 30.21: Types of herringbone gears

thrust on the bearings, which is a disadvantage of single helical gears. The disadvantage of herringbone gears is that axially they must be aligned very accurately if each half is to take its part of the load. Figure 30.21a shows an ordinary herringbone gear with a relief groove for the cutting tool cut in the center; Figure 30.21b shows a Wuest gear, which differs from the ordinary gear only in that the teeth are staggered; Figure 30.21c shows a gear with continuous teeth cutting in a cast blank with precast reliefs for the cutting tool; in Fig. 30.21d is shown a Sykes gear with continuous teeth cut by special machines. The accuracy obtained by these machines gives herringbone gears which have an efficiency of 99 to 99.5 per cent.

All discussion of single helical gears except that pertaining to the thrust applies also to herringbone gears. The helix angle is made up to 35°, since there is no axial thrust. Pitch-line velocities as high as 60 m/s have been attained with herringbone gears.

Strength of gears: A modified Lewis equation may be used to determine the pressure on a tooth of a helical or herringbone gear. This equation is

$$F_t = \frac{S_o\, c\, b\, y\, \pi m\, \cos\psi}{C_w} \qquad \ldots (30.38)$$

If pitch diameters is not known, the following Lewis equation may be used.

$$m^3 c S_o = \frac{2T}{k\, y\, \pi^2 z \cos\psi} \qquad \ldots (30.38a)$$

$$k = \frac{b}{p_c}$$

in which S_o is the allowable static stress and c is the velocity factor. Values of S_o are given in Table 30.10. Values of c may be found as follows: For low-angle helical gears when v_m is under 5 m/s, by equation 30.15; for all helical and herringbone gears when v_m is 5 to 10 m/s, by equation 30.16; for $v_m = 10$ to 20 m/s, using Barth's formula[13]

$$c = \frac{15}{15 + v_m} \qquad \ldots(30.39)$$

For precision gears with v_m greater than 20 m/s the velocity factor may be found by equation 30.17; and for nonmetallic gears, by equation 30.18.

The active width of the gear face b does not include the groove around the center of a herringbone gear, Fig. 30.21a or Fig. 30.21b. The equation for the minimum value of b for herringbone gears given by AGMA is

$$b \geq \frac{2.3\pi m}{\tan\psi} = \frac{2.3\pi m_n}{\sin\psi} \qquad \ldots (30.40)$$

There is no definite limit for the maximum value of b. Some designers suggest for herringbone gears the relation $b < \dfrac{30m}{\tan\psi}$ or $< \dfrac{30m_n}{\sin\psi}$. However, much greater values of b are used satisfactorily.

Table 30.10: Design data for helical and herringbone gears

Material	Elastic limit N/mm²	Allowable stress N/mm²
Cast iron, ordinary	84	56
Cast iron, better grade	110	80
Laminated phenolic materials	70	45
Bronze, SAE 65	165	84
Cast steel, ASTM class B, medium	240	140
0.4 to 0.5% carbon steel, not treated	290	230
0.4 to 0.5% carbon steel, heat-treated	370	270
High-carbon or alloy steels, heat-treated	620	450

[13] C. D. Albert, *Machine Design Drawing Room Problems*, 4th ed. (New York: John Wiley & Sons, Inc., 1948), p. 381.

The Lewis factor y in equation 30.40 is taken from Table 30.2 for the pressure angle in the plane of rotation and the number of teeth. There are no standard proportions for helical teeth, although they are usually made as stub teeth. For herringbone gears, the standard pressure angle[14] β in the plane of rotation is 20°. However, the AGMA gives for the pressure angle β the limits of 15° 23' and 25°, and for the helix angle ψ the limits of 20° and 45°. The relation between the pressure angle β_n in the normal plane and β in the plane of rotation is

$$\tan \beta_n = \tan \beta \cos \psi \qquad \ldots (30.41)$$

In equation 30.40, C_w is a wear and lubrication factor. For enclosed gears continuously lubricated with oil of the proper viscosity and character, $C_w = 1.15$; for scant lubrication but regular, frequent inspection, $C_w = 1.25$; for indifferent lubrication, $C_w = 1.35$. To prevent oil from being thrown from the tooth surfaces, it should be introduced at the point where the teeth are beginning the engagement.

Equation 30.40 is based on tooth proportions which give 1.5 or more teeth in contact on the line of action in the plane of rotation. If the contact is less than 1.5 teeth, the tooth load should be reduced proportionally.

Some gear manufacturers use their own formulas for the tooth load and power of herringbone gears of their type. In most case these data are more conservative than the values obtained by using equation 30.40.

Dynamic load: Helical and herringbone gears are also subject to dynamic loads caused by inaccuracies in generating the teeth, although to a lesser degree than straight-tooth spur gears. For metal gears, the total dynamic load may be determined by equation 30.20, where the increment load may be computed by the relation[15]

$$F_i = \frac{21v_m \left(Cb \cos^2 \psi + F_t \right) \cos \psi}{21v_m + \sqrt{Cb \cos^2 \psi + F_t}} \qquad \ldots (30.42)$$

As with spur gears, the total dynamic tooth load F_d must be smaller than the endurance strength F_{en} computed by equation 30.24, in which the value of S_{en} is taken from Table 30.6.

Wear resistance may be computed by the equation

$$F_w = \frac{kbD_p Q}{\cos^2 \psi} \qquad \ldots (30.43)$$

This differs from equation 30.25 only because of the term $\cos^2 \psi$. The value of F_w must be greater than that of F_d found by equation 30.20. In most cases it will be found that a pair of gears properly designed by equation 30.40 will have both F_{en} and F_w greater than F_d.

The safety margin should be the same as for spur gears, or 25, 35, or 50 per cent, the value depending on the uniformity of the load.

Design procedure: Basically, the design procedure for helical or herringbone gears is the same as for straight-tooth spur gears. First it is necessary to select the pressure and helix angles, the materials, and tentatively the pitch velocity. The diametral pitch and the face b of the pinion are found by using equation 30.40, in which the safe static stress is taken from Table 30.10. The next step is to determine the actual module m of the cutter to be used, the number of teeth in the pinion, and the number of teeth in the gear. After this the face of the gear is checked and, if necessary, both faces are increased to a safe value. Unlike for straight-tooth spur gears, there is no definite limit for the maximum face width b; but the minimum value of b is given by equation 30.38 or equation 30.42, as the case may be.

The final steps are to check the designed pinion and gear for dynamic load and resistance to wear. If the check gives unsatisfactory results, the design must be improved by using either a wider face or better materials, or by adopting both methods.

If the diameters of pinion and gear are unknown, the steps are the same as for spur gears and equation 30.38a is to be used.

---

[14] Brown & Sharpe Mfg. Co., *Practical Treatise on Gearing*, 25th ed. (Providence: 1951), p. 133.
[15] W. P. Schmitter, "Determining Capacity of Helical and Herringbone Gears," *Machine Design*, Vol. 6 (June and July, 1934), p. 40 and p. 33.

The detailed procedure given for spur gear including steps for computer programming are same for helical gear.

Example 30.6: Design a herringbone drive from a 2.25 kW steam turbine, running at 30,000 rpm, to a speed reducer that should run at 2,500 rpm.

Although the pinion speed is very high, the power is small and a pitch velocity 28 m/s may be selected. The pitch diameter of the pinion is then

$$D_p = \frac{28 \times 60 \times 1000}{\pi \times 30,000} = 17.8 \text{ mm say 18 mm}$$

The useful tooth pressure is found by equation 30.19, in which $C_s = 1$ from Table 30.4, for 10-hr service and steady load.

Thus,

$$F_t = \frac{2.25 \times 1000}{28} = 80 \text{ N}$$

The velocity factor c, by equation 33.17, is

$$c = \frac{5.6}{5.6 + \sqrt{28}} = 0.514$$

If the number of teeth in the pinion is taken as $z_p = 16$, the module is

$$m = \frac{18}{16} = 1.125 \text{ say } 1.25$$

$$D_p = 16 \times 1.25 = 20$$

and the number of teeth in the gear is

$$z_g = \frac{16 \times 30,000}{2,500} = 192$$

The helix ψ may be selected as 20° and $\beta = 20°$

$$z_f = \frac{16}{0.9396^3} = 19.23 \text{ say } 20$$

For 20 teeth, the Lewis factor, from Table 30.2, is $y = 0.102$.

The material for the pinion may be taken as alloy steel, for which $S_o = 345. \text{ N/mm}^2$.

At $v_m = 28$ m/s the gears cannot run is an oil bath. Therefore the factor C_w should be taken conservatively as 1.35. With these figures the face, from equation 30.40, is

$$b = \frac{80 \times 1.35}{345 \times 0.514 \times 0.102 \times \pi \times 1.25 \times 0.9396} = 1.6 \text{mm}$$

By equation 30.42, the minimum width should be

$$b = \frac{2.3 \pi m}{\tan \psi} = \frac{2.3 \pi \times 1.25}{0.3639} = 24.8$$

Taking $b = 35$ mm.

The design will be checked for dynamic load capacity. By equation 30.44, in which the permissible and probable error is $e = 0.02$ and $C = 158$, from Table 30.5,

$$F_i = \frac{21 \times 28 \left(158 \times 35 \times 0.9396^2 + 80\right)0.9396}{21 \times 28 + \sqrt{158 \times 35 \times 0.9396^2 + 80}}$$

$$= 4164 \text{ N}$$

Then $F_d = 4164 + 80 = 4244$ N

The endurance strength, with $S_{en} = 690$ from Table 30.6 is

$$F_{en} = 690 \times 35 \times \pi \times 1.25 \times 0.102 \times 0.9396 = 9089 \text{ N}$$

This corresponds to a safety margin of $(6016/3585 - 1) \times 100 = 114$ per cent.

The design must be checked for wear resistance by equation 30.45. From Table 30.7 for a pinion and gear with a Bhn of 300, $k = 1.344$. Also the term Q, by equation 30.27, is

$$Q = \frac{2 \times 192}{192 + 16} = 1.85$$

Then the wear resistance becomes

$$F_w = \frac{1.344 \times 35 \times 20 \times 1.85}{0.9396^2} = 1971.5 \text{N}$$

Since F_w is less than F_d, the pinion and gear materials are not hard enough. The necessary hardness can be found from equation 30.45 by solving it for K. Thus,

$$K = \frac{4244 \times 0.9396^2}{35 \times 20 \times 1.85} = 2.89$$

Table 30.7 shows that a pinion with a Bhn of 500 and a gear with a Bhn of 450 will give $K = 3.413$.

To complete the design, the size of the gear blanks should be determined. The outside diameter of the pinion with the addendum a taken from Table 30.1, is

$$D_p' = (20 + 2 \times 1.25) = 22.5 \text{ mm}$$

and that of the gear is

$$D_g' = 192 \times 1.25 + 2 \times 1.25 = 242.5 \text{ mm}$$

Example 30.7: A pair of helical gears used as reduction gears has to transmit 5 kW at 10,000 rpm of the pinion. The velocity of the gear is 2500 rpm. Both the gears are made of 20C8 heat treated steel with $S_o = 193$ MN/m^2. The gears are 20° stub and pinion is to be have a minimum of 24 teeth. Design the gears with helix angle of 23°.

As the material is same for both pinion and gear, the pinion is weaker of the two.

Torque

$$T = \frac{5 \times 1000 \times 60}{2\pi \times 10,000} = 4.77 \text{ Nm}$$

Assuming

$$k = 4, Z_p = 24,$$

$$z_f = \frac{z_f}{\cos^3 23} = \frac{24}{0.92^3} = 30.8 \text{ say } 31$$

$$y = 0.14$$

$$m^3 S = \frac{2T}{k\pi^2 yZ \cos \psi}$$

$$= \frac{2 \times 4.77 \times 1000}{4 \times \pi^2 \times 0.14 \times 24 \times 0.92} = 78.17$$

Assuming $c = 0.5$ for the first trial, allowable stress $= 0.5 \times 193 = 92.5$ N/mm^2

$$m = \left[\frac{78.17}{92.5}\right]^{1/3} = 0.945$$

Taking a standard module of 1, $D_p = 24$ mm,

$$v = \frac{\pi \times 24 \times 10,000}{1000 \times 60} = 12.57 \text{ m/s}$$

$$c = \frac{5.6}{5.6 + \sqrt{12.57}} = 0.613,$$

$$S_a = 0.613 \times 193 = 118 \text{N/mm}^2.$$

Induced stress $= \dfrac{78.17}{1^3} = 78.13$ N/mm^2.

As the induced stress is less than the allowable stress, the module 1 is satisfactory. However, the face width may be reduced to $\pi \times 1 \times 4 \times \dfrac{78.13}{118} = 8.32$ mm or say 10 mm.

Checking for dynamic load, for an error of 0.01, the value of $C = 118.7$

$$F_t = \frac{2 \times 4.77 \times 1000}{24} = 397.5 \text{ N}$$

$$Cb \cos^2 \psi + F_t = 118.7 \times 10 \times 0.92^2 + 397.5 = 1402$$

$$F_i = \frac{21 \times 12.57 \times 1402}{21 \times 12.57 + \sqrt{1402}} = 1228 \text{ N}$$

$$F_d = 1228 + 397.5 = 1628 \text{ N}$$

S_{en} from Tables, 4 14 N/mm^2.

Endurance strength, $F_{en} = 414 \times 10 \times \pi \times 1 \times 0.14 \times 0.92 = 1675$ N which $> F_d$.

Check for wear, $S_{enc} = 618$ N/mm^2 and $K = 0.902$

$$Q = \frac{2 \times z_g}{z_g + z_p} = \frac{2 \times 60}{24 + 60} = 1.42$$

Wear load

$$F_w = \frac{24 \times 10 \times 1.42 \times 0.902}{0.92^2} = 367 \text{ N}$$

which is unsafe. By successive trials by increasing the module to 3, $b = 30$, $D = 72$, $F_d = 3575$ N and by increasing the hardness of pinion to 300, $k = 1.108$

$$F_w = 4014 \text{ N}$$

F_e will be much higher at 15076 N

Alternately, the same gear dimensions will have materials of pinion and gear with each $Bhn = 500$, $k = 4.04$.

$$F_w = \frac{24 \times 10 \times 1.42 \times 4.04}{0.92^2} = 1626 \text{ N which}$$

is almost equal to $F_d = 1628$ N and much higher value of $S_e = 1305$ N/mm^2.

Example 30.8: A power of 18 kW at 10,000 rpm is to be transmitted through pair of helical gears with a velocity ratio of 3.5:1. The gears are 20° stub with a helix angle of 23°. The pinion diameter is 90 mm. Both gears are made of cast steel with a static stress of 110 N/mm^2. Find the module and facewidth. Find the facewidth of these have to be herringbone gears.

As material for pinion and gear is same, pinion is weaker.

Assume $z_p = 20$, $z_{pf} = \dfrac{20}{\cos^3 23} = 25.6$ say 26

$$y = 0.175 - \frac{0.95}{26} = 0.1385$$

$$T = \frac{18 \times 1000 \times 60}{2\pi \times 10,000} = 17.18 \; Nm$$

$$v = \frac{\pi \times 90 \times 10,000}{1000 \times 60} = 47.12 \; m/s.$$

$$c = \frac{5.6}{5.6 + \sqrt{43.12}} = 0.449$$

$$S_a = 0.449 \times 110 = 49.39 \; N/mm^2$$

$$m^3 S_a = \frac{2 \times 17.18 \times 1000}{3.5 \times \pi^2 \times 20 \times 0.1385 \times \cos 23} = 390.31$$

$$m = \left[\frac{390.31}{49.39} \right]^{1/3} = 1.99 \; \text{say} \; 2$$

With $z_p = \dfrac{90}{2} = 45,$

$$z_{pf} = \frac{45}{0.7786} = 57.8 \; \text{say} \; 58$$

$$y = 0.175 - \frac{0.95}{58} = 0.1587$$

$$S_i = \frac{390.31 \times 0.1385}{0.1587 \times 2^3} = 42.5 \; N/mm^2 \; OK$$

But k can be reduced to $\dfrac{3.5 \times 42.5}{49.39} = 3.01$

$b = 3.01 \times \pi \times 2 = 18.9 \; \text{say} \; 20 \; mm.$

with herringbone gear $b = 20 + 5 = 25 \; mm.$

$$z_g = 45 \times 3.5 = 157.5 \; \text{say} \; 158$$

30.15 GEAR BOX

The characteristics of a piston engine which is usually used in passenger cars and also goods and transport vehicles are such that it is required to make available very high torque to start the vehicle. The requirement of torque decreases as the speed increases. When it is moving at constant high speed, the force required is very less as there is no acceleration and it has to meet only frictional and other resistance. At the start, it has to meet very large frictional resistances of all moving parts including wheels, air resistances and also large force/torque to produce high acceleration

and also enough force to go up the steepest inclines. In contrast, the engine torque-speed characteristic shown in Fig. 30.22 shows that it produces low torque at low speed, increases to a maximum value as speed increases and falls to lower value with further increase in speed.

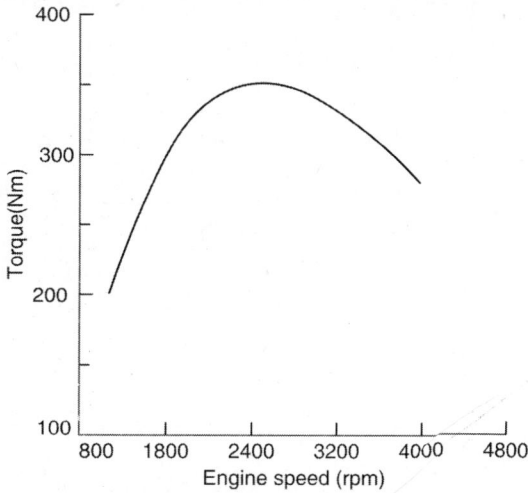

Fig. 30.22: Engine output

To make the lower torque at low speed compatible with the requirement of large torque, speed is reduced by about 4.25:1 with first gear multiplied by constant reduction of 3 to 6 at the differential depending on the vehicle. The combined effect is to increase the torque 12 to 25 times the available torque. As the speed increases, the torque requirement decreases. Even at the top gear (1:1) the gear box does not reduce the speed but the differential still multiples the torque by 3 to 6 and the speed reduces at the wheels.

Design procedure for gearbox:

1. Decide the gear arrangement: Sliding gear, constant mesh with sliding dog clutch with or without synchromesh or epicyclic gear box.

2. Conceive the layout of gears and make a sketch.

3. Decide the gear ratios for various speeds with first reduction

$\simeq \sqrt{\text{Maximum reduction}}$ for optimising the space.

4. With the approximate center distance, decide the number of teeth to obtain various gear ratios so that the maximum variation in speed is limited to ±2% with minimum number of teeth 18 plus and same module for all gear trains.

5. Assume various lengths of clutches (10–15), space for shifting (1–5 mm), operational clearances (1–2 mm), width of bearings (20–30 mm). Find the length of each shaft from the centers of bearings, i.e. half of bearing width on each side.

6. Decide the facewidth of each gear by taking $K = \dfrac{b}{p}$ as ≤ 3 and standard module.

   Calculate exact length of all shafts.

7. For the gear subjected to maximum torque in forward drive, find the static stress using Lewis equation $S_o = \dfrac{F_t}{c\,b\,\pi\,m\,y\cos\psi}$,

   Ultimate tensile stress as $S_u = (2.5 \text{ to } 3) \times S_o$. Consider helical gears

8. For this S_u, select a suitable material and check for dynamic load and wear load. If required, change the material.

9. Draw the bending moment diagram for each shaft and for each velocity ratio and find the maximum B.M in each case.

   Find the diameter of each shaft by ̣onsidering maximum B.M, maximum ̣rque and shock and endurance factors ̣ng ASTM code of shafting.

   ̣ssuming a life of 3000 hours to 4000 ̣ for each bearing, and calculated ̣ter for that shaft, find reactions at ̣ds. Select a suitable ball or roller ̣ from bearing catalogue/hand

   ̣uitable size of operating

   ̣nal elevation of the gear box ̣tails.

Example 30.9: Design a gear box for an automobile with the following data:

Maximum torque	350 Nm
At proper speed	1300 rpm
Maximum speed	1800 rpm
First speed reduction	4.28
Second speed reduction	2.56
Third speed reduction	1.62
Top speed (fourth)	1.00
Reverse speed reduction	5.97
Approximate centre distance	120 mm

Deciding number of teeth of gears: Consider first gear reduction of 4.28. To make compact gear box assembly, the first reduction

$$= \sqrt{\text{Maximum reduction}}$$

$$= \sqrt{4.28} = 2.06$$

$$\frac{z_2}{z_1} \times \frac{z_4}{z_3} = 4.28 \text{ and } \frac{z_2}{z_1} = \frac{z_4}{z_3} = 2.06$$

if $\quad z_1 = 20,\ z_2 = 20 \times 2.06 = 41.2$ say 41

i.e. $\quad \dfrac{z_2}{z_1} = \dfrac{41}{20}\quad \therefore\ \dfrac{z_4}{z_3} = \dfrac{4.28 \times 20}{41} = 2.08$

As sum of $z_2 + z_1 = z_4 + z_3$

Taking $\dfrac{z_4}{z_3} = \dfrac{41}{20}$

% age arror in reduction of speed

$$= \frac{\dfrac{41}{20} \times \dfrac{41}{20} - 4.28}{4.28} \times 100$$

$$= \frac{4.20 - 4.28}{4.28} \times 100 = -1.8\%$$

module $= \dfrac{120 \times 2}{61} = 3.93$, standard module $= 4$.

Exact centre distance

$$= \frac{4 \times 61}{2} = 122 \text{ mm}$$

Second speed reduction = 2.56

Reduction required $\dfrac{z_6}{z_5} = \dfrac{2.56}{2.05} = 1.24$

As $z_5 + z_6 = 61$ and $z_6 = 1.24\, z_5$

$$z_5 = \frac{61}{2.24} = 27 \text{ and}$$

$$z_6 = 1.24 \times 27 = 33.48 = 34$$

Actual reduction

$$= \frac{34}{27} \times \frac{41}{20} = 2.58$$

% error in second reduction

$$= \frac{2.58 - 2.56}{2.56} \times 100 = 0.78\%$$

Third gear reduction = 1.62.
Actual reduction required

$$\frac{z_8}{z_7} = \frac{1.62}{2.05} = 0.79$$

As $z_7 + z_8 = 61$ and $z_8 = 0.79\, z_7$

$$z_7 = \frac{61}{1.79} = 34$$

$$z_7 = 0.79 \times 34 = 27$$

Actual reduction

$$= \frac{34}{27} \times \frac{41}{20} = 1.627$$

% error in third reduction

$$= \frac{1.627 - 1.62}{1.62} = 0.43\%$$

Reverse speed reduction = 5.87
Actual reduction

$$= \frac{5.97}{2.05} = 2.91$$

As the gears z_9 and z_{11} should not touch at the addendum circle, the actual center distance for designing them should be as:

Sum of p.c.d of two gears = 61 × 4 = 244
Module size   = 4 × 2 = 8
Actual size   = 244 – 8 = 236

The distance of two shafts so that teeth of gears do not touch, is

$$< \frac{244 - 8}{2} = \frac{236}{2} = 118$$

Let it be 110. So sum of two addendum diameters will be

55 × 4 + 2 × 4 = 220 + 8 = 228 mm, i.e. there will be clearance between gears.

whereas sum of maximum diameters can be 122 × 2 = 244.

Considering sum of teeth as 55

$$z_9 + z_{11} = 55, \frac{z_{11}}{z_9} = 2.91$$

$$z_{11} = \frac{55}{3.91} = 14$$

$$z_9 = 14 \times 2.91 = 41$$

Actual reduction $= \frac{41}{14} \times \frac{41}{20} = 6.003$

% age error in reduction of reverse gear

$$= \frac{6.003 - 5.97}{5.97} \times 100 = 0.55\%$$

All the gear reduction have error of less than ±2%. The diameters are:

$D_1 = 4 \times 20 = 80$ mm
$D_2 = 4 \times 41 = 164$ mm
$D_3 = 4 \times 20 = 80$ mm
$D_4 = 4 \times 41 = 164$ mm
$D_5 = 27 \times 4 = 108$ mm
$D_6 = 34 \times 4 = 136$ mm
$D_7 = 4 \times 27 = 108$ mm
$D_8 = 4 \times 36 = 136$ mm
$D_9 = 4 \times 14 = 56$ mm
$D_{11} = 4 \times 41 = 164$ mm

The idle gear between D_9 and D_{11}, i.e. D_{10} will have a diameter of $= \frac{164 + 56}{2} = \frac{220}{2} = 110$ mm

having number of teeth $= \frac{110}{4} = 27$.

Designing the highly stressed gear no 3 in the forward motion: To reduce size of gear box. $k \le 3$.

$z_3 = 20$ teeth, diamter $D_3 = 80$ mm, $b = k\pi m. = 2.5 \times \pi \times 4 = 31.4$ say 32 mm, $\phi = 20°$ and take helix angle $\phi = 23°$.

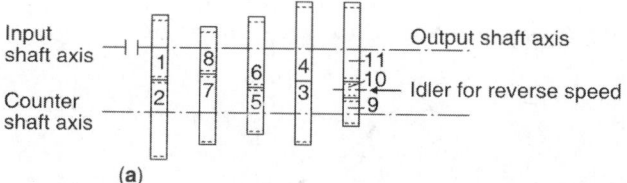

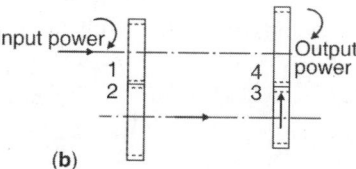

Fig. 30.23: (a) Layout of gears and (b) Transfer of motion

$$z_{3f} = \frac{20}{\cos^3 23} = \frac{20}{0.7799} = 25.6 \text{ say } 26$$

$$y = 0.154 - \frac{0.912}{26} = 0.119$$

Using Lewis equation

$$S_3 = \frac{2T}{m^3 k \pi^2 yz \cos \psi}$$

$$= \frac{2 \times 350 \times 1000 \times 2.05}{4^3 \times 2.5 \times \pi^2 \times 0.119 \times 20 \times 0.9205}$$

$$= 414.8 \text{ N/mm}^2$$

$$V_3 = \frac{\pi \times 1800 \times 80}{1000 \times 2.05 \times 60} = 3.68 \text{ m/s}$$

$$C = \frac{6}{6 + 3.68} = 0.6198$$

$$S_o = \frac{414.8}{0.6198} = 669.25 \text{ N/mm}^2$$

$S_u = 2.5 \times 669.25 = 1673 \text{ N/mm}^2$ very high stress
Taking module = 5, $z_3 = 20$, $D_3 = 100$, $b = 2.5 \times \pi \times 5 = 39.6$ Say 40 mm
The center distance

$$= \frac{100 + 205}{2} = 152.5 \text{ mm}$$

$$S_3 = \frac{2 \times T}{m^3 k \pi^2 yz \cos \psi}$$

$$\frac{2 \times 350 \times 1000 \times 2.05}{5^3 \times 2.5 \times \pi^2 \times 0.119 \times 0.9205}$$

$$= 212.4 \text{ N/mm}^2$$

$$V_3 = \frac{\pi \times 1800 \times 100}{1000 \times 2.05 \times 60}$$

$$= 4.597 \text{ say } 4.6$$

$$C = \frac{6}{6 + 4.6} = 0.566$$

$$S_0 = \frac{212.4}{0.566} = 375.26 \text{ N/mm}^2$$

$$S_u = 3 \times 375.26 = 1125.8 \text{ N/mm}^2$$

For this value of UTS, a suitable material from tables is case hardened 35 Nil Cr 18 with UTS 981–1460.

Taking UTS = 1200 N/mm² with $S_e = 620$ N/mm²

Tangential force

$$= \frac{2.05 \times 350 \times 100}{50} = 14350 \text{ N}$$

For module 5, maximum error 0.05, value c of dynamic load = 580 kN/m

$$F_d = F_t + \frac{21V \left[bc \cos^2 \psi + F_t \right] \cos \psi}{21V + \sqrt{bc \cos^2 \psi + F_t}}$$

$$= 14350 +$$

$$\frac{21 \times 4.6 [0.04 \times 580 \times 1000 \times 0.8437 + 14350] 0.9205}{21 \times 4.6 + [0.04 \times 580 \times 1000 \times 0.8437 + 14350]^{\frac{1}{2}}}$$

$$= 14350 + \frac{21 \times 4.6 \times 34007.36 \times 757 \times 0.9205}{96.6 + 184.4}$$

$$= 14350 + 10761.4 = 25111.4 \text{ N}.$$

Allowable endurance load

$$= S_e y b p \cos \psi$$

$$= 620 \times 0.119 \times 40 \times \pi \times 5 \times 0.9205$$

$$= 42671.9 \text{ N}$$

$$\geq 1.5 \times 25111.4 \geq 37667 \text{ O.K.}$$

i.e. quite safe for dynamic load.
Check for wear load

$$F_w = \frac{DbQK}{\cos^2 \varphi}$$

$$Q = \frac{2D_4}{D_3 + D_4} = \frac{2 \times 205}{100 \times 205} = 1.344$$

Load stress factor

$$K = 2553 \text{ kN/m}^2$$

$$F_w = \frac{0.1 \times 0.04 \times 1.344 \times 2.553 \times 10^3}{0.8473}$$

$$= 16.2 \times 10^3 = 16200 \text{ as it is less than } F_d \text{ so}$$
not suitable.

Suitable, k is found by equating

$$F_w = F_d$$

$$K = \frac{25111.4 \cos^2 \varphi}{DbQ} = \frac{25111.4 \times 0.8473}{0.1 \times 0.04 \times 1.344}$$

$$= 3957754 = 3957 \text{ kN/m}^2 \text{ or } 3.957 \text{ MN/m}^2$$

This will correspond to a BHN = 500 for both pinion and gear

$$K = \left(S_{enc}^2 \times \frac{\sin \beta}{1.4} \right) \left(\frac{1}{E_p} + \frac{1}{E_s} \right)$$

$$S_{enc} = \left[\frac{1.4 \, KE}{2 \sin \beta} \right]^{\frac{1}{2}}$$

$$S_{enc} = \left[\frac{1.4 \times 3957 \times 1000 \times 200 \times 10^9}{0.3420 \times 2} \right]^{\frac{1}{2}}$$

$$= 1273 \times 10^6 \, \text{N/m}^2$$

$$= 1273 \, \text{MN/m}^2 = 2.75 \, \text{H} - 70$$

$$H = 488$$

The pinion should have a hardens of 488, then it will be safe for wear. The table shows BHN of 500 for

$$S_{es} = 1305 \, \text{N/mm}^2.$$

With module 5, various diameters are $D_1 = 100$, $D_2 = 205$, $D_3 = 100$, $D_4 = 205$, $D_5 = 135$, $D_6 = 170$, $D_7 = 135$, $D_8 = 170$, $D_9 = 70$, $D_{11} = 200$ and $D_{10} = 135$.

There are three shafts: Input shaft I of length l_1, over hanging beyond bearing. Output shaft II of length l_2 supported inside the gear 1 and in the bearing at the right side in the gearbox body. Lay shaft supported in the left end and right end bearing.

Consider width of each dog clutch 15 mm, width of overlap for making contact 5 mm, clearance of 2 mm.

Shaft I: Taking each bearing width = 30 mm length of shaft $1 = l_1$ = half bearing width + clearance + width of gear 1 + clearance + width of dog clutch.

$$l_1 = 15 + 2 + 40 + 2 + 5 = 64.$$

As the load is transmitted by the width of gear, the length of shaft be taken

$$l_1 = 64 - \left(5 + 2 + \frac{40}{2} \right) = 37 \, \text{mm}$$

The length of the shaft II = 27 + 2 + 15 + 2 + 5 +5 + 2 + 40 + 2 + 15 + 2 + 5 + 2 + 40 + 2 + 40 + 2 + 5 + 15 + 2 + 5 + 2 + 40 + 2 + 15 = 292 say $300 = l_2$

Length of shaft III = 37 + 300 = 337

Design of shaft I

Tangential force

$$F_i = \frac{2 \times 350 \times 1000}{100} = 7000 \, \text{N}$$

Maximum horizontal bending moment

$$= 7000 \times 37$$

$$= 259000 \, \text{Nmm}$$

Radial force $F_r = 7000 \tan 20 = 2548 \, \text{N}$

Maximum bending moment due to F_r

$$= 2548 \times 37 = 94268 \, \text{Nmm}$$

Rasultant bending moment

$$= \sqrt{259000^2 + 94268^2} = 275.6 \times 10^3 \, \text{N/mm}$$

Equivalent torque T_e with $k_m = 2, k_t = 1.5$

$$= \sqrt{\left[2 \times 275.6 \times 10^3 \right]^2 + \left[1.5 \times 350 \times 10^3 \right]^2}$$

Taking $S_s = 90 \, \text{N/mm}^2$

Shaft diameter d_1

$$= \left[\frac{16 \times 761.2 \times 1000}{\pi \times 90 \times 0.75} \right]^{\frac{1}{3}} = 38.6 \, \text{mm}$$

Taking this diameter = 45 mm as it will be supporting the output shaft also and axial thrust has been neglected.

Design of shaft II

Considering the forces in the first gear on the output shaft, the force is transmitted at the center of gear 4.

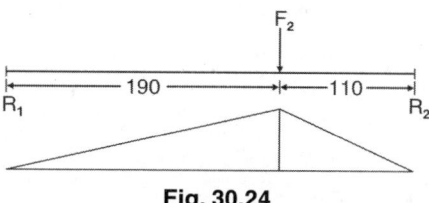

Fig. 30.24

Torque at gear 3 = 350 × 2.05 = 717.5 Nm.

Tangential force at gear 3

$$= \frac{350 \times 2.05 \times 1000}{50} = 14350 \, \text{N}$$

Radial force at gear 3

$$= 14350 \tan 20 = 5223 \, \text{N}$$

Resultant force F_r

$$= \sqrt{14350^2 + 5223^2} = 15.3 \, \text{kN}$$

Reaction

$$R_1 = \frac{15.3 \times 110}{300} = 5.61 \, \text{kN}$$

$$R_2 = 15.3 - 5.61 = 9.69 \, \text{kN}$$

Maximum bending moment

$$= 5.61 \times 190 = 1065.9 \, \text{kN/mm}$$

Equivalent torque

$$T_e = \sqrt{(1065.9 \times 2)^2 + (1.5 \times 717.5)^2}$$

$$= 2388 \, \text{kN/mm}$$

Shaft diameter

$$= \left[\frac{16 \times 2389 \times 1000}{\pi \times 180 \times 0.75} \right]^{\frac{1}{3}} = 44.8$$

say 45 mm

Design of shaft III

Forces due to gears 1 and 2 in mesh and gears 3 and 4 in mesh.

Tangential force

$F_{1t} = 7000$ N
$F_{1r} = 2548$ N
$F_{3h} = 14350$ N
$F_{3r} = 5223$ N

Horizontal plane Fig. 30.25.

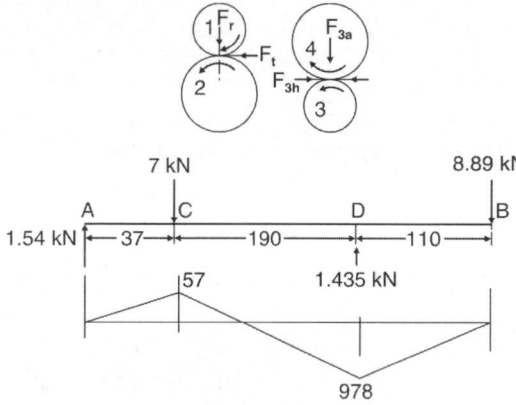

Fig. 30.25

Reaction

$$R_A = \frac{14350 \times 110 - 7000 \times 300}{300} = 1540 \text{ N or } 1.54 \text{ kN}$$

$R_B = 8.89$ kN

B. M. at C = $5.4 \times 37 = 57$ kN mm
B. M. at D = $8.89 \times 110 = 978$ kN mm
In the vertical plan Fig. 30.26.

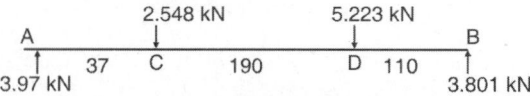

Fig. 30.26

$$R_A = \frac{2.548 \times 300 + 5.223 \times 110}{337} = 3.97 \text{ kN}$$

$R_B = 2.548 + 5.223 - 3.97 = 3.801$ kN mm
B. M. at C = $3.97 \times 37 = 146.89$ kN mm
B. M. at D = $3.801 \times 110 = 418.11$ kN mm
As maximum B.M is at D.
Resultant

B. M. at D = $\sqrt{978^2 + 418.11^2} = 1064$ kN mm.
= 1064 Nm

Torque on the shaft III

= $350 \times 2.05 = 717.5$ Nm

Equivalent torque

$$= \sqrt{(1064 \times 2)^2 + (717.5 \times 1.5)^2}$$
$$= 2384.6 \text{ Nm}$$

Shaft diameter

$$d_3 = \left[\frac{16 \times 2384.6 \times 1000}{\pi \times 180 \times 0.75} \right]^{\frac{1}{3}} = 44.8 \text{ mm}$$

Nearest standard diameter = 45 mm

Selection of bearings

Life = 3000 hrs

Maximum rpm = 1800

Life, $\quad L = \dfrac{3000 \times 1800 \times 60}{10^6} = 324$ mR

From catalogue of bearing for this life,

$$\frac{C}{F} = 6.87 \text{ for ball bearing}$$
$$= 5.67 \text{ for roller bearings}$$

Shaft I, $F_{1t} = 7000$ N

$F_{1r} = 2548$ N

$F_r = \sqrt{7000^2 + 2548^2} = 7449$ N

Axial compound $F_a = F_t \tan \psi = 7000 \tan 23 = 2971$ N

Ratio $e = \dfrac{F_a}{F_r} = \dfrac{2971}{7449} = 0.398 = 0.4$

For $e = 0.4$, $\quad X = 1 \quad Y = 1.04$

Equivalent bearing load = $7449 + 1.04 \times 2971$

= 10538.84 say 10539 N

$C = 6.87 \times 10539 = 72403$ N

$d = 45$

For these values, SKF, single row deep groove ball bearing 6408, with $d = 40$ mm, $D = 110$ mm, $B = 27$ mm with $C = 76.1$ kN is suitable.

Shaft II

The left end of shaft II will be reduced in diameter to 20 mm from 45 mm. Shaft I will be provided with a bore of 26 mm, shaft II end is inserted inside shaft I and provided with needle bearings of 3 mm diameter (Fig. 30.27).

The right end of shaft II
Radial force = 9.69 kN
Axial thrust = 14350 tan 23 = 6091 N

$$e = \frac{6091}{9690} = 0.62$$

For $e = 0.62$, $X = 1$, $Y = 1$

Equivalent bearing load $F = 9.69 + 6.091 = 15.781$ kN

$C = 6.87 \times 15.781 = 108.41$ kN

$D = 45$ mm

Single row deep groove bearing is not suitable

$\dfrac{C}{P}$ for roller bearing = 5.64

$C = 5.64 \times 15.781 = 89.0$ kN

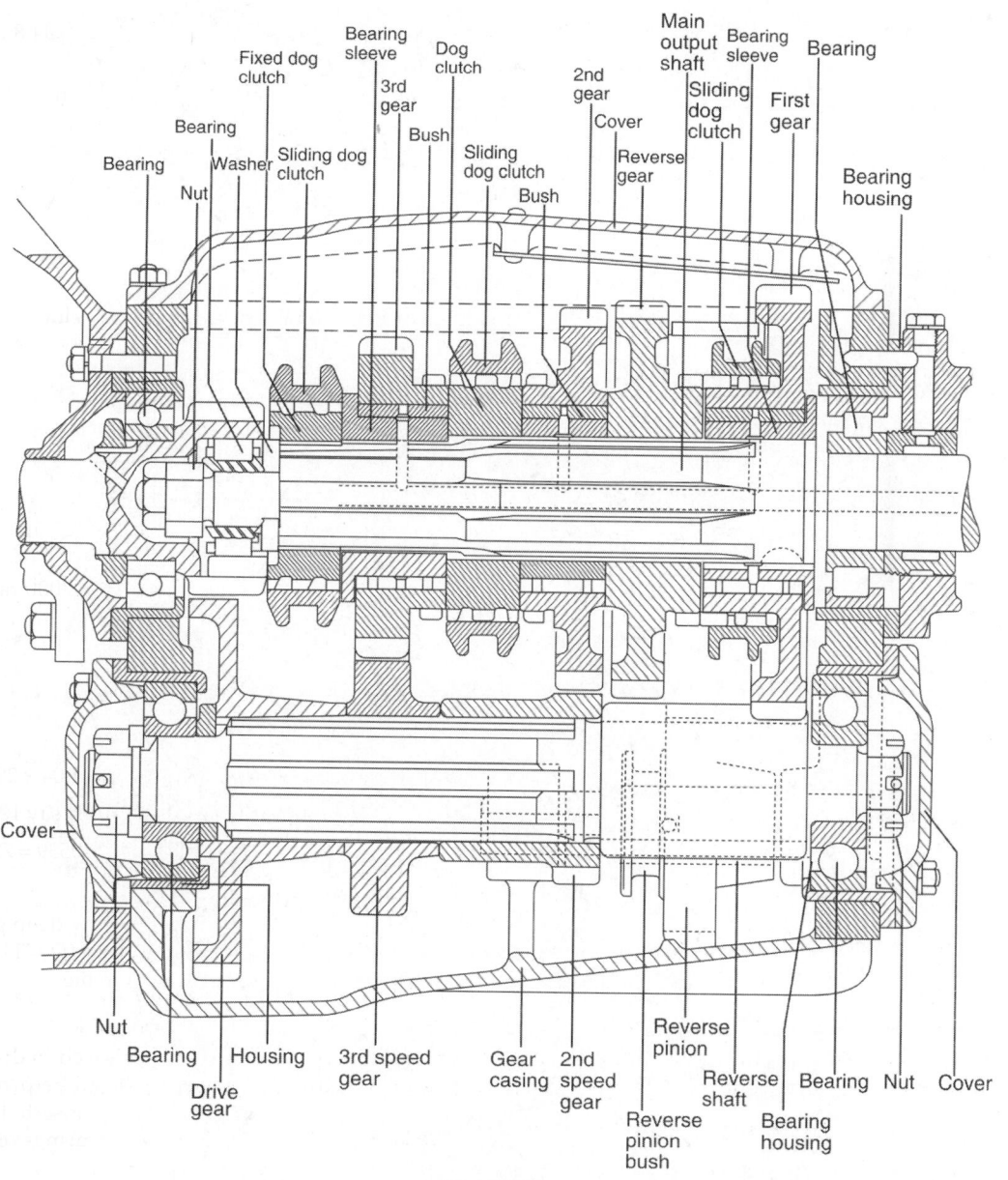

Fig. 30.27

For this value of C, SKF single row cylindrical roller bearing NU 308 ECP having $d = 40$ mm, $D = 90$ mm, $B = 23$ mm and $C = 93$ kN is suitable.

Shaft III

Maximum speed of this shaft

$$= \frac{1800}{2.05} = 878 \text{ say } 880 \text{ rpm}$$

Life $\qquad L = \dfrac{3000 \times 880 \times 60}{10^6} = 158.4 \text{ mR}$

$$\frac{C}{P} = 5.42$$

Left end bearing

Resultant reaction $F_r = \sqrt{3.97^2 + 1.54^2} = 4.26$ kN

This end, end thrust is considered because the right end bearing will take up end thrust

Equivalent bearing load = 4.26 kN

$$C = 4.26 \times 5.42 = 23.08 \text{ kN}$$

$$d = 45 \text{ mm}$$

SKF single row deep groove ball bearing 6206 having $d = 30$ mm, $D = 62$ mm and $B = 16$ mm with $C = 29.6$ kN is suitable.

Right end bearing

Reaction $\qquad F_r = \sqrt{8.89^2 + 3.801^2} = 9.67$ kN

Axial thrust $F_a = 14350 \tan 23° - 7000 \tan 23°$
$\qquad\qquad = 3119$ N

$$e = \frac{3119}{9670} = 0.32 \text{ for which } X = 1, Y = 1.15$$

Equivalent bearing load

$$F = 9670 + 3119 \times 1.15 = 13257.85 \text{ N}$$
$$= 13.257 \text{ kN}$$
$$C = 5.42 \times 13.257 = 71.85 \text{ kN}$$

For these values SKF single row deep groove ball bearing 6409 having $d = 45$ mm, $D = 120$ mm, $B = 29$ mm with $C = 76.1$ kN is suitable.

The complete design is shown in Fig. 30.28.

Epicyclic transmissive for automobiles: Epicyclic gear boxes of various designs and arrangements are being widely used. These are also used in automatic transmissions because of ease of operation. There is no loss of power or possibility of jerk in change of speeds. The scheme of gears in epicyclic gear train is shown

in Fig. 30.28. The Wilson gear box is shown in Fig. 30.29. The input is the engine side and output is the propeller shaft. There are four epicyclic trains of gears to get three velocity reductions and one direct speed (1:1) in the forward direction and one reverse reduction velocity is obtained.

Operation: The input shaft is on the engine side, i.e. left and output side is the propeller shaft side. Four trains are numbered 1, 2, 3, 4, and R shows the arrangement for reverse gear.[16]

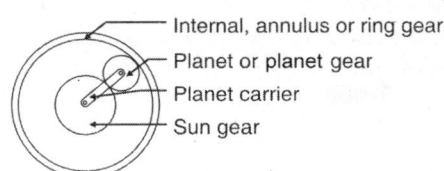

Fig. 30.28

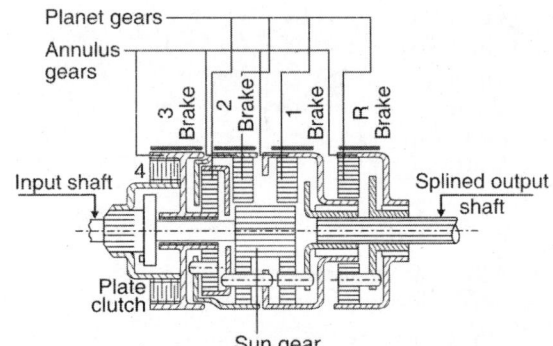

Fig. 30.29: Epicyclic gear box

First speed reduction is obtained by stopping the annulus of train 1. The engine in rotating the sun gear of train 1, so the planet gears will be rolling round it. This rolling of the carries is fixed to the output shaft through splines.

Second speed reduction is obtained by stopping the second train 2 annulus. The sun gear of this train is turning the planets and planet carrier which is connected to the first gear annulus. The annulus speeds up the rotation of its planet carrier. The output shaft

[16] Computer aided design of Wilson automatic epicycle transmission journal of institution of engineers (India) vol. 54 ME4 March 1974 J. Chakraborty, O. P. Grover.

runs faster than the first speed reduction, i.e. less reduction of speed is obtained.

Third speed reduction is obtained by holding the third train gear 3 annulus. This annulus is integral with planet gear of second train. The planet carrier of train 3 is connected to the annulus of train 2 and drives it faster. So the drive is taken through second gear planet carrier to the first gear annulus. The planet gear of train 1 speeds up and increases the speed of output shaft.

Top gear to obtain (1:1) top gear, all the trains are locked together so they revolve as one solid cylinder, driving the output shaft at the engine speed. This is brought about by releasing all the brakes and actuating the clutch which locks the sun gear of third train to the driving shaft, i.e. all the sun gears with annuli run simultaneously.

Reverse speed reduction: The first gear annulus is connected to the sun gear of reverse gear train R so the direction of rotation of sun gear is opposite to the rotation of other sun gears. When annulus R is stopped, the planets of this train with planet carrier which is connected to output shaft is rotating in the opposite direction.

Analysis: For calculating the number of teeth of various gears in all the trains, there is a need to consider the constraints given below Fig. 30.30.

(a) *Co-axiality constraint*: For an epicyclic gear train all the gears: sun, planets, planet carrier and annulus gears must be coaxial. This is required for all the trains.

(b) *Neighbourhood conditions:* The load carrying capacity is maximum when the number of planets is maximum and all the planet gears need to be accommodated in the available space, equally spaced without interference. The following condition is to be satisfied:

$$\sin\frac{\pi}{Q_i} \geqslant \frac{P_i+2}{P_i+S_i}$$

where, S_i and P_i is the number of teeth of sun and planet gears, Q is the number of planets.

(c) *Assembly condition:* The number of teeth in the sun and annulus must satisfy the assembly condition, i.e.

$$\frac{S_i+A_i}{Q_i} = \text{Integer}$$

(d) *Interference* (i) to avoid interference, the minimum number of teeth of pinion according to Buckingham for epicyclic gears should preferably be $\geqslant 16$ (ii) The sun and planet should be meshed at a center distance cosponsoring to $(S+P+2)$ teeth (iii) To avoid interference between annuli and planets, a correction coefficient of 0.5 be given to annuli teeth and size of planet be sufficiently smaller.

Size of annuli: The diameters of annulus of first, second and reverse train used drum for the third gear trains and top gear train should not have much variation to get a compact gear box as follows:

$1.05\,D_1 \geqslant D_2 \geqslant 0.95D_i$

$0.95\,D_1 \geqslant D_3 \geqslant 0.85D_i$

$1.05\,D_1 \geqslant D_4 \geqslant 0.95D_i$

(e) *Velocity constraint*: The final velocity ratio can have a maximum variation of 2% from the specified value.

(f) *Facewidth restriction:* According to BSI – 3681–1966, the ratio of facewidth and circular pitch should be between 2 and 5. The best value from expierience has been as 3.5

(g) *Planet teeth*: The number of teeth on the planets p should be equal to $\dfrac{A-S}{2}$ and $(A–S)$ should be an even number.

(h) *Stress constraint*: The gears are designed using Lewis equation. To ensure that the designed gears are safe for dynamic load in the running condition, Buckingham's equation is used as a check, i.e.

Dynamic load

$$F_d = F_t + \frac{21v\left(bc+F_t\right)}{21v+\sqrt{\left(bc+F_t\right)}}$$

The margins of safety is based on the consideration of static beam strength of teeth based on endurance stress, i.e. $F_e = S_e b \, py$ and $F_e \geqslant 1.35 F_d$.

Most of the time gears run at low torques and low speeds and in top gear, gears are not transmitting torque.

Example 30.10: Design a gear box with the following data for a bus:

Maximum torque	$= 621$ Nm at 1100 rpm
First speed reduction	$= 1 : 4.28 \, (R_1)$

Second speed reduction	$= 1 : 2.43 \, (R_2)$
Third speed reduction	$= 1 : 1.59 \, (R_3)$
Top speed	$= 1 : 1$
Reverse speed reduction	$= 1 : 5.97 \, (R_4)$

Considering the above speeds, tabulation method is used for calculating number of teeth

(i) $A_1 = (R_1 - 1) \, S_1$

$$P_1 = \frac{A_1 - S_1}{2} = \left(\frac{R_1 - 2}{2} \right) S_1$$

(ii) $S_2 = S_1$

$$A_2 = x S_2 \, (R_2) - S_1$$

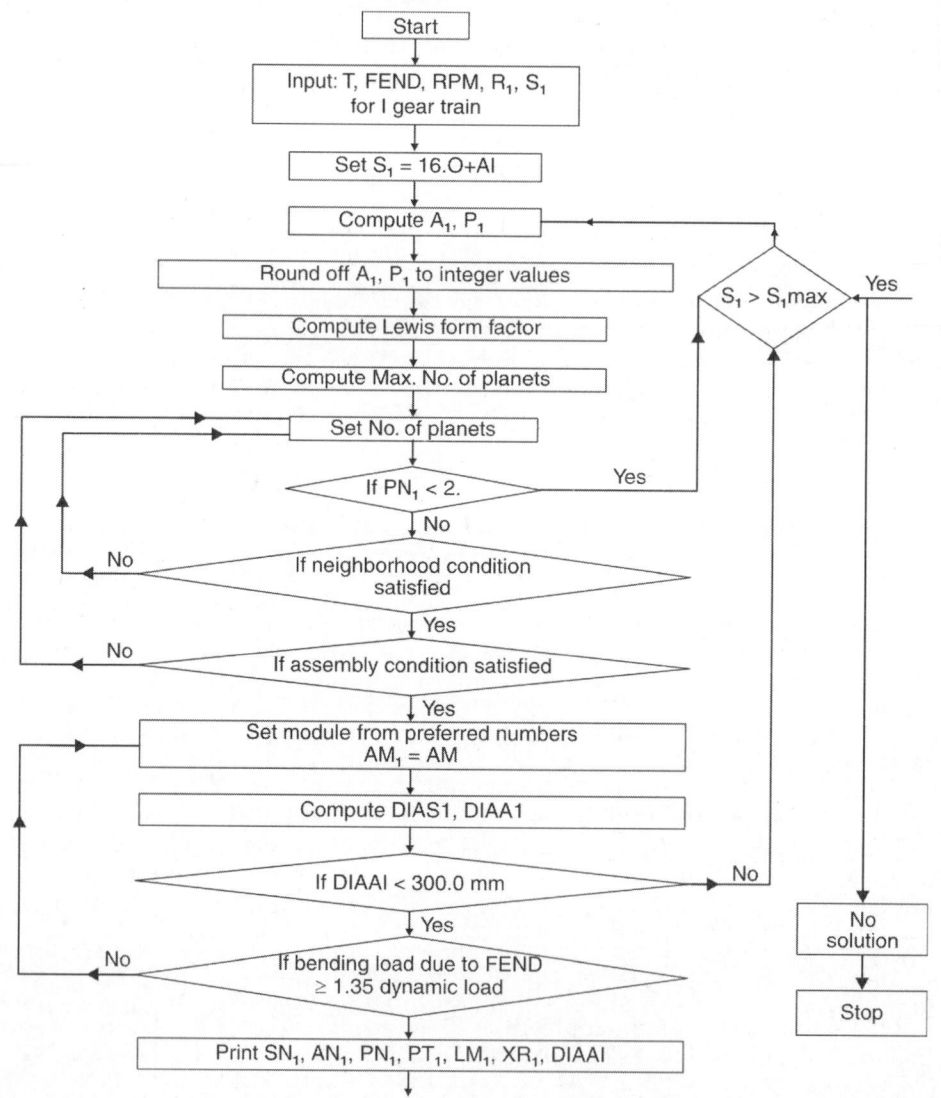

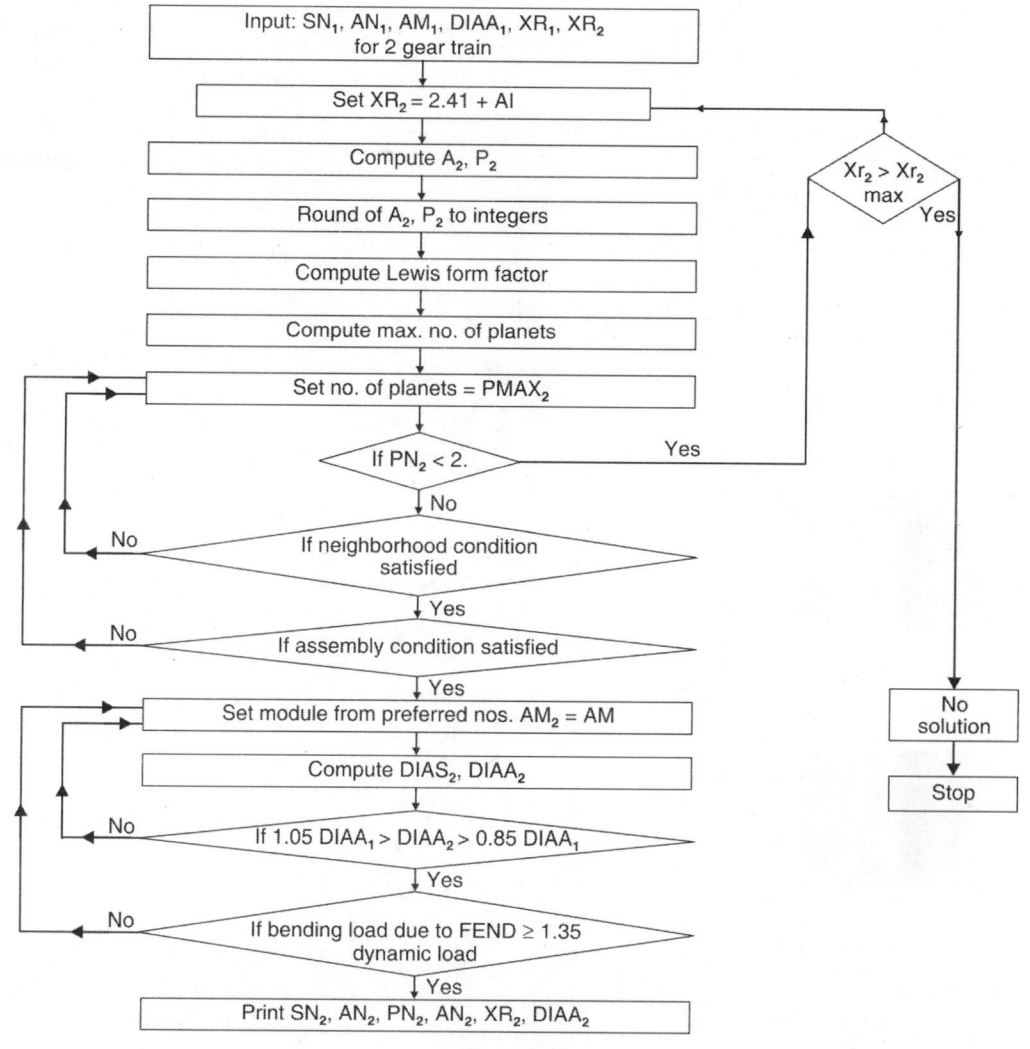

Fig. 30.30

$$x = \left(\frac{A_2 + S_2}{A_1 + S_1}\right)\frac{S_1}{S_2} + \frac{A_1}{A_1 + S_1}$$

$$P_2 = \frac{A_2 - S_2}{2}$$

(iii) $G_3 =$

$$\frac{A_2(A_1 + S_1) - A_2 S_1 R_3}{(A_1 + S_1)(A_2 + S_2) - R_3 A_1 S_1 - R_3 S_1 A_2 - R_1 S_1^2}$$

$$A_3 = \frac{S_3}{G_3 - 1}$$

$$P_3 = \frac{A_3 - S_3}{2}$$

(iv) $A_4 = \frac{(R_4 - 1)S_4}{(R_1 - 1)}$

Sun gear teeth of all the trains are taken as $\geqslant 16$ and after applying all the constraints is the values of teeth are calculated.

Solution: The design of the epicyclic gear trains to transmit a specified torque, the overall size of the gear trains should be as small as possible and none of the constraints given should be violated. If b_i is the facewidth of gears in the i th train, the objective function $f(x)$ to be minimised for the most compact box is given by

$$f(x) = \sum_{i=1}^{4} \frac{\pi}{4} D_i^2 \pi k m_i = K \sum_{i=1}^{4} D_i^2 m_i$$

where, k is constant, an m_i is the module of the gear of the i th train. It is assumed that the annulus drum of all the gear trains are of the same diameter. For the programme, the variation in diameter allowed to the annuli of second and reverse gear train is restricted to 5 per cent of the annulus of first gear train while the annulus of third gear train is 5% to 15% smaller as the third gear annulus is enveloped by second gear train annulus.

The following constraints are used for evaluation

1. Interference constraint

   $S_i \geqslant 16$

2. Neighborhood constraint

   $$\sin \frac{\pi}{Q_i} \geqslant \frac{(z_1)_i + 2}{(z_1)_i + (z_3)_i}$$

3. Stress constraint

   $(F_e)_i = 1.35(F_d)_i$

   and

   $$(S_a)_i \geqslant \frac{2T}{k \pi^2 m_i^3 y_i z}$$

   $$F_d = w_i + \frac{21 v_i (b_i c_i + w_i)}{21 v_i + \sqrt{b_i c_i + w_i}}$$

   $$F_e = S_e b_i \pi m_i y_i$$

4. Size constraint

   $D_2 \geqslant 0.95 D_1$

   $\qquad \leqslant 1.05 D_1$

   $D_3 \geqslant 0.85 D_1$

   $\qquad \leqslant 0.95 D_1$

   $D_4 \geqslant 0.95 D_1$

   $\qquad \leqslant 1.05 D_1$

5. Assembly constraint

   $$\frac{A_i + S_i}{Q_i} = \text{Integer}$$

6. Velocity ratio constraint

   $1.02 \, (XR)_i > XR_i > 0.98 \times R_i$

7. Interference constraint for planet gears

   $Z_p \geqslant 16$

8. Module constraint

   $m_i \geqslant 1.0$

The programme for the design has four stages. The output of the first gear train forms the input of the second gear train and reverse gear train. The output of second gear train forms the input for third gear train. The flow charts is shown in Fig. 30.30. The final results are number of teeth of trains

	Sun gear S_i	Annulus gear A_i	Planet gear Q_i	No. of planets Q_i	Module m_i	Face-width b_i
First	21	69	24	5	3	38.5
Second	21	69	24	3	3	33.5
Third	24	64	20	4	3	33.5
Reverse	28	59	16	3	3.5	33.5

The maximum annulus diameter is 260 mm for first, second and reverse gear trains as $pcd = 241.5$, minimum rim thickness $0.65p = 7.2$, minimum diameter $= 241.5 + 7.2 = 248.7$.

PROBLEMS

30.1 Explain why conjugate curves must be used for tooth profiles rather than just any curved surfaces which might be arbitrarily selected.

30.2 What is an involute curve and how is it developed?

30.3 State important reasons why involute curves are preferable to cycloidal profile in general.

30.4 Which type of gears use cycloidal profile and enumerate their advantages?

30.5 What is hobbing operation for cutting gear teeth and state the property of the involute profile which makes it possible?

30.6 What is the purpose of having standard modules, pressure angles and tooth proportions?

30.7 What is the significance of a parabola commonly used in the derivation of the Lewis equation?

30.8 What surface treatment could be used to increase the bending fatigue life and surface fatigue life of gear teeth?

30.9 What is a hunting tooth?

30.10 What is interference in involute gears?

30.11 What is the effect of increasing centre distance on the pressure angle of gears of involute profile?

30.12 What is the advantage of stub tooth over full depth tooth?

30.13 What type of stress is set up in the shaft of a gear drive? [**Ans.** Spur gear produces torque and bending moment. Helical gear produces torque, bending moment and axial thrust.]

30.14 What is the effect of increasing the helix angle of a helical gear drive? [**Ans** increase of torque capacity.]

30.15 What factors cause dynamic load on the gears?

30.16 What is law of gearing?

30.17 In a gear drive, the direction of motion at output is same as that of driving gears. How many idle gears should be used?

30.18 What is shape of surfaces making contact in involute and cycloidal gear drives?

30.19 A gear having a pitch diameter of 150 mm must transmit 3 kW at 100 rpm. The service is intermittent. Determine the pitch, the number of teeth, and the width of the gear face, using cast iron for (a) cast teeth and (b) cut teeth.

30.20 An 30C8 or SAE 1030 steel pinion has 18 teeth of 14.5° full-depth type with a 56 mm pitch and a 112 mm face. At 600 rpm it transmits a torque of 370 Nm to a cast-iron gear with 60 teeth. Determine the stresses in the teeth of the pinion and of the gear, the margin of safety for dynamic loads, and whether the pinion and the gear are suitable for continuous service if both are ordinary commercial products. State the changes which may be necessary to obtain a sufficient margin of safety for dynamic loads and to resist wear in continuous service.

30.21 A 40-tooth phosphor-bronze gear runs with a 20-tooth steel pinion of 175 Bhn. The helical gears are cut on the 20° full-depth involute system and are of

6 module with a 65 mm face. The pinion speed is 1,000 rpm and gear runs at 200 rpm. Determine the power that the gears can transmit, based on (a) static strength, (b) dynamic load, and (c) wear resistance.

30.22 A pair of 20° involute stub-tooth gears must transmit 90 kW at 720 rpm of the pinion, with a 3:1 speed reduction. The pinion has 24 teeth, and the gears have a module of 5 and 65 mm faces. Find the minimum necessary hardness of the teeth of the pinion and of the gear, based on wear resistance.

30.23 A 20 kW electric motor runs at 1,475 rpm and drives through a train of spur gears a shaft whose velocity is about 225 rpm. Use a silent-material pinion. Compute the main dimensions of the gears, using 14.5° full-depth involute teeth.

30.24 With the data of problem 30.23, determine all main dimensions of a drive with herringbone gears. The pinion is of hardened steel, and the gear is of cast iron and helix angle is 30°.

30.25 A pair of spur gears must transmit 36 kW from a shaft running at 300 rpm to another shaft with a speed reduction of 3.5:1. The center distance of the shafts is 400 mm. Determine (a) the module and the number of teeth of the gears, (b) the face of the gears, taking into account dynamic load and wear, and (c) the materials that must be used for the pinion and the gear.

30.26 A reciprocating compressor must be driven by an 970 rpm electric motor through a pair of straight spur gears. The compressor should run at about 200 rpm and requires a torque of 280 Nm. Assume a starting overload of 25 per cent, determine (a) the necessary power of the motor, (b) the module and face of the gears using 20° stub teeth, and (c) the number of teeth and the pitch diameter of each gear (d) specify the materials for the pinion and the gear.

30.27 A 12 kW motor running at 1,170 rpm drives a fan through a pair of helical gears with a reduction ratio of about 3.9:1. A micarta pinion and a cast-iron gear are specified. Determine (a) module and the number of teeth in the gears, (b) the gear face, and (c) the pitch diameters and the center distance. The helix angle is 23°.

30.28 The gate of a sluice valve weighting 60 kN is raised by means of a cast-iron rack and pinion. Design a train of gears, including the rack, so that the gate may be raised by two men working on 380 mm crank handles and exerting a pressure of 160 N each. Give also the linear speed of the rack motion, assuming that the hand crank makes 25 rpm.

30.29 A punch press running at 42 rpm is driven by a 10 kW motor running at 1,200 rpm. Using a double-gear reduction, select all materials and design the gear train, stating all other assumptions. Give a sketch of the gear train.

30.30 The drum diameter of a hoist is 380 mm, and the drum is to be bolted by a flange to a gear approximately 0.66 m in diameter. The capacity of the hoist is 20 kN and the motor speed is 700 rpm. A hoisting speed of about 0.6 m/s is desired. Select the number of gear pairs in this train, and materials for the gears. Determine the module using stub teeth; the faces; the number of teeth; and the center distances of the whole mechanism. Illustrate the results by sketches.

30.31 Design a train of gears to transmit 600 kW from a shaft running at 3,600 rpm to one running at 200 rpm. Use herringbone gears, for both gears, the static stress in 500 N/mm², $\phi = 20°$, $\psi = 23°$.

Toothed Bevel Gearing

31.1 GENERAL CONSIDERATIONS

Bevel gears are used to connect two intersecting shafts with any given speed ratio. Two types of bevel gearing are in general use—straight-tooth gears and spiral-tooth gears. In the straight-tooth bevel gears, called *straight bevel gears*, several types of which are shown in Fig. 31.1, the elements of the teeth converge to a common point O, called the apex, which is the point of intersection of the gear axes. The form of tooth used for bevel gears is the involute. Spiral bevel gears are made with curved teeth, as shown in Fig. 31.6. Spiral bevel gears compare with straight bevel gears much as helical gears on parallel shafts compare with straight-tooth spur gears. Their advantages are smoother tooth engagement, quiet operation, greater strength, and higher permissible velocities.

Bevel gears are not interchangeable and are designed in pairs. In the majority of cases the axes of the shafts form a right angle, but they may intersect at any desired angle, as shown in Fig. 31.1.

The names of bevel gears, based on the angles between the shafts and the pitch angles, are indicated in Fig. 31.1

Definitions: As may be seen from Fig. 31.2, any group of tooth elements lie in the surface of an imaginary cone. Thus, lines containing the pitch elements of the teeth are elements of the *pitch cone*. The apex O of the pitch cone is called the *cone center*. The length l of a pitch-cone element is called the *cone distance*, or

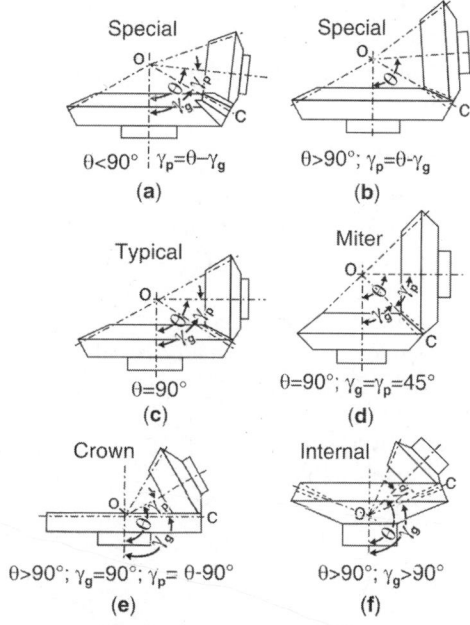

Fig. 31.1: Types of bevel gearing

pitch-cone radius. The angle γ that the pitch line makes with the axis is called the *pitch angle* or *center angle*. The angle α is called the *addendum angle*, and the *face angle* evidently is equal to $\gamma + \alpha$. The angle δ is called the *dedendum angle*, and the *root angle*, also called the *cutting angle*, is equal to $\gamma - \delta$. In speaking of the pitch or module of a bevel gear, the pitch of the large end is meant. The *diameter* of the gear is the diameter D of the largest pitch circle. The addendum a and dedendum d are measured at the large end of the tooth. The outside diameter is designated D_o.

559

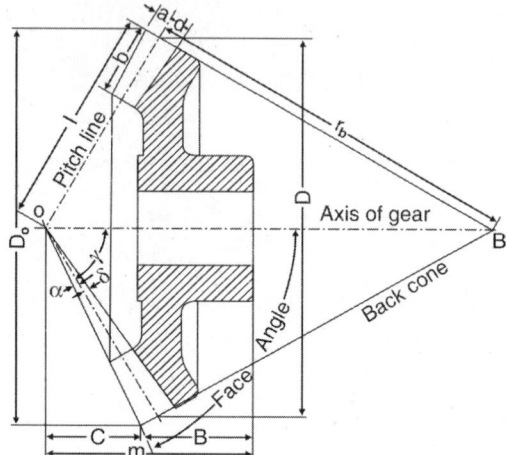

Fig. 31.2: Definitions and dimensions relating to bevel gears

The back cone is an imaginary cone the elements of which are perpendicular to the elements of the pitch cone at the large end of the tooth. The length r_b of a back-cone element is called the back-cone radius.

The formative number of teeth z_f is the number of teeth of the given pitch which would be contained in a spur gear having a radius equal to the back cone radius r_b (Fig. 31.4).

The distance c is called the crown height, B is the backing, and m is the mounting distance.

The subscripts p and g will refer to the pinion and the gear, respectively.

Method of manufacture: Bevel gears are either cast or cut. The casting process is similar to that used for making spur gears, but the cutting is made much more difficult by the continuously changing size and shape of the tooth tapering from the large end toward the apex. There are several different methods of cutting the teeth, some of which produce only approximately correct forms and require hand filing for better results. The methods by which the teeth are formed with theoretical accuracy require special and rather complicated machines. The output of these machines for generating

bevel gear teeth is high and the cost of production is comparatively low. Therefore most bevel gears have generated teeth.

Efficiency: The efficiency of properly cut and well-lubricated bevel gears equipped with antifriction bearings is in general slightly higher than that of spur gears and runs up to 99 per cent.[1]

31.2 ANGLE RELATIONS

The shaft angle θ (Fig. 31.1) between the axes of the shafts may have any value up to 180° but is commonly 90°. It is always measured to include the pitch-cone element O_c, which is common to both the pinion and the gear.

Acute-angle bevel gears: If the number of teeth are denoted by z_p and z_g, it follows from the geometry of Fig. 31.3 that

$$\tan \gamma_p = \frac{D_p}{2(m+n)}$$

where

$$m = \frac{D_g}{2 \sin \theta}$$

and

$$n = \frac{D_p}{2 \tan \theta}$$

Therefore,

$$\tan \gamma_p = \frac{D_p \sin \theta}{D_g + D_p \cos \theta} = \frac{\sin \theta}{\dfrac{z_g}{z_p} + \cos \theta} \qquad \dots (31.1)$$

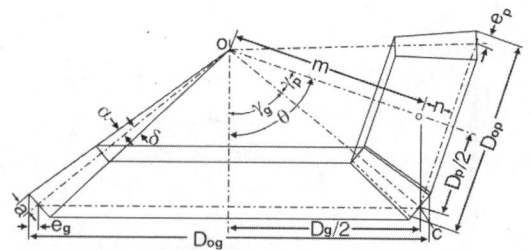

Fig. 31.3: Acute-angle bevel gears

[1] W.H. Kennersion, "Investigation of Efficiency of Worm Gearing for Automobile Transmissions," *Transactions of the American Society of Mechanical Enginners*, Vol. 34 (1912), p. 931; C. M. Allen and F. W. Roys, "Efficiency of Gear Drives," *Trans. ASME*, Vol.40 (1918), pp. 106–9.

The center angle of the gear is $\gamma_g = \theta - \gamma_p$. By reasoning in the same manner as for equation 31.1, we can get

$$\tan \gamma_g = \frac{\sin\theta}{\dfrac{z_p}{z_g} + \cos\theta} \qquad \dots (31.2)$$

The addendum angle α, also called the *angle increment*, is found from the relation

$$\tan \alpha = \frac{2a\sin\gamma_p}{D_p} = \frac{2a\sin\gamma_g}{D_g} \qquad \dots (31.3)$$

The dedendum angle δ, also called the *angle decrement*, is found from the relation

$$\tan \delta = \frac{2d\sin\gamma_p}{D_p} = \frac{2d\sin\gamma_g}{D_g} \qquad \dots (31.4)$$

For turning the blanks it is necessary to know the outside diameters of the pinion and the gear. These diameters are equal to the pitch diameters plus twice the diameter increment. The diameter increment for the pinion is

$$e_p = a \cos\gamma_p \qquad \dots (31.5)$$

and the outside diameter of the pinion is

$$D_{op} = D_p + 2a \cos\gamma_p \qquad \dots (31.6)$$

For the gear, similarly, the diameter increment is

$$e_g = a \cos\gamma_p \qquad \dots (31.7)$$

and the outside diameter is

$$D_{og} = D_g + 2a \cos\gamma_g \qquad \dots (31.8)$$

Right-angle gears: When $\theta = 90°$, equations 31.1 and 31.2 reduce to

$$\tan \gamma_p = \frac{z_p}{z_p} \qquad \dots (31.9)$$

and

$$\tan \gamma_g = \frac{z_g}{z_p} \qquad \dots (31.10)$$

All other relations remain unchanged.

Obtuse-angle gears: In obtuse-angle gears, θ is greater than 90°. The three possible arrangements are illustrated in Figs 31.1b,

31.1e, and 31.1f. A derivation similar to that just used gives for the pinion,

$$\tan \gamma_p = \frac{\sin(180° - \theta)}{\dfrac{z_g}{z_p} - \cos(180° - \theta)} \qquad \dots (31.11)$$

For the gear,

$$\tan \gamma_g = \frac{\sin(180° - \theta)}{\dfrac{z_p}{z_g} - \cos(180° - \theta)} \qquad \dots (31.12)$$

The remaining calculations are made by equations 31.3 to 31.8.

31.3 STRENGTH OF CUT TEETH

The load on a bevel-gear tooth varies, from the large end, along the face. The relation between the tooth strength and the tangential force corresponding to the power transmitted at a given speed may be found by considering an infinitesimal tooth section at a distance x from the apex O, as indicated in Fig. 31.4. The infinitesimal face of this section is dx, and the force which acts upon it is dF_x. Applying equation 30.12 to this section gives

$$dF_x = S_o cyp_{cx}\, dx \qquad \dots (31.13)$$

where, p_{cx} is the circular pitch at the distance x from the apex. Multiplying both sides of equation 31.13 by the gear radius r_x at this point, and integrating both sides we get

$$\int r_x dF_x = \int S_o cyp_{cx} r_x\, dx \qquad \dots (31.14)$$

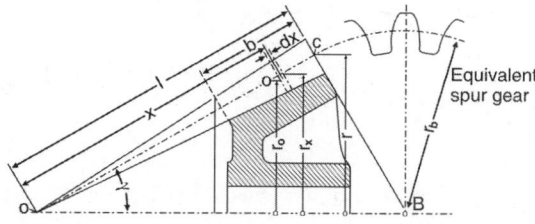

Fig. 31.4: Strength determination of a bevel-gear tooth

The left side represents a summation of the products of all elemental forces by their radii and is equal to the torque T transmitted by the gear.

If all constant terms are put before the integral sign, the result is

$$T = S_o cy \int p_{cx} r_x \, dx \qquad \text{... (31.15)}$$

Since all tooth elements on a straight-tooth bevel gear converge to the cone center O, the circular pitch is proportional to the distance from O. Thus

$$p_{cx} = \frac{p_c x}{l} . \qquad \text{... (31.16)}$$

Also, from the similar triangles in Fig. 31.4,

$$r_x = \frac{rx}{l} \qquad \text{... (31.17)}$$

Substituting values of p_{cx} and r_x from equations 31.16 and 31.17 in equation 31.15 gives

$$T = \left(\frac{S_o cy p_c r}{l^2}\right) \int x^2 dx \qquad \text{... (31.18)}$$

In order to include the whole face b, the limits of integration for the integral in equation 31.18 must be $(l-b)$ and l. Integration then gives

$$T = \frac{S_o cy p_c r}{l^2}\left[\frac{x^3}{3}\right]_{l-b}^{l}$$

$$= \frac{S_o cy p_c r}{l^2}\left[\frac{l^3 - l^3 + 3l^2 b - 3lb^2 + b^3}{3}\right]$$

$$= S_o cy p_c br\left(1 - \frac{b}{l} + \frac{b^2}{3l^2}\right) \qquad \text{... (31.19)}$$

Dividing both sides of equation 31.19 by the radius r, and noticing that $T/r = F_t$, and substituting for p_c its equivalent πm, we get

$$F_t = S_o cyb\pi m\left(1 - \frac{b}{l} + \frac{b^2}{3l^2}\right) \qquad \text{... (31.20)}$$

In actual gears the relative length of the face, $b/l \leq \frac{1}{3}$, and the term $b^2/3l^2$ in the parentheses is usually neglected. The American Gear Manufactures Association sanctions this omission, which is on the safe side. Equation 31.20 then becomes

$$F_t = S_o cyb\pi m\left(1 - \frac{b}{l}\right) \qquad \text{...(31.21)}$$

It should be pointed out that F_t is not the actual load at the large end of the tooth but is simply an imaginary safe load to be compared with the transmitted load obtained from the power equation when the speed at the large end is used.

In terms of torque transmitted

$$m^2 c S_o = \frac{2T}{\pi by z}\left(\frac{l}{l-b}\right) \qquad \text{... (31.22)}$$

The static stress S_o may be taken from Table 30.3. The velocity factor c should be computed by equation 30.6 if the teeth are cut by form cutters, and by equation 30.16 if they are generated with precision machine.

The form factor y is obtained from Table 30.2, but it is based on the *formative number of teeth* z_f (see section 31.1) and not on the actual number of teeth in the gear. It follows from the definition of the module that

$$z_f = \frac{2\pi r_b}{p} = \frac{2\pi r_b}{\pi m} = \frac{2r_b}{m} \qquad \text{... (31.23)}$$

In Fig. 31.5 the triangles OcB and OcE are similar. Hence, for the pinion, $r_b/l = 0.5D_p/0.5D_g$, or

$$r_b = \frac{lD_p}{D_g}$$

Substituting this value of r_b in equation 31.23 and noticing that $\dfrac{D_p}{m} = z_p$, we get

$$z_{fp} = z_p\left(\frac{2l}{D_g}\right) = \frac{z_p}{\cos\gamma_p} \qquad \text{... (31.24)}$$

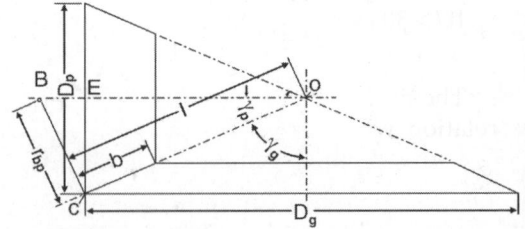

Fig. 31.5: Elements of a pair of bevel gears

Similarly, the formative number of teeth for the gear is

$$z_{fg} = z_g \left(\frac{2l}{D_p} \right) = \frac{z_g}{\cos \gamma_g} \qquad \ldots (31.25)$$

As can readily be seen,

$$\frac{D_g}{2l} = \cos \gamma_p \qquad \ldots (31.26)$$

and

$$\frac{D_p}{2l} = \cos \gamma_g \qquad \ldots (31.27)$$

Thus the relations in equations 31.24 and 31.25 may be given in the following general form:

$$z_f = \frac{z}{\cos \gamma} \qquad \ldots (31.28)$$

Equation 31.28 applies to any shaft angle θ and gives the formative number of teeth for acute-angle and obtuse-angle bevel gears as well.

For θ = 90°, the formative number of teeth can also be written as

$$z_{fp} = \frac{z_p \sqrt{z_p^2 + z_g^2}}{z_g} \text{ and}$$

$$z_{fg} = \frac{z_g \sqrt{z_p^2 + z_g^2}}{z_p} \qquad \ldots (31.28a)$$

Width of gear face: The maximum width b of the face of bevel gears should not be over one-third of the cone distance, or $b/l \leq \dfrac{1}{3}$. In addition, the practice of the Gleason Works is as follows:

If $l < 30$ m

$$b = 6 \text{ m to } 7 \text{ m} \qquad \ldots (31.29)$$

If $l > 30$ m

$$b = 7 \text{ m to } 10 \text{ m} \qquad \ldots (31.30)$$

The simplest way to find l is by the following relation, which is based on Fig. 31.5:

$$l = \sqrt{\left(0.5 D_g\right)^2 + \left(0.5 D_p\right)^2}$$

$$= 0.5 \sqrt{D_g^2 + D_p^2} \qquad \ldots (31.31)$$

Tooth proportions of ordinary bevel gears, at the large end, are made according to the AGMA composite $14\dfrac{1}{2}^\circ$ standard as given in Table 30.1 and also for the gears made with 20° and 20° stub teeth.

31.4 DESIGN OF SERVICE

The first step in the design of bevel gears should be to find a tooth size suitable for strength. Then if the drive is for continuous service, its wearing qualities should be investigated. In regard to assumptions, if there are space limitations, the maximum gear diameter may be assumed and the pinion diameter may be found from the prescribed speed ratio. For straight bevel gears the minimum number of teeth may be from 12, for a velocity ratio of about 4 up to 18, for a velocity ratio of 1. For spiral gears, the minimum number of teeth may be slightly lower, ranging from 10 to 15 for the above velocity ratios. However, smoother and quieter operation is obtained by using more teeth.

When there are no space limitations, the starting assumption may be that of the pitch-line speed at the larger tooth end. This speed may be about 5 to 6.5 m/s for ordinary cut teeth and about 10 m/s for generated teeth.

Dynamic load: Until specific data for bevel gears are available, equations 30.20 and 30.21 may be applied for the dynamic load. In computing the load increase from equation 30.21, the velocity of a point on the largest pitch circle must be used for v_m, and the transmitted load F_d must also be based on this velocity. The rest of the procedure is the same as for spur gears.

Check for wear: The limit load for wear is given by the equation

$$F_w = \frac{K b D_p Q}{\cos \gamma_p} \qquad \ldots (31.32)$$

The factor K is the same as for spur gears and is given in Table 30.7. The value of Q used is

Table 31.1: Service factors for Gleason gears

Character of power source	Character of load on driven machine		
	Uniform	Moderate shocks	Heavy shocks
Uniform	1.00	1.25	1.75
Light shocks	1.10	1.35	1.80
Medium shocks	1.25	1.50	1.85
Data from an AGMA (which may serve as a guide)			
Air compressor	1.35	Pneumatic tools	1.35
Airplane propeller	0.7–1.0	Pulverizers, coal or cement	1.0
Blowers, fans	1.0	Reciprocating pumps	1.5
Centrifugal extractors	1.0	Railway motor cars:	
Centrifugal pumps	1.0	(a) Based on starting torque	0.5
Coal dryers, rotary	1.0	(b) Based on normal running load	2.0
Coal and rock crushers	2.0	Road-building machinery	1.0–1.5
Conveyors	1.0–1.5	Rolling mills	2.0
Dredging machinery	1.5	Screens, coal or rock	1.0
Electric tools, portable	1.35	Speed reducers	1.0
Glass manufacturing machinery	1.0	Textile and woodworking machinery	1.0
Hoisting machinery	0.75–1.0	Washing machines	1.0
Machine tools:		Well-drilling machinery	1.35
(a) Motor-driven	1.5	Wire-drawing machinery	1.0
(b) Belt drive, direct	1.0		
(c) Belt drive, transmission	0.8		
Mining machinery	1.35		

Table 31.2: Material factors for bevel gears

	Pinion				Gear			
Material	Hardness		Material		Hardness		Durability
	Bhn	R-C			Bhn	R-C	C_m
Cast iron or soft steel	160–200	–	Cast iron		160–180	–	0.30
Heat-treated steel	245–280	24–29	Heat-treated steel		245–280	24–29	0.35
Surface-hardened steel	480	50*	Cast iron		160–180	–	0.40
Case hardened steel	550	55*	Cast iron		160–180	–	0.40
Case hardened or oil-hardened steel	550	55*	Soft or cast steel		180–200	–	0.45
Case hardened steel	–	55*	Heat-treated steel		245–280	24–29	0.50
Oil-hardened steel	–	–	Oil-hardened steel		–	–	0.65
Surface-hardened steel	–	50*	Surface-hardened steel		–	50*	1.00
Case hardened steel	–	55*	Surface-hardened steel		–	50*	1.00
Case hardened steel	–	55*	Case hardened steel		–	55*	1.00

*Minimum values.

determined for the formative number of teeth by the relation

$$Q = \frac{2z_{fg}}{z_{fp} + z_{fg}} \qquad \ldots (31.33)$$

Peak load rating: The peak load power rating recommended by AGMA standards is

Power in kW,

$$P = \frac{\pi m S n D_p b y}{19100}\left(\frac{L - 0.5b}{L}\right)\left(\frac{5.6}{5.6 + \sqrt{v_m}}\right) \qquad \ldots (31.34)$$

S = 1.7 Bhn of the weaker gear for hardened and for unhardened gears after cutting (MN/m³)

= 2 Bhn of the weaker gear if the gear is case hardened (MN/m³)

n = speed of pinion rpm

m = module in m

b = facewidth in m

The wear rating recommended by AGMA standards is

Power in kW = $0.8\, C_m\, C_B b$ for straight bevel gears

= $C_m C_b b$ for spiral bevel gears, C_m is material factor Table 31.2

$$C_b = \frac{D_p^{1.5} n}{0.032}\left(\frac{5.6}{5.6 + \sqrt{v_m}}\right)$$

The minimum number of teeth in a straight bevel pinion is 10 and for spiral bevel gear can be as low as 5.

Design procedure: The procedure in the design of bevel gears is the same as for spur gears. If an assumed pair of bevel gears with a certain combination of materials is found to be too small, as indicated by the fact that b is greater than $l/3$, or if the gears are found to be too big, because b is much less than $l/3$, the design is adjusted simply by changing the pitch diameters and, through them, changing the peripheral velocity v_m in the required direction. The number of teeth remains the same, and the module m changes automatically. Thus the preliminary design of a pair of bevel gears is simpler than that of a pair of spur gears.

Design steps for bevel gears: A generalized method as for bevel gears is:

1. Assume number of teeth for pinion and gear z_p and z_g depending on velocity ratio and find formative number of teeth to find y for each.

2. Assume suitable materials for pinion and gear and their stresses S_{op} and S_{og} if not given.

3. Find the strength factor for pinion and gear, i.e. $S_o y$ and find the weaker element.

4. Design the weaker element by using equation (31.22)

   i.e. $\quad m^2 c S_o = \dfrac{2T}{\pi b y z}\left(\dfrac{l}{l-b}\right) = A$ (say)

   Pitch length $l = 0.5m\sqrt{z_g^2 + z_p^2}$

   $$b = \frac{l}{3}$$

   Assume $c = 0.5$ as the first trial and calculate module m. Take the nearest standard module.

5. Check for allowable stress and induced stress as:

   With selected module m find,

   $$v = \frac{\pi D N}{60 \times 1000}$$

   $$c = \frac{3}{3 + v}$$

   Allowable stress $S_a = c S_o$

   Induced stress $S_i = \dfrac{A}{m^3}$

   If $S_i > S_a$ then increase module and repeat till $S_i \leqslant S_a$.

   In the diameter of any gear in specified, find v, c and allowable stress $c\, S_o = \sigma_a$

   The module is found as $m = \left(\dfrac{A}{\sigma_a}\right)^{\frac{1}{3}}$

   After taking the nearest standard module,

   the number of teeth $z = \dfrac{D}{m}$. After finding

the induced stress, $\sigma_i = \dfrac{2T}{m^2 \pi byz}\left(\dfrac{l}{l-b}\right)$

This has to be less than S_a, otherwise increase module m, and repeat the process.

31.5 GLEASON SYSTEMS OF BEVEL GEARS

The Gleason Works have developed a system for generating bevel gears which combines the following qualities, in the order of their importance: quietness, strength, and durability. The Gleason system for bevel gears, combined with high-grade workmanship made possible by the use of special automatic tooth-generating machines, has given such excellent results that it has been adopted as the recommended practice by the AGMA. This system takes full advantage of the fact that bevel gears are not interchangeable. The pressure angles and the addenda are varied in accordance with the ratios of the numbers of teeth in order to obtain the best result. The Gleason system is used for straight bevel gears and Zerol spiral bevel gears.

The difference between Gleason straight bevel gears and ordinary bevel gears is in the addenda and tooth depth. Zerol bevel gears, like spiral gears, have curved teeth, but the spiral angle α is 0°. Gleason spiral bevel gears have curved teeth with a spiral angle of 30° to 35°.

Although the same general principles are used in designing different Gleason gears, there are several practical differences which it is better to discuss separately.[2]

31.6 SPIRAL BEVEL GEARS

A spiral bevel gear and pinion are shown in Fig. 31.6. The teeth of the gear are curved on the

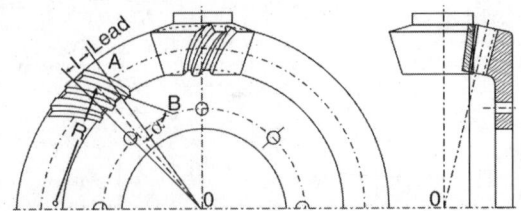

Fig. 31.6: Spiral-tooth bevel gears

arc of a circle with radius R. The teeth of the pinion are cut to mesh with those of the gear, but they are gradually relieved toward the ends to obtain a localized tooth contact, as mentioned in connection with straight bevel gears.

The angle α, which a tangent to the tooth at the middle point of the gear face makes with the element OA of the pitch cone, is called the *spiral angle* of the teeth. All latest data for spiral gears are based on a usual $\alpha = 35°$.

The spirals may be left-hand and right-hand. A right-hand spiral pinion meshes with a left-hand spiral gear, as shown in Fig. 31.6, and a left-hand spiral pinion meshes with a right-hand spiral gear. The hand, or incline, of the spiral of a pinion is the same as the incline of a screw. The main advantages of spiral gears were discussed in section 31.1. This system permits the use of pinions with as few as five teeth. However, Gleason data are worked out only for applications where the minimum number of teeth in the pinion is 12, the pinion is the driver, and the shaft angle θ is 90°. Pressure angles of $14\dfrac{1°}{2}$, 16°, $17\dfrac{1°}{2}$, and 20° are used, the choice depending on the numbers of teeth in the pinion and in the gear.

Spiral gears are designed to operate with high pitch velocities, 5 m/s and higher. Gears operating with speeds in excess of 4 m/s should have ground teeth.

Design: Because the tooth proportions of spiral bevel gears depend on the method of generating the teeth, no formulas are given for the load capacity or size of spiral gears. Such information may be obtained from manufacturers of this type of gearing.

31.7 HYPOID GEARS

Hypoid gears, shown in Fig. 31.7, are also a development of the Gleason Works. They are similar in appearance to spiral bevel gears, but their axes do not intersect. The main feature of hypoid gears is that the shafts of the pinion and the gear may continue past each other.

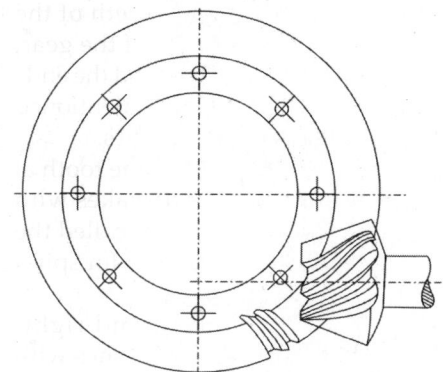

Fig. 31.7: Hypoid gearing

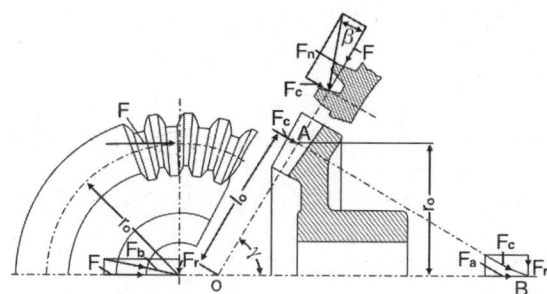

Fig. 31.8: Tooth load application

These gears are so called because the correct pitch surface for such gearing is a hyperboloid of revolution. Their design is based largely on empirical data, and their manufacture requires the use of the same precision machines that are used for generating spiral bevel gears.

31.8 BEARING LOADS AND THRUSTS

As has been stated, the formulas used in designing the teeth of a bevel gear do not use the actual or tangential load on the tooth, but an equivalent load at the large end of the tooth. The effective normal tooth load and its point of application must be determined, in order to analyze the bearing loads and thrusts caused by the action of bevel gears.

Effective tooth load: By a method similar to that used in deriving equation 31.20, it can be shown that the radius r_o (Fig. 31.8) of the application of the effective tooth load on a bevel gear is

$$r_o = D \frac{1 - \frac{b}{l} + \frac{1}{3}\left(\frac{b}{l}\right)^2}{2 - \frac{b}{l}} \qquad \dots (31.35)$$

However, the distance to the midpoint of the face is

$$r_o' = D \frac{(l - 0.5b)}{2l} \qquad \dots (31.36)$$

Since the distance differs from r_o by less than 1 per cent, it is usually taken as the radius of application of the effective tooth load.

The effective tooth load F may be obtained by simply dividing the torque T by the radius of application. Thus

$$F = \frac{T}{r_o} \qquad \dots (31.37)$$

If the tangential load F_t at the large end is known, however, the effective tooth load can be found from the relation

$$F = \frac{F_t l}{l - 0.5b} \qquad \dots (31.38)$$

Bearing loads in straight bevel gears: If the effective normal tooth load F_n, Fig. 31.8, is resolved into components along and perpendicular to the tangent to the pitch circle, the tangential component is

$$F = F_n \cos \beta \qquad \dots (31.39)$$

The component of F_n at right angles to the element of the pitch cone, namely that along the cone line AB, is

$$F_c = F_n \sin \beta = F \tan \beta \qquad \dots (31.40)$$

The component F_c can be resolved into a lateral or radial load F_r and an axial thrusts F_a, the magnitudes of which are

$$F_r = F_c \cos \gamma = F \tan \beta \cos \gamma \qquad \dots (31.41)$$

$$F_a = F_c \sin \gamma = F \tan \beta \sin \gamma \qquad \dots (31.42)$$

The force F produces only a lateral load upon the supporting bearings. As shown in Fig. 31.8, the total lateral load F_b on the bearings is the resultant of F and F_r.

Spiral bevel gears: The following analysis is based on the assumption that a tooth of a spiral bevel gear may be treated in a manner similar to a straight tooth having the same spiral angle α.[3]

For spiral gears the bearing loads and thrusts depend on the direction of rotation. If the driving member, which is nearly always the pinion, except when the speed ratio is 1.0, has a right-hand spiral and rotates the gear clockwise, as shown in Fig. 31.9, or if the pinion has a left-hand spiral and rotates the gear counterclockwise, the direction of rotation will be called *direct*. If the pinion has a right-hand spiral and the gear rotates counterclockwise, or if the pinion has a left-hand spiral and the gear rotates clockwise, the direction of rotation will be termed *reversed*. Also, the thrust will be considered *positive* if it acts away from the cone center, as in straight bevel gears, and the thrust will be considered *negative* if it acts toward the cone center.

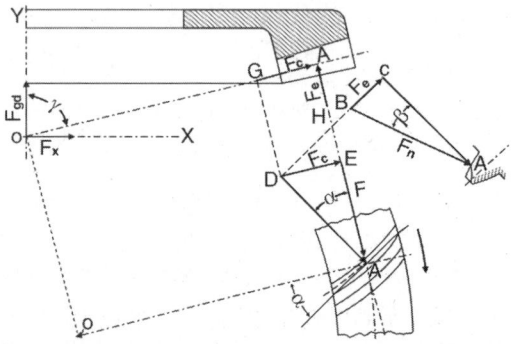

Fig. 31.9: Forces acting on a spiral-tooth gear

Since it is easier to consider the various forces acting upon the tooth of a driven gear, the analysis will be conducted by using Fig. 31.9, for illustration. The pinion has a right-hand spiral, so the gear must be a left-hand spiral. For direct rotation the pinion rotates counterclockwise and the gear rotates clockwise, as shown in Fig. 31.9. The gear has a spiral angle α, a pressure angle β, and a pitch angle γ. It should be noted that in all following equations the angle γ always refers

to the gear, even if the force is applied to the pinion.

Direct rotation: The effective normal tooth load F_n acting at the middle point A of a tooth may be resolved into the three components F, F_c, and F_e.

The component F, which is the tangential force acting on the gear at A, is given by the equation

$$F = F_n \cos \alpha \cos \beta \qquad \ldots (31.43)$$

The magnitude of the component F_c acting along element of the pitch cone is

$$F_c = F \tan \alpha \qquad \ldots (31.44)$$

The component F_e acting at right angle to the element of the pitch cone is

$$F_e = F_n \sin \beta = F \frac{\tan \beta}{\cos \alpha} \qquad \ldots (31.45)$$

If the three forces F, F_c, and F_e are resolved into components whose lines of action are long the center line OY of the shaft and at right angles to it, or along OX, the thrust along the shaft of the gear is

$$F_{gd} = F_c \cos \gamma + F_e \sin \gamma$$

$$= F \frac{(\sin \alpha \cos \gamma + \tan \beta \sin \gamma)}{\cos \alpha} \qquad \ldots (31.46)$$

Also, the thrust along the line OX is

$$F_x = F_c \sin \gamma - F_e \cos \gamma$$

$$= F \frac{(\sin \alpha \sin \gamma - \tan \beta \cos \gamma)}{\cos \alpha} \qquad \ldots (31.47)$$

The thrust exerted by the pinion upon its shaft is numerically equal to F_x, but will be in the opposite direction. Therefore,

$$F_{pd} = F \frac{(\tan \beta \cos \gamma - \sin \alpha \sin \gamma)}{\cos \alpha} \qquad \ldots (31.48)$$

Reversed rotation: If the direction of rotation is reversed, the component F_c reverses its direction and will act toward the apex O. Thus

$$F_c = -F \tan \alpha$$

The component F_e remains unchanged. Noticing that the force F again does not have

[3]O.A. Leutwiler, Elements of Machine Design (New York: McGraw-Hill Book Company, Inc., 1917), p. 346.

components along OX and OY, resolving the forces F_c and F_e, and combining their components, we find that the thrust along the shaft of the gear is

$$F_{gr} = F \frac{(\tan\beta \sin\gamma - \sin\alpha\cos\gamma)}{\cos\alpha} \qquad \ldots (31.49)$$

Also, the thrust along the shaft of the pinion, directly, is

$$F_{pr} = F \frac{(\tan\beta \cos\gamma + \sin\alpha\sin\gamma)}{\cos\alpha} \qquad \ldots (31.50)$$

Experimental results: Tests conducted by the Gleason Works in regard to the thrust of various types of bevel gears showed a good agreement with theoretical values calculated by the foregoing equation.

31.9 CONSTRUCTION DETAILS

Cast blanks for bevel gears have either solid webs or T-shaped arms, Fig. 31.10. The T-shape is particularly well adapted for resisting the stress due to the thrust load and is used extensively in larger gears. The strengthening rib is not taken into account when computing the width b which resists bending in the plane of rotation. The general procedure used in the design of spur-gear arms may be followed in the case of bevel gears. However, because of an eccentric force application, it is advisable to assume that only half of the arms carry the tangential load. With the designations of Fig. 31.10, equating the external moment to the resisting moment of one-half of the number j of the arms gives

$$Fr_0 = \frac{1}{12} Shb^2 j \qquad \ldots (31.51)$$

Solving equation 31.51 for the only unknown value gives

$$b = \sqrt{\frac{12Fr_0}{Shj}} \qquad \ldots (31.52)$$

The rim, bead, and hub dimensions may be made to conform to Fig. 31.10 and to the rules given for spur gears. However, the hub must be long enough to give a positive backing B, Fig. 31.10. A practical rule is to determine the backing of the pinion by the equation

$$B_p = \frac{0.25D_g D_P}{D_g + D_P} \qquad \ldots (31.53)$$

and to compute the backing of the gear by the relation

$$B_g = 0.25D_g - B_p \qquad \ldots (31.54)$$

Cast iron and soft-steel bevel gears are used extensively. However, heat-treated alloy-steel gears are used when higher strength and wearing qualities are desired and the gears are not too large for heat treatment.

Case hardened gears made of low-carbon steel, such as SAE 2315 (15Ni3), combine a very hard wearing surface with a tough core and therefore are much used. Gears made of heat-treated alloy steels are usually designed in the shape of a separate ring bolted to a hub or center, as in Fig. 31.12b.

Bevel-gear mounting should have some provision, such as spacing washers, for aligning the gears with respect to the apex. They should also provide for proper lubrication and, if the gears are not enclosed, for protective guards for the teeth. The bearing supports must be rigid, as in Fig. 31.11 which shows both the gear and the pinon mounted overhung. A better mounting is to support the shaft on two bearings Fig. 31.12a. A still better mounting, called *straddle mounting* because there are bearing on both sides of both the gear and the pinion, is shown in Fig. 31.12b.

Ball or roller bearings should always be preferred, since they are not subjected to wear and they keep the proper gear alignment. Shafts carrying bevel gears should be sufficiently rigid. Also, the gears should be

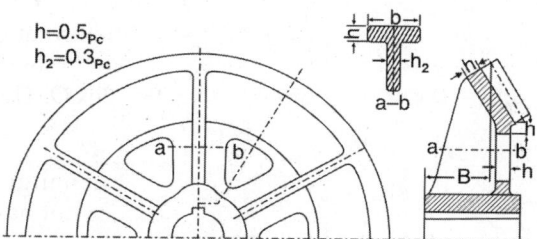

h=0.5$_{Pc}$
h$_2$=0.3$_{Pc}$

Fig. 31.10: Cast-iron bevel gear

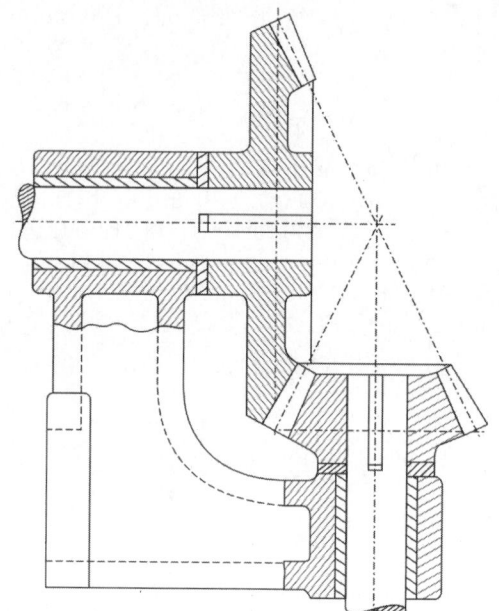

Fig. 31.11: Bevel-gear mounting with plain bearings

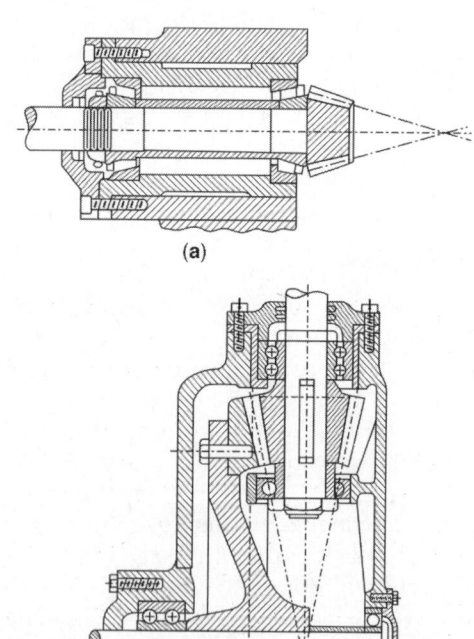

(a)

(b)

Fig. 31.12: Bevel-gear mountings with roller and ball bearings

located where the deflections are a minimum, because even slight deflection results in noise and is much more noticeable wear than in spur gears.

Example 31.1: Design a pair of bevel gears used to transmit 10 kW at 960 rpm of pinion. The gear runs at 320 rpm. The pinion has 20 tooth and is 30C8 (1030) with s_y of 220 N/mm². The gear is made of 20C8 (1020) with S_s of 140 N/mm². The gears are made with 20° involute full depth. Take $C = 456$ kN/m and $k = 919$ kN/m²

Assume,

$$z_p = 20 \text{ as } \quad V.R = \frac{960}{320} = 3 \ ; \quad z_g = 20 \times 3 = 60$$

$$z_{pf} = \frac{20\sqrt{20^2 + 60^2}}{60} = 21 \text{ and}$$

$$z_{gf} = \frac{60\sqrt{20^2 + 60^2}}{20} = 189.7 \text{ say } 190$$

$$y_p = 0.154 - \frac{0.912}{21} = 0.1106$$

$$z_g = 0.154 - \frac{0.912}{190} = 0.1492$$

$$y_p S_{sp} = 0.1106 \times 220 = 24.33$$
$$y_g S_{sg} = 0.1492 \times 140 = 20.88$$

As gear is weak, the gear is designed.

$$\text{Gear torque } T = \frac{10 \times 1000 \times 60}{2\pi \times 320} = 298.4 \text{ Nm}$$

$$l = \frac{m}{2}\sqrt{20^2 + 60^2} = 31.6 \text{ m}$$

$$b = \frac{l}{3} = 10.5 \text{ m}; \ \frac{l}{l-b} = 1.5$$

$$m^2 S = \frac{2 \times 298.4 \times 1000 \times 1.5}{\pi \times 10.5m \times 0.1492 \times 60}$$

or $\qquad m^3 S = 3031.5$

As first trial $c = 0.5$, $S_a = 0.5 \times 140 = 70$ N/mm²

$$m = \left[\frac{3031.5}{70}\right]^{\frac{1}{3}} = 3.504 \text{ take } m = 3.5$$

$$D_g = 3.5 \times 60 = 210 \text{ mm}, v = \frac{\pi \times 210 \times 320}{60 \times 1000} = 3.52 \text{ m/s}$$

$$c = \frac{3}{3+3.52} = 0.46$$

$$S_a = 0.46 \times 140 = 64.4 \, \text{N/mm}^2$$

$$S_i = \frac{3031.5}{3.5^3} = 70.7 \, \text{N/mm}^2$$

As $S_i > S_a$, Next higher module $m = 4$ is used

$$D_g = 240; \quad v = 4.02; \quad c = \frac{3}{3+4.02} = 0.427$$

$$S_a = 0.427 \times 140 = 59.78 \, \text{N/mm}^2$$

$$s_i = \frac{3031.5}{4^3} = 47.36 \, \text{N/mm}^2 \, \text{o.k}$$

$$b = 10.5 \times 4 = 42 \, \text{mm}, l = 31.6 \times 4 = 126.4$$

$$\cos \gamma_p = \frac{120}{126.4} = 0.9493$$

Check for dynamic load.

$$F_t = \frac{298.4 \times 1000}{120} = 2486.6 \text{ say } 2487 \text{N}$$

$$F_d = 2487 + \frac{21 \times 4.02 [456000 \times 0.042 + 2487]}{21 \times 4.02 + \sqrt{456000 \times 0.042 + 2487}}$$

$$= 10327 \, \text{N}$$

Endurance load

$$F_e = 180 \times 42 \times \pi \times 4 \times 0.1492 \times \frac{2}{3} = 9449$$

As it is less than F_d, increase b to 45

$$F_e = 11249 \, \text{N and } F_d = 10714 \, \text{N o.k.}$$

New wear load

$$F_w = \frac{D_p b K Q}{\cos^2 p} = \frac{0.080 \times 0.045 \times 919 \times 1.5 \times 1000}{0.9483}$$

$$= 5808 \, \text{N}$$

$$Q = \frac{2 \times 60}{60 \times 20} = 1.5$$

F_w is lower than F_e so a better material with $Bhn = 400$ and $K = 2189 \, \text{kN/mm}^2$. Then $F_w = 12465 \, \text{N}$.

PROBLEMS

31.1 What is miter and crown bevel gear?

31.2 Why bevel gear assemblies must be constructed so that the position of the gear and pinion can be adjusted along their axes?

31.3 A straight bevel gear and pinion provide a 4:1 ratio and have a perpendicular axes.

Could these gears be interchangeable with gears to provide 3:1 gear ratio, also having perpendicular axes?

31.4 Explain why there is no differential in a cycle rickshaw. How does it take a turn? Why is it not preferable to have the same arrangement in an automobile?

31.5 How do the carriages of a railway train take turn?

31.6 Determine the pitch, the width of face, the number of teeth, the outside diameter, and pitch angles—pitch, face, and cutting angle—for a pair of cast-iron straight bevel gears with normal tooth proportions to transmit 15 kW. The speed of the driving shaft is 600 rpm, and that of the driven shaft is 190 rpm. Check for dynamic load and wear load. Assume continuous operation and very light shock.

31.7 A pair of straight bevel gears must transmit 15 kW at 1,250 rpm of the 18-tooth pinion, The speed-reduction ratio is 3.5:1. Use 14-1/2° full-depth teeth. Select the materials to obtain a compact design. Determine the module, the gear face, the pitch diameters, and the pitch-cone angles for both gears.

31.8 A pair of straight bevel gears, with a module of 5 and 14-1/2° machine-cut teeth, are made of SAE 3240 steel and have a 2:1 reduction. The pitch diameter of the driver is 125 mm, and the gear face is 45 mm. Determine (a) the pitch angles of the pinion and gear, (b) the face angles of the pinion and gear, (c) the cutting angles of the pinion and gear, (d) the maximum diameters of both gears, (e) the formative numbers of teeth, (f) the equivalent tangential load at the large end of the teeth at 300 rpm of the driver. (g) the pitch diameter of the effective tooth load, (h) the effective tooth load, and (i) the power transmitted at 300 rpm of the driver.

31.9 Design a pair of straight cast-iron bevel gears to transmit 90 kW from a shaft running at 235 rpm to another running at 75 rpm. Make a sketch of the gear with all dimensions pertaining to the rim, arms, and hub. Check for continuous operation.

31.10 The ring gear of a truck differential has 50 straight teeth of module 6 and is made of SAE 2340 steel hardened to 240 Bhn. The pinion has 13 teeth with a 35 mm face and is made of SAE 2340 steel hardened to 300 Bhn. Determine (a) the power that can be transmitted at 1,100 rpm of the pinion, (b) the effective tooth load and the radius of its application, and (c) the magnitude of the axial thrust.

31.11 The straight-tooth bevel pinion driving the differential of an automobile has 15 teeth with 5 module and a 32 mm face. The pinion transmits 32 kW at 2,500 rpm. The pinion is supported on two bearings placed 35 mm and 135 mm behind the large pitch circle. The gear has 60 teeth. Determine (a) the beam strength of the teeth, (b) the effective tooth load and the radius of its application, (c) the axial thrust, and (d) the radial load on each bearing.

31.12 Using data from problem 31.11 but substituting spiral bevel teeth, a 20° pressure angle, and a 35° right-hand spiral (on the gear), determine (a) the effective tooth load and the radius of its application; (b) the axial thrusts on the pinion and gear; (c) the radial load on the bearings.

31.13 (a) Determine the axial thrusts for direct and reverse rotation of a pair of spiral gears. The number of teeth in the pinion is 14, the number of teeth in the gear is 42, the spiral angle is 35°, angle $\beta = 20°$, the pinion speed is 1,000 rpm, and it transmits 14 kW. (b) Compare these thrusts with those produced in the pinion and gear shaft of a straight bevel gear. The pinion diameter is 48 mm, and the faces are 42 mm.

32

Worm Gearing

32.1 GENERAL CONSIDERATIONS

Worm gearing is a type of screw gearing used for transmitting power between non-intersecting shafts which are at right angle to each other. *By this means higher speed reductions up to 100, even up to 500 may be obtained in a minimum of space in a single step (32.2).* There are two classes of worm gearing in common use, each of which has its advantages. One is a straight, or cylindrical, worm; the other is a worm with a hollow shape similar to that of an hourglass. Owing to its nature worm gearing is used mostly as a speed reducer, the worm being the driving member.

Definitions: The *linear pitch* of a worm is the distance p_c, Fig. 32.1, measured axially from a point on one thread to the corresponding point on the adjacent thread. Evidently the linear pitch of the worm is equal to the circular pitch of the worm gear.

The *lead* is the distance l that a thread advances in one turn of the worm. Thus

$$l = z_w p_c \qquad \dots (32.1)$$

where, z_w is the number of threads of the worm. In Fig. 32.1a is shown a single-thread worm; in Fig. 32.1b, a double-thread worm; and in Fig. 32.1c, a triple-thread worm.

The *lead angle* λ is the angle between a tangent to the thread and a plane normal to the worm axis.

The *velocity ratio* r_v is equal to the pitch circumference of the worm wheel divided by

the lead of the worm; or since the pitch can be canceled,

$$r_v = \frac{p_c z_g}{z_w p_c} = \frac{z_g}{z_w} \qquad \dots (32.2)$$

Where, z_g is the number of teeth in the gear.

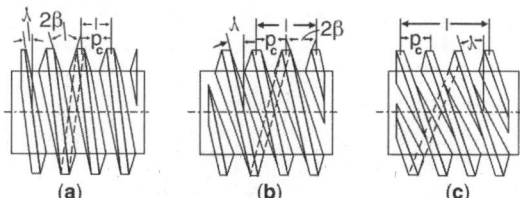

Fig. 32.1: Pitch and lead of worms

32.2 STRAIGHT WORM GEARS

The threads of a straight, or cylindrical worm, have an axial pitch that is constant for all points between the tops and the roots of the threads. The gear teeth are of the involute form. There exist two methods of cutting the worm-gear teeth. By the first method, which is used for ordinary worm gearing, the cutting hob has a constant diameter and is fed radially to the proper depth into the gear blank, both hob and blank being rotated in the required relation to each other. The teeth produced by this method are not theoretically correct but are sufficiently accurate for single-thread worm gears with great number of teeth.

When a higher efficiency and better service are desired, the teeth are cut with a tapered hob fed into the gear blank longitudinally at right angles to the axis of the blank.

Proportions of worm gears: Tables 32.1 and 32.2, with Fig. 32.2, give the proportions recommended by the American Gear Manufacturers Association for Worms and Worm gears for *industrial* use. *The reason for using a larger pressure angle for triple and quadruple threads is that such threads have a large lead angle, and it is difficult to cut threads with a small pressure* *and large lead angles because of undercutting by the hob.*

Pressure angles larger than 20° are often used. It is better to make the recommended pressure angle a function of the lead angle rather than of the number of threads. Thus, β may be 20° for values of λ up to 25°; 25° for values of λ up to 35°; and 30° for values of λ up to 45°. *For automotive gears a large angle λ is desirable, and a value of β of 30° is recommended in*

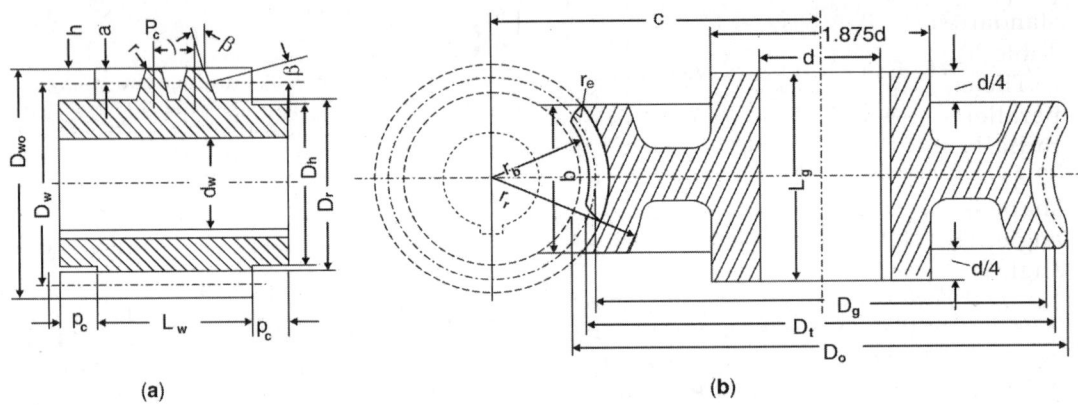

Fig. 32.2: Proportions of worms and worm gears

Table 32.1: Proportions of worms recommended

Dimension	Symbol (Fig. 32.2a)	Single and double threads	Triple and quadruple threads
Normal pressure angle (deg)	β	$14\frac{1}{2}$	20
Pitch diameter, bored for shaft mm (higher value)	D_w	$2.4\, p_c + 28$ $7.5\, m + 28$ $\dfrac{c^{0.875}}{3.48} \leq D_w \leq \dfrac{c^{0.875}}{1.7}$	$2.4\, p_c + 28$ $7.5\, m + 28$
Pitch diameter, integral with shaft mm (lower value)	D_w	$2.35\, p_c + 10$ $7.4\, m + 10$	$2.35\, p_c + 10$ $7.4\, m + 10$
Face length	L_w	$(4.5 + 0.02\, z_w)\, p_c$	$(4.5 + 0.02\, z_w)\, p_c$
Depth of tooth	h	$0.686\, p_c = 2.16\, m$	$0.623\, p_c = 1.96\, m$
Addendum	a	$0.318\, p_c = m$	$0.286\, p_c = 0.9\, m$
Top radius	r	$0.05\, p_c + 0.157\, m$	$0.05\, p_c + 0.1z57\, m$
Hub diameter mm	D_h	$1.66\, p_c + 25$	$1.726\, p_c + 25$
Maximum bore for shaft mm	d_w	$p_c + 16$ $3.14\, m + 16$	$p_c + 16$ $3.14\, m + 16$
Outer diamter of worm mm	D_{wo}	$D_w + 0.636\, p_c$ $d_w + 2\, m$	$D_w + 0.572\, p_c$ $D_w + 18\, m$

order to obtain a high efficiency and to permit overhauling (which means to permit the worm to be turned by the wheel) (32.1).

The teeth of single-thread and double-thread gears with a $14\frac{1}{2}^{\circ}$ pressure angel are made with an addendum of the standard full-height tooth. Therefore, when the number of gear teeth is less than 32, they must be undercut. The teeth for the triple and quadruple threads are stubbed, but the proportions are smaller than those of the standard interchangeable stub tooth given in Table 30.1.

The ends of the gear teeth are cut either parallel to the gear axis, as in Fig. 32.2b, or radially toward the worm axis, as in Fig. 32.3a, with a *face angle* 2δ or 60° to 75°. This shape is used for worms with a small lead angle. The recommended value for the face angle is[1]

$$\tan\delta \leq \frac{\tan\beta}{\tan\lambda} \qquad \ldots (32.3)$$

Straight teeth (Fig. 32.3b) cut with a form cutter are not efficient and are used only for intermittent service and small power transmitted.

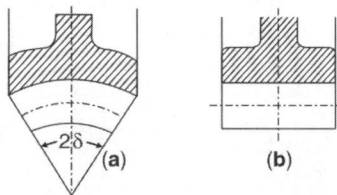

Fig. 32.3: Types of worm gears

Pitch: All the standard modules given for spur gears are in use for worm gears.

The lead angle may vary from 9° to 45°. Its magnitude may be found from Fig. 32.4, which shows the development of the worm. From Fig. 32.4 and equation 32.1, we can obtain the relation

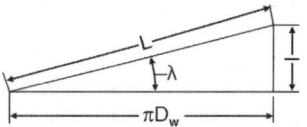

Fig. 32.4: Lead angle

$$\tan\lambda = \frac{z_w p_c}{\pi D_w} \qquad \ldots (32.4)$$

The values of pressures angle and lead angles are given below:

Pressure angle β, degree	14.5°	20°	25°	30°
Lead angle λ degree	0–16°	16–25°	25–35°	35–45°

Experience shows that an angle λ less than 9° results in rapid wear, and a safe value[2] is

$$\lambda \geq 12\frac{1}{2}^{\circ}.$$

For a compact design the angle λ may be selected approximately from the relations[3]

$$\tan\lambda = \sqrt[3]{\frac{n_g}{n_w}} \qquad \ldots (32.5)$$

Strength of worm-gear teeth: The teeth of a worm gear are weaker than the threads on the worm and should be checked by the Lewis formula (equation 30.13). Since both module and circular pitches are used for worm gearing, this equation may by given in the form

$$F = S_o cb\pi m y = S_o cbyp \qquad \ldots (32.6)$$

Equation 32.6 assumes conservatively that the entire load is carried by one tooth. If it is desired to make allowance for the fact that the load is distributed, the allowable load F may be multiplied by the number of teeth in actual contact with the worm.

For the single-thread and double-thread worm gears having the proportions in Table 32.2 with 14–1/2° and 20° pressure angles, the values of y may be taken from the

[1] O. F. Shepard, "Worm Gear Proportions," *Transactions of the American Gear Manufactures Association*, Vol. 11 (1927), p. 201.

[2] F. A Halsey, *Handbook for Machine Designers*, 2nd ed. (New York: McGraw-Hill Book Company, Inc., 1916), p. 133.

[3] C. D. Albert, *Machine Design Drawing Room Problems*, 4th ed. (New York: John Wiley & Sons, Inc., 1949), p. 384.

first two columns of Table 30.2. If the number of teeth in the gear plus the number of threads in 25 mm length in the worm is greater than 40, $Y = \pi y$ may be determined with safety for gears with any number of threads in the worm by the equation

$$Y = 0.314 + 0.0151 (\beta - 14.5°) \quad ... (32.7)$$

The Barth's velocity factor c, which takes into account the dynamic load, may be computed by equation 30.16, in which v_m is the pitch-line speed of the gear.

For the allowable static stress S_o the values of Table 32.3 should be used.

The permissible tooth load F found by equation 32.6 must be greater than the actual tooth load F_t, which may be determined from the effective transmitted torque T_e and the selected suitable gear diameter D_g by the relation

$$F_t = \frac{2T_e}{D_g} \quad ... (32.8)$$

In this case the effective torque T_e is equal to the nominal torque T multiplied by a load factor K_l based on data in Table 20.3.

Example 32.1: Design worm gearing to transmit 15 kW from an electric motor running at 1,165 rpm to a compressor which should run at 150 rpm.

In order to be sure that the gear teeth will not be undercut, assume that $z_g = 32$. By equation 32.2 the number of worm threads is

$$z_w = \frac{32 \times 150}{1165} = 4.12 \text{ or } 4$$

The corrected number of teeth in the gear, by equation 32.2, is

$$z_g = \frac{4 \times 1,165}{150} = 31.2, \text{ or } 32$$

The corresponding compressor speed is

$$= \frac{1165}{8} = 145.625 \text{ rpm}$$

The gear diameter D_g of 384 mm is taken, the module is

$$m = \frac{384}{32} = 12 \text{ mm}$$

which is a standard module, and circular pitch

$$p = \pi \times 12 = 37.7 \text{ mm}$$

The corresponding pitch velocity is

$$v_m = \frac{\pi \times 384 \times 145.625}{1000 \times 60} = 2.927 \text{ m/s}$$

which seems to be statisfactory

Table 32.2: Proportions of worms gears

Dimension	Symbol (Fig. 32.2b)	Single and double threads	Triple and quadruple threads
Normal pressure angle (deg)	β	$14\frac{1}{2}$	20
Outside diameter	D_o	$D_g + 1.0135\, p_c$	$D_g + 0.8903\, p_c$
Throat diameter	D_t	$D_g + 0.636\, p_c$	$D_g + 0.572\, p_c$
Facewidth (mm)	b	$2.38\, p_c + 6.35$	$2.15\, p_c + 5$
Radius of gear face (mm)	r_b	$0.882\, p_c + 14$	$0.914\, p_c + 14$
Radius of gear rim (mm)	r_r	$2.2\, p_c + 14$	$2.1\, p_c + 14$
Radius of edge (mm)	r_e	$0.25\, p_c$	$0.25\, p_c$

Table 32.3: Allowable static stresses for worm gears

Material	S_o	Material	S_o
Ordinary cast iron	70	Leaded gun metal	55
High-grade cast iron or semisteel	105	Manganese bronze	140
Bakelite, textolite, rawhide, etc.	42	Phosphor bronze	105

The effective torque on the gear shaft is, by equation 2.21, in which $K_1 = 1.75$ (from Table 20.3).

$$T_c = \frac{15 \times 60 \times 1000 \times 1.75}{2\pi \times 145.625} = 1721.3 \text{ Nm}$$

The tooth load, by equation 32.8, is

$$F_t = \frac{2 \times 1721.3 \times 1000}{384} = 8966 \text{ N}$$

The speed factor, by equation 30.16, is

$$c = \frac{6}{6 + 2.927} = 0.672 \text{ N}$$

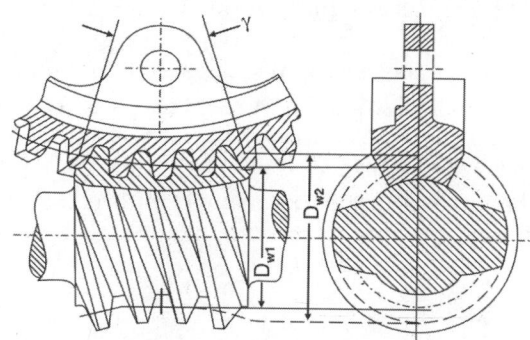

Fig. 32.5: Cone worm gearing

The facewidth, based on data from Table 32.2, is

$$2.15 \times \pi \times 12 + 5.00 = 86.06 \text{ mm say } 87 \text{ mm}$$

The Lewis factor y in equation 32.6, found from Table 30.2 by interpolation, is 0.116. If the gear is assumed to be of high-grade cast iron, $S_o = 105 \text{ N/mm}^2$ from Table 32.3. The maximum allowable load, by equation 32.6, is then

$$105 \times 0.672 \times 87 \times \pi \times 12 \times 0.116 = 26845 \text{ N}$$

which is considerably greater than the tooth load of 8966 N. This indicates that the circular pitch $p_c = 37.7$ may be somewhat too great. However, the deciding factor in worm-gear drives is wear, not strength, and wear will be checked later.

With p_c, z_g, and z_w determined, all other dimensions can be found by means of Tables 31.1 and 32.2. Thus, the pitch diameter of the worm is

$$2.35 \times \pi \times 12 + 10 = 98.6 \text{ say } 100 \text{ mm}$$

The face of the worm is $L_w = (4.5 + 0.02 \times 4) \pi \times 12 = 172.6$ say 175 mm.

The lead angle of the worm, found from equation 32.4, is

$$\tan \lambda = \frac{4 \times 12 \times \pi}{\pi \times 100} = 0.48$$

or $\lambda = 26°$, which seems satisfactory

32.3 CONE GEARING

Cone gearing, named after its inventor, Samuel L. Cone, is also called *double-enveloping worm gearing* because both the gear teeth and the worm teeth follow the shape of the other member. The gear has straight-sided teeth with a pressure angle of 20°. In general it resembles an ordinary worm gear, Figs 32.2b and 32.3a. The main difference is in the shape of the worm, Fig. 32.5, which has teeth cut to conform to the shape of the gear.

The hourglass-shaped worms were first introduced by Hindley for the purpose of decreasing the wear of the worm. However, only the method of cutting the teeth in both the worm and the gear developed by Cone resulted in a practical type of gearing with a high load capacity and small wear. Since cone gearing permits the use of shorter center distances, it needs only about two-thirds of the space and has about one-third of the weight of conventional worm gearing.[4]

Advantages: Cone drives that are properly designed and machined, and carefully assembled, show small wear and a high efficiency, combined with small size and weight. Their main disadvantage is the requirement of almost absolute accuracy in assembling and aligning. A small deviation from the correct center distance or the correct relative positions of the worm and gear results in the loss of the theoretical area of contact. However, wear does not affect a correctly assembled drive, as both the worm and the gear are regenerative and tend to correct themselves in case of a slight misalignment.

Design: The design procedure for a cone drive is not complicated. The necessary formulas have been established by the manufacturer, but they require the use of special diagrams. Both

F. E. Birth, "Double Enveloping Right Angle Gear Drives," *Product Engineering*, Vol 19 (August, 1948), p. 85.

the formulas and the diagrams are presented in a clear way in publications issued by the builder of the special machines for generating these gears.[5]

32.4 FORCE ANALYSIS

The forces acting between a worm thread and a gear tooth are represented in Fig. 32.6.

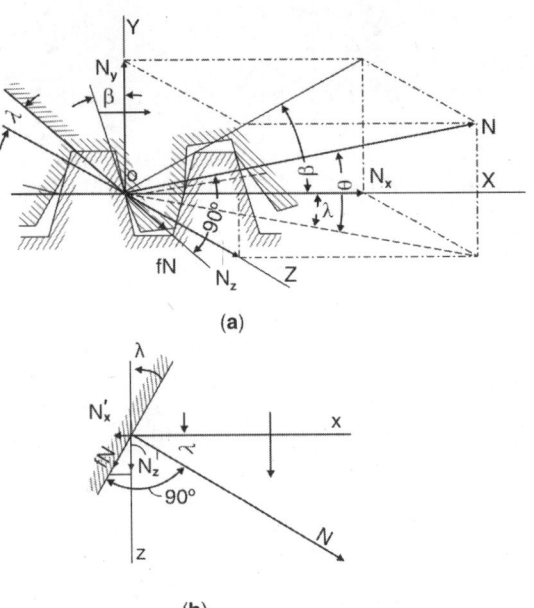

(a)

(b)

Fig. 32.6: Forces acting between the worm thread and the gear tooth

Turning force and gear load: The relation between the turning force Q on a worm and the tangential load F_t on the worm gear may be found by using a perspective picture of the forces in a three-dimensional system of coordinates, as in Fig. 32.6a. The X axis is parallel to the axis of the worm, the Y axis is in the radial direction on the worm, and the Z axis is in the tangential direction for the worm normal to the plane $X - Y$. The vector N represents the normal reaction between the worm and a gear tooth at the point of contact O. If the frictional resistances are disregarded

at first, the components of N along the axes OX, OY, and OZ are, respectively,

$$N_x = N \cos \theta \cos \lambda$$

$$N_y = N \sin \theta$$

$$N_z = N \cos \theta \sin \lambda$$

where, the angle θ between N and the plane X–Z may be found from the relation

$$\tan \theta = \tan \beta \cos \lambda \qquad \ldots (32.9)$$

If it is assumed that the worm rotates as indicated in Fig. 32.6b, the force $N' = fN$ due to friction between the worm and the gear teeth acts along the tangent to the helix. The components of this force along the axes OX, OY, and OZ are

$$N'_x = fN \sin \lambda \qquad N'_y = 0 \qquad N'_z = fN \cos \lambda$$

When the components along the same lines of action are combined, the following results are obtained: The magnitude of the tangential driving force exerted by the worm upon the worm-gear teeth is[6]

$$F_t = N'_x - N_x = N (\cos \theta \cos \lambda - f \sin \lambda) \ldots (32.10)$$
$$= \text{Axial force on the gear}$$

The magnitude of the turning force Q required at the pitch radius of the worm is

$$Q = N_z + N'_z = N (\cos \theta \sin \lambda + f \cos \lambda) \ldots (32.11)$$

The component along the axis OY represents the magnitude of either the downward pressure R upon the worm shaft or the upward pressure upon the worm-gear shaft. Thus,

$$R = N_y = N \sin \theta \qquad \ldots (32.12)$$

Elimination of N by dividing equation 32.11 by equation 32.10 gives

$$Q = F_t \frac{\cos\theta \sin\lambda + f \cos\lambda}{\cos\theta\cos\lambda - f \sin\lambda} \quad \ldots (32.13)$$

Efficiency: If there were no friction, f would be 0 and the turning force would be reduced to

$$Q' = F_t \frac{\cos\theta \sin\lambda}{\cos\theta\cos\lambda} = F_t \tan\lambda \ldots (32.14)$$

[5]Michigan Machine Tool Company, *Cone Drive Gears*, Catalog No. 50 (Detroit: 1951).
[6]W. Lewis, "Investigation of Worm Gear Drives," *Trans*, ASME, Vol. 7 (1885), p. 297.

The efficiency is the ratio of the ideal effort without friction to the actual effort with friction, or $e = Q'/Q$. When the values from equations 32.14 and 32.13 are substituted for Q' and Q, and a few simplifications are made, the result is

$$e = \frac{\cos\theta - f\tan\lambda}{\cos\theta + f\cot\lambda} \quad \cdots (32.15)$$

Equation 32.15 differs slightly from equation 11.18, because the latter was deduced by using the approximate equation 11.15: that is, by assuming that $\cos\lambda = 1$. This is sufficiently accurate for single-thread screws. The efficiency of a worm gear is influenced greatly by the helix angle λ. The dotted curve 7 (Fig. 32.7), shows values of e computed by equation 32.15, in which $\beta = 14\frac{1°}{2}$ and $f = 0.05$. The other curves, 1 to 6, were found experimentally for different values of the rubbing velocity v from 0.025 to 1 m/s. The influence of v is indirect, as there is a gradual change of the friction coefficient from about $f = 0.15$ at $v = 0.025$ m/s to $f = 0.02$ at 1 m/s. This variation is in agreement with the theory of lubrication. For single-thread worms of the irreversible or self-locking type, f may be assumed as 0.05. For multiple-thread worms operating at rubbing speeds from 0.2 to 2.75 m/s, with good workmanship and proper running conditions, including adequate lubrications, a safe value of f may be estimated by the equation[7]

$$f = \frac{0.042}{v^{0.28}} \quad \cdots (32.16)$$

For rubbing speeds v greater than 2.75 m/s

$$f = 0.025 + \frac{v}{300} \quad \cdots (32.17)$$

The friction coefficient f depends on the finish of the tooth surfaces and on the lubricant used. With the best materials, for the worm and the gear, with precision machining, and with good lubrication, it is possible to obtain values of f as low as 0.025; with low speeds and indifferent lubrication, f may be as high as 0.10.

Rubbing velocity: It should be noted that the rubbing velocity v is different from the pitch velocity v_m. If the notation in Fig. 32.2 is used, the rubbing velocity may be computed by the relation

$$v = \frac{\pi D_w n_w}{60 \times 1000\cos\lambda} = \frac{v_w}{\cos\lambda} \quad \cdots (32.18)$$

For ordinary industrial worms, the rubbing velocity should not exceed 6 m/s; for well-designed, precision-machined, hardened, and ground worms, it should not exceed 15 m/s.

The curves in Fig. 32.7 show clearly how desirable it is to use a large angle λ. This angle should be 30° or higher, and such a large angle is obtained only with multiple-thread worms. Tests with steel worms and bronze worm gears and with values of λ ranging from 38° to 45° showed efficiencies as high as 98 per cent.[8] For values of the speed ratio greater than 8, the De Laval Company uses the following empirical formula:

$$e = 1 - 0.005r_v \quad \cdots (32.19)$$

where, r_v is the speed ratio of the worm and the gear. This formula presumes that the best helix

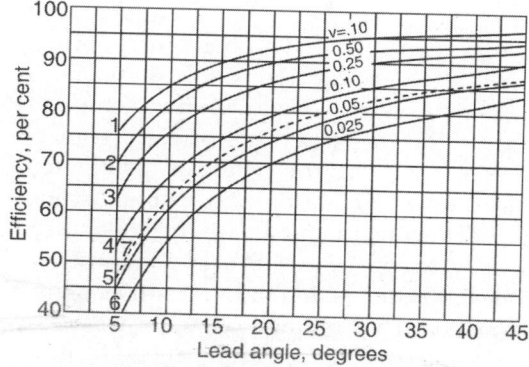

Fig. 32.7: Efficiencies of worm drives

[7]V. M. Faires, *Design of Machine Elements*, rev. ed. (New York: The Macmillan Company, 1942), p. 273. Modified on the basis of data from Lewis, *loc. cit.*
[8]C. M. Allen and F. W. Roys, "Efficiency of Gear Drives," *Trans.* (ASME, Vol. 40, 1918), pp. 110–11.

angle, the lowest friction coefficient, the best thrust bearing, and the best lubrication are used.

Generally speaking self locking occurs if lead angle is below 6° and it may occur with as much as 10° lead angle.

Example 32.2: Find the efficiency of the worm gear of example 32.1.

By equation 32.18, in which $\lambda = 26°$ and $\cos \lambda = 0.899$, the rubbing velocity is

$$v = \frac{\pi \times 100 \times 1165}{1000 \times 0.899 \times 60} = 6.78 \text{ m/s}$$

The coefficient of fiction, by equation 32.17, is

$$f = 0.025 + \frac{6.78}{300} = 0.0476$$

Since $\beta = 20°$ and $\tan \beta = 0.364$, it is found from equation 32.9 that

$$\tan \theta = 0.364 \times 0.899 = 0.3275 = \tan 18° \, 8'$$

The efficiency found by equation 32.15, in which $\cos \beta = 0.950$ and $\cot \lambda = 2.05$, is

$$e = \frac{0.950 - 0.0476 \times 0.438}{0.950 + 0.0476 \times 2.05} = 0.879$$

32.5 DESIGN OF WORM GEARING

Worm gearing should be designed to resist wear and provide for dissipation of the heat generated.

Wear: The determination of proper proportions of worm gearing for wear is a rather involved problem. The best proportions depend to a great extent on the capacity of the gear assembly for radiating the heat generated by friction. With correct alignment, excessive wear is the result of insufficient lubrication. Lubrication may be poor even if the worm runs in an oil bath, for it is caused either by excessive pressure or by a low viscosity of the lubricant resulting from insufficient heat dissipation. The limiting load F_w for wear may be computed by the relation[9]

$$F_w = D_g bK \qquad \ldots (32.20)$$

where, D_g is the pitch diameter of the gear; b is its face, in m; and K is a constant which depends on the materials of the worm and gear.

The values of K for various materials of the worm gear and worm are given in Table 32.4.

Table 32.4: Values of K

Worm	Gear	$K (kN/m^2)$
Hardened steel	Cast iron	345
Steel, 250 Bhn	Phosphor bronze	415
Hardened steel	Phosphor bronze	550
Hardened steel	Chilled phosphor bronze	830
Hardened steel	Antimony bronze	830
Cast iron	Phosphor bronze	1035

These values are given for lead angles up to 10°. For lead angles between 10° to 25°, increase K by 25% and for lead angles greater than 25°, increase K by 50%.

Another formula for F_w, which takes into account the various gear data but assumes the use of a proper grade of lubricant, is[10]

$$F_w = \frac{A \cos \lambda \; p_s c}{c_s} \qquad \ldots (32.21)$$

where, A is the projected tooth area of contact, in mm^2.

p_s is the allowable surface pressure, in N/mm^2, as given in Table 32.5;

c is Barth's velocity factor, from equation 30.6, for the worm gear;

c_s is a service factor, which may be found by increasing the value in Table 30.4 by 25 per cent.

The projected tooth area may be calculated from the equation

$$A = \frac{h D_w \delta}{57.3} \qquad \ldots (32.22)$$

where, h is the tooth depth, in mm, as given in Table 32.1;

δ is one-half of the face angle, in degrees (*see* Fig. 32.3).

[9]R. T. Kent, *Mechanical Engineers' Handbook*, 12th ed., Vol. II, *Design and production*, ed. by Colin Carmichael (New York: John Wiley and Sons, Inc., 1950), p. 14 – 43.
[10]Alex Vallance and V. L. Doughtie, *Design of Machine Members* (New York: McGraw-Hill Book Company, Inc., 1943), p 417.

Table 32.5: Allowable surface pressures p_s = N/mm^2 × 10^{-2} in equation 32.21

Material		Number of teeth in gear							
Worm	Gear	10	20	30	40	50	60	70	80 and more
20C8 (SAE) 1020 steel*	Cast iron	52	155	290	520	620	750	860	930
40C8 (SAE) 1040 steel*	(SAE 63) bronze[†]	80	240	430	750	930	1130	1320	1380
40C8 (SAE) 1040 steel[‡]	(SAE 63) bronze[†]	120	350	650	1110	1380	1680	1960	2070
0.10C alloy steel ⎤	(SAE 65) bronze[†]	160	470	860	1550	1860	2250	2620	2760
carburized	(SAE 65) bronze[§]	220	640	1190	2070	2550	3100	3620	4620
hardened, and	Ni bronze[†]	260	780	1480	2480	3100	3760	4370	4620
ground ⎦	Ni bronze[§]	310	1730	2960	3730	4480	–	5140	5510

* Untreated. [†]Sand-cast. [‡]Heat-treated, ground. [§]Chill-cast.

Values for allowable pressures in Table 32.5 are given for a $14\frac{1°}{2}$ pressure angle. For $\beta = 20°$ they can be increased by 5 per cent; and for $\beta = 30°$, by 10 per cent.

As may be seen from example 32.3, equation 32.21 gives considerably lower values than does equation 32.20. Which of these formulas give more accurate information can be decided only by additional research work.

Heat dissipation: In order to prevent overheating of the lubricating oil because of low efficiency, the work of friction, which may be calculated from the efficiency of the gear drive, must be dissipated chiefly by radiation (32.3). The heat-dissipating capacity depends on the size and surface of the housing and on the velocity of the air surrounding the housing. For average conditions the dissipating capacity Q is given by

$$Q = \frac{0.407}{1000} A(t_2 - t_1) \qquad \dots (32.23)$$

where, the difference between the gear temperature t_2 and the room temperature t_1 should be less than 55°C. A is the projected surface area of worm and gear.

$$A_w = L_w \times D_w ; A_g = \frac{\pi}{4} D_g^2$$

Whether the worm runs in an oil bath or not, the gears should be entirely enclosed to prevent oil leakage and to protect them from dust. Ball bearings or roller bearings should be used in order to increase the efficiency and to maintain proper alignment.

Materials: It should be remembered that unlike metals are more satisfactory for sliding contact than are like metals. Thus, for light loads and low speeds the worm may be made of steel, such as SAE 1040/40C8 and the worm gear may be made of cast iron or leaded gun metal. For medium service conditions the worm may be made of SAE 2320 (20 Ni3) or SAE 3120 (20 Ni 1 Cr 60) steel that is case hardened to a Brinell hardness number of at least 250, and the gear may be of phosphor bronze. For high speeds and heavy loads with shock action, the worm is made of molybdenum steel or chrome-vanadium steel and is hardened, and the gear is made of phosphor bronze, which may be chilled for hardness and refinement of grain structure. Often, in order to reduce the cost of a large gear, the rim alone is made of bronze, as in Fig. 32.5, and it is bolted to a cast-iron flange with arms and a hub.

Design Procedure Steps

1. For the given velocity ratio, select the number of starts of worm as given below:

Velocity ratio r_v	36 and above	12–36	8–12	6–12	4–10
Number of starts z_w	1	2	3	4	5

2. Number of teeth on worm gear $z_g = r_v z_w$.
3. If space dimensions are specified or center distance c is given, the diameter of worm is (AGMA):

$$\frac{C^{0.875}}{3.48} \le D_w \le \frac{C^{0.875}}{1.7}$$

4. Decide the pitch $p_c = \frac{D_w}{3}$.

5. Decide the module $m = \frac{p_c}{\pi}$ and take a standard value.

6. Worm wheel diameter $D_g = mz_g$

7. Make a check for centre distance

8. For compact design select a lead angle as

$$\lambda = \sqrt[3]{\frac{z_g}{z_w}}$$

Also $\qquad \lambda = \tan^{-1} \frac{p_c}{\pi D_w}$

To reduce wear $\lambda \ge 12.5°$

Decide the corresponding pressure angle of gear.

9. Decide the facewidth $b = 0.73\, p_c$ (AGMA)

Also b is given in Table 32.2.

As a first trial take the lower value.

10. Calculate velocity of gear $v = \frac{\pi D_g n}{60 \times 1000}$

find $\qquad c = \frac{3.05}{3.05 + v}$ or $\frac{6}{6 + v}$

and find allowable stress $S_a = CS_o$

11. Find torque T and F_t and estimate induced stress

$$S_i = \frac{F_t}{b p_c y}$$

If $S_i > S_a$ increase module m and repeat till $S_i \le S_a$

12. Find $e = \frac{\cos\theta - f \tan\lambda}{\cos\theta + f \cot\lambda}$

where $\tan\theta = \tan\beta \tan\lambda$ and f depends on velocity.

Find heat generated $H_g = P\,(1 - e) \times 1000$

and heat dissipated $H_d = \frac{0.407}{1000} A(t_2 - t_1)$

$$A = L_w D_w + \frac{\pi}{4} D_g^2$$

and $(t_2 - t_1) < 55°C$ if values are not given

13. Check for wear load $F_w = k\, b\, D_g$

and it should be greater than $F_d = \frac{F_t}{c}$

or provide more velocity of air surrounding the housing.

14. Specifications of worm and worm gear are:

$$z_w / z_g / q / m / c$$

where q is diameter quotient

$$= \frac{\text{Pitch diameter of worm}}{\text{Module}} = \frac{D_w}{m}$$

and $\tan\lambda = \dfrac{p_c z_w}{\pi D_w} = \dfrac{\pi m z_w}{\pi D_w} = \dfrac{m z_w}{D_w} = \dfrac{z_w}{q}$

Example 32.3: Check the worm drive discussed in examples 32.1 and 32.2 for wear and overheating.

The pitch diameter of the gear is $D_g = 384$ mm, and the face is $b = 87$ mm. If it is assumed that two teeth are in mesh at one time, the limiting load for wear for a cast-iron gear is, by equation 32.20,

$$F_w = 384 \times 87 \times 0.345 \times 2 = 23051 \text{ N}.$$

This is amply safe, since the transmitted tooth load is 8966 N.

The procedure in using equation 31.21 as a check is as follows. For a quadruple-thread worm, the tooth depth is

$$h = 1.96\, m = 1.96 \times 12 = 24 \text{ mm}$$

Heat to be dissipated $= 1000 \times P\,(1 - e)$ watts

The angle δ, Fig. 32.3, can be found approximately from the relation

$$\sin\delta = \frac{0.5b}{0.5(D_w + 2a)}$$

$$= \frac{87}{100 + 2 \times 0.286 \times 37.7} = 0.7156$$

Then $\delta = 46°$, and by equation 32.22 the projected tooth area, with two teeth in contact, is

$$A = \frac{24 \times 100 \times 46 \times 2}{57.3} = 3853 \text{ mm}^2$$

Since $\lambda = 26°$, $\cos\lambda = 0.899$. For a cast iron gear with 32 teeth, a worm of SAE 1020 (20C8) steel, and $\beta = 20°$ the allowable surface pressure, found with the aid of Table 32.4, is for 32 teeth, and increased by 5% for 20°.

$$p_s = \left[2.90 + \frac{2}{10}(5.2 - 2.90)\right] \times 1.05 = 3.53 \text{ N/mm}^2$$

The pitch-line velocity of the worm is

$$v_m = \frac{\pi \times 100 \times 1165}{60 \times 1000} = 6.1 \text{ m/s}$$

and the velocity factor is

$$\frac{3.05}{3.05 + 6.1} = 0.333$$

The service factor for an air-compressor drive, taken from Table 30.4, is $C_s = 1.25 \times 1.25 = 1.56$.

Hence, by equation 32.21,

$$F_w = \frac{3853 \times 0.899 \times 3.53 \times 0.333}{1.56} = 2610 \text{ N}$$

This is considerably less than the tooth load of 8966 N. For continuous operation the drive should therefore, be redesigned. Changing the material of the worm to hardened 0.10 C alloy steel and changing the material of the gear to *Ni* bronze chilled cast will increase F_w to 22055 N which is satisfactory.

The heat that must be dissipated per second is evidently

$$Q = 1000P (1 - e) = 1000 \times 15(1 - 0.879) = 1815 \text{ W}$$

The projected area of the gear is

$$A_g = \frac{\pi}{4} \times 384^2 = 115812 \text{ mm}^2$$

and the projected area of the worm is

$$A_w = 100 \times 177 = 1770 \text{ mm}^2$$

Solving equation 32.23 for $t_2 - t_1$ gives

$$t_2 - t_1 = \frac{1815 \times 1000}{0.407(115812 + 1770)} = 38°C$$

A temperature rise of 38°C is entirely satisfactory.

32.6 MOUNTINGS

The *worm shaft* is usually supported by two bearings. As indicated in Fig. 32.8a, the bearing *a* and *b* must take the pressures coming from the forces Q and R, and the thrust coming from the force F_t. If the shaft is mounted on ball bearings, a radial bearing can be installed to take care of the lateral and thrust loads simultaneously. Otherwise, a special thrust bearing is necessary.

Bearing loads: If it is assumed that the turning effort Q and the upward reaction R are applied midway between the bearings *a* and *b*, which are *c* mm apart, each of these bearings must take a pressure equal to one-half of these forces. The forces Q and R being at right angles, their components at the bearing are also at right angles to each other. The tangential force F_t causes an end thrust upon the bearing *a* and also exerts a pressure equal to $F_t D_w/2c$ upon each bearing, the pressure at *a* acting downward and that at *b* upward. Therefore the resultant radial, or lateral, load upon the bearing *a* is

$$A = \sqrt{\frac{Q^2}{4} + \left(\frac{R}{2} - \frac{F_t D_w}{2c}\right)^2} \qquad \dots (32.24)$$

and the resultant lateral load upon the bearing *b* is

$$B = \sqrt{\frac{Q^2}{4} + \left(\frac{R}{2} + \frac{F_t D_w}{2c}\right)^2} \qquad \dots (32.25)$$

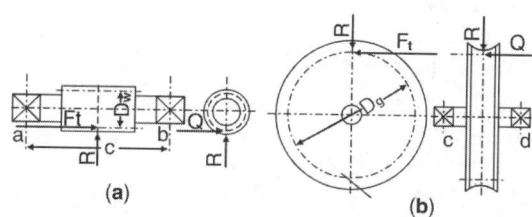

Fig. 32.8: Forces on worm and worm-gear shafts

Worm-gear shaft: If the bearing *c* and *d*, Fig. 32.8b, are located symmetrically with respect to the middle plane of the gear, the loads coming upon each bearing from the forces F_t and R will be equal to one-half of these forces. The force Q exerts a thrust on bearing *c* and also introduces lateral loads upon both bearings. The load upon bearing *c* is equal to $QD_g/2c$ and acts downward; the load upon bearing *d* is $QD_g/2c$ and acts upward. Since the components of F_t and those of R and Q act at right angles, the resultant radial load on the bearing *c* is

$$C = \sqrt{\frac{F_t^2}{4} + \left(\frac{R}{2} + \frac{QD_g}{2c}\right)^2} \qquad \dots (32.26)$$

and the resultant radial load on the bearing d is

$$D = \sqrt{\frac{F_t^2}{4} + \left(\frac{R}{2} - \frac{QD_g}{2c}\right)^2} \qquad \ldots (32.27)$$

Mountings: A worm-gear drive should always be mounted in a dustproof casing which permits either the worm or the gear to run in an oil bath. The worm shaft is usually mounted on ball bearings, and there is often a double-row radial ball bearing on the side, where an axial thrust load exists. The gear shaft is supported either by ball bearings or by adjustable roller bearings.

If a worm is mounted as in Fig. 32.9, the axial thrust creates tension in the shaft, regardless of the direction in which the worm rotates.

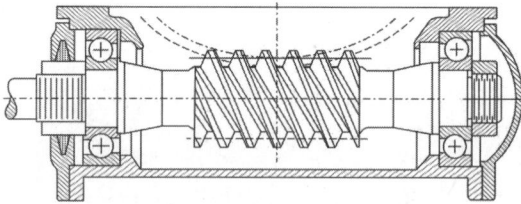

Fig. 32.9: Mounting of a worm

Example 32.4: A speed reducer is to designed for transmitting 40 kW at 480 rpm with a transmission ratio of 14 using worm and worm wheel. The pressure angle is 20° and static stress of manganese bronze gear is 140 N/mm². Take $f = 0.05$

Torque transmitted by gear

$$T = \frac{40 \times 1000 \times 60 \times 14}{2\pi \times 480}$$

$$= 11140.8 \text{ say } 11141 \text{ Nm}$$

For a velocity ratio of 14, a double start thread is specified, i.e. $z_w = 2, z_g = 2 \times 14 = 28$

As centre distance is not specified a pitch of $p_c = 30$ mm is assumed.

$$m = \frac{p}{\pi} = \frac{30}{\pi} = 9.5 \text{ say } 10$$

$$D_g = 10 \times 28 = 280 \text{ mm}$$

$$D_w = 3 \times 30 = 90 \text{ mm}$$

Also $\qquad D_w = 2.4 \times 30 + 28 = 100 \text{ mm}$

Adopting lower value of $D_w = 90$ mm.

$$b = 0.73 \times 90 = 65.7 \text{ say } 66$$

Also $\quad b = 2.38 \times 30 + 6.35 = 77.7 \text{ say } 78$

Adopting lower value of say 70 mm.

$$F_t = \frac{11141 \times 1000}{140} = 79578.5 \text{ N}$$

$$y = 0.154 - \frac{0.912}{28} = 0.1215$$

$$\upsilon = \frac{\pi \times 280 \times 480}{1000 \times 60 \times 14} = 0.502 \text{ m/s}$$

$$c = \frac{3.05}{3.05 + 0.502} = 0.858$$

$$S_a = 0.858 \times 140 = 120.1 \text{ N/mm}^2$$

$$S_i = \frac{79578.5}{70 \times 30 \times 0.1215} = 311.8 \text{ N/mm}^2$$

As induced stress is much higher than allowable stress, module is increased to 12

$$m = 12, p = 12 \times \pi = 37.69 \text{ say } 38 \text{ mm}$$

$$D_g = 12 \times 28 = 336 \text{ mm}$$

$$b = 2.38 \times 38 + 26 = 116.4 \text{ say } 120 \text{ mm}$$

$$F_t = \frac{11141 \times 1000}{168} = 66315.5 \text{ N}$$

$$\upsilon = \frac{\pi \times 336 \times 480}{1000 \times 60 \times 14} = 0.603 \text{ m/s}$$

$$c = \frac{3.05}{3.05 + 0.603} = 0.835,$$

$$S_a = 0.835 \times 140 = 116.9 \text{ N/mm}^2$$

$$S_i = \frac{66315.5}{120 \times 38 \times 0.1215} = 119.7 \text{ N/mm}^2$$

This value is slightly higher than allowable stress, a slightly higher value of $b = 125$ mm will be preferable to increasing the module which will be more costly.

$$S_i = \frac{66315.5}{125 \times 38 \times 0.1215} = 114.9 \quad \text{OK}$$

$$D_w = 3 \times 38 = 114 \text{ mm}$$

$$\tan\lambda = \frac{2 \times 38}{\pi \times 114} = 0.2122 \text{ or } \lambda = 12°$$

It is slightly lower than 12.5° but it is accepted. For the velocity of 0.603 m/s, $f = 0.02$

$$e = \frac{\cos 20 - 0.02 \tan 12}{\cos 20 + 0.02 \cot 12} = 0.905$$

$$H_g = 40 (1 - 0.905) \times 1000 = 3800 \text{ W}$$

$L_w = (4.5 + 0.02 \times 2) \times 38 = 172.2$ say 180 mm

$A_w = 180 \times 114 = 20520$ mm^2

$A_g = \dfrac{\pi}{4} \times 336^2 = 88668.5$ mm^2

$H_d = \dfrac{0.407}{1000}(20520 + 88668.5) \times 50 = 2221.9$ W

Assuming a temperature difference of 50°C.

The heat dissipated is lower than heat generated so artificial cooling is required to dissipate 3800 – 2222 = 1578 W of heat by providing movement of air and better lubrications.

Example 32.5: Design a worm and worm gear to transmit 2.8 kW through a worm rotating at 1750 rpm to the gear which is to rotate at 65 rpm. The involute has 14.5° profile. The centre distance may be 0.1 m, $S_s = 65$ N/mm^2.

The velocity ratio $= \dfrac{1750}{65} = 26.92$ say 27

Accordingly the number of starts of worm $z_w = 2$, $z_g = 2 \times 27 = 54$.

Taking smallest worm diameter

$$D_w = \dfrac{c^{0.875}}{3.48} = \dfrac{0.1^{0.875}}{3.48} = 0.429 \text{ m}$$

or $D_w = 45$ mm

$$p_c = \dfrac{45}{3} = 15 \text{ mm and } m = \dfrac{15}{\pi} = 4.77 \text{ say } 5$$

Then revised $p_c = 5 \times \pi = 15.7$ mm

$D_g = 5 \times 54 = 270$ mm

$$\tan\lambda = \dfrac{2 \times 15.7}{\pi \times 45} = 0.2221, \ \lambda = 12.52° \text{ ok}$$

Revised $D_w = 15.7 \times 3 = 47.1$ say 47 mm, $\lambda = 12°$

$b = 0.73 \times 50 = 36.5$ say 37 mm

Also $D_w = 2.4 \times 15.7 + 28 = 65.68$ say 66 mm

Taking lower value of $D_w = 50$ mm.

$$y = 0.124 - \dfrac{0.684}{54} = 0.1114$$

Torque $T = \dfrac{2.8 \times 1000 \times 60}{2\pi \times 65} = 411.35$ Nm

$$F_t = \dfrac{411.35 \times 1000}{135} = 3047 \text{ N}$$

$$v = \dfrac{\pi \times 270 \times 65}{1000 \times 60} = 0.9189 \text{ m/s say } 0.92 \text{ m/s}$$

$$c = \dfrac{3.05}{3.05 + 0.92} = 0.768$$

$S_a = 0.768 \times 65 = 49.92$ N/mm^2

$$S_i = \dfrac{3047}{37 \times 15.7 \times 0.1114} = 47.08 \text{ N/mm}^2 \text{ O.K.}$$

Taking $f = 0.1$ for $\beta = 14.5°$

$\tan\theta = \tan 14.5 \cos 12.00 = 0.2529$ or $\theta = 14.2°$

$$\eta = \dfrac{\cos 14.2 - 0.1 \times \tan 12.00}{\cos 14.2 + 0.1 \times \cot 12.00} = 0.658$$

Heat generated $H_g = 2.8 (1 - 0.658) \times 1000 = 957.6$ W

$L_w = (4.5 + 0.02 \times 2) \ 15.7 = 71.27$ say 72 mm

$A_w = 72 \times 47 = 3384$ mm^2

$A_g = \dfrac{\pi}{4} \times 270^2 = 57255.6$ say 572.56 mm^2

Heat dissipated

$$H_d = \dfrac{0.407}{1000}[3384 + 57256] \times 50 = 1234.0 \text{ W}$$

Design is OK as $H_d > H_g$

Checking for wear.

Wear load $F_w = k \, b \, D_g$.

Taking hardened steels worm with phosphor bronze gear $k = 555$ kN/m^2

$F_w = 555000 \times 0.037 \times 0.270 = 5544.4$ N

Dynamic load $F_d = \dfrac{3047}{0.768} = 3967.4$ N

As wear load is higher than Dynamic load, design is safe.

PROBLEMS

32.1 Which lead angles of worm give an efficiency for reversibility?

32.2 What are possible reductions of speed in a worm gear drive?

32.3 Why should worm gear drive requires to be checked for heat dissipations?

32.4 Make a sketch to show the mounting of a worm.

32.5 Design worm drive for a speed reducer, to transmit 30 kW at a worm speed of 600 rpm. The desired velocity ratio is 25:1, and an efficiency of at least 87 per cent is desired. Assume that the worm is made of hardened steel, and select the material of the gear.

32.6 The motor of a truck develops its maximum power at 1,200 rpm, and 91 per cent of this power is transmitted to the worm of the worm-gear drive on the rear axle. The speed reduction in the transmission is 3.33:1, and that in the worm and gear is 12.5:1. The worm has a quadruple thread of 95 mm lead, a pitch diameter of 67 mm and a pressure angle of 30°. Determine the maximum power of the motor based on the load which the worm-gear drive can stand, considering (a) strength alone, (b) permissible wear, and (c) heat-dissipating capacity. The worm is made of hardened steel, and the gear is of phosphor bronze and has a face 54 mm wide.

32.7 If the motor of problem 32.6 develops 26 kW at 1,200 rpm and all other data are the same, determine (a) the magnitude of the forces F, N, Q, and R; (b) the efficiency of the worm and gear; (c) the heat generated, at the surfaces in contact; and (d) the necessary area of the housing surface to dissipate that heat.

32.8 Design a worm and worm gear to transmit 45 kW at a worm speed of 480 rpm. The desired velocity ratio is 14:1. The efficiency must not be less than 92 per cent. Use a worm with three threads.

32.9 An elevator cage is lifted at the rate of 1.4 m/s. The elevator drum diameter is 0.6 m, and the load to be lifted is 18 kN. A worm gear is keyed to the drum shaft.

Assuming a speed of the driving motor of 900 rpm, determine (a) the worm and gear proportions, (b) the efficiency of the worm driver, and (c) the required power of the motor if the efficiency of the hoist itself is 94 per cent.

32.10 In a worm gear of 20 mm pitch the double-threaded worm is proportioned according to Table 32.2. Assuming a worm speed of 1,075 rpm and 30 teeth in the gear, determine the efficiency of the drive.

32.11 Determine the safe power which the drive of problem 32.10 can transmit if the worm is of 0.10 C alloy steel (Table 32.5) and the gear is of chilled-cast bronze. Consider (a) the strength of the gear teeth; (b) the limiting load for wear; (c) the heat-radiating capacity; (d) the method of changing the materials which would increase the allowable power without changing the diameters or speed, and the obtainable gain by this method for continuous operation.

32.12 A triple-thread worm and its gear have a pitch of 16 mm and a velocity ratio of 28:1. The material of the worm is steel, and that of the gear is Bakelite. The worm speed is 1,750 rpm. (a) Determine the strength of the worm-gear teeth. (b) Find the limiting load for wear. (c) Check the limiting load based on heat-radiating capacity, using equation 32.23. (d) Determine the power which the gear can transmit safely. (e) Describe the method of changing the specifications or proportions which would increase the allowable power without changing the diameters or speed. (f) Compute the possible power increase for continuous operation by this method.

33

Screw Gearing

33.1 GENERAL CONSIDERATIONS

Screw gears, or helical gears, often miscalled spiral gears, are used to connect shafts which are not intersecting. Their axes may be parallel, at right angles, or inclined at any angle to each other. Parallel gears are always called *helical gears* and are discussed in section 30.12. *Worm gears*, discussed in Chapter 32, are special types of gears with their axes at right angles. In order to preclude confusion with helical spur gears having parallel axes, a better name for those with nonparallel axes is *screw gears. Screw gears resemble helical spur gears as far as the shape of the teeth are concerned, but their action is different (33.1).* The engaging teeth slide over each other, instead of moving in the same direction at the pitch point. Such gears are suitable only for the transmission of light power at moderate speeds. They are used to convert rapid rotary motion into slow rotary motion or, when using a rack, into slow linear motion. When used as a speed increaser, they are subject to rapid wear.

Definitions: The two mating screw gears, which will be designated as the driver and the follower in order to differentiate them, may have different helix or pitch angles λ_d and λ_f and different circular pitches p_{cd} and p_{cf} measured on the pitch planes normal to the gear axes. However, the normal circular pitch p_{cn}, or the distance from the face of one tooth to the corresponding face of the next tooth, measured on the pitch cylinder normal to

the tooth face, is the same for both gears, as shown in Fig. 33.1 (33.2). The normal module is related to p_{cn} by the equation $p_{cn} = \pi\, m_n$. Any two screw gears having the same tooth system and the same normal pitch will work together, regardless of tangential pitches. Thus, from Fig. 33.1.

$$p_{cn} = p_{cd} \cos \lambda_d = p_{cf} \cos \lambda_f \qquad \ldots (33.1)$$

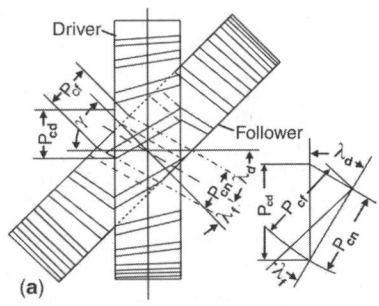

(a)

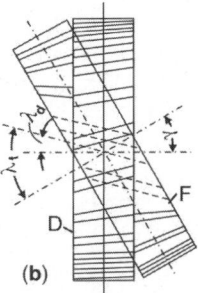

(b)

Fig. 33.1: Pitches and angles in helical gears

33.2 GEOMETRIC RELATIONS

If the pitch diameter of mating screw gears are designated by D_d and D_f, and the numbers of teeth are designated by, z_d and z_f

$$D_d = z_d \frac{p_{cd}}{\pi} = \frac{z_d p_{cn}}{\pi \cos \lambda_d} \qquad \dots (33.2)$$

and

$$D_f = \frac{z_f p_{cn}}{\pi \cos \lambda_f} \qquad \dots (33.3)$$

The velocity ratio r_v of any kind of tooth gearing is always equal to the inverse ratio of the number of teeth in the driver and follower, or

$$r_v = \frac{n_d}{n_f} = \frac{z_f}{z_d} = \frac{D_f \cos \lambda_f}{D_d \cos \lambda_d} \qquad \dots (33.4)$$

where, n_d and n_f are the respective rotative speeds. Equation 33.4 shows that when it is desired to use screw gears for connecting two shafts so as to produce a desired velocity ratio, and any one of a number of pairs of gears with various pitch angles and diameter ratios will give a satisfactory result.

By changing the pitch angles it is also possible to change the direction of rotation of the driven shaft. The pitch angle may be either right-hand or left-hand, the same as for screw threads, and the two helices of a pair of gears may be of the same hand or of opposite hands. Comparison of Figs. 33.1a and 33.1b shows that (1) when both helices are of the same hand, the shaft angle is $\gamma = \lambda_d + \lambda_f$; (2) when the helices are of opposite hands, $\gamma = \lambda_f - \lambda_d$.

Dimensions and angles for screw gears may be computed by using the formulas given in Table 33.1. The notations not shown in Fig. 33.1 are as follows; l_d and l_f are the leads of the tooth helix of the driver and the follower, respectively; a is the addendum of the normal pitch, which for the $14\frac{1}{2}°$ standard is equal to m; and C is the center distance.

33.3 DESIGN

Since screw gears have only a point contact, their wear is considerable and the efficiency cannot be very high. Therefore, they are not used for large tooth loads. This makes possible the selection of the pitch and the face width from geometrical considerations, rather than from strength considerations. Generally m is made from 2 to 5, and the face width b is made from $2 p_c$ to $4 p_c$. In order that the teeth may be cut with standard cutters, the normal module m_n must be a simple number, and the center distance C of the gears should be allowed to vary somewhat.[1] If a comparatively high efficiency is desired, the two pitch angles λ_d and λ_f should be approximately equal.

If the velocity ratio r_v, the shaft angle γ, and the approximate center distance C are given, the preliminary values for D_d and D_f are found by solving simultaneously equation 33.4, in which λ_f and λ_d are assumed to be equal, and equation 33.17 in Table 33.1. Then a suitable value for m_n is assumed, the corresponding value of p_{cn} is found, and the numbers of teeth z_d and z_f are calculated by equations 33.7 and 33.14, respectively. The nearest whole numbers are taken for z_d and z_f, and either the preliminary values of D_d and D_f are corrected or new values are found for λ_d and λ_f. The tooth proportions are based on the relations in Table 30.1 for a pressure angle $\beta = 14\frac{1}{2}°$ and for the values of the normal circular pitch and module.

Example 33.1: Design a pair of screw gears to have a velocity ratio of 1: 3, with the shafts at an angle of 60° and approximately 128 mm apart. The teeth are to be $14\frac{1}{2}°$ involute with full-depth proportion.

Assume that $\lambda_d = \lambda_f$. Then by equation 33.13, $\lambda = \gamma/2 = 30°$

$$D_f = 2 \times 128 - D_d$$

Now by equation 33.4, with $\lambda_d = \lambda_f$,

$$r_v = \frac{256 - D_d}{D_d} = \frac{1}{3}$$

---

[1]Brown & Sharpe Mfg. Co., *Practial Treatise on Gearing*, 17th ed. (Providence, R. I.: 1935), p. 139.

Table 33.1: Formulas for screw-gear calculations

Driver			Follower		
To find	Formula		To find	Formula	
λ_d	$\cos\lambda_d = \dfrac{p_{cn}}{p_{cd}} = \dfrac{z_d m_n}{D_d}$	... (33.5)	λ_f	$\cos\lambda_f = \dfrac{p_{cn}}{p_{cf}}$	... (33.12)
	$\tan\lambda_d = \dfrac{\pi D_d}{l_d}$	... (33.6)	λ_f	$\lambda_f = \lambda - \lambda_d$	... (33.13)
p_{cn}	$p_{cn} = \dfrac{\pi D_d \cos\lambda_d}{z_d}$	... (33.7)	p_{cn}	$p_{cn} = \dfrac{\pi D_f \cos\lambda_f}{z_f}$	... (33.14)
p_{cd}	$p_{cd} = \dfrac{\pi D_d}{z_d}$	... (33.8)	p_{cf}	$p_{cf} = \dfrac{\pi D_f}{z_f}$	... (33.15)
l_d	$l_d = \pi D_d \tan\lambda_f = p_{cf} z_d$	... (33.9)	l_f	$l_f = \pi D_f \tan\lambda_d = p_{cd} z_f$	... (33.16)
D_d	$D_d = \dfrac{2C}{\dfrac{z_f \cos\lambda_d}{z_d \cos\lambda_f} + 1}$	... (33.10)	D_f	$D_f = 2C - D_d$	... (33.17)
	$D_d = 0.3183\, z_d p_{cd}$	... (33.11)		$D_f = 0.3183\, z_f p_{cf}$	... (33.18)

$$C = \frac{z_d m_n}{2\cos\lambda_d} + \frac{z_f m_n}{2\cos\lambda_f} \qquad \text{... (33.19)}$$

From this relation, $D_d = 192$ mm; and $D_f = 256 - 192 = 64$ mm.

Take $m_n = 3$. Then $p_{cn} = \pi \times 3 = 9.42$ and, by equation 33.14, the number of teeth on the follower is

$$z_f = \frac{\pi D_f \cos\lambda_f}{p_{cn}} = \frac{\pi \times 64 \times \cos 30°}{\pi \times 3}$$

$$= \frac{64 \times 0.8660}{3} = 18.47 \text{ or } 19$$

By equation 33.4, the number of teeth on the driver is

$$z_d = \frac{z_f}{r_v} = 19 \times \frac{3}{1} = 57$$

The exact value of the pitch diameter D_d is, by equation 33.2,

$$D_d = \frac{57 \times 9.42}{\pi \times 0.866} = 197.35 \text{ mm say } 198 \text{ mm}$$

Then

$$D_f = 198 \times \frac{1}{3} = 66 \text{ mm}$$

The center distance is, from equation 33.17,

$$C = 0.5\,(198 + 66) = 132 \text{ mm}$$

By equation 33.8,

$$p_{cd} = \frac{\pi \times 198}{57} = 10.91$$

The face width b can be made between $2 \times 10.91 = 21.82$ and $4 \times 10.91 = 43.64$; select 40 mm

The addendum is

$$a = m = 3 \text{ mm}$$

The outside diameters of the gears will be

$$D_d = 198 + 2 \times 3 = 204 \text{ mm}$$

and

$$D_f = 66 + 2 \times 3 = 72 \text{ mm}$$

Example 33.2: Design a pair of screw gears for the conditions of example 33.1 but with the shafts exactly 125 mm apart.

Equation 33.19, in conjunction with equation 33.4, gives

$$\frac{2C}{z_d} = \sec\lambda_d + r_v \sec\lambda_f$$

For $C = 125$ mm, $r_v = \dfrac{1}{3}$, $m_n = 3$, $z_f = 19$ and $z_d = 57$, this equation becomes

$$\frac{2 \times 125}{57 \times 3} = 1.462 = \sec \lambda_d + \frac{1}{3}\sec \lambda_f$$

This equation can be solved by the trial method. For easier handling it can be rewritten in the form

$$3 \sec \lambda_d + \sec \lambda_f = 4.39$$

For the first try, it will be assumed that $\lambda_d = \lambda_f = 30°$ then

$$3 \times 1.1547 + 1.1547 = 4.6188$$

A check by using eqs 33.7 and 33.14 follows:

$$D_d = \frac{57 \times 3}{0.866} = 204.25; \quad D_f = \frac{19 \times 3}{0.866} = 65.8$$

$C = 0.5\,(204.25 + 65.8) = 135$ mm which is O.K.

33.4 SHAFTS AT RIGHT ANGLES

In the most common case, the shafts of a pair of screw gears are at right angles to each other, as in Fig. 33.2. Such gears are used either as a speed reducer, as in a drive from a crankshaft to a camshaft in a four-stroke gas engine with a speed ratio of 2 to 1, or as a speed increaser, as in a drive from a crankshaft to the governor with a speed ratio of 1 to 2 or higher but always carrying a very small load.

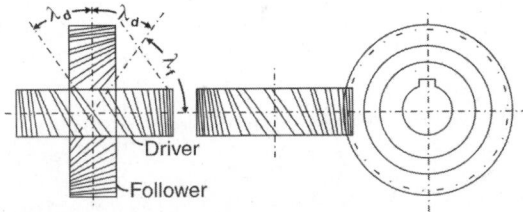

Fig. 33.2: Helical gears on shafts at right angles

In principle this screw gearing is simply straight-tooth worm gearing having a short worm with a long lead so that only short segments of the thread appear on the surface.

Efficiency: Since screw gearing with the shafts at right angles is geometrically identical with worm gearing, equation 32.15 applies to it, with a slight difference in notation. In deriving equation 32.15 the worm was taken as the driver and the gear was the follower. A comparison of Figs 32.1 and 33.2 shows that $\lambda = \lambda_f$. Also, in Fig. 33.2, $\lambda_f + \lambda_d = 90°$. Equation 32.14 then becomes

$$e = \frac{\cos\theta - f \tan \lambda_f}{\cos\theta + f \tan \lambda_d} \qquad \ldots (33.20)$$

The coefficient of friction f ranges from 0.05 to 0.10.

Design: The formulas of Table 33.1 become somewhat simpler when $\lambda_d + \lambda_f = \gamma = 90°$. *For the sake of efficiency and of small wear caused by sliding, the angles λ_d and λ_f should be about equal, and they must be kept between 20° and 70°*[2] *(33.3).* The magnitude of the tangential effort F_t, Fig. 32.8, is found from the torque T_d that must be transmitted. Thus

$$F_t = \frac{2T_d}{D_d} \qquad \ldots (33.21)$$

Strength of teeth: The strength of the gear teeth may be checked by the Lewis formula (equation 30.40), where the normal module m_n should be used and the value of y taken from Table 30.2 should be based on the formative number of teeth z_f. This number is determined by the relation

$$z_f = \frac{z}{\cos^3 \lambda} \qquad \ldots (33.22)$$

The velocity factor c should be computed by equation 30.15. The value of the allowable stress S_o may be taken from Table 30.10. The effective width b of the teeth is rather uncertain in this type of gearing, but it may be taken safely as $2p_{cn}$. The values of the factor C_w are the same as those given in section 30.14.

The value of F_t found by equation 30.40 is the safe normal load on a tooth and should be greater than F_t found by equation 33.21.

Limit load: Because of the point contact of screw gears, even light tooth loads set up very high compressive stresses at the point of contact. Tooth loads smaller than those that

---

[2]C. A. Norman, Principles of *Machine Design* (New York: The Macmillan Company, 1925), p. 389.

may break the teeth may produce excessive wear caused by pitting and abrasion.

Wear can be considerably reduced, and the load capacity can be increased, if the gears are first broken in by operating them in their actual working positions under a light load until a polished line or narrow band appears on the teeth. A further increase of the load capacity— about 150 to 250 per cent—can be obtained if the gears, after being broken in as just described, are run carefully under a gradually increasing load until the narrow bands on the contact surfaces become appreciably wider. The load capacity of gears broken in by the described method may be determined by the relation

$$F_w = KQD_d^2 \qquad \ldots (33.23)$$

Values for the factor K in this equation are given in Table 33.2, and the term Q is computed from the relation[3]

$$Q = \left(\frac{2D_f}{D_f + D_d}\right)^2 \qquad \ldots (33.24)$$

To prevent excessively fast wear, the limit load F_w for wear should be greater than the maximum or dynamic load determined by the equation

$$F_d = \frac{F_t}{C} \qquad \ldots (33.25)$$

where F_t is determined by equation 33.21 and c is found by equation 30.15.

In general, to avoid scuffing, soft steel and cast iron should not be used if the pitch-line velocity of either gear is considerably over 5 m/s. A bronze gear should not be used in combination with a nonmetallic gear, because of an excessive lapping in, or abrasion, of the bronze gear.

Bearing load: The load on bearing for screw gears may be found from equations 32.24 to 32.27 by substituting D_a for D_w and D_f for D_g.

The magnitude of the axial thrust Q on the follower is, by equation 32.13,

$$Q = F_t \frac{\cos\theta \tan\lambda_f + f}{\cos\theta - f \tan\lambda_f} \qquad \ldots (33.26)$$

By eliminating N from equation 32.12 by means of equation 32.10, the downward pressure on the shaft of the follower is found to be

$$R = F_t \frac{\sin\theta}{\cos\theta\cos\lambda_f - f\sin\lambda_f} \qquad \ldots (33.27)$$

Lubrication: Because the power transmitted by screw gears is usually small the problem of lubrication is often neglected, and as a result the screw gears wear out very rapidly. Screw gears, like worm gearing, should run in oil enclosed in a housing if proper efficiency and reasonably long life are desired. Since screw gears have comparatively low efficiency (not over 85 per cent), the heat generated by friction is comparatively great and provision should be made to dissipate it.

Table 33.2: Values of *K* for screw gears

One gear of pair		Meshing gear		K after polishing run	
Material	*Bhn*	*Material*	*Bhn*	*Short*	*Careful*
Steel	250	Steel	250	2	5
Steel	250	Bronze	100	4	12
Steel	500	Bronze	120	5	20
Steel	500	Cast iron	180	6	20
Steel	500	Steel	500	7	15
Cast iron	180	Cast iron	180	8	20
Nonmetallic	–	Steel or cast iron	--	10	25

[3]R. T. Kent, *Mechanical Engineers' Handbook*, 12th ed., Vol. II, *Design and Production*, ed. by Colin Carmichael (New York: John Wiley & Sons, Inc., 1950), p. 14–16.

PROBLEMS

33.1 How far are screw gears similar to helical gears?

33.2 Can the mating screw gears have different helix angles and circular pitches?

33.3 What relation of pitch angles is required to have a high efficiency?

33.4 Design a pair of screw gears. The speed of the driver is 520 rpm, and that of the follower is 200 rpm; the shaft angle is 72° and the center distance must be 116 mm; and the teeth are to be 14.5° involute with standard full depth proportions.

33.5 Assuming that the coefficient of friction is $f = 0.07$, determine the axial thrusts for the gears of problem 33.4. The driving torque is 3 Nm, and the helix angle of the driver is approximately 32°.

33.6 Design a pair of screw gears to transmit 0.2 kN. The speed of the driver is 600 rpm, and a speed reduction of 4:1 must be obtained. The shafts are at right angles to each other. Select the materials and probable coefficients of friction, and compute the efficiency of the drive.

33.7 Using data from problem 33.6 determine the bearing pressures on all four bearings. Assume that the distances from center to center of each pair of bearings is 1.27 m and that the gears are located centrally.

33.8 Design a pair of screw gears to drive a governor of an oil engine from the crank shaft, which makes 180 rpm and has a diameter of 180 mm. The governor speed increase should be about 1:4. The torsional resistance which the governor must overcome is 9 Nm. The shafts are at right angles.

33.9 Indicate what factors should be taken into account if it is desired to develop a formula for the design of screw gears based on values of allowable pressures as used in other machine parts and rubbing speeds between the teeth. Indicate how each factor should influence the design.

34

Choice of Transmission

Mechanical drive is a mechanism used to transmit power from the prime mover to the machine involving change of velocity, forces, torques, and character. In general, the machines are required to run at a speed lower than the speed of motors. By reducing velocity, the torque gets increased to high magnitudes.

A transmission is required for:

1. The velocity of machine does not match with the velocity of prime mover.
2. The machine needs a change in velocity at different times such as a lathe and automobile.
3. The requirement of torque is much higher than that of prime mover as in an automobile gear box.
4. One prime mover is required to run many mechanisms with different velocities.
5. The machine needs to perform rectilinear motion with varying velocities and required to stop at varying intervals.
6. For ease of maintenance and safety, it requires transmission.

Mechanical drives can be used along with hydraulic, electric, pneumatic or vacuum transmission. Preference of a drive is depending on specific requirement as per the following advantages:

Requirement	Types of transmission
1. Centralised power supply, transmission over large distance	Electric
2. High velocity ratio	Electric, pneumatic mesh
3. Step by step velocity change	Electric, friction, mesh
4. Accurate velocity ratio	Pneumatic, mesh
5. Easy accumulation of power	Hydraulic, mesh
6. No effect of ambient temperature	Electric, pneumatic, mesh
7. Automatic remote control	Electric

Types of drives: The following mechanical drives are available depending on mode of transmission.

Friction	Belt, rope, friction gears
Mesh	Gears: Spur, helical, bevel worm, chain

Constant velocity ratio: When the requirement is for constant velocity ratio, the following drives are available depending on the velocity ratio (Table 34.1).

Table 34.1: Transmissions for various velocity ratios

Velocity ratio	Drive
High velocity ratios 100 even up to 500	Single stage worm gear
4 to 6	Spur gear and bevel gears
6 to 20	Helical gears
6 to 15	Silent chain
6 to 10	Chain
8 to 15	V belt drive
Up to 10	Flat belt with idle gear
Up to 5	Flat belt
5 to 10	Friction drive

Depending on the maximum velocity, the following drives are preferable (Table 34.2).

Table 34.2: Drives of various velocities

Velocity m/s	Drive
Up to 50	Flat belt of special fibre
25 to 30	V belts
40	V belts with steel core
25–30	Chain drive
Up to 10	Spur gear
7 to 15	Bevel gears
Up to 30	Helical gears
Up to 20	Worm gears
Up to 20	Friction gears

Power transmission: Expected power transmission by various drives are specified in Table 34.3:

Table 34.3: For various power transmitted capacities

Power kW	Drive
37000	Gears: Spur, helical, bevel
550	Worm gear
35 × 22	V belt × 22
1750 to 3750	Leather, flat belt with special fiber
3500	Chains
150–200	Friction

Efficiency: The following Table 34.4 indicates the approximate maximum efficiency of various drives.

Table 34.4: Efficiencies of transmissions

Drive	Efficiency
Toothed gears	99%
Chain drive	98%
Flat belt	97.5%
Friction	96%
Worm	25–90

Size, weight and cost of drives: The size may be indicated by centre distance and width of gears and pulleys. The cost can be compared by taking the cost of V belt as 100 units and the relative values can be specified with respect to this cost. The comparative value are approximately given in Table 34.5.

These values are approximated by con-

Table 34.5: Comparison of cost of transmissions

Drives	Centre distance mm	Width mm	Approximate weight N	Cost
V belt	1800	130	5000	100
Flat belt	5000	350	5000	106
Flat belt with idler	2300	250	5000	125
Chain	850	350	5000	140
Toothed gears	280	160	6000	165
Worm	280	60	4500	125

sidering the power of 75 kW. From the table the weight range is quite narrow 4500–6000 N and compact size is obtained by using gears and the highest cost is of toothed gears.

PROBLEMS

34.1 Which drive can be used when power is to be transmitted over very large distance?

Ans: Flat belt drive.

34.2 Which is the most compact transmission for very large velocity ratio?

Ans: Toothed gears.

34.3 Which is the best transmission for very large power to be transmitted?

Ans: Gears.

34.4 Which transmission has the highest efficiency?

Ans: Gears.

34.5 For a velocity ratio of 4 to 8, which transmission has the least cost?

Ans: V belt.

35

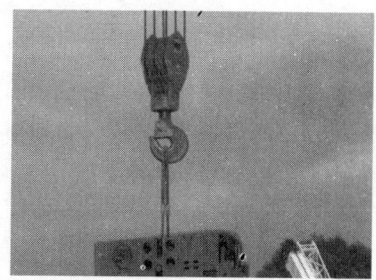

Crane Hook and Block

35.1 GENERAL CONSIDERATIONS

Cranes are an important equipment used for material handling. Cranes carry loads of various shapes and sizes. The crane is able to handle loads by means of chains or wire slings attached to hooks which are part of the snatch block. The block includes pulleys for wire ropes fixed in the side plates. On the lower side, the side plates hold the crosspiece or block which accommodates the shank of the hook. The shank is fixed to the thrust bearing. The upper end of shank is held by nut tightened on the block. A cover is fitted to the block to prevent dust entering the bearing. The assembly is shown in Fig. 35.1.

There are two possible designs of hooks: Single or Ramshorn, produced by flat die or closed die forging (Fig. 35.2) or made by series of shaped plates. One piece hooks are used for lifting loads up to 1 MN and laminated hooks can carry over 1 MN.

After forging: The hooks are carefully annealed and cleaned of scale. The inner diameter, i.e. bed diameter should be sufficient to accommodate two strands of chain/rope sling (Fig. 35.3) to carry loads. Loads are suspended on a four leg sling with two loops over the saddle of hook. *The throat width is made smaller than bed diameter to avoid ropes getting out easily (35.3). More often than not, hooks are having trapezoidal/made wider on the inner side to reduce the effect of bending stress and it approximates to a triangular form (35.1, 35.2).* The

hook is considered as a curved bar subjected to direct and bending stresses for simplifying the design calculations. The shank of the hook is subjected to a tensile stress. The core diameter of the threaded part is the basis of design and corresponding nominal diameter is the standard size.

Hook: The working load or proof load is always considered with 50 percent overload, say $P = (1.5 \times \text{Load } W \text{ to be lifted})$

The root diameter of d_c is given by

$$P = \frac{\pi}{4} d_c^2 s_1$$

s_1 being the allowable tensile stress < 50 N/mm².

By assuming a suitable pitch p; the outside diameter $d_1 = d_c + p$ and standard size is selected.

Considering a bearing stress/compressive stress of 30–35 N/mm², the length of threaded part is found as follows:

Number of threads in the nut

$$n = \frac{4P}{\pi \left[d_1^2 - d_c^2 \right] f_b}$$

f_b is taken 20–30 N/mm²
Length of nut $l = n \times p$.

Hook section: The bed diameter of hook c is given by:

$$C = x\sqrt{P}$$

$x = 12$ for mild steel
$\quad = 23.13$ for high tensile stress steel

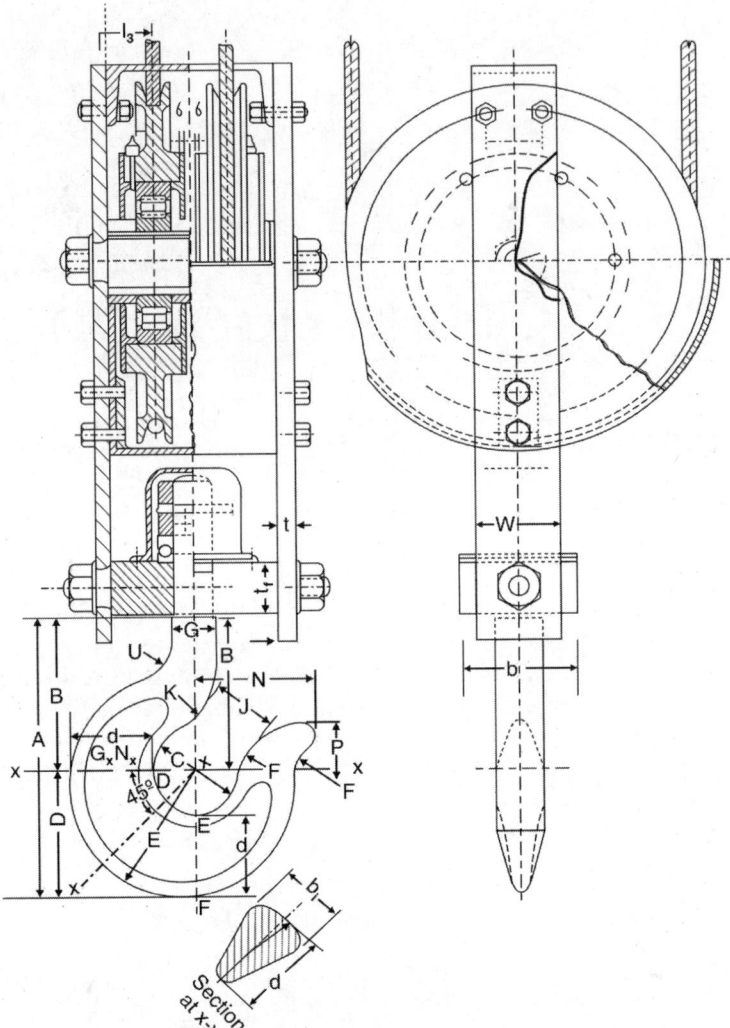

Fig. 35.1: Hook assembly with block and pulleys

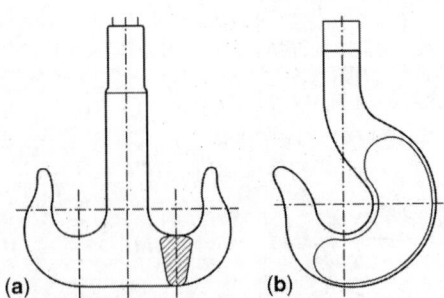

Fig. 35.2: Types of hooks

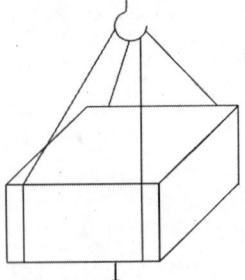

Fig. 35.3: Suspension of load from a hook

Depth of section $d = 0.93\,C$ as shown in Fig. 35.4.

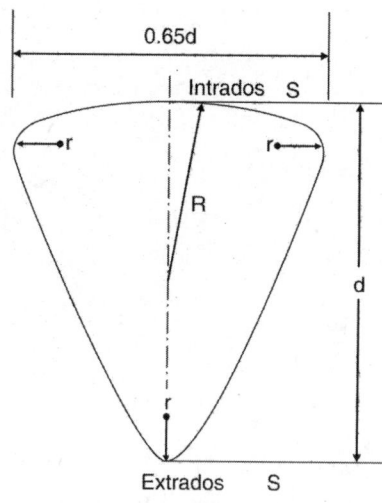

Fig. 35.4

This value is adopted at the horizontal and vertical section of the hook. *A section at 45°, i.e. in between vertical and horizontal sections is made 8% larger than d as the rope makes contact at this point, the contact stress is higher and to take care of wear (35.4).*

As the body curves to join the shank, the section, is reduced to d_1 as the stress gets reduced. The hook load is taken by a thrust bearing through the nut screwed on to the end of shank.

Stresses in the hook section: The stresses are checked at the critical section, i.e. the horizontal section which is a part of the curved surface of the hook and has direct tensile stress and bending stress. The maximum stress will be at the inner section—intrados (35.5).

$$\text{Direct tensile stress} = \frac{P}{A} \qquad \ldots (35.1)$$

Area

$$A = \left[\frac{0.65d + 0.25d}{2}\right]d \qquad \ldots (35.2)$$

For simplification to a triangle

$$A = \frac{0.65d \times d}{2}$$

The radius of curvature of the neutral axis from the central axis

$$r_n = \frac{\frac{1}{2}h(b_i + b_o)}{\frac{b_i r_o - b_o r_i}{h}\ln\left(\frac{r_o}{r_i}\right) - (b_i + b_o)} \qquad \ldots (35.3)$$

b_o is zero when considered as a triangle

r_g = distance of centroidal axis from the central axis

$$e = r_g - r_n$$

Stress at the intrados

$$S_i = \frac{M_b C_i}{Ael}$$

Stress at the extrados

$$= \frac{M_b C_o}{Aer_o} \qquad \ldots (35.4)$$

C_i = distance of inner fiber from N. A.

C_o = distance of outer fiber from N. A

$M_b = P \times r_n.$

Maximum stress at the intrados

$$= \frac{P}{A} + \frac{M_b C_o}{Aer_o} \qquad \ldots (35.5)$$

Maximum stress at the extrados

$$= \frac{P}{A} + \frac{M_b C_o}{A_e\, r_o} \qquad \ldots (35.6)$$

Other dimensions:

1. $C = 12\sqrt{P}$ for M. S.

2. $C = 23.17\sqrt{P}$ for high tensile steel.

3. $d = 0.93\,C$

4. $B = 1.31\,C$

5. $D = 1.44\,C$

6. $A = 1.75$

7. $b_i = 0.65\,d$

8. $N = 1.2\,C$

9. $U = 0.3\,C$

10. $K = 0.93\,C$

11. $G =$ Normal size

12. $P = 0.5\,C$

13. $F = C$

14. $E = 1.25\,C$

Thrust bearing: The rotation of the hook shank is very slow and intermittent, the load is considered stationary. For the load P and diameter d_1, a thrust bearing in selected from the catalogue. It will be necessary to select a bearing of a little higher capacity than P. The bearing specification will be d_1, d_2, D and B, Nut. A nut having outer diameter equal to or less than the outer diameter of bearing D is chosen. The nut needs to be locked by a 5 mm taper pin.

Block: The thrust bearing needs to be provided with a cover of thickness 2–3 mm and having a flange of about 15 mm wide to screw the cap on to the block. *The cover protects the bearing against dust (35.7).*

The width, l_1 of the block can be calculated as follows:

l_1 = Diameter of bearing + clearance of 2 to 3 mm + 2 × thickness of cap + 2 × flange width + clearance. The length of the block will have to be minimum l_1. The length will also be sufficient to allow for hubs of two pulleys in which a bearing will be fixed to reduce power loss due to friction between pulleys and axles when the load is lifted/lowered. This length is also the distance between two side plates.

Pins for block

The pins do not have any relative motion in the side plates. Considering shear of pins of diameter d_p

$$\frac{P}{2} = \frac{\pi}{4}\left(d_p\right)^2 \times s_s \qquad \ldots (35.7)$$

The diameter can be found by considering a low value of allowable shear stress in the pins, about 35–40 N/mm^2. By taking a standard thickness of plates t, the bearing stress/crushing stress can be calculated as load/projected area, i.e.

$$S_c = \frac{P/2}{t \times d_p}$$

Now considering the bending moment in the block by assuming points of supports at the mid section of pins, $l_2 = l_1 + t$

Maximum bending moment is $\dfrac{Pl_2}{4}$

Resisting moment

$$= \frac{\left(l_1 - d_1\right)t_1^2}{6}S_t$$

i.e. $\qquad \dfrac{\left(l_2 - d_1\right)t_1^2}{6}S_t = \dfrac{Pl_2}{4} \qquad \ldots (35.8)$

A suitable value of bending stress s_t of 70 N/mm^2 will give the value of t_1. Considering a groove is made of 4–5 mm to accommodate the bearing, the thickness at diameter D can be increased. The groove is at diameter D to accommodate the bearing. *As the bending stress at the sides is zero, the corners are cut at 45° reduce weight and its shape will be octagonal (35.6).* Now considering the pins to be supported at the outer edge of plate and loaded as cantilevers at the end.

The maximum bending moment

$$= P/2 \times t$$

The bending stress

$$= \frac{P/2 \times t}{\pi/32 \times d_p^3} \qquad \ldots (35.9)$$

This stress will be produced only when sufficient wear takes place.

The pins are reduced slightly in diameter beyond the plates and threaded, to be tightened by suitable nuts.

Plates: A standard reasonable thickness t of plate is already considered. To find the width of plate w, it is considered to be subjected to tension $\dfrac{P}{2}$, the effective width = $w - d_p$ or

$$\frac{P}{2} = \left(w - d_p\right)t \times S_t \qquad \ldots (35.10)$$

By considering a suitable allowable tensile stress S_t is the plates as 45–50 N/mm², w is calculated.

Ropes: Various loads to which wire ropes are subjected are: weight to be lifted + acceleration load + bending load + sudden/impact load to decide the wire rope diameter by equating the total load to the breaking load with a factor of safety of 6–8.

Sheaves: The sheave diameter D_1, is taken 25 to 35 times the rope diameter, the higher diameter is preferable to reduce the bending stress. The sheave will be cast iron with four arms of elliptical section. *The profile of groove for wire rope has a radius of 0.53 d. The walls of groove are tangential to the root arc and should not make contact with wire rope when the rope flattens under the load.* The spindle/axle is subjected to a B.M. due to two equal loads. The distance l_3 of center of sheave to end of axle up to half the plate thickness can be found.

The maximum bending moment,

$$M = \frac{P}{2} \times l_3$$

Axle diameter

$$d_2 = \frac{M}{\frac{\pi}{32} S_t} \qquad \ldots (35.11)$$

rounded off to next mm size

Hub of pulley $= 2.5 \times d_2$

Inner diameter of hub

$\qquad = 1.5\, d_2$ to accommodate bearing.

The bearings for sheave will have an internal diameter d_2 and outer race rotating. The bearing can have a life of 5 to 6 years and working for 12 hours/day at N rpm of the sheave. The race factor of 1.2 needs to be taken into account for calculating the equivalent bearing load.

Example 35.1: Design a snatch block for a crane to lift a load of 60 kN with 50 percent over load. Allowable tensile stress in shank, in 55 N/mm², bending stress in the pins 80 N/mm², shear stress in pins 35 N/mm², tensile stress in hook 130 N/mm².

The bearing pressure in threads is not to exceed 20 N/mm².

Hook section

Proof load $= 1.5 \times 60 = 90$ kN

Bed diameter $c = 12\sqrt{90} = 113.8$ say 115 mm

Depth of section $d = 0.93\, c = 106.95$ say 107 mm

Intrados $b_i = 0.65\, d = 69.55$ say 70 mm

Extrados $d_o = 0.25\, d = 26.75$ say 27 mm

Radius of curvature of intrados

$$r_i = \frac{c}{2} = \frac{107}{2} = 53.5 \text{ mm}$$

Radius of curvature of extrados

$$r_o = \frac{c}{2} + d = 160.5 \text{ mm}$$

Radius of curvature of centroidal axis

$$r_g = \frac{c}{2} + \frac{d}{3} = 89.1 \text{ mm}$$

Radius of curvature of neutral axis, taking triangular section:

$$r_n = \frac{\dfrac{1}{2} b_i\, h}{\dfrac{b_i r_o}{h} \ln \dfrac{r_o}{r_i} - b_i}$$

$$= \frac{70 \times 53.5}{70 \times 160.5/107 \ln \dfrac{160.5}{53.5} - 70}$$

$$= 82.6 \text{ mm}$$

$$e = 89.1 - 82.6 = 6.5 \text{ mm}$$

$$c_i = 82.6 - 53.5 = 29.1 \text{ mm}$$

$$A = \frac{70 \times 27}{2} \times 107 = 5189.5 \text{ mm}^2$$

$$c_o = 160.5 - 82.5 = 78 \text{ mm}$$

Direct tensile stress

$$S_d = \frac{90 \times 1000}{5189.5} = 17.34 \text{ N/mm}^2$$

Bending moment

$$M = P \times r_n = 90 \times 82.6 = 7434 \text{ kN/mm}$$

Bending stress at intrados

$$S_i = \frac{Mc_i}{Ae\, r_i} = \frac{7434 \times 1000 \times 29.1}{5189.5 \times 6.5 \times 53.5}$$

$$= 111.22 \text{ N/mm}^2$$

Bending stress at extrados

$$S_o = -\frac{Mc_o}{Aer_o} = \frac{7434 \times 1000 \times 78}{5189.5 \times 6.5 \times 160.5}$$

$$= -107.10 \text{ N/mm}^2$$

Total stress at intrados

$$= 17.34 + 111.22 = 128.56 \text{ N/mm}^2$$

Total stress at extrados

$$= 17.34 - 107.10 = -89.76 \text{ N/mm}^2$$

These stresses are within the prescribed limits.

Shank diameter and nut

Core diameter of shank

$$= \left[\frac{90 \times 1000 \times 4}{\pi \times 55}\right]^{1/2} = 45.64 \text{ mm}$$

Assuming a pitch of 9 mm, outside diameter

$$= 45.64 + 9 = 54.64 = \text{say } 55$$

Then core diameter = 46 mm
Number threads in the nut

$$\eta = \frac{90 \times 1000 \times 4}{\pi \left[55^2 - 46^2\right] \times 20} = 6.3 \text{ say } 7$$

Length of nut

$$l = 7 \times 9 = 63 \text{ mm.}$$

To compensate for the locking pin the diameter is increased to 63 + 5 = 68 say 70 mm.

The shank diameter at the lower end may be taken as:

$$G = 60 \text{ mm}$$

Bearing: For 60 mm shank diameter and a static load of 90 kN a thrust bearing with the following dimensions is selected.

SKF 51112, $d = 60$, $c = 62$, $D = 85$, $E = 85$, $B = 17$

It has a value of $c_o = 139{,}247$ N and maximum rpm of 3000 which is very safe.

Nut: A round nut of outer diameter equal to that of thrust bearing, i.e. 85 mm is used. The nut can be tightened by the pin hole provided for locking.

Block: The thrust bearing and nut are to be provided with suitable cover to avoid dust particles. The cover can have a thickness of 2 mm and a radial clearance of 5 mm. A flange of 20 mm radial width is to be provided for fixing it with screws on to the block.

So minimum width of block

$$= 85 + 5 \times 2 + 2 \times 2 + 2 \times 20$$

$$= 139 \text{ say } 140 \text{ mm}$$

To accommodate two sheaves with roller bearings, the distance between two side plates is taken as 210 mm. If the side plates are taken with 20 mm thickness, the center of supports of the block will be at a distance of 210 + 2 × 10 = 230 mm.

Maximum bending moment on the block

$$= \frac{90 \times 230}{4} = 5175 \text{ kN/mm}$$

Effective width of block

$$= 210 - 60 = 150 \text{ mm}$$

The thickness of the block t_1 is given by:

$$\frac{150 \times t_1^2 \times 75}{6} = 5175 \times 1000$$

$$t_1 = \left[\frac{5175 \times 1000 \times 6}{150 \times 75}\right]^{1/2} = 52.53 \text{ say } 55 \text{ mm}$$

Taking the allowable stress as 75 N/mm²

The block needs to be provided with a recess of 4 mm depth at diameter 85 mm to fit the lower race of the bearing.

Total thickness of the block = 66 + 4 = 70 mm

Pins for block and sheave spindle: The block is extended in the form of pin in the center of width which will pass through the plate for holding with nuts. The pins will be subjected to shear. The diameter of pins d_p subjected to a maximum shear stress of 35 N/mm² is given by:

$$\frac{\pi}{4} d_p^2 \times 35 = \frac{90 \times 1000}{2}$$

$$d_p = \left[\frac{90 \times 1000 \times 4}{\pi \times 35 \times 2}\right]^{1/2} = 40.46 \text{ mm}$$

Considering the bending stress in the pins, the diameter may be taken as 50 mm.

Bearing pressure/stress an each pin

$$= \frac{90 \times 1000}{2 \times 20 \times 50} = 45 \text{ N/mm}^2$$

As there is no relative motion between pins and plate, the stress is not high.

The bending stress in the pins

$$= \frac{90 \times 1000 \times 20 \times 32}{2 \times \pi \times 50^3} = 73.33 \text{ N/mm}^2$$

which is quite safe as there is no likelihood of clearance between the pins and the plates. The ends of pins are reduced in size and threads provided to fix the nuts, the size of nuts should be larger than 50 mm. So 50 M 3 hexagonal nuts are quite suitable.

Side plates: The plates are subjected to tension and have a width w. Each plate is subjected to half the total load

$$(w - 50) \times 20 \times 55 = \frac{90}{2} \times 1000$$

$$w = 41 + 50 = 91 \text{ mm}$$

Ropes: Considering two pulleys handling the load, the load on each rope

$$= \frac{90}{4} = 22.5 \text{ kN}$$

Assuming an acceleration of 3 m/s²
Accelerating force

$$= \frac{60 \times 3}{9.81} = 18.34 \text{ kN}$$

Force on each rope

$$= \frac{60}{4} = 15.00 \text{ kN}$$

Accelerating force on each rope

$$= \frac{18.34}{4} = 4.58 \text{ kN}$$

i.e. for lifting 60 kN load, force on each rope = 15 + 4.58 = 19.58 kN.

This is lower than proof load force so taking the value of 22.5 kN and a working factor of 6, breaking load = 22.5 × 6 = 135 kN. From tables, a suitable wire rope of breaking strength of 144.16 is 6 × 19 having $d_r = 16$ mm.

For cranes, sheave diameter = 30 d = 30 × 16 = 480 mm and *Sheaves* can be made of cast iron with suitable profile for wire ropes.

Roller bearing for sheaves: Considering the sheaves to be rotating at 50 rpm, and life of 20000 hrs which corresponds to 12 hours per day for 360 days for five years approximately.

Life in $mR = \dfrac{20000 \times 50 \times 60}{10^6} = 60 \text{ mR}$

For this life $\dfrac{c}{P}$ for roller bearing = 3.42

Radial load with shock factor of 1.5 and race factor of 1.2

$$= \frac{90}{2} \times 1.5 \times 1.2 = 81 \text{ kN}$$

Basis dynamic capacity

$$C = 81 \times 3.42 = 277 \text{ kN}$$

For this value of C, proper roller bearing is SKF: N 417, $d = 85$, $D = 210$, $B = 52$, $E = 177$ and $C = 240$ kN.

This is selected as for this bearing maximum speed is 50 rpm.

Spindle of the sheave

The spindle of the bearing will have a diameter of 85 mm and considered as a simply supported beam loaded by two equal loads at the center of the sheaves (Fig. 35.5).

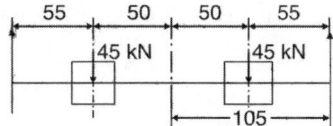

Fig. 35.5

Maximum bending moment

$$= 45 \times 55 = 2475 \text{ kNmm}$$

Bending stress

$$= \frac{2475 \times 1000 \times 32}{\pi \times 85^3}$$

$$= 41.05 \text{ N/mm}^2$$

which is quite low.

For proper retainment of bearings on the spindle, three sleeves of inner diameter 85 mm and outer diameter 100 mm are used. Length of outer sleeves will be 29 mm each and central sleeve will have a length of 48 mm.

The width of side plates can be increased to 100 mm each.

PROBLEMS

35.1 What shape of cross-section of a hook is preferable?

35.2 Why is it preferable to have a trapezoidal section for the crane hook?

35.3 Why is the throat width of a crane hook of smaller size than the bed diameter?

35.4 Why is the cross-section at 45° to vertical made larger than any other section?

35.5 Which is the most critical section of a hook?

35.6 Why is the cross-piece sometimes made of octagonal section?

35.7 Why is a cover required over the nut and bearing of a crane hook?

35.8 A workshop crane is to have a capacity of 80 kN. Design the cross-section, bed diameter for a standard trapezoidal section. Check for the maximum stress in the cross-section.

35.9 A crane hook has a trapezoidal section 56 mm deep. The intradoz and extrados have the width 37 mm and 14 mm respectively. Calculate the maximum stress at the critical section when a load of 12.5 kN is placed symmetrically in the hook.

35.10 Design the cross-section, bed diameter, throat diameter and shank diameter of a standard crane hook for a working load of 100 kN. Find also the dimensions of crosspiece, bearing, nut and side plates.

Index

603